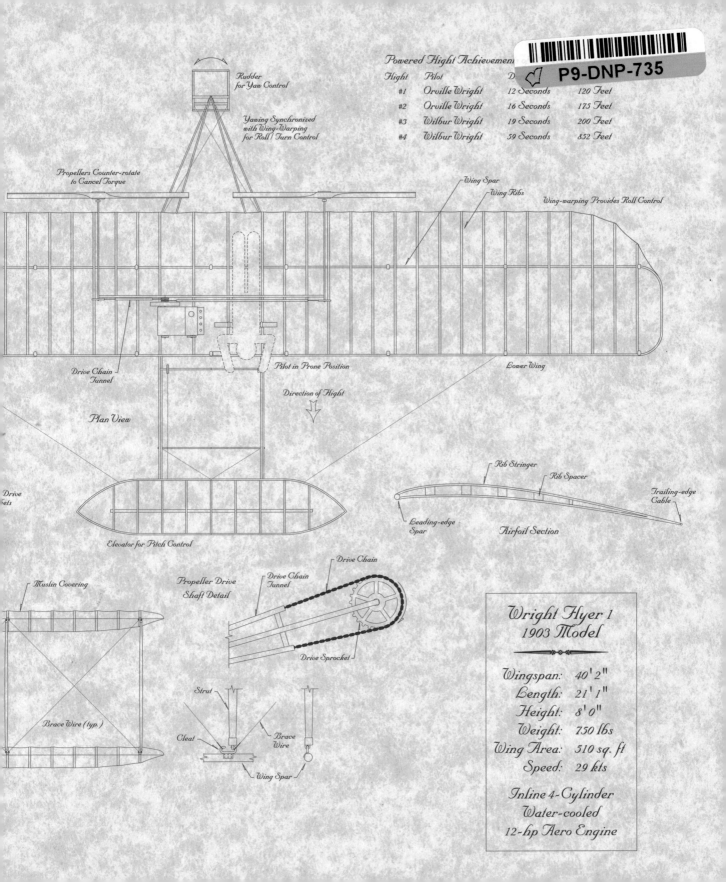

Rudder for Yaw Control

Yawing Synchronized with Wing-Warping for Roll / Turn Control

Powered Flight Achievement

Flight	Pilot	Duration	Distance
#1	Orville Wright	12 Seconds	120 Feet
#2	Orville Wright	16 Seconds	175 Feet
#3	Wilbur Wright	19 Seconds	200 Feet
#4	Wilbur Wright	59 Seconds	852 Feet

Propellers Counter-rotate to Cancel Torque

Wing Spar

Wing Ribs

Wing-warping Provides Roll Control

Drive Chain Tunnel

Pilot in Prone Position

Lower Wing

Direction of Flight

Plan View

Drive ...ets

Rib Stringer

Rib Spacer

Trailing-edge Cable

Leading-edge Spar

Airfoil Section

Elevator for Pitch Control

Muslin Covering

Drive Chain Tunnel

Drive Chain

Propeller Drive Shaft Detail

Brace Wire (typ.)

Drive Sprocket

Strut

Cleat

Brace Wire

Wing Spar

Wright Flyer 1
1903 Model

Wingspan: 40' 2"
Length: 21' 1"
Height: 8' 0"
Weight: 750 lbs
Wing Area: 510 sq. ft
Speed: 29 kts

Inline 4-Cylinder
Water-cooled
12-hp Aero Engine

ENGINEERING DESIGN GRAPHICS

AutoCAD® Release 14 • Ninth Edition

James H. Earle

 **ADDISON-WESLEY**

An imprint of Addison Wesley Longman, Inc.

Reading, Massachusetts • Menlo Park, California • New York • Harlow, England
Don Mills, Ontario • Sydney • Mexico City • Madrid • Amsterdam

Senior Acquisitions Editor: Denise Olson
Production Editor: Amy Willcutt
Production Assistant: Brooke D. Albright
Art Coordinator: Jennifer Brownlow Bagdigian
Technical Artist: John Sanderson/Horizon Design
Composition: Michael and Sigrid Wile
Copyeditor: Stephanie Argeros-Magean
Proofreader: Carmen Wheatcroft
Cover Designer: Diana Coe
Cover Illustrator: Chester Beals
Indexer: Nancy Fulton

Access the latest information about Addison-Wesley books from our World Wide Web site:
http://www.awl.com/cseng/cad

Many of the designations used by manufacturers and sellers to distinguish their products are claimed as trademarks. Where those designations appear in this book, and Addison-Wesley was aware of a trademark claim, the designations have been printed in initial caps or all caps.

The programs and applications presented in this book have been included for their instructional value. They have been tested with care, but are not guaranteed for any particular purpose. The publisher does not offer any warranties or representations, nor does it accept any liabilities with respect to the programs or applications.

Library of Congress Cataloging-in-Publication Data

Earle, James H.
    Engineering design graphics / James Earle. — 9th ed.
      p. cm.
    Includes bibliographical references and index.
    ISBN 0-201-82372-1
    1. Engineering design. 2. Engineering graphics. I. Title.
  TA174.E23   1999
  620'.0042'0222—dc21   98-18041
                                            CIP

This book was typeset in QuarkXPress 3.32 on a Power Macintosh 7500. The font used was Utopia. It was printed on New Era Matte.

1 2 3 4 5 6 7 8 9 10-DOC-0201009998

Dedicated to my father,
Hubert Lewis Earle,
October 25, 1900–October 22, 1967

Preface

The Best Yet

We are proud to say that this ninth edition of *Engineering Design Graphics* is the best edition since its introduction in all categories: content, format, readability, clarity, quality of illustrations, and economy.

Every revision has begun with the challenge, "How can the book be written and illustrated to make it easier for the student to learn and the teacher to teach?" Also asked is, "What should be the content for today's course that will fit tomorrow's needs?"

Comprehensive. Classically Modern. Competitively Priced.

Successfully meeting this goal is a difficult chore in any discipline, but it is especially daunting in the area of engineering design graphics. We believe we have met the challenge with the most comprehensive edition ever, with its 2,000 illustrations, 1,000 problems, and a multitude of topics compactly merged into a cohesive text. Every paragraph and illustration was evaluated for its contribution to the learning and teaching process before being revised for this edition. Despite its comprehensiveness, this text is value-priced and represents a cost savings over some previous editions.

Engineering Design Graphics, Ninth Edition continues the tradition of illustrating and presenting concepts in an understandable format to reduce the amount of classroom tutoring needed by the student. It also continues the tradition of presenting core concepts as well as timely, important topics such as CAD and the design process. As a result, this text can be used in a wide variety of programs.

Major content areas of the engineering design process that are covered in this text are: engineering drawing; descriptive geometry; problem solving; computer graphics; and design and creativity.

Design and Creativity

The eight chapters devoted to the introduction of design and creativity have been revised and improved. New case studies with examples of worksheets and design drawings guide the student through the design process.

Since the primary objective of design instruction is to teach the process of design, meaningful design assignments are given to make the process fun and to encourage the application of creativity and intuition. Most problems are at a challenge level that will encourage creative and inspirational solutions that may lead to patentable products. Care has been taken to offer realistic design problems that are within the grasp of beginning students rather than overwhelming them with problems beyond their capabilities.

Chapter 9 contains 115 design problems that can be used for quick problems, short assignments, or as semester-long design projects. Additional design exercises are included at the ends of the chapters throughout the book.

CAD Coverage

This text has been updated to cover AutoCAD Release 14, which is presented in a step-by-step format to aid the student in learning how to use this popular software. Computer Methods boxes illustrating the use of AutoCAD can be found throughout the text. In addition, Chapter 37 gives an introduction to two-dimensional computer graphics and Chapter 38 covers three-dimensional computer graphics, solid modeling, and rendering.

Other Revision Features Besides the updating of the engineering design process material, this text was revised for modern courses in other ways; specifically, the coverage of instruments, lettering and geometric construction has been compressed. Geometric tolerance was revised for current standards in Chapter 21. Additionally, new working drawing problems have been added to Chapter 23; perspectives have been added to Chapter 25; and new problems have been added to most chapters. The tables in the appendix have been revised to make them more readable and functional.

Updating and streamlining of some material created space for the inclusion of two specialty chapters of interest to many: Pipe Drafting and Electronics Drafting.

Reference Value

This text's numerous appendices make it an extremely valuable reference tool. In addition, some material in this book may not be formally covered in the course for which it is adopted due to time limitations or course emphasis. These lightly-covered topics may be the ones that will be needed in later courses or in practice. This book can thus be retained as a convenient reference for the engineer, technologist, or technician.

A Teaching System

Engineering Design Graphics used in combination with the supplements listed below form a complete teaching system.

Textbook Problems: Approximately 1,000 problems are offered to aid the student in mastering the principles of graphics and design.

Problem Manuals: Nineteen problem books and teacher's guides (with outlines, problem solutions, tests and test solutions) that are keyed to this book are available. New problem manuals are in development as well. A listing of these manuals is given inside the back cover of this book.

Visual Aids: Sixteen modules of SoftVisuals are available on disks from which multicolored overhead transparencies can be plotted on transparency film for classroom presentations. Transparency selection can be made from over 500 SoftVisuals keyed to this textbook that can be plotted or projected onto a screen with AutoCAD.

Acknowledgments

We are grateful for the assistance of many who have influenced the development of this volume. Many industries have furnished photographs, drawings, and applications that have been acknowledged in the accompanying legends. The Engineering Design Graphics staff of Texas A&M University have been helpful in making suggestions for the revision of this book.

Professor Tom Pollock provided valuable information on metallurgy for Chapter 19. Professor Leendert Kersten of the University of Nebraska, Lincoln, kindly provided his descriptive geometry computer programs for inclusion; his cooperation is greatly appreciated.

We are indebted to Jimm Meloy of Autodesk, Inc. for his assistance and cooperation. We are appreciative of the assistance of David Ratner of the Biomechanics Corporation Inc. for providing HUMANCAD® software.

We are appreciative of the fine editorial and production team assembled by Denise Olson at Addison-Wesley: Amy Willcutt, Jenny Bagdigian, and John Sanderson. Jenny worked closely with John in expertly managing 2,000 illustrations to see that they were properly reproduced and sized. Amy demonstrated professionalism in overseeing the project; she handled the minor details with the same dedication as the major issues. Thank you, Amy. It was a pleasure to work with Addison-Wesley's A-Team.

Above all, we appreciate the many institutions who have thought enough of our publications to adopt them for classroom use. This is the highest honor that can be paid an author. We are hopeful that this textbook will fill the needs of engineering and technology programs. As always, comments and suggestions for improvement and revision will be appreciated.

College Station, Texas   *Jim Earle*

Contents

Chapter 1 Engineering and Technology 2
1.1 Introduction 2
1.2 Engineering Graphics 2
1.3 Technological Milestones 3
1.4 The Technological and Design Team 4
1.5 Engineering Fields 6
1.6 Aerospace Engineering 6
1.7 Agricultural Engineering 7
1.8 Chemical Engineering 7
1.9 Civil Engineering 8
1.10 Electrical Engineering 9
1.11 Industrial Engineering 10
1.12 Mechanical Engineering 11
1.13 Mining and Metallurgical Engineering 12
1.14 Nuclear Engineering 12
1.15 Petroleum Engineering 13
1.16 Graphics/Drafting 13
Problems 14

Chapter 2 The Design Process 16
2.1 Introduction 16
2.2 Types of Design Problems 17
2.3 The Design Process 18
2.4 Problem Identification 19
2.5 Application of the Design Process 21
Problems 26

Chapter 3 Problem Identification 28
3.1 Introduction 28
3.2 Example: Ladder Attachment 28
3.3 The Identification Process 29
3.4 Design Worksheets 30
3.5 Example: Exercise Bench 30
3.6 Organization of Effort 32
3.7 Planning Design Activities 33
Problems 34

Chapter 4 Preliminary Ideas 36
4.1 Introduction 36
4.2 Individual Versus Team Methods 37
4.3 Plan of Action 38
4.4 Brainstorming 38
4.5 Sketching and Notes 39
4.6 Quickie Design 41

4.7	Background Information	42
4.8	Opinion Surveys	43
4.9	Preliminary Ideas: Exercise Bench	44
	Problems	46

Chapter 5 Refinement **48**

5.1	Introduction	48
5.2	Physical Properties	48
5.3	Application of Descriptive Geometry	49
5.4	Refinement Considerations	51
5.5	Refinement: Exercise Bench	51
5.6	Standard Parts	53
	Problems	53

Chapter 6 Design Analysis **56**

6.1	Introduction	56
6.2	Graphics and Analysis	56
6.3	Types of Analysis	56
6.4	Analysis: Exercise Bench	65
	Problems	68

Chapter 7 Decision **72**

7.1	Introduction	72
7.2	Types of Decisions	72
7.3	Decision: Exercise Bench	73
7.4	Types of Presentations	74
7.5	Organizing a Presentation	74
7.6	Visual Aids	75
7.7	Making a Presentation	78
7.8	Written Reports	79
	Problems	82

Chapter 8 Implementation **84**

8.1	Introduction	84
8.2	Working Drawings	84
8.3	Specifications	85
8.4	Assembly Drawings	86
8.5	Miscellaneous Considerations	86
8.6	Implementation: Exercise Bench	87
8.7	Patents	90
8.8	Patent Drawings	91
8.9	Patent Searches	93
8.10	Questions and Answers	93
	Problems	97

Chapter 9 Design Problems **98**

9.1	Introduction	98
9.2	The Individual Approach	98
9.3	The Team Approach	98
9.4	Selection of a Problem	99
9.5	Problem Specifications	99
9.6	Scheduling Team Activities	99
9.7	Short Design Problems	100
9.8	Systems Design Problems	104
9.9	Product Design Problems	106

Chapter 10 Drawing Instruments **112**

10.1	Introduction	112
10.2	Drawing Media	112
10.3	Drawing Equipment	114
10.4	Lines	117
10.5	Measurement	119
10.6	Presentation of Drawings	124
	Problems	125

Chapter 11 Lettering **128**

11.1	Introduction	128
11.2	Lettering Tools	128
11.3	Guidelines	129
11.4	Gothic Lettering	130
11.5	Computer Lettering	133
	Problems	135

Chapter 12 Geometric Construction **136**

12.1	Introduction	136
12.2	Angles	136
12.3	Polygons	136

12.4	Circles	137
12.5	Geometric Solids	137
12.6	Constructing Polygons	139
12.7	Bisecting Lines and Angles	140
12.8	Division of Lines	141
12.9	An Arc Through Three Points	142
12.10	Parallel Lines	143
12.11	Tangents	143
12.12	Conic Sections	149
12.13	Spirals	152
12.14	Helixes	152
	Problems	153

Chapter 13 Freehand Sketching   160

13.1	Introduction	160
13.2	Shape Description	160
13.3	Sketching Techniques	162
13.4	Six-View Sketching	163
13.5	Three-View Sketching	163
13.6	Circular Features	166
13.7	Oblique Pictorial Sketching	168
13.8	Isometric Pictorial Sketching	169
	Problems	173

Chapter 14 Orthographic Drawing with Instruments   178

14.1	Introduction	178
14.2	Orthographic Projection	178
14.3	Alphabet of Lines	180
14.4	Six-View Drawings	181
14.5	Three-View Drawings	182
14.6	Arrangement of Views	183
14.7	Selection of Views	183
14.8	Line Techniques	184
14.9	Point Numbering	185
14.10	Lines and Planes	185
14.11	Drawing with Triangles	185
14.12	Views by Subtraction	186
14.13	Three-View Drawing Layout	187
14.14	Views by Computer	189
14.15	Two-View Drawings	191

14.16	One-View Drawings	191
14.17	Simplified and Removed Views	191
14.18	Partial Views	192
14.19	Curve Plotting	192
14.20	Conventional Practices	193
14.21	Conventional Intersections	194
14.22	Fillets and Rounds	195
14.23	First-Angle Projection	199
	Problems	200

Chapter 15 Primary Auxiliary Views   208

15.1	Introduction	208
15.2	Folding-Line Principles	209
15.3	Auxiliaries from the Top View	209
15.4	Auxiliaries from the Top: Application	211
15.5	The Rules of Auxiliary View Construction	212
15.6	Auxiliaries from the Front View	213
15.7	Auxiliaries from the Profile View	216
15.8	Curved Shapes	218
15.9	Partial Views	219
15.10	Auxiliary Sections	219
15.11	Secondary Auxiliary Views	220
15.12	Elliptical Features	221
	Problems	221

Chapter 16 Sections   226

16.1	Introduction	226
16.2	The Basics of Sectioning	226
16.3	Sectioning Symbols	228
16.4	Sectioning Assemblies of Parts	229
16.5	Full Sections	229
16.6	Partial Views	232
16.7	Half-Sections	233
16.8	Offset Sections	234
16.9	Broken-Out Sections	234
16.10	Revolved Sections	234
16.11	Removed Sections	235
16.12	Conventional Revolutions	237
16.13	Conventional Breaks	239

16.14 Phantom (Ghost) Sections   240
16.15 Auxiliary Sections   240
      Problems 241

Chapter 17   Screws, Fasteners, and Springs   246
17.1   Introduction   246
17.2   Thread Terminology   246
17.3   English System Specifications   248
17.4   English Thread Notes   249
17.5   Metric Thread Notes   251
17.6   Drawing Threads   253
17.7   Detailed Symbols   254
17.8   Schematic Symbols   256
17.9   Simplified Symbols   258
17.10   Nuts and Bolts   258
17.11   Drawing Square Heads   261
17.12   Drawing Hexagon Heads   261
17.13   Types of Screws   263
17.14   Other Threaded Fasteners   265
17.15   Tapping a Hole   266
17.16   Washers, Lock Washers, and Pins   267
17.17   Pipe Threads and Fittings   268
17.18   Keys   269
17.19   Rivets   269
17.20   Springs   270
17.21   Drawing Springs   271
      Problems   273

Chapter 18   Gears and Cams   276
18.1   Introduction   276
18.2   Spur Gears   276
18.3   Gear Ratios   278
18.4   Drawing Spur Gears   279
18.5   Bevel Gears   280
18.6   Drawing Bevel Gears   282
18.7   Worm Gears   282
18.8   Drawing Worm Gears   284
18.9   Cams   284
18.10   Designing Plate Cams   286
      Problems   288

Chapter 19   Materials and Processes   290
19.1   Introduction   290
19.2   Commonly Used Metals   291
19.3   Properties of Metals   294
19.4   Forming Metal Shapes   294
19.5   Machining Operations   299
19.6   Surface Finishing   304
19.7   Plastics and Other Materials   305
      Review Questions   307

Chapter 20   Dimensioning   308
20.1   Introduction   308
20.2   Terminology   308
20.3   Units of Measurement   309
20.4   English/Metric Conversions   310
20.5   Dual Dimensioning   310
20.6   Metric Units   310
20.7   Numerals and Symbols   311
20.8   Dimensioning by Computer   313
20.9   Dimensioning Rules   314
20.10   Curved and Symmetrical Parts   323
20.11   Finished Surfaces   324
20.12   Location Dimensions   324
20.13   Outline Dimensioning   327
20.14   Machined Holes   328
20.15   Chamfers   330
20.16   Keyseats   330
20.17   Knurling   330
20.18   Necks and Undercuts   331
20.19   Tapers   332
20.20   Miscellaneous Notes   332
      Problems   332

Chapter 21   Tolerances   336
21.1   Introduction   336
21.2   Tolerance Dimensions   336
21.3   Mating Parts   338
21.4   Tolerancing Terms: English Units   339
21.5   Basic Hole System   340
21.6   Basic Shaft System   340

21.7	Cylindrical Fits	340
21.8	Tolerancing: Metric Units	341
21.9	Chain Versus Datum Dimensions	345
21.10	Tolerance Notes	346
21.11	General Tolerances: Metric Units	347
21.12	Geometric Tolerances	348
21.13	Rules of Tolerancing	350
21.14	Cylindrical Datum Features	351
21.15	Datum Targets	354
21.16	Location Tolerancing	354
21.17	Form Tolerancing	357
21.18	Profile Tolerancing	358
21.19	Orientation Tolerancing	358
21.20	Runout Tolerancing	360
21.21	Surface Texture	361
	Problems	363

Chapter 22   Welding   370

22.1	Introduction	370
22.2	Welding Processes	370
22.3	Weld Joints and Welds	372
22.4	Welding Symbols	373
22.5	Application of Symbols	374
22.6	Surface Contouring	377
22.7	Brazing	378
22.8	Soldering	378
	Problems	379

Chapter 23   Working Drawings   380

23.1	Introduction	380
23.2	Working Drawings as Legal Documents	380
23.3	Dimensions and Units	381
23.4	Laying Out a Detail Drawing	387
23.5	Notes and Other Information	389
23.6	Checking a Drawing	391
23.7	Drafter's Log	392
23.8	Assembly Drawings	392
23.9	Freehand Working Drawings	394

23.10	Working Drawings for Forged Parts and Castings	394
	Problems	396

Chapter 24   Reproduction of Drawings   432

24.1	Introduction	432
24.2	Computer Drawing Types	432
24.3	Types of Reproduction	432
24.4	Assembling Drawing Sets	435
24.5	Transmittal of Drawings	436

Chapter 25   Three-Dimensional Pictorials   438

25.1	Introduction	438
25.2	Oblique Drawings	439
25.3	Oblique Projection Theory	444
25.4	Isometric Pictorials	445
25.5	Isometric Drawings	446
25.6	Technical Illustration	453
25.7	Isometrics by Computer	455
25.8	Axonometric Projection	456
25.9	Perspective Pictorials	457
25.10	Three-Dimensional Modeling	460
25.11	The Human Figure	461
	Problems	462

Chapter 26   Points, Lines, and Planes   464

26.1	Introduction	464
26.2	Projection of Points	465
26.3	Lines	465
26.4	Visibility	467
26.5	Planes	468
26.6	Parallelism	470
26.7	Perpendicularity	472
	Problems	473

Chapter 27   Primary Auxiliary Views in Descriptive Geometry   476

27.1	Introduction	476
27.2	Geometry by Computer	476

27.3	True-Length Lines	478
27.4	Angles Between Lines and Principal Planes	480
27.5	Sloping Lines	480
27.6	Bearings and Azimuths of Lines	481
27.7	Application: Plot Plans	483
27.8	Contour Maps and Profiles	484
27.9	Plan-Profiles	485
27.10	Edge Views of Planes	486
27.11	Planes and Lines	488
27.12	Sloping Planes	490
27.13	Ore-Vein Applications	492
27.14	Intersections Between Planes	495
	Problems	496
	Review Questions	501

Chapter 28 Successive Auxiliary Views   504

28.1	Introduction	504
28.2	Point View of a Line	504
28.3	Dihedral Angles	505
28.4	True Size of a Plane	506
28.5	Shortest Distance from a Point to a Line: Line Method	508
28.6	Shortest Distance Between Skewed Lines: Line Method	509
28.7	Shortest Distance Between Skewed Lines: Plane Method	509
28.8	Shortest Level Distance Between Skewed Lines	510
28.9	Shortest Grade Distance Between Skewed Lines	510
28.10	Angular Distance to a Line	511
28.11	Angle Between a Line and a Plane: Plane Method	512
28.12	Angle Between a Line and a Plane: Line Method	513
	Problems	514

Chapter 29 Revolution   518

29.1	Introduction	518
29.2	True-Length Lines: Front View	518
29.3	True Size of a Plane	520
29.4	Angle Between Planes	522
29.5	Determining Direction	523
29.6	Revolution: Point About an Axis	524
29.7	A Line at Specified Angles	526
	Problems	527

Chapter 30 Vector Graphics   530

30.1	Introduction	530
30.2	Definitions	530
30.3	Coplanar, Concurrent Forces	531
30.4	Noncoplanar, Concurrent Forces	532
30.5	Forces in Equilibrium	533
30.6	Coplanar Truss Analysis	534
30.7	Noncoplanar Vector Analysis	537
30.8	Resultant of Parallel, Nonconcurrent Forces	539
	Problems	540

Chapter 31 Intersections and Development   544

31.1	Introduction	544
31.2	Intersections of Lines and Planes	544
31.3	Intersections Between Prisms	546
31.4	Intersections Between Planes and Cylinders	547
31.5	Intersections Between Cylinders and Prisms	549
31.6	Intersections Between Cylinders	549
31.7	Intersections Between Planes and Cones	551
31.8	Intersections Between Cones and Prisms	552
31.9	Intersections Between Prisms and Pyramids	552
31.10	Principles of Developments	554
31.11	Development of Rectangular Prisms	556
31.12	Development of Oblique Prisms	557
31.13	Development of Cylinders	558
31.14	Development of Oblique Cylinders	559

31.15	Development of Pyramids	561
31.16	Development of Cones	562
31.17	Development of Transition Pieces	563
	Problems	563

Chapter 32   Graphs **568**

32.1	Introduction	568
32.2	Pie Graphs	569
32.3	Bar Graphs	570
32.4	Linear Coordinate Graphs	571
32.5	Semilogarithmic-Coordinate Graphs	576
32.6	Schematics	578
32.7	Graphs by Computer	579
	Problems	580

Chapter 33   Nomography **584**

33.1	Introduction	584
33.2	Alignment Graph Scales	585
33.3	Concurrent Scales	586
33.4	Alignment Graphs: Three Variables	588
33.5	Parallel Scale Graphs: Linear Scales	588
33.6	N or Z Nomographs	590
	Problems	592

Chapter 34   Empirical Equations and Calculus **594**

34.1	Introduction	594
34.2	Linear Equation: $Y = MX + B$	595
34.3	Power Equation: $Y = BXM$	595
34.4	Exponential Equation: $Y = BM^X$	596
34.5	Graphical Calculus	597
34.6	Graphical Differentiation	598
34.7	Graphical Integration	602
	Problems	602

Chapter 35   Pipe Drafting **606**

35.1	Introduction	606
35.2	Welded and Seamless Steel Pipe	606
35.3	Cast-Iron Pipe	607
35.4	Copper, Brass, and Bronze Piping	608
35.5	Miscellaneous Piping Materials	608
35.6	Pipe Joints	608
35.7	Pipe Fittings	609
35.8	Screwed Fittings	611
35.9	Flanged Fittings	612
35.10	Welded Fittings	613
35.11	Valves	613
35.12	Fittings in Orthographic Views	614
35.13	Piping Systems in Pictorial	615
35.14	Dimensioned Isometrics	616
35.15	Vessel Detailing	618
35.16	Computer Drawings	619
	Problems	620

Chapter 36   Electric/Electronics Graphics **624**

36.1	Introduction	624
36.2	Types of Diagrams	624
36.3	Schematic Diagram Connecting Symbols	626
36.4	Graphic Symbols	628
36.5	Terminals	630
36.6	Separation of Parts	633
36.7	Reference Designations	634
36.8	Numerical Units of Function	635
36.9	Functional Identification of Parts	636
36.10	Printed Circuits	637
36.11	Shortcut Symbols	638
36.12	Installation Drawings	638
	Problems	639

Chapter 37   AutoCAD Computer Graphics **644**

37.1	Introduction	644
37.2	Computer Graphics Overview	644
37.3	Computer Hardware	645
37.4	Your First AutoCAD Session	647
37.5	Introduction to Windows	651
37.6	Format of Presentation	651
37.7	Using Dialogue Boxes	652
37.8	Drawing Aids	654

37.9	General Assistance Commands	655
37.10	Drawing Layers	656
37.11	Toolbars	658
37.12	Creating a New Drawing	659
37.13	Drawing Scale	662
37.14	Saving and Exiting	663
37.15	Plotting Parameters	663
37.16	Readying the Plotter	667
37.17	Two-Dimensional Lines (Draw Toolbar)	667
37.18	Circles (Draw Toolbar)	668
37.19	Arcs (Draw Toolbar)	669
37.20	Polygon (Draw Toolbar)	670
37.21	Ellipse (Draw Toolbar)	670
37.22	Fillet (Modify Toolbar)	671
37.23	Chamfer (Modify Toolbar)	671
37.24	Trim (Modify Toolbar)	672
37.25	Extend (Modify Toolbar)	672
37.26	Trace (Command Line)	673
37.27	Zoom and Pan (Standard Toolbar)	673
37.28	Selecting Objects	674
37.29	Erase and Break (Modify Toolbar)	675
37.30	Move and Copy (Modify Toolbar)	676
37.31	Undo (Standard Toolbar)	677
37.32	Change (Modify Toolbar)	677
37.33	Grips (Tools Menu)	678
37.34	Polyline (Draw Toolbar)	680
37.35	Pedit (Draw Toolbar)	681
37.36	Spline (Draw Toolbar)	683
37.37	Hatching (Draw Toolbar)	684
37.38	Text and Numerals	685
37.39	Text Style (Format Menu)	686
37.40	Multiline (Mtext): Draw Toolbar	686
37.41	Mirror (Modify Toolbar)	687
37.42	Osnap (Object Snap Toolbar)	688
37.43	Array (Modify Toolbar)	689
37.44	Donut (Draw Menu)	689
37.45	Scale (Modify Toolbar)	689
37.46	Stretch (Modify Toolbar)	690
37.47	Rotate (Modify Toolbar)	690
37.48	Setvar (Command Line)	690
37.49	Divide (Draw Menu)	691
37.50	Measure (Draw Menu)	691
37.51	Offset (Modify Toolbar)	691
37.52	Blocks (Draw Toolbar)	692
37.53	Transparent Commands (Command Line)	693
37.54	View (Command Line)	693
37.55	Inquiry Commands (Tools)	694
37.56	Dimensioning	695
37.57	Dimension Style (Dimstyle) Variables	696
37.58	Linear (Dimension Toolbar)	697
37.59	Angular (Dimension Toolbar)	698
37.60	Diameter (Dimension Toolbar)	698
37.61	Radius (Dimension Toolbar)	699
37.62	Dimension Style (Dimension Toolbar)	700
37.63	Saving Dimension Styles	704
37.64	Dimension Style Override (Dimension Toolbar)	704
37.65	Editing Dimensions	704
37.66	Toleranced Dimensions	706
37.67	Geometric Tolerances (Dimension Toolbar)	706
37.68	A Custom Title Block	707
37.69	Digitizing with the Tablet	708
37.70	Sketch (Miscellaneous Toolbar)	708
37.71	Oblique Pictorials	709
37.72	Isometric Pictorials	709
	Problems	710

Chapter 38 Three-Dimensional Modeling 712

38.1	Introduction	712
38.2	Paper Space and Model Space: An Overview	712
38.3	Paper Space Versus Model Space	714
38.4	Fundamentals of 3D Drawing	714
38.5	Elementary Extrusions	714
38.6	Coordinate Systems	716
38.7	Setting Viewpoints (VPoints)	718
38.8	Application of Extrusions	719
38.9	Dynamic View (Dview)	719
38.10	Basic 3D Shapes (Surfaces)	721
38.11	Surface Modeling	723
38.12	Line, Pline, and 3Dpoly	725

38.13	The 3Dface Command	725
38.14	XYZ Filters	726
38.15	Solid Modeling: Introduction	727
38.16	Extrusion Example: Tilemode=0	728
38.17	Solid Primitives	729
38.18	Modifying Solids	732
38.19	Section	733
38.20	Slice	734
38.21	A Solid Model Example	734
38.22	Views of a Solid	735
38.23	Mass Properties (Massprop)	737
38.24	Paper Space and Model Space: Tilemode=0	738

38.25	Dimensioning in 3D	740
38.26	Render	741
38.27	Lights	742
38.28	Working with Scenes	746
38.29	Materials	747
	Problems	747

| **Appendix** | **A-1** |

| **Index** | **I-1** |

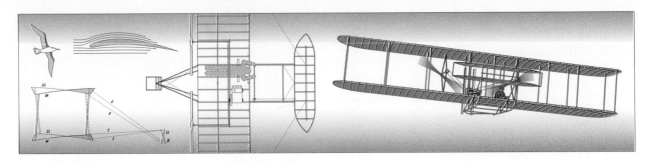

Engineering and Technology

1.1 Introduction

This book deals with the field of engineering design graphics and its application to the design process. Engineering graphics is the primary medium for developing and communicating design concepts. The solution of most engineering problems requires a combination of organization, analysis, problem-solving principles, graphics, skill, and communication (**Fig. 1.1**).

This book will help you use your creativity and develop your imagination because innova-

tion is essential to a successful career in engineering and technology. Albert Einstein said, "Imagination is more important than knowledge, for knowledge is limited, whereas imagination embraces the entire world . . . stimulating progress, or, giving birth to evolution."

1.2 Engineering Graphics

Engineering graphics covers the total field of graphical problem solving within two major areas of specialization: **descriptive geometry** and **documentation drawings**. Other areas of application are nomography, graphical mathematics, empirical equations, technical illustration, vector graphics, data analysis, and computer graphics. Graphics is one of the designer's most effective tools for developing design concepts and solving three-dimensional problems. It is also the designer's best means of communicating ideas to others.

Descriptive Geometry

Gaspard Monge (1746–1818), the "father of descriptive geometry," used graphical meth-

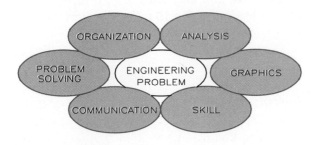

1.1 A visual depiction of the total approach to engineering, with the engineering problem as the focal point.

1900	Vacuum cleaner	1950	A—bomb tests
	Airplane		Optical fibers
	Dial telephone		Soviet satellite
	Light bulb		Microchip
	Model T Ford	1960	Commun. satellite
1910	Washing machine		Indus. robot
	Refrigerator		Nuclear reactor
	Wireless phone		Heart transplant
1920	Radio broadcasts		Man on moon
	Telephone service	1970	Silicon chip
	35mm camera		Personal computer
	Cartoons & sound		Videocassette recorder
1930	Tape recorder		Supersonic jet
	Atom split		Neutron bomb
	Jet engine	1980	Stealth bomber
	Television		Space shuttle
1940	Elect. computer		Artificial heart
	Missile		Soviet space station
	Transistor	1990	Computer voice
	Microwave		recognition
	Polaroid camera		Artificial intelligence
			Space—based
			assembly plant

1.2 A chronology of significant technological advances of the twentieth century.

1.3 Technological and design team members, with their varying experiences and areas of expertise, must communicate and interact with each other. (Courtesy of the Mitre Corporation.)

ods to solve design problems related to fortifications and battlements while a military student in France. His headmaster scolded him for not using the usual long, tedious mathematical process. Only after lengthy explanations and demonstrations of his technique was he able to convince the faculty that graphical methods (now called descriptive geometry) produced solutions in less time.

Descriptive geometry was such an improvement over mathematical methods that it was kept as a military secret for 15 years before the authorities allowed it to be taught as part of the civilian curriculum. Monge went on to become a scientific and mathematical aide to Napoleon.

Descriptive geometry is the projection of three-dimensional figures on a two-dimensional plane of paper in a manner that allows geometric manipulations to determine lengths, angles, shapes, and other geometric information about the figures.

1.3 Technological Milestones

Many of the technological advancements of the twentieth century are engineering achieve-ments. Since 1900, technology has taken us from the horse-drawn carriage to the moon and back, and more advancements are certain in the future.

Figure 1.2 shows a few of the many technological mileposts since 1900. It identifies products and processes that have provided millions of jobs and a better way of life for all. Other significant achievements were building a railroad from Nebraska to California that met at Promontory Point, Utah, in 1869 in less than four years; constructing the Empire State Building with 102 floors in a mere 13½ months in 1931; and retooling industry in 1942 for World War II to produce 4.5 naval vessels, 3.7 cargo ships, 203 airplanes, and 6 tanks each day while supporting 15 million Americans in the armed forces.

One of the "miracle projects" of the 1990s was the construction of the 31-mile "Chunnel" that connects England and France under the English Channel for high-speed shuttle trains. It consists of three tunnels drilled 131 feet under the channel floor; two of the tunnels are 24 feet in diameter. The trip from London to Paris can be made in 3½ hours.

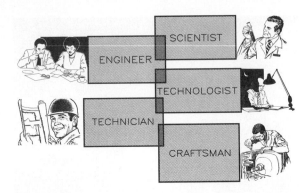

1.4 A ranking of the typical technological team, from the most theoretical level (scientists) to the least technical level (craftsperson).

1.4 The Technological and Design Team

Technology and design have become so broad and complex that teams of specialists rather than individuals undertake most projects (**Fig. 1.3**). Such teams usually consist of one or more scientists, engineers, technologists, technicians, and craftspeople, and may include designers and stylists (**Fig. 1.4**).

Scientists

Scientists are researchers who seek to discover new laws and principles of nature through experimentation and scientific testing (**Fig. 1.5**). They are more concerned with the discovery of scientific principles than with the application of those principles to products and systems. Their discoveries may not find applications until years later.

Engineers

Engineers receive training in science, mathematics, and industrial processes to prepare them to apply the findings of the scientists (**Fig. 1.6**). Thus engineers are concerned with converting raw materials and power sources into needed products and services. Creatively applying scientific principles to develop new products and systems is the design process, the engineer's primary function. In general, engineers use known principles and available

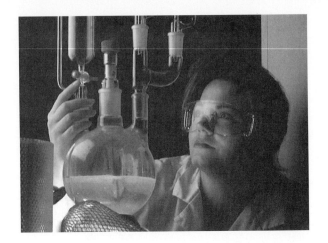

1.5 This scientist is working to develop a new drug to protect the body from contracting AIDS. (Courtesy of FMC Corporation.)

1.6 An engineer and a technologist combine their knowledge to work on an avionics modification. (Courtesy of Cessna Aircraft Company.)

resources to achieve a practical end at a reasonable cost.

Technologists

Technologists have backgrounds in science, mathematics, and industrial processes. Whereas engineers are responsible for analysis, overall design, and research, technologists are concerned with the application of engineering principles to planning, detail design, and production (see Fig. 1.6). Technologists apply their knowledge of engineering principles, manufacturing, and testing to assist in the implementation of projects and production. They also provide support and act as liaisons between engineers and technicians.

Technicians

Technicians assist engineers and technologists at a less theoretical level than technologists and act as liaisons between technologists and craftspeople (**Fig. 1.7**). They have backgrounds in mathematics, drafting, computer programming, and materials testing. Their work varies from conducting routine laboratory experiments to supervising craftspeople in manufacturing or construction.

Craftspeople

Craftspeople are responsible for implementing designs by fabricating them according to engineers' specifications. They may be machinists who make product parts or electricians who assemble electrical components. Their ability to produce a part according to design specifications is as necessary to the success of a project as the engineer's ability to design it. Craftspeople include electricians, welders, machinists, fabricators, drafters, and members of many other occupational groups (**Fig. 1.8**).

1.8 A craftsperson assembles a portion of an experimental aircraft. (Courtesy of Cessna Aircraft Company.)

Designers

Designers may be engineers, technologists, inventors, or industrial designers who have special talents for devising creative solutions. Designers do not necessarily have engineering backgrounds, especially in newer technologies where there is little design precedent. Thomas Edison (**Fig. 1.9**), for example, had little formal education, but he created some of the world's most significant inventions.

Stylists

Stylists are concerned with the appearance and market appeal of a product rather than its fundamental design (**Fig. 1.10**). They may design an automobile body or the exterior of an electric iron. Automobile stylists, for example, consider the car's appearance, driver's vision, passengers' enclosure, power unit's space requirement, and so on. However, they are not involved with the design of the car's internal mechanical functions, such as the engine, steering linkage, and brakes. Stylists must have a high degree of aesthetic awareness and an instinct for styling that will appeal to the consumer.

1.7 An engineering technician works on a phased-array radar antenna. (Courtesy of Northrop Grumman Corporation.)

1.9 Thomas Edison had little formal education, but he changed the world with his knowledge and creativity.

1.10 The stylist develops illustrations such as this one to present a product's outward appearance in a marketable and attractive style. (Courtesy of General Motors Corporation.)

1.5 Engineering Fields

Recent changes in engineering include the emergence of technologists and technicians and the growing number of women pursuing engineering careers. More than 15 percent of today's freshman engineering students are women; 15 percent of master's degrees and 10 percent of doctor's degrees in engineering are awarded to women.

1.6 Aerospace Engineering

Aerospace engineering has progressed from the Wright brothers' first flight at Kitty Hawk, North Carolina, in 1903 to the penetration of outer space. It deals with all aspects (speeds and altitudes) of flight. Aerospace engineering assignments range from developing complex vehicles capable of traveling millions of miles into space to hover aircraft that can transport and position large construction components. In the space exploration branch of this profession, aerospace engineers work on all types of aircraft and spacecraft—missiles, rockets, propeller-driven planes, and jet-powered planes (**Fig. 1.11**).

Second only to the auto industry in sales, the aerospace industry contributes immeasurably to national defense and the economy. Specialized areas include aerodynamics, structural design, instrumentation, propulsion systems, materials, reliability testing, and production methods.

Aerospace engineers specialize in one of two major areas: research engineering and design engineering. Research engineers investigate known principles in search of new ideas and concepts; design engineers translate them into workable applications. The

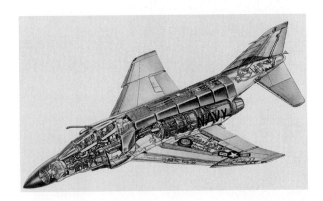

1.11 Aerospace engineering deals with all aspects of aircraft and spacecraft design. (Courtesy of McDonnell Douglas Corporation.)

professional society for aerospace engineers is the American Institute of Aeronautics and Astronautics (AIAA).

1.7 Agricultural Engineering

Agricultural engineers are trained to serve the world's largest industry—agriculture—in which they deal with the production, processing, and handling of food and fiber.

Mechanical Power Agricultural engineers who work with manufacturers of farm equipment are concerned with gasoline and diesel engine equipment, including pumps, irrigation machinery, and tractors. Machinery must be designed for the electrical curing of hay, milk and fruit processing, and heating environments for livestock and poultry (**Fig. 1.12**). Farm machinery designed by agricultural engineers has been largely responsible for the vastly increased productivity in U.S. agriculture. Today's farmer produces enough food for 120 people, whereas 100 years ago a farmer was able to feed only four people.

1.12 Agricultural engineers design food processing systems such as this one for oranges in Florida. (Courtesy of FMC Corporation.)

Farm Structures The construction of barns, shelters, silos, granaries, processing centers, and other agricultural buildings requires specialists in agricultural engineering. They must understand heating, ventilation, and chemical changes that might affect the storage of crops.

Electrical Power Agricultural engineers design electrical systems and select equipment that will operate efficiently and meet the requirements of many situations. They may serve as consultants or designers for manufacturers or processors of agricultural products.

Soil and Water-Control Agricultural engineers are responsible for devising systems to improve drainage and irrigation systems, resurface fields, and construct water reservoirs. They may perform activities in conjunction with the U.S. Department of Agriculture, the U.S. Department of the Interior, state agricultural universities, consulting engineering firms, or irrigation companies.

Characteristics

Most agricultural engineers are employed in private industry, especially by manufacturers of heavy farm equipment and specialized lines of field, barnyard, and household equipment; by electrical service companies; and by distributors of farm equipment and supplies. Although few agricultural engineers live on farms, they need to understand agricultural problems—farming, crops, animals, and farmers themselves. The professional society for this group of engineers is the American Society for Agricultural Engineers (ASAE).

1.8 Chemical Engineering

Chemical engineering involves the design and selection of equipment used to process and manufacture large quantities of chemicals

(**Fig. 1.13**). Chemical engineers develop and design methods of transporting fluids through ducts and pipelines, transporting solid material through pipes or conveyors, transferring heat from one fluid or substance to another through plate or tube walls, absorbing gases by bubbling them through liquids, evaporating liquids to increase concentration of solutions, distilling mixed liquids to separate them, and handling many other chemical processes. Process control and instrumentation are important specialties in chemical engineering. With the measurement of quality and quantity by instrumentation, process control is fully automatic.

Chemical engineers often utilize chemical reactions of raw products, such as oxidation, hydrogenation, reduction, chlorination, nitration, sulfonation, pyrolysis, and polymerization in their work. They develop and process chemicals such as acids, alkalies, salts, coal-tar products, dyes, synthetic chemicals, plastics, insecticides, and fungicides for industrial and domestic uses. Chemical engineers help develop drugs and medicines, cosmetics, explosives, ceramics, cements, paints, petroleum products, lubricants, synthetic fibers, rubber, and detergents. They also design equipment for food preparation and canning plants.

Approximately 80 percent of chemical engineers work in manufacturing industries, primarily the chemical industry. The other 20 percent work for government agencies, independent research institutes, and as independent consultants. New fields requiring chemical engineers are nuclear sciences, rocket fuel development, and environmental pollution control. The professional society for chemical engineers is the American Institute of Chemical Engineers (AIChE).

1.9 Civil Engineering

Civil engineering, the oldest branch of engineering, is closely related to virtually all our daily activities. The buildings we live and work in, the transportation we use, the water we drink, and the drainage and sewage systems we rely on are all the results of civil engineering.

Construction Civil engineers manage the workers, finances, and materials used on construction projects. Structural engineers design and supervise the construction of buildings, harbors, airfields, tunnels, bridges, stadiums, and other types of structures.

City Planning Civil engineers working as city planners develop plans for the growth of cities and systems related to their operation. They are involved in street planning, zoning, residential subdivisions, and industrial site development (**Fig. 1.14**).

Hydraulics Civil engineers work with the behavior of water and other fluids from their conservation to their transportation. Civil engineers design wells, canals, dams, pipelines, flood control and drainage systems, and other methods of controlling and using those resources.

1.13 Chemical engineers developed this closed-system container for insecticides that is the industry standard. (Courtesy of FMC Corporation.)

1.14 Civil engineers may be involved in a wide range of projects—from city planning and highway design to utility systems.

Transportation Civil engineers design and supervise construction, modification, and maintenance of railroad and mass transit systems. They design and supervise construction of airport runways, control towers, passenger and freight stations, and aircraft hangars. They are involved in all phases of developing the national systems and local networks of highways and interchanges for moving automobile traffic including the design of tunnels, culverts, and traffic control systems.

Sanitary Engineering Civil engineers maintain public health by designing pipelines, treatment plants, and other facilities for water purification and water and air pollution control. They also are involved in solid waste disposal activities.

Other Areas of Specialization Civil engineers are involved in geotechnical engineering, which deals with the study of soils. They also work in environmental engineering, which relates to all aspects of the environment and methods of preserving it.

Characteristics
Many civil engineers hold positions in administration and municipal management and are associated with federal, state, and local government agencies and the construction industry. Many work as consulting engineers for architectural firms and independent consultants. The remainder work for public utilities, railroads, educational institutions, and various manufacturing industries. The professional society for these engineers is the American Society of Civil Engineers (ASCE). Founded in 1852, it is the oldest engineering society in the United States.

1.10 Electrical Engineering

Electrical engineers are concerned with power, which deals with providing the large amounts of energy required by cities and large industries, and electronics, which deals with the small amounts of power used for communications and automated operations that have become part of everyday life.

Power Power generation, transmission, and distribution pose many electrical engineering problems—from the design of generators for producing electricity to the development of transmission equipment. Power applications in homes are quite numerous—from computers to washers, dryers, and vacuum cleaners. Only about one-quarter of electric energy is consumed in the home. About half is used by industry for metal refining, heating, motor drives, welding, machinery controls, chemical processes, plating, and electrolysis. Illumination (lighting) is required in nearly every phase of modern life. Improving its efficiency is a challenging area for electrical engineers.

Electronics Computers have contributed to the development of a gigantic industry that is the domain of electrical engineers. Used with industrial electronics (such as robotics),

1.15 The electrical engineer in conjunction with the computer programmer have revolutionized work with computers and numerous other electronic devices. (Courtesy of Cessna Aircraft Company.)

computers have changed industry's manufacturing and production processes, resulting in greater precision and less manual labor (**Fig. 1.15**).

Communications Systems These engineering applications are devoted to the improvement of radio, telephone, telegraph, and television systems, the nerve centers of most industrial operations.

Instrumentation These engineering applications involve systems of electronic instruments used in industrial processes. Electrical engineers make extensive use of the cathode-ray tube and the electronic amplifier in industry and nuclear reactors. Increasingly, engineers are using instrumentation for applications in medical diagnosis and therapy.

Military Electronics This field encompasses most weapons and tactical systems ranging from the walkie-talkie to radar networks for detecting enemy aircraft. Remote-controlled electronic systems are used for navigation and interception of guided missiles.

Characteristics
There are currently more electrical engineers than any other type of engineer. The increas-

ing need for electrical equipment, automation, and computerized systems is expected to sustain the rapid growth of this field. The professional society for electrical engineers is the Institute of Electrical and Electronic Engineers (IEEE). Founded in 1884, the IEEE is the world's largest professional society.

1.11 Industrial Engineering

Industrial engineering, one of the newer engineering fields, differs from other branches of engineering in that it relates primarily to people, their performance, and working conditions. Industrial engineers often manage people, machines, materials, methods, and money in the production and marketing of goods.

Industrial engineers may be responsible for plant layout, development of plant processes, or determination of operating standards that will improve the efficiency of a plant operation. They also design and supervise systems for improved safety of personnel and increased production at lower costs (**Fig. 1.16**).

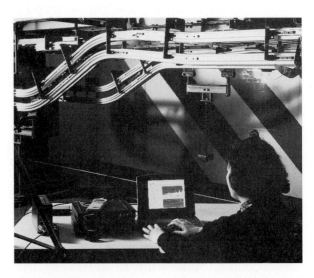

1.16 Industrial engineers are concerned with productivity and safety in the workplace. This engineer is conducting a noise-reduction analysis study of a new conveyor product. (Courtesy of Jervis B. Webb Company.)

Areas of industrial engineering include management, plant design and engineering, electronic data processing, systems analysis and design, control of production and quality, performance standards and measurements, and research. Industrial engineers are increasingly involved in implementing automated production systems.

People-oriented areas include the development of wage incentive systems, job evaluation, work measurement, and environmental system design. Industrial engineers are often involved in management-labor agreements that affect operations and production.

More than two-thirds of all industrial engineers are employed in manufacturing industries. Others work for insurance companies, construction and mining firms, public utilities, large businesses, and government agencies. The professional society for industrial engineers is the American Institute of Industrial Engineers (AIIE), which was organized in 1948.

1.12 Mechanical Engineering

Mechanical engineering's major areas of specialization are power generation, transportation, manufacturing, power services, and atomic energy. The professional society for mechanical engineers is the American Society of Mechanical Engineers (ASME).

Power Generation Mechanical engineers develop and design prime movers (machines that convert natural energy into work) to power electric generators that produce electricity. They are involved in the design and operation of steam engines, turbines, internal combustion engines, and other prime movers.

Transportation Mechanical engineers participate in the design of trucks, buses, automobiles, locomotives, marine vessels, and aircraft. In aeronautics, they develop aircraft engines, controls, and internal environmental systems. Mechanical engineers design marine vessels powered by steam, diesel, or gas-turbine engines, as well as power services throughout the vessels, such as lighting, water, refrigeration, and ventilation (**Fig. 1.17**).

Manufacturing Mechanical engineers design new products and the equipment and factories needed to manufacture them economically and at a uniformly high level of quality. The professional society for manufacturing engineers is the Society of Manufacturing Engineers (SME).

Power Services Mechanical engineers in this area must have a knowledge of pumps, ventilation equipment, fans, and compressors. They apply this knowledge to the development of methods for moving liquids and gases through pipelines and the design and installation of refrigeration systems, elevators, and escalators.

1.17 Mechanical engineers design unique traffic systems such as this Airtrans system for the Dallas-Fort Worth Airport. (Courtesy of Vought Aeronautics.)

1.18 Metallurgical engineers develop new alloys for special applications and new products. (Courtesy of Jones & Laughlin Steel Corporation.)

Nuclear Energy Mechanical engineers develop and handle protective equipment and materials and assist in constructing nuclear reactors.

1.13 Mining and Metallurgical Engineering

Mining engineers are responsible for developing methods of extracting minerals from the earth and preparing them for use by manufacturing industries (**Fig. 1.18**). Working with geologists to locate ore deposits, which are exploited through the construction of tunnels and underground operations or surface strip mining, mining engineers must understand safety, ventilation, water supply, and communications. Mining engineers who work at mining sites usually are employed near small, out-of-the-way communities, whereas those in research and consulting most often work in large urban areas.

The two main areas of metallurgical engineering are extractive metallurgy (the extraction of metal from raw ores to form pure metals) and physical metallurgy (the development of new products and alloys). Many metallurgical engineers work on the development of machinery for electrical equipment and in the aircraft industry. The need to develop new lightweight, high-strength materials for spacecraft, jet aircraft, missiles, and satellites will require more metallurgical engineers in the future.

The professional society for mining and metallurgical engineers is the American Institute of Mining, Metallurgical, and Petroleum Engineering (AIME).

1.14 Nuclear Engineering

The earliest work in nuclear engineering involved military applications. Nuclear power for domestic needs was developed for a time, but the new emphasis in nuclear engineering is for medical applications. Peaceful applications of nuclear engineering fall into two major areas: radiation and nuclear power reactors. Radiation is the propagation of energy through matter or space in the form of waves. In atomic physics, radiation includes fast-moving particles (alpha and beta rays, free neutrons, and so on), gamma rays, and X-rays. Nuclear science is closely allied with botany, chemistry, medicine, and biology.

The use of nuclear energy to produce mechanical or electric power is a major peaceful application of nuclear engineering (**Fig. 1.19**). In the production of electric power, nuclear energy is the fuel used to produce steam to drive turbine generators.

Most nuclear engineering training focuses on the design, construction, and operation of nuclear reactors. Other areas include the processing of nuclear fuels, thermonuclear engineering, and the use of various nuclear by-products. The professional society for nuclear engineers is the American Nuclear Society.

1.19 Nuclear engineers will design the nuclear reactors to economically produce much of the electric power of the future.

1.15 Petroleum Engineering

The recovery of petroleum and natural gas is the primary concern of petroleum engineers, but they also develop methods for transporting and separating various petroleum products. They are also responsible for improving drilling equipment and ensuring its economical operation. In exploring for petroleum, petroleum engineers are assisted by geologists and by instruments such as the airborne magnetometer, which indicates uplifts in the earth's subsurfaces that could hold oil or gas.

Petroleum engineers develop equipment to remove oil from the ground most efficiently and supervise oil-well drilling (**Fig. 1.20**). They also design systems of pipes and pumps to transport oil to shipping points or to refineries. Development, design, and operation of petroleum processing facilities are done jointly with chemical engineers.

The Society of Petroleum Engineers (SPE) is a branch of AIME, which includes mining and metallurgical engineers and geologists.

1.16 Graphics/Drafting

Drafters help engineers and designers select materials. They also prepare construction documents and specifications and translate designs into working drawings and technical illustrations. Working drawings are visual instructions for fabricating products and erecting structures in all fields of engineering. Technical illustration, the most artistic area of engineering design graphics, visually depicts projects and products for operations manuals and presentations.

In addition, drafters prepare maps, geological sections, and plats from data given them by engineers, geologists, and surveyors. These drawings show the locations of property lines, physical features, strata, rights-of-way, building sites, bridges, dams, mines, and utility lines.

The three levels of certification for drafters are drafters, design drafters, and engineering designers:

- Drafters are graduates of a two-year,

1.20 Petroleum engineers will be responsible for projects of this type which will extract gas from 5,300 feet below sea level in the Gulf of Mexico. (Courtesy FMC Corporation.)

1.21 The drafter of today may use an Apollo workstation such as this one for documentation and design needs. (Courtesy of Hewlett Packard Company.)

post–high school curriculum in engineering design graphics.

- Design drafters complete two-year programs at an approved junior college or technical institute.

- Engineering designers are graduates of a four-year college course in engineering design graphics who can become certified as technologists.

Industry has embraced computer graphics as a way to improve drafters' productivity. However, the use of computer graphics systems does not lessen the need for a knowledge of graphics. The principles of graphics remain the same; only the medium—the computer—is different (**Fig. 1.21**).

Problems

1. Write a report that outlines the specific duties of and relationships among the scientist, engineer, technologist, technician, craftsperson, designer, and stylist in an engineering field of your choice. For example, explain this relationship for an engineering team involved in an aspect of civil engineering. Your report should be supported by factual information obtained from interviews, brochures, or library references.

2. Investigate and write a report on the employment opportunities, job requirements, professional challenges, and activities of your chosen branch of engineering or technology. Illustrate this report with charts and graphs, where possible, for easy interpretation. Compare your personal abilities and interests with those required by the profession.

3. Arrange a personal interview with a practicing engineer, technologist, or technician in your field of interest. Discuss with that person the general duties and responsibilities of the position to gain a better understanding of this field. Summarize your interview in a written report.

4. Write to the professional society in your field of study for information about it. Prepare a notebook of these materials for easy reference. Include in the notebook a list of books that provide career information for that field.

5. Write a one-page report that gives your reasons for selecting engineering or technology as your field of study. List the type of work that you visualize yourself doing upon graduation, where you will work, and for whom. Also, list your personal attributes that you believe will make you successful in this career. Summarize by explaining why your career selection is a good one.

6. Have you ever considered self-employment? There are many rewards for the person who "invents a new mousetrap" and has the commitment to independently bring it into being, as Dr. Land did with the Land Polaroid camera. Write a one-page report listing the advantages and disadvantages of being self-employed with employees on your payroll. Summarize by assessing the "fit" of your personal attributes to being your own boss.

7. If you were to graduate with an engineering or technology degree today, you would probably look for a job where most graduates go—manufacturing companies, government branches, and large, well-established companies in industry. Make a list of the various options that you visualize yourself as having because of your technical training, and give a brief explanation of each. Try to think of as many unique and nonconventional opportunities as you can. Limit your best list to no more than one page.

Addresses of Professional Societies
Publications and information from these societies were used in preparing this chapter.

American Ceramic Society
65 Ceramic Drive, Columbus, OH 43214

American Institute of Aeronautics and Astronautics
1290 Avenue of the Americas, New York, NY 10019

American Institute of Chemical Engineers
345 East 47th Street, New York, NY 10017

American Institute for Design and Drafting
3119 Price Road, Bartlesville, OK 74003

American Institute of Industrial Engineers
345 East 47th Street, New York, NY 10017

American Institute of Mining, Metallurgical and Petroleum Engineering
345 East 47th Street, New York, NY 10017

American Nuclear Society
555 N. Kensington Avenue
LaGrange Park, IL 60526

American Society of Agricultural Engineers
2950 Niles Road, St. Joseph, MI 49085

American Society of Civil Engineers
1801 Alexander Bell Drive
Reston, VA 20191

American Society for Engineering Education
11 DuPont Circle, Suite 200, Washington, DC 20036

American Society of Mechanical Engineers
345 East 47th Street, New York, NY 10017

Institute of Electrical and Electronic Engineers
444 Hoes Lane, PO Box 459
Piscataway, NJ 08855

National Society of Professional Engineers
1420 King Street, Alexandria, VA 22314

Society of Petroleum Engineers
P.O. Box 833836, Richardson, TX 75083

Society of Women Engineers
120 Wall Street, 11th Floor, New York, NY 10005

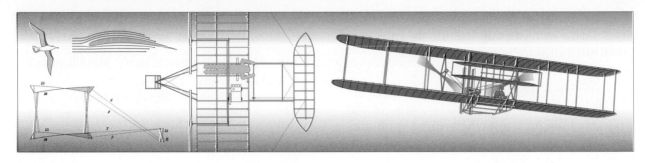

The Design Process

2.1 Introduction

The design process is the method of devising innovative solutions to problems that will result in new products or systems. Engineering graphics is the primary medium of design that is used for developing designs from initial concepts to final working drawings. Initially, a design consists of sketches that are refined,

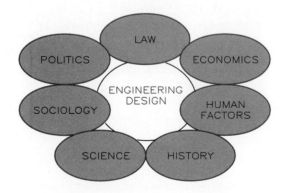

2.1 An engineering project may involve the interaction of people representing many professions and interests, with engineering design as the central function.

analyzed, and developed into precise detail drawings and specifications. They, in turn, become part of the contract documents for the parties involved in funding and implementing a project.

At first glance, the solution of a design problem may appear to involve merely the recognition of a need and the application of effort toward its solution, but most engineering designs are more complex than that. The engineering and design efforts may be the easiest parts of a project.

For example, engineers who develop roadway systems must deal with constraints such as ordinances, historical data, human factors, social considerations, scientific principles, budgeting, and politics (**Fig. 2.1**). Engineers can readily design driving surfaces, drainage systems, overpasses, and other components of the system. However, adherence to budgetary limitations is essential, and funding is closely related to politics on public projects.

Traffic laws, zoning ordinances, environmental impact statements, right-of-way acquisition, and liability clearances are legal aspects of roadway design that engineers must consider. Past trends, historical data, human factors (including driver characteristics), and safety features affecting the function of the traffic system must be analyzed. Social problems may arise if proposed roadways will be heavily traveled and will attract commercial development such as shopping centers, fast-food outlets, and service stations. Finally, designers must apply engineering principles developed through research and experience to obtain durable roads, economical bridges, and fully functional systems.

2.2 Types of Design Problems

Most design problems fall into one of two categories: **product design** and **systems design**.

Product Design

Product design is the creation, testing, and manufacture of an item that usually will be mass produced, such as an appliance, a tool, or a toy, or larger products such as automobiles (**Fig. 2.2**). In general, a product must have sufficiently broad appeal for meeting a specific need and performing an independent function to warrant its production in quantity. Designers of products must respond to current market needs, production costs, func-

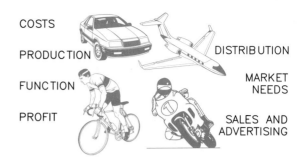

2.3   These are some of the major factors that must be considered when developing a product design.

tion, sales, distribution methods, and profit predictions (**Fig. 2.3**).

Products can perform one or many functions. For instance, the primary function of an automobile is to provide transportation, but it also contains products that provide communications, illumination, comfort, entertainment, and safety. Because it is mass produced for a large consumer market and can be purchased as a unit, the automobile is regarded as a product. However, because it consists of many products that perform various functions, the automobile also is a system.

Systems Design

Systems design combines products and their components into a unique arrangement and provides a method for their operation. A residential building is a system of products consisting of heating and cooling, plumbing, natural gas, electrical power, sewage, appliances, entertainment, and other features that form the overall system, as shown in **Fig. 2.4**.

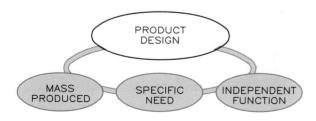

2.2   Product design seeks to develop a product that meets a specific need, that can function independently, and that can be mass-produced.

2.4   The typical residence is a system composed of many components and products.

Systems Design Example

Suppose that you were carrying luggage to a faraway gate in an airport terminal. It would be easy for you to recognize the need for a luggage cart that you could use and then leave behind for others to use. If you have this need, then others do too. The identification of this need could prompt you to design a cart like the one shown in **Fig. 2.5** to hold luggage, and even a child, and to make it available to travelers. The cart is a product.

How could you earn a profit from providing such a cart? First you would need a device to hold the carts and dispense them to customers, as shown in **Fig. 2.6**. You would also need a method for users to pay for cart rental, so you would need to design a coin-operated gate for releasing them (**Fig. 2.7**).

As an added customer convenience, you could provide a vending machine that would take bills, make change, and issue cards entitling customers to multiple use of carts at this and other terminals (see Fig. 2.7). A mechanism to encourage customers to return carts to another, conveniently located dispensing unit would be helpful and efficient. A coin dispenser (see Fig. 2.6) that gives a partial refund on the rental fee to customers returning carts to a dispensing unit at their destination might work.

The combination of these products and the method of using them is a system design.

2.6 Luggage carts are dispensed from coin-operated centers at major entrances of airport terminals. (Courtesy of Smarte Carte.)

Such a system of products is more valuable than the sum of the products alone.

2.3 The Design Process

Design is the process of creating a product or system to satisfy a set of requirements that has multiple solutions by using any available resources. In essentially all cases, the final design must be completed at a profit or within a budget.

2.5 The luggage cart in an airport terminal is a product and part of a system. (Courtesy of Smarte Carte.)

2.7 The coin-operated gate (left) releases a cart from the cart dispenser. A vending machine (right) makes change for $5, $10, and $20 bills and issues baggage cards that can be used at any other airport terminal having the same cart system. (Courtesy of Smarte Carte.)

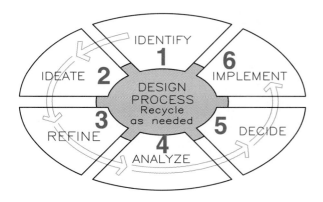

2.8 The design process consists of six steps, each of which can be repeated as necessary.

There are six steps in the design process, as shown in **Fig. 2.8**:

1. Problem identification
2. Preliminary ideas (ideation)
3. Refinement
4. Analysis
5. Decision
6. Implementation

Designers should work sequentially from step to step but should review previous steps periodically and rework them if a new approach comes to mind during the process.

2.4 Problem Identification

Most engineering problems are not clearly defined at the outset and require identification before an attempt is made to solve them (**Fig. 2.9**). For example, air pollution is a concern, but we must identify its causes before we can solve the problem. Is it caused by automobiles, factories, atmospheric conditions that harbor impurities, or geographic features that trap impure atmospheres?

Another example is traffic congestion. When you enter a street where traffic is unusually congested, can you identify the reasons for the congestion? Are there too many cars? Are the traffic signals poorly synchro-

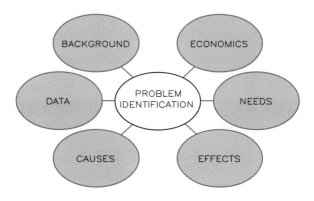

2.9 Problem identification requires that the designer accumulate as much information about a problem as possible before attempting a solution. The designer also should keep product marketing in mind at all times.

nized? Are there visual obstructions? Has an accident blocked traffic?

Problem identification involves much more than simply stating, "We need to eliminate air pollution." We need data of several types: opinion surveys, historical records, personal observations, experimental data, physical measurements from the field, and more. It is important that the designer resist the temptation to begin developing a solution before the identification step has been completed.

Preliminary Ideas

The second step of the design process is the development of as many ideas for problem solution as possible (**Fig. 2.10**). A brainstorming

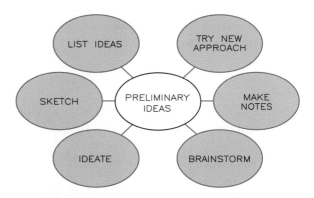

2.10 The designer gathers ideas from a brainstorming session and develops preliminary ideas for problem solution. Ideas should be listed, sketched, and noted.

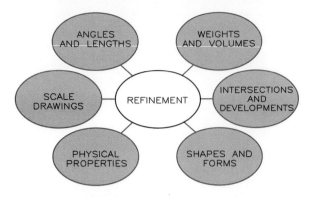

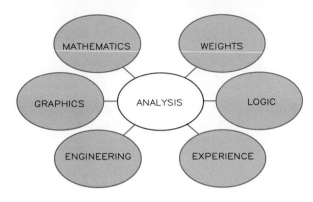

2.11 Refinement begins with the construction of scale drawings of the best preliminary ideas. Descriptive geometry and graphics are used to describe geometric characteristics.

2.12 All available methods, from science to technology to graphics to experience, should be used to analyze a design.

session is a good way at the outset to collect ideas that are highly creative, revolutionary, and even wild. Rough sketches, notes, and comments can capture and preserve preliminary ideas for further refinement. The more ideas, the better at this stage.

Refinement

Several of the better preliminary ideas are selected for refinement to determine their merits. The rough preliminary sketches are converted into scale drawings for spatial analysis, determination of critical measurements, and the calculation of areas and volumes affecting the design (**Fig. 2.11**). Descriptive geometry aids in determining spatial relationships, angles between planes, lengths of structural members, intersections of surfaces and planes, and other geometric relationships.

Analysis

Analysis is the step during which engineering and scientific principles are used most intensively to evaluate the best designs and compare their merits with respect to function, strength, safety, cost, and optimization (**Fig. 2.12**). Graphical methods also play an important role in analysis. Data can be analyzed graphically, forces analyzed as graphical vec-

tors, and empirical data analyzed, integrated, and differentiated by other graphical methods. Analysis is less creative than the previous steps.

Decision

After analysis, a single design, which may be a compromise among several designs, is decided upon as the solution to the problem (**Fig. 2.13**). The designer alone, or a team, may make the decision. The outstanding aspects of each design usually lend themselves to graphical comparisons of manufacturing costs, weights, operational characteristics, and other data essential in decision making.

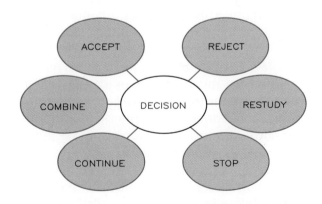

2.13 Decision involves the selection of the best design or design features to implement. This step may require an acceptance, a rejection, or a compromise of the proposed solution.

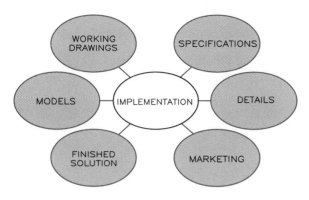

2.14 Implementation involves the preparation of drawings, specifications, and documentation from which the product can be made. The product is produced and marketing is begun.

Implementation

The final design must be described and detailed in working drawings and specifications from which the project will be built, whether it is a computer chip or a suspension bridge (**Fig. 2.14**). Workers must have precise instructions for the manufacture of each component, often measured within thousandths of an inch to ensure proper fabrication assembly. Working drawings must be sufficiently explicit to serve as part of the legal contract with the successful bidder on the job.

2.5 Application of the Design Process

The following example illustrates the application of the design process to a simple problem.

Hanger Bracket Problem

A bracket is needed to support a 2-inch diameter hot-water pipe from a beam or column 9 inches from the mounting surface. This design project is typical of an in-house assignment by an engineer of a company specializing in pipe hangers and supports.

Problem Identification

First, write a statement of the problem and a statement of need (**Fig. 2.15**). List limitations

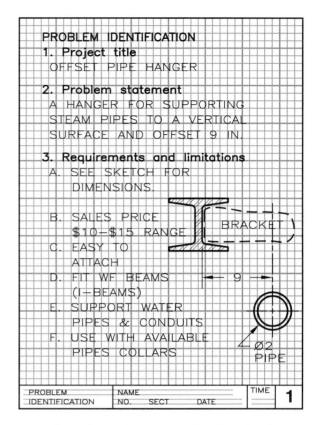

2.15 This worksheet shows aspects of problem identification, the first step of the design process.

and desirable features, and make descriptive sketches to better identify the requirements (**Fig. 2.16**). Even if much of this information may be obvious, writing statements and making sketches about the problem will help you "warm up" to the problem and begin the creative process. Also, you must begin thinking about sales outlets and marketing methods.

Preliminary Ideas

Brainstorm the problem for possible solutions with others, or alone, if necessary. List the ideas obtained on a worksheet (**Fig. 2.17**), and summarize the best ideas and design features on a separate worksheet (**Fig. 2.18**). Then translate and expand these verbal ideas into rapidly drawn freehand sketches (**Fig. 2.19** and **Fig. 2.20**). You should develop as many ideas as possible during this step because a

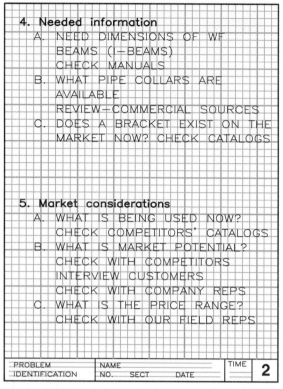

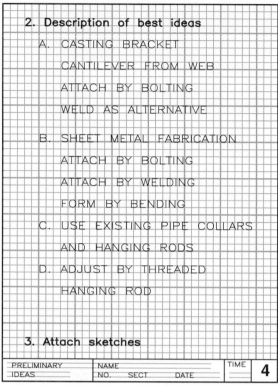

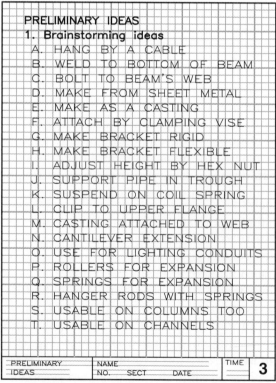

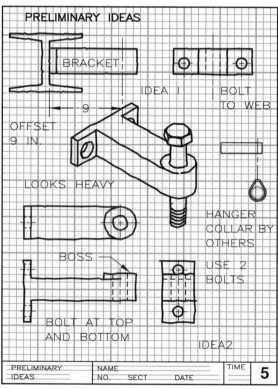

Card 2 — Problem Identification

4. Needed information
   A. NEED DIMENSIONS OF WF BEAMS (I-BEAMS) CHECK MANUALS
   B. WHAT PIPE COLLARS ARE AVAILABLE REVIEW-COMMERCIAL SOURCES
   C. DOES A BRACKET EXIST ON THE MARKET NOW? CHECK CATALOGS

5. Market considerations
   A. WHAT IS BEING USED NOW? CHECK COMPETITORS' CATALOGS
   B. WHAT IS MARKET POTENTIAL? CHECK WITH COMPETITORS INTERVIEW CUSTOMERS CHECK WITH COMPANY REPS
   C. WHAT IS THE PRICE RANGE? CHECK WITH OUR FIELD REPS

PROBLEM IDENTIFICATION · NAME · NO. · SECT · DATE · TIME · **2**

Card 4 — Preliminary Ideas

2. Description of best ideas
   A. CASTING BRACKET
      CANTILEVER FROM WEB
      ATTACH BY BOLTING
      WELD AS ALTERNATIVE
   B. SHEET METAL FABRICATION
      ATTACH BY BOLTING
      ATTACH BY WELDING
      FORM BY BENDING
   C. USE EXISTING PIPE COLLARS
      AND HANGING RODS
   D. ADJUST BY THREADED
      HANGING ROD

3. Attach sketches

PRELIMINARY IDEAS · NAME · NO. · SECT · DATE · TIME · **4**

Card 3 — Preliminary Ideas

PRELIMINARY IDEAS
1. Brainstorming ideas
   A. HANG BY A CABLE
   B. WELD TO BOTTOM OF BEAM
   C. BOLT TO BEAM'S WEB
   D. MAKE FROM SHEET METAL
   E. MAKE AS A CASTING
   F. ATTACH BY CLAMPING VISE
   G. MAKE BRACKET RIGID
   H. MAKE BRACKET FLEXIBLE
   I. ADJUST HEIGHT BY HEX NUT
   J. SUPPORT PIPE IN TROUGH
   K. SUSPEND ON COIL SPRING
   L. CLIP TO UPPER FLANGE
   M. CASTING ATTACHED TO WEB
   N. CANTILEVER EXTENSION
   O. USE FOR LIGHTING CONDUITS
   P. ROLLERS FOR EXPANSION
   Q. SPRINGS FOR EXPANSION
   R. HANGER RODS WITH SPRINGS
   S. USABLE ON COLUMNS TOO
   T. USABLE ON CHANNELS

PRELIMINARY IDEAS · NAME · NO. · SECT · DATE · TIME · **3**

Card 5 — Preliminary Ideas

PRELIMINARY IDEAS

BRACKET · IDEA 1 · BOLT TO WEB
OFFSET 9 IN. · 9
LOOKS HEAVY
BOSS
BOLT AT TOP AND BOTTOM
HANGER COLLAR BY OTHERS
USE 2 BOLTS
IDEA 2

PRELIMINARY IDEAS · NAME · NO. · SECT · DATE · TIME · **5**

2.16 (Sheet 2) This worksheet shows information needed before the designer can proceed, including sources from which it can be obtained.

2.17 (Sheet 3) A member of the brainstorming team records the ideas from the session.

2.18 (Sheet 4) The best ideas are selected from the brainstorming session to be developed as preliminary ideas.

2.19 (Sheet 5) Preliminary ideas are sketched and noted for further development. This is the most creative step of the design process.

large number of ideas represents a high level of creativity. This is the most creative step of the design process.

Problem Refinement

Describe the design features of one or more preliminary ideas on a worksheet for comparison (**Fig. 2.21**). Draw the better designs to scale in preparation for analysis; you need to

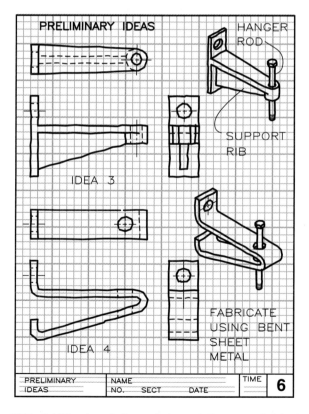

2.20 Additional preliminary design solutions are sketched here.

```
REFINEMENT
1. Description of best ideas
   A. BRACKET WITH RIB
      1. BRACKET MADE AS CASTING
         10-12 EXTENSION FROM BEAM
      2. ATTACHED WITH A SINGLE
         BOLT ON UPPER SIDE
      3. SUPPORTS PIPE WITH HANGER
         ROD—STANDARD PART
      4. STURDY AND STABLE

   B. BENT STRAP BRACKET
      1. MADE OF BENT SHEET METAL
      2. ATTACHED BY A SINGLE
         BOLT OR BY WELDING
      3. SUPPORTS HANGER ROD
         SAME AS CAST BRACKET
      4. ABOUT 2 LBS IN WEIGHT

2. Attach scale drawings
DESIGN          NAME              TIME   7
REFINEMENT      NO.  SECT  DATE
```

2.21 Refinement of preliminary ideas begins with written descriptions of the better ideas.

show only a few dimensions at this stage (**Fig. 2.22**).

Use instrument-drawn orthographic projections, computer drawings, and descriptive geometry to refine the designs and ensure precision. Let's say that you select ideas 3 and 5 for analysis. Figure 2.22 depicts orthographic and auxiliary views of the two designs.

Analysis

Use an analysis worksheet to analyze the cast-iron bracket design (**Figs. 2.23–2.25**). If you are considering more than one solution, analyze each design.

By using the maximum load of 200 lbs and the geometry of the bracket, it is possible to graphically determine the angle of the reaction that the bolt must carry—456 lbs—when the bracket is fully loaded (**Fig. 2.26**). Again, graphics is used as an important design tool.

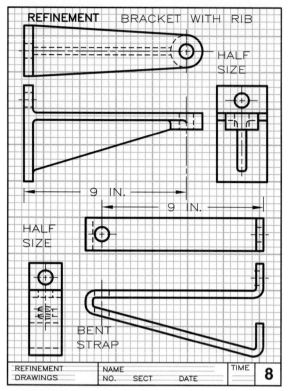

REFINEMENT   BRACKET WITH RIB

HALF
SIZE

9 IN.

9 IN.

HALF
SIZE

BENT
STRAP

| REFINEMENT DRAWINGS | NAME NO.   SECT   DATE | TIME | **8** |

4. **Physical description**
   A. BRACKET: 10–11 IN. LONG
      WITH RIB FOR STRENGTH
   B. ATTACHED WITH A SINGLE
      BOLT AT UPPER SIDE
   C. WEIGHT: ABOUT 3–4 LBS
   D. BOSS FOR EXTRA STRENGTH
      WHERE HANGER ROD ATTACHES
   E. FORMED AS A CASTING
      CAST IRON

5. **Strength**
   A. WILL SUPPORT ABOUT 200 LBS
      WITH A SAFETY FACTOR OF 5
   B. 1/2 BOLT SUFFICIENT FOR
      ATTACHMENT TO BEAM OR
      OR COLUMN
   C. 3/8 DIA HANGER ROD MORE
      THAN ADEQUATE TO SUPPORT
      200 LBS

| DESIGN ANALYSIS | NAME NO.   SECT   DATE | TIME | **10** |

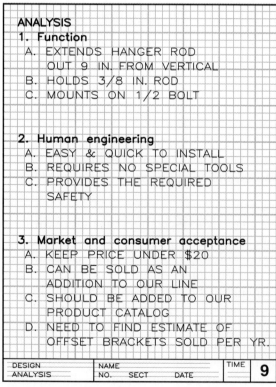

ANALYSIS
1. **Function**
   A. EXTENDS HANGER ROD
      OUT 9 IN. FROM VERTICAL
   B. HOLDS 3/8 IN. ROD
   C. MOUNTS ON 1/2 BOLT

2. **Human engineering**
   A. EASY & QUICK TO INSTALL
   B. REQUIRES NO SPECIAL TOOLS
   C. PROVIDES THE REQUIRED
      SAFETY

3. **Market and consumer acceptance**
   A. KEEP PRICE UNDER $20
   B. CAN BE SOLD AS AN
      ADDITION TO OUR LINE
   C. SHOULD BE ADDED TO OUR
      PRODUCT CATALOG
   D. NEED TO FIND ESTIMATE OF
      OFFSET BRACKETS SOLD PER YR.

| DESIGN ANALYSIS | NAME NO.   SECT   DATE | TIME | **9** |

6. **Production procedures**
   A. FABRICATE AS A CASTING
      USING CAST IRON
   B. DRILL .50 DIA HOLE FOR
      .375 DIA HANGER ROD
   C. DRILL .625 DIA HOLE FOR
      .50 DIA BOLT TO ATTACH
      BRACKET
   D. PAINT TO GIVE RUST–PROOF
      COATING

7. **Economic analysis**
   A. COSTS
      1. DEVELOPMENT $0.20
      2. CAST IRON   .50
      3. CASTING COST   1.40   } 2.50
      4. DRILL 2 HOLES   .20
      5. PAINTING   .20
   B. LABOR   .70
   C. PACKAGING   .30
   D. PROFIT   2.00
   E. WHOLESALE PRICE   6.50
   F. RETAIL PRICE $12.00

| DESIGN ANALYSIS | NAME NO.   SECT   DATE | TIME | **11** |

2.22 (Sheet 8) Scale drawings of two designs are made to describe two designs. Almost no dimensions are needed.

2.23 (Sheet 9) A continuation of the analysis step.

2.24 (Sheet 10) The next two steps of the analysis.

2.25 (Sheet 11) This portion of the analysis focuses on the production and economic considerations.

Decision

The decision table (**Fig. 2.27**) compares two designs: the bracket with a rib and the bent strap. Assign weight factors to be analyzed by assigning points to them that add up to 10. You can then rank the designs by their overall scores from highest to lowest.

Draw conclusions and summarize the features of the recommended design, along with a projection of its marketability (**Fig. 2.28**). In this case, we decide to implement the bracket with a rib.

Implementation

Detail the bracket with a rib design in a working drawing which graphically describes and dimensions each individual part (**Fig. 2.29**).

Use notes to specify standard parts—the nuts, bolts, and hanger rod—but it will be unnecessary to draw them because they will be bought as standard parts. This working drawing shows how the bracket will be made. Now, you need to build a prototype or model of the product and test it for function.

Figure 2.30 shows the final product, the cast-iron bracket with a rib. After determining how it will be packaged and distributed, your next task will be to add it to your company's product line, list it in your catalog, and introduce it to the marketplace.

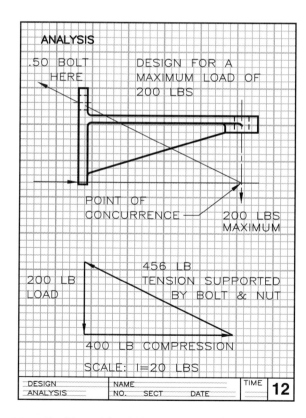

2.26 Graphics and descriptive geometry can be used to determine the load carried by the support bolt.

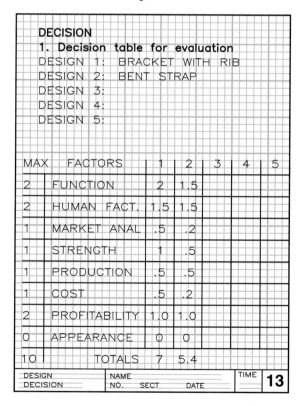

2.27 A decision table is used to evaluate alternative designs in arriving at the final selection.

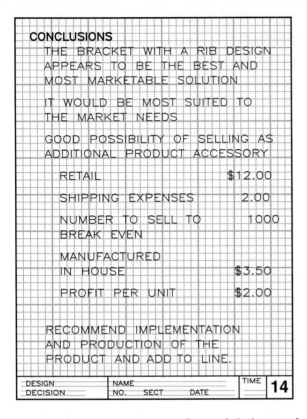

CONCLUSIONS

THE BRACKET WITH A RIB DESIGN
APPEARS TO BE THE BEST AND
MOST MARKETABLE SOLUTION

IT WOULD BE MOST SUITED TO
THE MARKET NEEDS

GOOD POSSIBILITY OF SELLING AS
ADDITIONAL PRODUCT ACCESSORY

RETAIL	$12.00
SHIPPING EXPENSES	2.00
NUMBER TO SELL TO BREAK EVEN	1000
MANUFACTURED IN HOUSE	$3.50
PROFIT PER UNIT	$2.00

RECOMMEND IMPLEMENTATION
AND PRODUCTION OF THE
PRODUCT AND ADD TO LINE.

DESIGN DECISION   NAME   NO.   SECT   DATE   TIME   **14**

2.28 The decision step is summarized to conclude this step of the design process.

Problems

Most of the following problems are to be solved on 8½ × 11 inch paper, using instruments or by sketching freehand as assigned. Use a paper with a 0.20-inch printed grid or plan paper and an engineer's scale to lay out the problems.

Endorse each problem sheet with your name and file number, date, and problem number. Letter all points, lines, and planes with ⅛-inch letters, using guidelines. Letter the answers to essay problems with approved single-stroke Gothic lettering (see Chapter 11).

1. Outline your plan of activities for the weekend. Indicate aspects of your plans that you feel display creativity or imagination. Explain why.

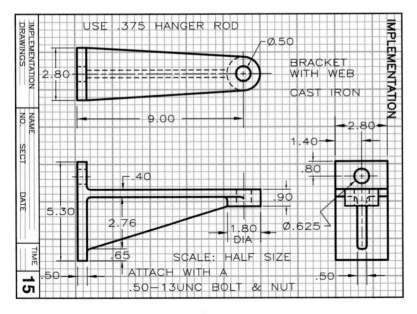

IMPLEMENTATION DRAWINGS   NAME   NO.   SECT   DATE   TIME   **15**

USE .375 HANGER ROD

Ø.50

BRACKET WITH WEB

CAST IRON

2.80

9.00

2.80

1.40

.80

.40

.90

5.30

2.76

.65

1.80 DIA

Ø.625

SCALE: HALF SIZE

.50

.50

ATTACH WITH A
.50—13UNC BOLT & NUT

IMPLEMENTATION

2.29 The completed swing set anchor ready to market.

2.30 The completed bracket is shown here ready to be marketed. (Courtesy of Grinnell Company, Inc.)

2. Write a short report (not to exceed two pages) on the engineer or engineering achievement that you believe exhibits a high degree of creativity. Justify your selection by outlining the creative aspects of your choice.

3. Test your ability to recognize the need for new designs. List as many improvements as you can think of for the typical automobile. Make suggestions for implementing these improvements. Follow the same procedure for another product of your choice.

4. List as many systems as you can that affect your daily life. Separate several of these systems into their components (subsystems).

5. Subdivide the following items into their individual components: (a) a classroom, (b) a wristwatch, (c) a movie theater, (d) an electric motor, (e) a coffee percolator, (f) a golf course, (g) a service station, and (h) a bridge.

6. Indicate which of the items in Problem 5 are systems and which are products. Explain your answers.

7. Make a list of products and systems that you believe would be necessary for life on the moon.

8. You have been assigned the responsibility for organizing and designing a skateboard installation that will be self-supporting. Write a paragraph on each of the six steps of the design process to explain how you would apply each step to the problem. For example, explain the action you would take to identify the problem.

9. Suppose that you are responsible for designing a motorized wheelbarrow to be marketed for home use. Write a paragraph on each of the six steps of the design process, explaining how you would apply each step to the problem. For example, what action would you take to identify the problem?

10. List and explain a sequence of steps that you believe would be adequate for, yet different from, the design process given in this chapter. Your version of the design process may contain any of the steps discussed.

11. Design a simple device for holding a fishing pole in a fishing position while the person fishing rows the boat. Make sketches and notes to describe your design.

12. Design a doorstop to keep a door from slamming into the wall behind it. Make rapid freehand sketches and notes using the six design steps. Do not spend more than 30 minutes on this problem. Indicate any information you would need at the decision and implementation steps that you may not have now.

13. List factors to consider during the problem-identification step for designing (a) a skillet, (b) a bicycle lock, (c) a handle for a piece of luggage, (d) improving your grades, (e) a child's toy, (f) a stadium seat, (g) a desk lamp, (h) an umbrella, and (i) a hot dog stand.

14. Hanger Bracket identification: Make a worksheet that could follow Sheet 2 (Fig. 2.16) to provide additional identification information. For example, how much per linear foot will a 2" XXX pipe weigh (refer to Chapter 35) when it is full of water (refer to the Appendix for the weight of water)? Show your calculations on the worksheet. How far apart should hangers be spaced if each is to carry no more than 200 lbs. with a safety factor of 5 (the capacity to carry 5 times the design load)? Make sketches to clarify this information.

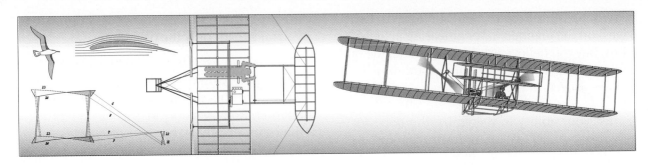

Problem Identification

3.1 Introduction

Problem identification is the initial step that a designer takes to solve a problem. It first involves recognizing a need and then proposing design criteria (**Fig. 3.1**).

Recognizing a need may begin with observation of a problem or a defect that needs to be corrected in an existing product or a sys-

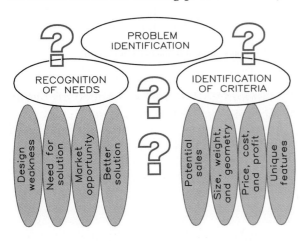

3.1 The two aspects of problem identification.

tem. Or, you may suddenly think about a device or a gadget that no one else has thought about—maybe an improved automobile safety belt, a system allowing full wheelchair access to public buildings, or a new exercise apparatus. The solution may be a product or a system improvement that will increase reliability and profits.

Identifying design criteria follows need recognition. Here, the designer proposes the specifications a new product or system must meet.

3.2 Example: Ladder Attachment

When you use a ladder for house repairs, it works well enough when placed against the roof or fascia at the eaves of the house (**Fig. 3.2**). However, if a gutter is attached to the fascia, the ladder damages the gutter when placed against it. The need is to prevent gutter damage by the ladder.

Now, you must propose design criteria before proceeding with a solution. You need

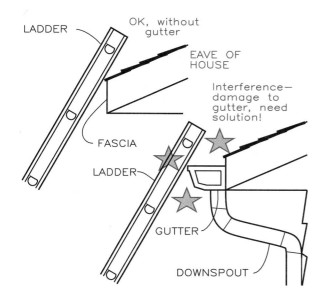

3.2 A ladder placed against the fascia or roof works fine, but when a gutter is attached to the fascia, the ladder damages the gutter. This problem needs a solution.

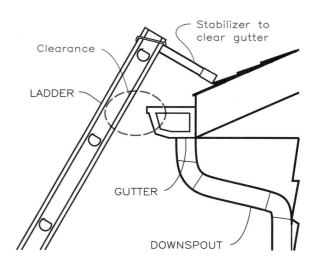

3.3 The problem recognized in Fig. 3.2 can be solved with a stabilizer attached to the ladder, which helps the ladder clear the gutter.

to know the geometry and dimensions of typical gutters, pitches of roofs, angles of ladder placement, and the potential market for the solution. Later in the design process, you might design a product that attaches to the ladder, allowing it to clear the gutter (**Fig. 3.3**). Other solutions are possible, but they all begin with recognizing the need.

3.3 The Identification Process

Problem identification requires the designer to determine requirements, limitations, and other background information before becoming involved in problem solution. The designer should take the following steps during problem identification (**Fig. 3.4**).

1. *Problem statement.* Describe the problem to begin the thinking process.

2. *Requirements.* List the conditions that the design must meet. Most will be questions to be answered after data are gathered.

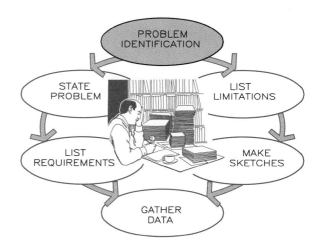

3.4 The major steps of problem identification.

3. *Limitations.* List the factors affecting design specifications, such as maximum weight or size.

4. *Sketches.* Make sketches and add notes and dimensions to identify geometric and physical characteristics.

5. *Data collection.* Gather data on population trends, related designs, physical characteristics, sales records, and market studies.

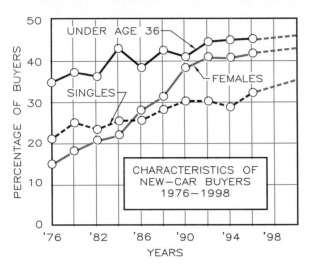

3.5 Data gathered to identify the changing characteristics of new-car buyers can be interpreted more easily when it has been graphed.

Graph the data, as in **Fig. 3.5**, for easier interpretation of raw numbers and data tabulation.

3.4 Design Worksheets

Designers must make numerous notes and sketches on worksheets throughout the design process. Worksheets serve to document what has been done and allow periodic review of earlier ideas to avoid overlooking previously identified concepts. Moreover, a written and visual record of design work helps establish ownership of patentable ideas.

The following materials aid in maintaining permanent records of design activities:

1. *Worksheets* (8½ by 11 inches). Sheets can be either grid-lined or plain and should be three-hole punched for a notebook or binder.

2. *Pencils.* A medium-grade pencil (F or HB) is adequate for most purposes.

3. *Binder* or *envelope.* Keep worksheets in a binder or envelope for reference.

PROBLEM IDENTIFICATION

1. Project title
AN EXERCISE BENCH FOR WEIGHT LIFTERS

2. Problem statement
THERE ARE MANY PEOPLE WHO LIFT WEIGHTS IN ORDER TO KEEP PHYSICAL FITNESS. AN EXERCISE BENCH FOR HOME USE IS NEEDED THAT WILL HELP THEM WITH THEIR EXERCISE REGIMEN.

3. Requirements and Limitations

A. SHOULD BE AFFORDABLE $100-$200 PRICE RANGE
B. SIZE: SHOULD NOT TAKE MORE THAN 3 FT X 5 FT FLOOR SPACE
C. WEIGHT: LESS THAN 50 LBS FOR EASY MOVING
D. COLLAPSIBLE: WOULD BE GOOD IF IT WOULD FOLD UP FOR EASY STORAGE
E. MUST BE HELPFUL IN PERFORMING THE MOST COMMON WEIGHT EXERCISES —MUST DETERMINE FROM LIBRARY OR EXERCISE GYMS
F. MUST BE USABLE WITH THE TYPES OF WEIGHTS USED IN MOST EXERCISES —VISIT EXERCISE GYMS
G. MUST SUPPORT THE EXERCISER AND THE POUNDAGE OF THE WEIGHTS USED —OBSERVE AT EXERCISE GYMS
H. ATTRACTIVE AND COMFORTABLE TO APPEAL TO THE CUSTOMER
I. SUPPORT 275-LB PERSON

| PROBLEM IDENTIFICATION | NAME | | | GRADE | 1 |
| | FILE | SEC | DATE | | |

3.6 The problem-identification step gives the project a title, describes what the product will be used for, and lists basic requirements and limitations for the design.

3.5 Example: Exercise Bench

Design an exercise bench that can be used by those who lift weights for body fitness. This apparatus should be as versatile as possible and at the low end of the price range of exercise equipment. The contents of worksheets shown in **Figs. 3.6–3.9** identify the problem in a typical manner.

First, give the title of the project and a brief problem statement to describe the problem better. Then list requirements and limitations and add sketches as necessary. You will have to list some requirements as questions for the time being, but in all cases make estimates and give sources for the answers (**Fig. 3.6**).

4. Needed information

A. NUMBER OF PEOPLE WHO EXERCISE? —LIBRARY, INTERVIEW GYM OWNERS

B. AGES OF EXERCISERS? OBSERVE GYM; HEALTH MAGAZINES

C. WHAT IS THE POUNDAGE OF THE WEIGHTS USED BY MOST EXERCISERS

D. WHERE DO MOST PEOPLE EXERCISE? HOME OR GYM? —LIBRARY AND INTER—VIEW GYM OPERATORS

E. WHAT FEATURES WOULD BE DESIRABLE ON AN EXERCISE BENCH? —INTERVIEW USERS AT GYM

F. IS HOME EXERCISING DONE INSIDE OR OUTSIDE? —INTERVIEW USERS AT GYM

G. HOW OFTEN DO PEOPLE EXERCISE? —INTERVIEW USERS AT GYM

5. Market considerations

A. SUITABLE MARKET PRICE? —CHECK RETAIL OUTLETS AND CATALOGS FOR COMPETING PRICES

B. HOW MUCH IS SPENT ON EXERCISING AT GYMS? —CHECK YELLOW PAGES TO DETERMINE THE NUMBER OF GYMS AND RELATE POPULATION SIZE

C. BRISKNESS OF SALES FOR EXERCISING BENCHES? —INTERVIEW SALESMEN AT SPORTING GOODS STORES

D. WHAT FEATURES DO BUYERS OF EQUIPMENT LIKE? —QUESTION SALESMEN AT SPORTING STORES

E. IS THERE A VOID IN THE EXERCISE APPARATUS MARKET? —QUESTION SALES—PEOPLE AND INTERVIEW USERS AT GYMS

F. HOW MANY PEOPLE HAVE MEMBERSHIPS AT GYMS? SPOT CHECK SEVERAL GYMS.

PROBLEM IDENTIFICATION | NAME | FILE | SEC | DATE | GRADE | 2

3.7 Continuation of problem identification for the exercise bench includes gathering information on market potential.

PROBLEM IDENTIFICATION

A. CHECK YELLOW PAGES FOR GYMS
—8 GYMS FOR POPULATION OF 100,000
—ONE GYM PER 12,500 PEOPLE

B. GYM MEMBERSHIP FOR STILLMAN'S GYM

YEAR	MEN	WOMEN
1988	75	20
1990	70	36
1992	110	70
1994	175	82
1996	180	120
1998	210	135

C. INTERVIEW RETAILERS OF EQUIPMENT
—3 DEALERS POSITIVE ABOUT BENCH
—2 DEALERS NEUTRAL
—1 DEALER NEGATIVE ABOUT PROSPECTS

D. AGES OF EXERCISERS (BY OBSERVATION)
—20% UNDER 20
—40% BETWEEN 20 AND 30
—30% BETWEEN 30 AND 50
—10% OVER 50

E. PRICES AT STORES

1. COMPLEX MULTI—USE	$1000
2. MEDIUM—RANGE EQUIPMENT	700
3. LIGHTWEIGHT BENCHES	130
4. WEIGHTS	100

F. TYPICAL BRANDS OF EQUIPMENT ON THE MARKET?
1. WEIDER
2. NAUTILUS
3. BODY WONDERFUL
4. HEALTH PLUS

PROBLEM IDENTIFICATION | NAME | FILE | SEC | DATE | GRADE | 3

3.8 This worksheet shows the data collected for use in designing the exercise bench.

Make estimates as you go along; for example, it must cost between \$100 and \$200; it must weigh less than 50 pounds; and it must at least support a person weighing 275 pounds. Give estimates as ranges of prices or weights rather than exact numbers. Use catalogs offering similar products as sources for prices, weights, and sizes.

Next, make a list of the questions that need answers. Follow each question with a source for its answer and give a preliminary answer. How many people exercise? What are their ages? Do they buy exercise equipment? What are the most popular weight exercises? You may obtain this type of information from interviews, product catalogs, the library, and sporting-goods stores. Market considerations include the average income of a typical exerciser. How much does he or she spend on physical fitness per year? The opinions of sporting-goods dealers are helpful, and they should be able to direct you to other sources of information (**Fig. 3.7**).

Record the data that you gather by interviewing gym managers, looking at catalogs, and visiting sporting-goods stores to learn about the people who exercise and the equipment they use (**Fig. 3.8**). Determine the most popular weight exercises of those for whom this bench is to be designed by surveying users, coaches, and sporting-goods outlets. Sketch those exercises on a worksheet (**Fig. 3.9**).

Think about costs and pricing the product even during problem identification. **Figure 3.10** shows how products may be priced from

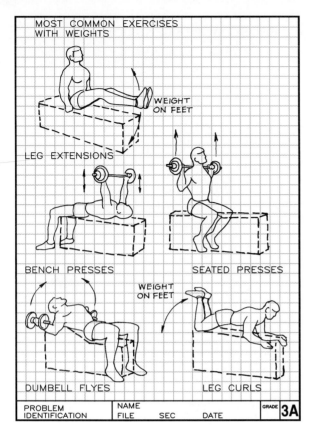

PROBLEM IDENTIFICATION | NAME FILE SEC DATE | GRADE **3A**

3.9 Sketches of the most popular weight exercises.

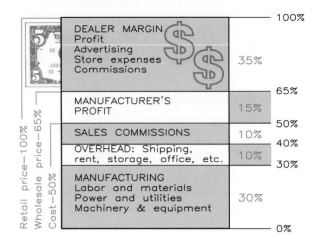

3.10 The breakdown of costs involved in the retail price of a product.

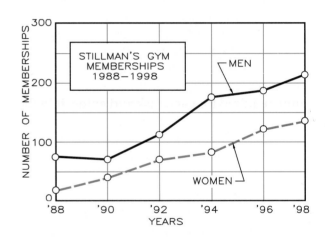

3.11 This graph describes visually the trends in potential customers for the exercise apparatus.

wholesale to retail, but these percentages vary by product. For instance, profit margins are smaller for food sales than furniture sales. If an item retails for $50, the production and overhead cost cannot be more than about $20 to maintain the necessary margins.

Data are easier to interpret if presented graphically (**Fig. 3.11**). Thorough problem identification includes graphs, sketches, and schematics that improve the communication of your findings.

Problem identification is not complete at this point. However, this example should give you a basic understanding of the process.

3.6 Organization of Effort

The designer should prepare a schedule of required design activities after completing problem identification. The **project evaluation and review technique** (**PERT**), developed by managers for projects requiring coordination of many activities, aids in scheduling tasks. PERT is a means of scheduling activities in sequence and reviewing progress made toward their completion.

The **critical path method** (**CPM**) of scheduling evolved from PERT and is used with it. The tasks that must be completed before others can begin are critical and should be given priority. The critical path is the sequence of tasks requiring the longest time from start to finish and has the least flexibility. Activities not in the critical path receive less emphasis.

3.7 Planning Design Activities

Figure 3.12 illustrates the three steps involved in planning a job (project): (1) list the tasks on a form called a Design Schedule and Progress Record; (2) prepare an Activities Network Diagram, arranging the tasks in sequence; and (3) prepare an Activity Sequence Chart that graphs the tasks in the sequence shown in the network.

Design Schedule and Progress Record

The designer separates the job into tasks, numbers them, and lists them in the **Design Schedule and Progress Record** (**DS&PR**) without concern for their sequence (**Fig. 3.13**). Next, the designer estimates the amount of time required for each task and enters it in the third column. The designer adjusts the amount of time for each task so that the total matches the time allotted for the job.

Activities Network

The designer prepares an **Activities Network** (**AN**) by arranging the tasks from the DS&PR in their proper sequence in either note or symbol form (**Fig. 3.14**). The note form identifies activities by task name; the symbol form iden-

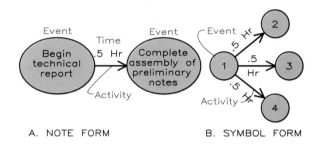

		Est.	Act.	PERCENT COMPLETE
JOB	ASSIGNMENT	Hrs.	Hrs.	0 20 40 60 80
7	WRITE REPORT—ALL	15		
6	BRAINSTORM—ALL	.5	1	
3	WRITE MFGRS—JHE	2		
10	GRAPH DATA—HLE	2		
5	MARKET ANALYSIS—DL	3	3	
12	COLLECT DATA—TT	2		

Design Schedule & Progress Record

TEAM __4__ PROJECT __PRODUCT DESIGN__

WORK PERIODS __II__ MAN HOURS __70__ FINISHING DATE __4−15__

3.13 The Design Schedule and Progress Record (DS&PR) shows typical entries for a product design project.

3.14 Two ways of preparing an Activities Network are (A) the note form and (B) the symbol form, both of which graphically arrange activities in sequence.

tifies activities by task number. Arrows connect activities, showing their sequence. Labels on the arrows indicate the amount of time needed to complete the activities. For example, the first activity in preparing the technical report is to assemble preliminary notes, which requires 0.5 hour (Fig. 3.14A).

Dummy activities simply indicate connections between activities that involve no work and no time, because activities must be connected in the network. In **Fig. 3.15** two dummy activities (dashed lines) show sequential connections but no expenditure of time. In other words, the project data and the preliminary sketches and diagrams that were assembled were not needed to complete the draft of Chapter 1 but are available for later use in the project.

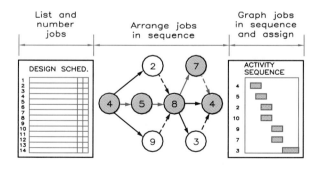

3.12 Planning and scheduling a project involves three steps.

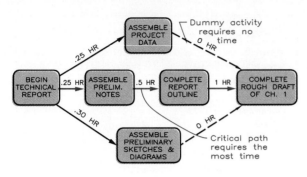

3.15 The critical path is the path from start to end of the project that requires the most time; a portion of the product design project is shown here.

Activity Sequence Chart

JOB	MEMBER	PROJECT HOURS
		0 2 4 6 8 10 12 14 16
7	ALL	
6	SMITH	
3	PRISK	
10	REED	
5	ALL	
12	FLYNN	

3.16 The activity sequence chart (ASC) contains tasks, assignments, and initial time estimates from the DS&PR and activity sequence from the Activity Network.

The critical path identified in the Activities Network (Fig. 3.15) is the longest time path from the project's beginning to its end. For the partial Activities Network shown, 1.75 hours are required to complete the draft of Chapter 1 of the technical report.

Activity Sequence Chart

The designer next lists the activities on the Activity Sequence Chart (ASC) (**Fig. 3.16**), which shows task numbers and team member assignments from the DS&PR. A bar graph shows the sequence of and the time allocated to each task. For example, 0.5 hour is scheduled for brainstorming, or task 2, from the

DS&PR and is to be done first. The ASC also is the overall project schedule.

As work progresses, the designer graphs the status of each activity on the DS&PR (Fig. 3.13). The actual hours required to complete tasks appear in column 4, for easy comparison with the original time estimates. When the extra time required becomes greater than that scheduled, the designer has to make adjustments in the ASC.

Problems

Problems should be presented on 8½-by-11-inch paper, grid or plain. All notes, sketches, drawings, and graphs should be neat and accurate. Written matter should be typed or lettered on paper with ⅛-in. guidelines.

General

1. Identify a need for a design solution as a short class assignment requiring less than three hours to complete. Submit a proposal outlining this need and your general plan for a design solution. Limit the written proposal to two pages.

2. You are marooned on an uninhabited island with no tools or supplies. Identify the major problems that you would have to solve. List the factors that you would have to consider before attempting a solution for each problem. For example: There is the need for food. Determine (a) available sources of food on the island, (b) methods of gathering/catching, (c) methods of storing a food supply, and (d) method of cooking.

3. While you were walking to class today, what irritants or discomforts did you recognize? Were the sidewalks too narrow? Were the entrances to the buildings inconvenient?

Using worksheets, identify the cause of these problems, write a problem statement (including need recognition), and list the requirements for and limitations on solutions.

4. Repeat Problem 3, but use irritants or discomforts found in your (a) living quarters, (b) classroom, (c) recreation facilities, (d) dining facilities, or (e) another environment with which you are familiar.

Product Design Problems

5. Reconstruct the designer's approach to development of the tab-opening can. Even though the problem has been solved and its solution marketed, identify the need for the product and propose specifications for it. Is the existing solution the most appropriate one, or does your problem identification statement suggest others? Explain your answer.

6. Repeat Problem 5 for a travel iron used to press clothes.

7. List the problems involved in designing a motorized wheelbarrow.

8. Suppose that you recognize the need for a device that could be attached to a bicycle to allow it to be ridden over street curbs to sidewalk level. Identify the problems involved in determining the marketability of such a device.

9. Identify the problems you might encounter in developing a portable engineering travel kit to give engineers the capability of making engineering calculations, notes, sketches, and drawings. The kit might include a carrying case, calculator, computer, drawing instruments, paper, reference material, and other accessories.

10. Use Section 3.7 as a guide and prepare a Design Schedule & Progress Record, an Activities Network, and an Activity Sequence Chart for the portion of your design project as assigned below:

   a. An overview of the entire project as you anticipate it at the present.

   b. The first three steps of the design process.

   c. The problem identification step only. Use as many sheets as necessary.

11. Develop your ability to make educated guesses and estimates regarding your world. For example, can you guess how many McDonald's hamburger stores there are in a neighboring city? If you looked in the yellow pages of your town, counted the McDonald's there, divided by the local population to find the population per store, would this factor help you with your estimate? Check a phone book from the neighboring city to see.

Using this approach, go to your library, look in the yellow pages, and determine the business outlets per 1000 of population for any of the following categories: Banks, gas stations, movie theaters, bookstores, fitness centers, restaurants, department stores, or other businesses that are of interest to you.

Learning to make intelligent guesses will help you be a better engineer and entrepreneur.

12. Pricing products is similar to making estimates covered in Problem 11. Use common sense and instincts (that you may not realize you have) to explain why mark-up and profit margins vary among various types of products. Give your explanation in a brief outline form that will make your key points easy to process by the reader.

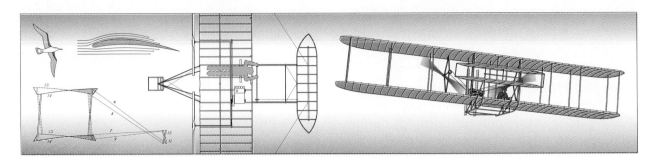

4

Preliminary Ideas

4.1 Introduction

Creativity is highest during the preliminary idea step of the design process because there are no limitations on being innovative, experimental, and daring. During subsequent steps of the design process, freedom of creativity diminishes, and the need for information and facts increases. **Figure 4.1** shows this relationship between creativity and information accumulation during the design process.

An example of a creative solution to a problem is the wearable data terminal (**Fig. 4.2**), which utilizes electronic technology in a unique manner. The device is an optical scanner that reads bar codes, worn on the forearm of the user. A mirror on the shoulder-mounted model reflects the data from screen to user.

Figure 4.3 illustrates a less technical product—the instant coffee spoon—with a unique design that prompts the question, "Why didn't I think of that?" A notable example of a very simple but highly successful design is the Post-it, the "sticky" removable note that was developed by the 3M Corporation.

Another advanced design is the prototype voice/data/fax mobile terminal (**Fig. 4.4**), which is being developed for trucking and transportation companies. A short time ago, these ideas were considered futuristic

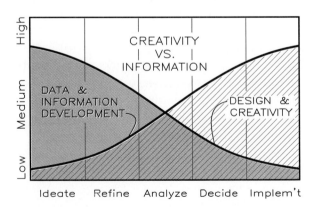

4.1 Creativity is highest during the early stages of the design process (excluding problem identification), and information accumulation increases as the process continues.

36

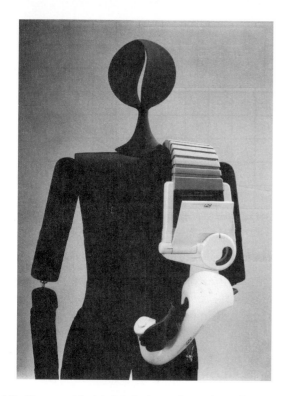

4.2 The wearable data terminal scans bar codes and transmits data to the screen of the shoulder unit and is reflected onto a mirror, permitting the user to see the reading. (Courtesy of NEC U.S. Communications Office.)

fantasies. Progress in the future will be limited only by our imagination in applying technology.

4.2 Individual Versus Team Methods

Designers work both as individuals and as members of design teams during the process of developing products. Both methods have their advantages and disadvantages.

Individual Approach

Designers working alone must make sketches and notes to communicate first with themselves and then with others. Their primary goal is to generate as many ideas as possible because better ideas are more likely to come from long lists than from short lists. Rapidly drawn sketches can capture fleeting thoughts that might otherwise be lost during ideation.

4.3 A simple but unique design that solves several problems is the coffee spoon: it serves as a container for the instant coffee and also functions as a spoon. (Courtesy of Aluminum Company of America.)

4.4 This voice/data/fax mobile terminal that operates from satellite signals is being developed for trucking and transportation, utility, and resource companies. (Courtesy of BCE, Inc.)

Team Approach

The team approach brings diversity and a broad range of ideas to the design process, but along with it comes problems of management and coordination. Groups perform better with a leader with the authority to guide their activities and make assignments.

Teams should alternate between individual and group work to take advantage of both approaches. For example, each team member could individually develop preliminary ideas, bring them to a team meeting, compare

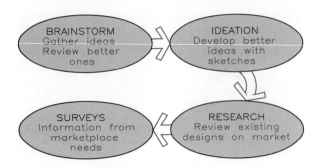

4.5 A plan of action is needed for the preliminary ideas step of the design process, which will probably involve most of the steps shown here.

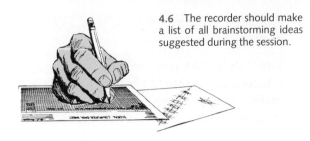

4.6 The recorder should make a list of all brainstorming ideas suggested during the session.

solutions, merge ideas, and return to individual work with a renewed outlook.

4.3 Plan of Action

Figure 4.5 illustrates the four steps for completing the preliminary ideas step of the design process: (1) hold brainstorming sessions, (2) begin ideation through sketches and notes, (3) research background data, and (4) conduct surveys. Periodically reviewing notes and worksheets made during problem identification ensures that efforts will stay focused on the design objectives.

4.4 Brainstorming

Brainstorming is a problem-solving technique in which members of a group spontaneously contribute ideas. The best session is one that adheres to the established rules.

Rules of Brainstorming

The following guidelines can be used for any brainstorming session:*

1. *Criticism is ruled out.* Judgment of ideas must be withheld until later.

2. *Freewheeling is encouraged.* The wilder the idea, the better; it is easier to tame down than to think up. A good solution may

*From Alex Osborn, *Applied Imagination.* New York: Scribner, 1963.

emerge from a suggestion that was made as a joke.

3. *Quantity is wanted.* The larger the number of ideas, the greater will be the likelihood of useful ideas.

4. *Combination and improvement are sought.* Participants should seek ways of improving the ideas of others.

Organization of a Brainstorming Session

The organization of a brainstorming session involves selecting the panel, becoming familiar with the problem, selecting a moderator and recorder, holding the session, and following up.

1. *Panel selection.* The optimum number of participants in a brainstorming session is 12 people. For diversity, they should be people both with and without knowledge of the subject. Because supervisors may restrict the flow of ideas, panels should be composed of nonsupervisory professionals of similar status.

2. *Preliminary work.* An information sheet about the session should be given to panel members a couple of days before the session to allow ideas to incubate.

3. *The problem.* The problem to be brainstormed should be defined concisely. Instead of presenting the problem as "how to improve our campus," it should be presented by category, such as "how to

improve student parking," or "how to improve food services."

4. *The moderator and recorder.* The panel selects a moderator to be in charge of the session and a recorder to keep track of the ideas presented (**Fig. 4.6**).

5. *The session.* The moderator introduces the problem and recognizes the first member holding up a hand to respond. The person responding should state an idea as briefly as possible. The moderator then recognizes the next person holding up a hand, and the process continues. A suggestion made by one member often stimulates ideas in others, who snap their fingers to signify that they want to "hitchhike" on the previous idea. This interaction is central to a brainstorming session. The moderator's most important job is to keep the ideas flowing and to prevent participants from giving long, drawn-out responses that dampen the spontaneous and "fun" aspects of the session.

6. *Length.* A session should move at a brisk pace and end when ideas slow to an unproductive rate. An effective session can last from a few minutes to an hour, but 20 minutes is considered about the best length.

7. *Follow-up.* The recorder should reproduce the list of ideas gathered during the session and distribute them to the participants. As many as a hundred ideas may be gathered during a 20-minute session. The designer should pare them down to those having the most merit.

4.5 Sketching and Notes

Sketching is an effective medium of design, its development, and its communication whether it is a simple part or a complex product (**Fig.**

4.7 Designers use sketches and notes to develop preliminary ideas and communicate with themselves and others. (Courtesy of Ford Motor Company.)

4.7). There are many instances where modification of designs and their descriptions are almost impossible without the ability to make simple and rapid freehand sketches. Sketching allows the designer's ideas to take form as three-dimensional pictorials or as two-dimensional views that are visual extensions of the thinking process. The sketch shown in **Fig. 4.8** was one of many used to develop the styling features of an automobile, an advanced application. **Figure 4.9** shows one of many sketches used in the design of the various components and systems of the space program, one of the most advanced design programs in history.

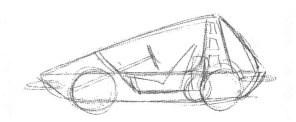

4.8 This simple sketch was made within a couple of minutes by a designer as a preliminary idea for the development of an automobile design. (Courtesy of American Motors Corporation.)

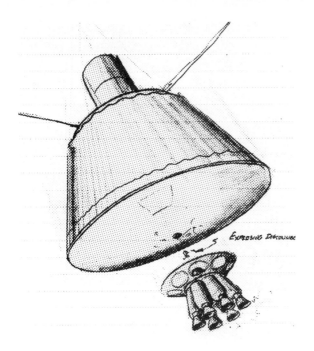

4.9 Sketches are important to the development of solutions to the most complex of engineering and scientific designs. This sketch illustrates a preliminary idea for a manned space capsule. (Courtesy of the National Aeronautics and Space Administration.)

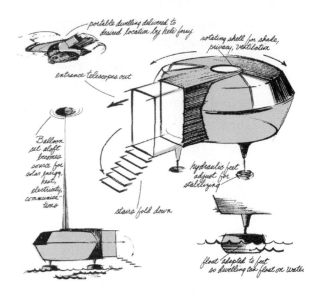

4.10 These preliminary sketches and notes depict a transportable uni-lodge, a mobile dwelling of the future. (Courtesy of Lippincott and Margulies, Inc., and Charles Bruning Company.)

The preliminary ideas for the transportable uni-lodge (**Fig. 4.10**) illustrate the importance of sketches in developing and communicating ideas. These ideas propose that the uni-lodge be transported by helicopter to previously unreachable areas and lowered onto the site or the water. Its retractable legs can be equipped with pontoons, allowing it to float. Additional features are noted on the sketches.

Another concept is a self-contained pipelayer (**Fig. 4.11**) for laying irrigation pipe in desert areas. The tractor has a cab, sleeping accommodations, radio equipment, power plant, and bulk storage tanks for plastic. The van consists of an extrusion machine, a refrigeration unit, and a control station. The pipelayer transports bulk plastic and machinery that can extrude and lay approximately two miles of pipe from each pair of storage tanks. Empty tanks are discarded and

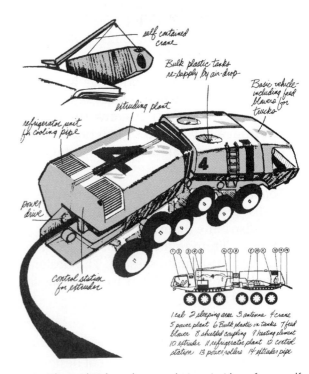

4.11 These sketches show a designer's ideas for a self-contained pipelayer for laying irrigation pipe in large tracts of desert areas. (Courtesy of Donald Desky Associates, Inc., and Charles Bruning Company.)

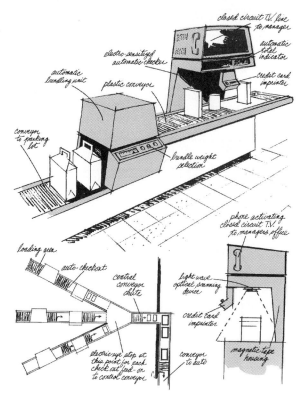

4.12 These sketches describe a concept for an automatic checkout/packaging unit and for a supermarket. (Courtesy of Lester Beall, Inc., and Charles Bruning Company.)

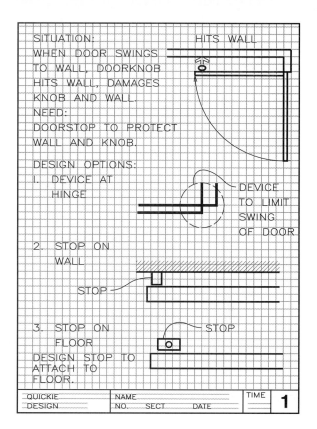

4.13 This worksheet of sketches and notes identifies the need for a device to prevent damage caused by a door swinging into a wall.

replaced by full tanks that are air-dropped to the crew.

Figure 4.12 shows a sketch of a checkout system for a grocery store, a type that is presently in use. The shopper sets the machine in operation by inserting a credit card in a slot. After the card is scanned, the customer receives an order number tag, and the conveyor moves the items under a scanner and totals the prices. If a question arises, the customer can stop the machine and talk to a store employee by lifting the phone. The conveyor moves the items to a unit where they are packaged in plastic containers and marked with the customer's number. A central conveyor transports large orders to an exterior pickup point near the customer's parking space.

The concepts in these examples could not have been developed without the use of sketching as a medium of design and as a means of thinking. Design sketches are fascinating to study whether done by Leonardo or a student in a classroom. Design sketches capture a person's thought process, imagination, and creativity in a manner that can be shared by others.

4.6 Quickie Design

Let's assume that we recognize the need for a product to protect walls and doorknobs where doors swing into walls. First, we make sketches on a worksheet to identify the problem and its geometry (**Fig. 4.13**). If we decide to develop a floor-mounted doorstop, we

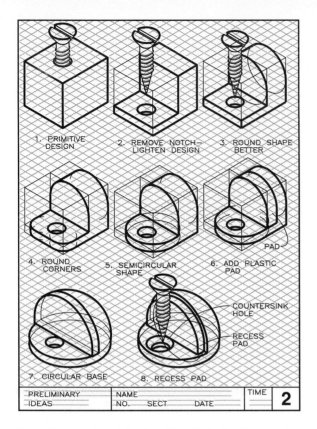

4.14 A series of worksheet sketches illustrating the evolution of a doorstop design.

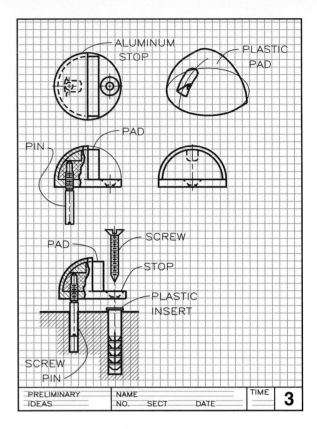

4.15 We used detailed sketches to refine the design and make it easily understandable for others.

might begin with sketches (made by hand or by computer) of a design that begins as basic block, which will do the job (**Fig. 4.14**). However, we can do better than that; we make additional sketches to refine and develop the design until we find a suitable and marketable solution.

The sketches in **Fig. 4.15** further develop the doorstop's details and can be shown to others for consultation and evaluation. **Figure 4.16** shows solutions to this problem that are products already being marketed. These products were developed in a similar manner by designers using graphics as the medium of design.

4.7 Background Information

One way of gathering preliminary ideas is to look for existing products and designs that are similar to the one being considered. Possible sources of background information include magazines, patents, and consultants (**Fig. 4.17**).

Magazines

Articles in both general and technical magazines often present unique designs, complete with drawings and photographs. Advertisements in these magazines may give helpful information on materials and innovative devices. There are various magazines that specialize in most product areas that your librarian can help you find as a reference.

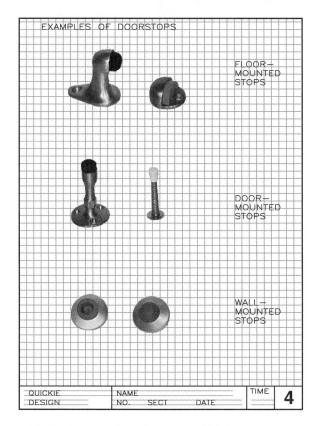

FLOOR—MOUNTED STOPS

DOOR—MOUNTED STOPS

WALL—MOUNTED STOPS

QUICKIE DESIGN	NAME			TIME	4
	NO.	SECT	DATE		

4.16 The doorstops shown here are available in today's market.

Consultants

Background data

Questionnaires

Surveys

Patents

Magazines

Interviews

4.17 The gathering of background information is helpful and necessary to the development of preliminary ideas.

Patents

Patents from the U.S. Patent Office (at $1.50 each) illustrate the details of all designs that were granted a patent. The designer can use them to understand competing products better and to ensure that there is no infringement on existing patents.

Consultants

Complex projects may require specialists in manufacturing, electronics, materials, and instrumentation. Manufacturers' representatives also provide valuable assistance with projects related to their companies' products.

4.8 Opinion Surveys

Designers need to know the attitudes of consumers about a new product at the prelimi-

nary design stage. Is there a need for the product? Are consumers excited about the prospects for a particular product being designed? Would they buy the product? What features do they like or dislike? What price range would be acceptable to them? What do retailers think it will sell for? Does size, weight, or color matter?

For a survey to be of value, the population for whom the product is being developed should be identified. Are they homeowners, high-school-age males, or single women? A selected sample of the population can be surveyed by a personal interview, telephone survey, or mail questionnaire.

The Personal Interview

A personal interview survey should be organized to obtain and summarize reliable

```
PRELIMINARY IDEAS

1. Brainstorming Ideas
   A. FEATURES
      1. HANG FROM CEILING
      2. CONVERT INTO A PIECE OF FURNITURE
      3. PADDED BENCH—INFLATED PAD—FOAM
         PAD, WATER—BED PAD
      4. USE SPRINGS, NOT WEIGHTS
      5. USE WATER FOR WEIGHTS
      6. FOLD OUT FROM WALL
      7. ADJUSTABLE HEIGHTS
      8. SAFETY DEVICE TO KEEP WEIGHTS FROM
         FALLING ON EXERCISER
      9. MOUNT ON WHEELS—MOUNT ON PADS
     10. MAKE WEIGHTS VARIABLE

   B. WHERE TO USE
      1. USE IN GARAGE
      2. USE IN BEDROOM, STORE IN CLOSET
      3. USE WHILE WATCHING TV
      4. MAKE FUN TO USE—ADD SOUND—USE
         WITH TV TAPE—RECORD ON TV TAPE
      5. USE IN DORM

   C. WHERE TO SELL
      1. DOOR TO DOOR
      2. SELL BY MAIL
      3. AT RETAIL OUTLES
      4. SELL AT EXERCISE GYMS
      5. SELL BY SCOUTS AS FUND RAISER
      6. SELL BY TV MARKETING
      7. SELL BY RADIO ADS
      8. ADVERTISE IN FITNESS MAGAZINES
      9. ADVERTISE IN GENERAL MAGAZINES
     10. RECOMMENDATIONS BY DOCTORS

PRELIMINARY     NAME                           GRADE
IDEAS           FILE    SEC    DATE                  4A
```

4.18 Ideas gathered during a brainstorming session are listed on a worksheet for future reference.

responses quickly and easily. For this reason, questions should be true–false or multiple choice for ease of tabulation.

Interviewers should introduce themselves, explain the purpose of the interview, and ask for permission to proceed. They should make the interview brief, record responses, and thank the interviewee for participating.

The Telephone Interview

If the opinions of the general public are desired, interviewers can talk to people selected from the telephone book. If opinions from a certain group—say, sporting-goods retailers—are needed, the Yellow Pages provide prospective interviewees.

The Mail Questionnaire

An economical method of contacting a large number of people in many locations is by mail questionnaire. To test the suitability of questions to be asked, the questionnaire can be mailed to a small group of people for their response as a test. The final questionnaire should be mailed to at least three times as many people as the number of responses desired. Inserting a self-addressed, stamped envelope will increase responses.

4.9 Preliminary Ideas: Exercise Bench

We introduced the design of an exercise bench in Chapter 3, where it was taken through the problem identification step. Recall that the bench is to be used by those who lift weights to maintain body fitness. This apparatus should be versatile and at the low end of the price range for exercise equipment.

Now, to apply the preliminary ideas step of the design process, begin holding a brainstorming session with team members to generate and record ideas on a worksheet (**Fig. 4.18**). You can then select the better ideas, list their features, and summarize them on a worksheet (**Fig. 4.19**), even if you have more features than you could possibly use in a single design. The reason is to make sure that you do not forget or lose any concepts.

Using rapid freehand drawing techniques, such as orthographic views and three-dimensional pictorials, sketch ideas on worksheets. Note on the drawings any ideas or questions that come to mind while you sketch. Do not erase and modify your sketches; instead, make new sketches that incorporate revisions and modifications. By so doing, you will be able to review your thinking process from idea to idea. You should occasionally go back to previous steps to be sure that usable concepts have not been overlooked.

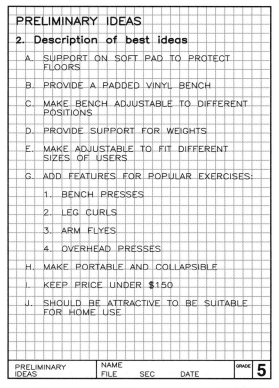

PRELIMINARY IDEAS

2. Description of best ideas

A. SUPPORT ON SOFT PAD TO PROTECT FLOORS

B. PROVIDE A PADDED VINYL BENCH

C. MAKE BENCH ADJUSTABLE TO DIFFERENT POSITIONS

D. PROVIDE SUPPORT FOR WEIGHTS

E. MAKE ADJUSTABLE TO FIT DIFFERENT SIZES OF USERS

G. ADD FEATURES FOR POPULAR EXERCISES:

    1. BENCH PRESSES

    2. LEG CURLS

    3. ARM FLYES

    4. OVERHEAD PRESSES

H. MAKE PORTABLE AND COLLAPSIBLE

I. KEEP PRICE UNDER $150

J. SHOULD BE ATTRACTIVE TO BE SUITABLE FOR HOME USE

PRELIMINARY IDEAS	NAME FILE SEC DATE	GRADE **5**

4.19 (Sheet 5) The best of the brainstorming ideas are selected and summarized for further development.

4.20 (Sheet 6) Preliminary concepts for the exercise bench are shown as sketches and notes.

4.21 (Sheet 7) Sketches and notes describing additional ideas about the exercise bench.

In **Fig. 4.20**, note the adaptation of ideas from various types of benches and exercise techniques. A number identifies each idea. Another worksheet (**Fig. 4.21**) shows other ideas and modifications of previous ideas. Additional sketches can be made of specific details of each design: connections, fabrication details, padding, and so forth.

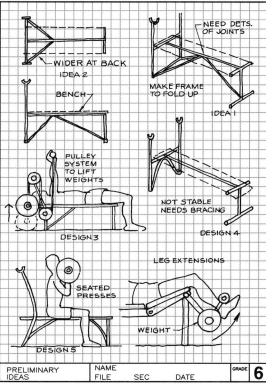

PRELIMINARY IDEAS	NAME FILE SEC DATE	GRADE **6**

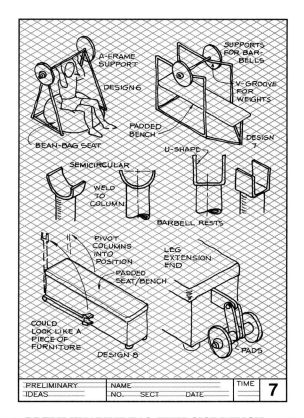

PRELIMINARY IDEAS	NAME NO. SECT DATE	TIME **7**

Problems should be presented on 8½-by-11-inch paper, grid or plain. All notes, sketches, drawings, and graphs should be neat and accurate. Written matter should be legibly lettered, using ⅛-in. guidelines.

1. Select one or more of the following items and list as many of their uses as possible: (a) empty vegetable cans (3 by 5 in. in diameter), (b) 2,000 sheets of 8½-by-11-in. bond paper, (c) 1 cubic yd. of dirt, (d) three empty oil drums (24-in. diameter by 36 in.), (e) a load of egg cartons, (f) 25 bamboo poles (10 ft. long), (g) 10 old tires, or (h) old newspapers.

2. If you were going to select an ideal team to develop an engineering solution to a problem, what characteristics would you look for? Explain.

3. What are the advantages and disadvantages of working independently on a project? Of working as a member of a design team? Explain and give examples of instances in which each approach would have the advantage.

4. Gather background information on one of the following design problems or on one of your own selection: (a) a one-person canoe, (b) a built-in car jack, (c) an automatic blackboard eraser, (d) a built-in coffee maker for an automobile, (e) a self-opening door to permit a pet to leave or enter the house, (f) an emergency fire escape for a two-story building, (g) a rain protector for people attending outdoor spectator activities, (h) a new household appliance, or (i) an at home exerciser. You are concerned with costs, methods of construction, dimensions, existing products, estimates of need, and other information that will assist you in understanding the problem and its feasibility as a project. List your references and present your findings.

5. List and describe the type of consulting services required for the following design projects: (a) a zoning system for a city of 20,000 people, (b) a shopping center, (c) a go-cart operation, (d) a water-purification facility, (e) a hydroelectric system, (f) a nuclear-fallout disaster plan, (g) a processing plant for refining petroleum products, and (h) a drainage system for residential and rural areas.

6. Develop a questionnaire to determine the public's attitude toward a particular product of your selection. Explain how you would tabulate the responses to the questionnaire.

7. Organize a group of classmates, and hold a brainstorming session to identify problems in need of solutions. Make a list of the ideas.

8. With a team of classmates, make a list of items that are in need of "invention." If you can't think of any, make a list of problems and inconveniences that you are unhappy with. Their solution might lead to a new product or system.

9. Make sketches of a toothbrush holder that can sit on a table surface or be attached to the wall. What options would make it most marketable?

10. Identify the features needed for a dog house that could be sold to the mass market. Make sketches of your better ideas. Are there other products for the pet market that you can think of that would be worthy of development?

11. A tensioner that keeps the chain taut that engages one side of the sprocket (the toothed wheel) is shown in **Fig. 4.22**. Tension is applied to the chain by moving the position of the 6 in. DIA sprocket along its shaft by turning the nut on the threaded screw. The entire apparatus must be mounted in a stationary position (either horizontal or vertical) with two bolts, one at each end.

Although some of the parts are not clearly illustrated, make a series of freehand orthographic sketches to describe the parts. Describe them as if you were the designer who is going to explain the design to your drafting department so they can make the finished working drawings of the parts. You will find it helpful to refer to Chapters 13 and 14.

4.23 (Problem 12) A V-belt tensioner with a 6 in. DIA V-pulley (not shown) mounted on a shaft in the movable arm. (Courtesy of the Universal Machine Company.)

4.22 (Problem 11) A chain tensioner with a 6 in. DIA sprocket mounted on a movable shaft. (Courtesy of Brewer Machine and Gear Company.)

14. Design a different device of your own that could be used to sharpen a chisel as well as the one described in Problem 13. Make freehand sketches as specified in Problem 13.

12. Another design for a belt tensioner is shown in **Fig. 4.23**, but the pulley (instead of a sprocket) is not shown. Notice that the base has a slot that is an arc of a circle for adjustment. An additional adjustment can be made by moving the arm to different positions. The V-pulley (6 in. DIA) for a flexible belt will be mounted on the shaft passing through the arm and held on by a collar with a setscrew.

Make freehand orthographic sketches of the individual parts to better explain how they are made and their relationships to each other. Refer to Chapters 13 and 14 as needed.

13. Problems 12 and 13 give different solutions to the same problems that are marketable products. Other products are on the market to serve this same need. Can you think of other solutions? Make freehand sketches of your ideas that could be used to describe your design to a classmate.

4.24 (Problem 13) A chisel-sharpening guide.

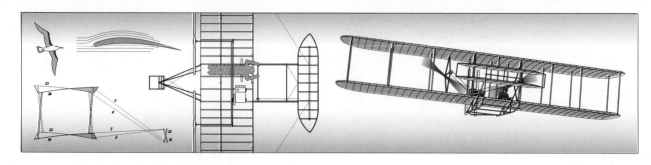

5

Refinement

5.1 Introduction

Refinement of preliminary ideas is the first departure from unrestricted creativity and imagination. The designer must now give primary consideration to function, costs, and practicality.

This step of the design process calls for the designer to make scale drawings with instruments to check dimensions and geometry that cannot be accurately measured in unscaled sketches. However, it is unnecessary to fully dimension these drawings. Descrip-

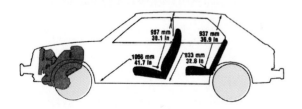

5.1 This scale drawing gives the dimensions, clearances, and seat positions necessary for a comfortable automobile interior. (Courtesy of Chrysler Corporation.)

tive geometry has its greatest application as a design tool in the refinement step of the design process.

5.2 Physical Properties

An important step in refining an idea is determining the product's physical properties. An example of a refinement drawing is the profile of an automobile, which shows its basic dimensions, clearances, and seat positions (**Fig. 5.1**). These dimensions are based on the dimensions of the ultimate user, the average-sized person.

Although most refinement drawings will be rendered in two dimensions that can be accurately scaled and compared, three-dimensional drawings can also be used. **Figure 5.2** is an example of a three-dimensional pictorial that is drawn to scale with only a few major dimensions given.

The configuration of an aircraft is refined in the same manner as simpler designs by

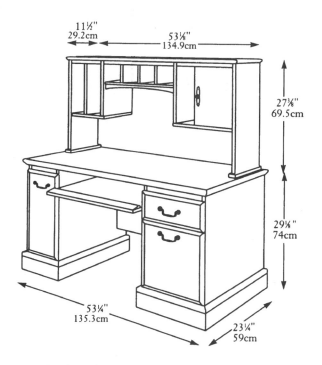

5.2 This three-dimensional pictorial of a computer station is drawn to scale with a few of its major dimensions shown. Most refinement drawings are two-dimensional orthographic views. (Courtesy of OfficeMax.)

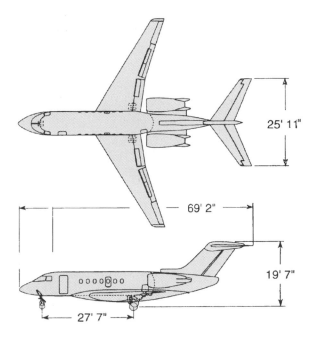

5.3 This two-dimensional, orthographic drawing shows the configuration of an aircraft, the Hawker Horizon. (Courtesy of Raytheon Aircraft Company.)

beginning with top and front views as shown in **Fig. 5.3**. The computer is a significant aid to the designer in taking the refinement a step further. In **Fig. 5.4**, the body of the plane is drawn by computer, which permits it to be viewed from any direction. Mechanisms with moving parts can be made to operate on the computer's monitor to test its operation. An additional example of a wire-frame model of a Volkswagen automobile, drawn with Auto-CAD, is illustrated in **Fig. 5.5**.

5.3 Application of Descriptive Geometry

Descriptive geometry is the study of points, lines, and surfaces in three-dimensional space, which are the geometric elements that comprise all forms. Before descriptive geometry can be applied, the designer must draw orthographic views to scale, from which auxiliary views can be projected. **Figure 5.6** shows

5.4 The body of the Raytheon Premier 1 is drawn to scale as a solid model by computer as a means of refinement. (Courtesy of Raytheon Aircraft Company.)

how descriptive geometry is used to determine the clearance between a hydraulic cylinder and the fender of an automobile where the cylinder is attached with a clip. This con-

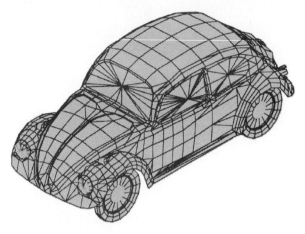

5.5 This computer-drawn wire-frame model of a Volkswagen enables the designer to develop and refine its design and view it from any direction. (Courtesy of AutoDesk Corporation.)

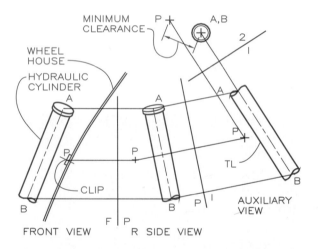

5.6 Descriptive geometry is an effective way to determine clearances between components, such as that between a hydraulic cylinder and a fender.

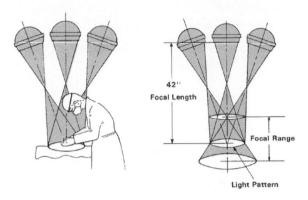

5.7 By using scale drawings developed in the refinement step of the design process, the designer can study the geometry of a surgical lamp. (Courtesy of Sybron Corporation.)

5.8 A refinement drawing shows the overall dimensions of the final design of the surgical lamp. (Courtesy of Sybron Corporation; Photo by Brad Bliss.)

struction can be performed either by pencil or by computer.

Another application of descriptive geometry can be seen in the design of a surgical light. The light fixture must provide maximum light on the operating area with the least obstruction, as shown in **Fig. 5.7**. This scale drawing depicts the converging beams of light emitted from the reflectors, its position above the operating area, and approximate positions of the surgeons. The beams are positioned so their narrowest rays are at shoulder level to minimize shadows cast by the surgeon's shoulders, arms, and hands.

From these scale drawings, lengths, angles, areas, and other geometric relations can be determined by the designer at the drawing board or at the computer (**Fig. 5.8**). When the three-dimensional geometrical relationships have been determined, the engineering details can be developed for more analysis and testing. The major dimensions of the sur-

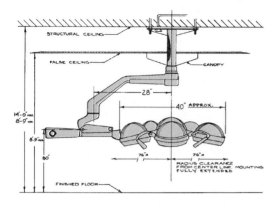

5.9 A well-designed surgical lamp emits light that passes around the surgeon's shoulders with the minimum of shadow. The focal range of this surgical lamp is between 30 and 60 inches. (Courtesy of Sybron Corporation.)

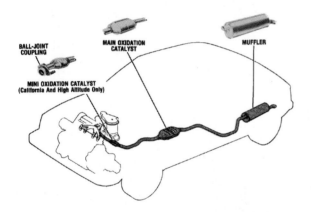

5.10 Descriptive geometry is useful in determining the lengths and angles of an automobile's exhaust system that are necessary to clear the structural members of the chassis. (Courtesy of Chrysler Corporation.)

gical lamp are shown in the refinement drawing in **Fig. 5.9**.

5.4 Refinement Considerations

In advanced designs, such as that of a new model automobile, numerous features must be refined. The exhaust system shown in **Fig. 5.10** is the result of the many refinement drawings needed to determine its geometry. Descriptive geometry was used to determine the exhaust pipe's bend angles, its length, and

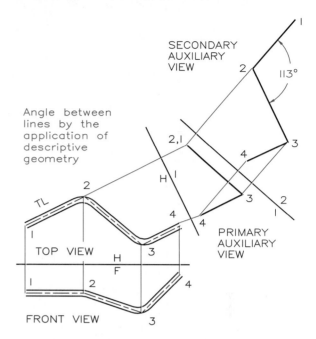

5.11 Descriptive geometry is applied to find the lengths and angles between exhaust pipe segments.

the clearances required for fitting it to the chassis without interference (**Fig. 5.11**).

A designer's refinement drawing of the fuselage of a business jet aircraft is shown as an orthographic section in **Fig. 5.12**. This drawing, although simple in concept, is effective in determining clearance, heights, and seat sizes for its passengers.

5.5 Refinement: Exercise Bench

In Chapter 3 the exercise bench design problem was identified and in Chapter 4 preliminary ideas for it were developed. Now, in this chapter, we refine the preliminary ideas for the exercise bench with instrument drawings.

First, list the features to be incorporated into the design on a worksheet (**Fig. 5.13**). Then refine a preliminary idea, say, idea 2 from Fig. 4.20 in an orthographic scale drawing of the seat (**Fig. 5.14**). Block in extruded parts, such as the framework members, to expedite the drawing process and omit unneeded hidden lines.

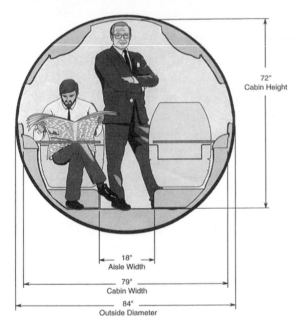

5.12 This refinement drawing is an orthographic section showing clearance, heights, and seat sizes of the interior of the fuselage of the Hawker Horizon, a business jet aircraft. (Courtesy of Raytheon Aircraft Corporation.)

Refinement drawings must be drawn to scale. The use of instruments is important to portray precisely the design from which angles, lengths, shapes, and other geometric elements will be obtained. The drawing shows only overall dimensions and several connecting joints are detailed to explain the design. **Figure 5.15** shows additional design features. These worksheets depict representative types of drawings required to refine a design; additional drawings would be required for a complete refinement of the design.

5.13 (Sheet 8) This list of desirable features is a refinement of the exercise bench.

5.14 (Sheet 9) This refinement drawing for the exercise bench is a scale drawing with several of the major dimensions shown.

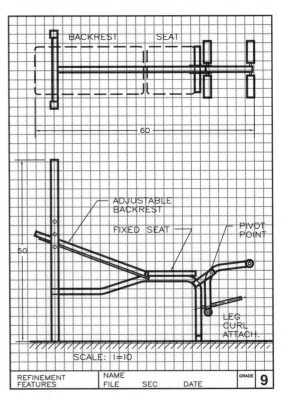

REFINEMENT

1. Description of design

   A. MUST SUPPORT 300 LB EXERCISER

   B. WEIGH UNDER 100 LBS

   C. PADDED BENCH

   D. ADJUSTABLE TO VARIOUS SIZE USERS

   E. FOLD UP FOR EASY STORAGE

   F. SHOULD BE ATTRACTIVE FOR HOME USE

   G. USE FOR

      1. OVERHEAD LIFTS & PRESSES

      2. LEG EXTENSIONS

      3. DUMBELL FLYES

      4. NONWEIGHT EXERCISES

   H. ECONOMICAL CONSTRUCTION— UNDER $200 RETAIL

REFINEMENT FEATURES | NAME | FILE | SEC | DATE | GRADE 8

BACKREST | SEAT
60
ADJUSTABLE BACKREST
FIXED SEAT
PIVOT POINT
50
LEG CURL ATTACH.
SCALE: 1=10

REFINEMENT FEATURES | NAME | FILE | SEC | DATE | GRADE 9

5.6 Standard Parts

When preparing refinement drawings, specify standard parts whenever possible because they are more economical and they are readily available. Merchandise catalogs, sales brochures, magazine advertisements, newspapers, and similar sources contain specifications for standard parts.

Make a practice of keeping files on stock items—such as leveling devices and casters (**Fig. 5.16**)—that can be used in developing products and specified in refinement drawings by referring to literature from manufacturers and vendors. You can become a better designer by observing how standard parts and devices are made and how they function. When you see a device that you are unfamiliar

A. LEVELING   B. CASTERS FOR MOVABLE
    DEVICES   EQUIPMENT

5.16 Leveling devices (A) are standard parts for balancing equipment on uneven floors. Casters (B) are attached to equipment that must be moved about a work area. They range from small sizes for TV sets to those that carry over a thousand pounds. (Courtesy of Vlier Industries and Hamilton Casters.)

with, ask yourself these questions: Why is it made the way it is? How is it used? What application can you think of for it in a design?

Problems

Problems should be presented on 8½-by-11-inch paper, grid or plain. All notes, sketches, drawings, and graphical work should be neatly presented. Written matter should be lettered legibly using ⅛-inch guidelines.

1. When refining a design for a folding lawn chair, what physical properties would a designer need to determine? What physical properties would be needed for a (a) TV set base, (b) golf cart, (c) child's swing set, (d) portable typewriter, (e) shortwave radio, (f) portable camping tent, and (g) warehouse dolly used for moving heavy boxes?

2. Why should scale drawings rather than freehand sketches be used in the refinement of a design?

3. List five examples of problems involving spatial relationships that could be solved by the application of descriptive geometry. Explain your answers.

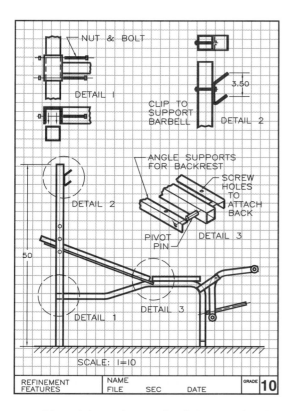

5.15 Additional design features for the exercise bench are shown in this refinement drawing.

4. In the refinement step, how many preliminary designs should be refined? Why?

5. Make a list of refinement drawings that would be needed to develop the installation and design of a 100-foot radio antenna. Make rough sketches of the types of drawings needed, with notes to explain their purposes.

6. If a design is eliminated as a possible solution after refinement drawings have been made, what should be the designer's next step? Explain.

7. For the exercise bench discussed in Section 5.5, what refinement drawings are necessary in addition to those presented? Make freehand sketches of the necessary drawings, with notes to explain what they should show.

Refinement Problems

8. Students often draw wheels as disks with holes at their centers, but even the simplest wheels have more sophistication in their design. Make a refinement drawing of one of the wheels shown in the sketches in **Fig. 5.17.** Why do wheels have raised hubs at their centers? Why do they have bushings in the holes through them?

9. Preliminary sketches for a pipe hanger bracket, a device used to support steam pipes from overhead beams, are given in **Fig. 5.18.** The sketches are sufficient for you to understand the concepts, but they need further refinement. Make refinement drawings of

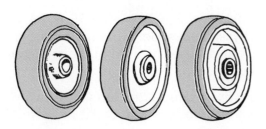

5.17 Problem 9: Why do wheels have raised hubs and bushings?

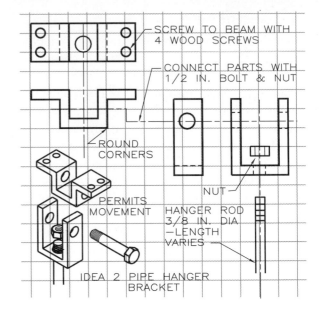

5.18 Problem 10: Preliminary sketches of a pipe hanger bracket for supporting steam pipes from an overhead beam on a hanger rod.

these sketches, incorporating whatever modifications in the design that you think would improve their design.

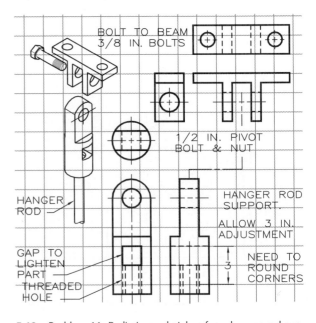

5.19 Problem 11: Preliminary sketches for a hanger rod support for suspending steam pipes.

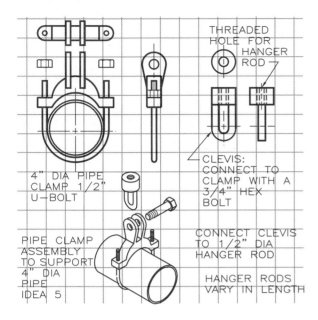

5.20 Problem 12: Preliminary sketches of a pipe clamp assembly for holding steam pipes and connecting to an overhead hanger bracket.

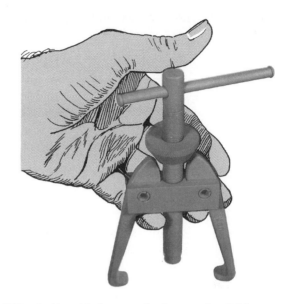

5.22 Problem 14: A gear puller for removing tightly assembled gears from their shafts.

10. The preliminary sketches in **Fig. 5.19** illustrate another concept for a pipe hanger bracket. Solve this problem by following the steps in Problem 9.

11. The preliminary sketches in **Fig. 5.20** illustrate a pipe clamp assembly that attaches to a hanger bracket of the type shown in the two previous figures. Solve this problem by following the steps in Problem 9.

12. Make orthographic refinement drawings of the parts of a metal-to-water discharge assembly (**Fig. 5.21**) to better understand their relationship.

13. Make orthographic refinement drawings of the parts of a gear puller (**Fig. 5.22**). Make separate drawings of each part, and make an assembly drawing showing how the parts fit together.

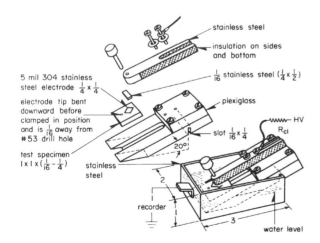

5.21 Problem 13: A metal-to-water discharge assembly. (Courtesy of Westinghouse Corporation.)

14. Make orthographic refinement drawings of the casters shown in Fig. 5.16.

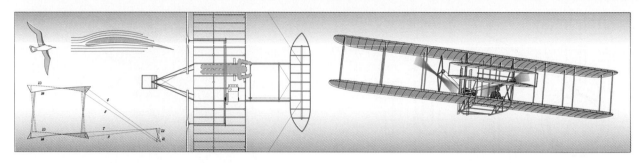

6 Design Analysis

6.1 Introduction

Design analysis involves the development and evaluation of a proposed design by objective thinking and the application of engineering and technology. For example, bridge designs are analyzed for loads, stresses, travel loads, dimensions, materials, sizes, function, economy, and much more. Less creativity is needed during analysis than during the previous steps of the design process. Analysis is the step in the design process most thoroughly covered in engineering courses.

6.2 Graphics and Analysis

Descriptive geometry and graphics are effective tools for analyzing a design in addition to the numerical methods normally used in engineering. Empirical data obtained from laboratory experiments and field observation can be transformed into formats suitable for graphical analysis and evaluation (**Fig. 6.1**).

Figure 6.2 shows an example of computer graphics applied to the analysis of a linkage system to determine the clearances, limits, and velocities of its members. Graphics is also an effective way to present and analyze technical information, background data, market surveys, population trends, and sales projections. The graph shown in **Fig. 6.3** enables the designer to quickly select the appropriate conveyor for transporting raw material at the desired rate.

6.3 Types of Analysis

Analysis includes evaluation of the following attributes:

1. Function
2. Human factors
3. Product market
4. Physical specifications
5. Strength
6. Economic factors
7. Models

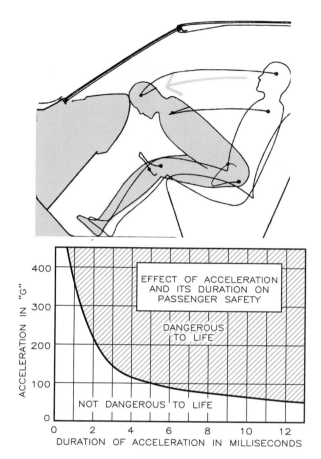

6.1 Designers analyze experimental data and human factors to determine comfortable and safe automobile designs. Graphics is a helpful tool in this step of the design process. (Courtesy of General Motors Corporation.)

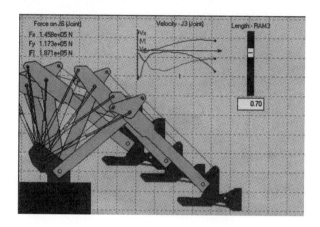

6.2 Clearance between functional parts and linkage systems can be analyzed efficiently with graphical methods by computer. (Courtesy of Knowledge Revolution, Inc.)

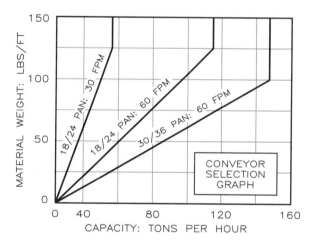

6.3 Graphics can be used to organize and present laboratory and field data to make it much easier to analyze for design applications.

Function

Function is the most important characteristic of a design because a product that does not function properly is a failure regardless of its other desirable features (**Fig. 6.4**). Many products serve a narrow, utilitarian purpose, such as the hand-operated clamping device used for holding a part while it is being machined (**Fig. 6.5**). In those cases, the designer is concerned with function to a much greater extent. Functional analysis usually involves the optimization of several aspects of the design, including safety, economics, durability, appearance, and marketability. For example, despite high fuel costs, most consumers will not accept cars designed to get good mileage at the expense of comfort, safety, and styling. Buyers may be willing to give up some features, but few would give up air conditioning, tape decks, radios, comfortable seats, and safety features for improved gasoline mileage.

Human Factors

Human engineering (ergonomics) is the design of products and workplaces suited to

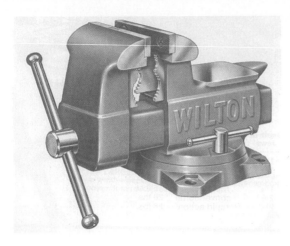

6.4 This vise must function properly to be acceptable. No other features can offset a design that operates poorly. (Courtesy of Wilton Tool Division.)

6.6 Human dimensions.

A Leonardo da Vinci analyzed body dimensions and proportions with graphics in the fifteenth century.

B Engineers used his techniques in the twentieth century to analyze motions by astronauts when restricted by radiation protection vests. (Courtesy of General Dynamics Corporation.)

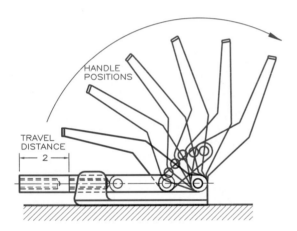

6.5 Graphical analysis is an efficient way to determine operating limits for a product such as this hand-operated clamping device.

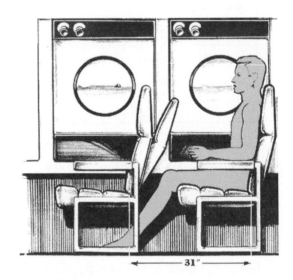

6.7 Analysis of human factors and living environments was part of designing the interior of this aircraft. (Courtesy of Bell Helicopter Corporation.)

the humans who use and occupy them. Safety and comfort are essential for efficiency, productivity, and profitability. Therefore, the designer must consider the physical, mental, safety, and emotional needs of the user and how to best satisfy them.

In the fifteenth century, Leonardo da Vinci was analyzing body dimensions in meticulously executed drawings such as the one shown in **Fig. 6.6A**. Nearly five hundred years later, NASA performed similar analyses to determine the range of mobility permitted by

a radiation protection garment used by astronauts (**Fig. 6.6B**). Analysis of human factors is crucial in the space program because even the simplest, most familiar tasks require training and adaptation when astronauts perform them while in a weightless state. It is also important in the configuration of aircraft cabins, which share many similarities with

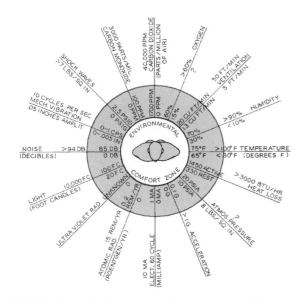

6.8 The inner circle represents the environmental comfort zone and the outer circle the bearable limit zone of the human environment. (Courtesy of Henry Dreyfuss, *The Measure of Man.*)

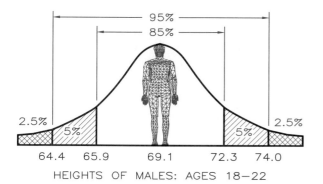

HEIGHTS OF MALES: AGES 18–22

6.9 The distribution of average heights in inches of American men from 18 to 22 years of age. Fifty percent of American men in this age range are taller than 69.1 in., and 50 percent are shorter. (Courtesy of HumanCAD.)

automobile interiors (**Fig. 6.7**). Both must provide comfort and space in which to function, but the aircraft has the added restriction of less space to work within.

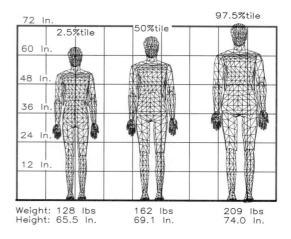

Weight: 128 lbs   162 lbs   209 lbs
Height: 65.5 In.   69.1 In.   74.0 In.

6.10 Men of average build have the body measurements shown here. (Courtesy of HumanCAD.)

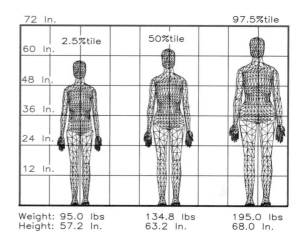

Weight: 95.0 lbs   134.8 lbs   195.0 lbs
Height: 57.2 In.   63.2 In.   68.0 In.

6.11 Women of average build have the body measurements shown here. (Courtesy of HumanCAD.)

Dimensions and Ranges A design must take into account the sizes, ranges of movement, senses, and comfort zones of the people using the finished product (**Fig. 6.8**). Variations in people's physical characteristics conform to the normal distribution curve shown in **Fig. 6.9**. Designers must use the body dimensions of the average American man (**Fig. 6.10**) and the average American woman (**Fig. 6.11**) as the basis for industrial designs. It is a challenging assignment to design accommodations that will be comfortable for the smallest

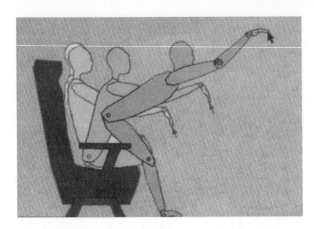

6.12 Working Model, a computer program, is used here to analyze body movements through a range of movements. (Courtesy of Knowledge Revolution, Inc.)

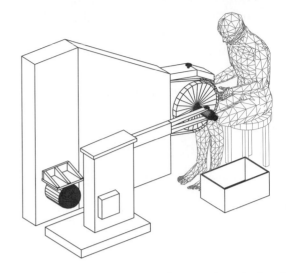

6.13 Three-dimensional analysis of this grinding-wheel workstation helps ensure that it is well-adapted to the operator. (Courtesy of HumanCAD.)

6.14 The design of an efficient, safe automobile instrument panel involves the analysis of human factors. (Courtesy of Ford Motor Company.)

as well as the largest subjects. **Figure 6.12** shows a computer-generated drawing analyzing the range of movement for a person in relationship to his seat.

Motion The study of body motion begins with the understanding of the amount of space required for a person to function comfortably, safely, and efficiently. **Figure 6.13** shows a computer-generated analysis of a grinding wheel workstation for the purpose of adapting it to the operator's needs.

Vision Designs that include gauges and controls must provide the most visually effective means of aiding the operator. The automobile dashboard configuration shown in **Fig. 6.14** is a futuristic design, but it is similar enough to traditional dashboards to make the driver feel at home with it. Lights allow the driver to operate the controls easily and to obtain information from the instruments quickly.

Sound Sound must be within specified frequencies so as not to adversely affect a person's stress level and productivity. Many types of sound cause stress and contribute to an unsafe work environment.

Environment Working environments may include an entire industrial plant, a particular workstation, or a specialized location, such as the cockpit of a farm machine. Important environmental factors are temperature, lighting, color, sound, and comfort.

Product Market

Designers study the market for a product during all stages of product development (as discussed in Chapters 3–5) and review it more formally during the analysis step. Areas of product analysis are market prospects, retail outlets, sales features, and advertising.

Market Prospects   Market information should be collected to learn about the age groups, income brackets, and geographical locations of prospective purchasers of a product. This information is helpful in planning advertising campaigns to reach potential customers.

Retail Outlets   A product may be marketed through existing wholesale and retail channels or newly established dealerships, or by the manufacturer directly. For example, mainframe computers are not suitable for distribution through retail outlets, so manufacturers' technical representatives work with clients individually. However, an exercise machine can be sold effectively through department stores, television commercials, direct mail, and sporting-goods outlets.

Sales Features   Designers should itemize the unique features of a new design that would stimulate interest in the product and attract consumers. They must continually ask questions, such as "What features make this design better than my competitor's?"

Advertising   Manufacturers, wholesalers, and retailers use several media, including personal contact, direct mail, radio, TV, newspapers, and periodicals, to advertise their products to potential customers. Advertising costs vary widely, and each medium should be analyzed for suitability before one or more is selected.

Physical Specifications

During the refinement step, the designer specifies various measurements, such as lengths, areas, shapes, weight, and angles for the product. During the analysis step, the

SPECIFICATIONS

CAPACITY: 1000 lb.
HYDRAULIC CYLINDER: Heavy duty, 3000 lb.
MAXIMUM LIFT HEIGHT: 8'
REACH: 4'
FLOOR-TO-HOOK HEIGHT,
   BOOM LOWERED: 32"
MAST HEIGHT: 58"
BOOM LENGTH: 54"
OVERALL WIDTH: 44"
WIDTH INSIDE WHEELBASE: 34"
OVERALL LENGTH: 57"
SHIPPING WEIGHT: 150 lb. (approx.)

6.15   Designs must be analyzed to determine their physical properties, including weights, ranges, geometries, and capacities. (Courtesy of Air Technical Industries.)

designer uses the product's geometry and measurements to calculate member sizes and dimensions, weights, volumes, capacities, velocities, operating ranges, packaging and shipping requirements, and similar information (**Fig. 6.15**).

Sizes and Dimensions   The designer must evaluate product sizes and dimensions to ensure that they meet any standards specified, such as permissible widths, lengths, and weights in automobile design. For products that have moving parts, such as a construction crane, the designer must analyze the size of the product when extended, contracted, or positioned differently, as well as weight and balance requirements.

Ranges   Many products have ranges of operation, capacities, and speeds that the designer

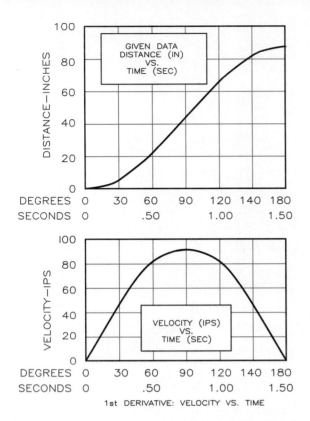

6.16 This graph shows the distance traveled versus the time required for a part on a conveyor. The designer differentiates the plot of the data given to obtain a graph of velocity versus time as the first step in designing parts of sufficient strength for the conveyor.

must analyze before finalizing a design. For example, the designer must determine ranges and maximum limits such as seating capacity, miles per gallon, pounds of laundry per cycle, flows in gallons per minute, or power required.

Packaging and Shipping The designer must also be concerned with product packaging and shipping. Packaging relates both to product protection and consumer appeal: how the product is to be shipped—air, rail, mail, or truck—and whether it is to be shipped one at a time or in quantity are important considerations. Shipping and marketing a product assembled, partially assembled, or disassem-

6.17 An economic model that can be used as a guide in the economic analysis and pricing of a product.

bled requires design attention and analysis, as does the cost of each method.

Strength Much of engineering is devoted to analyzing a product's strength to support maximum design loads, withstand specified shocks, and endure necessary repetitive motions. **Figure 6.16** illustrates motion analysis for a moving part on a conveyor. The designer plots the data obtained and then uses graphical calculus to find the conveyor's velocity versus time profile as the first step in determining the strength needed by the part.

Economic Factors

Designs must be economically competitive to have a chance of being successful. Therefore, before releasing a product for production, the designer must analyze its cost and expected profit margin. Two methods of pricing a product are **itemizing** and **comparative pricing**.

Itemizing The process of totaling the costs of each part and its related overhead to determine its final cost is the first step in itemizing a product's price. From the working drawings, the designer (or an estimator) can estimate the costs of materials, manufacturing, labor, overhead, and other items to arrive at the total production expense. The wholesale price is production cost plus profit. Dealer margin plus the wholesale cost gives the retail price. One example of an economic model is shown in **Fig. 6.17**; these percentages vary for different areas of manufacturing marketing, and retailing.

Comparative Pricing   The other method used to estimate the price of a proposed product is to compare it with the prices of similar products. For example, the power tools shown in **Fig. 6.18** are priced at approximately $100 each. These tools are similar: All use the same type of power source, are made from the same materials, and have the same styling. Most importantly, these products will have about the same market size. Approximately the same number of drills, sanders, and saws are sold; consequently, production costs and retail prices are similar for each.

Another example of comparative pricing are the prices of the hunting seat (**Fig. 6.19**) and the exercise bench (**Fig. 6.20**), which have

6.19   This hunting seat and the exercise bench in Fig. 6.20 can be comparatively priced at between $90 and $100 because both have similar manufacturing requirements and market volume potential. (Courtesy of Baker Manufacturing Company, Valdosta, Georgia.)

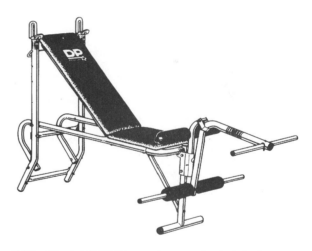

6.20   Priced at $100, this exercise apparatus is similar in manufacture and market appeal to the hunting seat shown in Fig. 6.19. (Courtesy of Diversified Products Corporation.)

6.18   Each of these three products retails for approximately $100. They can be priced comparatively because they are similar in design and have essentially the same market size. (Courtesy of Sears, Roebuck and Company.)

somewhat equal market sizes. Since the manufacturing requirements and market volumes of both products are similar, both sell for about the same price of $90 to $100. However, the baby stroller in **Fig. 6.21** sells for about $60

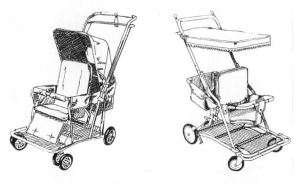

6.21 These baby strollers are priced from $55 to $75, or considerably less than the hunting seat and exercise apparatus, because the stroller market is larger and competition for customers is greater. (Courtesy of Strolee of California.)

6.22 A mock-up of a product is useful for testing its resistance to air flow in the laboratory prior to a decision about its final configuration. (Courtesy of General Motors Corporation.)

because of its larger and more competitive market, even though it is very similar in configuration to the hunting seat and the exercise bench.

Manufacturers also use comparative pricing to estimate cost per square foot, per mile, per cubic foot, or per day. These factors yield rough cost estimates as a basis for doing more detailed studies.

Miscellaneous Expenses Various expenses incurred in the development of new products can be easily overlooked and thereby affect a product's profitability projections. For example, warehousing and storage costs for finished products must be included in their price. Associated with warehousing are the costs of insurance, temperature control, shelving, forklifts, and employees.

Models

Models are effective aids for analyzing a design in the final stages of its development. Designers use three-dimensional models to study a product's proportion, operation, size, function, and efficiency. Types of models often used are **conceptual models**, **mock-ups**, **prototypes**, and **system layout models**.

Conceptual Models Designers use rough models to analyze a preliminary design or feature concept.

Mock-ups Designers use full-size dummies of the finished design to demonstrate the product's size, appearance, and component relationships (**Fig. 6.22**). Mock-ups present a visual impression, not the operation, of a product.

Prototypes Designers use full-size working models to demonstrate the operation of a final product. Because prototypes are made mostly by hand, materials that are easy to fabricate are used instead of those to be used in final production.

System Layout Models Designers use detailed scale models that show the relationships among components of large manufacturing systems, building complexes, and traffic layouts. Designers usually construct system layout models of refineries to supplement working drawings for contractors and construction supervisors during construction (**Fig. 6.23**).

Model Materials Designers commonly use balsa wood, cardboard, and clay in model construction because they are easy to shape

6.23 This system layout model is used to analyze the details of construction of a refinery. (Courtesy of E. I. du Pont de Nemours and Company.)

6.24 A student's model demonstrates how a portable home caddy will fold flat for ease of storage.

6.25 Computer graphics can run tests on the computer screen much more economically than real models. (Courtesy of Knowledge Revolution, Inc.)

and require few tools. Standard parts such as wheels, tubing, figures, dowels, and other structural shapes can be purchased, rather than made, to save time and effort. Plexiglass can be used to construct models that illustrate both interior and exterior design features. Finished models should give a realistic impression of the design, especially when they are used for sales presentations and displays.

Model Scale A model should be large enough to show the function of the smallest significant moving parts. For example, the student model of a portable home caddy shown in **Fig. 6.24** (made of balsa wood) demonstrates a linkage system that permits the wheels to be collapsed for storage. The model's scale is large enough to permit the linkage system to operate as it will in the final product.

Model Testing Using models to test performance is helpful in determining how well a design meets requirements. Aerodynamic characteristics of the rear styling of an automobile can be evaluated by wind tunnel tests. Physical relationships and the functional work-

ings of movable components, as for the hatch and storage area of a car, can be tested in a prototype. Products can be tested on the computer screen to economically obtain results, as illustrated in **Fig. 6.25**, where a crash test is performed. Designers also use models to test consumer reactions to new products before releasing the design for production (**Fig. 6.26**).

6.4 Analysis: Exercise Bench

To illustrate a method of analyzing a product design, we return to our problem example, the exercise bench, which was carried through

6.26 Prototypes are full-size models for demonstrating and testing a design's operation. (Courtesy of General Motors Corporation.)

6.27 A worksheet containing an analysis of function, human engineering, and market considerations for the exercise bench.

ANALYSIS

1. Function

  A. PROVIDES SUPPORT FOR BASIC EXERCISES

  B. SUPPORTS BAR BELLS

  C. ADJUSTABLE TO SUIT INDIVIDUAL

2. Human engineering

  A. PADDED BENCH

  B. ADJUSTABLE BACKREST

  C. CONFORMS TO BODY MOTIONS

  D. BARBELL BRACKETS FOR SAFETY

  E. FOAM PADS FOR COMFORT

3. Market and consumer acceptance

  A. POTENTIAL MARKET
    1 STATE—56,000
    2 NATION—7,000,000

  B. MORE EFFECTIVE EXERCISING

  C. AFFORDABLE AT $120—$150 RANGE

  D. USABLE FOR NONWEIGHT EXERCISES ALSO

DESIGN ANALYSIS	NAME		GRADE **11**
	FILE	SEC	DATE	

4. Physical description

  A. BENCH COMPOSED OF BACKREST & SEAT

  B. BACKREST ADJUSTABLE FROM 0 DEG. TO +30 DEG. WITH HORIZONTAL

  C. 2 BAR HOLDER BRACKETS FOR BARBELLS

  D. 2 BUTTERFLY EXERCISE ATTACHMENTS

  E. LEG CURL ATTACHMENT

  F. WEIGHT: 45 LB

  G. SIZE: 52" LONG X 31" WIDE X 50" TALL

5. Strength

  A. RECOMMENDED WEIGHT SET: 160 LB

  B. SUPPORT PERSON WEIGTHING UP TO 300 LB

  C. CROSS BRACED TO PROVIDE STABILITY

  D. REPLACEABLE PLASTIC SLEEVES FOR BUTTERFLY ATTACHMENTS

  E. MAX. LEG CURL WEIGHT: 100 LB.

  F. MAX RECOMMENDED BUTTERFLY WEIGHTS: 50 LB EACH

DESIGN ANALYSIS	NAME		GRADE **12**
	FILE	SEC	DATE	

6. Production procedures

  A. STRUCTURAL MEMBERS HOLLOW REC-TANGULAR SECTIONS—STEEL, BENT TO SHAPE

  B. PARTS WELDED OR BOLTED TOGETHER

  C. VINYL SEAT COVERS STAPLED TO PLYWOOD SEAT AND BACKREST

  D. PLASTIC CAPS AT ENDS OF OPEN SUPPORT MEMBERS

  E. METAL PARTS NICKEL PLATED

7. Economic analysis

MATERIALS	$15	
LABOR	16	
SHIPPING	7	
WAREHOUSING	1	
TOTAL	$39	
SALES COMMISSION	$ 4	
PROFIT	$20	
WHOLESALE PRICE	$63	
RETAIL PRICE	$90	

DESIGN ANALYSIS	NAME		GRADE **13**
	FILE	SEC	DATE	

6.28 (Sheet 12) A worksheet giving the physical description and strength analysis for the exercise bench.

6.29 (Sheet 13) A worksheet containing an analysis of the production procedures for and economics of the exercise bench.

the first three steps of the design process in Chapters 3–5. The main areas of analysis listed on the worksheets in **Figs. 6.27–6.30** will assist you in analyzing the design. Additional worksheets and large sheet sizes for analysis drawings may be used if more space is needed.

Figure 6.30 shows how graphics is used to determine the range of positions of the backrest. Those positions affect the design of the angle-iron supports for the backrest and the locations of the semicircular holes in the angle irons for a range of settings of 30°. The leg-exercising attachment at the end of the bench is designed to move through a 90° arc, which is sufficient for leg extensions. By determining the maximum loads on the backrest and the leg exerciser, you can select the member sizes and materials that provide the strength required. For further analysis, construct a model and test the design for suitability. A product that must support body weight plus weights that are being lifted should be rigorously tested to ensure that it is adequately sturdy.

Figure 6.31 illustrates a model of the exercise bench for testing its functional features. The catalog description of the bench shown in **Fig. 6.32** lists the physical properties of the Weider exercise bench to help the consumer understand its features. You should keep a list of descriptive characteristics of your design for the final catalog specifications. These points become very important to consumers as they come closer to buying a product.

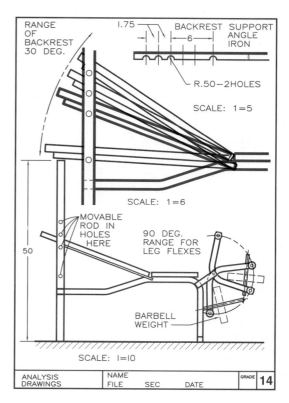

6.30 A worksheet showing graphical analysis of the range of movements for the adjustable parts of the exercise bench.

6.31 A full-size model of the exercise bench is tested for function and acceptability.

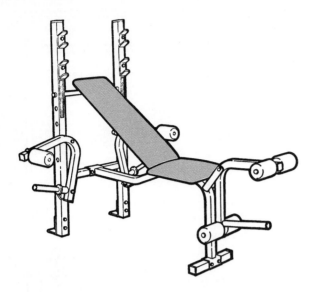

6.32 This catalog description gives the key features of the exercise bench: Weider bench with butterfly attachment. It features no-pinch supports, multiposition padded back and leg lift, and tubular steel frame. Total weight capacity is 1000 lb; butterfly capacity, 50 lb; leg lift capacity, 65 lb; overall size, 58" × 45" × 41"; weight, 48 lb; and price, $89.99. (Courtesy of Sears, Roebuck and Company.)

Problems

The following problems should be solved on 8½-by-11-inch paper and the solution presented in drawing, note, and text forms. Answers to essay problems can be typed or lettered. All sheets should be placed in a binder or folder.

General

1. Make a list of human factors that must be considered in designing the following items: (a) canoe, (b) hairbrush, (c) water cooler, (d) automobile, (e) wheelbarrow, (f) drawing table, (g) study desk, (h) pair of binoculars, (i) baby stroller, (j) golf course, (k) stadium seats, (l) coffee table, (m) exercise apparatus, and (n) lunch box.

2. What physical quantities have to be determined for the designs in Problem 1?

3. Select one of the items in Problem 1 and outline the steps required to analyze the following: (a) function, (b) human factors, (c) product market, (d) physical specifications, (e) strength, (f) economic factors, and (g) a prototype model.

Human Engineering

4. Design a computer graphics table for your body size to meet your own working and comfort needs. Make a drawing indicating the optimum working areas and tilt angle for the computer when you sit at the station. The drawing should also show the most efficient positioning of supplies, materials, and manuals.

5. Using the dimensions for the average man and woman given in Figs. 6.10 and 6.11, design stadium benches to meet the optimum needs of spectators. A primary consideration is slope of the stadium seating to allow an adequate view of the playing field. Spectator comfort and provision for traffic along aisles in front of the benches must also be considered.

6. Compare the measurements of the male and female students in your class with the averages given in Figs. 6.10 and 6.11. Tabulate the results and compare them with the percentiles given in the example in Fig. 6.9.

7. Design a backpack for use on a week-long camping trip. Determine the minimum number of articles campers should carry; use their weights and volumes in establishing design criteria. Make sketches of the pack and the method of attaching it to the body to provide mobility, comfort, and capacity.

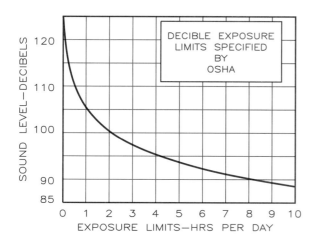

6.33 This graph shows the maximum sound levels for various periods of time permitted by OSHA

8. Sound levels (measured in decibels, dB) are harmful to workers and affect their productivity. Refer to the comfort level for noise shown in Fig. 6.8 and the maximum noise levels permitted by the Occupational Safety and Health Act (OSHA) shown in the graph in **Fig. 6.33** to answer the following questions.

    a. For an 8-hour work day, what is the maximum continuous noise level that a worker can be exposed to?

    b. How long during a work day can a worker be safely exposed to noise levels of 97 decibels?

9. State the dimensions, facilities, and provisions needed for a one-person storm shelter to provide protection for 48 hours. Make sketches of the interior in relationship to a person and the supplies.

10. Design a manhole access to an underground facility. Determine the diameter of the manhole required to permit a person to climb a ladder for a distance of 10 ft with freedom of movement. Make a sketch of your design and explain your method of solving the problem.

11. Analyze the needs for an observation facility for temporary service in the Arctic. This facility is to be as compact as possible, but it must provide for the needs of one person during a 72-hour duty watch. Make sketches of your design and explain the items considered essential to human survival in that harsh climate.

12. Design an automobile steering wheel that is different from current designs but that is just as functional. Base your design on human factors such as arm position, grip, and vision. Make sketches of your design and list the factors that you considered.

13. Make sketches to indicate safety features that you would build into your automobile to reduce the severity of personal injury in case of a bad accident. Explain your ideas and the advantages of your designs.

14. Assume that you prefer to alternate between sitting and standing when working at a study desk. Determine the ideal height of the table top for working in each position. Indicate how you would devise the table to permit instant conversion from the height for standing to the height for sitting.

15. Identify some human engineering problems that you believe need to be solved. Present several to your instructor for approval. Solve the approved problems. Make a series of sketches and notes to explain your approach.

Market Analysis

16. Conduct a market analysis for the drill shown in Fig. 6.18, covering the areas mentioned in the text. Assume that this power tool has never been introduced before. Outline the steps you would take in conducting a product market analysis.

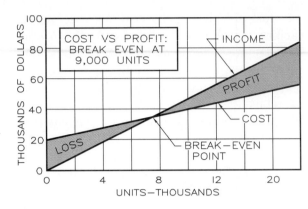

6.34 This break-even graph can be used to determine profit and loss for various quantities of units produced.

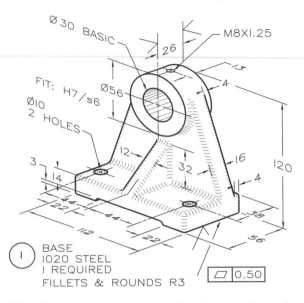

6.35 This base needs to be analyzed to determine its weight and mode of shipping to customers.

17. The production department has analyzed the manufacturing costs for a product and illustrated this data in the form of a break-even graph (**Fig. 6.34**). Use this graph to determine the following.

  a. What would be the loss if only 4,000 units were sold?
  b. What would be the profit if 20,000 units were sold?
  c. What was the fixed start-up cost before any units were sold?

18. Make a market analysis of the hunting seat shown in Fig. 6.19, following the steps suggested in the text. Determine a reasonable price, potential outlets, and other marketing information for the product.

19. Assume that the costs of producing hunting seats are estimated as follows: 100 seats, $35 each; 200 seats, $20 each; 400 seats, $10 each; 1000 seats, $8.50 each. Using these figures, determine the price at which you could introduce the seats to consumers on a trial basis and still make some money. Explain your plan.

20. Packing and shipping modes must be considered to determine the optimum way of delivering the manufactured products to the customer. Make the following determinations regarding the base shown in **Fig. 6.35:**

  a. What is the weight of the part? Refer to Chapter 19 and Appendix 45 for the weight of 1020 steel.
  b. What size box and stuffing material would be best for shipping this product?
  c. Compare various costs for shipping the packaged part 1000 miles by UPS, Federal Express, U.S. Post Office, or by motor freight.

21. List the unique features of the hunting seat that would be important to a sales campaign, including advertising. Make sketches and notes to explain these features.

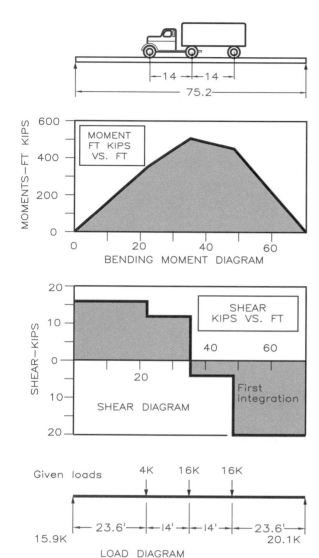

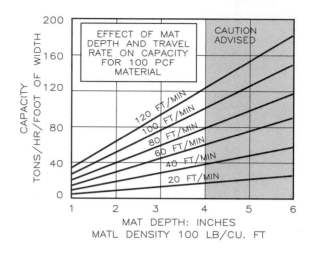

6.36 The loads imposed on the bridge by the truck have been analyzed graphically as integral curves to show shear and bending moments. This is an example of how graphics can be used as an analysis tool in addition to the usual numerical methods.

Strength and Function Analysis

22. Refer to the integral curves plotted in **Fig. 6.36** and answer the following questions. (The term *kips* represents 1000 pounds.) At what point on the bridge is shear the greatest?

6.37 These straight-line curves represent conveyor speeds used for transporting materials of 100 lb per cubic foot. The capacity in tons per hour per foot of width of the conveyor is given on the *y* axis and mat depth is shown on the *x* axis. Mat depth is the thickness of the material applied to the conveyor.

Where is shear the least? What is the greatest value of shear? Where is the bending moment the greatest and least? What is the value of the greatest bending moment?

23. The graph in **Fig. 6.37** can be used to estimate the capacities of material conveyors at various speeds and mat depths (thickness of the material on the conveyor). By referring to this graph, answer the following questions: (a) for mat depth of 4 in., and a speed of 120 ft per minute, how many tons per hour are transported; (b) if you wish to have 100 tons per hour transported, what is the slowest speed of the conveyor that would be safe (would avoid the "caution advised" area); (c) for a mat depth of 5 in. and a speed of 20 ft per minute, what would be the capacity?

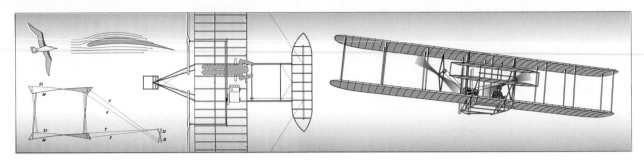

Decision

7.1 Introduction

After the designer has conceived, developed, refined, and analyzed several designs, one must be selected for implementation. The decision process begins with a presentation by the designer (or design team) of all significant findings, features, estimates, and recommendations. The presentation should be organized in an easy-to-follow form, and it must communicate the designer's conclusions and recommendations because it is the means of gaining support for the project in order for it to become a reality. A committee usually makes the decision when funding must be obtained. Although decision making is aided by facts, data, and analyses, it is subjective at best.

7.2 Types of Decisions

The purpose of oral presentations and written reports is to present the findings of a project so that a decision can be made whether to implement it. One of three types of decisions may be made:

1. *Acceptance.* A design may be accepted in its entirety, which indicates success by the designer.

2. *Rejection.* A design may be rejected in its entirety, which does not necessarily mean that the designer has failed. Changes in the economic climate, moves by competitors, or other factors beyond the designer's control may make the design obsolete, premature, or unprofitable.

3. *Compromise.* Parts of a design may have weaknesses, and compromises may be suggested. For example, the initial production run might be increased or decreased, or various features might be eliminated, modified, merged, or added.

7.3 Decision: Exercise Bench

We have used the exercise bench problem introduced in Chapter 3 to illustrate the first four steps of the design process. We continue to use it here to help explain the decision step.

Decision Table

A decision table like the one shown in **Fig. 7.1** can be used to compare designs, where each idea is listed and given a number for identification. Assign maximum values for each factor of analysis, based on your best judgment, so they total to 10 points. Rate each factor for the competing designs by entering points for each.

Sum the columns of numbers to determine the total for each design and compare the scores of each design. If your instincts disagree with the outcome of this numerical analysis, have enough faith in your judgment to go with your intuition. The scores from the decision table are meant to be a guide for you, not the final word in your decision.

Conclusion

After making a decision, state it and the reasons for it clearly (**Fig. 7.2**). Record any additional information, such as number to be produced initially, selling price per unit, profit per unit, estimated sales during the first year, break-even number, and the product's most

---

DECISION

1. Decision for evaluation

DESIGN 1   A-FRAME
DESIGN 2   U-FRAME
DESIGN 3   2-COLUMN FRAME
DESIGN 4
DESIGN 5

MAX	FACTORS	1	2	3	4
3.0	FUNCTION	2.0	2.3	2.5	
2.0	HUMAN FACTORS	1.6	1.4	1.7	
0.5	MARKET ANALYSIS	0.4	0.4	0.4	
1.0	STRENGTH	1.0	1.0	1.0	
0.5	PRODUCTION EASE	0.3	0.2	0.4	
1.0	COST	0.7	0.6	0.8	
1.5	PROFITABILITY	1.1	1.0	1.3	
0.5	APPEARANCE	0.3	0.4	0.4	
10	TOTALS	7.4	7.3	8.5	

DESIGN DECISION	NAME FILE SEC DATE	GRADE **16**

7.1 A worksheet showing the decision table used to evaluate the design alternatives for the exercise bench.

CONCLUSIONS

THE 2-COLUMN FRAME IS THE BEST SOLUTION FOR IMPLEMENTATION BECAUSE

1. GOOD MARKET POTENTIAL
2. AIDS WELL IN EXERCISING
3. FULFILLS PRODUCT REQUIREMENTS
4. ATTRACTIVE PRICE
5. MANUFACTURED EASILY

RECOMMEND IMPLEMENTATION AND PRODUCTION OF THE DESIGN

ECONOMIC FORECAST

SALES PRICE	$ 90
SHIPPING EXPENSES	$ 12
CUSTOMER PRICE	$102
MANUFACTURER'S PROFIT	$ 20 EACH
BREAK-EVEN AT	500 UNITS

PRODUCTION: RECOMMEND THAT BENCHES BE MADE BY QUALIFIED MANUFACTURER ON A CONTRACT BASIS

PROFITABILITY: BENCH SHOULD YIELD AN ATTRACTIVE RETURN ON INVESTMENT

DESIGN CONCLUSION	NAME FILE SEC DATE	GRADE **17**

7.2 This worksheet summarizes the designer's conclusions and recommendations for implementing that design.

marketable features, that will help you prepare your presentation.

If you believe that none of your designs are satisfactory, you should recommend that they not be implemented. A negative recommendation is not a failure of the design process; it means only that your solutions developed so far are not feasible. Going forward with an inadequate solution could waste both money and effort.

Presentation

Until now, your efforts have been self-directed and mostly free from supervision. The work is your own (or that of your team), you have solved the problem to the best of your ability, and you are ready to make recommendations regarding its implementation.

At this point, the project usually involves the input of others besides the designers. These outsiders may be other engineers, managers, administrators, salespeople, company shareholders, and investors or bankers who will loan money for the project. You must prepare a presentation suitable for your audience in order to communicate the important features of your design, the data you gathered and analyzed, and the benefits to be gained by implementing your design. Present your findings, conclusions, and recommendations as objectively as possible so that the group can make a valid decision. At no time should your enthusiasm for the project outweigh an impartial presentation of the facts.

7.4 Types of Presentations

Presentations may be made to groups ranging from a few knowledgeable design associates to a large number of laypeople unfamiliar with the project and its objectives. Presentations of the first type are usually informal; the second type of presentation is formal.

Informal Presentations

Informal presentations are made to several associates and, perhaps, a supervisor. Although formally prepared visual aids are unnecessary for presentations to a small group, the designer nevertheless needs to graph data, draw pictorials, sketch schematics, and build models to explain design concepts. Ideas and concepts may be sketched on a blackboard or informally discussed in a one-on-one situation (**Fig. 7.3**).

Formal Presentations

Formal presentations usually involve large groups that may include various combinations of associates, administrators, and/or laypeople. They may be clients for whom the project is designed, potential investors, or politicians who will vote to approve or disapprove the design. Function and acceptability of a design are the primary concerns of engineering associates, and profitability is most important to investors.

7.5 Organizing a Presentation

An effective method of planning oral and written reports is to use 3-by-5-inch index cards

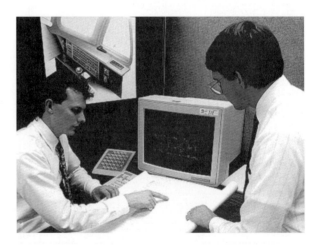

7.3 A decision may be the outcome of an informal presentation to an associate where ideas and designs are discussed one-on-one. (Courtesy of the Cessna Aircraft Company.)

7.4 Planning cards are useful in preparing the sequence of a presentation by arranging the cards on a planning board (as shown here) or on a table top. (Courtesy of Eastman Kodak.)

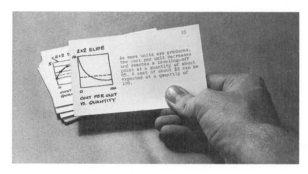

7.5 This 3-by-5-inch card, showing a sketch of the visual and its accompanying text, illustrates sound planning.

for the ideas to be presented. Placing the cards on a table or tacking them to a bulletin board (**Fig. 7.4**) allows an easy choice of sequence and rearrangement as needed. Each card (**Fig. 7.5**) should contain the following information:

1. *Number.* The card's position in the presentation sequence.

2. *Illustration.* A sketch of any visual material.

3. *Text.* A brief outline of the points to be covered for that idea.

7.6 Visual Aids

Visual aids commonly used in presentations are flip charts, photographic slides, overhead visuals, models, computer images, and videotapes. The following suggestions apply to the preparation of visual aids:

1. Limit each visual to a single concept or point.

2. Reduce statements to key points to communicate thoughts clearly and concisely.

3. Make any text accompanying visuals large enough to be readable.

4. Use illustrations, color, and attention-getting devices.

5. Prepare enough visuals so that notes are unnecessary.

Flip Charts

Flip charts consist of bold illustrations drawn on sheets (usually 30 by 36 inches) for presentation to groups no larger than about 30 people (**Fig. 7.6**).

Paper Flip charts may be drawn on brown wrapping paper or white newsprint paper attached to a cardboard backing board. A stand or easel is needed to support the cardboard-backed set of sheets.

7.6 The flip chart is an effective method for presentations to small groups.

Lettering Felt-tipped markers, ink, tempera, or sign paints are fine for lettering. When used correctly, felt-tipped markers can yield bold, visible lines in a variety of colors and with sophisticated effects (**Fig. 7.7**). India ink is an effective medium for lettering and for adding emphasis to a chart.

Color Construction paper cutouts mounted with rubber cement are especially effective for adding color to bar graphs. The use of felt-tipped markers and tempera colors also adds color and interest to a chart.

Assembly Flip chart sheets should be arranged in order of presentation with a title page covered by a blank sheet of paper on top to control audience attention. The sheets are fastened at the top to the backing board.

Presentation Each sheet is flipped in sequence after the presenter has covered the points on it. A pointer should be used to direct the audience's attention to specific items on the sheet. Well-prepared flip charts should require no additional notes.

Photographic Slides

Slides are effective for larger audiences and for showing actual scenes or examples of hardware.

Artwork The method for proportionally sizing artwork for a 35-mm slide is shown in **Fig. 7.8**. An 8-by-12-inch size is suitable for most slides. The artwork should contain color to make the slides more attractive and effective in maintaining audience interest. Colored construction paper, mat board, and other poster materials should be used in preparing slides.

Allow at least an inch margin on all artwork, so the edges will not show when photographed. Uppercase letters are best for slides, with the space between lines of text equal to the height of the letters. Do not use a white background for slide artwork, because it is tiring to the eyes.

Photographing Layouts To make slides, a camera, copy stand, and lights are required. A 35-mm reflex camera with through-the-lens viewfinder is best because the photographer sees exactly what is being photographed. The copy stand holds the camera steady. If all the artwork is uniform in size, the camera can be left in the same position during photography. Small illustrations can be photographed with a close-up lens. The finished slides are reviewed, arranged in sequence, numbered, and loaded in a tray for showing.

7.7 This flip chart sheet is brown wrapping paper on which India ink was applied with a brush. Overlays of colored construction paper or colored felt-tipped markers can be used to highlight topics.

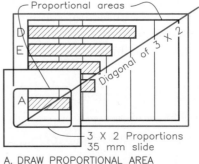

7.8 Use of this method of proportionally sizing artwork ensures that the subject will properly fill a photographic slide.

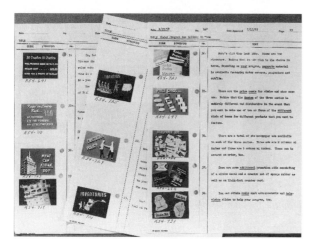

7.9 A slide script is useful for lengthy slide presentations and those that will be given repetitively. (Courtesy of Eastman Kodak.)

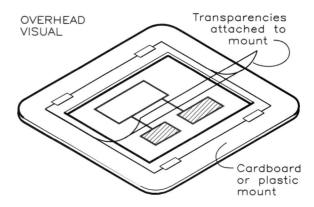

7.10 A transparency used on an overhead projector consists of an 8½ × 11 in. transparency attached to a 10 × 12 in. mount. The projection area within the mount is about 7½ × 9½ in. To show information sequentially, different-colored overlays can be flipped over one at a time.

Slide Scripts A slide script is useful when a presentation will be made repeatedly. Photographic copies of the slides attached to the left side of the script serve as prompts for the presenter (**Fig. 7.9**).

Overhead Projector Transparencies

Overhead projector transparencies are reproduced on 8-by-10-inch plastic sheets by the heat-transfer or diazo processes, or plotted by the computer. Tracing paper is the most commonly used drawing surface for preparing art from which transparencies are made. Tracing paper can be used in both the heat-transfer and diazo processes (opaque paper cannot be used in the diazo process). Computer plotting can be done directly onto plastic film with special pens. Diazo transparencies are reproduced on plastic film in the same manner in which blue-line prints are made.

Drawings should be made in black India ink. Stick-on shapes and graphing tapes can give the drawing a professional appearance. Lettering should be at least 0.20 in. high.

Color Overlays Several color overlays can be hinged to the transparency mount for presentations with multiple steps (**Fig. 7.10**). Each color overlay requires a separate piece of artwork, which is drawn on tracing paper placed over the basic layout.

Computer Plotted Transparencies Computer-generated art and text can be plotted directly on plastic film with fiber-tipped pens, which come in many colors and match the film. The Romand font of AutoCAD is better suited for large lettering on transparencies than is the Romans font.

Presentation The presenter stands or sits near the projector in a semilighted room and refers to the transparencies while facing the audience. The presenter can emphasize items on the stage of the projector with a pointer. In the same manner that multiple overlays are hinged to a mount, paper overlays can be attached to mounts in order to block out parts of the transparency to control audience attention.

Models

A model is the most realistic visual aid for showing a final design. Models should be large enough to be seen by the entire audience. A series of photographic close-ups of

7.11 A model is an effective visual aid to use in a presentation to a small group. (Courtesy of Cessna Aircraft Company.)

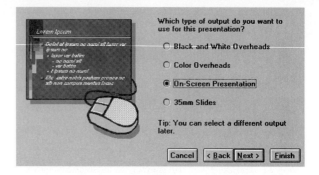

7.12 The dialogue screen illustrates a computer visual and the prompts for composing it. (Courtesy of Power Point.)

the model, taken from different angles, can be used to supplement the presentation. Obviously, a full-sized prototype of the completed design (**Fig. 7.11**) provides the most accurate description of the product and demonstrates its operation.

Computer Visuals

Software programs such as Microsoft's Power Point can be used for making slides that can be viewed on the computer monitor for small groups or projected onto a large screen connected to a computer for large groups. **Figure 7.12** shows a dialogue screen used in composing a computer visual with Power Point. Any drawing or photograph that can be seen on the computer monitor can be used as a visual.

When slides are made by the computer, there is no need for special art supplies, storage problems are eliminated, and the entire slide show can be shown on a lap-top computer. Full-color artwork, special effects, and animation can be incorporated into these presentations.

Videotapes

Videotaping presentations or visuals supplemented by a voice-over narration is an effective and sophisticated method. It may include

various special effects such as music, sound effects, close-ups, motion, and realism. Formal presentations in the future will be multimedia shows using video and the computer in combination.

7.7 Making a Presentation

You should inspect the room in which the presentation is to be made in advance of the meeting. You should also view projected visuals from various audience locations. Projectors and visual-aid equipment must be positioned and focused before the audience arrives, and remote controls for slide projectors should be readied for use. You should determine where to stand so that you do not block anyone's view.

Delivery

You should move through the presentation at a moderate pace while emphasizing information on the visual aids with a pointer (**Fig. 7.13**). A positive approach in selling ideas should not become deceptive high-pressure salesmanship. The presenter should be frank in pointing out weaknesses in a design and show alternatives that compensate for them.

Conclusions and recommendations should be supported by data and analyses. A recom-

7.13 A presentation should include graphical aids and models to help the speaker communicate with the audience. (Courtesy of Hewlett Packard Corporation.)

The table below should be completed jointly by the team with only the grade column completed by the instructor who will use the chart on page 4 and the factor "F" that was computed for each member.

Oral Report

Team No. **5** Project: **TOY MANUFACTURING**

Names	No. (N= 7)	%Contribution (C)	F=NC	GRADE
1. Brown, George		17	119	91
2. Prisk, Helen		14.3	100	87
3. Smith, Roger		20	140	95
4. Reed, Ralph		5.7	40	63
5. Potter, Joyce		14.3	100	87
6. Flynn, Errol		14.3	100	87
7. Ross, Lawrence		14.4	100	87
8.				

Evaluation by instructor	Max.		Comments:
1. Introduction of team members	2	2	Good introduction to project
2. Proper dress of team members	2	2	
3. Statement of purpose of presentation	5	4	
4. Use of visuals—point to important points, do not block screen, do not fumble, etc.	10	8	Several visuals too complex to see very well
5. Adequate number of visual aids	9	9	
6. Quality of visual aids	15	12	
7. Clear presentation of recommended design	10	8	Economic analysis could use a little more study
8. Presentation of alternate solutions considered	2	2	
9. Consideration of human factors	5	5	
10. Coverage of economics (manufacturing, shipping, packing, overhead, mark-up, etc.)	10	7	Very good professional manner in giving the presentation
11. Presentation of an effective conclusion	5	4	
12. Continuity of presentation	3	3	
13. Poise and professionalism	2	2	Good conclusion and proposed solution
14. Team participation (perfect score if all participate)	10	10	
15. Use of allotted time	10	9	
TOTAL 100		87	

Instructor comments on back of this sheet.

7.14 An evaluation form for grading oral reports.

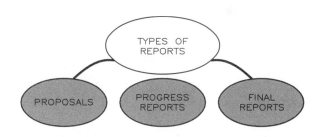

7.15 The three basic types of written reports.

mendation to reject or accept a design should be supported by reasons. A period for questions and answers should follow the presentation for clarification purposes. If available, a technical report should be given to the audience.

Critique

In a classroom setting, when a team gives a presentation to the class, both the design recommendations and presentation skills of the group will be evaluated and critiqued. The evaluation form shown in **Fig. 7.14** is typical of the type that may be used for the critique. It can also be used as a guide in preparing the presentation. The names of the team members are listed at the top of the sheet. As a group, the team must agree on the percent contribution of each member to the project prior to presentation. The sum of the percent contributions of all team members must equal 100 percent. The F-factor for each member is the number of members (N) times the contribution of each (C). The chart in Appendix XX illustrates how an individual's contribution to the project is translated into his or her individual grade by using the F-factor.

7.8 Written Reports

Engineers, technologists, and technicians must know how to prepare a well-written report. The three basic types of written reports are proposals, progress reports, and final reports (**Fig. 7.15**).

Proposals

Proposals are written to establish the need for projects and to obtain authorization of funds and support to pursue them. A proposal outlines data, costs, specifications, time schedules, personnel requirements, completion dates, and other information concerning the project. The purpose of the project is stated, with emphasis on its benefit to the client or organization.

Proposals should reflect the interests and language of the reader. For instance, investors are interested in profits and returns, whereas engineers are more concerned with function and feasibility. Typical elements of a proposal are the following:

- *Statement of the problem.* Identify the problem and its purpose.

- *Method of approach.* Outline procedures for attacking the problem.

- *Personnel needs and facilities.* Itemize requirements for equipment, space, and personnel.

- *Schedule.* Give estimated completion dates for each phase of the project.

- *Budget.* Itemize the funds required for each phase of the project.

- *Summary.* Review the important points made.

Progress Reports

Progress reports are periodic reports on the status of a project. Usually, they are brief and may take the form of a letter or memo. They generally review progress and project the outlook for further developments, including the need for increases or decreases in expenditures or time.

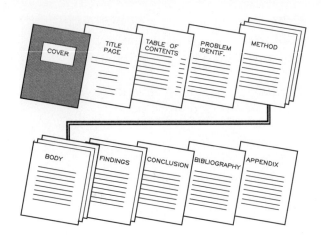

7.16   The elements of a typical technical report.

Final Reports

The most comprehensive type of written reports are final reports, which contain five sections:

1. Problem identification.

2. Method.

3. Body.

4. Findings.

5. Conclusions and recommendations.

Some reports present the conclusions at the beginning and others at the end. The order of presentation will depend on the requirements of your instructor or employer.

Report Format   The elements of the typical technical report (**Fig. 7.16**) are as follows:

1. *Cover.* The report should be inserted in a binder, with its title and author indicated on the cover.

2. *Letter of transmittal* (optional). The second page may be a letter of transmittal, describing briefly the contents of the report and the reasons for the project.

3. *Title page.* The title page contains the title of the report, the name of the person or team that prepared it, and the date.

4. *Table of contents.* The major headings in the report and their page numbers are listed.

5. *Table of illustrations* (optional). A list of the illustrations contained in the report may be included, especially in long, formal reports having many illustrations.

6. *Problem identification.* (Actual heading may vary.) This section explains the importance of and need for a solution to the problem by outlining background information on the problem.

7. *Method.* (Actual heading may vary.) This section should cover the general method used in solving the problem.

8. *Body.* (Actual heading may vary.) This section is the main part of the report. It describes the data collected and analyzed and the steps taken to solve the problem. Subheadings should be used to emphasize the various parts of the report's body.

9. *Findings.* (Actual heading may vary.) Project findings should relate clearly to the data gathered, analyzed, and described in the preceding section.

10. *Conclusions.* This section should summarize the findings and recommendations made.

11. *Bibliography.* The references—books, magazines, brochures, interviews—used in the preparation of the report are listed alphabetically by author.

12. *Appendix.* The appendix includes information (such as drawings, sketches, raw data, brochures, and letters) that supplements (often in more detail) the main sections of the report.

Common Omissions Topics that are often overlooked and omitted from reports by students are sales estimates, advertising methods, shipping costs, product packaging, miscellaneous overhead expenses, and recommendation summaries. Each of these topics must be adequately covered to provide a complete understanding of the total project.

Illustrations Technical reports should be liberally illustrated. Drawings and other illustrations should be numbered and given captions that describe them and relate them to the text. The text should refer to each figure by number and explain the figure's contents. For example, "Figure 6 shows the number of boats sold between 1988 and 1998" identifies the figure being referred to and what it depicts.

Illustrations for the report should be drawn or reproduced on opaque paper rather than tracing paper; ink illustrations are preferred. Drawings should be positioned on the sheet so that they can be read from the bottom or from the right. Large drawings should be folded to 8 1/2 by 11 in. to fit the binder, and they should be easy to unfold for reading.

Evaluating a Written Report The form shown in **Fig. 7.17** is an example of the type that may be used by instructors to grade reports. The team determines the contribution of each team member and fills in the blanks at the top of the form. The teacher grades the report, and each member's grade is found by using the table in Appendix 44. You will have a good

Written Report

TEAM NO. __5__ PROJECT __TOY MFGR__

	NAMES	NO. (N = 7)	% CONTRIBU-TION (C)	F = NC	GRADE (G)
1.	BROWN, G.	14.3	100	92
2.	PRISK, A.	14.3	100	92
3.	SMITH, L.	12.0	84	87
4.	REED, T.	16.3	114	93
5.	POTTER, M.	10.0	70	82
6.	FLYNN, O.	8.8	62	79
7.	ROSS, N.	14.3	100	92
8.				

EVALUATION BY INSTRUCTOR	100%	Max. Value	Points Earned
1. Use of an appropriate cover		2	2
2. Inclusion of an evaluation sheet		2	2
3. Inculsion of a proper letter of transmittal		2	2
4. Correct title page		2	2
5. Proper table of contents		2	2
6. Sufficient introduction to the the report		5	4
7. Thoroughness in identifying the problem		10	8
8. Continuity and quality of the body the report		10	9
9. Collection and presentation of background data		5	4.5
10. Justification of major decisions		5	5
11. Review of costs, overhead expenses, shipping costs and similar expenses		5	5
12. Arrival at strong conclusion and recommendation		5	4.5
13. Sufficient number of graphs and graphics		10	9
14. Quality of graphics		10	9
15. Bibliography—form and content		5	5
16. Use of footnotes		5	5
17. Appendix—content and form		5	5
18. Form and appearance of report (spelling, punctuation, margins, typing, neatness)		10	9
		100	92

7.17 An evaluation form for grading an oral report by a design team.

idea about the guidelines for preparing a written report if you review this form before beginning your report.

Problems

1. Prepare a checklist for evaluating an oral presentation by one of your classmates. List items to consider and develop a point scale for them. Devise a rating system to arrive at an overall evaluation.

2. Use 3-by-5-in. cards to plan a flip-chart presentation that will last no more than 5 minutes. The subject of your presentation may be of your choosing or one assigned by your instructor. Here are some examples: (a) your career plans for the first two years after graduation; (b) the role of this course in your overall educational program; (c) the importance of effective communication; (d) the need for a design project that you are proposing; and (e) a comparison of engineering with another profession.

3. Prepare graphical aids for an oral presentation using the methods and materials covered in this chapter.

4. Using the planning cards developed in Problem 2, prepare a 5-minute briefing on a technique that you choose or is assigned by your instructor. Present this briefing to your class.

5. Assume that you are an engineer responsible for developing a proposal for a project that could result in a sizable contract for your firm. Make a list of instructions to give to your assistants for their help in preparing a presentation for a group of 20 people, ranging in background from bankers to engineers.

Your instructions should outline the materials needed, types and number of graphical aids required, method of projection or presentation, assistance needed during the presentation, room seating arrangements, and other factors. Your outline should cover the entire presentation for the length of time you think most desirable. Select a topic or use one assigned by your instructor.

6. When giving a presentation, be sensitive to what the most important and significant points are. Important recommendations and findings must be stressed above all others. Do not get bogged down in ideas and approaches that were attempted but discarded at the

expense of the points that are most helpful in attaining acceptance of your recommendations.

Make a list of the major points that should receive emphasis if you were giving one of the following reports: (a) an application for a summer job; (b) a request for a loan from your parents; (c) a proposal to a banker for opening a hotdog stand or a business of your choice; (d) a résumé of introduction to a prospective girl or boy friend; (e) a reason for talking your instructor into an excused absence; (f) a list of your qualifications for being elected class president.

7. Many designs and products come onto the market that are mistakes from the beginning—they don't sell and often they may not function well. Poor decisions were made when the "go ahead" was given. Make a list of as many of these "mistakes" as you can think of. For starters, the Susan B. Anthony dollar is one mistake that has been discontinued by the government.

8. One of the best-selling cars of all time was the Ford T-Model touring car (**Fig 7.18**) Make a list of the T-Model's attributes that made it the success that it was at the time it was introduced to market in 1908. Make another list of the disadvantages that it would have today.

7.18 The Ford T-Model was one of the best-selling and most popular cars of all time.

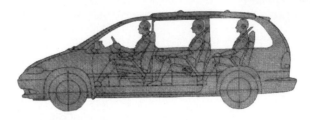

7.19 The Dodge Caravan is a popular vehicle concept that seats seven passengers comfortably. (Courtesy of Chrysler Corporation.)

9. A best selling car of today (**Fig. 7.19**) is the seven-passenger minivan, a new concept vehicle that has outsold the traditional sedan. Make two lists; one that lists the advantages of the minivan over the sedan, and one that lists the disadvantages of the minivan over the sedan.

10. The sales generated by advertisement spots on TV must be sufficiently effective in order to pay for all other programs, shows, and sports. Therefore, TV commercials are good examples of extremely short reports that must get the viewer's attention, tell a story in a matter of seconds, and be persuasive in selling a concept or product.

Pay close attention to several TV commercials and tabulate your analysis of the following points:

    a. What introduction technique got your attention?

    b. What "plot" was used to direct your thinking?

    c. What method was used to make you receptive to the advertised product?

    d. Was the commercial effective?

    e. List negative factors.

Compare your findings with those of a classmate to determine the level of agreement. Make mental notes of these findings and use them in preparing future presentations.

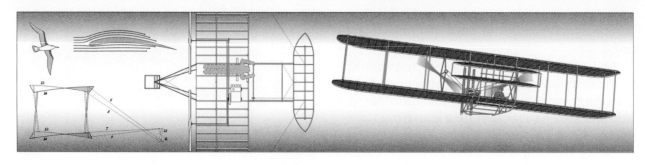

Implementation

8.1 Introduction

Implementation is the final step of the design process, in which the design becomes a reality. The designer details the product in working drawings with specifications and notes for its fabrication. Graphical methods are particularly important during implementation, because all products are manufactured from working drawings and specifications. Implementation also involves the packaging, warehousing, distribution, and sales of the manufactured product.

8.2 Working Drawings

Working drawings, with orthographic views, dimensions, and notes, describe how to make the individual parts of a product. The blank hanger in **Fig. 8.1** is drawn by computer as a working drawing for implementation in **Fig. 8.2**. Properly executed working drawings ensure that the resulting products will be

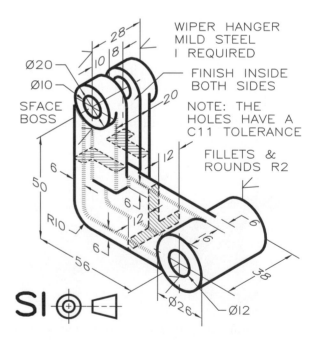

8.1 This hanger wiper is shown in a dimensioned working drawing in Fig. 8.2 with three orthographic views.

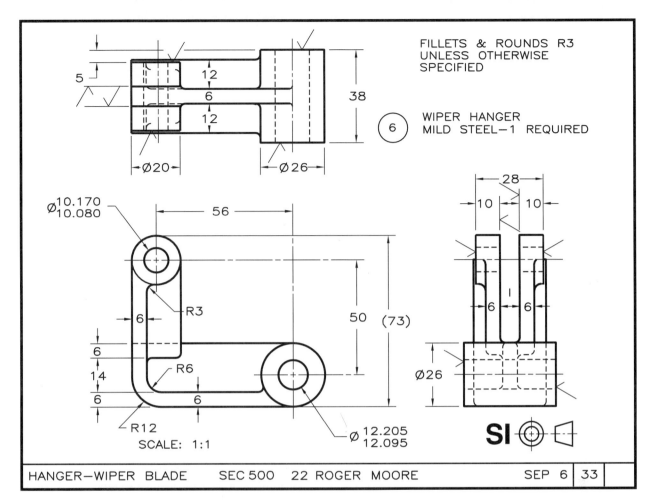

FILLETS & ROUNDS R3
UNLESS OTHERWISE
SPECIFIED

WIPER HANGER
(6) MILD STEEL—1 REQUIRED

HANGER—WIPER BLADE   SEC 500   22 ROGER MOORE   SEP 6 | 33

8.2 This computer-drawn working drawing enables the design for this part to be implemented as a final product.

identical when the instructions on the drawings are followed, regardless of the shop in which they are made.

When making working drawings, designers draw several parts on the same sheet without attempting to arrange them in relationship to each other or in order of their assembly. The names of the parts, their identifying numbers, the quantity required, and the materials to be used in making them are noted near the views.

8.3 Specifications

Specifications are written notes and instructions that supplement the information shown in drawings. Specifications may be prepared as separate typed documents that accompany drawings or that stand alone when graphical representation is unnecessary. Instructional notes such as the following are adequate as written specifications without drawings:

> METALLURGICAL INSPECTION IS
> REQUIRED BEFORE MACHINING.

or

> PAINT WITH TWO COATS OF FLAT BLACK
> PAINT (NO. 780) AFTER FINISHING.

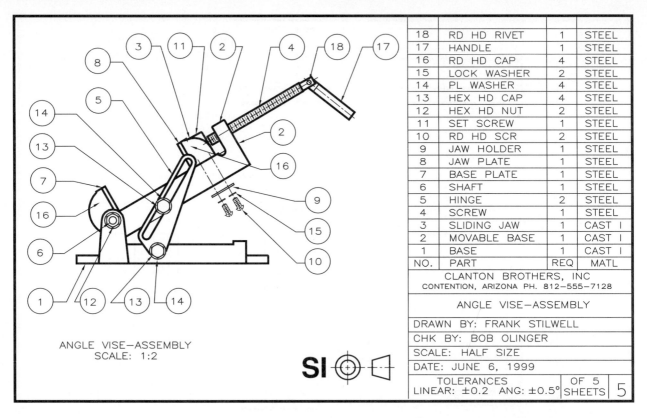

18	RD HD RIVET	1	STEEL
17	HANDLE	1	STEEL
16	RD HD CAP	4	STEEL
15	LOCK WASHER	2	STEEL
14	PL WASHER	4	STEEL
13	HEX HD CAP	4	STEEL
12	HEX HD NUT	2	STEEL
11	SET SCREW	1	STEEL
10	RD HD SCR	2	STEEL
9	JAW HOLDER	1	STEEL
8	JAW PLATE	1	STEEL
7	BASE PLATE	1	STEEL
6	SHAFT	1	STEEL
5	HINGE	2	STEEL
4	SCREW	1	STEEL
3	SLIDING JAW	1	CAST I
2	MOVABLE BASE	1	CAST I
1	BASE	1	CAST I
NO.	PART	REQ	MATL

CLANTON BROTHERS, INC
CONTENTION, ARIZONA PH. 812-555-7128

ANGLE VISE—ASSEMBLY

DRAWN BY: FRANK STILWELL
CHK BY: BOB OLINGER
SCALE: HALF SIZE
DATE: JUNE 6, 1999
TOLERANCES
LINEAR: ±0.2 ANG: ±0.5° | OF 5 SHEETS | 5

ANGLE VISE—ASSEMBLY
SCALE: 1:2

SI

8.3 This partially exploded orthographic assembly drawing of a vise shows how the individual parts fit together.

When space permits, specifications should be given on the working drawing rather than in a separate document.

8.4 Assembly Drawings

Assembly drawings illustrate how individual parts are to be put together to become the final product. They can be drawn as three-dimensional pictorials or orthographic views that are fully assembled, fully exploded, or partially exploded. **Figure 8.3** shows a partially exploded orthographic view of an assembly with part numbers in balloons. An assembly drawing usually contains a parts list for easy reference.

8.5 Miscellaneous Considerations

After preparing drawings and specifications, designers must consider other aspects of implementation: product packaging, storage, shipping, and marketing.

Packaging

In some industries, packaging is elaborate and may be as expensive as the product. Designers in the toy industry, for example, must be aware of packaging problems as they develop a design because a product that is difficult to package will cost more. Many products are shipped partially disassembled to make packaging easier and cheaper.

Storage

Most manufacturers maintain an inventory of products for shipment. Therefore, warehousing costs must be figured into the product's final selling price.

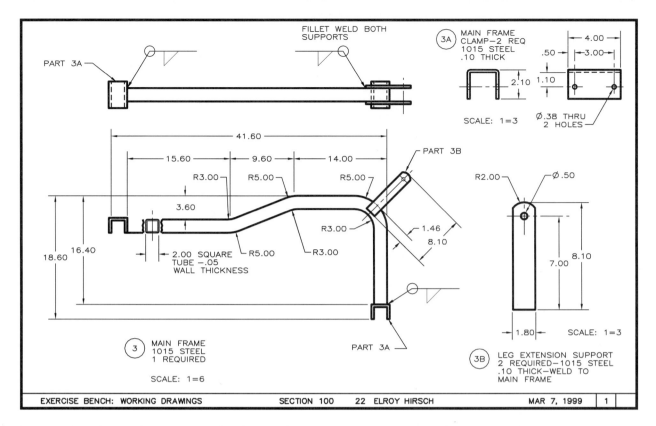

8.4 A working drawing depicting exercise bench parts (sheet 1 of 5).

Shipping

Industries that locate warehouse facilities in the middle of their market areas have lower shipping costs than those with warehouses at the peripheries of their market areas.

Marketing

Designers must be concerned with all aspects of a product after it enters the marketplace, including its marketability and consumer acceptance. Complaints about a product's reliability and function are important to designers, alerting them to design or manufacturing defects that must be overcome in future versions of the product.

8.6 Implementation: Exercise Bench

To illustrate the implementation of a product design, we return to the exercise bench, which was introduced in Chapter 3 and has been used to demonstrate the application of each step in the design process.

Working Drawings

The two working drawings shown in **Figs. 8.4** and **8.5** depict some details of the exercise bench design. Additional working drawings are required to show the other parts of the bench, which are dimensioned in decimal inches. Standard parts to be purchased from suppliers are not drawn but are itemized on the drawing, given part numbers, and listed in the parts list on the assembly drawing.*

*This design was developed, patented, and marketed by Weider Health and Fitness, 2100 Erwin Street, Woodland Hills, CA 91367.

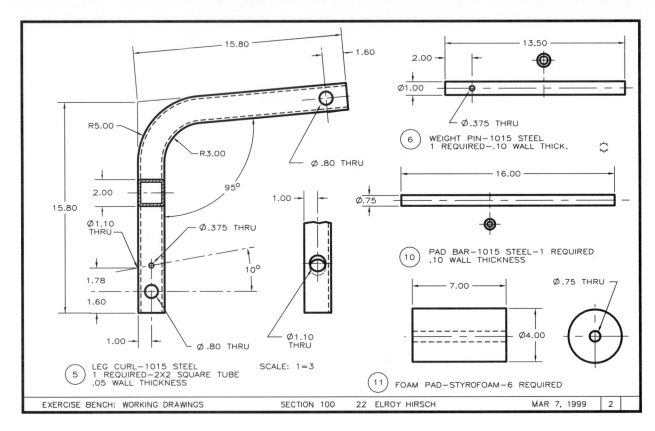

8.5 Another working drawing depicting exercise bench parts (sheet 2 of 5).

Assembly Drawing

Figure 8.6 shows an assembly drawing that illustrates how the parts are to be assembled after they have been made. The assembly is shown pictorially, with the different parts identified by numbered balloons attached to leaders. The parts list identifies each part by number and describes it generally.

Packaging

The Weider exercise bench is packaged in a corrugated cardboard box and weighs approximately 40 pounds. It is shipped unassembled so that it will fit into a smaller carton for ease of handling during shipment (**Fig. 8.7**).

Storage

An inventory of benches must be maintained to meet retailer demand. The need to hold inventory increases overhead costs for interest payments, warehouse rent, warehouse personnel, and loading equipment.

Shipping

Shipping costs for all types of carriers (rail, motor freight, air delivery, and mail services) must be evaluated. The shipping cost for a Weider bench with its accessories is $10 to $15, depending on distance, when shipped one at a time by United Parcel Service. The cost per unit is about 50 percent less when units are shipped in bundles of ten to the same destination.

Accessories

Examples of accessories, or add-ons, are the butterfly attachments for arm exercises.

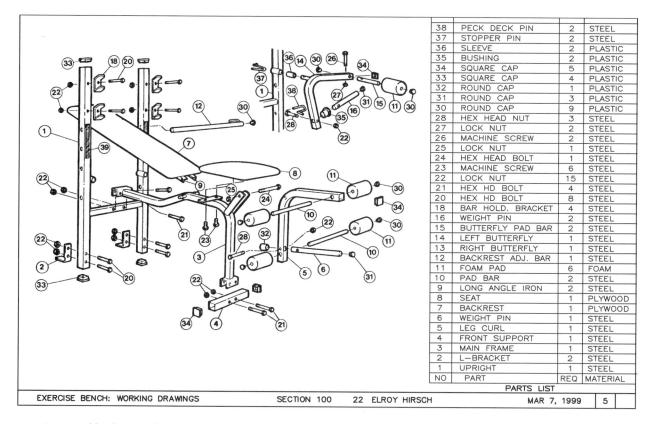

38	PECK DECK PIN	2	STEEL
37	STOPPER PIN	2	STEEL
36	SLEEVE	2	PLASTIC
35	BUSHING	2	PLASTIC
34	SQUARE CAP	5	PLASTIC
33	SQUARE CAP	4	PLASTIC
32	ROUND CAP	1	PLASTIC
31	ROUND CAP	3	PLASTIC
30	ROUND CAP	9	PLASTIC
28	HEX HEAD NUT	3	STEEL
27	LOCK NUT	2	STEEL
26	MACHINE SCREW	2	STEEL
25	LOCK NUT	1	STEEL
24	HEX HEAD BOLT	1	STEEL
23	MACHINE SCREW	6	STEEL
22	LOCK NUT	15	STEEL
21	HEX HD BOLT	4	STEEL
20	HEX HD BOLT	8	STEEL
18	BAR HOLD. BRACKET	4	STEEL
16	WEIGHT PIN	2	STEEL
15	BUTTERFLY PAD BAR	2	STEEL
14	LEFT BUTTERFLY	1	STEEL
13	RIGHT BUTTERFLY	1	STEEL
12	BACKREST ADJ. BAR	1	STEEL
11	FOAM PAD	6	FOAM
10	PAD BAR	2	STEEL
9	LONG ANGLE IRON	2	STEEL
8	SEAT	1	PLYWOOD
7	BACKREST	1	PLYWOOD
6	WEIGHT PIN	1	STEEL
5	LEG CURL	1	STEEL
4	FRONT SUPPORT	1	STEEL
3	MAIN FRAME	1	STEEL
2	L—BRACKET	2	STEEL
1	UPRIGHT	1	STEEL
NO	PART	REQ	MATERIAL

PARTS LIST

EXERCISE BENCH: WORKING DRAWINGS | SECTION 100 | 22 ELROY HIRSCH | MAR 7, 1999 | 5

8.6 An assembly drawing for the Weider exercise bench (sheet 5 of 5). (Courtesy of Weider Health and Fitness.)

Accessories enable buyers to upgrade the basic product in stages, which can increase product marketability and sales.

Prices

The retail price of the Weider bench is about $100. This type of product generally retails for about five or six times the cost of manufacturing them (materials and labor). Retailers receive approximately a 40-percent margin, distributors earn about 10 percent, and the remainder of the price represents advertising costs and the other miscellaneous costs mentioned previously. The consumer pays all of these costs (prorated to each exercise bench) as part of the purchase price.

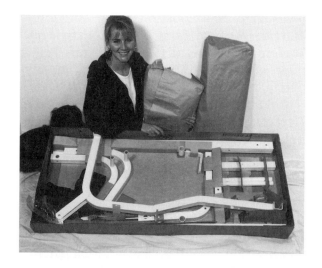

8.7 The exercise bench is packaged unassembled and flat for ease of packaging and handling during shipment.

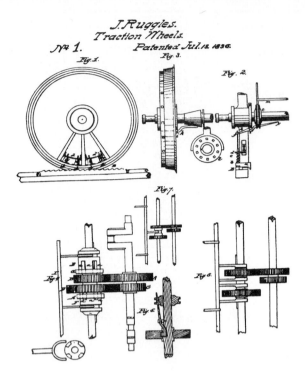

J. Ruggles.
Traction Wheels.
№ 1. Patented Jul. 12. 1836.

T. A. EDISON.
Electric-Lamp.
No. 223,898. Patented Jan. 27, 1880.

8.8 The U.S. Patent and Trademark Office issued this first patent in 1836.

8.9 Thomas Edison received this patent for the electric lamp in 1880, thus beginning the electrical/electronics revolution.

8.7 Patents

Inventors of processes or products should investigate the possibility of obtaining patents on them from the U.S. Patent and Trademark Office (PTO) before disclosing their inventions. The PTO issued its first patent in 1836, a patent for traction wheels (**Fig. 8.8**). The patent procedure is outlined in *General Information Concerning Patents,* a publication available from the PTO from which the material in the following sections was extracted.

What May Be Patented?

Any person who "invents or discovers any new and useful process, machine, manufacture, or composition of matter, may obtain a patent," subject to the conditions and requirements of law. These categories include everything made by humans and the processes for making them. **Figure 8.9** illustrates the cover page of Edison's 1880 patent for the electric lamp.

Inventions used for the development of nuclear and atomic weapons for warfare are not patentable because they are not considered "useful." Also, a design for a mechanism that will not operate as described is not patentable. An idea or concept for a new invention is not patentable; it must be designed and described in detail before it can be considered for patent registration.

Who May Apply for a Patent?

Only the inventor may apply for a patent. A patent given to a person who was not the inventor would be void and the recipient subject to prosecution for perjury. However, the executor of a deceased inventor's estate may apply for a patent, and two or more people may apply for a patent as joint inventors.

Patent Rights

An inventor granted a patent has the right to exclude others from making, using, or selling the invention throughout the United States for 20 years from the time of application. At the end of that time, anyone may make, use, or sell the invention without authorization from the patent holder.

Application for a Patent

An inventor applying for a patent must provide the following:

1. A completed form that includes a petition, specification (description and claims), and oath or declaration;

2. A drawing, if a drawing is possible; and

3. The filing fee.

Petition and Oath   In the petition and oath (usually on one form), the inventor asks to be given a patent on the invention and declares that he or she is the original inventor of the device described in the application.

Specification   The inventor must submit a written specification, describing the invention in detail so that a person skilled in the field to which the invention pertains can produce the item. Drawings should carry figure numbers and contain part numbers for text references (**Fig. 8.10**).

Claims   The inventor's claims are brief descriptions of the invention's features that distinguish it from already patented items. The PTO studies claims to judge the novelty and patentability of an invention.

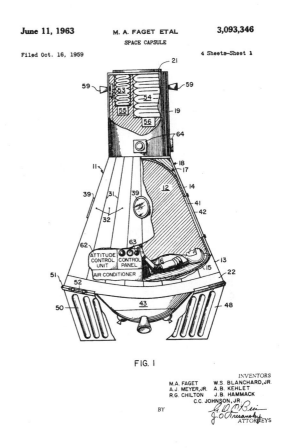

8.10   This patent drawing of a space capsule was developed by the National Aeronautics and Space Administration (NASA) in 1963.

Fee   As part of the application for a patent, the inventor must submit a $790 filing fee. Additional fees can be charged based on additional specifications and claims. An issue fee of $1320 is payable when the PTO grants the patent. As a general rule, patents cost the inventor about $4000–$6000, excluding attorney's fees.

8.8 Patent Drawings

A booklet, *Guide for Patent Draftsmen* (available from the U.S. Government Printing Office), outlines the required format for patent drawings. If the inventor cannot furnish drawings, the PTO will recommend a drafter who can prepare them at the inventor's expense.

Patent Drawing Standards

Patent drawings must meet the following standards.

Paper and Ink Drawings must be on pure white paper of the thickness of a two- or three-ply Bristol board with a surface that is calendared and smooth to permit erasure and correction. India ink is required for permanence and solid black lines. The use of white pigment to cover errors is not allowed.

Sheet Size and Margins Sheet size must be 8 1/2 by 14 in. (21.6 by 35.6 cm) or 21.0 by 29.7 cm. All sheets in a particular application must be the same size. One of the shorter sides is regarded as the top of the sheet. On 8 1/2-by-14-in. sheets, the top margin is 2 in. and the side and bottom margins are 1/4 in. Margin border lines cannot be drawn on the sheets, but all work must be included within the margins. Sheets may be punched with two 1/4-in. holes, with their centerlines 11/16 in. below the top edge and 23/4 inches apart and centered from the sides of the sheet. The margins for 21.0 by 29.7 cm sheets are 2.5 cm from the top, 2.5 cm from the left, 1.5 cm from the right, and 1 cm from the bottom.

Character of Lines All lines and lettering must be absolutely black, regardless of line thickness. Freehand work is to be avoided.

Hatching and Shading Hatching lines used to shade the surface of an object should be parallel and at least 1/20 in. apart (**Fig. 8.11**). Heavy lines are used on the shade side of the views if they do not cause confusion in the drawing. The light is assumed to come from the upper left-hand corner at an angle of 45°. **Figure 8.12** depicts several types of surface delineation.

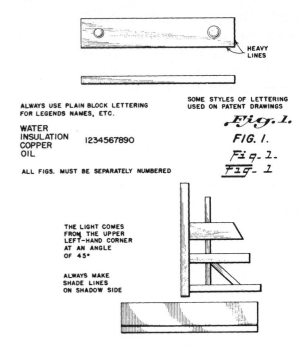

8.11 These are typical examples of lines and lettering recommended for patent drawings.

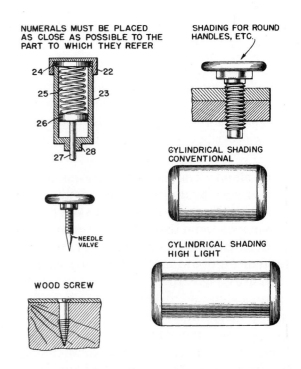

8.12 Several techniques of representing surfaces and beveled planes may be used on patent drawings.

Scale The scale must be large enough to show the mechanism without crowding when the drawing is reduced for reproduction. Portions of the mechanism may be drawn at a larger scale to show details.

Reference Characters The drafter should identify different views of a mechanism by consecutive plain, legible numerals at least 1/8-in. high figure numbers, not encircled, and placed close to their parts. A blank space should be provided on hatched surfaces if numbers are to be placed on them. The same part appearing in more than one view on the drawing should be labeled with the same numeral.

Symbols Symbols used to represent various materials in sections, electrical components, and mechanical devices are recommended by the PTO and conform to engineering drawing standards.

Signature and Names The signature or name of the applicant and the signature of the attorney or agent are placed in the lower right-hand corner of each sheet within the marginal lines or below the lower marginal line.

Views Figures should be numbered consecutively in order of their appearance. Figures may be plan, elevation, section, perspective, or detail views. Exploded views may be used to describe an assembly of multiple parts. Large parts may be broken into sections and drawn on several sheets if this approach is not confusing. Removed sections may be used if the cutting plane is labeled to indicate the section by number. All sheet headings and signatures are to be placed in the same position on the sheet whether the drawing is read from the bottom or the right of the sheet. Completed drawings should be sent flat, protected by heavy board, or rolled in a suitable mailing tube.

8.9 Patent Searches

A patent can be granted only after PTO examiners have searched existing patents to verify that the invention has not been patented previously. With more than 6,000,000 patents on record, a search is the most time-consuming part of obtaining a patent. Most inventors employ patent attorneys or agents to do preliminary searches for possible infringement on other patents.

8.10 Questions and Answers

We used the PTO's pamphlet, *Questions and Answers About Patents,* to answer the following questions.

Nature and Duration of Patents

1. **Q.** *What is a patent?*

   A. A patent is a grant issued by the U.S. Government, giving an inventor the right to exclude all others from making, using, or selling his or her invention within the United States, its territories, and possessions.

2. **Q.** *For how long is a patent granted?*

   A. Twenty years from the date on which an application is filed; except for patents on ornamental designs, which are granted for terms of 3-1/2, 7, or 14 years.

3. **Q.** *May the term of a patent be extended?*

   A. Only by special act of Congress, which occurs rarely and only under exceptional circumstances.

4. **Q.** *Does the person granted the patent have any control over the patent after it expires?*

   A. No. Anyone has the right to use an invention covered in an expired patent so long as they do not use features covered in other unexpired patents.

5. **Q.** *On what subject matter may a patent be granted?*

A. A patent may be granted to the inventor or discoverer of any new and useful process, machine, manufacture, or composition of matter, or any new and useful improvement thereof, or on any distinct and new variety of plant, or on any new, original, and ornamental design for an article of manufacture.

6. **Q.** *What may not be patented?*

A. A patent may not be granted on a useless device, on printed matter, on a method of doing business, on an improvement in a device that would be obvious to a person skilled in the art, or on a machine that will not operate, particularly on alleged perpetual motion machines.

7. **Q.** *What do "patent pending" and "patent applied for" mean?*

A. They are used by a manufacturer or seller of an article to indicate that a patent application for that article is on file with the U.S. Patent and Trademark Office. Those using these terms falsely to deceive the public can be fined.

8. **Q.** *I have made some changes and improvements in my invention after my patent application was filed with the PTO. May I amend my patent application by adding a description or illustration of these features?*

A. No. The law provides that new matter shall not be introduced into a patent application. You should call to the attention of your patent agent any such changes you may make, or plan to make, so steps may be taken for your protection.

Thomas Alva Edison, the most important inventor in history, was granted more than 1000 patents.

9. **Q.** *How does someone apply for a patent?*

A. By making application to the Commissioner of Patents, Patent and Trademark Office, Washington, DC, 20231.

10. **Q.** *What are the PTO's fees in connection with filing of an application for patent and issuance of the patent?*

A. A filing fee of $790 plus certain additional charges for claims, depending on their number and the manner of their presentation, are required when the application is filed. An issue fee of $1320 plus certain printing charges are required if the patent is to be granted.

11. **Q.** *Are models required as a part of the application?*

A. Only in exceptional cases. The PTO has the authority to require that a model be submitted, but rarely exercises it.

12. **Q.** *Is it necessary for me to go to the PTO in Washington to transact business concerning patent matters?*

A. No. Most business is conducted by correspondence. Interviews regarding pending applications can be arranged with examiners if necessary and often are helpful.

13. **Q.** *Can the PTO give me advice about whether to apply for a patent?*

    A. No. It can only consider the patentability of an invention when an application comes before it.

14. **Q.** *Is there any danger that the PTO will give others information contained in my application while it is pending?*

    A. No. All patent applications are kept secret until the patent is issued. After the patent is issued, the PTO file containing the application and all correspondence leading to its issuance is made available in the Patent Office Search Room to anyone, and copies may be purchased from the PTO.

15. **Q.** *May I write to the PTO about my application after it is filed?*

    A. The PTO will answer your inquiries about the status of the application and indicate whether the application has been rejected, allowed, or is awaiting action. However, you should forward correspondence through your patent attorney or agent.

16. **Q.** *What happens when two inventors apply separately for a patent on the same invention?*

    A. The PTO declares an "interference" and requires that testimony be submitted to determine which inventor is entitled to the patent.

17. **Q.** *May applications be examined out of their regular order?*

    A. No. All applications are examined in the order in which they are filed, except under special conditions.

When to Apply for a Patent

18. **Q.** *I have been making and selling my invention for the past 13 months and have not filed any patent application. Is it too late for me to apply?*

    A. Yes. A patent may not be obtained if the invention has been in public use or for sale in this country for more than a year prior to application. Your own use and sale of it for more than a year before filing will bar your right to a patent as though someone else had done so.

19. **Q.** *I published an article describing my invention in a magazine 13 months ago. Is it too late to apply for a patent?*

    A. Yes. The inventor is not entitled to a patent if the invention has been described in a printed publication anywhere in the world more than a year before filing an application.

20. **Q.** *If two or more people work together on an invention, to whom will the patent be granted?*

    A. If each had a share in the ideas forming the invention, they are joint inventors and a patent will be issued to them jointly if an application is filed by them jointly. If one person provided all the ideas and the other has only followed instructions in making the device, the person contributing the ideas is the sole inventor and the patent application and patent should be in his or her name only.

21. **Q.** *If one person furnishes all the ideas for an invention and someone else employs that person or furnishes the money for building and testing the invention, should the patent application be filed by them jointly?*

A. No. The application must be signed, executed, sworn to, and filed in the name of the inventor, who is the person furnishing the ideas, not the employer or the person furnishing the money.

22. **Q.** *May a patent be granted if an inventor dies before filing an application?*

A. Yes. The application may be filed by the executor or administrator of the inventor's estate.

23. **Q.** *While in England this summer, I found an ingenious article that has not been introduced into the United States or patented. May I obtain a U.S. patent on it?*

A. No. A U.S. patent may be obtained only by the inventor, not by someone learning of someone else's invention.

24. **Q.** *May the inventor sell or otherwise transfer the right to the patent or patent application to someone else?*

A. Yes. The inventor may sell all or part of the interest in the patent application or patent to anyone by a properly worded legal assignment. However, the application for a patent must be filed in the name of the inventor, not in the name of the purchaser.

25. **Q.** *Is it advisable to conduct a search of patents and other records before applying for a patent?*

A. Yes. If the device has been patented previously, making an application is useless. A patent search avoids the expense of filing a needless application.

Technical Knowledge Available from Patents

26. **Q.** *May I obtain information through patents of what has been done by others to solve a particular problem?*

A. The patents in the Patent Office Search Room in Washington contain a wealth of technical information. The information is organized so that you can easily find and review previous work related to your problem or general field of interest. You may review these patents personally, or hire a patent practitioner to do so and send you copies of patents related to your problem.

27. **Q.** *Can I obtain information about patents and the patent process on the World Wide Web?*

A. Yes. You may contact the Patent Office at WWW.USPTO.GOV to obtain information about patents and trademarks plus most answers to questions that you will have about patents. Although the patents are accessible by computer, the patent drawings are not available by computer.

28. **Q.** *If I obtain a patent on my invention, will that protect me against the claims of others who assert that I am infringing on their patents when I make, use, or sell my own invention?*

A. No. There may be a patent of a more basic nature on which your invention is an improvement. If your invention is a detailed refinement or feature of a basically protected invention, you may not use it without consent of the patentee, just as no one will have the right to use your patented improvement without your consent.

29. **Q**. *Will the PTO help me prosecute others if they infringe on the rights granted me by my patent?*

A. No. The PTO has no jurisdiction over questions relating to the infringement of patent rights. If your patent is infringed upon, you may sue the infringer at your own expense.

Patent Protection in Foreign Countries

30. **Q**. *Does a U.S. patent give protection in foreign countries?*

A. No. The U.S. patent protects your invention only in this country. If you want to protect your invention in foreign countries, you must file an application in the patent office of each such country within the time required by law.

Problems

1. Prepare working and assembly drawings as the implementation step of the design process for one of the refinement problems at the end of Chapter 5 on 11-by-17-in. sheets of tracing vellum or film.

2. Prepare working and assembly drawings for one of the problems assigned or selected from those at the end of Chapter 23. Produce the drawings on 11-by-17-in. sheets of tracing vellum or film.

Patents

3. Write for a copy of a patent that is of interest to you. List the features used as a basis for the patent.

4. Suggest modifications to the patent obtained in Problem 3. Sketch innovations that would improve the patented mechanism.

5. Write to the U.S. Patent and Trademark Office for patent application forms. Prepare a patent application for a simple invention that has been previously patented, such as a fountain pen, drafting instrument, or similar item. Determine the drawings and materials needed to complete your application.

6. Make a list of ideas for products that you believe to be patentable.

7. Write a technical report on the history and significance of the patent system and its role in our industrial society. Consult your library and available government publications on patents.

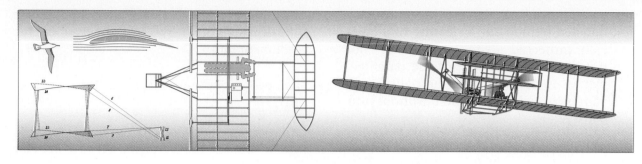

Design Problems

9.1 Introduction

This chapter offers problems that are suitable for both individual assignments and team projects to provide experience in applying the methods of creative problem solving presented in this textbook. Graphics has many applications during the design process: All new products begin with sketches at the preliminary idea step and end with working and assembly drawings at the implementation step.

9.2 The Individual Approach

The solution of short problems (one to two hours) is best suited to students working alone. Although simple design problems may involve fewer details and less depth than comprehensive problems do, the same design steps are involved.

9.3 The Team Approach

An effectively organized team working on a problem represents more talent than the typical individual possesses. However, management of talent becomes as much of the process as solving the problem.

Team Size

Student design teams should have from three to eight members. Three is the minimum number needed for a valid team experience, and four is the number needed to minimize the possibility of domination by one or two members.

Team Composition

In practice, an engineering team often consists of representatives of different departments, or even different firms, who may be unacquainted. This situation can be advantageous because it reduces the impact of preconceived notions about individuals.

Team Leader
A leader is necessary for teams to function effectively. The leader is responsible for making assignments, ensuring that deadlines are met, and mediating disagreements.

9.4 Selection of a Problem
The best problem for a student design project is one that involves familiar and accessible conditions that can be observed, measured, and inspected. A design for a water-ski rack for an automobile is more feasible than is a design for a support bracket for an airplane. When a student design team selects a design problem, the team should prepare a written proposal identifying the problem and outlining its limits. Assignment of problems by the instructor in the classroom is analogous to assignments by a supervisor in the workplace.

9.5 Problem Specifications
An individual or team may be expected to complete any or all of the following tasks.

Short Problems (One or Two Hours)
1. Worksheets that record development of a design procedure (Chapter 2).
2. Freehand sketches of the design for implementation (Chapters 8 and 13).
3. Instrument drawings of the solution (Chapters 8 and 23).
4. Pictorial sketches (or drawings made with instruments or computer) illustrating the design (Chapters 4, 13, and 25).

5. Visual aids, flip charts, or other media for presentation to a group (Chapter 7).

Comprehensive Problems (40 to 100 Hours)
1. A proposal identifying the problem and outlining an approach for solving it (Chapters 2, 3, and 7).
2. Worksheets documenting the preliminary ideas for a solution (Chapter 4).
3. Schematic diagrams, flowcharts, or other graphics to illustrate refinements of the design (Chapter 5).
4. A market survey evaluating the product's possible acceptance and estimated profit (Chapters 3 and 6).
5. A model or prototype for analysis and/or presentation (Chapter 6).
6. Pictorials to illustrate features of the final design solution (Chapters 7 and 25).
7. Dimensioned working drawings and assembly drawings to give details and specifications (Chapters 8 and 23).
8. A written or oral report, illustrated with graphs and diagrams, to explain the method of solution and present conclusions and recommendations (Chapter 7).

9.6 Scheduling Team Activities
The semester schedule shown in **Fig. 9.1** is suggested for a comprehensive design project. Spreading design projects over the semester allows time for thinking about the problem, gathering information, and working on the solution. Refer to the exercise bench example in Chapters 3 through 8 as a guide for carrying out your project.

	MONDAY	WEDNESDAY	FRIDAY
1			
2			
3		Assign teams	
4		Identify problem	
5		Identify problem	
6		Brainstorm	
7		Preliminary ideas	
8		Refinement	
9		Refinement	
10		Analysis	
11		Decision	
12		Implementation	
13		Prepare	
14		Present	
15			
16			

9.1 The semester schedule for the integration of a design project.

9.2 Problem 6. Base redesign.

9.7 Short Design Problems

The following short design problems can be completed in less than two hours.

1. **Lamp bracket**. Design a bracket to attach a desk lamp to a vertical wall for reading in bed. It should be removable for use as a conventional desk lamp.

2. **Towel bar**. Design a towel bar for a kitchen or bathroom. Determine optimum size and consider styling, ease of use, and method of attachment.

3. **Pipe aligner for welding**. The initial requirement for joining pipes with a butt weld is to align the pipes. Design a device for aligning 2-to-4-in. diameter pipes for on-the-job welding.

4. **Boot puller**. Design a device for helping you remove cowboy boots from your feet.

5. **Side-mounted mirror**. Design an improved side-mounted rearview mirror for an automobile. Consider aerodynamics, protection from inclement weather, visibility, and other factors.

6. **Part modification** (**Fig. 9.2**). Modify the base that supports the 2-in. diameter shaft by changing the square base to a circular base with six holes instead of four. Modify the ribs accordingly.

7. **Pipe column support**. Design a base that can be attached to a concrete slab with bolts that would provide a base for 3-in. diameter pipe columns.

8. **Motor bracket** (**Fig. 9.3**). Design a bracket to support a motor. The plate should be the upper part of the finished bracket.

9. **Pulley bracket clamp** (**Fig. 9.4**). Design a clamp that can be attached to an overhead I-beam to support a pulley without welding or drilling holes in the beam.

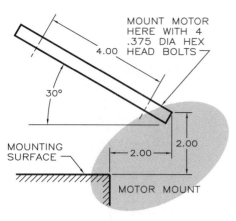

9.3 Problem 8. Motor bracket.

9.5 Problem 10. Pipe roll stand. (Courtesy of Grinnell Company, Inc.)

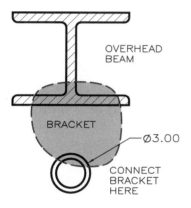

9.4 Problem 9. Pulley bracket clamp.

10. **Pipe roll stand** (**Fig. 9.5**) Design an alternative design to this roll stand that supports up to a 10 in. diameter pipe with a roller that allows expansion and contraction.

11. **Paint can holder**. Paint cans held by their wire bails are difficult to hold and get paint brushes into. Design a holding device that can be attached and removed easily from a gallon-size paint can. Consider weight, grip, balance, and function.

12. **Foot scraper**. Design a device that can be attached to the sidewalk for scraping mud from your shoes.

13. **Audiocassette storage unit**. Design a storage unit for an automobile that will hold several audiocassettes, making them accessible to the driver but not to a thief.

14. **Slide projector elevator**. Design a device for raising a slide projector to the proper angle for projection on a screen. It may be part of the original projector or an accessory to be attached to existing projectors.

15. **Book holder 1**. Design a holder to support a book while reading in bed.

16. **Book holder 2**. Design a holder to support a textbook or reference book at a workstation for ease of reading and accessibility.

17. **Table leg design**. Do-it-yourselfers build a variety of tables using hollow doors or plywood for the tops and commercially available legs. Determine standard heights for various types of tables and design a family of legs that can be attached to table tops with screws.

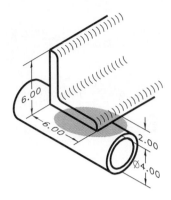

9.6 Problem 19. Pipe clamp.

9.7 Problem 21. Swivel caster.

18. **Leaf cart**. Design a cart or a container that leaves can be raked into while working in the yard and then transported to their final destination. Make it portable, light-weight, and economical.

19. **Pipe clamp** (**Fig. 9.6**). A pipe with a 4-in. diameter must be supported by angles that are spaced 8 ft apart. Design a clamp that will support the pipe without drilling holes in the angles.

20. **Toothbrush holder**. Design a toothbrush holder for a cup and two toothbrushes that can be attached to a bathroom wall.

21. **Caster lock** (**Fig. 9.7**). Design a device that will permit the caster to be locked in non-rolling mode when desired.

22. **Clothes hook**. Design a clothes hook that can be attached to a closet door for hanging clothes.

23. **Step stool**. Design a step stool that can be used for standing on to reach various heights as needed for home chores such as changing light bulbs.

24. **Hammock support**. Design a hammock support that will fold up and that will require minimal storage space.

25. **Doorstop**. Design a doorstop that can be attached to a wall or floor to prevent a doorknob from hitting the wall.

26. **Basketball goal**. Design a basketball goal that is easy to install for the 8–10 age range.

27. **Roller brackets** (**Fig. 9.8**). The rollers are to be positioned as shown to support a conveyor belt that carries bulk material. Design brackets to support the rollers.

28. **Jib-crane brackets**. Design the brackets at the joints indicated (**Fig. 9.9** A, B, and C) to form a jib crane made of a 8 in. × 4 in. × 8 ft long I-beam and a steel connecting rod. The crane should have at least a 180° swing.

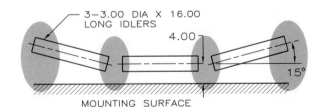

9.8 Problem 27. Roller brackets.

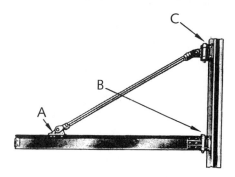

9.9 Problem 28. Jib crane bracket.

9.10
Problem 34.
Sit-up bench.

29. **Drawer handle**. Design a handle for a standard file cabinet drawer.

30. **Paper dispenser**. Design a dispenser that will hold a 6 × 24-in. roll of wrapping paper.

31. **Handrail bracket**. Design a bracket that will support a tubular handrail to be used on a staircase.

32. **TV yoke**. Design a yoke to hang from a classroom ceiling that will support a TV set and that will permit it to be adjusted for viewing from various parts of the room.

33. **Flagpole socket**. Design a flagpole socket that is to be attached to a vertical wall.

34. **Sit-up bench** (**Fig. 9.10**). Design a sit-up bench for exercising. Can you make it serve multiple purposes?

35. **Trash-can cover** (**Fig. 9.11**). Design a functional lid with an appropriate opening through which to put garbage.

36. **Conduit connector hanger** (**Fig. 9.12**). Design an attachment for a 3/4-in. conduit to support a channel used as a raceway for electrical wiring.

9.11 Problem 35.
Trash-can cover.

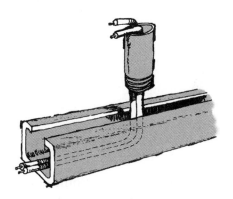

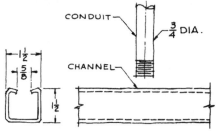

CONDUIT

$\frac{3}{4}$ DIA.

CHANNEL

$1\frac{1}{2}$

$\frac{5}{8}$

$1\frac{1}{2}$

9.12 Problem 36. Conduit connector hanger.

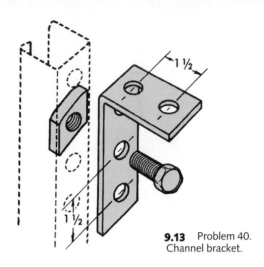

9.13 Problem 40. Channel bracket.

37. **Hose spool**. Design a spool/rack on which a garden hose can be wound and left neatly near the outside faucet.

38. **Gate hinge**. Design a hinge that can be attached to a 3-in. diameter tubular post to support a 3-ft-wide wooden gate.

39. **Cup holder**. Design a holder that will support a soft-drink can or bottle in an automobile.

40. **Channel bracket** (**Fig. 9.13**) The bracket shown is designed to fit on the flat side of the channel. Design a method of attaching the bracket to the slotted side so the inside nut will not drop down inside the channel.

9.8 Systems Design Problems

Systems problems require analysis of the interrelationship of various components as well the application of design principles.

41. **Multipurpose utility meter**. Residences have meters for electricity, water, and gas, which are checked monthly by separate utility companies. Consider the feasibility of combining all the meters into a single unit that one meter-reading service could read. Describe briefly how to organize and implement such a system.

42. **Portable bleachers**. Design portable bleachers that can be easily assembled, disassembled, and stored. Identify the uses that would justify their production for the general market.

43. **Archery range**. Determine the feasibility of providing an archery range that can be operated profitably. Investigate the potential market for such a facility and factors such as location, equipment needed, method of operation, utilities, concessions, parking, costs, and fees.

44. **Bicycle rental system**. Investigate the feasibility of a student-operated bicycle rental system. Determine student interest, cost factors, number of bikes needed, prices, personnel needs, storage, maintenance, and so on. Summarize your location, operating cost, and profitability conclusions.

45. **Model airplane field**. Investigate the need for a model-airplane field, including space requirements, types of surface needed, sound control, safety factors, and method of operation. Select a site on or near your campus that is adequate for this facility, and evaluate the equipment, utilities, and site preparation required.

46. **Overnight campsite**. Analyze the feasibility of converting a vacant tract of land near a major highway into sites for overnight campers. Determine the facilities required by the campers and the venture's profitability.

47. **Skateboard facility study**. Determine the cost of building and operating a skateboard facility on your campus. Consider where it could be located, how many students would use it, and the amounts of

equipment and labor necessary to operate it. Would it be financially feasible?

48. **Hot water supply**. Your weekend cottage does not have a hot-water supply, but cold water is available from a private well. Design a system that uses the sun's energy in the summer to heat water for bathing and kitchen use. Devise a system for heating the water in the winter by some other source. Determine whether one or both of these systems could be made portable for showers on camping trips and, if so, how.

49. **Information center**. Design a drive-by information center to help campus visitors find their way around. Determine the best location for it and the informational material needed, such as slides, photographs, maps, sound, and other audio-visual aids.

50. **Golf driving range ball-return system**. Balls at golf driving ranges are usually retrieved by hand or with a specially designed vehicle. Design a system capable of automatically returning balls to the tee area.

51. **Car wash**. Design a car wash facility for your campus that would be self-supporting and would provide the basics needs for washing cars. Think simple and economical.

52. **Instant motel**. Many communities need temporary housing for celebrations and sporting events. Investigate methods of providing an "instant motel" involving the use of tents, vans, trailers, train cars, or other temporary accommodations. Estimate profitability.

53. **Mountain lodge**. Determine the food supply and other provisions needed for a mountain lodge to accommodate six people who might be snowed-in for two

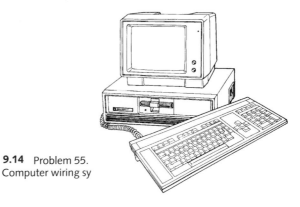

9.14 Problem 55.
Computer wiring sy

weeks. Identify and explain the features that should be included in the design of the lodge.

54. **Drive-through trash disposal system**. Design a system for leaving wrappers, boxes, and napkins left over after eating in your car at a fast-food drive-through. Preferably, this should be done without getting out of your car.

55. **Computer wiring system** (**Fig. 9.14**). Computer installations become cluttered with wiring as accessories are attached and access to connections becomes difficult. Design a system whereby the electrical wiring can be more organized, convenient, and accessible.

56. **Bonfire**. Develop a plan for building a bonfire to be burned the night before a major football game at your school. A few questions must first be answered: Where will logs be obtained? How many people will be required to build the bonfire? Where can it be located? How can it be funded? What hazards must be controlled?

57. **Stadium expansion**. Study the attendance figures for your school's stadium (football, basketball, or baseball) and predict future attendance. Recommend whether to enlarge the stadium. If you recommend that it be enlarged, determine the quantity

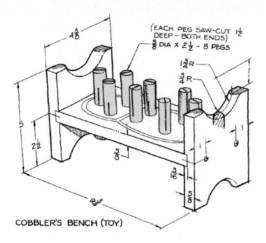

COBBLER'S BENCH (TOY)

9.15 Problem 62. Educational toy production.

of seating needed and prepare a schedule for adding it.

58. **Shopping checkout system**. A problem for grocery stores and shopping malls is check out, bagging, payment, and delivery of purchases to customers' automobiles. Develop a system that will improve these conditions.

59. **Injury-proof playground**. Design a playground that permits the greatest degree of participation by children with the least risk of injury.

60. **Modification of an existing facility**. Select a facility on your campus or in your community that is inadequate, such as a street intersection, parking lot, recreational area, or classroom. Identify its deficiencies and propose improvements to it.

61. **Recreational facility**. Analyze the various recreational activities on your campus that could be improved with the construction of a multipurpose facility to accommodate activities such as movies, plays, sports, meetings, and dances. The facility should be designed as an outdoor installation with the minimum of structures.

62. **Educational toy production** (Fig. 9.15). You are responsible for establishing the production system for manufacturing the educational toy shown. Do an analysis on what would be needed to produce 500 units per month—space, raw materials, office facilities, manufacturing facilities, people, and warehousing; the workstations required; and so forth. Calculate the cost of producing the item and the selling price necessary to provide a profit.

63. **Patio table production**. Proceed as in Problem 62 for production of a plywood patio table. Determine the number to be manufactured per month to break even and the selling price. Establish quantity breaks for selling prices for quantities that exceed the break-even level.

64. **Simple product**. Select a simple product and determine the production system and requirements for its manufacture, proceeding as in Problem 60. Your instructor should approve the product before you go ahead.

9.9 Product Design Problems

Product design involves developing a device that will perform a specific function, be mass-produced, and be sold to a large number of consumers.

65. **Hunting blind**. Design a portable hunting blind adequate for two geese or duck hunters that can be easily carried to a site by the hunters.

66. **Mailbox** (Fig. 9.16). Design a residential mailbox that either attaches to the house or is supported on a pole at the street.

67. **Writing table arm**. Design a writing table arm that can be attached to a folding chair.

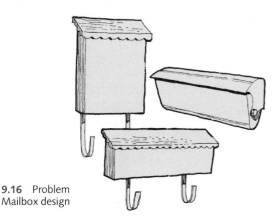

9.16 Problem Mailbox design

68. **Firewood rack** (**Fig. 9.17**). Design a rack for holding firewood inside near the fireplace. Can you give it multiple uses?

69. **Computer mount**. Design a device that can be clamped to a desktop for holding a computer, permitting it to be adjusted to various positions while leaving the desktop free to work on.

70. **Workers' stilts**. Design stilts to give workers access to an 8-ft high ceiling, permitting them to nail 4 × 8-ft ceiling panels into position.

71. **Pole-vault uprights**. Pole-vault uprights must be adjusted for each vaulter by moving them forward or backward 18 in. The crossbar must be replaced at heights of over 18 ft by using poles and ladders. Develop a more efficient set of uprights that can be readily adjusted and allow the crossbar to be replaced easily.

72. **Sports chair**. Design a sports chair that can be used for camping, for fishing from a bank or boat, at sporting events, and for other purposes.

73. **Bicycle child carrier**. Design a seat that can be used to carry a small child as a passenger on a bicycle.

74. **Car washer**. Design a garden-hose attachment that can apply water and agitation to wash a car. Suggest other applications for this device.

75. **Yard helper** (**Fig. 9.18**). Design a movable container on wheels that can be used for gardening and yard work.

76. **Drum truck** (**Fig. 9.19**) Design a truck that can be used for handling 55-gal drums of turpentine (7.28 lb per gallon) one at a time. Drums are stored in a vertical position and used in a horizontal position. The truck should be useful in tipping a drum into a horizontal position (as shown), as well as for moving the drum.

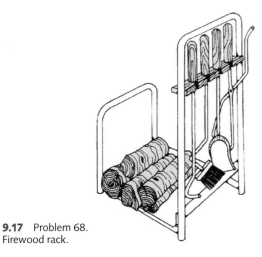

9.17 Problem 68. Firewood rack.

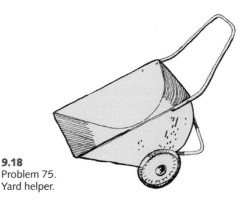

9.18 Problem 75. Yard helper.

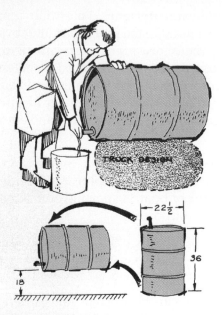

9.19   Problem 76. Drum truck.

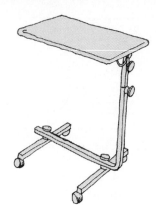

9.20   Problem 78. Bed table.

9.21   Problem 80. Monitor support.

77. **Power lawn-fertilizer attachment**. The rotary power lawn mower emits a force caused by the rotating blades that might be used to distribute fertilizer during mowing. Design an attachment for a lawn mower that can be used in this manner.

78. **Bed table** (**Fig. 9.20**). Modify the design of a hospital table to permit it to be used for studying, computing, writing, and eating in bed.

79. **Sawhorse**. Design a portable sawhorse for use in carpentry projects that folds up for easy storage. It should be about 36 in. long and 30 in. high.

80. **Monitor support** (**Fig. 9.21**). Design an arm support to position a computer monitor for ease of use.

81. **Projector cabinet**. Design a cabinet to serve as an end table or some other function while housing a slide projector and slide trays ready for use.

82. **Heavy appliance mover**. Design a device for moving large appliances—stoves, refrigerators, and washers—about the house for the purposes of rearranging, cleaning, and servicing them.

83. **Car jack**. The average car jack does not attach itself adequately to the automobile's frame or bumper, which causes a safety problem. Design one that would employ a different method of lifting a car on various types of terrain.

84. **Map holder**. Design a map holder to give the driver a view of the map in a convenient location in the car while driving.

9.22 Problem 86. Can crusher.

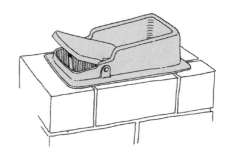

9.23 Problem 92. Chimney cover.

9.24 Problem 94.
Hand truck.

Provide a method of lighting the map that will not distract the driver.

85. **Stump remover**. Design an apparatus that can be attached to a car bumper to remove dead stumps by pushing or pulling.

86. **Can crusher** (**Fig. 9.22**). Design a device for flattening aluminum cans.

87. **Gate opener**. An annoyance to farmers and ranchers is the necessity of opening and closing gates. Design a manually operated gate that could be opened and closed by the driver from his vehicle.

88. **Paint mixer**. Design a product for use at paint stores or by paint contractors to mix paint quickly in the store or on the job.

89. **Automobile coffee maker**. Design a device that will provide hot coffee from the dashboard of an automobile. Consider the method of changing and adding water, the spigot system, and similar details.

90. **Portable hauler**. Design a portable, collapsible hauler that can be used for various home applications.

91. **Baby seat** (**cantilever**). Design a chair to support a child. It should attach to and be cantilevered from a standard table top. The chair should also be collapsible for ease of storage.

92. **Chimney cover** (**Fig. 9.23**). Design a chimney cover that can be closed from inside the house and will repel rain and reduce temperature loss.

93. **Miniature-TV support**. Design a device that would support TV sets ranging in size from 6×6 in. to 7×7 in. for viewing from a bed. Provide adjustments on the device to allow positioning of the set.

94. **Hand truck** (**Fig. 9.24**). Design a hand truck that can be used to carry a variety of items in the warehouse.

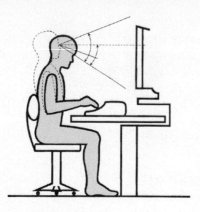

9.25 Problem 96. Computer workstation.

95. **Panel holder.** A worker applying 4 × 8-ft plasterboard to a ceiling needs a helper to hold the panel in place while she nails it. Design a device to hold the panel and eliminate the need for an assistant.

96. **Computer workstation** (**Fig. 9.25**). Design a computer workstation that reduces user discomfort and fatigue.

97. **Backpack.** Design a backpack that can be used for carrying camping supplies. Adapt the backpack to the human body for maximum comfort for extended periods of time. Suggest other uses for your design.

98. **Automobile controls.** Design a control system that can be easily attached to the standard automobile to permit driving without the use of the legs.

99. **Bathing apparatus.** Design an apparatus that would help a wheelchair-bound person to get in and out of a bathtub without assistance.

100. **Nutcracker** (**Fig. 9.26**). Design a nutcracker similar to the one shown.

101. **Adjustable TV base.** Design a base to support full-sized TV sets and allow maximum adjustment up and down and rotation about vertical and horizontal axes.

102. **Log splitter.** Design a device to aid in splitting logs for firewood.

103. **Projector cabinet.** Design a portable cabinet that can remain permanently in a classroom to house a slide projector and/or a movie projector in a ready-to-use position. It should provide both convenience and security from theft.

104. **Projector eraser.** Design a device to erase grease-pencil markings from the acetate roll of a specially equipped overhead projector as the acetate is cranked past the stage of the projector.

105. **Cement mixer.** Design a portable cement mixer that a home owner can operate manually. Such a mixer would be used only occasionally, so it should also be affordable to make it marketable.

9.26 Problem 100. Nutcracker.

106. **Boat trailer**. Design a trailer from which a boat hangs rather than riding on top of it so that the boat can be launched in very shallow water.

107. **Washing machine**. Design a manually operated washing machine. It may be considered an "undesign" of an electrically powered washing machine.

108. **Pickup truck hoist**. Design a manual lift that can be attached to the tailgate of a pickup truck for raising and lowering loads from the bed of the truck.

109. **Display booth**. Design a portable display booth for use behind or on an 8-ft long table for displaying your company's name and product information. It must be collapsible so that it can be carried by one person as airplane luggage.

110. **Toy Design** (**Fig. 9.27**). Design a toy similar to the one shown. Do a production and marketing analysis of it as covered in Problem 62.

111. **Shop bench**. Design an adjustable shop bench that is collapsible for easy storage and will accommodate accessories such as vises, anvils, and electrical tools.

112. **Deer-hunting seat**. Design a deer-hunting seat that can be carried to the field and attached to a tree trunk.

113. **Punching bag platform**. Design a portable platform for a speed punching bag that is adjustable to various heights and ships in a flat box.

114. **Shopping caddy**. Design a portable lightweight caddy that a shopper can use for carrying parcels.

115. **Patio grill**. Design a portable charcoal grill for cooking on the patio. Consider how it would be cleaned, stored, and used. Study competing products already on the market.

9.27 Problem 110. Toy design.

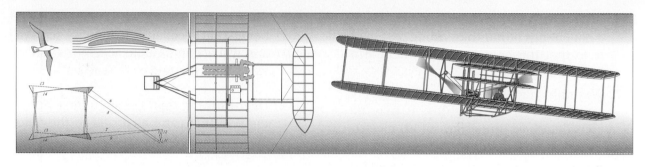

10

Drawing Instruments

10.1 Introduction

The preparation of technical drawings requires the ability to use drawing instruments. Even people with little artistic ability can produce professional technical drawings when they learn to use drawing instruments properly.

The drawing instruments covered in this chapter are traditional ones that are used by hand. Such instruments will always have an application in the development of drawings, but to a lesser degree than in the past.

Electronic drawing instruments—computers, plotters, scanners, and similar equipment—are covered in Chapter 35. The ability to produce computer graphics and to use the associated hardware is a necessary skill for engineers today.

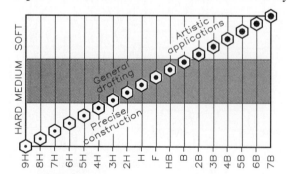

10.1 The hardest pencil lead is 9H, and the softest is 7B. The diameters of the hard leads are smaller than those of the soft leads.

10.2 Drawing Media

Pencils

A good drawing begins with the correct pencil and its proper use. Pencil grades range from the hardest, 9H, to the softest, 7B (**Fig. 10.1**). The pencils in the medium-grade range, 4H-B, are used most often for drafting work of the type covered in this textbook.

Figure 10.2 shows three standard pencils used for drawing. The leads used in the lead

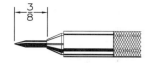

A. LEAD HOLDER

Holds any size lead; point must be sharpened.

B. FINE−LINE HOLDER

Must use different size holder for different lead sizes; does not need to be sharpened.

C. WOOD PENCIL

Wood must be trimmed and lead must be pointed.

D. THE PENCIL POINT

Sharpen point to a conical point with a lead pointer or a sandpaper pad.

10.2 Sharpen the drafting pencil to a tapered conical point (not a needle point) with a sandpaper pad or other type of sharpener.

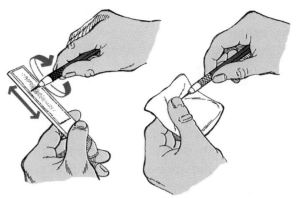

10.3 Revolve the drafting pencil about its axis as you stroke the sandpaper pad to form a conical point. Wipe away the graphite from the sharpened point with a cloth or tissue.

holder shown in Fig. 10.2A (the best all-around pencil of the three) are marked in white at their ends.

The fine-line leads used in the lead holder in Fig. 10.2B are more difficult to identify because their sizes are smaller and are not marked. A different fine-line holder must be used for each size of lead. The common sizes are 0.3 mm, 0.5 mm, and .007 mm, which are the diameters of the leads. A disadvantage of the fine-line pencil is the tendency of the lead to snap off when you apply pressure to it.

Although you have to sharpen it and point its lead, the wood pencil shown in Fig. 10.2C is a very satisfactory pencil. The grade of the lead is marked on one end of the pencil; therefore, the opposite end should be sharpened so the identity of the grade of lead will be retained. The wood can be sharpened with a knife or a drafter's pencil sharpener to leave about 3/8 inch of lead exposed.

You must sharpen pencil lead properly to obtain the conical point shown in Fig. 10.2D. To obtain this point, stroke the pencil lead against a sandpaper pad and revolve the pencil about its axis in the process (**Fig. 10.3**). Wipe excess graphite from the point with a cloth or tissue.

Although sharpening a pencil point with a sandpaper pad may seem outdated, it is a very suitable and practical way to sharpen pencil points and about the only way to sharpen compass points. However, there are a multitude of pencil pointers available, some of which work well while others do not.

Papers and Films

Sizes Sheet sizes are specified by the letters A–E. These sizes are multiples of either the standard 8 1/2 × 11-inch sheet (used by engineers) or the 9 × 12-inch sheet (used by architects) (**Fig. 10.4**). The metric sizes (A4–A0) are equivalent to the 8 1/2 × 11-inch modular sizes.

Detail Paper When drawings are not to be reproduced by the diazo or blue-line process, an opaque paper, called **detail paper,** can be

	ENGINEERS'	ARCHITECTS'		METRIC		
A	11" X 8.5"	12" X 9"	A4	297	X	210
B	17" X 11"	18" X 12"	A3	420	X	297
C	22" X 17"	24" X 18"	A2	594	X	420
D	34" X 22"	36" X 24"	A1	841	X	594
E	44" X 34"	48" X 36"	A0	1189	X	841

10.4 The three types of standard sheet sizes.

used as the drawing surface. The higher the rag content (cotton additive) of the paper, the better its quality and durability. You may draw preliminary layouts on detail paper and then trace them onto the final surface.

Tracing Paper   Tracing paper, or tracing vellum, is a thin, translucent paper that permits light to pass through it, allowing reproduction by the blue-line process. Tracing papers that yield the best reproductions are the most translucent ones. Tracing vellum is chemically treated to improve its translucency, but it does not retain its original quality as long as high-quality, untreated tracing papers do.

Tracing Cloth   Tracing cloth is a permanent drafting medium used for both ink and pencil drawings. It is made of cotton fabric and is coated with a starch compound to provide a tough, erasable drafting surface that yields excellent blue-line reproductions. Tracing cloth does not change shape as much as tracing paper with variations in temperature and humidity. Repeated erasures do not damage the surface of tracing cloth.

Polyester Film   An excellent drafting surface, polyester film is available under trade names such as Mylar. It is more transparent, stable, and tougher than paper or cloth, and it is waterproof. Mylar film is used for both pencil and ink drawings. A plastic-lead pencil must be used with some films, whereas standard lead pencils may be used with others.

10.3 Drawing Equipment

The Drafting Machine

Most professional drafters prefer the mechanical drafting machine (**Fig. 10.5**), which is attached to the drawing table top and has fingertip controls for drawing lines at any angle. A traditional, fully equipped drafting station is

10.5 The drafting machine is used for drawings made by hand. (Courtesy of Keuffel & Esser Company.)

10.6 The professional drafter or engineer may work in this type of environment. (Courtesy of Martin Instrument Company.)

shown in **Fig. 10.6.** Today's state-of-the-art offices are equipped with computer graphics stations, which have replaced much of the manual equipment (**Fig. 10.7**).

Triangles

The two types of triangles used most often are the 45° triangle and the 30°–60° triangle. The size of a 30°–60° triangle is specified by the

10.7 The computer graphics station is an integral part of today's classroom.

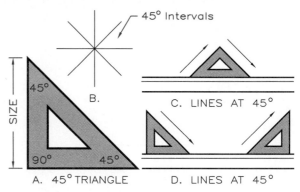

10.9 Use the 45° triangle to draw lines at 45° angles throughout 360°.

A. 45° TRIANGLE
B.
C. LINES AT 45°
D. LINES AT 45°

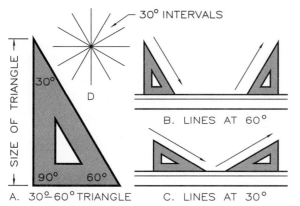

A. 30°–60° TRIANGLE
B. LINES AT 60°
C. LINES AT 30°
D

10.8 The 30°–60° triangle is used to draw lines at 30° intervals throughout 360°.

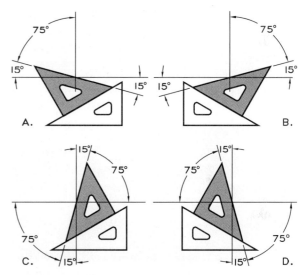

10.10 By using a 30°–60° triangle in combination with a 45° triangle, angles can be drawn at 15° intervals.

longer of the two sides adjacent to the 90° angle (**Fig. 10.8**). Standard sizes of 30°–60° triangles range in 2-inch intervals from 4 to 24 inches.

The size of a 45° triangle is specified by the length of the sides adjacent to the 90° angle. These range in 2-inch intervals from 4 to 24 inches, but the 6-inch and 10-inch sizes are adequate for most classroom applications. **Figure 10.9** shows the various angles that you may draw with this triangle. By using the 45° and 30°–60° triangles in combination, you may draw angles at 15° intervals throughout 360° (**Fig. 10.10**).

The Protractor

When drawing or measuring lines at angles other than multiples of 15°, a protractor is used (**Fig. 10.11**). Protractors are available as semicircles (180°) or circles (360°). Adjustable triangles with movable edges that can be set at different angles with thumbscrews are also available.

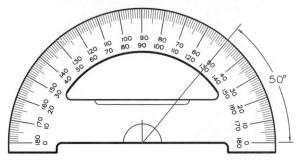

10.11 This semicircular protractor is used to measure angles.

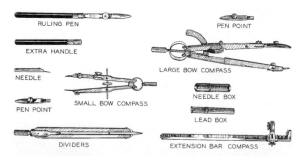

10.12 Drawing instruments can be purchased individually or as a set.

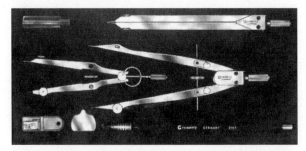

10.13 This is a typical cased set of drawing instruments. (Courtesy of Gramercy Guild.)

The Instrument Set

Figure 10.12 shows some of the basic drawing instruments. They are available individually or in cased sets, such as the one shown in **Fig. 10.13**.

Compass Use the compass to draw circles and arcs in pencil or ink (**Fig. 10.14**). To draw circles well with a pencil compass, sharpen the lead on its outside with a sandpaper pad (**Fig. 10.15**). A bevel cut of this type gives the best point for

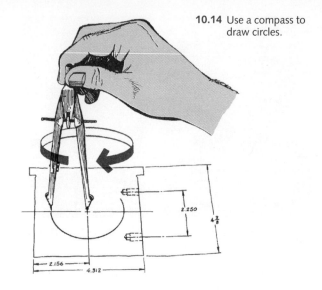

10.14 Use a compass to draw circles.

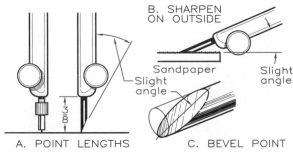

10.15 Sharpening the compass.
A Adjust the pencil point to be the same length as the compass point.
B&C Sharpen the lead from the outside with a sandpaper pad.

drawing circles. You cannot draw a thick arc in pencil with a single sweep; draw a series of thin concentric circles by adjusting the radius of the compass slightly.

When setting the compass pivot point in the drawing surface, insert it just enough for a firm set, not to the shoulder of the point. When the tabletop has a hard covering, place several sheets of paper under the drawing to provide a seat for the compass point.

Use a small bow compass (**Fig. 10.16**) to draw circles of up to a 2-inch radius. For larger circles, use an extension bar to extend the range of the large bow compass (**Fig. 10.17**).

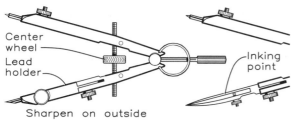

Center wheel
Lead holder

Inking point

Sharpen on outside

A. LEAD COMPASS   B. INK COMPASS

10.16 Use a small bow compass to draw circles of up to a 2-inch radius in pencil or in ink.

10.17 Use an extension bar with a bow compass to draw large arcs.

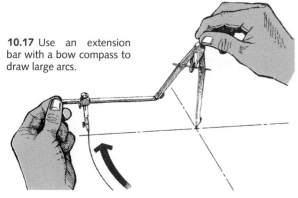

You may draw small circles conveniently with a circle template aligned with the centerlines of the circles (**Fig. 10.18**).

Divider The divider looks like a compass without a drawing point. It is used for laying off and transferring dimensions onto a drawing. For example, you can step off equal divisions rapidly and accurately along a line (**Fig. 10.19**). As you make each measurement, the divider's points make a slight impression in the drawing surface.

10.4 Lines

The type of line produced by a pencil depends on the hardness of its lead, the drawing surface used, and your drawing technique. You must experiment in order to achieve the ideal combination.

Dividers are also used to transfer dimensions from a scale to a drawing (**Fig. 10.20**) or to divide a line into a number of equal parts. Bow

10.18 Circle templates are convenient for drawing small circles without a compass.

10.19 Use a divider to step off measurements.

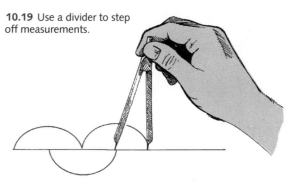

10.20 Use dividers to transfer dimensions from a scale to a drawing.

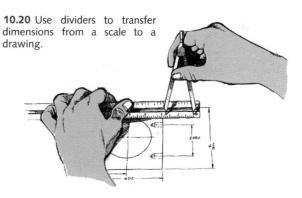

dividers (**Fig. 10.21**) are useful for transferring smaller dimensions, such as the spacing between lettering guidelines.

Various templates are available for use in drawing nuts and bolts, circles and ellipses, architectural symbols, and many other shapes.

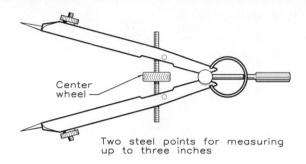

Center wheel

Two steel points for measuring up to three inches

10.21 Bow dividers are used to transfer small dimensions such as spacing for guidelines for lettering.

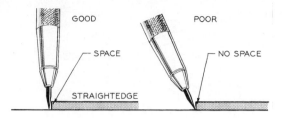

GOOD

POOR

SPACE

NO SPACE

STRAIGHTEDGE

10.24 To draw the best line, hold the pencil or pen point in a plane perpendicular to the paper, leaving a space between the point and the straightedge.

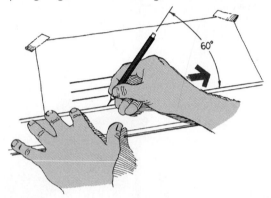

60°

10.22 You may use a straightedge and a 45° triangle to draw a series of parallel lines. Hold the straightedge firmly in position and move the triangle into position to draw line CD.

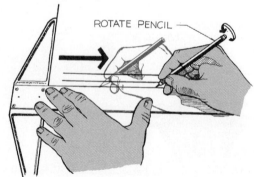

ROTATE PENCIL

10.23 Rotate the pencil about its axis while drawing so that its point will wear evenly.

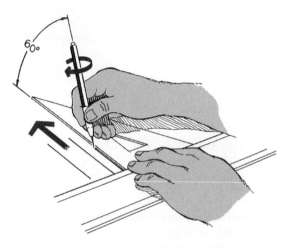

60°

10.25 Draw vertical lines along the left side of a triangle (if you are right-handed) in an upward direction, holding the pencil or pen in a plane perpendicular to the paper and at 60° to the surface.

Horizontal Lines

To draw a horizontal line, use the upper edge of your horizontal straightedge and make strokes from left to right if you are right-handed (**Fig. 10.22**) and from right to left if you are left-

handed. Rotate the pencil about its axis so that its point will wear evenly (**Fig. 10.23**). Darken pencil lines by drawing over them with multiple strokes. For drawing the best line, leave a small space between the straightedge and the pencil or pen point (**Fig. 10.24**).

Vertical Lines

Use a triangle and a straightedge to draw vertical lines. Hold the straightedge firmly with one hand and position the triangle where needed and draw the vertical lines with the other hand (**Fig. 10.25**). Draw vertical lines upward along the left side of the triangle if you are right-handed and upward along the right side of the triangle if you are left-handed.

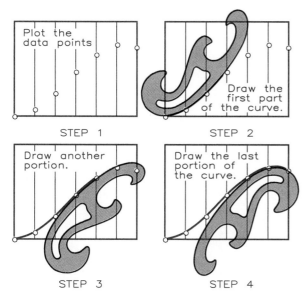

| STEP 1 | STEP 2 |
| Plot the data points | Draw the first part of the curve. |

| STEP 3 | STEP 4 |
| Draw another portion. | Draw the last portion of the curve. |

10.26 Using the irregular curve.

Step 1 Plot the data points.

Step 2 Position the irregular curve to pass through as many points as possible and draw that portion of the curve.

Step 3 Reposition the irregular curve and draw another portion of the curve.

Step 4 Draw the last portion to complete the smooth curve.

Irregular Curves

Curves that are not arcs must be drawn with an irregular curve (sometimes called French curves). These plastic curves come in a variety of sizes and shapes, but the one shown in **Fig. 10.26** is typical. Here, we used the irregular curve to connect a series of points to form a smooth curve.

Erasing Lines

Always use the softest eraser that will do a particular job. For example, if you are erasing pencil lines, do not use ink erasers because they are coarse and may damage the surface of the paper. When working in small areas, use an erasing shield to avoid accidentally removing adjacent lines (**Fig. 10.27**). Follow this step by brushing away the "crumbs" with a dusting brush. Wiping the crumbs away with your hands will smudge the drawing. A typical

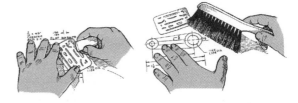

10.27 Use an erasing shield in tight spots. Use a brush, not your hand, to brush away the erasure crumbs.

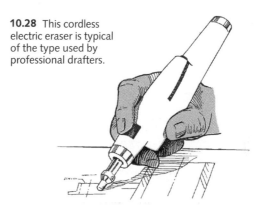

10.28 This cordless electric eraser is typical of the type used by professional drafters.

cordless electric eraser, shown in **Fig 10.28**, can be used with several grades of erasers to meet various needs.

10.5 Measurement

Scales

All engineering drawings require the use of scales for measuring lengths and sizes. Scales are flat or triangular and are made of wood, plastic, or metal. **Figure 10.29** shows triangular architects', engineers', and metric scales. Most scales are either 6 or 12 inches long.

Architects' Scale

Drafters use architects' scales to dimension and scale features such as room-size, cabinets, plumbing, and electrical layouts. Most indoor measurements are made in feet and inches with the architects' scale. **Figure 10.30** shows how to indicate the scale you are using on a drawing. Place this scale designation in the title block or in a prominent location on the drawing. Because dimensions measured

A. ARCHITECTS' SCALE

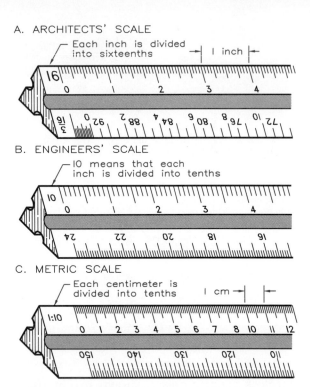

B. ENGINEERS' SCALE

C. METRIC SCALE

10.29 The architects' scale (A) measures in feet and inches. The engineers' scale (B) and the metric scale (C) are calibrated in decimal units.

with the architects' scale are in feet and inches, you must convert all dimensions to decimal equivalents (all feet or all inches) before making calculations.

Use the 16-scale for measuring full-size lines (**Fig. 10.31A**). An inch on the 16-scale is divided into sixteenths to match the ruler used by carpenters. The measurement shown is 3 1/8 in. When the measurement is less than 1 ft, a zero may precede the inch measurements, with inch marks omitted, or 0′–3-1/2.

Figure 10.31B shows the use of the 1 = 1′-0 scale to measure a line. Read the nearest whole foot (2 ft, in this case) and then the remainder in inches from the end of the scale (3 1/2 in.) for a total of 2′–3-1/2. Note that, at the end of each scale, a foot is divided into inches for use in measuring fractional parts of feet in inches. The scale 1 in. = 1′-0 is the same

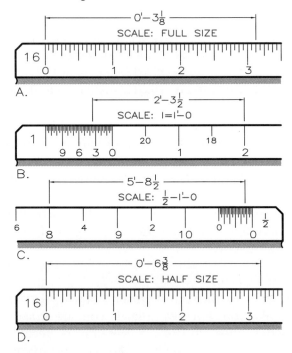

10.30 Use this basic form to indicate the scale on a drawing made with an engineers' scale.

10.31 Lines measured with an architects' scale.

as saying that 1 in. is equal to 12 in. or that the drawing is 1/12 the actual size of the object.

When you use the 1/2 = 1′-0 scale, 1/2 in. represents 12 in. on a drawing. Thus, the line in **Fig. 10.31C** measures 5′-8 1/2.

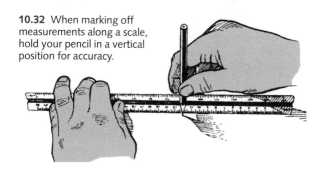

10.32 When marking off measurements along a scale, hold your pencil in a vertical position for accuracy.

2'-7½ Omit inch marks

4'-0¼ Zero here

0'-1½ Zero optional

10.33 Omit inch marks but show foot marks (according to current standards). When the inch measurement is less than a whole inch, use a leading zero.

To obtain a half-size measurement, divide the full-size measurement by 2 and measure it with the 16-scale. Half-size is sometimes specified as SCALE: 6 = 12 (inch marks omitted). The line in **Fig. 10.31D** measures to be 0'-6 3/8.

When marking measurements, hold your pencil or pen vertical for the greatest accuracy (**Fig. 10.32**). Letter dimensions in feet and inches are shown in **Fig. 10.33,** with fractions twice as tall as whole numerals.

Engineers' Scale

On the engineers' scale, each inch is divided into multiples of 10. Because it is used for making drawings of outdoor projects—streets, structures, tracts of land, and other topographical features—it is sometimes called the civil engineers' scale.

Figure 10.34 shows the form for specifying scales when using the engineers' scale. For example, scale: 1 = 10'. With measurements already in decimal form, performing calculations is easy; there is no need to convert from one unit to another as there is when you use the architects' scale.

FROM END OF SCALE

BASIC FORM SCALE: /= XX

EXAMPLE SCALES

10	SCALE: /=/'	SCALE: /=1,000
20	SCALE: /=200'	SCALE: /=20 LB
30	SCALE: /=3'	SCALE: /=3,000'
40	SCALE: /=4'	SCALE: /=40'
50	SCALE: /=50'	SCALE: /=500'
60	SCALE: /=6	SCALE: /=0.6'

10.34 Use this basic form to indicate the scale on a drawing made with an engineers' scale.

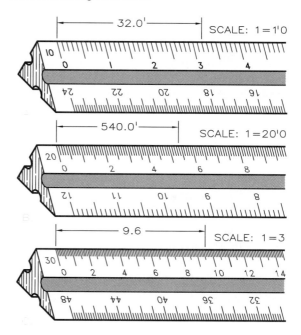

10.35 These lines are measured with engineer's scales.

Each end of the scale is labeled 10, 20, 30, and so on, which indicates the number of units per inch on the scale (**Fig. 10.35**). You may obtain many combinations simply by mentally moving the decimal places of a scale.

Figure 10.35A shows the use of the 10-scale to measure a line 32.0 ft long drawn at the scale of 1 = 10'. Figure 10.35B shows the use of the 20-scale to measure a line 540.0 ft long

.13 2.13 0.15' ~~0.13"~~

Omit zero; omit inch marks

GOOD
No zero in front of decimal

GOOD
Space for decimal

GOOD
Zero for fractional feet

POOR
Decimal point crowded

10.36 For decimal fractions in inches, omit leading zeros and inch marks. For feet, leave adequate space for decimal points between numbers and show foot marks.

Value		Prefix	Symbol	Pronunciation
1 000 000	= 10^6	= mega	M	megah
1 000	= 10^3	= kilo	k	keylow
100	= 10^2	= hecto	h	heck tow
10	= 10^1	= deka	da	dekah
1	=			
0.1	= 10^{-1}	= deci	d	des see
0.01	= 10^{-2}	= centi	c	cen tee
0.001	= 10^{-3}	= milli	m	mil lee
0.000 001	= 10^{-6}	= micro	μ	microw

10.37 These prefixes and abbreviations indicate decimal placement for SI measurements.

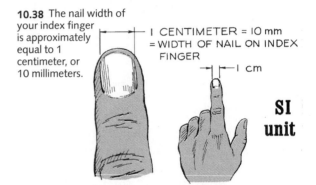

10.38 The nail width of your index finger is approximately equal to 1 centimeter, or 10 millimeters.

1 CENTIMETER = 10 mm = WIDTH OF NAIL ON INDEX FINGER

1 cm

SI unit

drawn at a scale of 1 = 200'. Figure 10.35C shows the use of the 30-scale to measure a line of 9.6 in. long at a scale of 1 = 3. **Figure 10.36** shows the proper format for indicating measurements in feet and inches.

English System of Units

The English (Imperial) system of units has been used in the United States, Britain (until recently), and Canada since it was established. This system is based on arbitrary units (of length) of the inch, foot, cubit, yard, and mile. Because there is no common relationship among these units, calculations are cumbersome. For example, finding the area of a rectangle that measures 25 in. × 6 3/4 yd first requires conversion of one unit to the other.

Metric System of Units

France proposed the metric system in the fifteenth century. In 1793, the French National Assembly agreed that the meter (m) would be one ten-millionth of the meridian quadrant of the earth and fractions of the meter would be expressed as decimal fractions. An international commission officially adopted the metric system in 1875.

The worldwide organization responsible for promoting the metric system is the **International Standards Organization (ISO).** It has endorsed the Système International d'Unites (International System of Units), abbreviated **SI.** Prefixes to SI units indicate placement of the decimal, as **Fig. 10.37** shows.

Metric Scales

The meter is 39.37 inches. The basic metric unit of measurement for an engineering drawing is the millimeter (mm), which is one-thousandth of a meter, or one-tenth of a centimeter. Dimensions on a metric drawing are understood to be in millimeters unless otherwise specified.

The width of the fingernail of your index finger is a convenient way to approximate the dimension of one centimeter, or ten millimeters (**Fig. 10.38**). **Figure 10.39** presents the basic form for indicating metric scales on a drawing.

Decimal fractions are unnecessary on most drawings dimensioned in millimeters. Thus, dimensions are rounded off to whole numbers except for those measurements dimensioned with specified tolerances. For metric measurements of less than 1, a zero goes in front of the decimal. In the English system, the zero is omitted from measurements of less than one inch (**Fig. 10.40**).

BASIC FORM *SCALE: I= XX* ←FROM END OF SCALE

EXAMPLE SCALES

SCALE: I:I (Imm=Imm; Icm=Icm)

SCALE: I:2 (Imm=2mm; Imm=20mm)

SCALE: I:3 (Imm=30mm; Imm=0.3mm)

SCALE: I:4 (Imm=4mm; Imm=40mm)

SCALE: I:5 (Imm=5mm; Imm=500mm)

SCALE: I:6 (Imm=6mm; Imm=60mm)

10.39 Use this basic form to indicate the scale of a drawing made with a metric scale.

22.0 — GOOD   0.15 — GOOD   .15 — POOR (Missing zero)

Too little room for decimal

146 — GOOD   14.6 — GOOD   14.6 — POOR

10.40 In the metric system, a zero precedes the decimal. Allow adequate space for it.

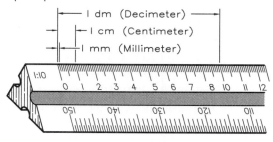

I dm (Decimeter)
I cm (Centimeter)
I mm (Millimeter)

$I\ dm=\frac{m}{I0}$ $I\ cm=\frac{m}{I00}$ $I\ mm=\frac{m}{I000}$ $I\ \mu m=\frac{m}{I\ 000\ 000}$

10.41 The decimeter is one-tenth of the meter, the centimeter is one-hundredth of a meter, a millimeter is one-thousandth of a meter, and a micrometer is one-millionth of a meter.

Metric scales are expressed as ratios: 1:20, 1:40, 1:100, 1:500, and so on. These ratios mean that one unit represents the number of units to the right of the colon. For example, 1:10 means that 1 mm equals 10 mm or 1 cm equals 10 cm or 1 m equals 10 m. The full-size metric scale (**Fig. 10.41**) shows the relationship between the metric units of the decimeter, centimeter, millimeter, and micrometer.

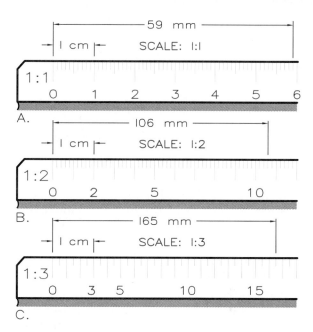

A. 59 mm — I cm — SCALE: I:I — 1:1 — 0 1 2 3 4 5 6

B. 106 mm — I cm — SCALE: I:2 — 1:2 — 0 2 5 10

C. 165 mm — I cm — SCALE: I:3 — 1:3 — 0 3 5 10 15

10.42 Measurements with metric scales.

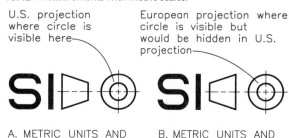

U.S. projection where circle is visible here

European projection where circle is visible but would be hidden in U.S. projection

A. METRIC UNITS AND THIRD ANGLE OF PROJECTION

B. METRIC UNITS AND FIRST ANGLE OF PROJECTION

10.43 The large SI indicates that measurements are in metric units. The partial cones indicate whether the views are drawn in (A) the third angle (U.S. system) or (B) the first angle of projection (European system).

The line shown in **Fig. 10.42A** measures 59 mm. Use the 1:2 scale when 1 mm represents 2 mm, 20 mm, 200 mm, and so on. The line shown in **Fig. 10.42B** measures 106 mm. **Figure 10.42C** shows a line measuring 165 mm, where 1 mm represents 3 mm.

Metric Symbols To indicate that drawings are in metric units, insert SI in or near the title block (**Fig. 10.43**). The two views of the partial cone denote whether the orthographic views were drawn in accordance with the U.S. system

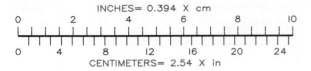

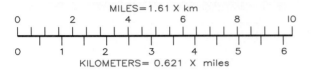

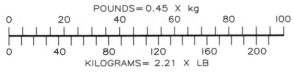

10.44 These scales show a comparison of the English system units with the metric system units.

Omit commas and group into threes	1 000 000	1,000,000
	GOOD	POOR
Use a raised dot for multiplication	N·M	NM
	GOOD	POOR
Precede decimals with zeros	0.72 mm	.72 mm
	GOOD	POOR
Methods of division	kg/m or kg·m^{-1}	
	GOOD	GOOD

10.45 Follow these general rules for expressing SI units.

(third-angle projection) or the European system (first-angle projection). **Figure 10.44** shows several comparisons of English and SI units.

Expression of Metric Units The general rules for expressing SI units are given in **Fig. 10.45.** Do not use commas to separate digits in large numbers; instead, leave a space between them, as shown.

Scale Conversion Appendix 2 gives factors for converting English to metric lengths and vice versa. For example, multiply decimal inches by 25.4 to obtain millimeters, and divide millimeters by 0.394 to obtain inches.

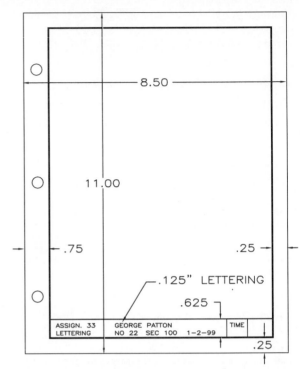

10.46 Use the format and title strip shown for a size AV sheet (8 1/2 × 11, vertical position) to present the solutions to problems at the end of each chapter.

Multiply an architects' scale by 12 to convert it to an approximate metric scale. For example, Scale: 1/8 = 1´-0 is the same as 1/8 in. = 12 in. or 1 in. = 96 in., which closely approximates the metric scale of 1:100. You cannot convert most metric scales exactly to English scales, but the metric scale of 1:60 does convert exactly to 1 = 5´ which is the same as 1 in. = 60 in.

10.6 Presentation of Drawings

The following formats are suggested for the presentation of drawings. Most problems can be drawn and solved on 8 1/2 × 11-inch sheets with a title strip, as **Fig. 10.46** shows. When it is used vertically, we call this 8 1/2 × 11-inch sheet size AV throughout the rest of this textbook.

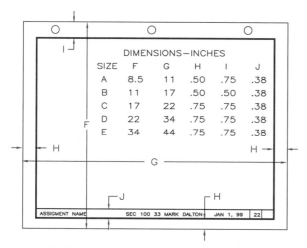

10.47 This format is for a size AH sheet (11 × 8 1/2, horizontal position) and other sheet sizes. The numbers in columns A–E are the various layout dimensions.

When this sheet is used horizontally, as **Fig. 10.47** shows, we call it size AH. This figure also shows the standard sizes of sheets, from sizes A through E and an alternative title strip for sizes B, C, D, and E. Always use guidelines for lettering title strips.

Problems

Problems 1–9 (Figs. 10.48–10.50): Solve these problems on size AH plain or paper with a printed grid, using the format shown in **Fig. 10.48.** You may be assigned to solve two half-size drawings per sheet on size AV paper; use the side of each square of a grid to represent .25 in., or 6 mm.

Problems 10–20 (Figs. 10.51–10.61): Draw full-size views on size AH sheets and omit the dimensions.

Design Application

Problem 21 (Figure 10.62): The knob has a diameter of 3 in. (flat-to-flat). Make an instrument drawing of the descriptive end of this part by estimating the dimensions as if you were its designer. (Courtesy of Balcrank Products Inc.)

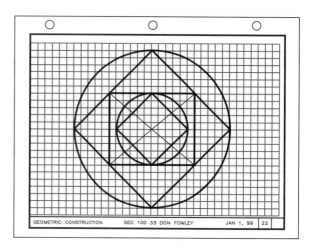

10.48 Problem 1 and sheet format.

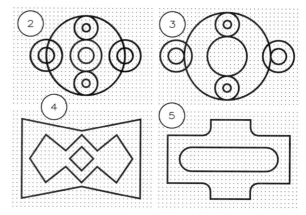

10.49 Problems 2–5.

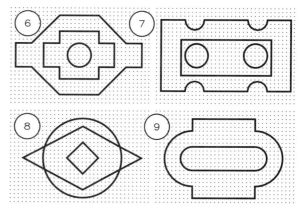

10.50 Problems 6–9.

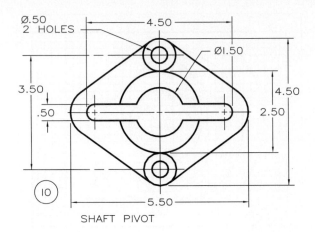

SHAFT PIVOT

10.51 Problem 10.

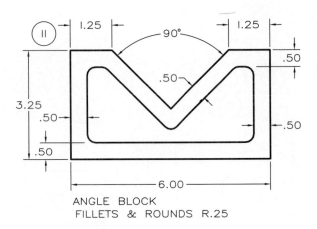

ANGLE BLOCK
FILLETS & ROUNDS R.25

10.52 Problem 11.

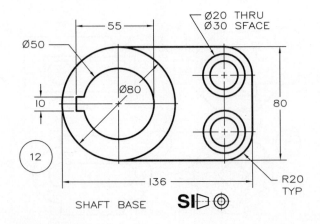

SHAFT BASE

10.53 Problem 12.

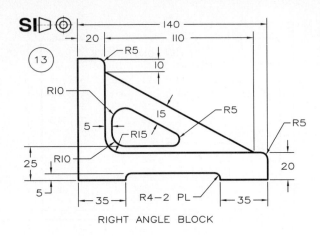

RIGHT ANGLE BLOCK

10.54 Problem 13.

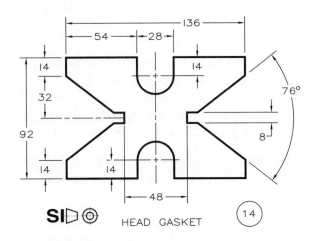

HEAD GASKET

10.55 Problem 14.

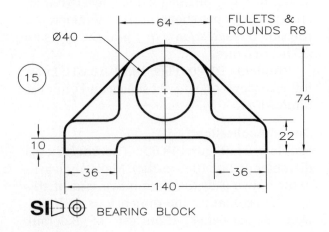

BEARING BLOCK

10.56 Problem 15.

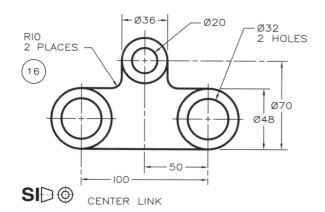

RIO
2 PLACES

Ø36 Ø20 Ø32
2 HOLES

Ø70
Ø48

50
100

SI⊳◉ CENTER LINK

10.57 Problem 16.

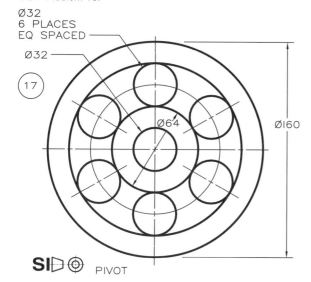

Ø32
6 PLACES
EQ SPACED

Ø32

Ø64

Ø160

SI⊳◉ PIVOT

10.58 Problem 17.

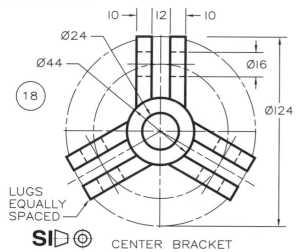

10 → 12 ← 10

Ø24
Ø44

Ø16

Ø124

LUGS
EQUALLY
SPACED

SI⊳◉ CENTER BRACKET

10.59 Problem 18.

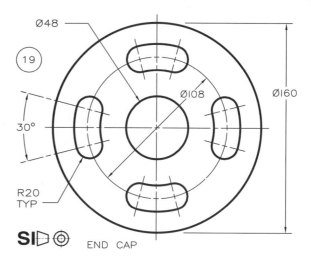

Ø48

Ø108
Ø160

30°

R20
TYP

SI⊳◉ END CAP

10.60 Problem 19.

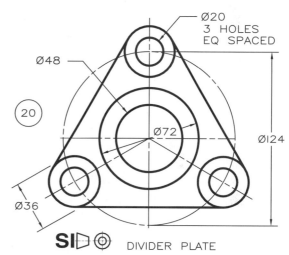

Ø20
3 HOLES
EQ SPACED

Ø48

Ø72

Ø124

Ø36

SI⊳◉ DIVIDER PLATE

10.61 Problem 20.

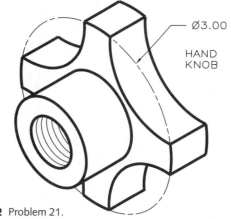

Ø3.00

HAND
KNOB

10.62 Problem 21.

PROBLEMS • 127

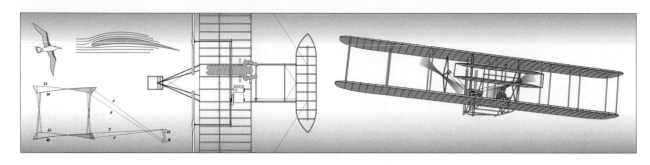

11

Lettering

11.1 Introduction

The notes, dimensions, and specifications that supplement all drawings must be lettered. The ability to letter freehand is thus an important skill to develop because it affects the use and interpretation of drawings. It also displays an engineer's skill with graphics and may be taken as an indication of professional competence.

11.2 Lettering Tools

The best pencils for lettering on most surfaces are the H, F, and HB grades, with an F grade pencil being the one most commonly used. Some papers and films are coarser than others and may require a harder pencil lead. To give the desired line width, round the point of the pencil slightly (**Fig. 11.1**) because a needle point will break off when you apply pressure.

Hold your pencil in the position shown in **Fig. 11.2**, revolving it slightly about its axis as you make strokes so that the lead will wear

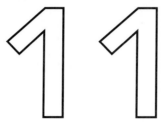

Sharpen your pencil point
to a slightly rounded point
for good lettering.

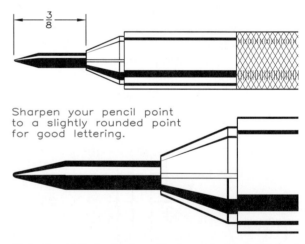

11.1 Good lettering begins with a properly sharpened pencil point. The F grade pencil is good for lettering.

evenly. For good reproduction, bear down firmly to make letters black and bright with a single stroke. Prevent smudging while lettering by placing a sheet of paper under your hand to protect the drawing (**Fig. 11.3**).

11.2 Hold your pencil in a plane perpendicular to the paper and at 60° to the surface while lettering.

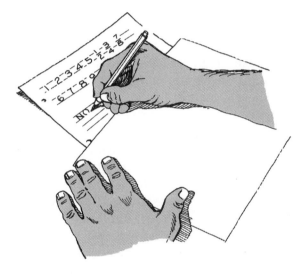

11.3 Place a protective sheet under your hand to prevent smudges; work from a comfortable position for natural strokes.

11.3 Guidelines

The most important rule of lettering is to use guidelines at all times, whether you are lettering a paragraph or a single letter. **Figure 11.4** shows how to draw and use guidelines. Use a sharp pencil in the 2H–4H grade range and draw light guidelines, just dark enough to be seen.

Most lettering on an engineering drawing is done with capital letters that are 1/8 inch (3 mm) high. The spacing between lines of lettering should be no closer than half the height of the capital letters, or 1/16 inch, in this case.

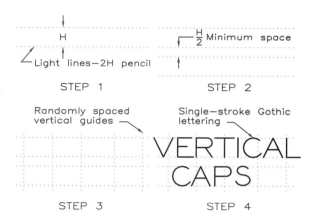

11.4 Method of using lettering guidelines

Step 1 Lay off letter heights, H, and draw light guidelines with a 2H pencil.

Step 2 Space lines no closer than H/2 apart.

Step 3 Draw vertical guidelines as light, thin, randomly spaced lines.

Step 4 Draw letters with single strokes using a medium-grade (H, F, or HB) pencil. Do not erase the lightly drawn guidelines.

Lettering Guides

Two instruments for drawing guidelines are the **Ames** lettering guide and the **Braddock-Rowe** lettering triangle.

The **Ames lettering guide** (**Fig. 11.5A**) is a device that has a circular dial for selecting guideline spacing. The numbers around the dial represent thirty-seconds of an inch. For example, the number 8 represents 8/32 inch, or guidelines for drawing capital letters that are 1/4 inch tall.

The **Braddock-Rowe triangle** contains sets of holes for spacing guidelines (**Fig. 11.5B**). The numbers under each set of holes represent thirty-seconds of an inch. For example, the numeral 4 represents 4/32 inch or 1/8 inch for making uppercase (capital) letters. Some triangles have millimeter markings. Intermediate holes provide guidelines for lowercase letters, which are not as tall as capital letters.

While holding a horizontal straightedge firmly in position, place the Braddock-Rowe

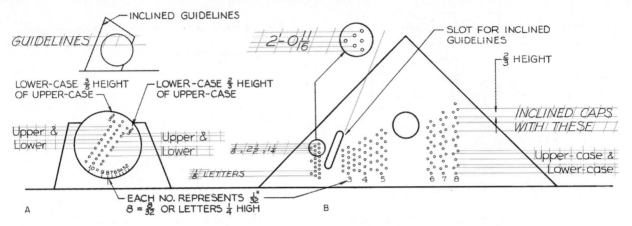

11.5   Lettering Guides

A The **Ames lettering guide** is used for drawing guidelines for lettering. Set the dial to the desired number for the height of uppercase letters.

B The **Braddock-Rowe triangle** is also used for drawing guidelines for lettering. Each number near the guideline holes represents 1/32 inch.

triangle against its upper edge. Insert a sharp 2H pencil point in the desired guideline hole to contact the drawing surface and guide the pencil point across the paper, drawing the guideline while the triangle slides along the straightedge. Repeat this procedure by moving the pencil point to each successive hole to draw other guidelines. Use the slanted slot in the triangle to draw guidelines, spaced randomly, for inclined lettering.

The **Ames lettering guide** (**Fig. 11.5A**) is a similar device, but it has a circular dial for selecting guideline spacing. Each number around the dial represents 1/32 inch. For example, the number 8 represents 8/32 inch, or guidelines for drawing capital letters that are 1/4 inch tall.

11.4 Gothic Lettering

The lettering recommended for engineering drawings is single-stroke Gothic lettering, so called because the letters are a variation of the Gothic style made with a series of single strokes. Gothic lettering may be vertical or

inclined (**Fig. 11.6**), but only one style or the other should be used on a single drawing.

Vertical Letters

Uppercase   **Figure 11.7** shows the alphabet in the single-stroke Gothic vertical uppercase letters. Each letter is drawn inside a square box of guidelines to show their correct proportions. Draw straight lines with a single stroke; for example, draw the letter A with three single strokes. Letters composed of curves can best be drawn in segments; for example, draw the letter O by joining two semicircles.

The shape (form) of each letter is important. Small wiggles in strokes will not detract from your lettering if the letter forms are correct. **Figure. 11.8** shows common errors in lettering that you should avoid.

11.6   Use either vertical or inclined single-stroke Gothic lettering on engineering drawings.

11.7 This alphabet shows the form of single-stroke Gothic vertical uppercase letters. Letters are drawn inside squares to show their proportions.

A. LETTERS POORLY DONE

LETTERS THIN

B. STROKES TOO THIN

TOO HEAVY

C. STROKES TO THICK

D. STROKES TOO LIGHT

11.8 Avoid these common errors when lettering.

Lowercase The alphabet of lowercase letters is shown in **Fig. 11.9**, which should be either two-thirds or three-fifths as tall as uppercase letters. Both lowercase ratios are labeled on the Ames guide, but only the two-thirds ratio is available on the Braddock-Rowe triangle.

Some lowercase letters, such as the letter b, have ascenders that extend above the body of the letter; some, such as the letter p, have descenders that extend below the body. Ascenders and descenders should be equal in length.

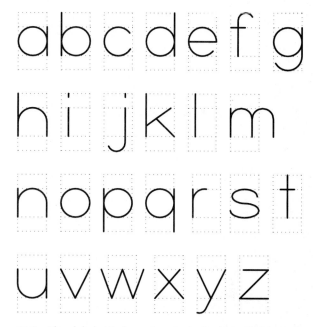

11.9 The alphabet is drawn here in single-stroke Gothic vertical lowercase letters.

The guidelines in Fig. 11.9 that form squares about the body of each letter are used to illustrate their proportions. Letters that have circular bodies may extend slightly beyond the sides of the guideline squares. **Figure 11.10** shows examples of capital and lowercase letters used together.

Numerals **Figure 11.11** illustrates vertical numerals for use with single-stroke Gothic lettering; each numeral is enclosed in a square of guidelines. Numbers should be the same height as the capital letters being used, usually 1/8 inch. The numeral 0 (zero) is an oval, whereas the letter O is a circle in vertical lettering.

Inclined Letters
Uppercase Inclined uppercase letters have the same heights and proportions as vertical letters; the only difference is their 68° inclination (**Fig. 11.12**). Guidelines for inclined lettering can be drawn with both the Braddock-Rowe triangle and the Ames guide.

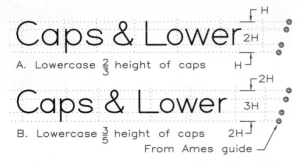

A. Lowercase ⅔ height of caps   H

B. Lowercase ⅗ height of caps   2H

From Ames guide

11.10 The ratio of lowercase letters to uppercase letters should be either two-thirds (A) or three-fifths (B). The Ames guide has both ratios, but the Braddock-Rowe triangle has only the two-thirds ratio.

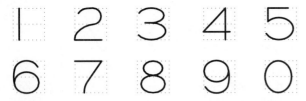

11.11 These numerals are used with single-stroke Gothic, vertical letters.

Lowercase Inclined lowercase letters are drawn in the same manner as vertical lowercase letters (**Fig. 11.13**), but circular features are drawn as ovals (ellipses). The angle of inclination is 68°, the same as for uppercase letters.

Numerals Examples of inclined numerals in **Fig. 11.14**; both inclined letters and numerals used in combination are shown in **Fig. 11.15**. The ratio of lowercase to the uppercase letters should be either two-thirds (A) or three-fifths (B). The Ames guide has both ratios, but the Braddock-Rowe triangle has only the two-thirds ratio. Guidelines are drawn using the Braddock-Rowe triangle or the Ames lettering guide shown in **Fig. 11.16**.

Spacing Numerals and Letters
Allow adequate space between numerals for the decimal point and fractions (**Fig. 11.17A–C**). Common fractions are twice as tall

11.12 This is an alphabet of single-stroke Gothic, inclined uppercase letters.

11.13 This is an alphabet of single-stroke Gothic, inclined lowercase letters.

as single numerals (**Fig. 11.17D–G**). Both the Braddock-Rowe triangle and the Ames guide have separate sets of holes spaced 1/16 inch apart for common fractions. The center guideline locates the fraction's crossbar.

When grouping letters to spell words, make the areas between the letters approximately equal for the most visually pleasing result (**Fig. 11.18**). **Figure 11.19** shows the incorrect use of guidelines and other

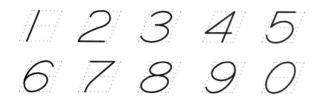

11.14 Single-stroke Gothic, inclined numerals.

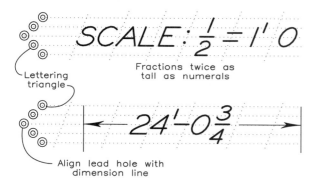

11.15 Inclined common fractions are twice as tall as single numerals. Omit inch marks in dimensions.

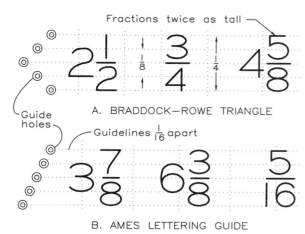

11.16 Guidelines are drawn for fractions so they can be twice as tall as single numerals.

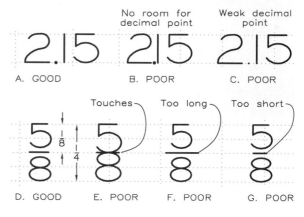

11.17 Avoid making these errors in lettering fractions.

11.18 Letters should be spaced so that the areas between them are about equal.

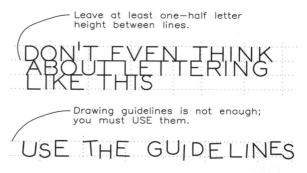

11.19 Always use guidelines (vertical and horizontal) whether you are lettering a paragraph or a single letter.

violations of good lettering practice that you should avoid.

11.5 Computer Lettering

Although AutoCAD offers a wide range of lettering (text) fonts, the ROMANS font shown in **Fig. 11.20** is the Gothic single-stroke text recommended for engineering drawings. The command used to apply text to a drawing is DRAW/TEXT/Single Line Text, but you can simply type DTEXT. The text will be displayed across the screen as it is typed:

Command: (Type) DTEXT (Enter)
Justify/Style/<Start point>: J (Enter)
Align/Fit/Center/Middle/Right/TL/TC/TR/ML/ MC/MR/BL/BC/BR:

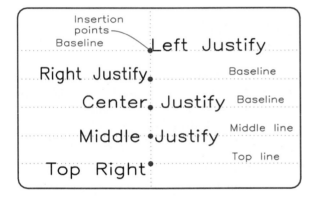

11.20 ROMANS is the text font that is most like Gothic lettering and is recommended for engineering drawings.

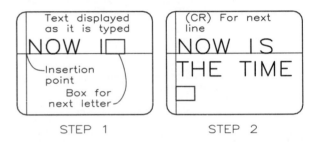

11.21 The options of DTEXT/JUSTIFY are available for variations in text lettering placement.

The abbreviations in the last line above are insertion points for lines of text defined in **Fig. 11.21.**

With the STYLE command, you can modify a single font, ROMANS (for example), to obtain a variety of letter forms with it.

Command: STYLE (Enter)
Text style name (or ?) <Standard>: NEWFONT (Enter)
Existing style <TXT>: ROMANS (Enter)
Height <0>: 0 (Enter)
Width factor <1>: 1.3 (Enter)
Obliquing angle <0>: (Enter)
Backwards? <N>: (Enter)
Upside-down? <N>: (Enter)
Vertical? N<N>: (Enter)

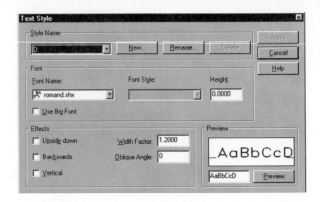

11.22 The TEXT STYLE menu (Format/Text Style) can be used to assign various characters to the letter form (text) being used. (See Chapter 36 for more details.)

11.23 The DTEXT command
Step 1 Command: DTEXT (Enter)
Justify/Style/<start points>:(cursor-select)
Height<.18>:0 (Enter)
Rotation angle <0>:0 (Enter)
Text: NOW IS (Enter)
Step 2 (Box moves to start of next line.)
Text: THE TIME. (Enter) (Box moves to start of next line.)
(Enter to save the text.)

Instead of typing STYLE at the command line, the TEXT STYLE dialogue box can be obtained under FORMAT to obtain the screen shown in **Fig. 11.22** for making the same selections.

NEWFONT is now the current style; it has a height of zero, and each letter has a width that is 130 percent of (1.3 times) the default width. The width of 1.3 matches the recommended letter proportions of Gothic engineering lettering. Assigning zero instead of a specific letter height permits you to change heights by responding when prompted by the DTEXT command. Had a height been specified, it would remain constant until you modified it

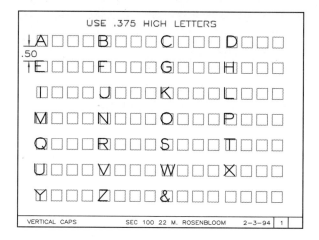

11.24 The sheet layout for uppercase lettering assignments.

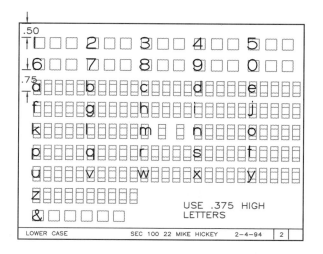

11.25 A typical sheet layout for lowercase lettering assignments

by using the STYLE command. Additional information on AutoCAD lettering is covered in Section 38 of Chapter 36.

Figure 11.23 illustrates how DTEXT is used in conjunction with TEXT STYLE to type a sequence of lines on the screen.

PROBLEMS

Complete lettering problem solutions on size AH (11 × 8-1/2-in.) paper, plain or with a grid using the format shown in **Fig. 11.24**.

1. Practice drawing the alphabet in vertical uppercase letters, as shown in Fig. 11.24. Construct each letter three times (three As, three Bs, and so on). Use a medium-weight pencil (H, F, or HB).

2. Practice drawing vertical numerals and the alphabet in lowercase letters, as shown in **Fig. 11.25**. Construct each letter and numeral two times: two 1s, two 2s, two a's, two b's, and so on. Use a medium-weight pencil (H, F, or HB).

3. Practice drawing the alphabet in inclined uppercase letters, as shown in Fig. 11.12. Construct each letter three times. Use a medium-weight pencil (H, F, or HB) and the layout shown in Fig. 11.24.

4. Practice lettering the vertical numerals and the alphabet in lowercase letters, as shown in Figs. 11.9 and 11.11. Construct each letter two times. Use a medium-weight pencil (H, F, or HB) and the layout shown in Fig. 11.25.

5. Construct guidelines for 1/8-in. capital letters starting 1/4 in. from the top border of a sheet similar to the one shown in Fig. 11.25. Each guideline should end 1/2 in. from the left and right borders. Using these guidelines, letter the first paragraph of the text of this chapter. Use all vertical capitals. Spacing between the lines should be 1/8 in.

6. Repeat Problem 5, but use inclined capital (uppercase) letters. Use inclined guidelines to help you slant letters uniformly.

7. Repeat Problem 5, but use vertical capital and lowercase letters in combination. Capitalize only those words that are capitalized in the text.

8. Repeat Problem 5, but use inclined capital and lowercase letters in combination. Capitalize only those letters that are capitalized in the text. Use single-stroke Gothic vertical lowercase letters.

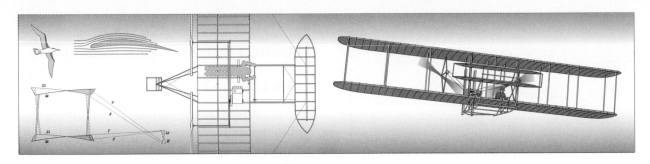

12

Geometric Construction

12.1 Introduction

The solution of many graphical problems requires the use of geometry and geometric construction. Because mathematics was an outgrowth of graphical construction, the two areas are closely related. The proofs of many principles of plane geometry and trigonome-

try can be developed by using graphics. Moreover, graphical methods can be applied to solve some types of problems in algebra and arithmetic and virtually all types of problems in analytical geometry.

12.2 Angles

A fundamental application of geometric construction involves drawing lines at specified angles to each other. **Figure 12.1** gives the names and definitions of various angles.

The unit of angular measurement is the degree; in a circle there are 360 degrees. A degree (°) can be divided into 60 parts called minutes ('), and a minute can be divided into 60 parts called seconds ("). An angle of 15°32'14" is an angle of 15 degrees, 32 minutes, and 14 seconds.

12.3 Polygons

A **polygon** is a multisided plane figure of any number of sides. If the sides of a polygon are

12.1 Standard angles and their names and definitions.

SQUARE
4 sides

PENTAGON
5 sides

HEXAGON
6 sides

OCTAGON
8 sides

12.2 Any regular polygon can be inscribed in a circle.

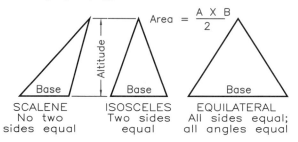

$$Area = \frac{A \times B}{2}$$

SCALENE
No two
sides equal

ISOSCELES
Two sides
equal

EQUILATERAL
All sides equal;
all angles equal

12.3 Types of triangles and their characteristics.

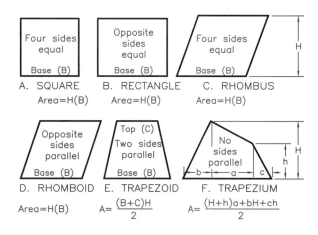

A. SQUARE
Four sides equal
Base (B)
Area=H(B)

B. RECTANGLE
Opposite sides equal
Base (B)
Area=H(B)

C. RHOMBUS
Four sides equal
Base (B)
Area=H(B)

D. RHOMBOID
Opposite sides parallel
Base (B)
Area=H(B)

E. TRAPEZOID
Top (C)
Two sides parallel
Base (B)
$A = \frac{(B+C)H}{2}$

F. TRAPEZIUM
No sides parallel
$A = \frac{(H+h)a + bH + ch}{2}$

12.4 Types of quadrilaterals, their characteristics, and how to calculate their areas.

equal in length, the polygon is a **regular polygon**. A regular polygon can be inscribed in a circle and all its corner points will lie on the circle (**Fig. 12.2**). Other regular polygons (not pictured here) are the **heptagon** (7 sides), the **nonagon** (9 sides), the **decagon** (10 sides), and the **dodecagon** (12 sides).

The sum of the angles inside any polygon (interior angles) is $S = (n - 2) \times 180°$, where n is the number of sides of the polygon.

Triangles

A **triangle** is a three-sided polygon. The four types of triangles are **scalene, isosceles, equilateral**, and **right** triangles (**Fig. 12.3**). The sum of the interior angles of a triangle is always 180°; that is: $[(3 - 2) \times 180°]$.

Quadrilaterals

A **quadrilateral** is a four-sided polygon of any shape. The sum of the interior angles of a quadrilateral is 360° $[(4 - 2) \times 180°]$. **Figure 12.4** shows the various types of quadrilaterals and the equations for their areas.

12.4 Circles

Figure 12.5 gives the names of the elements of a circle that are used throughout this book. A circle is constructed by swinging a radius from

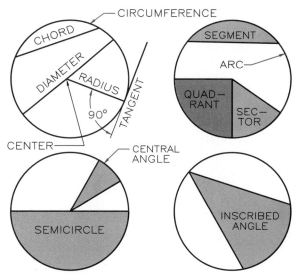

12.5 Elements of a circle and their definitions.

a fixed point through 360°. The area of a circle equals πR^2.

12.5 Geometric Solids

Figure 12.6 shows the various types of solid geometric shapes and their characteristics.

Polyhedra

A **polyhedron** is a multisided solid formed by intersecting planes. If its faces are regular polygons, it is a regular polyhedron. Five regu-

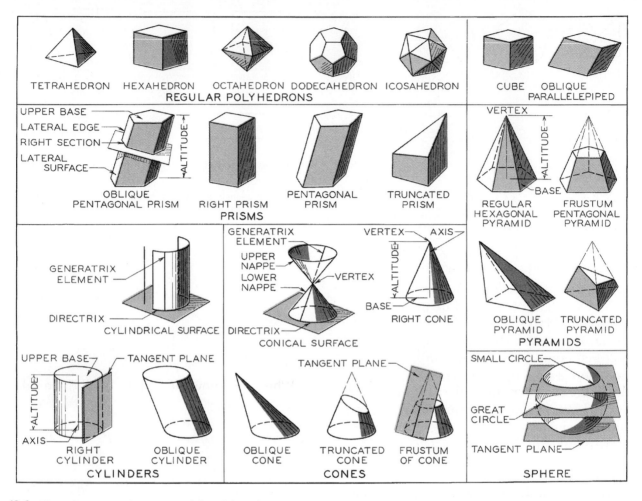

12.6 The various types of geometric solids and their elements.

lar polyhedra are the **tetrahedron** (4 sides), the **hexahedron** (6 sides), the **octahedron** (8 sides), the **dodecahedron** (12 sides), and the **icosahedron** (20 sides).

Prisms
A **prism** has two parallel bases of equal shape connected by sides that are parallelograms. The line from the center of one base to the center of the other is the axis. If its axis is perpendicular to the bases, the prism is a right prism. If its axis is not perpendicular to the

bases, the prism is an oblique prism. A prism that has been cut off to form a base not parallel to the other is a truncated prism. A **parallelepiped** is a prism with bases that are either rectangles or parallelograms.

Pyramids
A **pyramid** is a solid with a polygon as a base and triangular faces that converge at a vertex. The line from the vertex to the center of the base is the axis. If its axis is perpendicular to the base, the pyramid is a **right pyramid**. If its axis is not perpendicular to the base, the pyramid is an **oblique pyramid**. A truncated pyramid is called a **frustum** of a pyramid.

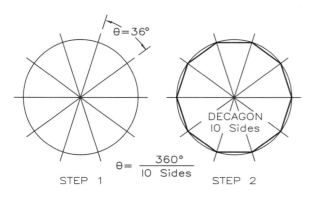

DECAGON
10 Sides

$\theta = 36°$

$$\theta = \frac{360°}{10 \text{ Sides}}$$

STEP 1 STEP 2

12.7 A regular polygon (sides of equal length).

Step 1 Divide circle into the required number of equal sectors.

Step 2 Connect the division points with straight lines where they intersect the circle.

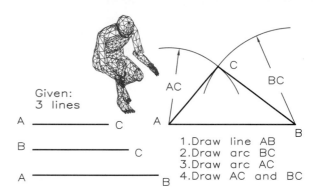

Given:
3 lines

A ——————— C
B ——————— C
A ——————— B

AC   C   BC

A   B

1. Draw line AB
2. Draw arc BC
3. Draw arc AC
4. Draw AC and BC

12.8 A triangle can be constructed by triangulation with a compass when three sides are given.

Cylinders

A **cylinder** is formed by a line or an element (called a *generatrix*) that moves about the circle while remaining parallel to its axis. The axis of a cylinder connects the centers of each end of a cylinder. If the axis is perpendicular to the bases, it is the altitude of a **right cylinder**. If the axis does not make a 90° angle with the base, the cylinder is an **oblique cylinder**.

Cones

A **cone** is formed by a generatrix, one end of which moves about the circular base while the other end remains at a fixed vertex. The line from the center of the base to the vertex is the axis. If its axis is perpendicular to the base, the cone is a **right cone**. A truncated cone is called a frustum of a cone.

Spheres

A **sphere** is generated by revolving a circle about one of its diameters to form a solid. The ends of the axis of revolution of the sphere are poles.

12.6 Constructing Polygons

A regular polygon (having equal sides) can be inscribed in or circumscribed about a circle.

When it is inscribed, all corner points will lie on the circle (**Fig. 12.7**). For example, constructing a 10-sided polygon involves dividing the circle into 10 sectors and connecting the points to form the polygon.

Triangles

When you know the lengths of all three sides of a triangle, you may construct it with a compass by triangulation, as **Fig. 12.8** shows. Any triangle inscribed in a semicircle, will be a right triangle (**Fig. 12.9**).

Hexagons

The **hexagon**, a six-sided regular polygon, can be inscribed in or circumscribed about a circle (**Fig. 12.10**). Use a 30°–60° triangle to draw the hexagon. The circle represents the distance from corner to corner for an inscribed hexagon and from flat to flat when the hexagon is circumscribed about a circle.

Octagons

The **octagon**, an eight-sided regular polygon, can be inscribed in or circumscribed about a circle (**Fig. 12.11**) or inscribed in a square (**Fig. 12.12**). Use a 45° triangle to draw the octagon in the first case and a compass and straightedge in the second case.

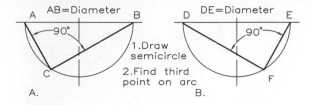

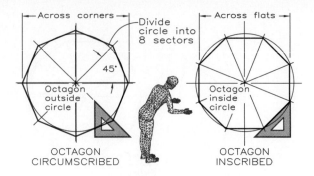

12.9 Any triangle inscribed in a semicircle will be a right triangle.

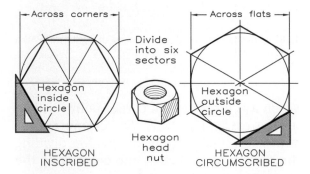

12.10 A hexagon can be inscribed in or circumscribed about a circle with a 30°–60° triangle.

12.11 An octagon can be inscribed in or circumscribed about a circle with a 45° triangle.

Pentagons

The **pentagon**, a five-sided regular polygon, can be inscribed in or circumscribed about a circle. **Figure 12.13** shows the steps of constructing a pentagon with a compass and straightedge where the vertices lie on the circle.

Computer Method You may use one of two POLYGON options from under the DRAW command to draw polygons. The CENTER option asks you to give the number of sides, select the center, specify the radius, and indicate whether the polygon is to be inscribed in or circumscribed about an imaginary circle (**Fig. 12.14**). The EDGE option allows you to specify the number of sides and specify the length and direction of one edge of the polygon before drawing it.

12.7 Bisecting Lines and Angles

Bisecting Lines

Two methods of finding the midpoint of a line with a perpendicular bisector are shown in

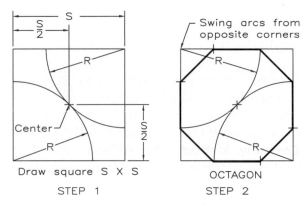

12.12 An octagon can be constructed in a square with a compass.

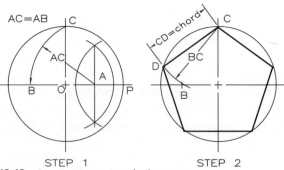

12.13 Constructing an inscribed pentagon.

Step 1 Bisect radius OP to locate point A. With A as the center and AC as the radius, locate point B on the diameter.

Step 2 With point C as the center and BC as the radius, locate point D. Use line CD as the chord to locate the other corners of the pentagon.

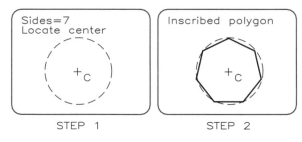

STEP 1   STEP 2

12.14 Drawing a polygon by computer (DRAW menu).

Step 1 Command: POLYGON *(Enter)*
Number of sides: 7 *(Enter)*
Edge <Center of polygon>: *(Enter)*
Inscribed in circle/Circumscribed about circle (I/C): I *(Enter)*

Step 2 Radius of circle: (Select radius with cursor.) (Polygon is inscribed inside the imaginary circle.)

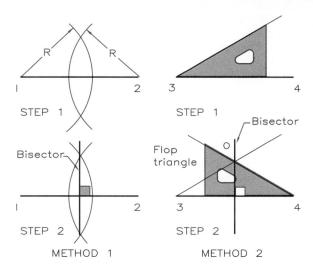

12.15 Bisecting a line.
Method 1 Use a compass and any radius.
Method 2 Use a standard triangle and a straightedge.

Fig. 12.15. In the first, a compass is used to construct the perpendicular bisector to the line. In the second, a standard triangle and a straightedge are used.

Computer Method The midpoint of a line may be found by using the MIDPOINT mode of OSNAP while drawing a line from any point, P, to the line (**Fig. 12.16**). The line will snap to the line's midpoint.

Bisecting Angles

You may bisect angles by using a compass and drawing three arcs, as shown in **Fig. 12.17**.

Computer Method Use the ARC command (with OSNAP set to Midpoint) and draw an arc of a convenient radius between the two lines, with its center at vertex A (**Fig. 12.18**). Use the LINE command (with OSNAP set to Midpoint) to draw a line from the vertex, A, to the arc's midpoint, D. Line AD is the bisector.

12.8 Division of Lines

Dividing a line into several equal parts is often necessary. **Figure 12.19** shows the method used to solve this type of problem where the

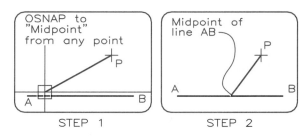

12.16 Finding the midpoint of a line by computer (DRAW menu).

Step 1 Command: LINE *(Enter)*
From point: P (Locate P anywhere.)
To point: OSNAP (Select Midpoint mode.)
(Select any point on line AB.)

Step 2 The line from point P is drawn to the midpoint of the line AB.

line AB is divided into five equal lengths by using a scale to lay off the five divisions.

The same principle applies to locating equally spaced lines on a graph (**Fig. 12.20**). Lay scales with the desired number of units (0 to 3 and 0 to 5, respectively) across the graph up and down and then left to right. Make marks at each whole unit and draw vertical and horizontal index lines through these points. These index lines are used to show data in a graph.

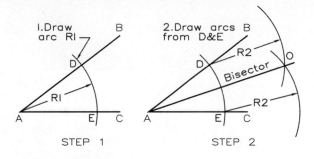

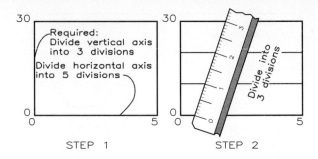

12.17 Bisecting an angle.

Step 1 Swing an arc of any radius R1 to locate points D and E.

Step 2 Draw equal arcs from D and E to locate point O. Line AO is the bisector of the angle.

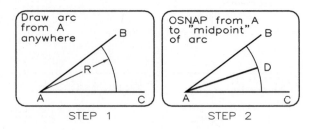

12.18 Bisecting an angle by computer (DRAW menu).

Step 1 Use the ARC command, any radius, and center A, to draw an arc that OSNAPs, nearest to AC and AB.

Step 2 Command: LINE (Enter)
From point: (Select vertex A.)
To point: OSNAP, Midpoint of (Select arc.) (The line AD is the bisector of the angle.)

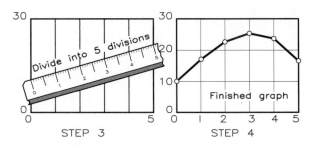

12.20 Dividing axes on a graph.

Step 1 Draw the outline of the graph.

Step 2 To divide the y axis into three equal segments, lay a scale so that three units of measurement span the graph, with the 0 and 3 located on the top and bottom lines. Mark points 1 and 2 and draw horizontal lines through them.

Step 3 To divide the x axis into five equal segments, lay a scale so that five units of measurement span the graph, with the 0 and 5 on the left- and right-hand vertical lines. Make marks at points 1, 2, 3, and 4 and draw vertical lines through them.

Step 4 Plot the data points and draw the curve to complete the graph.

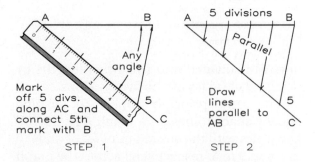

12.19 Dividing a line.

Step 1 To divide line AB into five equal lengths, lay off five equal divisions along line AC, and connect point 5 to end B with a construction line.

Step 2 Draw a series of five construction lines parallel to 5B to divide line AB.

12.9 An Arc through Three Points

An arc can be drawn through three points by connecting the points with two lines and drawing perpendicular bisectors through each line to locate the center of the circle at C (**Fig. 12.21**). Draw the arc; lines AB and BD will become chords of the arc.

To find the center of a circle or an arc, reverse this process by drawing two chords that intersect at a point on the circumference and then bisect them. The perpendicular bisectors intersect at the center of the circle.

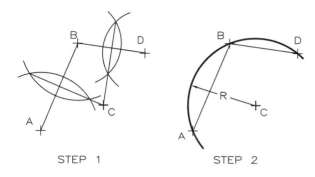

12.21 An arc through three points.

Step 1 Connect points A, B, and D with two lines and construct their perpendicular bisectors, intersecting at the center, C.

Step 2 Using the center C, and the distance to the points as the radius R, draw the arc through the points.

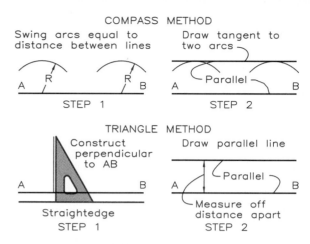

12.22 Drawing parallel lines.

Compass Method

Step 1 Swing two arcs from line AB.

Step 2 Draw the parallel line tangent to the arcs.

Triangle Method

Step 1 Draw a line perpendicular to AB.

Step 2 Measure the desired distance, R, along the perpendicular and draw the parallel line through it.

12.10 Parallel Lines

You may draw one line parallel to another by using either method shown in **Fig. 12.22**. In the compass method, draw two arcs having radius R to locate a parallel line at the desired distance (R) from the first line. In the triangle method, measure the desired perpendicular distance R from the first line, mark it, and draw the parallel line through it with your drafting machine.

12.11 Tangents

Marking Points of Tangency

A point of tangency is the theoretical point at which a line joins an arc or two arcs join without crossing. **Figure 12.23** shows how to find the point of tangency with triangles by constructing a perpendicular line to the tangent line from the center of the arc. **Figure 12.24** shows the conventional methods of marking points of tangency.

Line Tangent to an Arc

The point of tangency may be found by using a triangle and a straightedge, as shown in **Fig. 12.25**. The classical compass method of find-

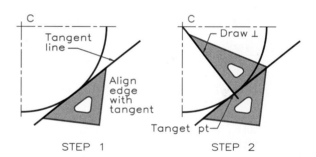

12.23 Locating a point of tangency.

Step 1 Align a triangle with the tangent line while holding it against a firmly held straightedge.

Step 2 Hold the triangle in position, locate a second triangle perpendicular to it, and draw a line from the center to locate the tangent point.

ing the point of tangency between a line and a point is shown in **Fig. 12.26**. Connect point A to the arc's center and bisect line AC (step 1); swing an arc from point M through point C locating tangent point T (step 2); draw the tangent to T (step 3); and mark the tangent point (step 4).

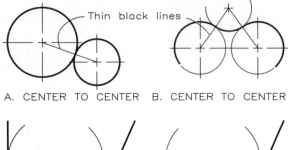

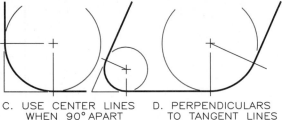

A. CENTER TO CENTER B. CENTER TO CENTER

C. USE CENTER LINES D. PERPENDICULARS
   WHEN 90° APART   TO TANGENT LINES

12.24 Use thin, black lines that extend from the centers slightly beyond the arcs to mark points of tangency.

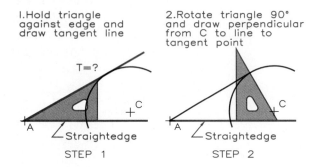

1.Hold triangle against edge and draw tangent line

2.Rotate triangle 90° and draw perpendicular from C to line to tangent point

STEP 1   STEP 2

12.25 A tangent to an arc from a point.

Step 1 Hold a triangle against a straightedge and draw a line from point A that is tangent to the arc.

Step 2 Rotate triangle 90° and locate the point of tangency by drawing a line through center C.

Computer Method A line can be drawn from a point tangent to an arc by using the OSNAP Tangent option (**Fig. 12.27**). When prompted for the second point, select a point on the arc near the tangent point and the line will be drawn to the true tangent point.

Arc Tangent to a Line from a Point

To construct an arc that is tangent to line DE at T and that passes through point P (**Fig. 12.28**), draw the perpendicular bisector of

144 • **CHAPTER 12   GEOMETRIC CONSTRUCTION**

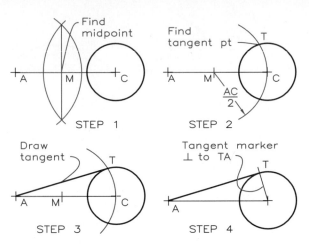

STEP 1   STEP 2

STEP 3   STEP 4

12.26 A line tangent to an arc from a point.

Step 1 Connect point A with center C and locate point M by bisecting AC.

Step 2 Using point M as the center and MC as the radius, locate point T on the arc.

Step 3 Draw a line from A to T that is tangent to the arc at point T.

Step 4 Draw the tangent marker perpendicular to TA from the center past the arc as a thin, dark line.

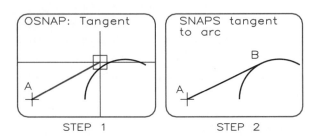

STEP 1   STEP 2

12.27 Using a computer to locate a line tangent to an arc.

Step 1 Command: LINE *(Enter)*
From point: Select point A
To point: (Type OSNAP/Tangent mode) *(Enter)*

Step 2 Tangent to: Select point on arc
Line AB is drawn tangent to the arc.

line TP. Draw a perpendicular to line DE at point T to locate the center at point C and swing an arc with radius OT. A similar problem (**Fig. 12.29**) requires drawing an arc of a given radius that is tangent to line AB and that passes through point P. An arc with its center at C is tangent to the line.

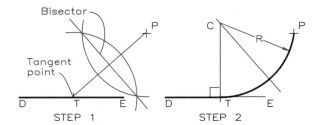

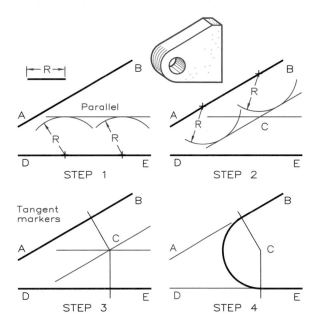

12.28 An arc through two points.

Step 1 To draw an arc through point P tangent to line DE at point T, draw the perpendicular bisector of TP.

Step 2 Construct a perpendicular to line DE at point T to intersect the bisector at point C and draw the arc from C with radius CT.

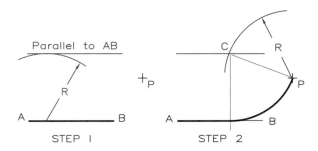

12.29 An arc tangent to a line through a point.

Step 1 When an arc is to be drawn tangent to line AB and through point P, first draw a line parallel to line AB at a distance of R from it.

Step 2 Draw an arc from point P with radius R to locate center C and draw the arc with radius R.

Arc Tangent to Two Lines

Figure 12.30 shows how to construct an arc of a given radius tangent to two nonparallel lines that form an acute angle. The same steps apply to constructing an arc tangent to two lines that form an obtuse angle (**Fig. 12.31**). In both cases, the points of tangency are located with thin, dark lines drawn from their centers perpendicular to and past the original lines. **Figure 12.32** shows a technique for finding an arc tangent to two lines that are perpendicular.

12.30 An arc tangent to two lines forming an acute angle.

Step 1 Construct a light construction line parallel to line DE with radius R.

Step 2 Draw a second light line parallel to and R distance from line AB to locate center C.

Step 3 Draw thin, dark lines from center C perpendicular to lines AB and DE to locate the tangency points.

Step 4 Draw the tangent arc and darken your lines.

Computer Method Draw an arc tangent to two nonparallel lines with the FILLET command (MODIFY menu) (**Fig. 12.33**). When the radius length has been set and a point selected on each line, the arc is drawn and the lines trimmed. To mark tangent points, snap to the center of the arc with the CENTER option of OSNAP and draw two lines perpendicular (use OSNAP's Perpend option) to lines AB and CD from the center.

Arc Tangent to an Arc and a Line

Figure 12.34 shows the steps for constructing an arc tangent to an arc and a line. **Figure 12.35** shows a variation of this technique for an arc drawn tangent to a given arc and line with the arc reversed.

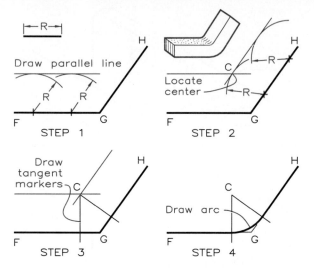

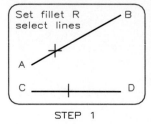

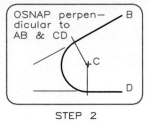

STEP 1 — STEP 2

12.31 An arc tangent to two lines forming an obtuse angle.

Step 1 Using radius R, draw a light line parallel to FG.

Step 2 Construct a light line parallel to line GH that is R distance from it to locate center C.

Step 3 Draw thin lines from center C perpendicular to lines FG and GH to locate the tangency points.

Step 4 Draw the tangent arc and darken your lines.

12.33 An arc tangent to two lines (by computer).

Step 1 Command: FILLET *(Enter)*
Polyline/Radius/<Select two objects>: R *(Enter)*
Enter fillet radius <0.0000>: .75 *(Enter)*
Command: *(Enter)*
FILLET Polyline/Radius/<Select two objects>: (Select points on AB and CD.)

Step 2 Command: LINE *(Enter)*
From point: CENTER *(Enter)* to (Select pt on the arc.)
To point: PERPEND *(Enter)* to (Select line AB; a perpendicular is drawn to the point of tangency. Locate the tangent point on CD in the same manner.)

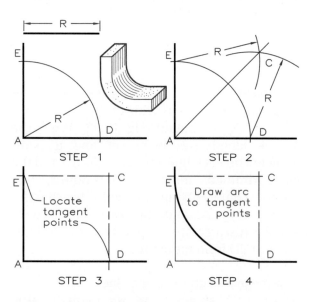

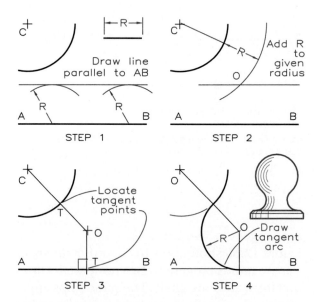

12.32 An arc tangent to perpendicular lines.

Step 1 Using radius R and center A, locate D and E.

Step 2 Find point C by swinging two arcs with radius R.

Step 3 Locate tangent points with perpendiculars CE and CD.

Step 4 Draw the tangent arc and darken your lines.

12.34 An arc tangent to an arc and a line.

Step 1 Draw a light line parallel to line AB that is R distance from it.

Step 2 Add radius R to the radius extended from center C. Swing the extended radius to find the center, O.

Step 3 Draw lines OC and OT to locate the tangency points.

Step 4 Draw the tangent arc between the points of tangency with radius R and center O.

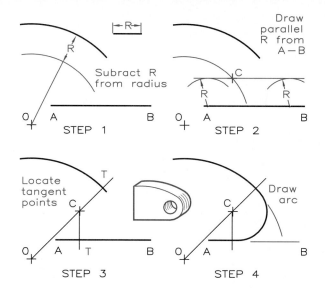

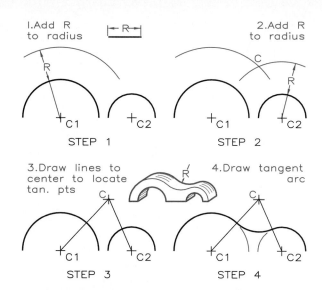

12.35 An arc tangent to an arc and a line:

Step 1 Subtract radius R from the radius through center O. Draw a concentric arc with this shortened radius.

Step 2 Draw a line parallel to line 1–2 and R distance from it to locate the center, C.

Step 3 Locate the tangency points with lines from O through C and from C perpendicular to line 1–2.

Step 4 Draw the tangent arc between the tangent points with radius R and center C.

12.36 A concave arc tangent to two arcs.

Step 1 Extend the radius of one circle by adding the radius R to it. Use the extended radius to draw a concentric arc.

Step 2 Extend the radius of the other circle by adding radius R to it. Use this extended radius to construct a concentric arc to locate center C.

Step 3 Connect center C with centers C1 and C2 with thin, dark lines to locate the tangency points.

Step 4 Draw the tangent arc between the points of tangency using radius R and center C.

Arc Tangent to Two Arcs

Figure 12.36 shows how to draw a concave arc tangent to two arcs. Lines drawn between the centers locate the points of tangency. The resulting tangent arc is concave from the top. Drawing a convex arc tangent to the given arcs requires that its radius be greater than the radius of either of the given arcs, as shown in **Fig. 12.37**.

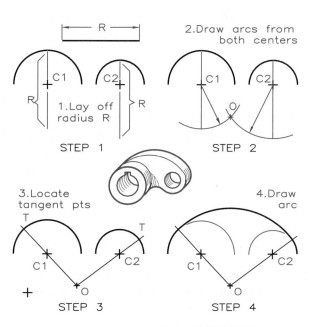

12.37 A convex arc tangent to two arcs.

Step 1 Extend the radius of each arc from the arc past its center by a distance of R, the radius, along these lines.

Step 2 Use the distance from each center to the ends of the extended radii to swing arcs to locate center O.

Step 3 Draw thin, dark lines from center O through centers C1 and C2 to locate the points of tangency.

Step 4 Draw the tangent arc between the points of tangency using radius R and center O.

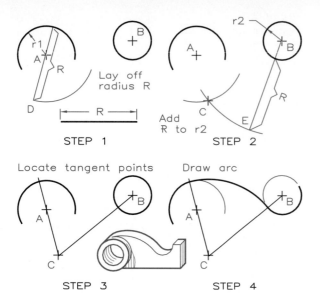

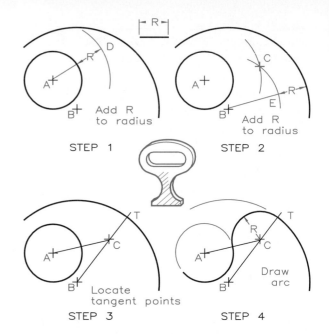

12.38 An arc tangent to two circles.

Step 1 Lay off radius R from the arc along an extended radius to locate point D.

Step 2 Extend the radius from center B and add radius R to it to point E. Use radius BE to locate center C.

Step 3 Draw thin lines from center C through centers A and B to locate the points of tangency.

Step 4 Draw the tangent arc between the tangent points using radius R and center C.

One variation of this problem (**Fig. 12.38**) is to draw an arc of a given radius tangent to the top of one arc and the bottom of the other. Another is to draw an arc tangent to a circle and a larger arc (**Fig. 12.39**).

Computer Method Use the FILLET command (**Fig. 12.40**) to draw an arc tangent to two arcs. After entering the command, specify the radius when prompted. Press the carriage return (Enter) twice to return to the COMMAND mode, and the FILLET command is ready for use.

Select the two arcs with your cursor; the tangent arc is drawn and the arcs are trimmed at the points of tangency. Mark the tangency points by drawing lines from centers C1 and C2.

12.39 An arc tangent to two arcs.

Step 1 Add radius R to the radius from A. Use radius AD to draw a concentric arc from center A.

Step 2 Subtract radius R from the radius through B. Use radius BE to draw an arc to locate center C.

Step 3 Draw thin lines to connect the centers and mark the points of tangency.

Step 4 Draw the tangent arc between the tangency points using radius R and center C.

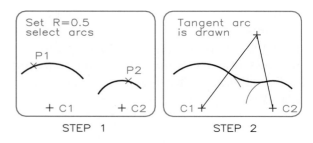

12.40 An arc tangent to two arcs by computer:

Step 1 Command: FILLET (Enter)
Polyline/Radius/<Select two lines>: R (Enter)
Enter fillet radius <current>: .5 (Enter)

Step 2 Command: FILLET (Enter)
Polyline/Radius/<Select two objects>: P1, P2 (Points on each arc.) (Draw the tangent arc, and TRIM the arcs. Tangent points are located by lines between centers.)

Ogee Curves

The **ogee curve** is an S curve formed by tangent arcs. The ogee curve shown in **Fig. 12.41** is the result of constructing two arcs tangent to three intersecting lines. **Figure 12.42** shows an unequal-arc ogee curve drawn to pass through points B, E, and C.

12.12 Conic Sections

Conic sections are plane figures that can be described both graphically and mathematically; formed by passing imaginary cutting planes through a right cone (**Fig. 12.43**). They are circles, ellipses, hyperbolas, and parabolas.

Ellipses

The ellipse is a conic section formed by passing a plane through a right cone at an angle (Fig. 12.43B). Mathematically, the ellipse is the path of a point that moves in such a way that the sum of the distances from two focal points is a constant. The largest diameter of an ellipse—the major diameter—is always the true length. The shortest diameter—the minor diameter—is perpendicular to the major diameter.

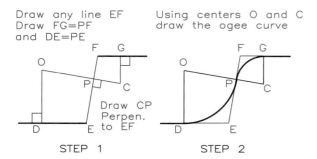

12.41　An ogee curve.

Step 1　To draw an ogee curve between two parallel lines, draw light line EF at any angle. Locate point P anywhere along EF. Find the tangent points by making FG equal to FP and DE equal to EP. Draw perpendiculars at G and D to intersect the perpendicular at O and C.

Step 2　Use radii CP and OP at centers O and C to draw two tangent arcs to complete the ogee curve.

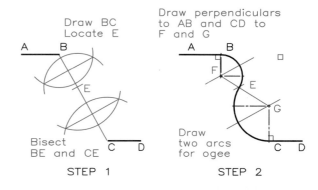

12.42　An unequal-arc ogee curve.

Step 1　Parallel lines are to be connected by an ogee curve passing through points B and C. Draw light line BC and select point E on it. Bisect BE and CE.

Step 2　Construct perpendiculars at points B and C to intersect the bisectors and locate centers F and G. Locate the points of tangency and draw the ogee curve using radii FB and GC.

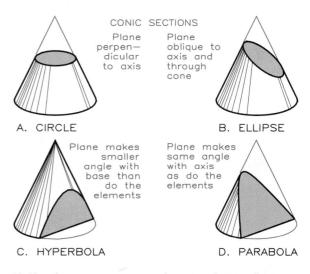

12.43　The conic sections are the (A) circle, (B) ellipse, (C) hyperbola, and (D) parabola. They are formed by passing cutting planes through a right cone.

Revolving the edge view of a circle yields an ellipse (**Fig. 12.44**). The ellipse template shown in **Fig. 12.45** is used to draw the same ellipse. The angle between the line of sight and the edge of the circle is the ellipse tem-

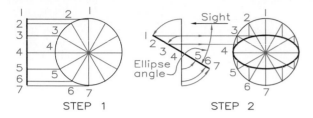

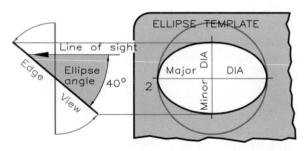

12.44 An ellipse by revolution.

Step 1 When the edge of a circle is perpendicular to the projectors between its adjacent view, it appears as a circle. Mark equally spaced points around the circle's circumference and project them to the edge.

Step 2 Revolve the edge of the circle and project the points to the circular view. Project the point vertically downward to obtain the elliptical view.

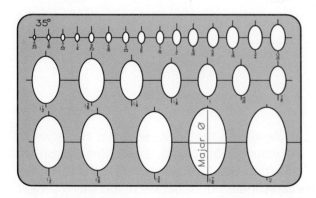

12.45 When the line of sight is not perpendicular to a circle's edge, it appears as an ellipse. The angle between the line of sight and the edge of the circle is the ellipse template angle.

plate angle (or the one closest to this size) that should be used. Ellipse templates are available in 5° intervals and in major diameter sizes that vary in increments of about 1/8 inch (**Fig. 12.46**).

You may construct an ellipse inside a rectangle or parallelogram by plotting a series of points to form the ellipse (**Fig. 12.47**).

An ellipse can be drawn on x and y axes by plotting x and y coordinates from the equation of an ellipse. The mathematical equation of an ellipse is

$$\frac{x^2 + y^2}{a^2 + b^2} = 1, \quad \text{where a, b} \neq 0.$$

Parabolas

The **parabola** is defined as a plane curve, each point of which is equidistant from a straight line (called a directrix) and a focal point. The parabola is the conic section formed when the cutting plane and an element on the cone's surface make the same angle with the cone's base, as shown in Fig. 12.43D.

Figure 12.48 shows the construction of a parabola by using its mathematical definition, as is done in analytical geometry. A second method of drawing a parabola, which involves

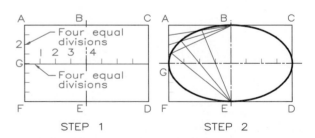

12.47 An ellipse by the parallelogram method.

Step 1 Draw an ellipse inside a rectangle or parallelogram by dividing the horizontal centerline and AF into the same number of equal divisions.

Step 2 The curve construction is shown for one quadrant. Sets of rays from E and B cross at points on the curve.

12.46 Ellipse templates are calibrated at 5° intervals from 15° to 60° and by their major diameter size.

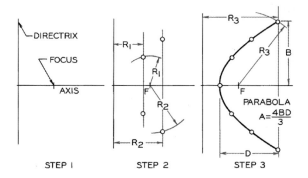

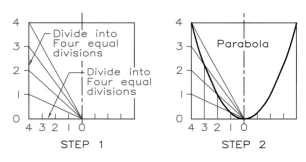

STEP I STEP 2 STEP 3 STEP 1 STEP 2

12.48 A parabola by the mathematical method.

Step 1 Draw an axis perpendicular to the directrix (a line). Choose a point for the focus, F.

Step 2 Use a series of selected radii to find points on the curve. For example, draw a line parallel to the directrix and R_2 from it. Swing R_2 from F to intersect the line and plot the point.

Step 3 Continue the process with a series of arcs of varying radii until you find an adequate number of points to complete the curve.

12.49 A parabola by the parallelogram method.

Step 1 Draw a parallelogram or rectangle to contain the parabola; draw its axis parallel to the sides through 0. Divide the sides into equal segments. Draw rays from 0.

Step 2 Draw lines parallel to the sides (vertical, in this case) to locate points along the rays from 0 and draw a smooth curve through them.

the use of a rectangle or parallelogram, is shown in **Fig. 12.49**. The mathematical equation of the parabola is

$$y = ax^2 + bx + c, \text{ where } a \uparrow 0.$$

Hyperbolas

The hyperbola is a two-part conic section defined as the path of a point that moves in such a way that the difference of its distances from two focal points is a constant (Fig. 12.44C). **Figure 12.50** shows construction of a hyperbola according to this definition. By selecting a series of radii until enough points have been located, the hyperbolic curve can be accurately drawn.

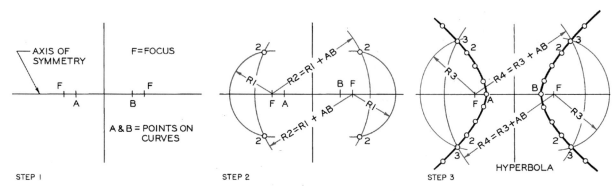

STEP I STEP 2 STEP 3

12.50 Constructing a hyperbola.

Step 1 Draw a perpendicular through the axis of symmetry. Locate focal points F equidistant from it on both sides. Locate points A and B equidistant from the perpendicular at a distance of your choosing, but between the focal points.

Step 2 Use radius R1 to draw arcs with focal points F as the centers. Add R1 to AB (the nearest distance between the hyperbolas) to find R2. Draw arcs using radius R2 and the focal points as centers. R1 and R2 cross at 2 on the curves.

Step 3 Select other radii and add them to AB to locate additional points as in step 2. Draw a smooth curve through the points.

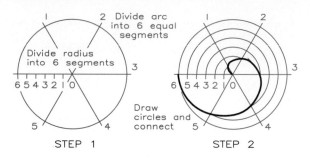

12.51 Constructing a spiral.

Step 1 Draw a circle and divide it into equal parts. Divide the radius into the same number of equal parts (six, in this case).

Step 2 Begin inside and draw arc 0–1 to intersect radius 0–1. Then swing arc 0–2 to radius 0–2, and continue to point 6 on the original circle. Connect the points.

12.13 Spirals

The **spiral** is a coil lying in a single plane that begins at a point and becomes larger as it travels around the origin. **Figure 12.51** shows the steps for constructing a spiral. The number of divisions selected in this construction depends upon the degree of accuracy desired.

12.14 Helixes

The **helix** is a three-dimensional curve that coils around a cylinder or cone at a constant angle of inclination. Applications of helixes are corkscrews and the threads on a screw. **Figure 12.52** shows how a helix is constructed about a cylinder, and **Fig. 12.53** shows how a helix is constructed about a cone.

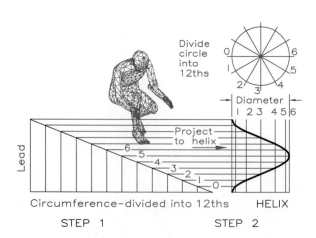

12.52 A cylindrical helix.

Step 1 Divide the top view of the cylinder into equal parts and project them to the front view. Lay out the circumference and the lead (pronounced *leed*) or height of the cylinder. Divide the circumference into the same number of equal parts transferred from the top view.

Step 2 Project the points along the inclined rise to their respective points on the diameter and connect them with a smooth curve.

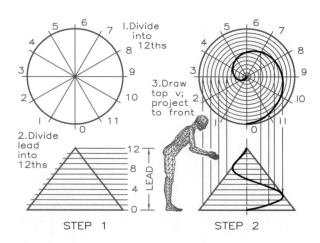

12.53 A conical helix.

Step 1 Divide the cone's base into equal parts. Pass a series of horizontal cutting planes through the front view of the cone. Use the same number as the number of divisions on the base (12, in this case).

Step 2 Project all the divisions along the front view of the cone to the line 3–9 and draw a series of arcs from the center to their respective radii in the top view and plot the points. Project the points to their respective cutting planes in the front view and connect them with a smooth curve.

Problems

Present your solutions to these problems on size AV (8 1/2 × 11 inch) paper similar in appearance to that shown in Sheet 1. The printed grid represents 0.20-in. intervals, so you can use your engineers' 10-scale to lay out the problems. By equating each grid interval to 5 mm, you also can use your full-sized metric scale to lay out and solve the problems. Show your construction and mark all points of tangency, as recommended in the chapter.

1. Basic constructions (Sheet 1):

    (A) Draw triangle ABC using the given sides.

    (B–C) Inscribe an angle in the semicircles with the vertexes at point P.

    (D) Inscribe a three-sided regular polygon inside the circle.

    (E) Circumscribe a four-sided regular polygon about the circle.

2. Construction of regular polygons (Sheet 2):

    (A) Circumscribe a hexagon about the circle.
    (B) Inscribe a hexagon in the circle.
    (C) Circumscribe an octagon about the circle.
    (D) Construct a pentagon inside the circle using the compass method.

3. Basic constructions (Sheet 3):

    (A) Bisect the lines.
    (B) Bisect the angles.
    (C) Bisect the sides of the triangle.

4. Line division and tangencies (Sheet 4):

    (A) Divide AB into seven equal parts using a construction line through point A.

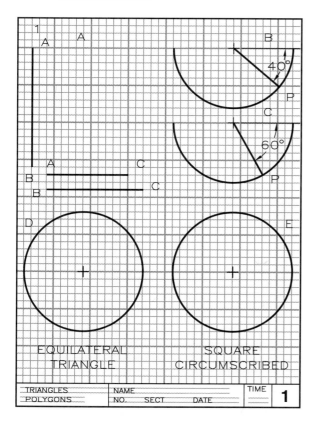

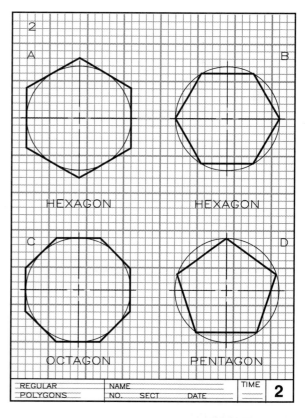

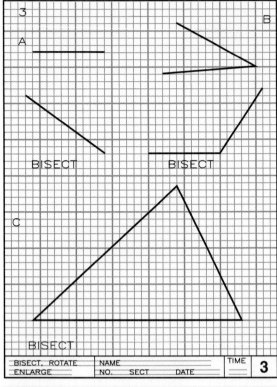

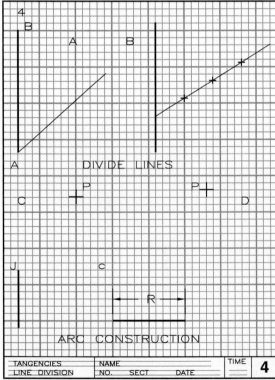

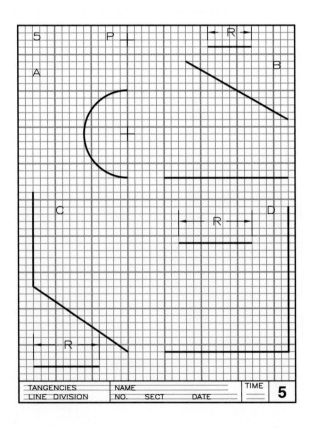

(B) Divide the space between the two vertical lines into four equal segments. Draw three vertical lines at the division points that are equal in length to the given lines.

(C) Construct an arc with radius R that is tangent to the line at J and that passes through point P.

(D) Construct an arc with radius R that is tangent to the line and passes through P.

5. Tangency construction (Sheet 5):

   (A) Using the compass method, draw a line from P tangent to the arc. Mark the points of tangency.

   (B–D) Construct arcs with the given radii tangent to the lines.

6. Tangency construction (Sheet 6):

(A–D) Construct arcs that are tangent to the arcs or lines shown. The radii are given for each problem.

7. Ogee curve construction (Sheet 7):

(A–D) Construct ogee curves that connect the ends of the given lines and pass through P.

8. Tangency construction (Sheet 8):

(A–B) Using the given radii, connect the circles with the tangent arcs as indicated in the diagrams.

9. Ellipse construction (Sheet 9):

(A–C) Construct ellipses inside the rectangles given. Use enough divisions to make it possible to draw accurate ellipses with your irregular curve.

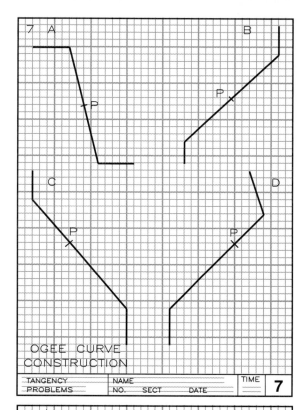

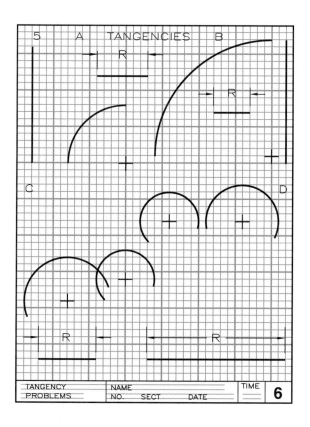

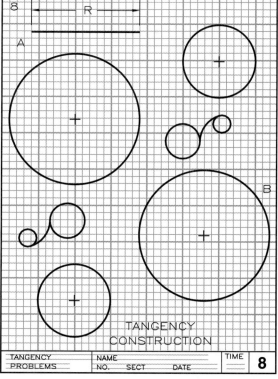

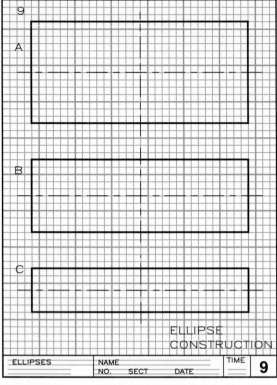

10. Ellipse and parabola construction (Sheet 10):

    (**A**) Construct an ellipse inside the circle when the edge view has been rotated as shown.

    (**B**) Using the focal point F and the directrix, plot and draw the parabola formed by these elements.

11. Hyperbola and spiral construction (Sheet 11):

    (**A**) Using the focal point F, points A and B on the curve, and the axis of symmetry, construct the hyperbola.

    (**B**) Construct a spiral by using the four divisions marked along the radius.

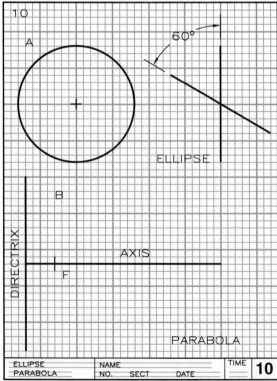

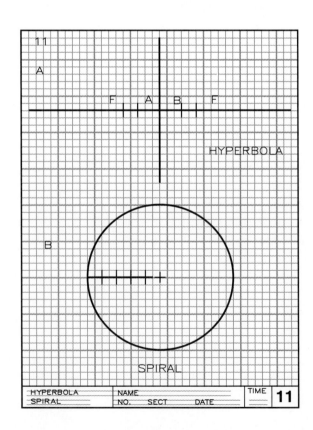

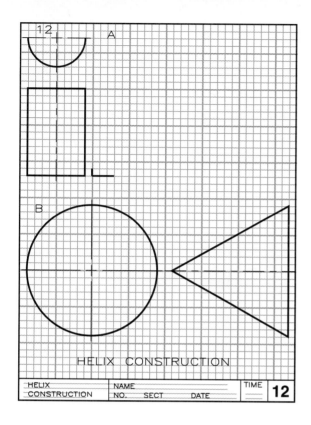

HELIX CONSTRUCTION

| HELIX
CONSTRUCTION | NAME
NO.   SECT   DATE | TIME | 12 |

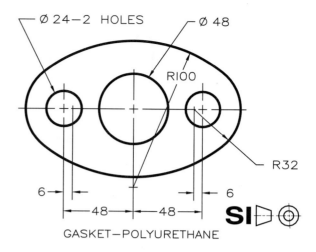

GASKET—POLYURETHANE

12.66 Problem 13.

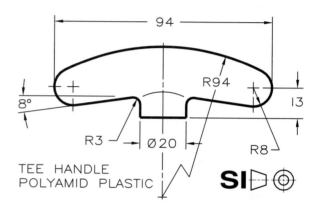

TEE HANDLE
POLYAMID PLASTIC

12.67 Problem 14.

12. Helix construction (Sheet 12):

(**A–B**) Construct helixes that have a rise equal to the heights of the cylinder and cone. Show construction and the curve in all views.

13–22. Practical applications (**Figs. 12.66–12.76**): Construct the given shapes on size A sheets, one problem per sheet. Select the scale that will best fit the problem to the sheet. Mark all points of tangency and strive for good line quality.

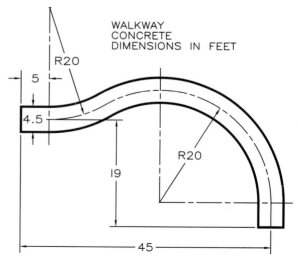

WALKWAY
CONCRETE
DIMENSIONS IN FEET

12.68 Problem 15.

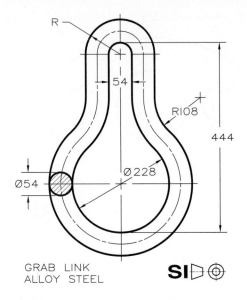

R

54

RI08

444

Ø228

Ø54

GRAB LINK
ALLOY STEEL

SI▷⊕

12.69 Problem 16.

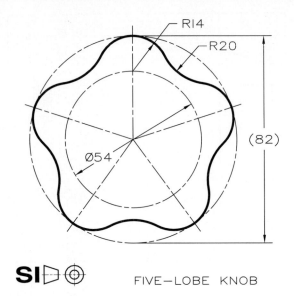

RI4

R20

Ø54

(82)

SI▷⊕ FIVE−LOBE KNOB

12.71 Problem 18.

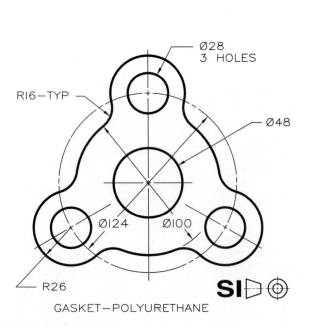

Ø28
3 HOLES

RI6−TYP

Ø48

Ø124 Ø100

R26

SI▷⊕

GASKET−POLYURETHANE

12.70 Problem 17.

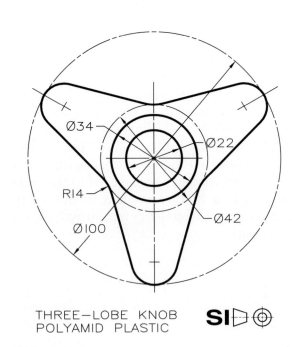

Ø34 Ø22

RI4

Ø42

Ø100

THREE−LOBE KNOB
POLYAMID PLASTIC

SI▷⊕

12.72 Problem 19.

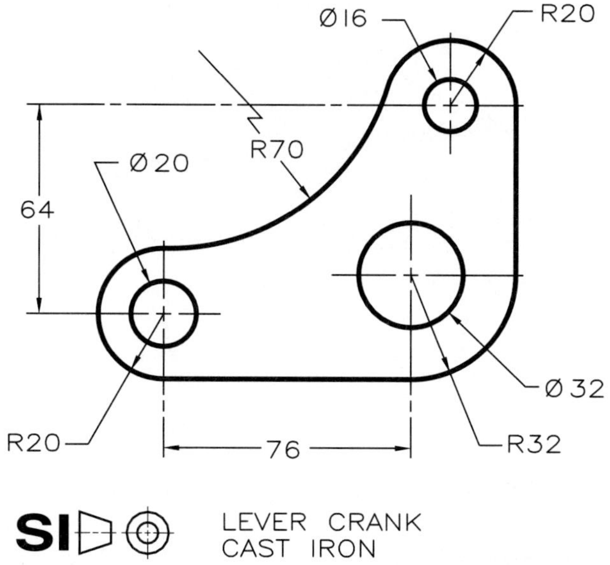
Ø16
R20
R70
Ø20
64
Ø32
R20
76
R32
SI
LEVER CRANK
CAST IRON

12.73 Problem 20.

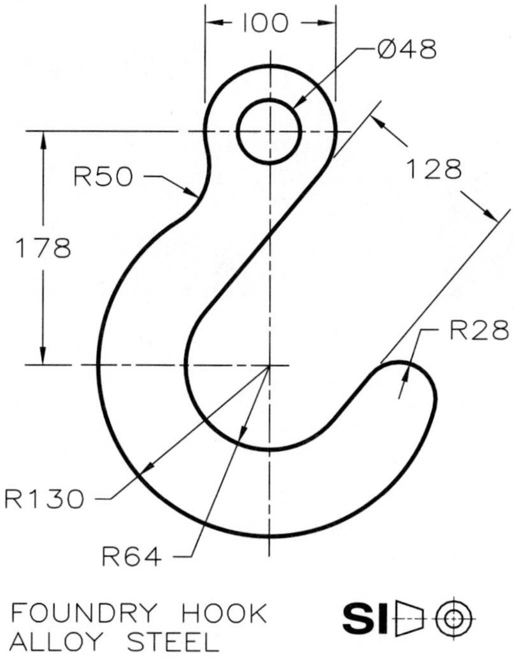
100
Ø48
128
R50
178
R28
R130
R64
FOUNDRY HOOK
ALLOY STEEL
SI

12.74 Problem 21.

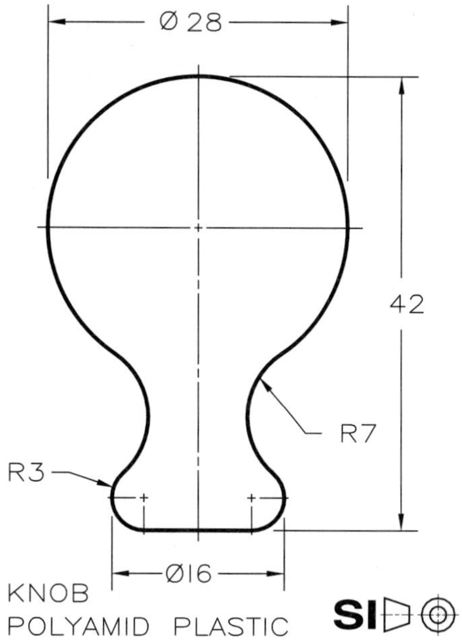
Ø28
42
R7
R3
Ø16
KNOB
POLYAMID PLASTIC
SI

12.75 Problem 22.

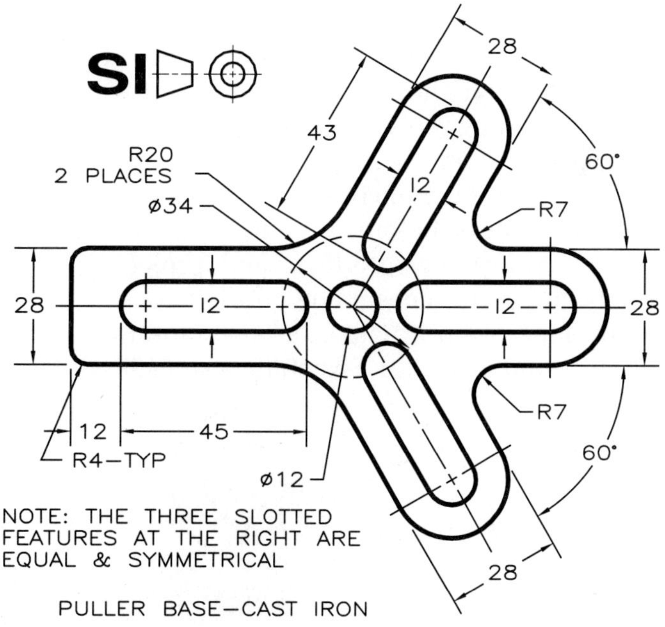
SI
28
43
60°
R20
2 PLACES
Ø34
12
R7
28
12
12
28
R7
12
45
R4–TYP
Ø12
60°
NOTE: THE THREE SLOTTED FEATURES AT THE RIGHT ARE EQUAL & SYMMETRICAL
28
PULLER BASE–CAST IRON

12.76 Problem 23.

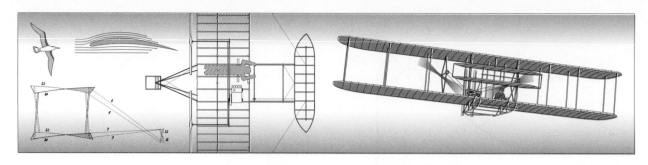

13

Freehand Sketching

13.1 Introduction

Sketching is a rapid, freehand method of drawing without the use of drawing instruments. Sketching is also a thinking process, a method of communication. Designers and engineers rely on sketches as a method of developing ideas before arriving at a final solution. Many new products and projects have begun as sketches made on the back of an envelope or on a napkin at a restaurant. (**Fig. 13.1** and **Fig. 13.2**). Sketching also facili-

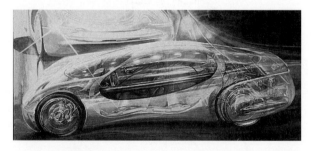

13.2 As sketching progresses, more sophisticated drawings can be made. (Courtesy of General Motors Corporation.)

tates communication when on-site problems are encountered.

The ability to communicate by any means is a great asset, and sketching is one of the best ways to transmit ideas. Engineers must use their sketching skills to explain their ideas before they can delegate assignments and obtain the assistance of their team members.

13.2 Shape Description

Although the angle bracket in **Fig. 13.3** is a simple three-dimensional object, describing

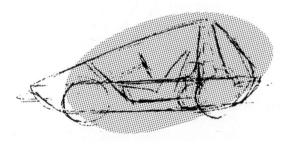

13.1 Designs begin with rough, freehand sketches as a means of developing concepts. (Courtesy of Chrysler Corporation.)

160

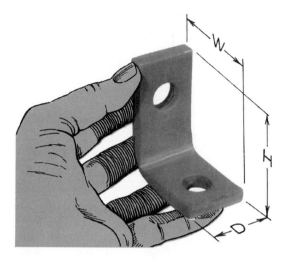

13.3 How can you sketch this angle bracket to convey its shape effectively?

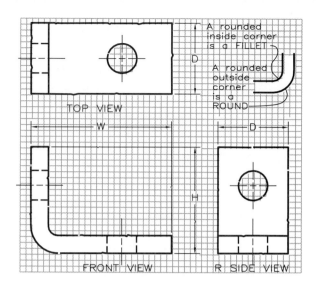

13.5 This sketch shows three orthographic views of the angle bracket: top, front, and right side.

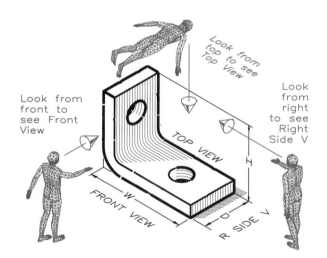

13.4 These positions give the viewpoints for three orthographic views of the angle bracket: top, front, and right side.

it with words is difficult. Most untrained people would think that drawing it as a three-dimensional pictorial would be a challenge. To make drawing such objects easier, engineers devised a standard system, called **orthographic projection**, for showing objects in different views.

In orthographic projection, separate views represent the object at 90° intervals as the viewer moves about it (**Fig. 13.4**). **Figure 13.5**

shows two-dimensional views of the bracket from the top, front, and right side. The top view is drawn above the front view, and both share the dimension of width. The right-side view is drawn to the right of the front view, and both share the dimension of height.

The views of the bracket is drawn with three types of lines: **visible lines, hidden lines**, and **centerlines**. Visible lines are the thickest. Thinner hidden lines (dashed lines) represent features that are invisible, or hidden, in a view. The thinnest lines are centerlines, which are imaginary lines composed of long and short dashes to show the centers of arcs and the axes of cylinders.

The space between views may vary, but the views must be positioned as shown here. This arrangement is logical, the views are easiest to interpret in this order, and the drawing process is most efficient because the views project from each other. **Figure 13.6** illustrates the lack of clarity when views are incorrectly positioned, even though each view is properly drawn.

13.2 SHAPE DESCRIPTION • 161

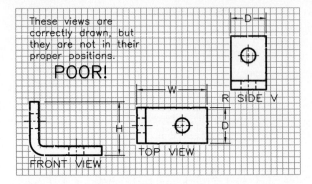

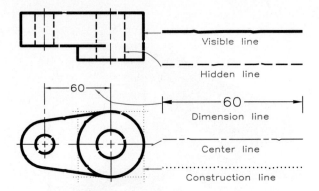

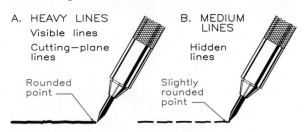

13.6 Views must be sketched in their standard orthographic positions. If they are incorrectly positioned, the object cannot be readily understood.

13.7 The alphabet of lines for sketching are shown here. The lines at the right are full size.

13.3 Sketching Techniques

You need to understand the application of line types used in sketching (freehand) orthographic views before continuing with the principles of projection. The "alphabet of lines" for sketching is presented in **Fig. 13.7**. All lines, except construction lines, should be black and dense. Construction lines are drawn lightly so that they need not be erased. The other lines are distinguished by their line widths (line thicknesses), but they are equal in darkness.

Medium-weight pencils, such as H, F, or HB grades, are best for sketching the lines shown in **Fig. 13.8**. By sharpening the pencil point to match the desired line width, you may use the same grade of pencil for all these lines. Lines sketched freehand should have a freehand appearance; do not attempt to make them appear mechanical. Using a printed grid or laying translucent paper over a printed grid can aid your sketching technique (**Fig. 13.9**).

When you make a freehand sketch, lines will be vertical, horizontal, angular, and/or circular. By not taping your drawing to the table top, you can position the sheet for the most comfortable strokes, usually from left to right (**Fig. 13.10**). Examples of correctly sketched lines are contrasted with incorrectly sketched ones in **Fig. 13.11**.

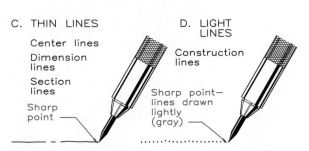

13.8 An F pencil is a good choice for sketching all lines if you sharpen it for varying line widths.

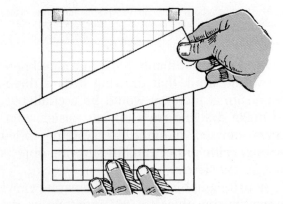

13.9 A grid placed under a sheet of tracing paper will provide guidelines as an aid in freehand sketching.

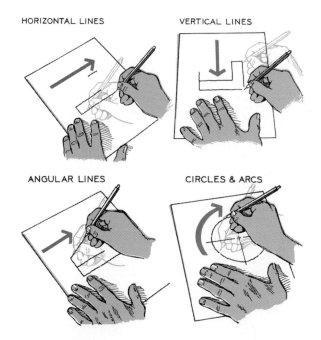

HORIZONTAL LINES VERTICAL LINES

ANGULAR LINES CIRCLES & ARCS

13.10 Sketch lines as shown here for the best results; rotate your drawing sheet for comfortable sketching positions.

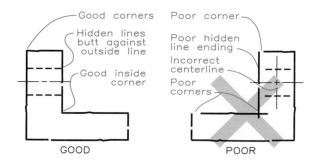

Good corners
Hidden lines butt against outside line
Good inside corner

GOOD

Poor corner
Poor hidden line ending
Incorrect centerline
Poor corners

POOR

13.11 Follow the examples of good technique. Try to avoid the common errors of poor technique shown.

13.4 Six-View Sketching

The maximum number of principal views that can be drawn in orthographic projection is six, as the viewer changes position at 90° intervals (**Fig. 13.12**). In each view, two of the three dimensions of height, width, and depth are seen.

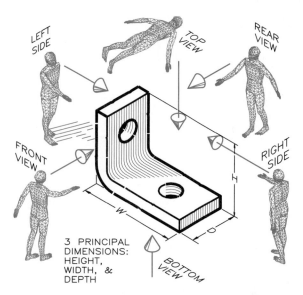

LEFT SIDE • TOP VIEW • REAR VIEW

FRONT VIEW • RIGHT SIDE

3 PRINCIPAL DIMENSIONS: HEIGHT, WIDTH, & DEPTH

BOTTOM VIEW

13.12 Six principal views of the angle bracket can be sketched from the viewpoints shown.

These views must be sketched in their standard positions (**Fig. 13.13**). The width dimension is shared by the top, front, and bottom views. The height dimension is shared with the right-side, front, left-side, and rear views. Note the simple and effective dimensioning of each view with two dimensions. Seldom is an object so complex that it requires six orthographic views.

13.5 Three-View Sketching

You can adequately describe most objects with three orthographic views—usually the top, front, and right-side views. **Figure 13.14** shows a typical three-view sketch of a jaw nut with height, width, and depth dimensions and the front, top, and right-side views labeled.

The fixture block shown in **Fig. 13.15** is represented by three orthographic views on a grid in **Fig. 13.16**. To obtain those views, first sketch the overall dimensions of the object, then sketch the slanted surface in the top view and project it to the other views. Finally, darken the lines, label the views, and letter the overall dimensions of height, width, and depth.

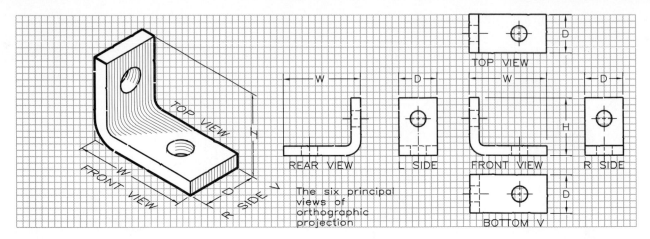

13.13 This six-view sketch of the angle bracket shows the principal views of orthographic projection. Note the placement of dimensions on the views.

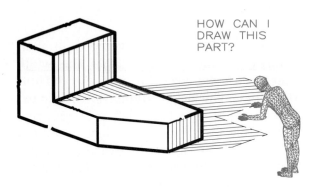

HOW CAN I DRAW THIS PART?

13.15 This fixture block is analyzed in the sketches of three orthographic views shown in Fig. 13.16.

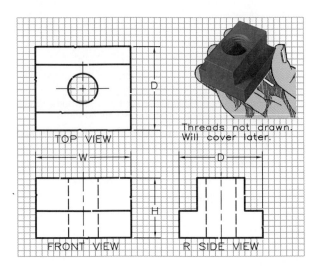

Threads not drawn. Will cover later.

13.14 The standard orthographic arrangement for three views of a jaw nut, with dimensions and labels.

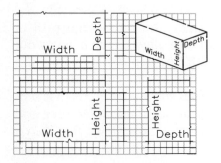

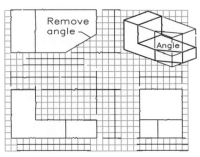

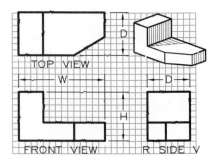

13.16 Three-view sketching.

Step 1 Block in the views with light construction lines. Allow proper spacing for labeling and dimensioning the views.

Step 2 Remove the notches and project from view to view.

Step 3 Check for correctness, darken the lines, and letter the labels and dimensions.

Slanted surfaces will appear as edges or foreshortened (not-true-size) planes in the principal views of orthographic projection (**Fig. 13.17**). In Fig. 13.17C, two intersecting planes of the object slope in two directions; thus, both appear foreshortened in the front, top, and right-side views.

A good way to learn orthographic projection is to construct a missing third view (the front view in **Fig. 13.18**) when two views are given. In **Fig. 13.19**, we construct the missing right-side view from the given top and front views. To obtain the depth dimension for the right-side view, transfer it from the top view with dividers; to obtain the height dimension, project it from the front view.

Figure 13.20 shows a rest pad sketched in three views. The pad has a finished surface,

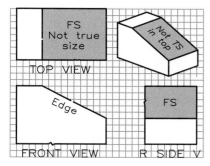 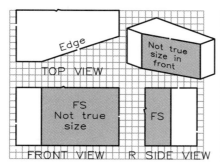 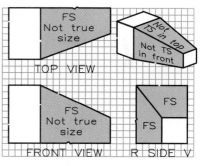

FORESHORTENED IN TOP   FORESHORTENED IN FRONT   FORESHORTENED IN ALL

13.17 Views of planes.

A The plane appearing as an angular edge in the front view is foreshortened in the top and side views.

B The plane appearing as an angular edge in the top view is foreshortened in the front and side views.

C Two sloping planes appear foreshortened in the side view and neither appears as an edge in either the top or front views.

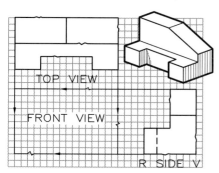 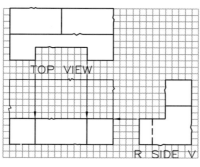 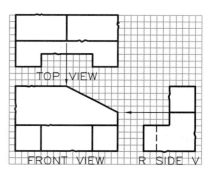

13.18 Sketching a missing front view.

Step 1 To sketch the front view, begin by blocking it in with light construction lines that will not need to be erased.

Step 2 Project the notch from the top view to the front view and sketch the lines as final lines.

Step 3 Project the ends of the angular notch from the top and right-side views, check the views, and darken the lines.

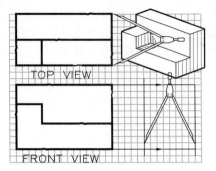
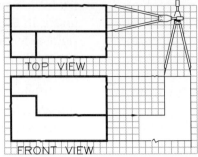
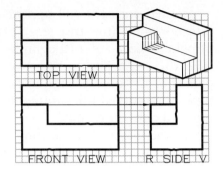

13.19 Sketching a missing side view.

Step 1 Transfer the depth with dividers and project the height from the front. Block in the view with construction lines.

Step 2 Locate the notch in the side view with your dividers and project its base from the front view. Use light construction lines.

Step 3 Project the top of the notch from the front view, check for correctness, darken the lines, and label the views.

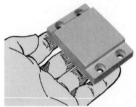

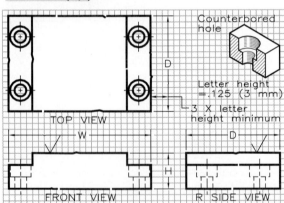

13.20 Three orthographic views adequately describe the rest pad. Space dimension lines at least three letter heights from the views. Finish marks (√ marks) indicate that the top surface has been machined to a smooth finish. Counterbored holes allow bolt heads to be recessed.

indicated by √ marks in the two views where the surface appears as edges, and four counterbored holes. Dimension lines for the height, width, and depth labels should be spaced at least three letter heights from the views. For example, when you use 1/8-inch

letters, position them at least 3/8 inch from the views.

Apply the finish mark symbol to the edge views of any finished surfaces, visible or hidden, to specify that the surface is to be machined to make it smoother. The surface in **Fig. 13.21** is being finished by grinding, which is one of many methods of smoothing a surface.

13.6 Circular Features

The pulley shaft depicted in **Fig. 13.22** in two views is composed of circular features.

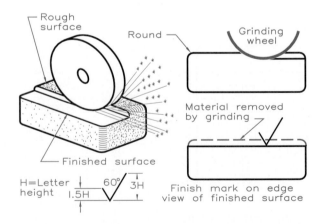

13.21 Place a finish mark on any edge view of a surface (visible or hidden) that has been (or is to be) smoothed by machining. Grinding is one of the methods used to finish a surface.

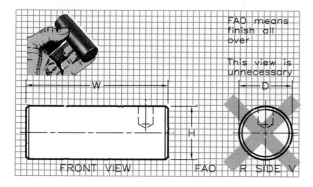

13.22 This pulley shaft is a typical cylindrical part that can be represented adequately by two views.

PRIORITIES
First: Visible lines
Second: Hidden lines
Third: Centerlines

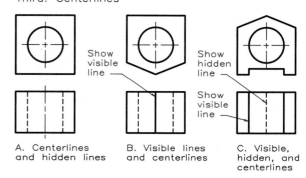

A. Centerlines and hidden lines   B. Visible lines and centerlines   C. Visible, hidden, and centerlines

13.24 When visible lines coincide with hidden lines, show the visible lines. When hidden lines coincide with centerlines, show the hidden lines.

Centerlines are added to better identify these cylindrical features. **Figure 13.23** shows how to apply centerlines to indicate the center of the circular ends of a cylinder and its vertical axis. Perpendicular centerlines cross in circular views to locate the center of the circle and extend beyond the arc by about 1/8 inch. Centerlines consist of alternating long and short dashes, about 1 inch and 1/8 inch in length, respectively.

When centerlines coincide with visible or hidden lines, the centerline should be omitted because object lines are more important and centerlines are imaginary lines. **Figure 13.24** shows the priorities of lines.

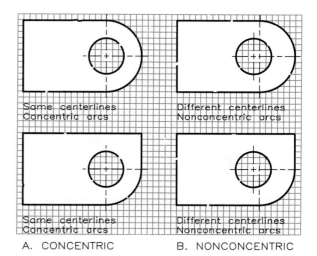

A. CONCENTRIC   B. NONCONCENTRIC

13.25 Centerlines.

A Extend centerlines beyond the last arc that has the same center.

B Sketch separate centerlines when the arcs are not concentric.

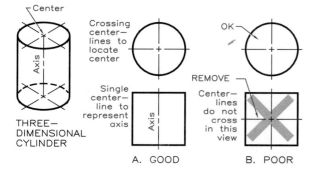

A. GOOD   B. POOR

13.23 Centerlines identify the centers of circles and axes of cylinders. Centerlines cross only in the circular view and extend about 1/8 inch beyond the outside lines.

The centerlines shown in **Fig. 13.25** clarify whether the circles and arcs are concentric (share the same centers). **Figure 13.26** shows the correct manner of applying centerlines to orthographic views of an object composed of concentric cylinders.

13.6 CIRCULAR FEATURES • 167

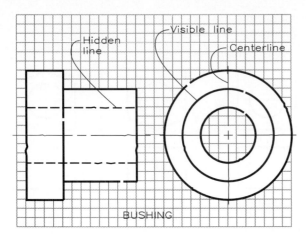

13.26 This orthographic sketch depicts the application of centerlines to concentric cylinders and the relative weights of various lines.

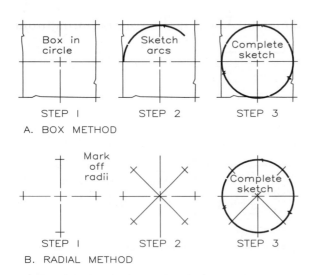

13.27 Sketching circles.

A Box Method

Step 1 Block in the diameter of the circle about the centerlines.

Step 2 Sketch an arc tangent through two tangent points.

Step 3 Complete the circle with other arcs.

B Radial Method

Step 1 Mark off radii on the centerlines.

Step 2 Mark off radii on two construction lines drawn at 45°.

Step 3 Sketch the circles with arcs passing through the marks.

Sketching Circles

Circles can be sketched by using either the box method or the radial method shown in **Fig. 13.27**. Use light guidelines and dark centerlines to block in the circle. Drawing a freehand circle in one continuous arc is difficult, so draw arcs in segments with the help of the guidelines.

A typical part having circular features is represented by the two sketched views in **Fig. 13.28**. Note the definitions of a round and a chamfer. **Figure 13.29** shows the steps involved in constructing three orthographic views of a part having circular features.

13.7 Oblique Pictorial Sketching

An oblique pictorial is a three-dimensional representation of an object's height, width, and depth. It approximates a photograph of an object, making the sketch easier to under-

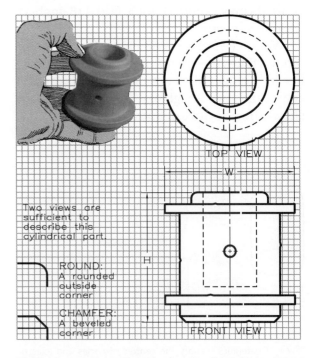

13.28 Two views adequately describe this cylindrical pivot base.

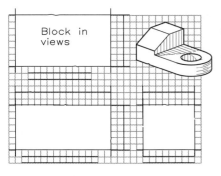
Block in views

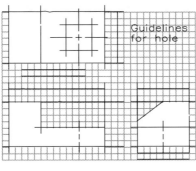
Guidelines for hole

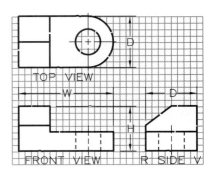
TOP VIEW

W

D

FRONT VIEW   R SIDE V

H

13.29 Sketching circular features.

Step 1 Begin by blocking in the overall dimensions with construction lines. Leave room for labels and dimensions.

Step 2 Draw the centerlines and the squares that block in the diameter of the circle. Find the slanted surface in the side view.

Step 3 Sketch the arcs, darken the lines, label the views, and show the dimensions W, D, and H between the views.

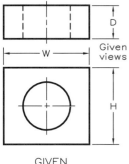
D

W

H

Given views

GIVEN

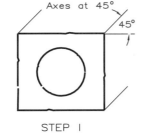
Axes at 45°

45°

STEP 1

Draw front view true size

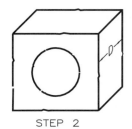
D

STEP 2

Measure depth along receding axes

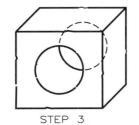
STEP 3

Draw visible portion of hole on back surface (omit hidden lines)

13.30 An oblique pictorial sketching.

Step 1 Sketch the front of the part as an orthographic front view and the receding lines at 45° to show the depth dimension.

Step 2 Measure the depth along the receding axes and sketch the back of the part.

Step 3 Locate the circle on the rear plane, show the visible portion of it, and omit the hidden lines.

stand at a glance than do orthographic views. Sketch the front of the object as a true-shape orthographic view (**Fig. 13.30**). Sketch the receding axes at an angle of between 20° and 60° oblique with the horizontal in the front view. Lay off the depth dimension as its true length along the receding axes. When the depth is true length, the oblique is a **cavalier oblique**.

The major advantage of an oblique pictorial is the ease of sketching circular features as circular arcs on the true-size front plane. **Figure 13.31** shows an oblique sketch of a shaft block. Circular features on the receding planes appear as ellipses, requiring slanted guidelines, as shown.

13.8 Isometric Pictorial Sketching

Another type of three-dimensional representation is the isometric pictorial, in which the axes make 12° angles with each other (**Fig. 13.32**). Specially printed isometric grids with lines intersecting at 60° angles make isometric sketching easier (**Fig. 13.33**). Simply transfer the dimensions from the squares in the orthographic views to the isometric grid. You cannot measure angles in isometric pictorials

13.8 ISOMETRIC PICTORIAL SKETCHING • 169

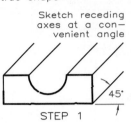

Sketch front view
true shape

Sketch receding
axes at a con—
venient angle

STEP 1

45°

Sketch rear of part

Sketch
guidelines
around holes

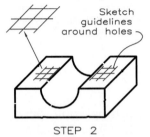

Sketch elliptical
views of holes
tangent to
guidelines

STEP 2   STEP 3

13.31 Sketching arcs in oblique pictorials.

Step 1 Sketch the front view of the mounting bracket saddle as a true front view. Sketch the receding axes from each corner.

Step 2 Sketch the rear of the part by measuring its depth along the receding axes. Sketch guidelines about the holes.

Step 3 Sketch the circular features as ellipses on the upper planes tangent to the guidelines.

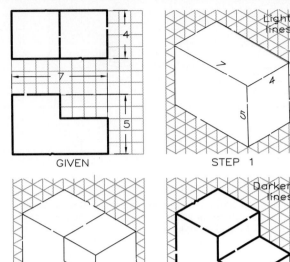

GIVEN   STEP 1

STEP 2   STEP 3

13.33 Sketching isometric pictorials.

Step 1 Use an isometric grid, transfer dimensions from the given views, and sketch a box having those dimensions.

Step 2 Locate the notch by measuring over four squares and down two squares, as shown in the orthographic views.

Step 3 Finish the notch and darken the lines.

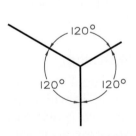

A. THE ISOMETRIC AXES   B. ISOMETRIC DRAWING

13.32 An isometric sketch.

A Begin an isometric pictorial by sketching three axes spaced 120° apart. One axis usually is vertical.

B Sketch the isometric shape parallel to the three axes and use its true measurements as the dimensions.

with a protractor; you must find them by connecting coordinates of the angle laid off along the isometric axes. In **Fig. 13.34**, the ends of the angular plane can be located by using the coordinates for width and height. When a part has two sloping planes that intersect (**Fig. 13.35**), you must sketch them one at a time to find point B. Line AB is found as the line of intersection between the planes. A more thorough coverage of isometric drawing is given in Chapter 25.

Circles in Isometric Pictorials

Circles appear as ellipses in isometric pictorials. When you sketch them, begin with their centerlines and construction lines enclosing

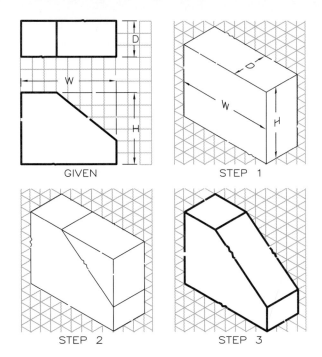

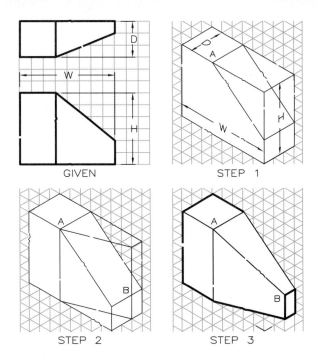

13.34 Sketching angles in isometric pictorials.

Step 1 Sketch a box from the overall dimensions given in the orthographic views.

Step 2 Angles cannot be measured with a protractor. Find each end of the angle with coordinates measured along the axes.

Step 3 Connect the ends of the angle and darken the lines.

13.35 Sketching double angles in isometric pictorials.

Step 1 This object has two sloping angles that intersect; begin by sketching the overall box and draw one of the angles.

Step 2 Find the second angle, which locates point B, the intersection line between the planes.

Step 3 Connect points A and B and darken the lines. Line AB is the line of intersection between the two sloping planes.

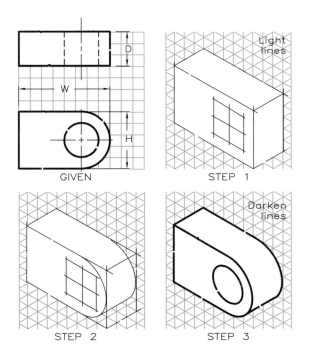

13.36 Sketching circles in isometric pictorials.

Step 1 Sketch a box using the overall dimensions given. Sketch the centerlines and a rhombus blocking in the circular hole.

Step 2 Sketch the isometric arcs tangent to the box. These arcs are elliptical rather than circular.

Step 3 Sketch the hole and darken the lines. Hidden lines usually are omitted in isometric pictorial sketches.

13.8 ISOMETRIC PICTORIAL SKETCHING • 171

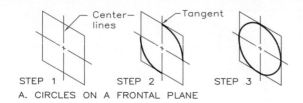

A. CIRCLES ON A FRONTAL PLANE

B. CIRCLES ON A HORIZONTAL PLANE

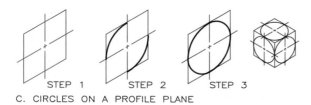

C. CIRCLES ON A PROFILE PLANE

13.37 Sketching circular features in isometric pictorials.
Step 1 Lay out centerlines and guidelines.
Step 2 Sketch two arcs.
Step 3 Connect the ends of the arcs to complete the ellipses.

their diameters, as shown in **Fig. 13.36**. The end of the block is semicircular in the front view, so its center must be equidistant from the top, bottom, and end of the front view. Circles and ellipses are easier to sketch if you use construction lines.

Figure 13.37 shows how to use centerlines and construction lines to draw ellipses in the three isometric planes: frontal, horizontal (top), and profile (side) views. This technique is used to sketch a cylinder in **Fig. 13.38** and an object having semicircular ends and circular holes in **Fig. 13.39**. Hidden lines are usually omitted in isometric drawings.

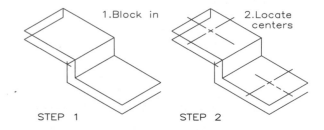

STEP 1   STEP 2

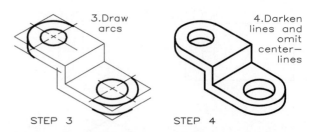

STEP 3   STEP 4

13.39 Sketching semicircular ends in isometric pictorials.
Step 1 Block in the isometric shape of the object with light lines.
Step 2 Locate the centerlines of the holes and the rounded ends.
Step 3 Sketch the semicircular ends of the part and the holes.
Step 4 Draw the bottoms of the holes and darken the lines.

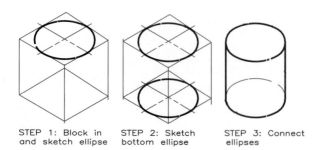

STEP 1: Block in and sketch ellipse   STEP 2: Sketch bottom ellipse   STEP 3: Connect ellipses

13.38 Sketching a cylinder as an isometric pictorial.
Step 1 Block in the cylinder and sketch the upper ellipse.
Step 2 Sketch the lower ellipse.
Step 3 Connect the ellipses with lines tangent to the elliptical ends and darken the lines.

Sketch your solutions to these problems (**Figs. 13.40–13.43**) on size A (8 1/2 × 11-inch) paper, with or without a printed grid. The format for this size sheet and a 0.20-in. grid (convertible to an approximate metric grid by equating each square to 5 mm) is shown in **Fig. 13.40**. Execute all sketches and lettering by applying the principles covered in this chapter and Chapter 11.

Problems 1–24: (A) Sketch the top, front, and right-side views of the problems assigned; supply lines that may be missing from all views. (B) Sketch oblique pictorials of the problems assigned. (C) Sketch isometric pictorials of the problems assigned.

Problems 25–64: Sketch the top, front, and right-side views of the problems assigned, two problems per sheet.

Design 2: The clevis (**Design 2**) has a 1.50 in. DIA pin that passes through the pair of holes. The single end is threaded for a 1.50 DIA threaded rod.

Design 1

Design Sketching
Design 1: The fixture block in **Design 1** has two holes at each end in the top view, and each is counterbored to half the height of the block for a bolt head. The holes in the front pass through the part and the small hole is drilled to intersect the horizontal hole. Sketch three orthographic views of the block.

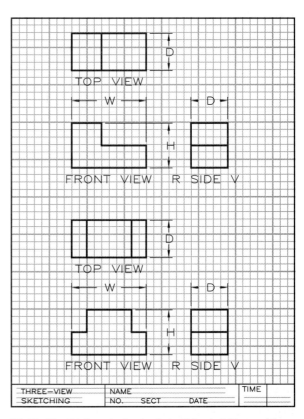

13.40   Use this layout of a size A sheet for sketching problems.

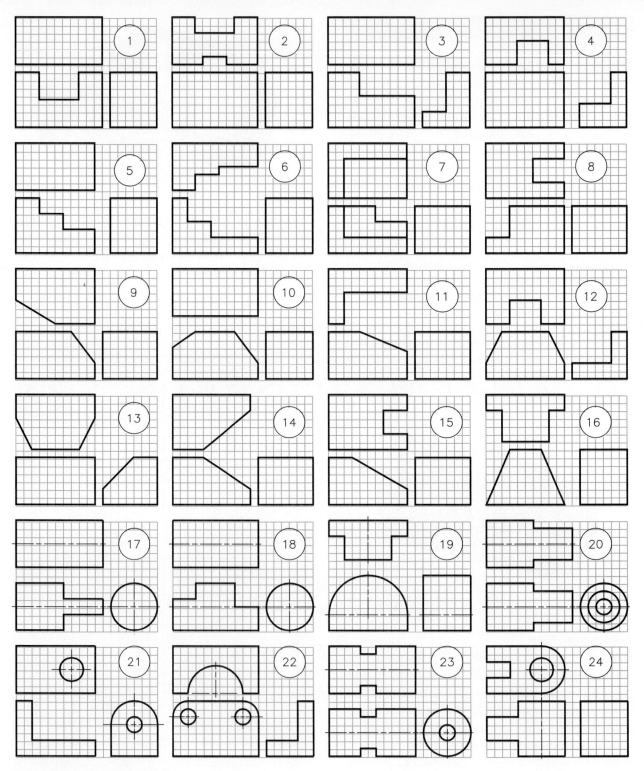

13.41 Sketching problems: 1–24

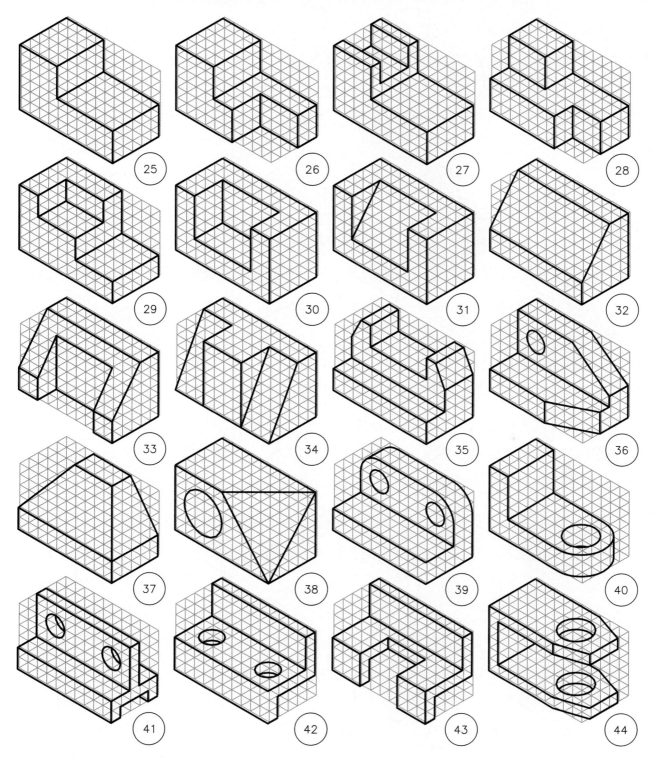

25 26 27 28 29 30 31 32 33 34 35 36 37 38 39 40 41 42 43 44

13.42   Sketching problems: 25–44

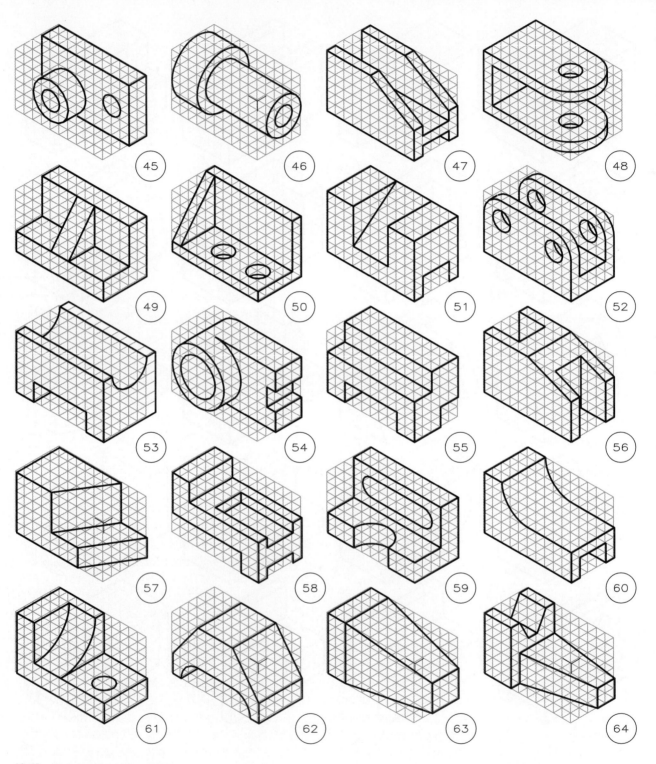

13.43 Sketching problems: 45–64

Design 3

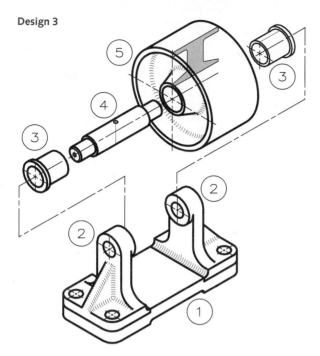

Design 4

Design Sketching

There are five parts in the pulley assembly in **Design 3**, which has a 1 in. diameter shaft (Part 4). Sketch one part per sheet following the instructions below.

Option 1: Make a two-view sketch of the shaft (Part 4). Do you know what the holes in the shaft are for?

Option 2: Make a two-view drawing of the bushing (Part 3). What are the bushings and why are they used?

Option 3: Make a two-view drawing of the base (Part 1). How are the brackets (Part 2) connected to the base?

Option 4: Make a three-view drawing of the shaft bracket (Part 2). How do the brackets support the shaft?

Option 5: Make a two-view drawing of the pulley (Part 5).

Option 6: Redesign the bracket (Part 2) and show your proposed modification in a three-view sketch.

Design 4: The two-part journal shown in **Design 4** supports the end of a 3 in. DIA shaft that rotates at a slow speed. Make three-view orthographic sketches of both parts. How are the parts held together? Use your common sense to determine what type of fasteners would be used and make sketches of them.

Design 5: The latching device in **Design 5** can be attached to a sheet-metal drawer (a file cabinet, for example) and turned 90° to lock the drawer. Make three-view sketches of the individual parts in order to better understand them yourself and to explain them to someone else.

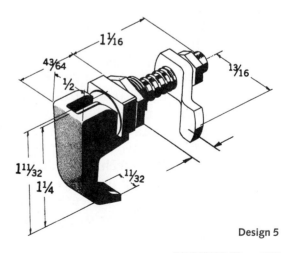

Design 5

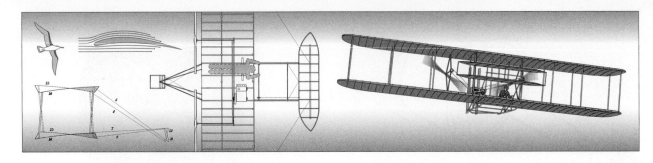

14

Orthographic Drawing with Instruments

14.1 Introduction

In Chapter 13, we discussed orthographic projection by freehand sketching, which is an excellent way to develop a design concept. Now, we must convert these sketches into orthographic views drawn to scale with instruments (or by computer) to more precisely define our designs. Afterward, we will add dimensions, notes, and specifications to convert these views into working drawings from which the design will become a reality.

Orthographic drawings represent three-dimensional objects in separate views arranged in a standard manner that will be readily understood by a technological team. Because multiview drawings usually are executed with instruments and drafting aids, they are often called **mechanical drawings**. They are called **working drawings**, or **detail drawings**, when sufficient dimensions, notes, and specifications are added to enable a prod-

uct to be manufactured or built from the drawings.

14.2 Orthographic Projection

An artist has the option of representing objects impressionistically, but an engineer must represent them precisely. Orthographic projection is used to prepare accurate, scaled, and clearly presented drawings from which an object can be built.

Orthographic projection is the system of drawing views of an object by projecting them perpendicularly onto projection planes with parallel projectors. **Figure 14.1** illustrates this concept of projection by imagining that the object is inside a glass box and three of its views are projected to planes of the box.

Figure 14.2 illustrates the principle of orthographic projection where the front view is projected perpendicularly onto a vertical projection plane, called the frontal plane,

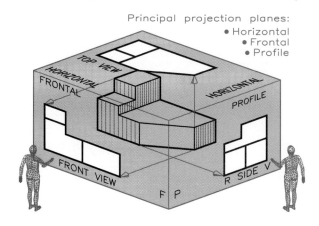

Principal projection planes:
• Horizontal
• Frontal
• Profile

14.1 Orthographic projection is the system of projecting views onto an imaginary glass box with parallel projectors to the three mutually perpendicular projection planes.

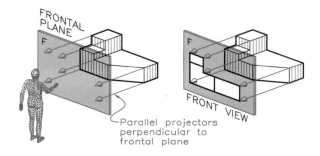

Parallel projectors perpendicular to frontal plane

14.2 An orthographic view is found by projecting from the object to a projection plane with parallel projectors that are perpendicular to the projection plane.

with parallel projectors. The projected front view is two-dimensional because it has only width and height and lies in a single plane. Similarly, the top view is projected onto a horizontal projection plane, and the side view is projected onto a second vertical projection plane.

Imagine that the box is opened into the plane of the drawing surface. **Figure 14.3A** illustrates how three planes of a glass box are opened into a single plane (**Fig. 14.3B**) to yield the standard positions for the three orthographic views. These views are the front, top, and right-side views.

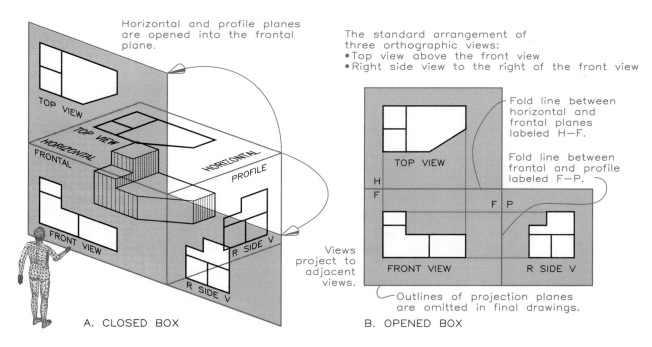

Horizontal and profile planes are opened into the frontal plane.

Views project to adjacent views.

A. CLOSED BOX

The standard arrangement of three orthographic views:
• Top view above the front view
• Right side view to the right of the front view

Fold line between horizontal and frontal planes labeled H—F.

Fold line between frontal and profile planes labeled F—P.

Outlines of projection planes are omitted in final drawings.

B. OPENED BOX

14.3 When the imaginary glass box is opened, the orthographic views and labeling are drawn in this format.

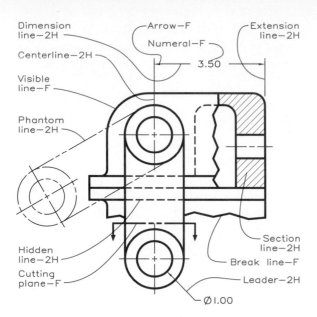

14.4 The alphabet of lines and recommended pencil grades for drawing orthographic views.

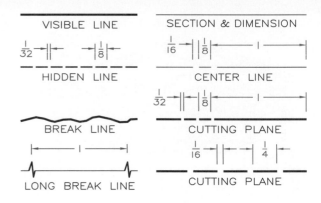

14.5 The full-size line weights recommended for drawing orthographic views.

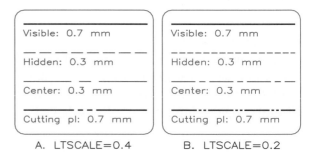

| A. LTSCALE=0.4 | B. LTSCALE=0.2 |

14.6 These lines were drawn using AutoCAD's standard LINETYPEs and two pens (P.3 and P.7 points).

A The LTSCALE factor of 0.4 gives relatively long dashes and spaces between dashes.

B In contrast, the LTSCALE factor of 0.2 produces dashes and spaces that are half as long as the factor of 0.4.

The principal projection planes of orthographic projection are the **horizontal (H)**, **frontal (F)**, and **profile (P)**. Views projected onto these principal planes are principal views. The dimensions that are used to give the sizes of principal views are **height (H)**, **width (W)**, and **depth (D)**.

14.3 Alphabet of Lines

All orthographic views should be drawn with dark and dense lines, as if done with ink. Only the line widths should vary except for guidelines and construction lines, which are drawn very lightly for layout and lettering. **Figure 14.4** presents the alphabet of lines used in orthographic projection and the recommended pencil grades to use. The lengths of dashes in hidden lines and centerlines are drawn longer as a drawing's size increases. **Figure 14.5** further describes these lines.

Computer Lines Computer graphics plotters use pens with points of varying widths, usually 0.7 mm and 0.3 mm wide, to draw lines of different thicknesses. To vary the lengths of dashes and the spaces between them in dashed lines, use LTSCALE (**Fig. 14.6**). When LTSCALE is used, ALL noncontinuous (centerlines and hidden lines) on the drawing are changed at the same time.

The laser and dot-matrix printers have come into widespread usage and have replaced many pen plotters. The ability to print with a high degree of sharpness and in

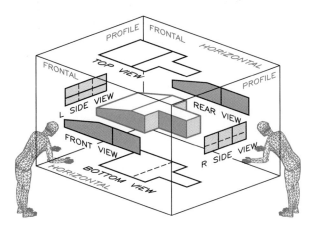

14.7 Six principal views of an object can be drawn in orthographic projection. Imagine that the object is in a glass box with the views projected onto its six planes.

color is the major advantage of these printers over the pen plotter. However, drawings in the 36-in. range are plotted mostly on the pen plotter.

14.4 Six-View Drawings

When you imagine that an object is inside a glass box, you will see two horizontal planes, two frontal planes, and two profile planes (**Fig. 14.7**). Therefore, the maximum number of principal views that can be used to represent an object is six. The top and bottom views are projected onto horizontal planes, the front and rear views onto frontal planes, and the right- and left-side views onto profile planes.

To draw the six views on a sheet of paper, imagine that the glass box is opened up into the plane of the drawing paper, as shown in **Fig. 14.8**. Place the top view over and the

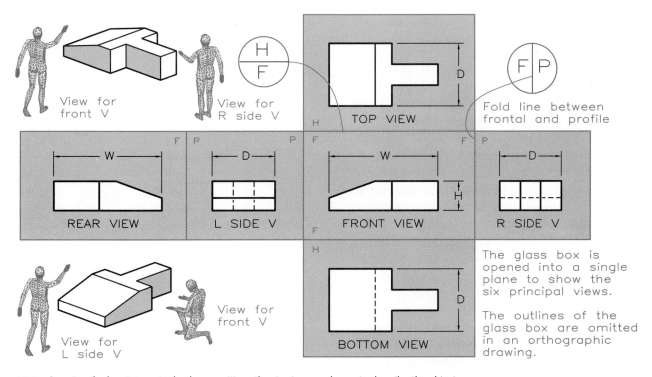

14.8 Opening the box into a single plane positions the six views as shown to describe the object.

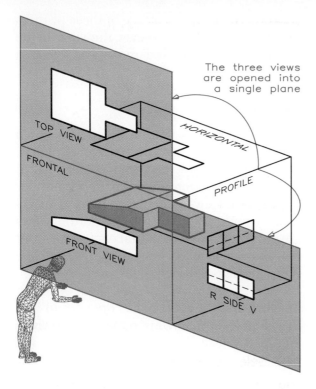

14.9 Three-view drawings are commonly used to describe small objects such as machine parts.

applies to the four horizontally aligned views. The width dimension is placed between the top and front views but applies to the bottom view also.

14.5 Three-View Drawings

The most commonly used orthographic arrangement of views is the three-view drawing, consisting of front, top, and right-side views. Imagine that the views of the object are projected onto the planes of the glass box (**Fig. 14.9**) and the three planes are opened into a single plane, the frontal plane. **Figure 14.10** shows the resulting three-view drawing where the views are labeled and dimensioned with H, W, and D.

In addition to using the glass-box approach for visualizing orthographic projection, imagine that the object is revolved until the desired view is obtained. For example, if the front view of the part in **Fig. 14.11** is revolved 90°, you will see its side view true shape (TS). Placed in its proper position, the side view aligns with the projectors from the front view at its right.

bottom view under the front view; place the right-side view to the right and the left-side view to the left of the front view; and place the rear view to the left of the left-side view.

Projectors align the views both horizontally and vertically about the front view. Each side of the fold lines of the glass box is labeled **H**, **F**, or **P** (horizontal, frontal, or profile) to identify the projection planes on each side of the imaginary fold lines (Fig. 14.8).

Height (H), width (W), and depth (D), the three dimensions necessary to dimension an object, are shown in their recommended positions in Fig. 14.8. Because of these standard arrangements, the six views can share dimensions by projection. For example, the height dimension, which is shown only once between the front and right-side views,

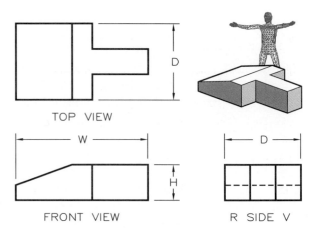

14.10 This three-view drawing depicts the object shown in Fig. 14.10.

182 • CHAPTER 14   ORTHOGRAPHIC DRAWING WITH INSTRUMENTS

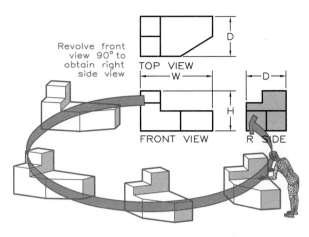

14.11 By rotating the front view 90° about a vertical axis, the side view is found.

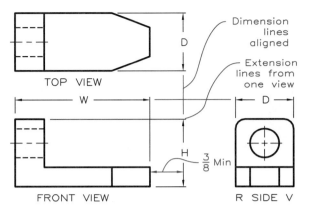

14.13 Dimension and extension lines used in three-view orthographic projection should be aligned. Draw extension lines from only one view when dimensions are placed between views.

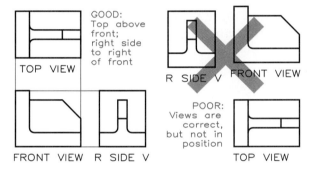

14.12 Orthographic views must be arranged in their proper positions in order for them to be interpreted correctly.

14.6 Arrangement of Views

Figure 14.12 shows the standard positions for a three-view drawing: The top and side views are projected from and aligned with the front view. Improperly arranged views that do not project from view to view are also shown. **Figure 14.13** illustrates the rules of projection and shows the proper alignment of dimensions. Orthographic projection shortens layout time, improves readability, and reduces the number of dimensions required because they are placed between and are shared by the views to which they apply.

14.7 Selection of Views

The best selection of orthographic views has the fewest hidden lines. **Figure 14.14A** shows that the right-side view is preferable to the left-side view because it has fewer hidden lines. Although the three-view arrangement of top, front, and right-side views is more commonly used, the top, front, and left-side view arrangement is acceptable (**Fig. 14.14B**) if the left-side view has fewer hidden lines than the right-side view. Use the right-side view when the left-side view is no better (**Fig. 14.14C**).

The most descriptive view usually is selected as the front view. If an object such as a chair has predefined views that people generally recognize as the front and top views, you should label the accepted front view as the orthographic front view.

Although the right-side view is usually placed to the right of the front view, the side view can be projected from the top view (**Fig. 14.15**). This alternative position is advisable when the object has a much larger depth than height.

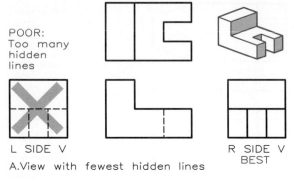

POOR:
Too many
hidden
lines

L SIDE V
A.View with fewest hidden lines

R SIDE V
BEST

BEST
L SIDE V
B.Left—side view can be used

POOR: Too
many
hidden
lines

R SIDE V

Same no.
of hidden
lines; Right—
Side View
preferred

L SIDE V
C.Right—side view preferred

R SIDE V
BEST

14.14 Selection of views.

A Select the sequence of views with the fewest hidden lines.

B Select the left-side view because it has fewer hidden lines than the right-side view.

C When both views have an equal number of hidden lines, select the right-side view.

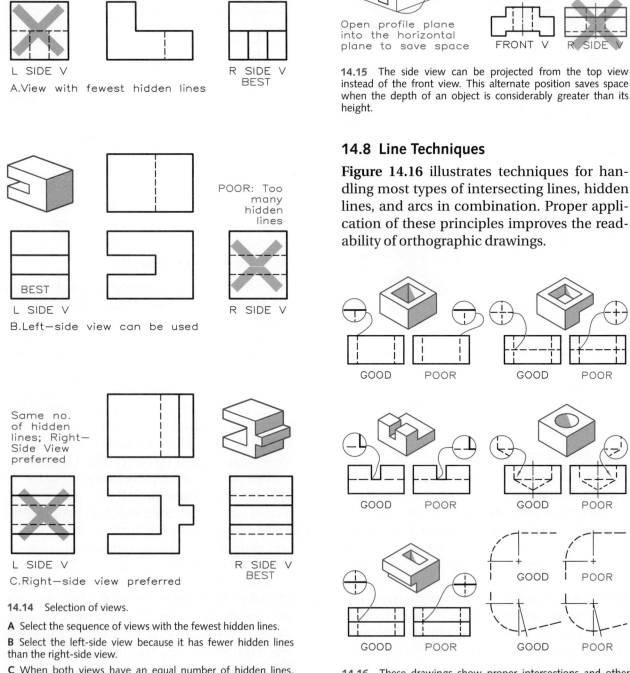

14.15 The side view can be projected from the top view instead of the front view. This alternate position saves space when the depth of an object is considerably greater than its height.

14.8 Line Techniques

Figure 14.16 illustrates techniques for handling most types of intersecting lines, hidden lines, and arcs in combination. Proper application of these principles improves the readability of orthographic drawings.

14.16 These drawings show proper intersections and other line techniques in orthographic views.

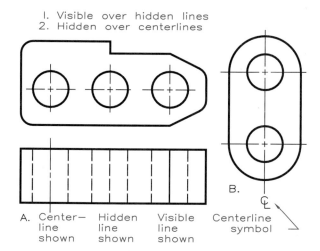

1. Visible over hidden lines
2. Hidden over centerlines

| A. Center-line shown | Hidden line shown | Visible line shown | Centerline symbol |

B.

14.17 When lines coincide with each other, the more important lines take precedence (cover up) the other lines. The order of importance is (1) visible lines, (2) hidden lines, and (3) centerlines.

Become familiar with the order of importance (precedence) of lines (**Fig. 14.17**). The most important line, the visible object line, is shown regardless of any other line lying behind it. Of next importance is the hidden line, which is more important than the centerline.

14.9 Point Numbering

Some orthographic views are difficult to draw due to their complexity. By numbering the endpoints of the lines of the parts in each view as you construct it (**Fig. 14.18**), the location of the object's features will be easier. For example, using numbers on the top and side views of this object aids in the construction of the missing front view. Projecting points from the top and side views to the intersections of the projectors locates the object's front view.

14.10 Lines and Planes

A line can appear **true length** (**TL**) or **foreshortened** (**FS**), or as a **point** (**PT**) in an orthographic projection (**Fig. 14.19**). A line that appears true length is parallel to the reference line in the adjacent view. The reference line in these examples represents the fold line between the horizontal and frontal planes and is therefore labeled H–F. A plane can appear true size (**TS**) or foreshortened, or as an edge in orthographic projection.

Lines and planes that are true length or true size can be measured with a scale. Foreshortened lines and planes are not true length and are less than full size.

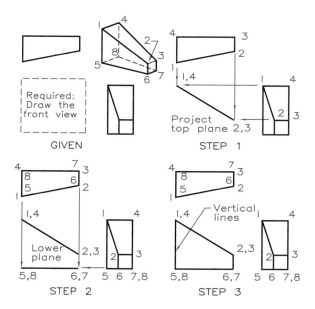

14.18 Point numbering.

Required: Draw the front view.

Step 1 Number the corners of plane 1–2–3–4 in the top and side views and project these points to the front view.

Step 2 Number the corners of plane 5–6–7–8 in the top and side views and project these points to the front view.

Step 3 Connect the numbered lines to complete the front view.

14.11 Drawing with Triangles

Two triangles can be effectively used to make instrument drawings on 8 1/2 × 11 sheets without taping the sheet to the drawing surface. It is better that the paper can be moved about to comfortably position the triangles.

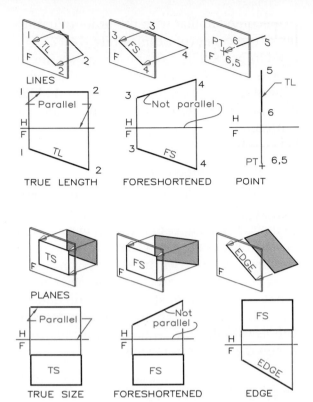

14.19 In orthographic projection a line will appear as true length or foreshortened or as a point. A plane projection will appear as true size or foreshortened or as an edge.

Parallel lines

The 45° triangle or the 30°–60° triangle can be used to draw parallel lines as shown in **Fig. 14.20**. One triangle or a straightedge is held in position while the other triangle is moved to where the parallel line is drawn.

Perpendiculars

Figure 14.21 illustrates the technique for drawing perpendiculars. To draw a line perpendicular to AB, align the 30°–60° triangle's hypotenuse with AB and against the lower triangle or straightedge. Holding the lower triangle in position, rotate it so that the hypotenuse is perpendicular to AB and line CD can be drawn.

Angles

To draw a line making 30° with AB, follow the steps shown in **Fig. 14.22**. Two triangles can be used in other combinations as an informal means of making instrument drawings executed with a sufficient degree of accuracy.

14.12 Views by Subtraction

Figure 14.23 illustrates how three views of a part are drawn by beginning with a block having the overall height, width, and depth of the

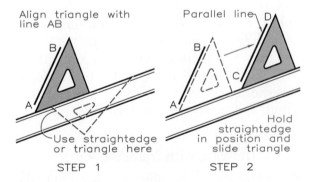

14.20 Drawing parallel lines.

Step 1 Align the upper triangle with AB and in contact with the lower triangle (or straightedge).

Step 2 Hold the lower triangle in position and slide the upper triangle to where CD is drawn parallel to AB.

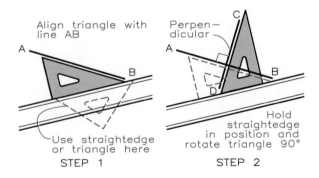

14.21 Drawing perpendiculars.

Step 1 Align your triangle with line AB and in contact with the lower triangle (or straightedge).

Step 2 Hold the lower triangle in position, rotate your triangle, and draw CD perpendicular to AB.

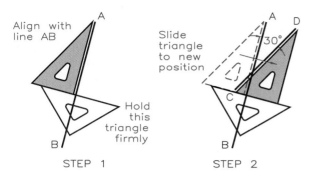

14.22 Drawing a 30° angle.

Step 1 Hold the 30°–60° triangle aligned with AB and in contact with the lower triangle (or straightedge).

Step 2 Hold the straightedge in position then slide the triangle and draw CD at 30° to AB.

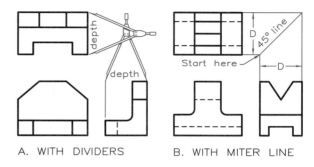

14.24 Transferring depth.

A Transfer the depth dimension to the side view from the top view with your dividers.

B Use a 45° miter line to transfer the depth dimension between the top and side views by projection.

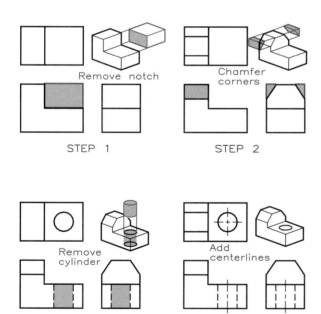

14.23 Views by subtraction.

Step 1 Block in the views of the object using overall dimensions of H, W, and D. Create the notch by removing the notch.

Step 2 Remove the triangular volumes at the corners.

Step 3 Form the hole by removing a cylindrical volume.

Step 4 Add centerlines to complete the views.

finished part and removing volumes from it. This drawing procedure is similar to the steps of making the part in the shop.

14.13 Three-View Drawing Layout

The depth dimension applies to both the top and side views, but these views usually are positioned where depth does not project between them (**Fig. 14.24**). The depth dimension can be transferred between the top and side views with dividers or by using a 45° miter line.

Layout Rules The basic rules of laying out orthographic drawings are summarized below. The examples in **Fig. 14.25** through **Fig. 14.29** illustrate how to apply these rules. Notice how the dimensions have been applied and how the views have been labeled.

1. Draw orthographic views in their proper positions.

2. Select the most descriptive view as the front view, if the object does not have a predefined front view.

3. Select the sequence of views with the fewest hidden lines.

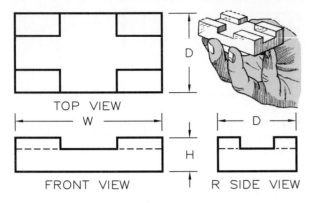

14.25 A three-view drawing depicting an object with horizontal and vertical planes.

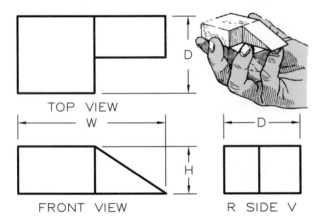

14.26 A three-view drawing depicting an object with a sloping plane.

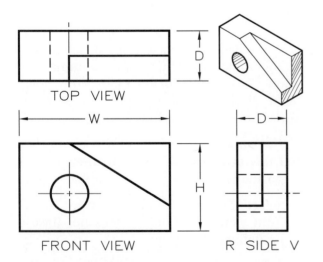

14.27 A three-view drawing depicting an object with a sloping plane with a cylindrical hole through it.

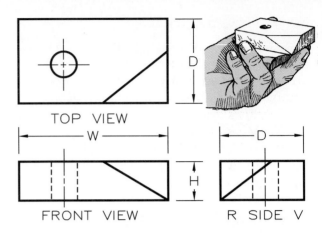

14.28 A three-view drawing depicting an object that has a plane with a compound slope.

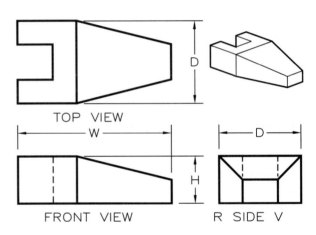

14.29 A three-view drawing depicting an object that has planes with compound slopes.

4. Label the views—for example, top view, front view, and right-side view.

5. Place dimensions between the views to which they apply.

6. Use the proper alphabet of lines.

7. Leave adequate room between the views for labels and dimensions.

8. Draw the views necessary to describe a part. Sometimes fewer and more views are required.

14.14 Views by Computer

The steps of drawing three orthographic views of an object are shown in **Fig. 14.30**, where the `LINE` command is used to draw the overall outlines of the three views. Other visible and hidden lines are added in by projecting from view to view. The `DTEXT` command is used to label the views if this is desired. Additional details of using the `LINE` command can be found in Section 37.17.

Three-dimensional (3D) Solids The object in **Fig. 14.31A** illustrates how a three-dimensional solid can be constructed and then rendered as shown in **Fig. 14.31B**. Once the part is drawn as a three-dimensional solid, it can be viewed from different directions to obtain the standard orthographic views.

In order to make a three-view orthographic layout, set `TILEMODE` to `0`, which

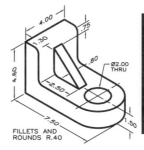

14.31 The given problem (left) was constructed as a three-dimensional solid and rendered as shown on the right.

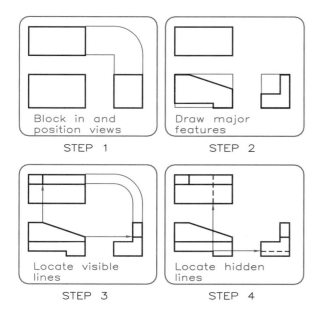

14.30 Three views by computer.

Step 1 Use the `LINE` command to draw the outlines of the views.

Step 2 Using orthographic projection, draw the major features.

Step 3 Draw the visible lines.

Step 4 Draw the hidden lines.

changes the screen to `Paper Space (PS)` and blanks the screen. Before the three-dimensional solid can be seen, you must create `Floating Viewports` with the `MVIEW` (make view) command and select `4` and the `Fit` option to create four viewports on the screen (**Fig. 14.32**). Type `MS` (model space) and activate the upper right port. Type `UCS` (user coordinate system) and select the option of `SAVE` and after desired UCS name type `ISO` to keep this isometric view (**Fig. 14.33**).

Select the upper-left port with the cursor to make it active where the X-Y icon is parallel to the top of the object. Type `UCS`, select `SAVE`, and name the view `TOP`. The X-Y icon must be rotated 90° to be parallel with the front of the part; type `UCS> X> 90`, then type `PLAN` to get front view. Type `UCS> SAVE> Name> FRONT`.

Select the lower-right port to make it active and make the X-Y icon parallel to the side view (**Fig. 14.34**): Type `UCS> Y> 90`. Type `PLAN` to get the side view; type `UCS> Save> Name> SIDE`.

The three views that have been found are probably shown at different sizes and are not aligned correctly as orthographic views should be. Select each view individually and restore the saved views. For example, if you select the top viewport, type `UCS> Restore> Name> TOP` and the X-Y icon will return parallel to the top view. Type `ZOOM> 1.2` to size the top view to appropriately fill the viewport.

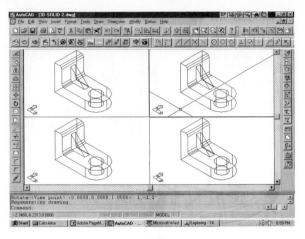

14.32 Creating floating viewports.

Step 1 Command: <u>TILEMODE</u>> 0. The screen is set to Paper Space (PS) with a triangle icon in the lower left corner.

Step 2 Type MVIEW> Select 4 to specify 4 viewports> Fit. The screen will be filled with four viewports showing the bracket in each one. An X-Y model-space icon appears in each viewport. Type MS (model space) and select and activate a viewport with your cursor.

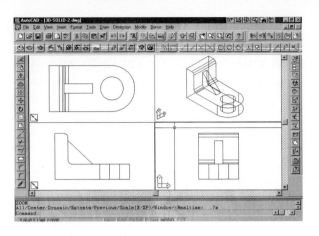

14.34 Creating a side view.

Step 1 Select the lower-right viewport and type UCS> Y> 90 to make the X-Y icon parallel to the side of the part. Type PLAN to obtain the side view.

Step 2 Type UCS> Save> Name> Side, to save the UCS named side for the side view. The three orthographic views should now appear in their respective positions.

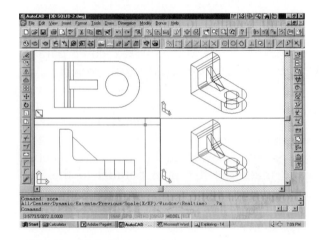

14.33 Creating top and front views.

Step 1 Select the upper right viewport, type UCS> Save> Desired UCS name: ISO.

Step 2 Select the upper left viewport and type PLAN to obtain a view parallel to the X-Y icon, the top view. Type UCS> Save>Desired UCS name: TOP. (This saves the top-view UCS.)

Step 3 Select the lower-left viewport, type UCS> X> Rotation> 90 to make the X-Y icon parallel to the front of the object. Type PLAN to get a front view; type UCS> Save> Name> FRONT to save the UCS.

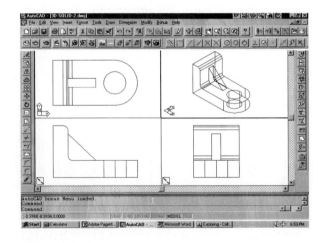

14.35 Scaling and aligning the views.

Step 1 Select the top view by cursor, type UCS> Restore> Name?> TOP, to restore the X-Y icon parallel to the top view. Type ZOOM> 1.2; this will size the top view to properly fill the viewport. Repeat this step for the front and side views.

Step 2 To align the equally sized views, type MVSETUP (multi-view setup), which changes the screen to paper space (PS). Select the Align option > Vertical> Base point> select a corner of the part in the top view> 0 point> select the same corner in the front view to align the top and front views. Repeat, but use the horizontal option to align the front and side views.

Select the front- and side-view ports, restore the views by name, and use ZOOM to 1.2 to make them equal in size to each other.

Type MVSETUP (model view setup), which changes the screen to paper space (PS), to align the views properly, as shown in **Fig. 14.35**. AutoCAD commands selected in sequence are separated by the > sign. MVSETUP> Align> Select base point> Select a corner point of the top view> Other point>. Pick the same point in the adjacent viewport and the views are automatically aligned. Now, you have three orthographic views drawn from a single three-dimensional solid. Refer to Chapter 37 for a more thorough coverage of solid modeling.

14.15 Two-View Drawings

Time and effort can be saved by drawing only the views and features that are necessary to describe a part. **Figure 14.36** shows typical objects that can be described in only two views. The fixture block in **Fig. 14.37** is another example of a part needing only two views. It is clear that the top and front views are the best for this part. The front and side views would not be as good.

14.16 One-View Drawings

Simple cylindrical parts and parts of a uniform thickness can be described by only one view, as shown in **Fig. 14.38**. Supplementary notes clarify features that would have been shown in the omitted views. Diameters are labeled with diameter signs and thicknesses are noted.

14.17 Simplified and Removed Views

The right- and left-side views of the part in **Fig. 14.39** would be harder to interpret if all hidden lines were drawn by rigorously following the rules of orthographic projection.

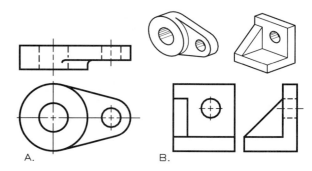

14.36 These objects can be adequately described with two orthographic views.

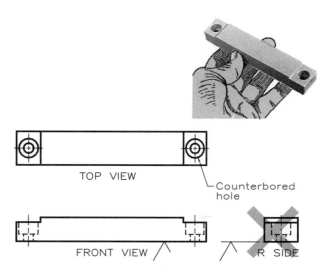

14.37 Two views adequately describe this fixture block.

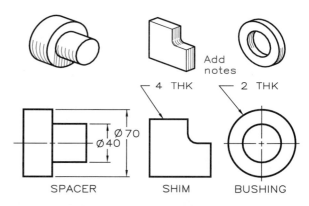

14.38 Describe objects that are cylindrical or of a uniform thickness with only one orthographic view and supplementary notes.

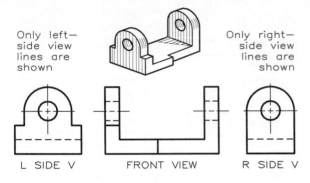

Only left—side view lines are shown

Only right—side view lines are shown

L SIDE V   FRONT VIEW   R SIDE V

14.39 Use simplified views with unnecessary and confusing hidden lines omitted to improve clarity.

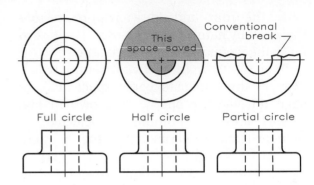

Conventional break

This space saved

Full circle   Half circle   Partial circle

14.41 Save space and time by drawing the circular view of a cylindrical part as a partial view.

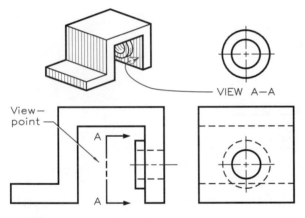

VIEW A—A

View—point

14.40 Use a removed view, indicated by the directional arrows, to show hard-to-see views in removed locations.

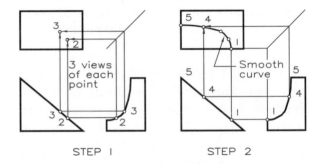

3

2

3 views of each point

3

2

3

2

5   4

1

Smooth curve

5   5

4   4

1   1

STEP 1   STEP 2

14.42 Curve plotting.

Step 1 Locate points 2 and 3 in the front and side views by projection. Project points 2 and 3 to the top view.

Step 2 Locate the remaining points in the three views and connect the points in the top view with a smooth curve.

Simplified views in which confusing and unnecessary lines have been omitted are better and more readable.

When it is difficult to show a feature with a standard orthographic view because of its location, a **removed view** can be drawn (**Fig. 14.40**). Such a view, indicated by directional arrows, is clearer when moved to an isolated position.

14.18 Partial Views

Partial views of symmetrical or cylindrical parts may be used to save time and space. Omitting the rear of the circular top view in **Fig. 14.41** saves space without sacrificing clar-

ity. To clarify that a part of the view has been omitted, a conventional break is used in the top view.

14.19 Curve Plotting

An irregular curve can be plotted by following the rules of orthographic projection, as shown in **Fig. 14.42**. Begin by numbering the points in the given front and side views along the curve. Next, project from the points having the same numbers in the front and side views to the top to where the projectors intersect. Continue projecting in this manner and connect the points in the top view with a smooth

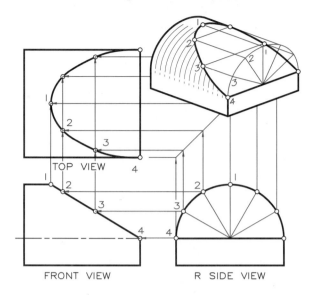

14.43 The ellipse in the top view was found by numbering points in the front and side views and projecting them to the top view.

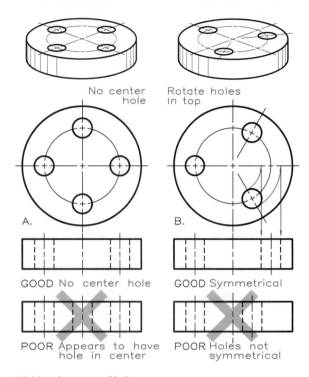

14.44 Placement of holes.

A Omit the center hole found by true projection that gives an impression that a hole passes through the center of the plate.

B Use a conventional view to show the holes located at their true radial distances from the center. They are imagined to be rotated to the centerline in the top view.

curve drawn with an irregular curve. **Figure 14.43** shows an ellipse plotted in the tcp view by projecting points from the front and side-views. It is best to number points one at a time as they are transferred to avoid losing them in your construction.

14.20 Conventional Practices

The readability of an orthographic view may be improved if the rules of projection are violated. Violations of rules customarily made for the sake of clarity are called **conventional practices**.

Symmetrically spaced holes in a circular plate (**Fig. 14.44**) are drawn at their true radial distance from the center of the plate in the front view as a conventional practice. Imagine that the holes are revolved to the centerline in the top view before projecting them to the front view.

This principle of revolution also applies to symmetrically positioned features such as

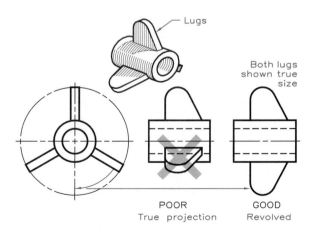

14.45 Symmetrically positioned external features, such as webs, ribs, and these lugs, are imagined to be revolved to their true-size positions for the best views.

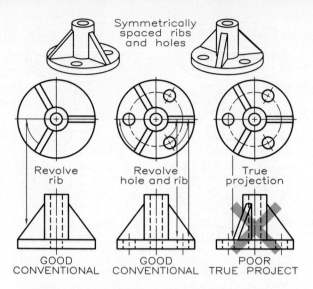

14.46 Conventional methods of revolving holes and ribs in combination improve clarity.

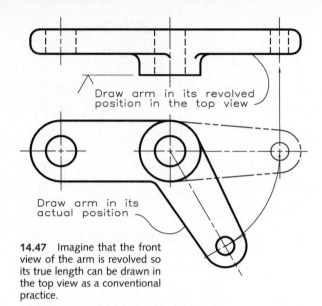

14.47 Imagine that the front view of the arm is revolved so its true length can be drawn in the top view as a conventional practice.

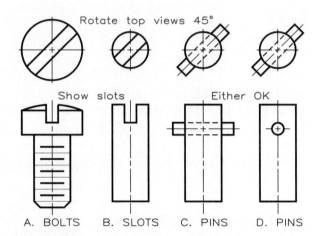

14.48 It is conventional practice to draw the slots and holes at 45° in the top view and true size in the front view.

ribs and webs as well as to the three lugs on the outside of the part shown in **Fig. 14.45**. **Figure 14.46** shows the applications of conventional practices to holes and ribs in combination.

Another conventional revolution is illustrated in **Fig. 14.47** where the front view of an inclined arm is revolved to a horizontal position so that it can be drawn true size in the top view. The revolved arm in the front view is not drawn because the revolution is imaginary. **Figure 14.48** shows how to improve views of parts by conventional revolution. By revolving the top views of these parts 45°, slots and holes no longer coincide with the centerlines and can be seen more clearly. Draw the front views of the slots and holes true size by imagining that they have been revolved 45°.

Another type of conventional view is the true-size development of a curved sheet-metal part drawn as a flattened-out view (**Fig. 14.49**). The top view shows the part's curvature.

14.21 Conventional Intersections

In orthographic projection, lines are drawn to represent the intersections (fold lines)

between the planes of an object. Wherever planes intersect, forming an edge, the line of intersection is projected to its adjacent view. **Figure 14.50** illustrates examples showing where lines are required.

Figure 14.51 shows how to draw intersections between cylinders rather than plotting more complex, orthographically correct lines of intersection. Parts A and C show conventional intersections, which means they are approximations drawn for ease of construction

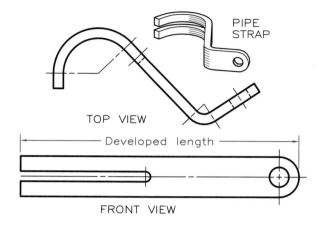

PIPE STRAP

TOP VIEW

Developed length

FRONT VIEW

14.49 It is conventional practice to use true-size developed (flattened-out) views of parts made of bent sheet metal.

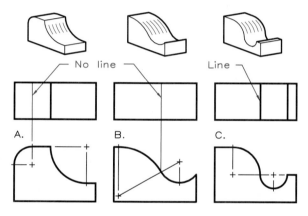

No line   Line

A.   B.   C.

14.50 Object lines are drawn only where there are sharp intersections or where arcs are tangent at their centerlines, as they are in Part C.

while being sufficiently representative of the object. Part B shows an easy-to-draw intersection between cylinders of equal diameters, and this is a true intersection as well. **Figures 14.52** and **14.53** show other cylindrical intersections, and **Fig. 14.54** shows conventional practices for depicting intersections formed by holes in cylinders.

14.22 Fillets and Rounds

Fillets and **rounds** are rounded intersections between the planes of a part that are used on castings, such as the body of the pillow block in **Fig. 14.55**. A **fillet** is an inside rounding and

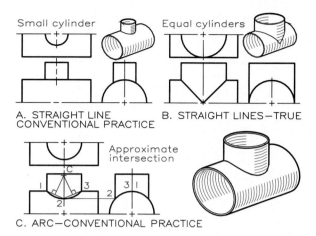

Small cylinder   Equal cylinders

A. STRAIGHT LINE
CONVENTIONAL PRACTICE   B. STRAIGHT LINES—TRUE

Approximate intersection

C. ARC—CONVENTIONAL PRACTICE

14.51 Intersections between cylinders.
A and **C** Use these methods of construction.
B Equal size cylinders have straight-line intersections.

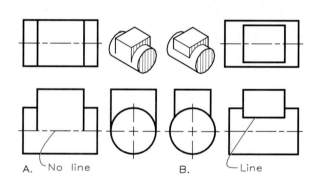

A. No line   B.   Line

14.52 True intersections between cylinders and prisms.

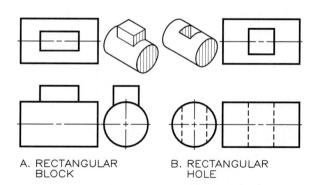

A. RECTANGULAR
BLOCK   B. RECTANGULAR
HOLE

14.53 Conventional intersections between cylinders and prisms.

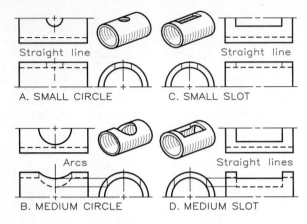

14.54 Conventional methods of depicting intersections formed by holes in cylinders.

A. SMALL CIRCLE — Straight line

C. SMALL SLOT — Straight line

B. MEDIUM CIRCLE — Arcs

D. MEDIUM SLOT — Straight lines

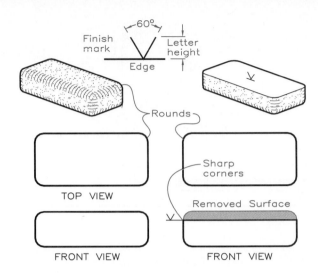

14.56 When a surface is finished (machined), the cut removes the rounded corners and leaves sharp corners. The finish mark placed on the edge of the surface indicates that it is to be finished.

14.55 The edges of this pillow block are rounded with fillets and rounds. The surface of the casting is rough except where it has been machined. (Courtesy of Dodge Manufacturing Company.)

a **round** is an external rounding on a part. The radii of fillets and rounds usually are small, about 1/4 inch. Fillets give added strength at inside corners, rounds improve appearance, and both remove sharp edges (**Fig. 14.56**).

The square corners that a casting has when its surface has been finished, are improved through the process of machining away part of the surface to a smooth finish (**Fig. 14.57**). Finished surfaces are indicated by placing a finish mark (V) on all edge views of finished surfaces whether the edges are visible or hidden. **Figure 14.58** shows four types of finish

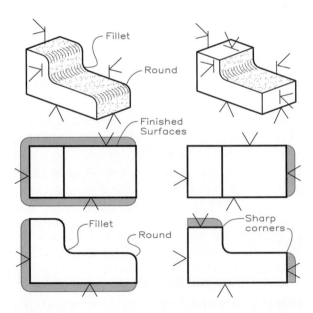

14.57 Fillets and rounds are rounded inside and outside corners respectively that are standard features on castings. When surfaces are finished, fillets and rounds are removed, as shown here.

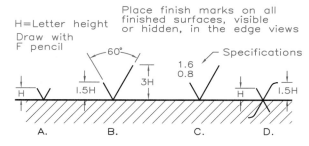

H=Letter height
Draw with
F pencil

Place finish marks on all
finished surfaces, visible
or hidden, in the edge views

60°

Specifications

1.6
0.8

3H

1.5H

H   1.5H

A.   B.   C.   D.

14.58 Four types of finish marks may be placed on all views of finished surfaces, visible or hidden.

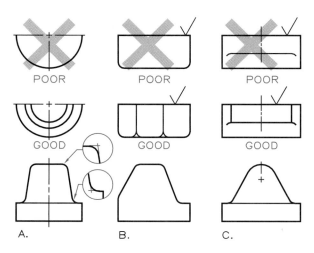

POOR   POOR   POOR

GOOD   GOOD   GOOD

A.   B.   C.

14.59 Poorly drawn and conventionally drawn fillets and rounds.

marks. A more detailed surface texture symbol is presented in Chapter 21. **Figure 14.59** illustrates three techniques for showing fillets and rounds on orthographic views drawn with a circle template.

Computer Method Fillets and rounds are drawn by computer as shown in **Fig. 14.60**. The object is first drawn with angular intersections, and the FILLET command is used to round the corners for both fillets and rounds.

Figure 14.61 compares the intersections and runouts of parts with and without fillets and rounds. Large runouts are constructed as

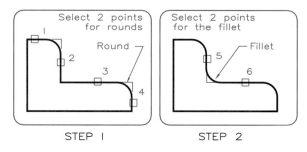

Select 2 points
for rounds

Select 2 points
for the fillet

Round

Fillet

STEP 1   STEP 2

14.60 Fillets and rounds drawn by computer.

Step 1 Command: FILLET (Enter)
Polyline-Radius-<Select-two-objects>: R (Enter)
Enter fillet radius <0.000>: 0.50 (Enter)
Command: FILLET Polyline Radius/<@Select two objects>: 1 and 2 (Repeat, selecting 3 and 4 to draw round.)

Step 2 Press Enter (Enter), and select points 5 and 6 on the inside lines to construct a fillet. The lines are trimmed as the fillets and rounds are applied to them.

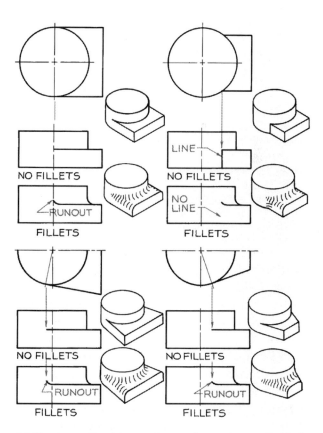

NO FILLETS   NO FILLETS

LINE

NO
LINE

RUNOUT

FILLETS   FILLETS

NO FILLETS   NO FILLETS

RUNOUT   RUNOUT

FILLETS   FILLETS

14.61 Conventional intersections and runouts on cylindrical features of parts. Runouts result from fillets and rounds intersecting cylinders.

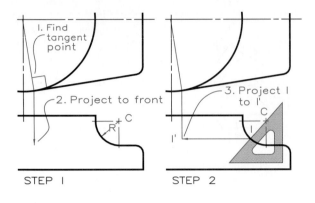

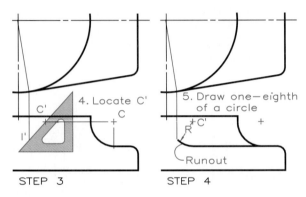

14.62 Plotting runouts.

Step 1 Find the tangency point in the top view and project it to the front view.

Step 2 Find point 1 with a 45° triangle and project it to 1′.

Step 3 Move the 45° triangle to locate point C′ on the horizontal projector from center C.

Step 4 Use the radius of the fillet to draw the runout with C′ as its center. The runout arc is equal to one-eighth of a circle.

one-eighth of a circle with a compass, as shown in **Fig. 14.62**. Small runouts are drawn with a circle template. Runouts on orthographic views reveal much about the details of an object. For example, in **Fig. 14.63**, the runout in the top view of Part A tells us that the rib has rounded corners, whereas the top view of Part B tells us the rib is completely round. **Figures 14.64** and **14.65** illustrate other types of filleted and rounded intersections.

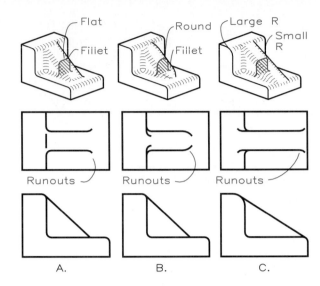

14.63 Typical runouts of edges with fillets and rounds.

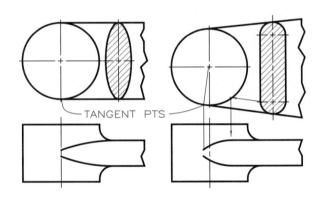

14.64 Conventional representations of runouts.

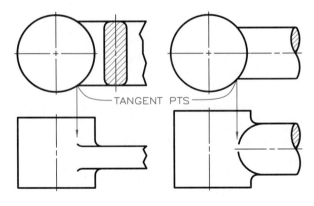

14.65 Conventional runouts for different cross-sections.

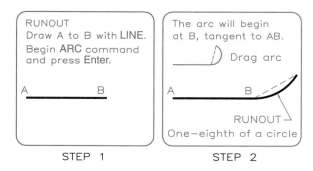

STEP 1   STEP 2

14.66 Runouts by computer.

Step 1 Command: LINE (Enter)
From point: (Select A) (Enter)
To point: (Select B) (Enter) (Enter)

Step 2 Command: ARC (Enter)
Center/<Start point>: (Enter)
End point: (Drag to end of runout.)

Computer Method Figure 14.66 shows a drawing of a part with runouts at its tangent points. The runouts are plotted by using the LINE command and then the ARC command, and finally pressing (Enter) to obtain the first point of an arc tangent to and connected to the end of the line. Locate the other end of the arc to complete the runout.

14.23 First-Angle Projection

The examples in this chapter are third-angle projections in which the top view is placed over the front view and the right-side view is placed to the right of the front view, as shown in **Fig. 14.67**. This method is used in the United States, Britain, and Canada. However, most of the world uses first-angle projection.

The first-angle system is illustrated in **Fig. 14.68**, in which an object is placed above the horizontal plane and in front of the frontal plane. When these projection planes are opened onto the surface of the drawing paper, the front view projects over the top view, and the right-side view projects to the left of the front view.

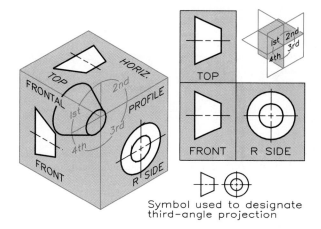

Symbol used to designate third-angle projection

14.67 Third-angle projection is used for drawing orthographic views in the United States, Britain, and Canada. The top view is placed over the front view and the right-side view is placed to the right of the front view. The truncated cone is the symbol used to designate third-angle projection.

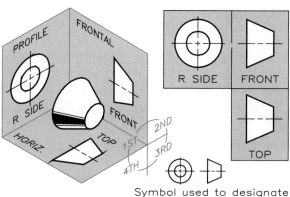

Symbol used to designate First-angle projection

14.68 First-angle projection is used in most of the world. It shows the right-side view to the left of the front view and the top view under the front view. The truncated cone designates first-angle projection.

The angle of projection used in making a drawing is indicated by placing the truncated cone in or near the title block (**Fig. 14.69**). When metric units of measurement are used, the SI symbol is given in combination with the cone on the drawing.

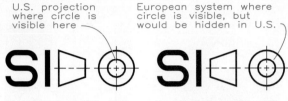

U.S. projection where circle is visible here

European system where circle is visible, but would be hidden in U.S.

A. METRIC UNITS AND THIRD—ANGLE PROJECTION

B. METRIC UNITS AND FIRST—ANGLE PROJECTION

14.69 These symbols are placed on drawings to specify first-angle or third-angle projection and metric units of measurement.

Problems

1–7. (**Figs. 14.70–14.76**) Draw the given views on size A sheets, two per sheet, using the dimensions given, and draw the missing top, front, or right-side views. Lines may be missing in the given views.

8–17. (**Figs. 14.77–14.86**) Draw three views of the objects. Each square grid is equal to 0.20 inches or 5 mm. Two problems can be placed on a size A sheet (AV format). Label the views and show the overall dimensions as W, D, and H.

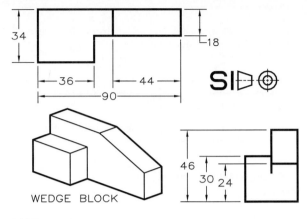

WEDGE BLOCK

14.71 Problem 2.

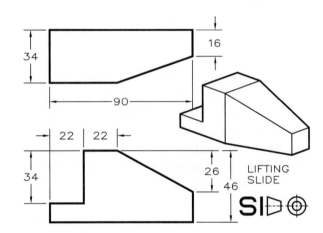

LIFTING SLIDE

14.72 Problem 3.

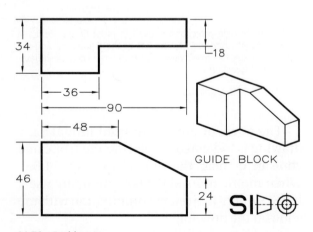

GUIDE BLOCK

14.70 Problem 1.

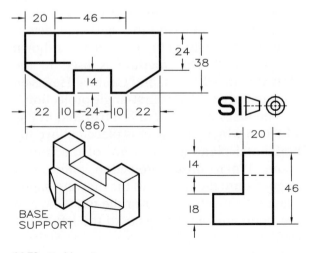

BASE SUPPORT

14.73 Problem 4.

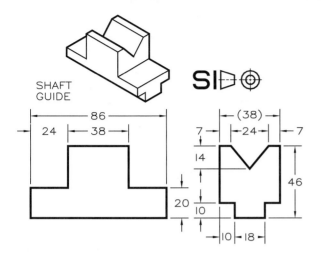

SHAFT
GUIDE

14.74 Problem 5.

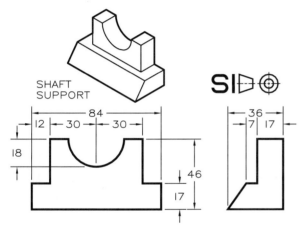

SHAFT
SUPPORT

14.75 Problem 6.

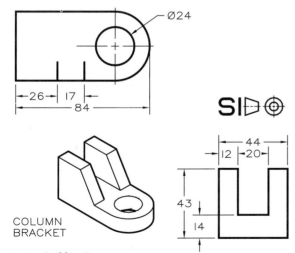

COLUMN
BRACKET

14.76 Problem 7.

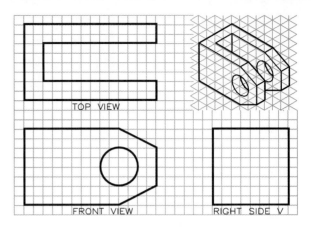

TOP VIEW

FRONT VIEW

RIGHT SIDE V

14.77 Problem 8.

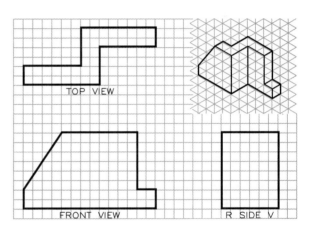

TOP VIEW

FRONT VIEW

R SIDE V

14.78 Problem 9.

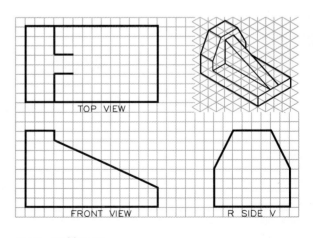

TOP VIEW

FRONT VIEW

R SIDE V

14.79 Problem 10.

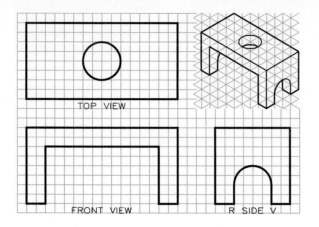

TOP VIEW

FRONT VIEW   R SIDE V

14.80 Problem 11.

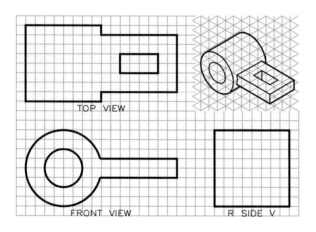

TOP VIEW

FRONT VIEW   R SIDE V

14.81 Problem 12.

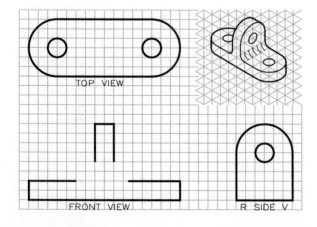

TOP VIEW

FRONT VIEW   R SIDE V

14.82 Problem 13.

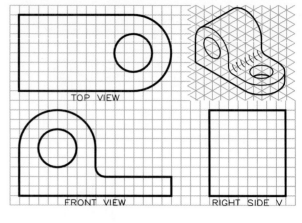

TOP VIEW

FRONT VIEW   RIGHT SIDE V

14.83 Problem 14.

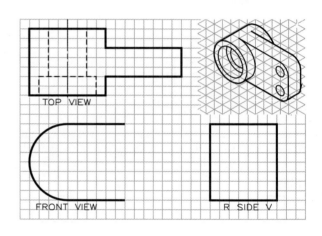

TOP VIEW

FRONT VIEW   R SIDE V

14.84 Problem 15.

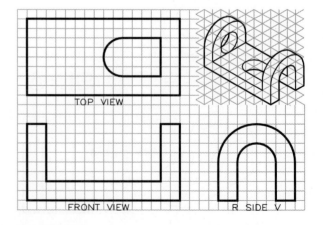

TOP VIEW

FRONT VIEW   R SIDE V

14.85 Problem 16.

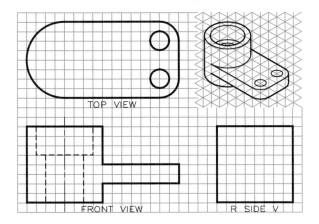

14.86 Problem 17.

18. (**Fig. 14.87**) Draw three full-size views of the tensioner base in millimeters on a size A sheet. The overall dimensions are height, 120 mm; width, 112 mm; and depth, 56 mm. Estimate the other dimensions.

20–42. (**Figs. 14.89–14.111**) Construct the necessary orthographic views to describe the objects on B-size sheets at an appropriate scale. Label the views and show the overall dimensions of W, D, and H.

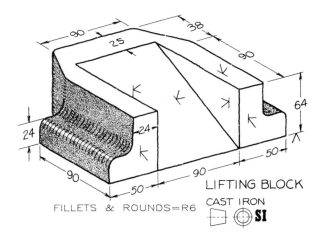

14.89 Problem 20.

19. (**Fig. 14.88**) Draw three full-size views of the base plate in millimeters on a size A sheet. The overall dimensions are height, 10 mm; width, 126 mm; and depth, 90 mm. Estimate the other dimensions.

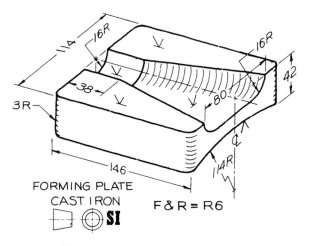

14.90 Problem 21.

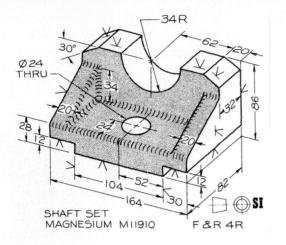

14.91

SHAFT SET
MAGNESIUM M11910
F & R 4R

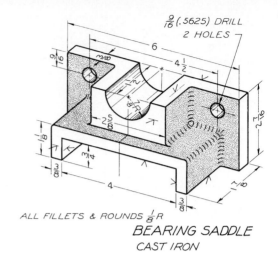

14.94

$\frac{9}{16}$ (.5625) DRILL
2 HOLES

ALL FILLETS & ROUNDS $\frac{1}{8}$ R

BEARING SADDLE
CAST IRON

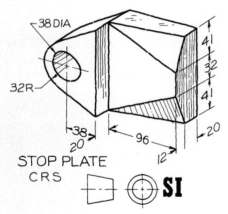

14.92

38 DIA

32R

STOP PLATE
CRS

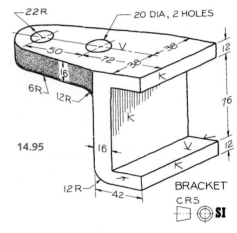

14.95

22R

20 DIA, 2 HOLES

6R

12R

12R

BRACKET
CRS

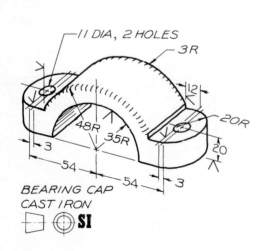

14.93

11 DIA, 2 HOLES

3R

48R

35R

20R

BEARING CAP
CAST IRON

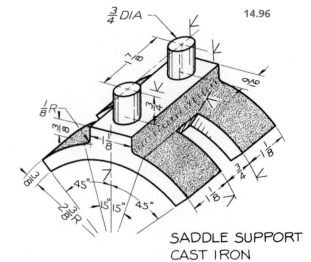

14.96

$\frac{3}{4}$ DIA

$\frac{1}{8}$ R

45°

15° 15°

45°

SADDLE SUPPORT
CAST IRON

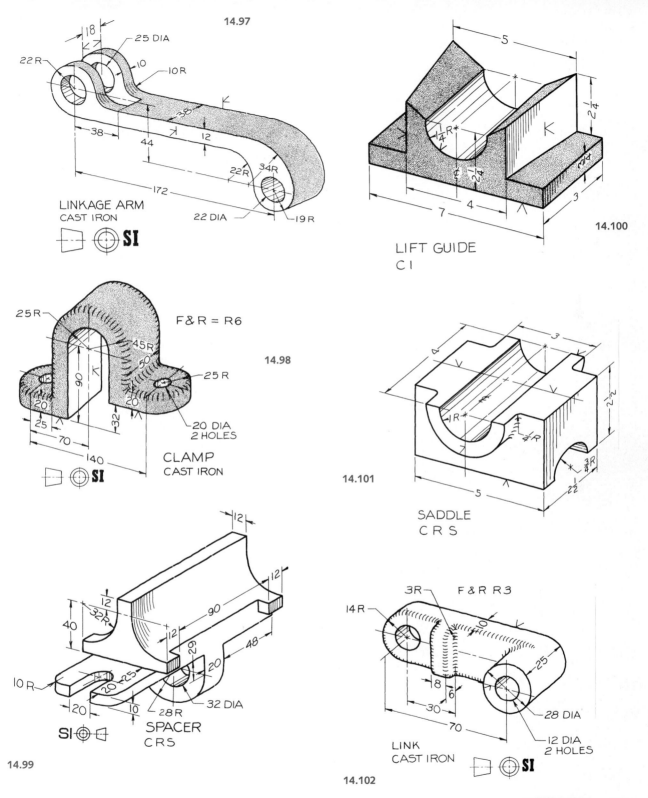

14.97

LINKAGE ARM
CAST IRON

25 DIA
22R
18
10
10R
38
38
44
12
172
22R
34R
22 DIA
19R

14.100

LIFT GUIDE
C I

5
2¼
4
7
3
1¼R*
2¼

14.98

CLAMP
CAST IRON

F & R = R6
25R
45R
50
25R
90
20
20
25
32
70
140
20 DIA
2 HOLES

14.101

SADDLE
C R S

4
3
1R*
¼R
5
2½
2½
¾R

14.99

SPACER
C R S

12
12
12
40
32R
90
12
29
48
20
10R
20
25
20
10
28R
32 DIA

14.102

LINK
CAST IRON

3R
F & R R3
14R
10
25
8
6
30
70
28 DIA
12 DIA
2 HOLES

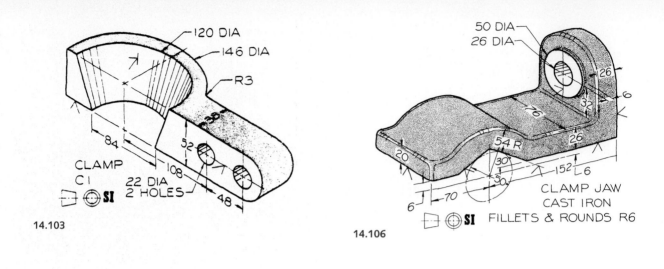

14.103

CLAMP
C1
SI

120 DIA
146 DIA
R3
84
52
108
22 DIA
2 HOLES
48

50 DIA
26 DIA
26
6
32
76
54 R
26
20
30
152 L6
6
70
50

CLAMP JAW
CAST IRON
SI FILLETS & ROUNDS R6

14.106

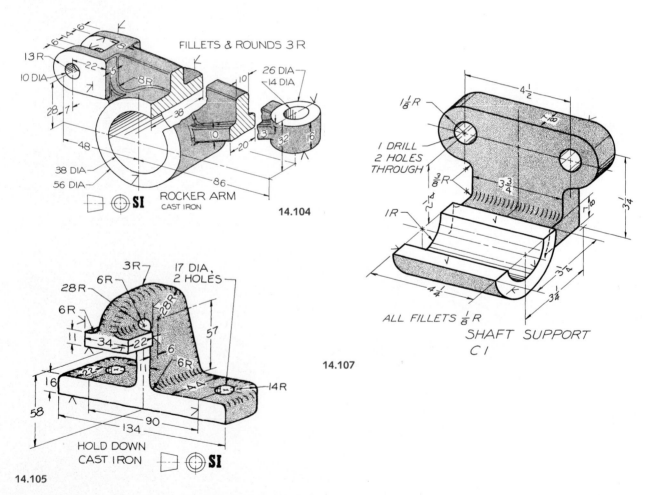

FILLETS & ROUNDS 3 R

13R
10 DIA
22
6 14 6
8R
28
7
38
48
38 DIA
56 DIA
86
10
10
20
26 DIA
14 DIA
3
32
16

ROCKER ARM
CAST IRON
SI

14.104

3R
6R
28R
28R
6R
11
34
22
6
6R
57
4R
16
58
134
90
17 DIA,
2 HOLES

HOLD DOWN
CAST IRON SI

14.105

$4\frac{1}{2}$
$1\frac{1}{8}R$
1 DRILL
2 HOLES
THROUGH
$\frac{3}{8}R$
$\frac{3}{4}$
1R
$\frac{2}{4}$
$3\frac{3}{4}$
$\frac{1}{8}$
$3\frac{1}{4}$
$3\frac{3}{4}$
$3\frac{1}{4}$
$4\frac{1}{4}$

ALL FILLETS $\frac{1}{8}$ R

SHAFT SUPPORT
C1

14.107

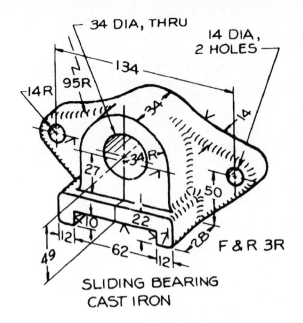

14.108

SLIDING BEARING
CAST IRON

- 34 DIA, THRU
- 14 DIA, 2 HOLES
- 134
- 95R
- 14R
- 34
- 14
- 34R
- 27
- 50
- 10
- 22
- 12
- 49
- 62
- 12
- 28
- F & R 3R

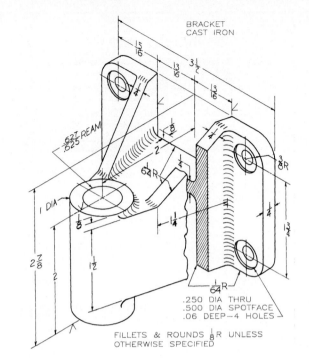

14.110

BRACKET
CAST IRON

- $\frac{15}{16}$
- $3\frac{1}{16}$
- $3\frac{1}{2}$
- $\frac{3}{16}$
- $\frac{3}{8}$R
- .627 REAM / .625
- 64R
- $\frac{1}{64}$R
- 2
- $1\frac{1}{4}$
- $\frac{1}{4}$
- 1 DIA
- $\frac{1}{8}$
- $1\frac{1}{2}$
- $2\frac{7}{8}$
- 2
- .250 DIA THRU
- .500 DIA SPOTFACE
- .06 DEEP—4 HOLES

FILLETS & ROUNDS $\frac{1}{8}$R UNLESS
OTHERWISE SPECIFIED

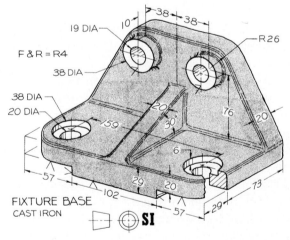

14.109

FIXTURE BASE
CAST IRON

- 19 DIA
- 10
- 38
- 38
- R26
- F & R = R4
- 38 DIA
- 38 DIA
- 20 DIA
- 159
- 20
- 30
- 76
- 20
- 6
- 57
- 29
- 20
- 73
- 102
- 57
- 29

SI

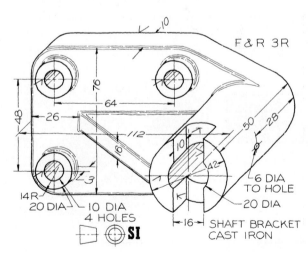

14.111

SHAFT BRACKET
CAST IRON

- F & R 3R
- 10
- 76
- 64
- 48
- 26
- 112
- 16
- 50
- 28
- 10
- 42
- 14R
- 20 DIA
- 10 DIA
- 4 HOLES
- 3
- 16
- 6 DIA TO HOLE
- 20 DIA

SI

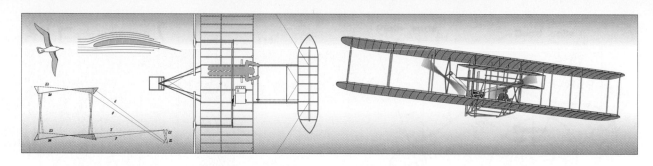

15

Primary Auxiliary Views

15.1 Introduction

Many objects are designed to have sloping or inclined surfaces that do not appear true size in principal orthographic views. A plane of this type is not parallel to a principal projection plane (horizontal, frontal, or profile) and

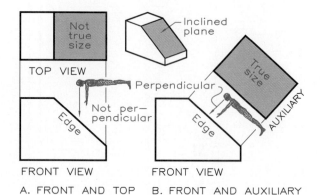

A. FRONT AND TOP   B. FRONT AND AUXILIARY

15.1 A surface that appears as an inclined edge in a principal view can be found true size by an auxiliary view. (A) The top view is foreshortened, but (B) the inclined plane is true size in the auxiliary view.

208

is, therefore, a **nonprincipal plane**. Its true shape must be projected onto a plane that is parallel to it in what is called an **auxiliary view**.

An auxiliary view projected from a primary view (principal view) is called a primary auxiliary view. An auxiliary view projected from a primary auxiliary view is a secondary auxiliary view. By the way, get out your dividers; you must use them all the time in drawing auxiliary views.

The inclined surface of the part shown in **Fig. 15.1A** does not appear true size in the top view because it is not parallel to the horizontal projection plane. However, the inclined surface will appear true size in an auxiliary view projected perpendicularly from its edge view in the front view (**Fig. 15.1B**).

The relationship between an auxiliary view and the view it was projected from is the same as that between any two adjacent orthographic views. **Figure 15.2A** shows an auxiliary view projected perpendicularly from the

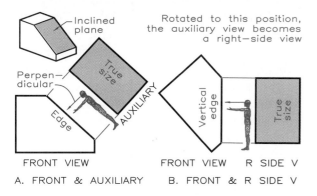

FRONT VIEW | FRONT VIEW | R SIDE V

A. FRONT & AUXILIARY | B. FRONT & R SIDE V

15.2 An auxiliary view has the same relationship with the view it is projected from as that of any two adjacent principal views.

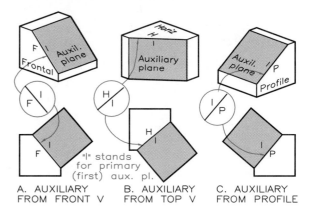

A. AUXILIARY FROM FRONT V | B. AUXILIARY FROM TOP V | C. AUXILIARY FROM PROFILE

15.3 A primary auxiliary plane can be folded from the frontal, horizontal, or profile planes. The fold lines are labeled F–1, H–1, and P–1, with 1 on the auxiliary plane side and P on the principal-plane side.

edge view of the sloping surface. By rotating these views (the front and auxiliary views) so that the projectors are horizontal, the views have the same relationship as regular front and right-side views (**Fig. 15.2B**).

15.2 Folding-Line Principles

The three principal orthographic planes are the **frontal** (F), **horizontal** (H), and **profile** (P) planes. An auxiliary view is projected from a principal orthographic view (a top, front, or side view), and a primary auxiliary plane is perpendicular to one of the principal planes and oblique to the other two.

Think of auxiliary planes as planes that fold into principal planes along a folding line (**Fig. 15.3**). The plane in Fig. 15.3A folds at a 90° angle with the frontal plane and is labeled F–1 where F is an abbreviation for frontal, and 1 represents first, or primary, auxiliary plane. Figures 15.3B and 15.3C illustrate the positions for auxiliary planes that fold from the horizontal and profile planes, labeled H–1 and P–1, respectively.

It is important that reference lines be labeled as shown in Fig. 15.3, with the numeral 1 placed on the auxiliary side and the letter H, F, or P on the principal-plane side.

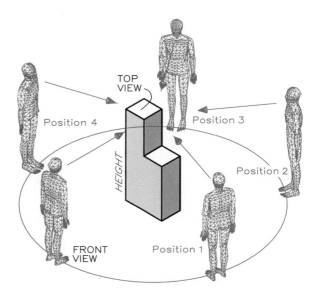

15.4 By moving your viewpoint around the top view of an object, you will see a series of auxiliary views in which the height dimension (H) is true length in all of them.

15.3 Auxiliaries from the Top View

By moving your position about the top view of a part as shown in **Fig. 15.4**, each line of sight is perpendicular to the height dimension. One of the views, the front view, is a principal view; the other positions are nonprincipal views called **auxiliary views**.

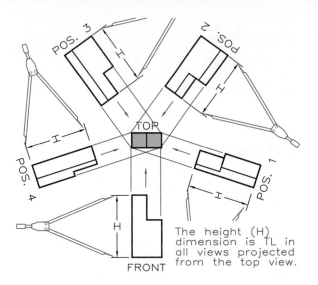

15.5 The views shown in Fig. 15.4 would be drawn as shown here with the same height dimensions common to each view.

Labels within the figure: POS. 3, POS. 2, POS. 1, POS. 4, TOP, FRONT, H. The height (H) dimension is TL in all views projected from the top view.

Figure 15.5 illustrates how these five views (one of which is a front view) are projected from the top view. The line of sight for each auxiliary view is parallel to the horizontal projection plane; therefore, the height dimension is true length in each view projected from a top view. You can transfer the height (H) dimensions from the front to each of the auxiliary views by using your dividers.

Folding-Line Method The inclined plane shown in **Fig. 15.6** is an edge in the top view and is perpendicular to the horizontal plane. If an auxiliary plane is drawn parallel to the inclined surface, the view projected onto it will be a true-size view of the inclined surface. **A surface must appear as an edge in a princi-**

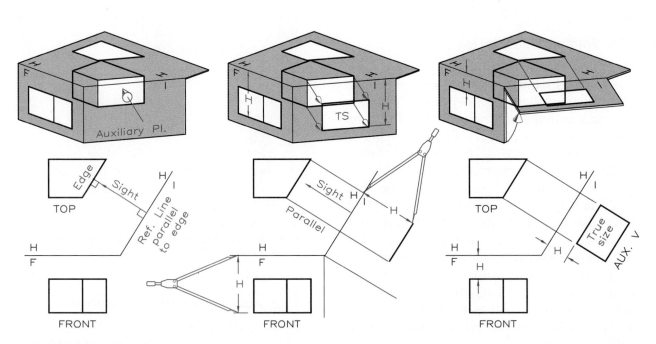

15.6 Auxiliary view from the top.

Step 1 Draw the line of sight perpendicular to the edge view of the inclined surface. Draw the H–1 line parallel to its edge and draw the H–F reference line between the top and front views.

Step 2 Project from the edge view of the inclined surface parallel to the line of sight. Transfer the H dimensions from the front to locate a line in the auxiliary view.

Step 3 Locate the other corners of the inclined surface by projecting to the auxiliary view and locating the points by transferring the height (H) from the front view.

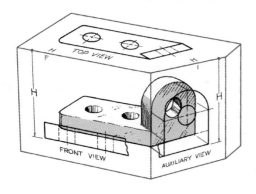

15.7 If you imagine that the object is inside a glass box, you can see the relationship of the auxiliary plane, on which the true-size view is projected, and the horizontal projection plane.

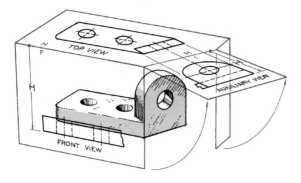

15.8 Fold the auxiliary plane into the horizontal projection plane by revolving it about the H–1 fold line.

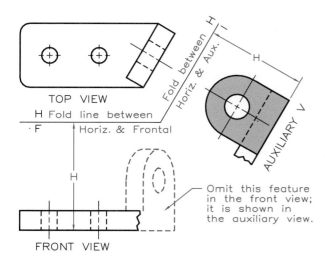

15.9 The front and auxiliary views are partial, to avoid drawing elliptical features that are shown in the auxiliary view as true arcs.

pal view in order for it to be found true size in a **primary auxiliary view.** When the auxiliary view is projected from the top view, transfer the height dimensions in the front view to the auxiliary view with your dividers.

15.4 Auxiliaries from the Top: Application

Figure 15.7 illustrates how the folding-line method is used to find an auxiliary view of a part that is imagined to be in a glass box. The semicircular end of the part does not appear true size in front or side views, making these views difficult to draw or to interpret if they were drawn. However, because the inclined surface appears as an edge in the top view, it can be found true size in a primary auxiliary view projected from the top view. The height dimension (H) in the frontal view will be the same as in the auxiliary plane, because both planes are perpendicular to the horizontal projection plane. Height is transferred from the front to the auxiliary view with dividers. In **Fig. 15.8**, the auxiliary plane is rotated about the H–1 fold line into the plane of the top view, the horizontal projection plane. This rotation illustrates how the placement of the views is arrived at when drawing the views.

When drawn on a sheet of paper, the views of this object appear as shown in **Fig. 15.9**. The top view is a complete view, but the front view is drawn as a partial view because the omitted portion would have been hard to draw and would not have been true size. The auxiliary view also is drawn as a partial view because the front view shows the omitted features better, which saves drawing time and space on a drawing.

Reference-Plane Method A second method of locating an auxiliary view uses reference planes instead of the folding-line method. **Figure 15.10A** shows a horizontal reference

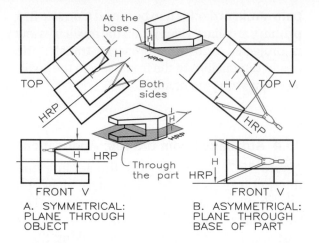

A. SYMMETRICAL:
PLANE THROUGH
OBJECT

B. ASYMMETRICAL:
PLANE THROUGH
BASE OF PART

15.10 A horizontal reference plane (HRP) can be positioned through the part or in contact with it. The dimension of height (H) is measured from the HRP and transferred to the auxiliary view with dividers.

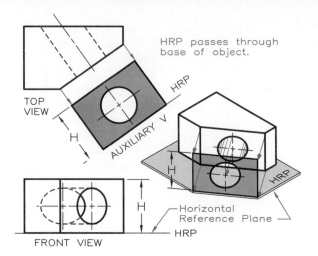

15.11 An auxiliary view projected from the top view is used to draw a true-size view of the inclined surface using a horizontal reference plane. The HRP is drawn through the bottom of the front view.

plane (HRP) drawn through the center of the front view. Because this view is symmetrical, equal height dimensions on both sides of the HRP can be conveniently transferred from the front view to the auxiliary view and laid off on both sides of the HRP.

The reference plane can be placed at the base of the front view, as shown in **Fig. 15.10B**. In this case, the height dimensions are measured upward from the HRP in both the front and auxiliary views. You may draw a reference plane (the HRP in this example) in any convenient position in the front view: through the part, above it, or below it.

A similar example of an auxiliary view drawn with a horizontal reference plane is shown in **Fig. 15.11**. In this example, the hole appears as a true circle instead of an ellipse in the auxiliary view.

15.5 The Rules of Auxiliary View Construction

Now that several examples of auxiliary views have been discussed, it would be helpful to summarize the general rules of their construction, which are summarized in **Fig. 15.12**.

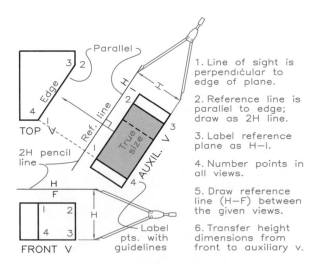

1. Line of sight is perpendicular to edge of plane.

2. Reference line is parallel to edge; draw as 2H line.

3. Label reference plane as H–I.

4. Number points in all views.

5. Draw reference line (H–F) between the given views.

6. Transfer height dimensions from front to auxiliary v.

15.12 The rules of auxiliary view construction.

Step 1 Draw a line of sight perpendicular to the edge of the inclined surface. Draw the H–1 fold line parallel to the edge of the inclined surface and draw an H–F fold line between the given views.

Step 2 Find points 1 and 2 by transferring the height (H) dimensions with your dividers from the front view to the auxiliary view.

Step 3 Find points 3 and 4 in the same manner by transferring the H dimensions.

1. An auxiliary view that shows a surface true size must be projected perpendicularly from the edge view of the surface. Usually, the inclined surface, or a partial view, is all that is needed in the auxiliary view, but the entire object can be drawn in the auxiliary view if desired as shown here.

2. Draw the sight line perpendicular to the inclined edge of the plane you wish to find true size (TS).

3. Draw the reference line (H–15, for example) parallel to the edge view of the inclined plane which will be perpendicular to the line of sight.

4. Draw a reference line between the given views (front and top in this example). Reference lines (fold lines) should be drawn as thin black lines with a 2H or 3H pencil.

5. If an auxiliary is projected from the front view, it will have an F–1 reference line; if it is projected from the horizontal view (top view), it will have an H–1 reference line; and if projected from the side view (profile view), it will have a P–1 reference plane.

6. Transfer measurements from the other given view with your dividers (not the view you are projecting from), height in the front view in this example.

7. Number each point one at a time in the primary views and the auxiliary views as they are plotted.

8. Do your lettering in a professional manner with guidelines.

9. Connect the points with light construction lines and use light gray projectors that do not have to be erased with a pencil in the 2H–4H range.

10. Draw the outlines of the auxiliary view as thick visible lines the same as visible lines in principal views, with an F or HB pencil.

15.6 Auxiliaries from the Front View

By moving about the front view of the part as shown in **Fig. 15.13**, you will be looking parallel to the edge view of the frontal plane. Therefore the depth dimension (D) will appear true size in each auxiliary view projected from the front view. One of the positions gives a principal view, the right-side view, and position 1 gives a true-size view of the inclined plane. **Figure 15.14** illustrates the relationship between the auxiliary views projected from the front view.

Folding-Line Method A plane of an object that appears as an edge in the front view (**Fig. 15.15**) is true size in an auxiliary view projected perpendicularly from it. Draw fold line F–1 parallel to the edge view of the inclined plane in the front view at a convenient location.

Draw the line of sight perpendicular to the edge view of the inclined plane in the front view. Observed from this direction, the frontal plane appears as an edge; therefore, measurements perpendicular to the frontal plane

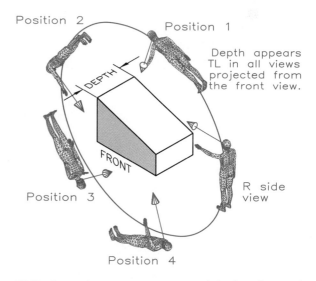

15.13 By moving your viewpoint around the frontal view of an object, you will see a series of auxiliary views in which the depth dimension (D) is true length in all views.

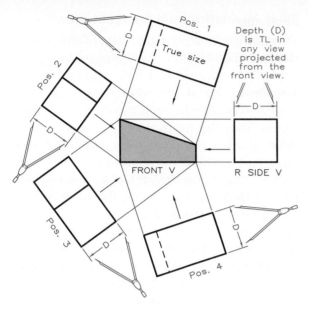

15.14 The auxiliary views shown in Fig. 15.15 would be seen in this arrangement when viewing the front view.

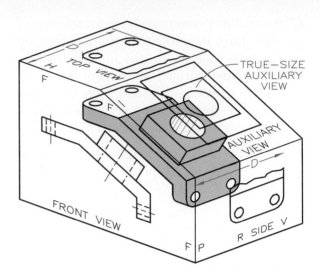

15.16 The imaginary glass box enclosing this part illustrates the relationship of the auxiliary plane, on which the true-size view of the inclined surface is projected, with the principal planes.

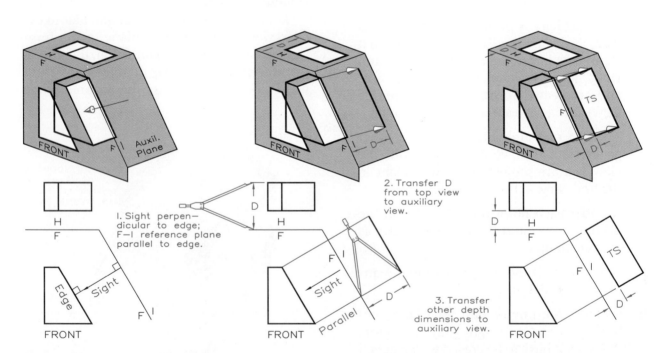

15.15 Auxiliary from the front: folding-line method.

Step 1 Draw the line of sight perpendicular to the edge of the plane and draw the F–1 line parallel to it. Draw the H–F fold line between the top and front views.

Step 2 Project perpendicularly from the edge view of the inclined surface and parallel to the line of sight. Transfer the depth dimensions (D) from the top to the auxiliary view with your dividers.

Step 3 Locate the other corners of the inclined surface by projecting to the auxiliary view. Locate the points by transferring the depth dimensions (D) from the top to the auxiliary view.

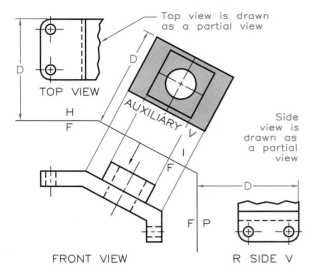

Top view is drawn as a partial view

Side view is drawn as a partial view

TOP VIEW

AUXILIARY

FRONT VIEW

R SIDE V

15.17 The layout and construction of an auxiliary view of the object shown in Fig. 15.18.

depth dimensions (D) will be seen true length. Transfer depth dimensions from the top view to the auxiliary view with your dividers.

The object in **Fig. 15.16** is imagined to be enclosed in a glass box and an auxiliary plane is folded from the frontal plane to be parallel to the inclined surface. When drawn on a sheet of paper, the views appear as shown in **Fig. 15.17**. The top and side views are drawn as partial views because the auxiliary view eliminates the need for drawing complete views. The auxiliary view, located by transferring the depth dimension measured perpendicularly from the edge view of the frontal plane in the top view and transferred to the auxiliary view, shows the true size of the surface.

Computer Method Find the true-size view of the inclined surface that appears as an edge in the front view (**Fig. 15.18**) by using the LISP

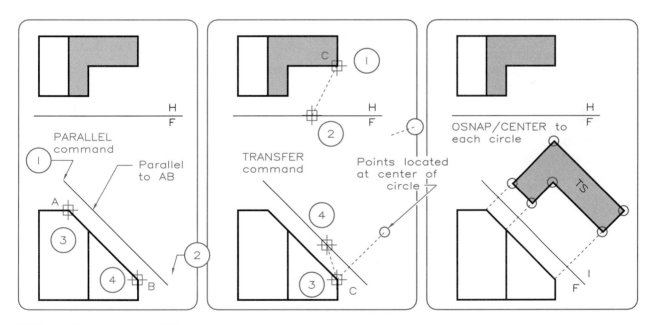

15.18 Auxiliary view by AutoCAD.

Step 1 Type PARALLEL to receive the prompts for the first end of the reference line (1) and its approximate second endpoint (2). You will then be prompted for the ends of line AB, to which the reference line is parallel.

Step 2 Type TRANSFER; you will be prompted for a point in the top view (1) and its distance from the H–F line (2) to be transferred. You will be prompted for the front view of the point to be projected (3) and the reference line (4). Point C is projected to the auxiliary view.

Step 3 Continue using TRANSFER to locate the other corner points of the inclined plane. Connect the points using the CENTER option of the OSNAP to snap to the centers of the circles. Use the ERASE command to remove the circles after connecting their centers.

15.6 AUXILIARIES FROM THE FRONT VIEW • 215

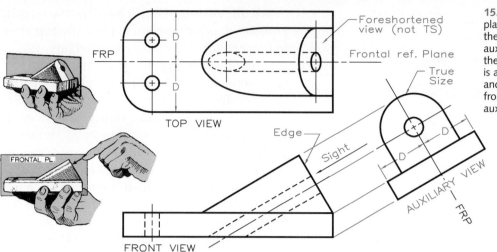

15.19 The frontal reference plane (FRP) is passed through the symmetrical object and the auxiliary view is projected from the edge of the plane. The FRP is an edge in the auxiliary view and depth dimensions are laid from it to find the true-size auxiliary view.

commands, PARALLEL and TRANSFER (see Section 27.2). (This program is not a regular part of AutoCAD, but it is an excellent addition to have available for solving auxiliary problems. It was developed by Professor Leendert Kersten of the University of Nebraska. Once copied as a LISP file, you can access it by typing (LOAD "ACAD") at the command line; be sure to use the parentheses.)

While in AutoCAD's drafting mode, type PARALLEL to draw the reference line parallel to edge AB. Type TRANSFER to obtain prompts for transferring measurements from the top view as if you were using your dividers. Connect the circular points to complete the auxiliary view.

Reference-Plane Method

The object shown in **Fig. 15.19** has an inclined surface that appears as an edge in the front view; therefore, this plane can be found true size in a primary auxiliary view. It is helpful to draw a reference plane through the center of the symmetrical top view because all depth dimensions can be located on each side of the frontal reference plane. Because the reference plane is a frontal plane, it is labeled FRP in the top and auxiliary views. In the auxiliary view, the FRP is drawn parallel to the edge view of

the inclined plane at a convenient distance from it. By transferring depth dimensions from the FRP in the top view to the FRP in the auxiliary view, the symmetrical view of the part is drawn.

15.7 Auxiliaries from the Profile View

By moving your position about the profile view (side view) of the part, as shown in **Fig. 15.20**, you will be parallel to the edge view of the profile plane. Therefore, the width dimension will appear true size in each auxiliary view projected from the side view. One of the positions gives a principal view, the front view, and position 1 gives a true-size view of the inclined plane. **Figure 15.21** illustrates the arrangement of the auxiliary views projected from the side view if they were drawn on a sheet of paper.

Folding-Line Method Because the inclined surface in **Fig. 15.22** appears as an edge in the profile plane, it can be found true-size in a primary auxiliary view projected from the side view. The auxiliary fold line, P–1, is drawn parallel to the edge view of the inclined surface. A line of sight perpendicular to the auxiliary plane shows the profile plane as an edge.

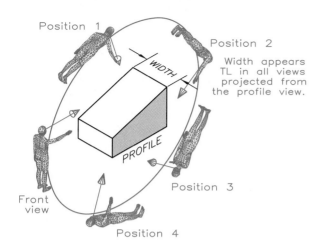

Position 1

Position 2

Width appears TL in all views projected from the profile view.

WIDTH

PROFILE

Front view

Position 3

Position 4

15.20 By moving your viewpoint around a profile (side) view of an object, you will obtain a series of auxiliary views in which the width dimension (W) is true length.

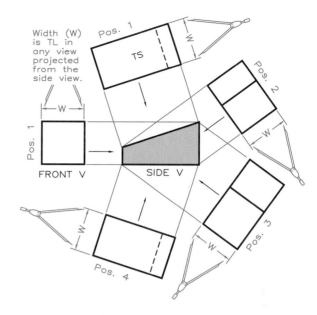

Width (W) is TL in any view projected from the side view.

Pos. 1

TS

W

Pos. 2

Pos. 1

W

W

FRONT V

SIDE V

Pos. 4

W

W

Pos. 3

15.21 The auxiliary views shown in Fig. 15.20 would be seen in this arrangement when projected from the side view.

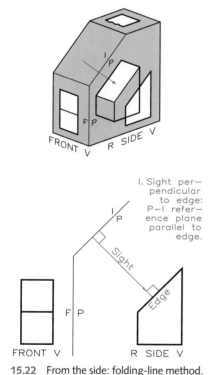

FRONT V   R SIDE V

1. Sight per—pendicular to edge: P—1 reference plane parallel to edge.

P

Sight

Edge

FRONT V   R SIDE V

15.22 From the side: folding-line method.

Step 1 Draw a line of sight perpendicular to the edge of the inclined surface. Draw the P–1 fold line parallel to the edge view, and draw the F–P fold line between the given views.

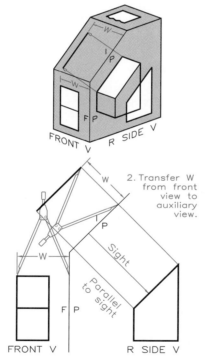

FRONT V   R SIDE V

2. Transfer W from front view to auxiliary view.

W

Sight

Parallel to sight

FRONT V   R SIDE V

Step 2 Project the corners of the edge view parallel to the line of sight. Transfer the width dimensions (W) from the front view to locate a line in the auxiliary view.

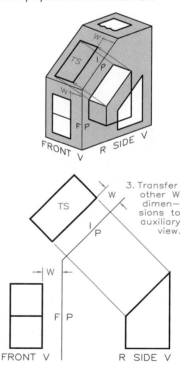

FRONT V   R SIDE V

3. Transfer other W dimen—sions to auxiliary view.

W

TS

W

FRONT V   R SIDE V

Step 3 Find the other corners of the inclined surface by projecting to the auxiliary view. Locate the points by transferring the width dimensions (W) from the front view to the auxiliary view.

Therefore width dimensions (W) transferred from the front view to the auxiliary view appear true length in the auxiliary view.

Reference-Plane Method The object shown in **Fig. 15.23** has an inclined surface that appears as an edge in the right-side view, the profile view. This inclined surface may be drawn true size in an auxiliary view by using a profile reference plane (PRP) that is a vertical edge in the front view. Draw the PRP through the center of the front view because the view is symmetrical. You can then find the true-size view of the inclined plane by transferring equal width dimensions (W) with your dividers from the edge view of the PRP in the front view to both sides of the PRP in the auxiliary view.

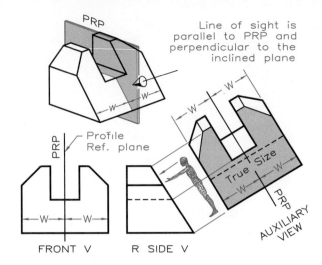

15.23 An auxiliary view is projected from the right-side view by using a profile reference plane (PRP) to show the true-size view of the inclined surface.

15.8 Curved Shapes

The cylinder shown in **Fig. 15.24** has an inclined surface that appears as an edge in the front view. The true-size view of this plane can be seen in an auxiliary view projected from the front view.

Because the cylinder is symmetrical, a frontal reference plane (FRP) is drawn through the center of the side view so that equal dimensions can be laid off on both sides of it. Points located about the circular side view are projected to its edge view in the front view.

In the auxiliary view, the FRP is drawn parallel to the edge view of the plane in the front view, and the points are projected perpendicularly from the edge view of the plane. Dimensions A and B are shown as examples of depth dimensions used for locating points in the auxiliary view. To construct a smooth elliptical curve, more points than shown are needed.

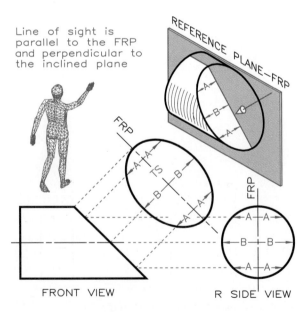

15.24 The auxiliary view of this elliptical surface was found by locating a series of points about its perimeter. The frontal reference plane (FRP) is drawn through its center in the side view since the object is symmetrical.

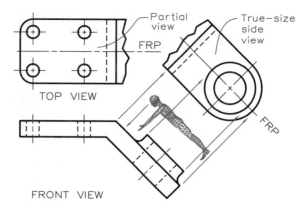

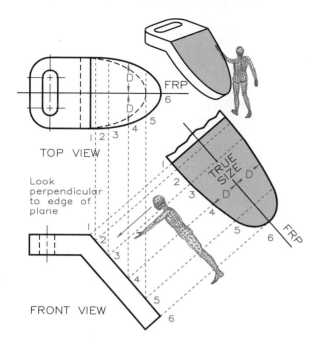

15.25 The auxiliary view of this curved surface required that a series of points be located in the top view and then projected to the front view, followed by the auxiliary view. The FRP passes through the top view.

A true-size auxiliary view of a surface bounded by an irregular curve is shown in **Fig. 15.25**. Project points from the curve in the top view to the front view. Locate these points in the auxiliary view by transferring depth dimensions (D) from the FRP in the top view to the auxiliary view.

15.9 Partial Views

Auxiliary views are used as supplementary views to clarify features that are difficult to depict with principal views alone. Consequently, portions of principal views and auxiliary views may be omitted, provided that the partial views adequately describe the part. The object shown in **Fig. 15.26** is composed of a complete front view, a partial auxiliary view, and a partial top view. These partial views are easier to draw and are more descriptive without sacrificing clarity.

15.26 Partial views with foreshortened portions omitted can be used to represent objects. The FRP reference line is drawn through the center of the object in the top view because the object is symmetrical to make point location easier.

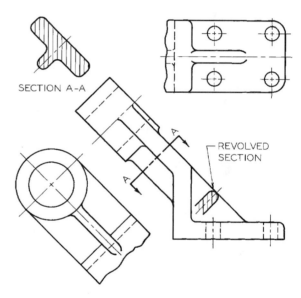

15.27 A cutting plane labeled A–A is passed through the object and the auxiliary section, section A–A, is drawn as a supplementary view to describe the part. The top and front views are drawn as partial views.

15.10 Auxiliary Sections

In **Fig. 15.27**, a cutting plane labeled A–A is passed through the part to obtain the auxiliary section labeled section A–A. The auxiliary

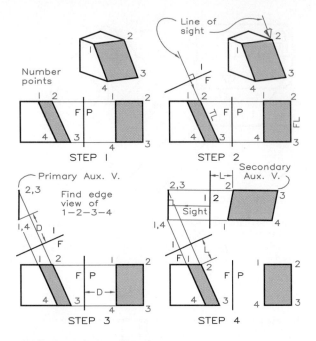

15.28 Secondary auxiliary views.

Step 1 Draw a fold line F–P between the front and side views. Label the corner points in both views.

Step 2 Line 2–3 is a true-length frontal line in the front view. Draw reference line F-1 perpendicular to line 2–3 at a convenient location with a line of sight parallel to line 2–3.

Step 3 Find the edge view of plane 1–2–3–4 by transferring depth dimensions (D) from the side view.

Step 4 Draw a line of sight perpendicular to the edge view of 1–2–3–4 and draw the 1–2 fold line parallel to the edge view. Find the true-size auxiliary view by transferring the dimensions (L) from the front view to the auxiliary view.

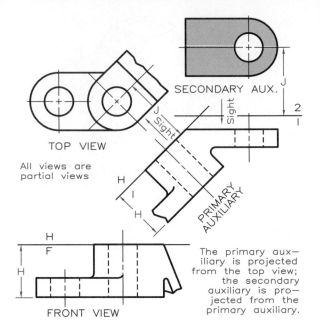

15.29 A secondary auxiliary view projected from a primary auxiliary view that was projected the top view is shown here. All views are drawn as partial views.

section provides a good and efficient way to describe features of the part that could not be as easily described by additional principal views.

15.11 Secondary Auxiliary Views

Figure 15.28 shows how to project a secondary auxiliary view from a primary auxiliary view. An edge view of the oblique plane is found in the primary auxiliary view by finding the point view of a true-length line (2–3) that lies on the oblique surface. A line of sight perpendicular to the edge view of the plane gives a secondary auxiliary view that shows the oblique plane as true size.

Note that the reference line between the primary auxiliary view and the secondary auxiliary view is labeled 1–2 to represent the fold line between the primary plane (1) and the secondary plane (2). The 1 label is placed on the primary side and the 2 label is placed on the secondary side.

Figure 15.29 illustrates the construction of a secondary auxiliary view that gives the true-size view of a surface on a part using these same principles and a combination of partial views. A secondary auxiliary view must be used in this case because the oblique plane does not appear as an edge in a principal view.

Find the point view of a line on the oblique plane to find the edge view of the plane in the primary auxiliary view. The secondary auxiliary view, projected perpendicularly from the edge view of the plane in the primary auxiliary view, gives a true-size view of the plane. In this example, all of the views are drawn as partial views.

15.12 Elliptical Features

Occasionally, circular shapes will project as ellipses, which must be drawn with an irregular curve or an ellipse template. The ellipse template (guide) is by far the most convenient method of drawing ellipses. The angle of the ellipse template is the angle the line of sight makes with the edge view of the circular feature. In **Fig. 15.30** the angle is found to be 45° where the curve is an edge in the front view, so the right-side view of the curve is drawn as a 45° ellipse.

Problems

1–13. (**Fig. 15.31**) Using the example layout, change the top and front views by substituting the top views given at the right in place of the one given in the example. The angle of incli-

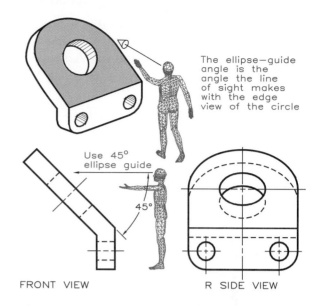

The ellipse–guide angle is the angle the line of sight makes with the edge view of the circle

Use 45° ellipse guide

45°

FRONT VIEW   R SIDE VIEW

15.30 The angle formed by the ellipse template is the angle that the line of sight makes with the edge view of the circular feature. The ellipse angle for the right-side view is 45°.

nation in the front view is 45° for all problems, and the height is 38 mm (1.5 inches) in the front view. Construct auxiliary views that show the inclined surface true size. Draw two problems per size A sheet.

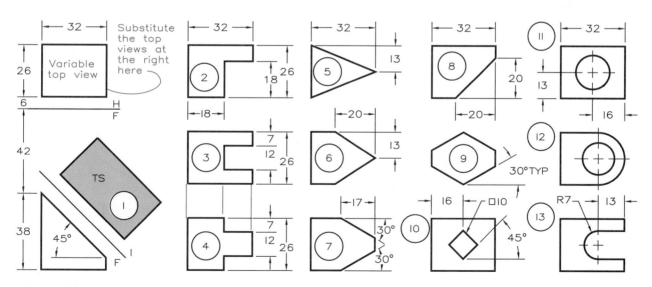

15.31 Problems 1–13. Primary auxiliary views.

14–29. (Figs. 15.32–15.50) Draw the necessary primary and auxiliary views to describe the objects assigned. Draw one per size A or size B sheet. Adjust the scale of each to utilize the space on the sheet.

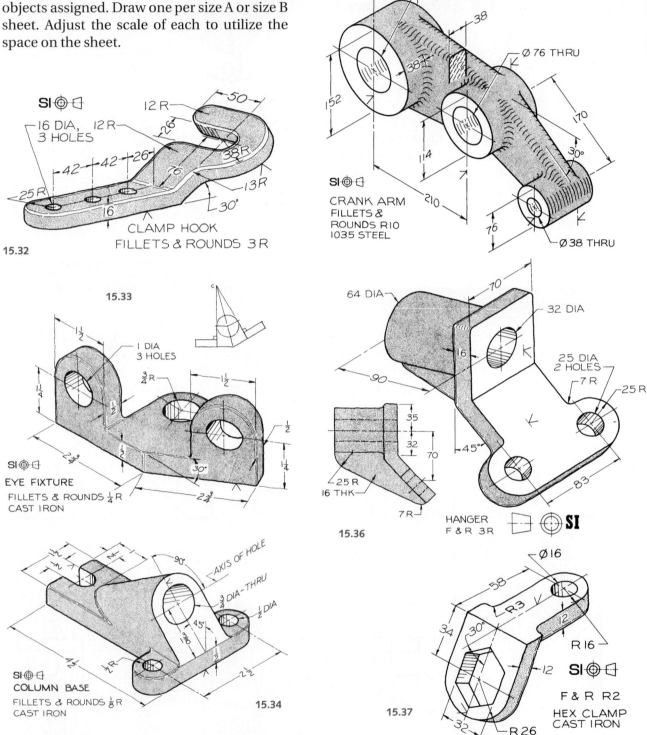

15.35

SI⊕⊟

16 DIA, 12 R
3 HOLES

12 R

50

26

38R

13R

30°

42   42   26

25R

16

CLAMP HOOK
FILLETS & ROUNDS 3 R

15.32

Ø 90 THRU

96

38

Ø 76 THRU

38

152

114

170

30°

210

SI⊕⊟

CRANK ARM
FILLETS &
ROUNDS R10
1035 STEEL

76

Ø 38 THRU

15.33

1 DIA
3 HOLES

1½

1¼

¾ R

1½

½

½

½

1¼

2¾

30°

2¾

SI⊕⊟

EYE FIXTURE

FILLETS & ROUNDS ¼ R
CAST IRON

64 DIA

70

32 DIA

16

25 DIA
2 HOLES

7 R

25 R

90

45°

83

35

32

70

25 R
16 THK

7 R

HANGER
F & R 3R   ⊟ ⊕ **SI**

15.36

90

AXIS OF HOLE

¾ DIA-THRU

½ DIA

45°

3/10

4½

2 R

2½

SI⊕⊟

COLUMN BASE

FILLETS & ROUNDS ⅛ R
CAST IRON

15.34

Ø16

58

R3

34

30°

12

R 16

12

SI⊕⊟

F & R R2

HEX CLAMP
CAST IRON

32

R 26

15.37

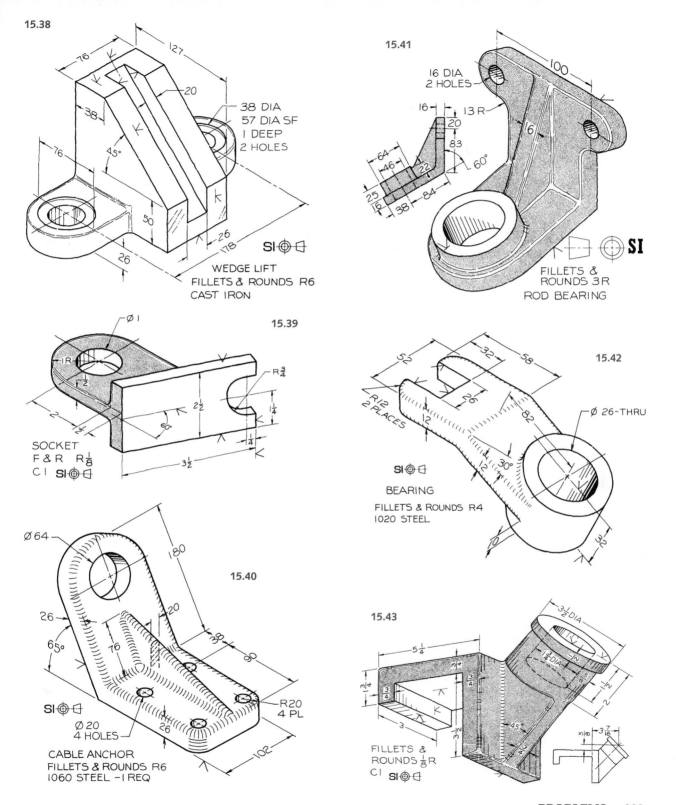

15.38

76
127
20
38 DIA
57 DIA SF
I DEEP
2 HOLES
38
45°
76
50
26
178
26
SI
WEDGE LIFT
FILLETS & ROUNDS R6
CAST IRON

15.41

16 DIA
2 HOLES
100
16
13 R
20
16
83
64
46
60°
22
25
16
38
84
SI
FILLETS &
ROUNDS 3 R
ROD BEARING

15.39

Ø1
IR
½
2
¼
6°
2½
1¼
4
3½
SOCKET
F & R R⅛
CI SI

15.42

52
32
58
R12
2 PLACES
26
82
Ø 26-THRU
12
30°
12
SI
BEARING
FILLETS & ROUNDS R4
1020 STEEL

15.40

Ø 64
180
26
20
65°
38
76
90
R20
4 PL
SI
Ø 20
4 HOLES
26
102
CABLE ANCHOR
FILLETS & ROUNDS R6
1060 STEEL – I REQ

15.43

3½ DIA
5¼
¾
¾
¾
¾
45°
3
3
3½
FILLETS &
ROUNDS ⅛R
CI SI

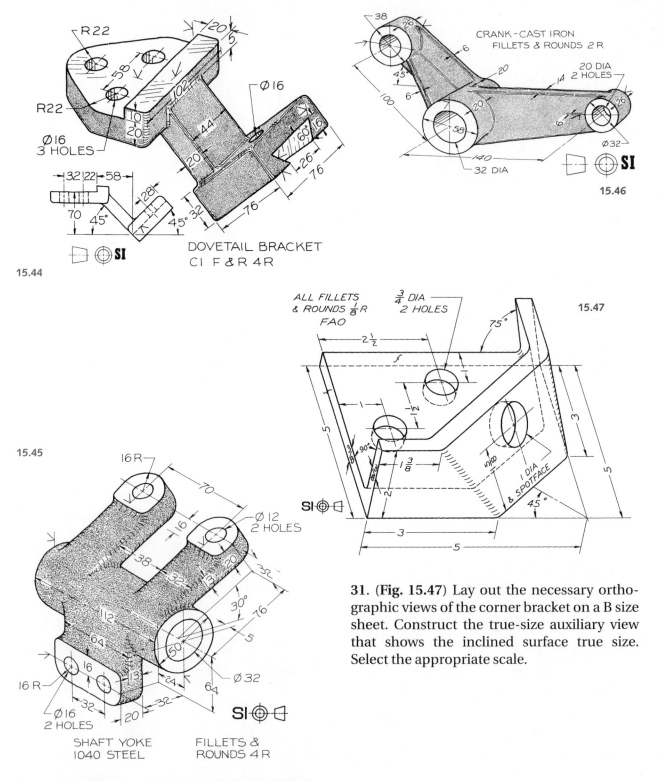

R22

58

102

R22

Ø16
3 HOLES

Ø16

44

10
20

20

60°

16
16

26

76

76

32° 45°

32 22 58

28

70   45°   45°   32

DOVETAIL BRACKET
CI F & R 4R

SI

15.44

CRANK-CAST IRON
FILLETS & ROUNDS 2 R

38   26

6

45°

20

20 DIA
2 HOLES

14

20

100   6

58   6

26

140

32 DIA

Ø32

SI

15.46

15.45

16 R

70

16

Ø 12
2 HOLES

38

32   13   20

32

112

30°

76

64   5

16

Ø 32

13   24   50

16 R

64

Ø16
2 HOLES

32   20   32

SI

SHAFT YOKE
1040 STEEL

FILLETS &
ROUNDS 4 R

ALL FILLETS
& ROUNDS $\frac{1}{8}$ R
FAO

$\frac{3}{4}$ DIA
2 HOLES

75°

15.47

2$\frac{1}{2}$

f

1

$\frac{1}{2}$

5

3
8

90°

60°

1$\frac{3}{8}$

5
100

3

1 DIA
& SPOTFACE

2

SI

3

45°

5

5

31. (**Fig. 15.47**) Lay out the necessary orthographic views of the corner bracket on a B size
sheet. Construct the true-size auxiliary view
that shows the inclined surface true size.
Select the appropriate scale.

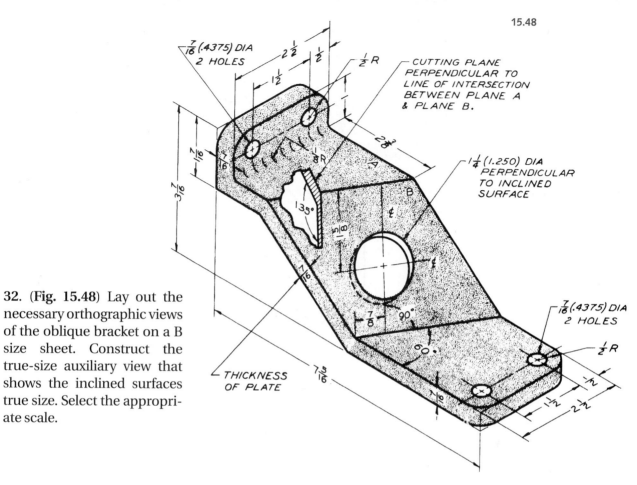

32. (**Fig. 15.48**) Lay out the necessary orthographic views of the oblique bracket on a B size sheet. Construct the true-size auxiliary view that shows the inclined surfaces true size. Select the appropriate scale.

15.49

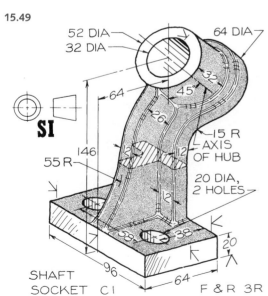

15.50

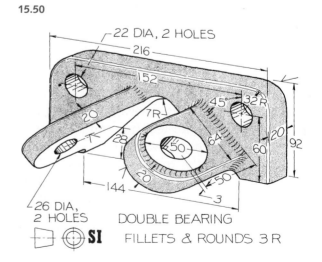

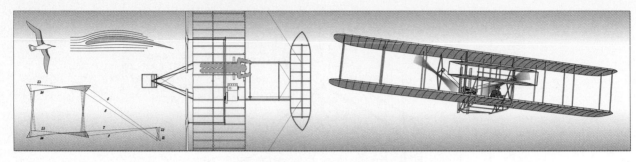

16

Sections

16.1 Introduction

Correctly drawn orthographic views that show all hidden lines may not clearly describe an object's internal details. The gear housing shown in **Fig. 16.1** is such an example: It is better understood when a section has been cut from it. The technique of constructing imaginary cross-sectional cuts through a drawing of a part results in an orthographic view called a **section**.

16.2 The Basics of Sectioning

In **Fig. 16.2A** standard views of a cylinder are shown as the interior features and are drawn as hidden lines. If you imagined a knife edge cutting through the top view, the front view would become a **section**. This section is a **full section** since the cutting plane passes fully through the part (**Fig. 16.2B**). The portion of the part that was cut by the imaginary plane is crosshatched to give an internal view.

Figure 16.3 shows two types of cutting planes. Both are acceptable, although the plane with pairs of short dashes is most often used. The spacing and proportions of the dashes depend on the size of the drawing. The

16.1 This gear housing has many internal features that cannot be described clearly in a standard orthographic view. Sections are used to clarify interior parts.

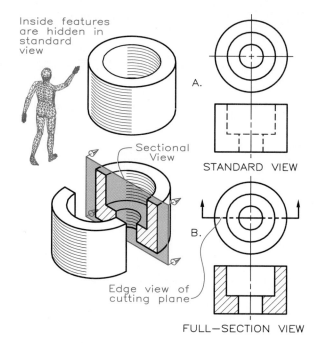

Inside features are hidden in standard view

A.

STANDARD VIEW

Sectional View

Edge view of cutting plane

B.

FULL—SECTION VIEW

16.2 A part with internal features can be better shown with sectional views than with standard views with hidden lines.

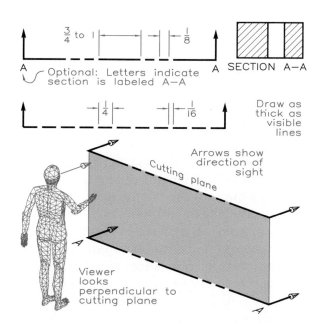

$\frac{3}{4}$ to 1

$\frac{1}{8}$

SECTION A—A

A — Optional: Letters indicate section is labeled A—A

$\frac{1}{4}$ $\frac{1}{16}$

Draw as thick as visible lines

Arrows show direction of sight

Cutting Plane

Viewer looks perpendicular to cutting plane

16.3 Cutting planes can be thought of as knife edges that pass through views to reveal interior features in sections. The cutting plane marked A—A results in a section labeled Section A—A.

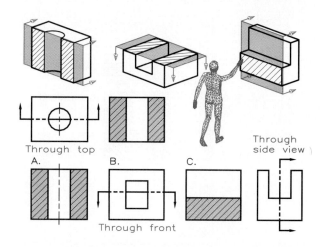

Through top
A.

Through front
B.

C.

Through side view

16.4 The three basic positions of sections and their respective cutting planes that pass through the principal views: (A) top, (B) front, and (C) side views. The arrows point in the direction of the line of sight for each section.

line thickness of the cutting plane is the same as the visible object line. Letters placed at each end of the cutting plane are used to label the sectional view (see as section A–A in Fig. 16.3).

The sight arrows at the ends of the cutting plane are always perpendicular to the cutting plane. In the sectional view, the observer is looking in the direction of the sight arrows, perpendicular to the surface of the cutting plane.

Figure 16.4 shows the three basic positions of sections and their respective cutting planes. In each case, perpendicular arrows point in the direction of the line of sight. For example, the cutting plane in Fig. 16.4A passes through and removes the front of the top view, and the line of sight is perpendicular to the remainder of the top view.

The top view appears as a section when the cutting plane passes through the front view and the line of sight is downward (Fig. 16.4B). When the cutting plane passes vertically through the side view (Fig. 16.4C), the front view becomes a section.

16.2 THE BASICS OF SECTIONING • 227

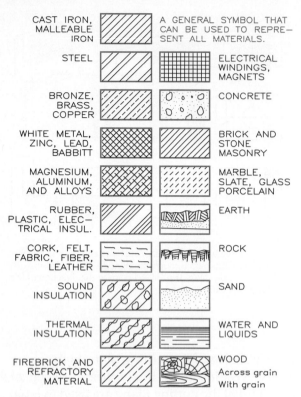

16.5 The symbols used to distinguish between different materials in a section. The cast-iron symbol may be used for any material.

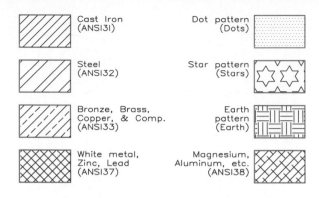

16.6 A few of the hatching symbols available in AutoCAD. The spacing and size of the symbols can be varied by adjusting the pattern scale factor.

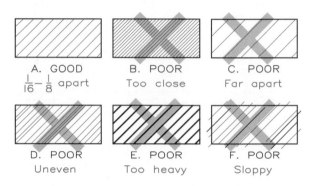

16.7 Hatching techniques. Correctly drawn section lines are thin black lines placed 1/16 to 1/8 in. apart. Common errors of section lining are shown in B–F.

16.3 Sectioning Symbols

The hatching symbols used to distinguish between different materials in sections are shown in **Fig. 16.5**. Although these symbols may be used to indicate the materials in a section, you should provide supplementary notes specifying the materials to ensure clarity.

The symbol that is used most often is the cast-iron symbol (evenly spaced section lines) which may be used to represent any material. Draw cast-iron symbols with a 2H pencil, slant the lines upward and to the right at 30°, 45°, or 60° angles, and space the lines about 1/16 inch apart (close together in small areas and farther apart in larger areas).

Computer Method A few of the many cross-sectional symbols available with AutoCAD are shown in **Fig. 16.6**. The spacing between the lines and the dash lengths can be changed by adjusting the pattern scale factor.

Properly drawn section lines—thin and evenly spaced—are shown in **Fig. 16.7**. Parts B–F show common errors of section lining.

Thin parts such as sheet metal, washers, and gaskets are sectioned by completely blacking in their areas because space does not permit the drawing of section lines (**Fig. 16.8**). The large parts in that figure are hatched with outline sectioning to save time and effort.

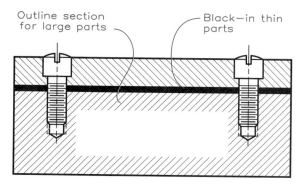

Outline section for large parts — Black—in thin parts

16.8 Large sectional areas are hatched with outline sectioning (around their edges) and thin parts are blacked in solid.

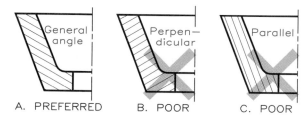

General angle — Perpen-dicular — Parallel

A. PREFERRED | B. POOR | C. POOR

16.9 Draw hatch lines that are neither parallel nor perpendicular to the outlines of the part, so that they are not confused with serrations or other machining treatments of the surface.

Sectioned areas should be hatched with lines that are neither parallel nor perpendicular to the outlines of the parts so that they will not be confused with serrations or other machining treatments of the surface (**Fig. 16.9**).

Computer Method The basic principles of using AutoCAD to apply hatching to an area are shown in **Fig. 16.10**. After assigning the proper hatch symbol with the HATCH command, select the area to be sectioned with a window and the hatch lines will be drawn automatically. To vary the spacing of the section lines, change the pattern scale factor of the HATCH command. BHATCH and other hatching commands are covered in Chapters 36 and 37.

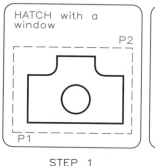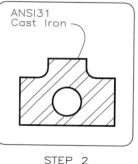

HATCH with a window — P2 — P1 | ANSI31 Cast Iron

STEP 1 | STEP 2

16.10 Hatching by computer.
Step 1 Command: HATCH (Enter)
Enter pattern name or (? /Solid/User defined) [ANSI31]: ANSI32 (Enter)
Scale for pattern [1.00]: .50 (Enter)
Angle for pattern [0]: 0 (Enter)
Select objects: W (window option)
Step 2 Press (Enter) to apply the hatching.

The lines that outline the areas to be hatched must intersect perfectly at each corner point; no T-joints are permitted (**Fig. 16.11**). Poor intersections may cause hatching symbols to fill the desired area improperly.

16.4 Sectioning Assemblies of Parts

When sectioning an assembly of several parts, draw section lines at varying angles to distinguish the parts from each other (**Fig. 16.12A**). Using different material symbols in an assembly also helps distinguish between the parts and their materials. Crosshatch the same part at the same angle and with the same symbol, even though portions of the part may be separated (**Fig. 16.12B**).

16.5 Full Sections

A cutting plane that passes fully through an object and removes half of it forms a **full section view**. **Figure 16.13** shows two orthographic views of an object with all its hidden lines. We can describe the part better by pass-

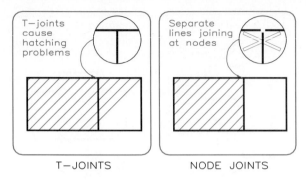

T—JOINTS   NODE JOINTS

16.11 For the HATCH command to work properly, the outlines of areas to be hatched must be drawn with perfect closing outlines. T-joints, overlaps, or gaps at intersections can result in irregular results.

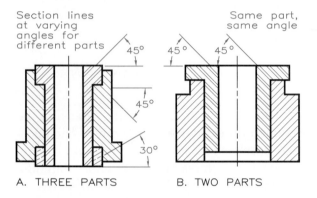

A. THREE PARTS   B. TWO PARTS

16.12 Hatching an assembly of parts.
A Draw section lines of different parts in an assembly at varying angles to distinguish the parts.
B Draw section lines on separated portions of the same part (both sides of a hole here) in the same direction.

ing a cutting plane through the top view to remove half of it. The arrows on the cutting plane indicate the direction of sight. The front view becomes a full section, showing the surfaces cut by the cutting plane. **Figure 16.14** shows a full section through a cylindrical part, with half the object removed. Figure 16.14A shows the correctly drawn sectional view. A common mistake—omitting the visible lines behind the cutting plane—is shown in Fig. 16.14B.

Omit hidden lines in sectional views unless you consider them necessary for a clear understanding of the view. Also, omit

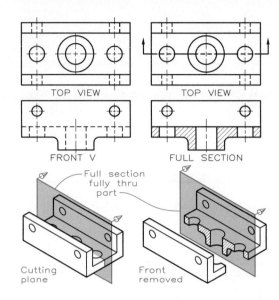

TOP VIEW   TOP VIEW

FRONT V   FULL SECTION

16.13 A full section is found by passing a cutting plane fully through the top view of this part, removing half of it. The arrows on the cutting plane gives the direction of your sight. The front sectional view shows the internal features clearly.

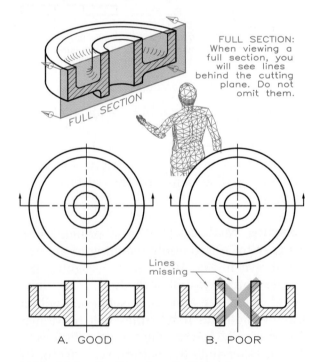

FULL SECTION: When viewing a full section, you will see lines behind the cutting plane. Do not omit them.

A. GOOD   B. POOR

16.14 A full section through a cylindrical part.
A When a front view of cylinder is shown as a full section, visible lines will be seen behind the cut surface.
B By showing only the lines of the cut surface, the view will be incomplete.

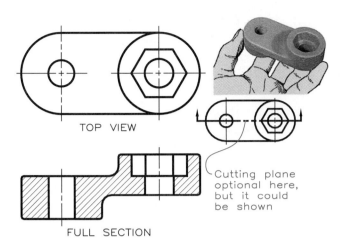

TOP VIEW

FULL SECTION

Cutting plane optional here, but it could be shown

16.15 The cutting plane of a section can be omitted if its location is obvious.

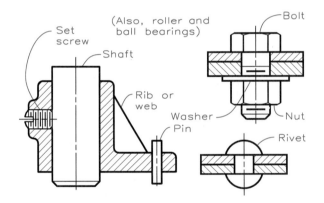

Set screw

Shaft

(Also, roller and ball bearings)

Rib or web

Pin

Bolt

Washer

Nut

Rivet

16.16 By conventional practice, these parts are not hatched even though cutting planes pass through them.

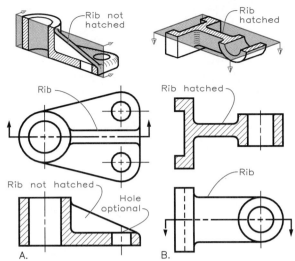

Rib not hatched

Rib hatched

Rib

Rib hatched

Rib not hatched

Hole optional

Rib

A.

B.

16.17 Ribs in section.

A Do not hatch ribs cut in a flatwise direction.

B Hatch ribs when cutting planes pass through them, showing their true thickness.

cutting planes if you consider them unnecessary. **Figure 16.15** shows a full section of a part from which the cutting plane was omitted because its path is obvious.

Parts Not Requiring Hatching

Many standard parts, such as nuts and bolts, rivets, shafts, and set screws, do not require section lining, even though the cutting plane passes through them (**Fig. 16.16**). These parts have no internal features, so sections through them would be of no value. Other parts not

requiring section lining are roller bearings, ball bearings, gear teeth, dowels, pins, and washers.

Ribs

Do not hatch ribs when the cutting plane passes flatwise through them (**Fig. 16.17A**), because to do so would give a misleading impression of the rib. But ribs do require hatching when the cutting plane passes perpendicularly through them and shows their true thickness (**Fig. 16.17B**).

By not hatching the ribs in **Fig. 16.18A**, we provide a more descriptive section view of the part. Had we hatched the ribs, the section would give the impression that the part was solid and conical (**Fig. 16.18B**).

Figure 16.19 shows an alternative method of hatching webs and ribs. The outside ribs in Fig. 16.19A do not require hatching because the cutting plane passes flatwise through them and they are well identified. As a rule, webs do not require hatching, but the webs shown in Fig. 16.19B are not well identified in

16.5 FULL SECTIONS • 231

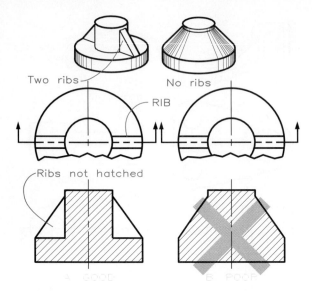

16.18 Ribs in section.

A Do not hatch ribs in some sections to better describe the part.

B If you hatch ribs in section, a misleading impression of the part will be given.

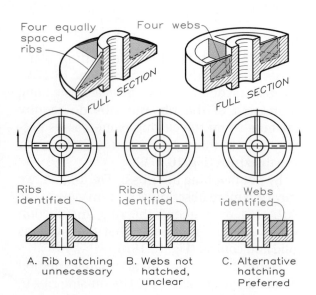

16.19 Ribs and webs in section.

A The well-defined ribs in this section are not hatched.

B These webs are poorly defined when not hatched in the sectional view.

C Alternate hatch lines clarify poorly defined webs.

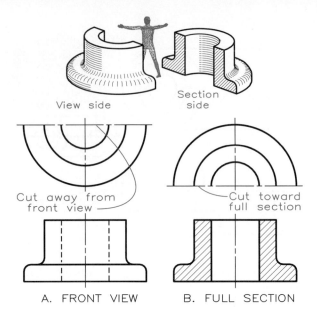

16.20 Half-views of symmetrical parts can be used to conserve space and drawing time as an approved conventional practice.

A The external portion of the half-view is toward the front view.

B The internal portion of the half-view is toward the front view when it is a section. In half-sections, the omitted half-view can be either adjacent to or away from the section.

the front section and could go unnoticed. Therefore, using alternate hatching as shown in Fig. 16.19C is better. Here, extending every other section line through the webs ensures that they can be identified easily.

16.6 Partial Views

A conventional method of representing symmetrical views is the half-view, which requires less space and less drawing time than the full view (**Fig. 16.20**). A half top view is sufficient when drawn adjacent to the section view or front view. For half-views (not sections), the removed half is away from the adjacent view (Fig. 16.20A). For full sections, the removed segment is the half nearest the section (Fig. 16.20B). When drawing partial views with

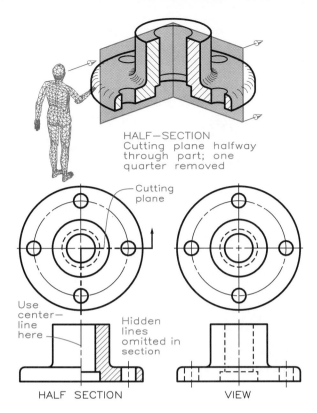

HALF—SECTION
Cutting plane halfway
through part; one
quarter removed

Cutting
plane

Use
center—
line
here

Hidden
lines
omitted in
section

HALF SECTION

VIEW

16.21 In a half-section, the cutting plane passes halfway through the object, removing a quarter of it, to show half the outside and half the inside of it. Omit hidden lines in sectional views unless they are needed for clarity.

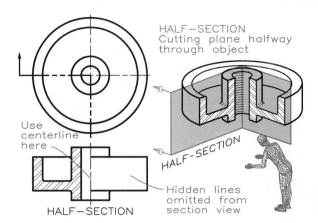

HALF—SECTION
Cutting plane halfway
through object

Use
centerline
here

HALF—SECTION

HALF—SECTION

Hidden lines
omitted from
section view

16.22 This half-section describes a pulley that is shown orthographically and pictorially.

the section. **Figure 16.22** shows a half-section of a pulley.

In **Fig. 16.23** the cutting plane from the half-section has been omitted because its location is obvious. Because the parting line of the half-section is not at a centerline, you may use a solid line or a centerline to separate the sectional half from the half that appears as an external view.

half-sections, you may omit either the near or the far halves of the partial views.

16.7 Half-Sections

A **half-section** is a view obtained by passing a cutting plane halfway through an object and removing a quarter of it to show both external and internal features. Half-sections are used with symmetrical parts and with cylinders, in particular, as shown in **Fig. 16.21**. By comparing the half-section with the standard front view, you can see that both internal and external features show more clearly in a half-section than in a partial view. Unnecessary hidden lines have been omitted to simplify

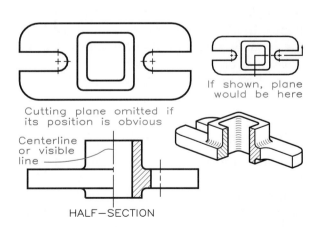

If shown, plane
would be here

Cutting plane omitted if
its position is obvious

Centerline
or visible
line

HALF—SECTION

16.23 The cutting plane can be omitted when its location is obvious. The parting line between the section and the view may be a visible line or a centerline if the part is not cylindrical.

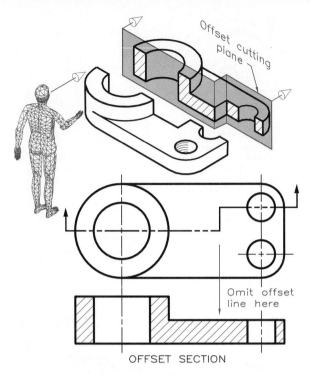

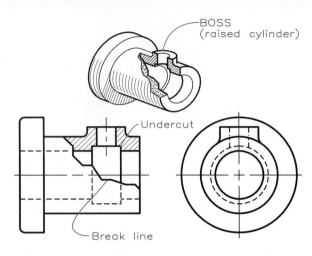

16.24 An offset section is formed by a cutting plane that must be offset to pass through features not in a single plane.

16.25 A broken-out section is one in which a part of the object has been broken away to show internal features.

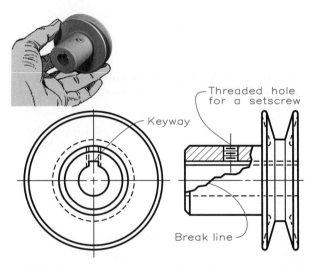

16.26 This broken-out section shows the keyway and the threaded hole for the setscrew in the pulley.

16.8 Offset Sections

An **offset section** is a full section in which the cutting plane is offset to pass through important features that do not lie in a single plane. **Figure 16.24** shows an offset section in which the plane is offset to pass through the large hole and one of the small holes. The cut formed by the offset is not shown in the section because it is imaginary.

16.9 Broken-Out Sections

A **broken-out section** shows a partial view of a part's interior features. The broken-out section of the part shown in **Fig. 16.25** reveals details of the wall thickness to describe the part better. The irregular lines representing the break are conventional breaks (discussed later in this chapter).

The broken-out section of the pulley in **Fig. 16.26** clearly depicts the keyway and threaded hole for a setscrew. This method shows the part efficiently, with the minimum of views.

16.10 Revolved Sections

In a **revolved section**, the cross-section of a part is revolved about an axis of revolution

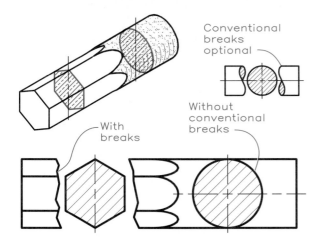

16.27 Revolved sections show cross-sectional features of a shaft and eliminate the need for additional orthographic views. Revolved sections may be superimposed on the given views, or conventional breaks can be used to separate them from the given view.

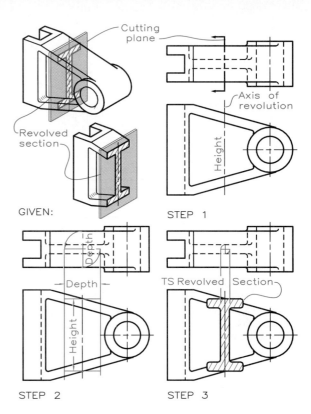

16.28 Revolved section construction.

Given: The part is shown pictorially with a cutting plane that shows the cross-section that will be revolved.

Step 1 An imaginary cutting plane is located in the top view and the axis of rotation is located in the front view.

Step 2 The depth in the top view is rotated in the top view and is projected to the front view. The height and depth in the front view give the overall dimensions of the revolved view.

Step 3 The revolved section is drawn, fillets and rounds are added, and the section lines are applied to finish the view.

and placed on the view where the revolution occurred. Note the use of revolved sections to explain two cross-sections of the shaft shown in **Fig. 16.27** (with and without conventional breaks). Conventional breaks are optional; you may draw a revolved section on the view without them.

A revolved section helps to describe the part shown in **Fig. 16.28**. Imagine passing a cutting plane through the top view of the part, as in Step 1. Then imagine revolving the cutting plane in the top view and projecting it to the front (Step 2). The true-size revolved section is completed in Step 3. Conventional breaks could be used on each side of the revolved section.

Figure 16.29 demonstrates how to use typical revolved sections to show cross-sections through parts without having to draw additional orthographic views.

16.11 Removed Sections

A **removed section** is a revolved section that is shown outside the view in which it was revolved (**Fig. 16.30**). Centerlines are used as

axes of rotation to show the locations from which the sections are taken. Where space does not permit revolution on the given view (**Fig. 16.31A**), removed sections must be used instead of revolved sections (**Fig. 16.31B**).

Removed sections do not have to position directly along an axis of revolution adjacent to the view from which they were revolved.

16.11 REMOVED SECTIONS • 235

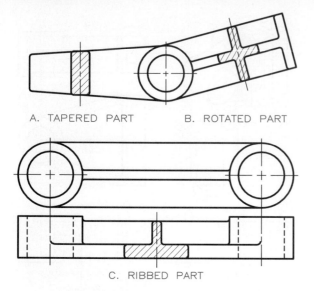

A. TAPERED PART   B. ROTATED PART

C. RIBBED PART

16.29 These revolved sections describe the cross-sections of the two parts that would be difficult to depict in supplementary orthographic views, such as a side view.

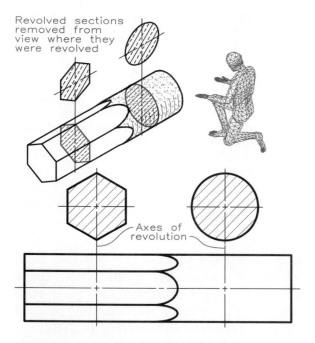

Revolved sections removed from view where they were revolved

Axes of revolution

16.30 Removed sections are revolved sections that are drawn outside the object and along their axes of revolution.

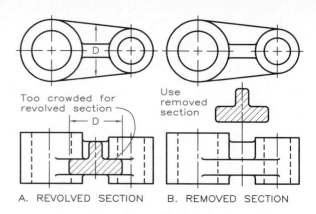

Too crowded for revolved section

Use removed section

A. REVOLVED SECTION   B. REMOVED SECTION

16.31 Removed sections are necessary when space does not permit the revolved section to be superimposed on a part.

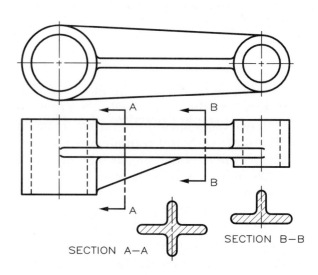

SECTION A–A

SECTION B–B

16.32 Lettering each end of a cutting plane (such as A–A) identifies the removed section labeled Section A–A located elsewhere on the drawing.

Instead, removed sections can be located elsewhere on a drawing if they are properly labeled (**Fig. 16.32**). For example, the plane labeled with an A at each end identifies the location of section A–A; the same applies to section B–B.

When a set of drawings consists of multiple sheets, removed sections and the views from which they are taken may appear on dif-

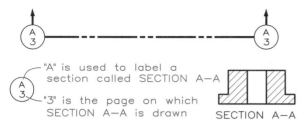

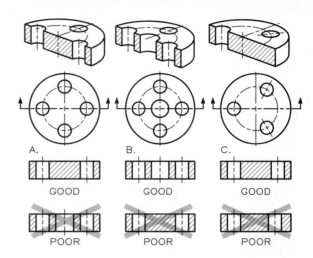

"A" is used to label a section called SECTION A—A

"3" is the page on which SECTION A—A is drawn

SECTION A—A

16.33 When you place a removed section on another page of a set of drawings, label each end of the cutting plane with a letter and a number. The letter identifies the section and the numbers indicate the page on which it is drawn.

A. GOOD   B. GOOD   C. GOOD

POOR   POOR   POOR

16.34 Symmetrically spaced holes are revolved to show their true radial distance from the center of a circular part in sectional views.

Step A Omit the middle holes; it is not at the center of the plate.

Step B Show the middle hole; it is at the center of the plate.

Step C Rotate the holes to the centerline and project them to make a symmetrical section.

ferent sheets. When this method of layout is necessary, label the cutting plane in the view from which the section was taken and the sheet on which the section appears (**Fig. 16.33**).

16.12 Conventional Revolutions

True projection can be improved by using standard convention practices of revolution. In **Fig. 16.34A**, the middle hole is omitted because it does not pass through the center of the circular plate. However, in **Fig. 16.34B**, the hole does pass through the plate's center and is shown in the section. Although the cutting plane does not pass through one of the symmetrically spaced holes in the top view (**Fig. 16.34C**), the hole is revolved to the cutting plane to show the full section.

When ribs are symmetrically spaced about a hub (**Fig. 16.35**), it is conventional practice to revolve them so that they appear true size in both their orthographic and sectional views. **Figure 16.36** illustrates the conventional practice of revolving both holes and ribs (or webs) of symmetrical parts. Revolution gives a better description of the parts in a manner that is easier to draw.

A cutting plane may be positioned in either of the two ways shown in **Fig. 16.37**. Even though the cutting plane does not pass through the ribs and holes in Fig. 16.37A, they

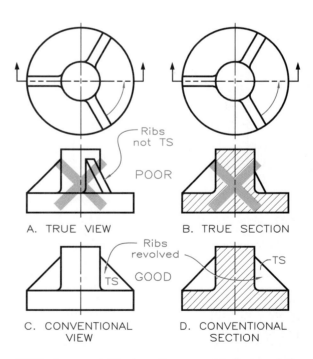

A. TRUE VIEW   B. TRUE SECTION

Ribs not TS   POOR

Ribs revolved   GOOD

C. CONVENTIONAL VIEW   D. CONVENTIONAL SECTION

TS

16.35 As a conventional practice, symmetrically spaced ribs are revolved and drawn true size in their orthographic and sectional views to show them true size.

16.12 CONVENTIONAL REVOLUTIONS • 237

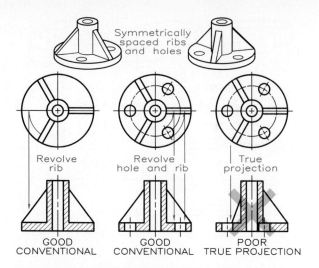

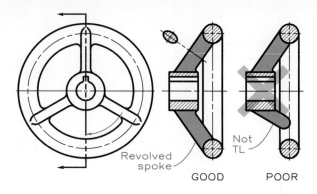

16.36 Symmetrically spaced ribs and holes should be shown in sections with the ribs rotated to show them true size and the holes rotated to show them at their true radial distance from the center.

16.38 Revolve spokes to show them true size in section. Do not hatch the spokes.

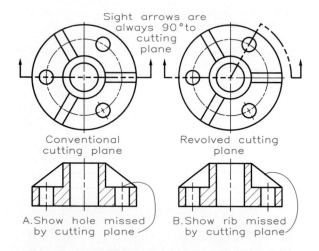

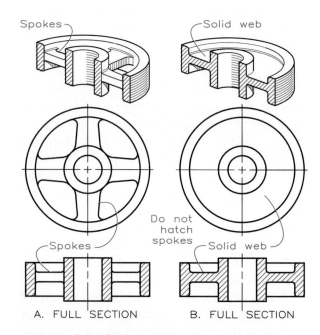

16.37 Show symmetrically spaced ribs true size, whether or not the cutting plane passes through them. As an alternative, the cutting plane can be revolved to pass through the ribs for clarity.

16.39 Spokes and webs in section.

A Spokes are not hatched even though they have been cut.

B Webs are hatched when cut by the cutting plane.

may appear in section as if the cutting plane passed through them. The path of the cutting plane may also be revolved, as shown in Fig. 16.37B. In this case, the ribs are revolved to their true-size position in the section view, although the plane does not cut through them.

The same principles apply to symmetrically spaced spokes (**Fig. 16.38**). Draw only the revolved, true-size spokes and do not hatch them. In **Fig. 16.39** if the spokes shown in part A were hatched, they could be misunderstood as a solid web, as shown in part B.

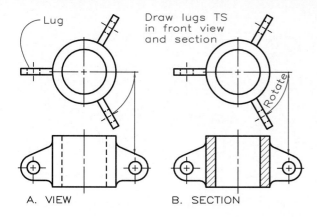

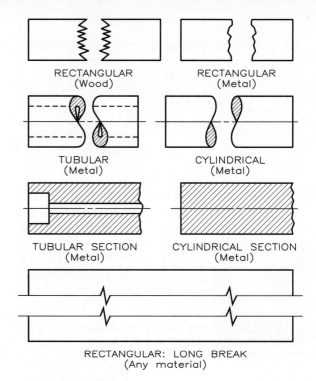

16.40 Lugs are revolved to show their true size in (A) the front view, and also in (B) the sectional view.

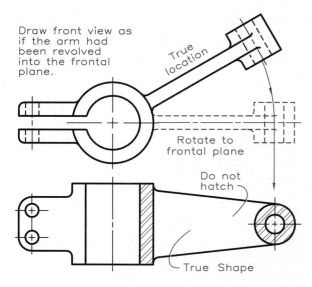

16.41 It is conventional practice to revolve parts with inclined features in order to show them true size in both sectional and regular orthographic views.

Revolving the symmetrically positioned lugs shown in **Fig. 16.40** gives their true size in both the front view and sectional view. The same principles of rotation apply to the part shown in **Fig. 16.41**, where the inclined arm appears in the section as if it had been revolved to the centerline in the top view and then projected to the sectional view. These

16.42 The conventional break indicates that a portion of an object has been omitted.

conventional practices save time and space on a drawing and also make the views more understandable to the reader.

16.13 Conventional Breaks

Conventional breaks indicate removed portions of parts and gives an indication of material and cross-sectional shape as well. **Figure 16.42** shows the types of conventional breaks to use when you remove portions of an object. You may draw the "figure-eight" breaks used for cylindrical and tubular parts freehand (**Fig. 16.43**) or with a compass when they are larger, as shown in **Fig. 16.44**.

Conventional breaks can be used to shorten a long piece by removing the portion between the breaks so that it may be drawn at a larger scale (**Fig. 16.45**). The dimension

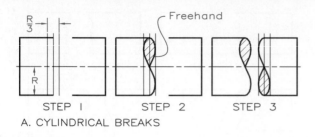

STEP 1　　　STEP 2　　　STEP 3

A. CYLINDRICAL BREAKS

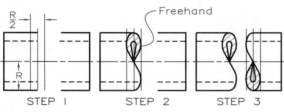

STEP 1　　　STEP 2　　　STEP 3

B. TUBULAR BREAKS

16.43 It is essential to use guidelines in drawing conventional breaks for both (A) solid cylinders and (B) tubular cylinders freehand. The radius R is used to determine the widths.

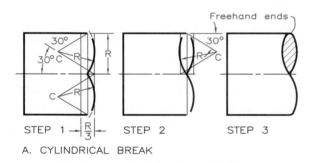

STEP 1 —|R/3|—　　STEP 2　　　STEP 3

A. CYLINDRICAL BREAK

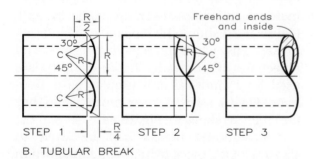

STEP 1 —|　|—R/4　　STEP 2　　　STEP 3

B. TUBULAR BREAK

16.44 These steps can be followed to draw conventional breaks with a compass.

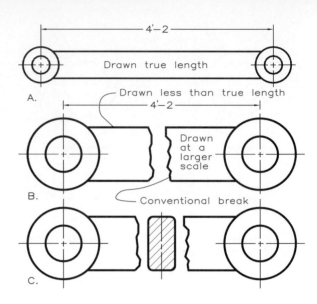

A.

B.

C.

16.45 By using conventional breaks, part of an object can be removed to save space or to draw the remaining part at a larger scale. A revolved section can also be inserted between the breaks to further describe the part.

specifies the true length of the part, and the breaks indicate that a portion of the length has been removed.

16.14 Phantom (Ghost) Sections

A **phantom** or **ghost section** depicts parts as if they were being X-rayed. In **Fig. 16.46**, the cutting plane is used in the normal manner, but the section lines are drawn as dashed lines. If the object were shown as a regular full section, the circular hole through the front surface could not be shown in the same view. A phantom section lets you show features on both sides of the cutting plane.

16.15 Auxiliary Sections

You may use **auxiliary sections** to supplement the principal views of orthographic projections (**Fig. 16.47**). Pass auxiliary cutting plane A–A through the front view and project the

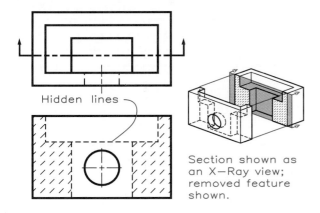

16.46 Phantom sections provide an X-ray view of a part to show features on both sides of the cutting plane. Section lines are drawn as dashed lines.

Hidden lines

Section shown as an X—Ray view; removed feature shown.

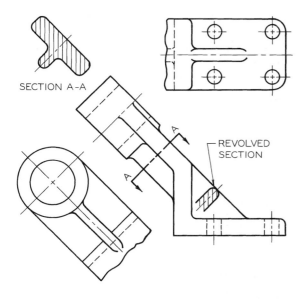

SECTION A-A

REVOLVED SECTION

16.47 Auxiliary sections are helpful in clarifying the details of the inclined features of a part.

auxiliary view from the cutting plane as indicated by the sight arrows. Section A–A gives a cross-sectional description of the part that would be difficult to depict by other principal orthographic views.

An auxiliary view that is also a phantom section is shown in **Fig. 16.48.** In this example the auxiliary view supplements the front view, and the top view is unneeded. A cutting plane indicates the path of a plane that passes

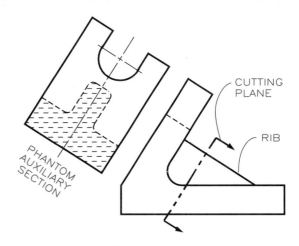

CUTTING PLANE

RIB

PHANTOM AUXILIARY SECTION

16.48 A phantom auxiliary section shows both the inclined surface true size and a cross section through the rib.

through the rib and the base which is shown as a phantom auxiliary section. This type of section permits the true-size inclined surface to be shown also.

Problems

1–14. (**Fig. 16.49**) Solve the problems shown in **Fig. 16.49** on size A sheets by drawing two solutions per sheet. Each grid space equals 0.20 in., or 5 mm.

25–29. (**Figs. 16.50–16.54**) Complete these drawings as full sections. Draw one problem per size AH sheet. Each grid space equals 0.20 in. or 5 mm. Show the cutting planes in each problem.

30–33. (**Figs. 16.55–16.58**) Complete the drawings as half-sections with one solution per size AH sheet. Each grid equals 0.20 in. or 5 mm. Show the cutting planes in each problem.

34–35. (**Figs. 16.59–16.62**) Complete the problems as offset sections. Draw one solution per size AH sheet. Each grid equals 0.20 in. or 5 mm. Show the cutting planes in each solution.

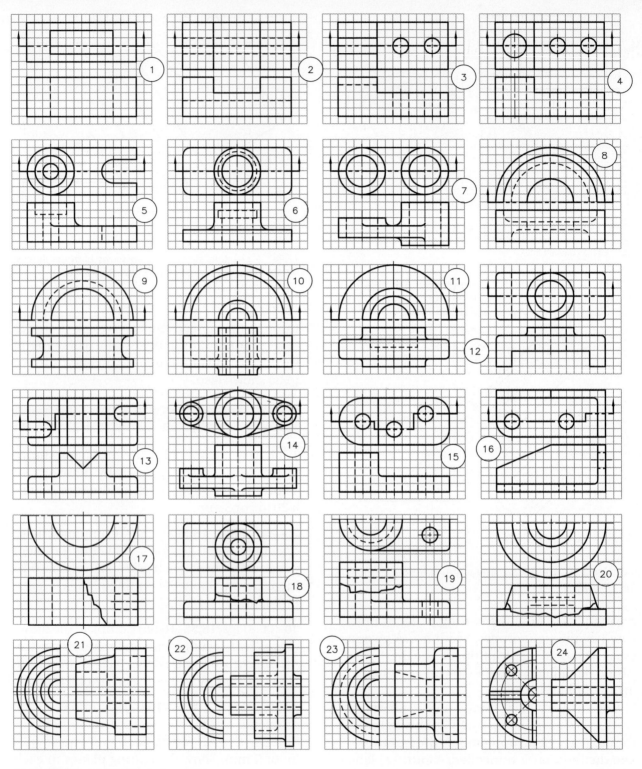

16.49 Problems 1–24. Introductory sections.

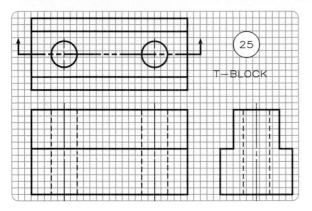

16.50 Problem 25. Full section.

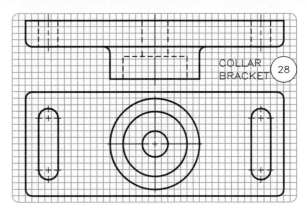

16.53 Problem 28. Full section.

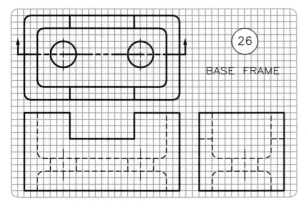

16.51 Problem 26. Full section.

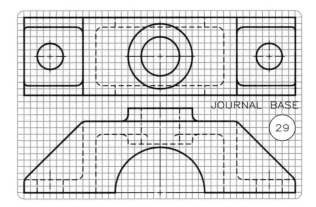

16.54 Problem 29. Full section.

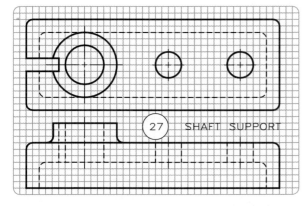

16.52 Problem 27. Full section.

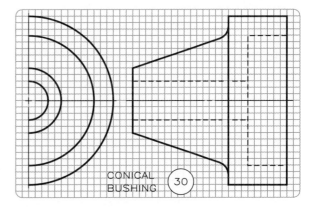

16.55 Problem 30. Half-section.

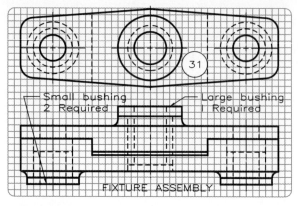

16.56 Problem 31. Half-section.

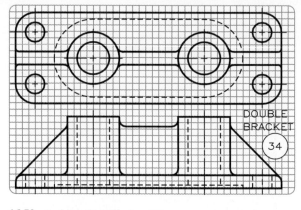

16.59 Problem 34. Offset section.

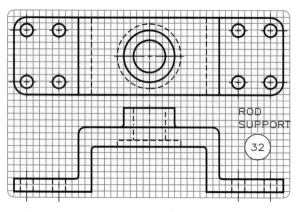

16.57 Problem 32. Half-section.

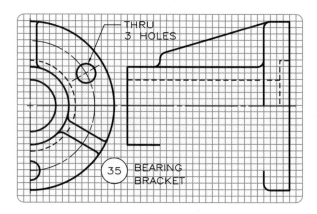

16.60 Problem 35. Offset section.

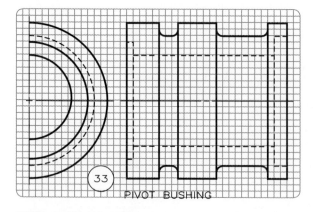

16.58 Problem 33. Half-section.

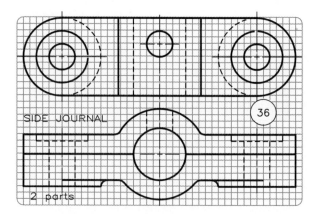

16.61 Problem 36. Offset section.

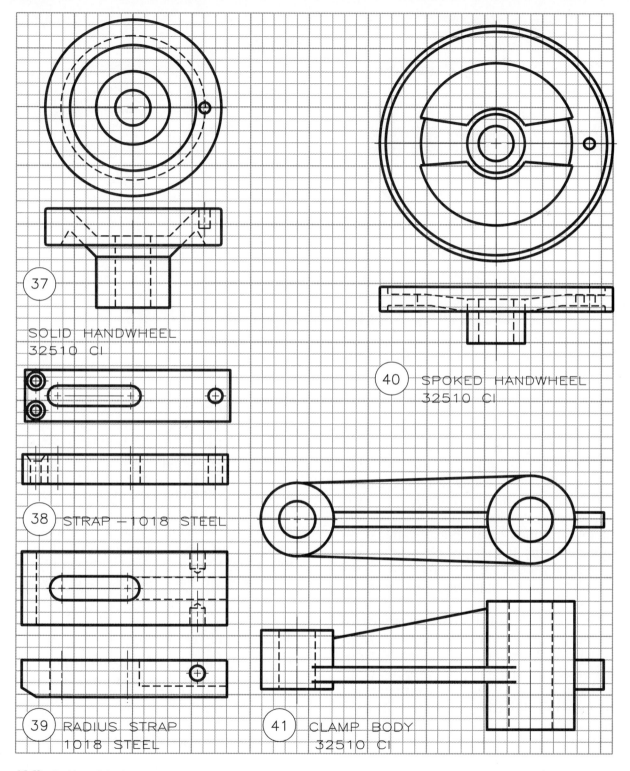

37 SOLID HANDWHEEL
32510 CI

40 SPOKED HANDWHEEL
32510 CI

38 STRAP – 1018 STEEL

39 RADIUS STRAP
1018 STEEL

41 CLAMP BODY
32510 CI

16.62 Problems 36–40.

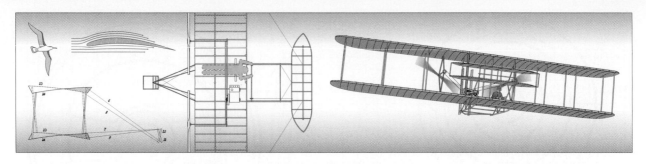

17

Screws, Fasteners, and Springs

17.1 Introduction

Screws provide a fast and easy method of fastening parts together, adjusting the position of parts, and transmitting power. Screws, sometimes called threaded fasteners, should be purchased rather than made as newly designed parts for each product. Screws are available through commercial catalogs in countless forms and shapes for various specialized and general applications. Such screws are cheap, interchangeable, and easy to replace.

The types of threaded parts most often used in industry are covered by current ANSI Standards and include both Unified National (UN) and International Organization for Standardization (ISO) threads. Adoption of the UN thread in 1948 by the United States, Britain, and Canada (sometimes called the ABC Standards), a modification of the American

Standard and the Whitworth thread, was a major step in standardizing threads. The ISO developed metric thread standards for even broader worldwide applications.

Other types of fasteners include **keys**, **pins**, and **rivets**. Other devices covered are **springs**, which resist and react to forces and have applications varying from pogo sticks to automobiles. Springs are also available in many forms and styles from specialty manufacturers who supply most of them to industry.

17.2 Thread Terminology

Understanding threaded parts begins with learning their terminology, which is used throughout this chapter.

External thread: a thread on the outside of a cylinder, such as a bolt (**Fig. 17.1**).

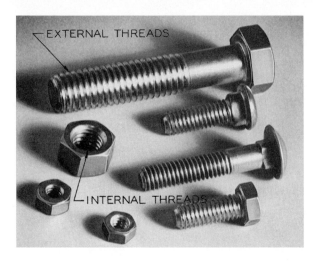

17.1 Examples of internal and external threads and three head types. (Courtesy of Russell, Burdsall and Ward Blot and Nut Company.)

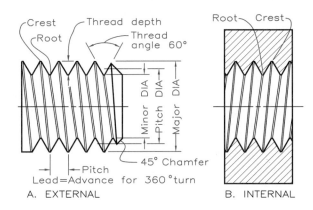

17.2 Most of the definitions of thread terminology are labeled for (A) external and (B) internal threads.

Internal thread: a thread cut on the inside of a part, such as a nut (Fig. 17.1).

Major diameter: the largest diameter on an internal or external thread (**Fig. 17.2**).

Minor diameter: the smallest diameter on an internal or external thread (Fig. 17.2).

Crest: the peak edge of a screw thread (Fig. 17.2).

Root: the bottom of the thread cut into a cylinder to form the minor diameter (Fig. 17.2).

Depth: the depth of the thread from the major diameter to the minor diameter; also measured as the root diameter (Fig. 17.2).

Thread angle: the angle between threads cut by the cutting tool, usually 60° (Fig. 17.2).

Pitch (thread width): the distance between crests of threads, found by dividing 1 inch by the number of threads per inch of a particular thread (Fig. 17.2).

Pitch diameter: the diameter of an imaginary cylinder passing through the threads at the points where the thread width is equal to the space between the threads (Fig. 17.2 and Fig. 17.4).

Lead (pronounced *leed*): the distance a screw will advance when turned 360°.

Form: the shape of the thread cut into a threaded part (**Fig. 17.3**).

Series: the number of threads per inch for a particular diameter, grouped into coarse, fine, extra fine, and eight constant-pitch thread series.

Class: the closeness of fit between two mating parts. Class 1 represents a loose fit and Class 3 a tight fit.

Right-hand thread: will assemble when turned clockwise. A right-hand external thread slopes downward to the right when its axis is horizontal and in the opposite direction on internal threads.

Left-hand thread: will assemble when turned counterclockwise. A left-hand external thread slopes downward to the left when its axis is horizontal and in the opposite direction on internal threads.

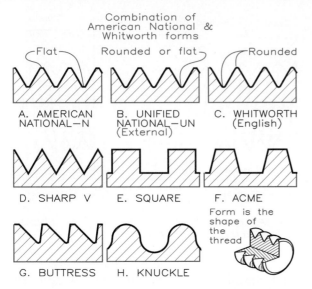

17.3 The various types of thread forms for external threads.

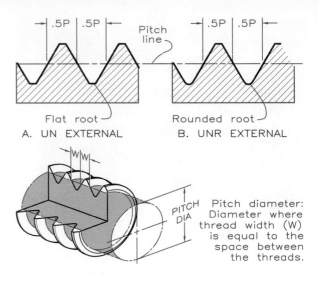

17.4 UN and UNR thread forms.

A The UN external thread has a flat root (a rounded root is optional) and a flat crest.

B The UNR thread has a rounded root formed by rolling rather than by cutting. The UNR form does not apply to internal threads.

17.3 English System Specifications

Form Thread form is the shape of the thread cut into a part (Fig. 17.3). The Unified National form, denoted by UN in thread notes, is the most widely used form in the United States. The American National form, denoted by N, appears occasionally on older drawings. The sharp V thread is used for set screws and in applications where friction in assembly is desired. Acme, square, and buttress threads are used in gearing and other machinery applications (Fig. 17.3).

The Unified National Rolled form (UNR) is used for external threads, never internal threads, because internal threads cannot be formed by rolling. The standard UN form has a flat root (a rounded root is optional) (**Fig. 17.4A**), and the UNR form (**Fig. 17.4B**) has a rounded root formed by rolling a cylinder across a die. The UNR form can be used instead of the UN form where precision of assembly is less critical.

Series The thread series designates the spacing of threads that vary with diameter. The

American National (N) and the Unified National (UN/UNR) forms are designated in three graded series: **coarse (C)**, **fine (F)**, and **extra fine (EF)**. They are also available in eight **constant-pitch series** (4, 6, 8, 12, 16, 20, 28, and 32 threads per inch).

Coarse Unified National forms are denoted UNC or UNRC, which is a combination of form and series designation. The coarse thread (UNC/UNRC or NC) has the largest pitch of any series and is suitable for bolts, screws, nuts, and general use with cast iron, soft metals, and plastics when rapid assembly is desired. An American National (N) form for a coarse thread is written NC.

Fine threads (NF or UNF/UNRF) are used for bolts, nuts, and screws when a high degree of tightening is required. Fine threads are closer together than coarse threads, and their pitch is graduated to be smaller on smaller diameters.

Extra-fine threads (UNEF/UNREF or NEF) are suitable for sheet metal screws and bolts, thin nuts, ferrules, and couplings when the length of engagement is limited and high stresses must be withstood.

Constant-pitch threads (4 UN, 6 UN, 8 UN, 12 UN, 16 UN, 20 UN, 28 UN, and 32 UN) are used on larger-diameter threads (beginning near the 1/2-inch size) and have the same pitch size regardless of the diameter size. The most commonly used constant-pitch threads are 8 UN, 12 UN, and 16 UN members of the series, which are used on threads of about 1 inch in diameter and larger. Constant-pitch threads may be specified as UNR or N thread forms. The ANSI table in Appendix 6 shows constant-pitch threads for larger thread diameters instead of graded pitches of coarse, fine, and extra fine.

Class of Fit The class of fit is the tightness between two mating threads as between a nut and bolt, and is indicated in the thread note by the numbers 1, 2, or 3 followed by the letters A or B. For UN forms, the letter A represents an external thread and the letter B represents an internal thread. The letters A and B do not appear in notes for the American National form (N).

   Class 1A and **1B** threads are used on parts that assemble with a minimum of binding and precision. **Class 2A** and **2B** threads are general-purpose threads for bolts, nuts, and screws used in general and mass-production applications. **Class 3A** and **3B** threads are used in precision assemblies where a close fit is required to withstand stresses and vibration.

Single and Multiple Threads A single thread (**Fig. 17.5A**) is a thread that advances the distance of its pitch in a revolution of 360°; that is, its pitch (P) is equal to its lead. The crest

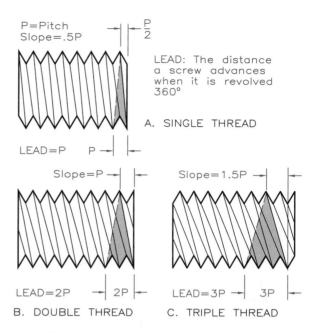

17.5 Threads can be (A) single, (B) double, or (C) triple. This represents the distance that each advances when turned 360°.

lines have a slope of 1/2 P since only 180° of the revolution is visible in the view.

   Multiple threads are used where quick assembly is required. A double thread is composed of two threads that advance a distance of 2P when turned 360° (**Fig. 17.5B**); that is, its lead is equal to 2P. The crest lines have a slope of P because only 180° of the revolution is visible in the view. A triple thread advances 3P in 360° with a crest line slope of 1 1/2 P in the view where 180° of the revolution is visible (**Fig. 17.5C**).

17.4 English Thread Notes

Drawings of threads are only symbolic representations and are inadequate to give the details of a thread unless accompanied by notes (**Fig. 17.6**). In a thread note, the major diameter is given first, then the number of threads per inch, the form, the series, the class of fit, and the letter A or B to denote external

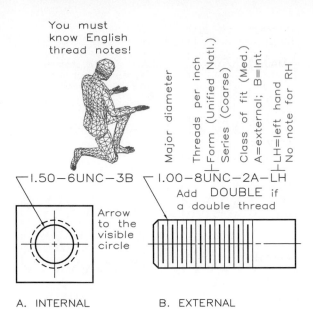

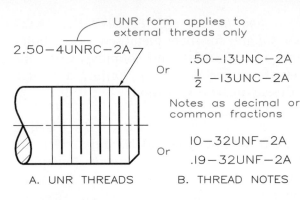

A. UNR THREADS

B. THREAD NOTES

17.7 Thread notes.

A The UNR thread note is applied to only external threads.

B Diameters in thread notes can be given as decimal fractions or common fractions.

A. INTERNAL

B. EXTERNAL

17.6 Thread notes of the English form that are applied to (A)internal and (B)external threads.

or internal threads, respectively. For a double or triple thread, include the word "double" or "triple" in the note, and for left-hand threads add the letters LH.

Figure 17.7 shows a UNR thread note for the external thread. (UNR does not apply to internal threads.) When inches are the units of measurement, fractions can be written as decimal or common fractions. The information for thread notes comes from ANSI tables in Appendix 5 and 6.

Using Thread Tables

A portion of the Appendix 5 is shown in **Fig. 17.8**, which gives the UN/UNR thread table from which specifications for standardized interchangeable threads can be selected. Note that a 1 1/2-inch-diameter bolt with fine thread (UNF) has 12 threads per inch, and its thread note is written as

1.500-12 UNF-2A   *or*   1-1/2-12 UNF-2A.

If the thread were internal (nut), the thread note would be the same but the letter B would

AMERICAN NATIONAL STANDARDS INSTITUTE
UNIFIED INCH SCREW THREADS (UN AND UNR)

Nominal Diameter	Basic Diameter	Coarse NC & UNC		Fine NF & UNF		Extra Fine NEF/UNEF	
		Thds per In.	Tap Drill DIA	Thds per In.	Tap Drill DIA	Thds per In.	Tap Drill DIA
1	1.000	8	.875	12	.922	20	.953
1-1/16	1.063	...	...	...	...	18	1.000
1-1/8	1.125	7	.904	12	1.046	18	1.070
1-3/16	1.188	...	...	...	...	18	1.141
1-1/4	1.250	7	1.109	12	1.172	18	1.188
1-5/16	1.313	...	...	...	...	18	1.266
1-3/8	1.375	6	1.219	12	1.297	18	1.313
1-7/16	1.438	...	...	...	...	18	1.375
1-1/2	1.500	6	1.344	12	1.422	18	1.438

17.8 This is a portion of the ANSI tables for UN and UNR threads, presented in Appendix 5.

be used instead of the letter A. For constant-pitch thread series, selected for larger diameters, write the thread note as

1.750-12 UN-2A   *or*   1-3/4-12 UN-2A.

For the UNR thread form (for external threads only), substitute UNR for UN in the last three columns—for example, UNREF for UNEF (extra fine). **Figure 17.9** shows the preferred placement of thread notes (with leaders) for external and internal threads.

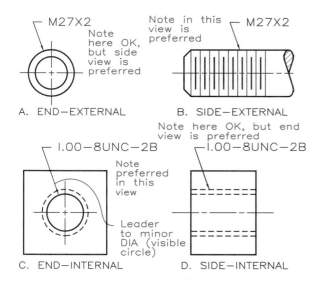

A. END—EXTERNAL   B. SIDE—EXTERNAL

C. END—INTERNAL   D. SIDE—INTERNAL

17.9 Notes for external threads are best if they are placed on the rectangular view of the threads. Notes for internal threads are best in the circular view, if space permits.

17.5 Metric Thread Notes

Metric thread notes usually given as a basic designation are suitable for general applications. However, for applications where the assembly of threaded parts is crucial, a complete thread designation should be noted. The ISO thread table in Appendix 7, a portion of which is shown in **Fig. 17.10**, contains specifications for metric thread notes.

Basic Designation **Figure 17.11** shows examples of metric screw thread notes. Each note begins with the letter M, designating the note as metric, followed by the major diameter size in millimeters and the pitch in millimeters separated by the multiplication sign and the number 3.

Complete Designation The first part of a complete designation note (**Fig. 17.12**) is identical to the basic designation, to which is added the tolerance class designation separated by a dash. (A tolerance is a specified

COARSE		FINE	
MAJ. DIA & THD PITCH	TAP DRILL	MAJ. DIA & THD PITCH	TAP DRILL
M20 X 2.5	17.5	M20 X 1.5	18.5
M22 X 2.5	19.5	M22 X 1.5	20.5
M24 X 3	21.0	M24 X 2	22.0
M27 X 3	24.0	M27 X 2	25.0
M30 X 3.5	26.5	M30 X 2	28.0
M33 X 3.5	29.5	M33 X 2	31.0
M36 X 4	32.0	M36 X 2	33.0
M39 X 4	35.0	M39 X 2	36.0
M42 X 4.5	37.5	M42 X 2	39.0

17.10 This portion of the ISO thread table in Appendix 7 shows specifications for metric thread notes.

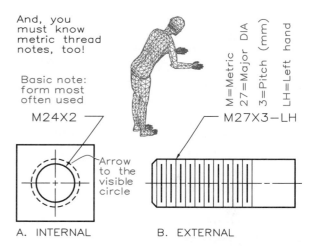

A. INTERNAL   B. EXTERNAL

17.11 This is the basic thread note that will be used for (A) internal and (B) external threads.

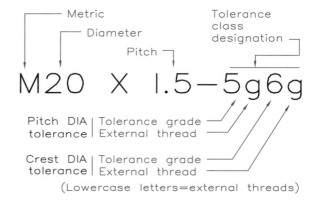

17.12 A complete designation note for metric threads adds tolerance specifications to the basic note.

External Thread			Internal Thread	
Major Diameter	Pitch Diameter		Minor Diameter	Pitch Diameter
—	3	FINE	—	—
4	4		4	4
—	5		5	5
6	6	MEDIUM	6	6
—	7		7	7
8	8	COARSE	8	8
—	9		—	—

17.13 Tolerance grades for major and minor diameters and pitch diameters of each range from fine to coarse. Tolerance grades combined with position symbols (such as 6g) become tolerance classes.

EXTERNAL THREADS (Lowercase letters)	INTERNAL THREADS (Uppercase letters)
e =Large allowance	G =Small allowance
g =Small allowance	H =No allowance
h =No allowance	

LENGTH OF ENGAGEMENT

S =Short   N =Normal   L =Long

EXAMPLE: Refer to Figure 17.13
Position   Lower—case=external threads
                Upper—case=internal threads
Allowance e, g, h, G, H=amount of allowance
Engagement S, N, & L Columns of Fig. 17.13

17.14 These symbols represent position, allowance, and engagement length. Position means either external or interior threads. A lowercase letter indicates an external thread, and an uppercase letter signifies an internal thread.

maximum variation in size that ensures assembly of the threaded parts.) The 5g represents the pitch diameter tolerance, and the 6g represents the crest diameter tolerance.

The numbers 5 and 6 are thread tolerance grades (variations in size from the basic diameter) from **Fig. 17.13**. These grades are for the pitch diameter and the major and minor diameters of medium general-purpose thread similar to class 2A and 2B threads in the UN system. Grades of less than 6 are best suited to fine-series fits and short lengths of engagement. Grades greater than 6 are best suited to coarse-series fits and long lengths of engagement.

The letters following the grade numbers denote tolerance positions, which means either external or internal threads. Lowercase letters designate external threads (bolts), as shown in **Fig. 17.14**. The letters e, g, and h represent large allowance, small allowance, and no allowance, respectively. (Allowance is the permitted variation in size from the basic diameter.) Uppercase letters designate internal threads (nuts). The letters G and H, placed after the tolerance grade number, denote small allowance and no allowance, respectively. For example, 5g designates a medium

tolerance with small allowance for the pitch diameter of an external thread, and 6H designates a medium tolerance with no allowance for the minor diameter of an internal thread.

Tolerance classes are designated by **fine**, **medium**, and **coarse**, as shown in **Fig. 17.15**. They represent combinations of tolerance grades, tolerance positions, and lengths of engagement—**short** (**S**), **normal** (**N**), and **long** (**L**). A table of lengths of engagement (the actual length of the assembled thread in mating parts) is given in the ANSI Standards but not in this text.

After deciding whether to use a fine, medium, or coarse class of fit, select the thread designation first from those in boxes (commercial threads), second from those in bold print, third from those in medium-size print, and fourth from those in small print. The 6H class is comparable to the 2B class of fit in the UN system for an interior thread, and the 6g class is similar to the 2A class of fit.

Figure 17.16 shows variations for complete thread notes. When the minor and pitch diameters have identical grades, the tolerance class symbol consists of one number and one letter, such as 6H (**Fig. 17.16A**). The uppercase H indicates that the position

Quality	External Threads (bolts)									Internal Threads (nuts)					
	Tolerance position e (large allowance)			Tolerance position g (small allowance)			Tolerance position h (no allowance)			Tolerance position G (small allowance)			Tolerance position H (no allowance)		
	Length of engagement			Length of engagement			Length of engagement			Length of engagement			Length of engagement		
	Group S	Group N	Group L	Group S	Group N	Group L	Group S	Group N	Group L	Group S	Group N	Group L	Group S	Group N	Group L
Fine							3h4h	4h	5h4h				4H	5H	6H
Medium		6e	7e6e	5g6g	6g	7g6g	5h6h	6h	7h6h	5G	6G	7G	5H	6H	7H
Coarse					8g	9g8g					7G	8G		7H	8H

*In selecting tolerance class, select first from the commercial classes in boxes, second from the bold print, third from the medium-size print, and fourth from the small-size print.

17.15 Tolerance classes for large and small allowances and for no allowance for internal and external threads are shown here. Select the most commonly used tolerance class, commercial threads (in boxes) first, then the classes in bold, medium-, and small-size print, respectively.

If pitch and crest DIA tolerances are equal, use only one tolerance symbol

Letters S, N, or L are used to indicate length of thread engagement

M22X1.5−6H   M24X3−7g6gL

A. NOTE WITH EQUAL TOLERANCE GRADES

B. LENGTH OF THREAD ENGAGEMENT INDICATED

17.16 Complete metric notes.

A When both pitch and major diameter tolerance grades are the same, show the tolerance class symbol only once.

B The letters S, N, and L indicate the length of the thread engagement.

g=External thds. and small allow.
6=Toler. grade

H=Internal thds. and no allowance
6=Toler. grade

g=External thds. and small allow. 6=Tol. grade for Crest DIA

g=External thds. and small allow. 5=Tol. grade for Pitch DIA

M6XI−6H/6g   M20X2−6H/5g6g

A.   B.

17.17 A slash mark is used to separate the tolerance class designations of mating internal and external threads.

is internal and no allowance is added to the basic thread note. If necessary, add the length of engagement symbol (S, N, L) to the tolerance class designation (**Fig. 17.16B**). For unknown lengths of thread engagement, use group N (normal).

Specify the fit between mating threads with notes as shown in **Fig. 17.17** with a slash separating the tolerance class designations of internal and external threads. Additional information about ISO threads may be obtained from *ISO Metric Screw Threads*, a booklet of standards published by ANSI that was used as the basis for most of this section.

17.6 Drawing Threads

Threads may be represented by **detailed**, **schematic**, and **simplified** symbols (**Fig. 17.18**). Detailed symbols represent a thread most realistically, simplified symbols represent a thread least realistically, and schematic symbols are a compromise between the two. Detailed and schematic symbols can be used for drawing larger threads on a drawing (1/2 inch in diameter and larger), and simplified symbols are best for smaller threads.

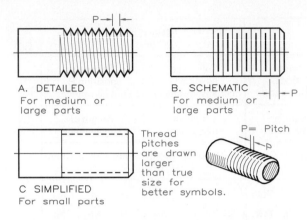

A. DETAILED
For medium or large parts

B. SCHEMATIC
For medium or large parts

C. SIMPLIFIED
For small parts

P = Pitch

Thread pitches are drawn larger than true size for better symbols.

17.18 The three types of thread symbols used for drawing threads are (A) detailed, (B) schematic, and (C) simplified.

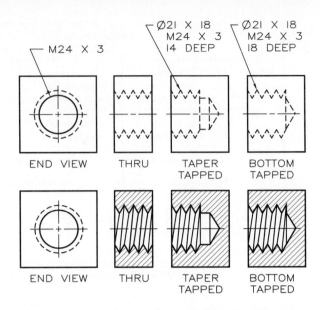

M24 X 3

Ø21 X 18
M24 X 3
14 DEEP

Ø21 X 18
M24 X 3
18 DEEP

END VIEW · THRU · TAPER TAPPED · BOTTOM TAPPED

END VIEW · THRU · TAPER TAPPED · BOTTOM TAPPED

17.20 These detailed symbols represent internal threads. Approximate the minor diameter as 75 percent of the major diameter. Tap drill diameters (21) are found in Appendix 7.

17.7 Detailed Symbols

UN/UNR–Threads Detailed thread symbols for external threads in view and in section are shown in **Fig. 17.19**. Instead of drawing helical curves, straight lines are used to depict crest and root lines. Variations of detailed thread symbols for internal threads drawn in views and sections are shown in **Fig. 17.20**.

The steps for drawing a detailed thread representation, whether English or metric threads, are shown in **Fig. 17.21**. When using the English tables, calculate the pitch by dividing 1 inch by the number of threads per inch. In the metric system, pitch is given in the tables. Draw the spacing between crest lines larger than the actual pitch size to avoid "clogged-up" lines.

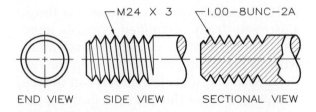

M24 X 3 · 1.00–8UNC–2A

END VIEW · SIDE VIEW · SECTIONAL VIEW

17.19 These detailed symbols represent external threads in view and section.

Computer Method You can draw detailed thread symbols by computer (**Fig. 17.22**) and duplicate them with the COPY commands and the MULTIPLE option. The program produces a typical set of threads in Step 1 and then copies it repetitively in Step 2.

Square Threads The steps needed to draw and note a detailed drawing of a square thread are shown in **Fig. 17.23**. Follow the same basic steps to draw views and sections of square internal threads (**Fig. 17.24**). In section, draw both the internal crest and root lines, but in a view draw only the outline of the threads. Place thread notes for internal threads in the circular view, whenever possible, with the leader pointing toward the center and stopping at the visible circle.

When a square thread is long, it can be represented by using phantom lines, without drawing all the threads (**Fig. 17.25**). This

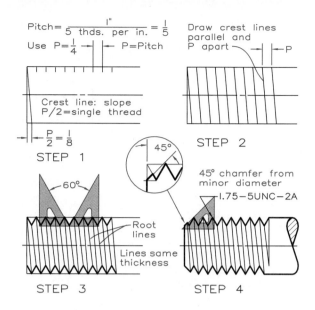

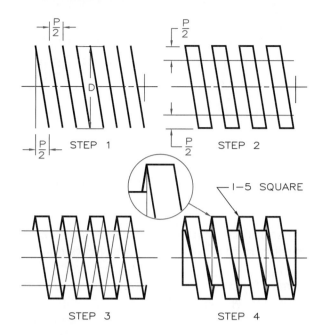

17.21 Detailed representation of threads.

Step 1 To draw a detailed drawing of a 1.75–5 UNC–2A thread, find the pitch by dividing 1 inch by the number of threads per inch, or 5 in this case. Use a pitch of 1/4 instead of 1/5 to space the threads apart. Lay off the pitch along the length of the thread, and draw a crest line at a slope of P/2, or 1/8 inch in this case.

Step 2 Draw the other crest lines as dark, visible lines parallel to the first crest line.

Step 3 Find the root lines by constructing 60° vees between the crest lines. Draw the root lines from the bottom of the vees. Root lines are parallel to each other but not to crest lines.

Step 4 Construct a 45° chamfer at the end of the thread from the minor diameter. Darken all lines, and add a thread note.

17.23 Drawing a square thread.

Step 1 Lay out the major diameter. Space the crest lines 1/2P apart and slope them downward to the right for right-hand threads.

Step 2 Connect every other pair of crest lines. Find the minor diameter by measuring 1/2P inward from the major diameter.

Step 3 Connect the opposite crest lines with light construction lines to establish the profile of the thread form.

Step 4 Connect the inside crest lines with light construction lines to locate the points on the minor diameter where the thread wraps around the minor diameter. Darken the final lines.

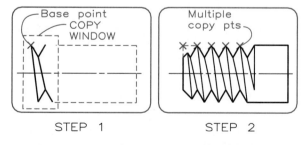

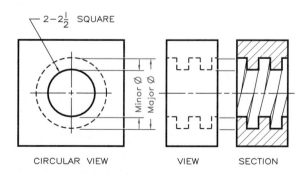

17.22 Detailed thread symbols by computer.

Step 1 Draw a typical detailed thread symbol at the end of the screw.

Step 2 Duplicate a typical set of threads with the `COPY` command and the `MULTIPLE` option along the predetermined snap points of the screw.

17.24 Square internal threads in view and section.

17.7 DETAILED SYMBOLS • 255

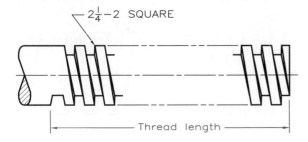

17.25 The conventional method of showing square threads is to draw sample threads at each end and to connect them with phantom lines.

conventional practice saves time and effort without reducing the drawing's effectiveness.

Acme Threads A modified version of the square thread is the Acme thread, which has tapered (15°) sides for easier engagement than square threads. The steps involved in drawing detailed Acme threads are shown in **Fig. 17.26.** Acme threads are heavy threads that are used to transmit force and power in mechanisms such as screw jacks, leveling devices, and lathes. Appendix 8 contains the table for Acme thread specifications and dimensions.

Internal Acme threads are shown in view and section in **Fig. 17.27.** Left-hand internal threads in section appear the same as right-hand external threads.

17.8 Schematic Symbols

Figure 17.28 shows schematic representations of external threads with metric notes. Because schematic symbols are easy to draw and adequately represent threads, it is the thread symbol used most often for medium-size threads. Draw schematic thread symbols by using thin parallel crest lines and thick root lines. Schematic drawings of left-hand and right-hand threads are identical; only the LH in the thread note indicates that a thread is left-handed. Right-hand threads are not marked RH but are understood to be right-hand threads.

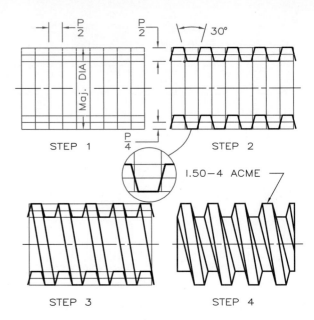

17.26 Drawing the Acme thread.

Step 1 Lay out the major diameter and thread length and divide the shaft into equal divisions 1/2P apart. Locate the minor and pitch diameters by using distances 1/2P and 1/4 P.

Step 2 Draw construction lines at 15° angles with the vertical along the pitch diameter to make a total angle of 30°.

Step 3 Draw the crest lines across the screw.

Step 4 Darken the lines, draw the root lines, and add the thread note to complete the drawing.

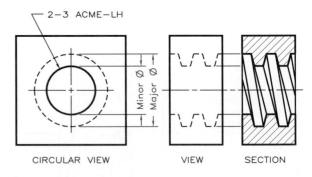

17.27 Internal Acme threads in view and section.

Figure 17.29 shows threaded holes in view and in section drawn with schematic symbols. The size of the tap drill diameter is approximately equal to the major diameter minus the pitch. However, the minor diameter is usually

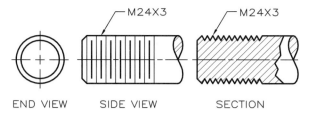

17.28 External threads in view and section.

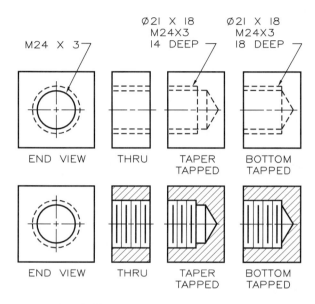

17.29 Threaded holes in view and section. Tap drill diameters are found in Appendix 7.

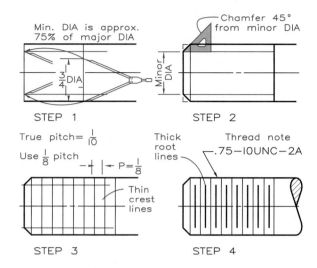

17.30 Schematic representation of threads.

Step 1 Lay out the major diameter and locate the minor diameter (about three-quarters of the major diameter). Draw the minor diameter with light construction lines.

Step 2 Chamfer the end of the threads with a 45° angle from the minor diameter.

Step 3 Find the pitch of a .75–10UNC–2A thread (0.1) by dividing 1 inch by the number of threads per inch (10). Use a larger pitch, 1/8 inch in this case, for spacing the thin crest lines.

Step 4 Draw root lines as thick as the visible lines between the crest lines to the construction lines representing the minor diameter. Add a thread note.

drawn a bit smaller to provide better separation between the lines representing the major and minor diameters.

Figure 17.30 outlines the steps needed to draw schematic threads using English specifications. Draw the minor diameter at approximately three-quarters of the major diameter and the chamfer (bevel) 45° from the minor diameter. Draw crest lines as thin lines and root lines as thick visible lines.

Computer Method You may draw schematic thread symbols by computer (**Fig. 17.31**) and duplicate them with the COPY command's MULTIPLE option. The program produces a typical set of threads in Step 1 and copies it repetitively in Step 2.

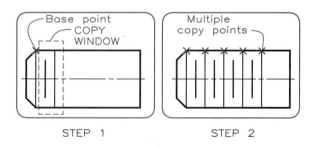

17.31 Schematic threads by computer.

Step 1 Draw the outline of the threaded shaft having a chamfered end. Produce typical minor and major diameters and window them for a MULTIPLE COPY command.

Step 2 Repetitively COPY the threads along the screw at the predetermined snap points.

17.8 SCHEMATIC SYMBOLS • 257

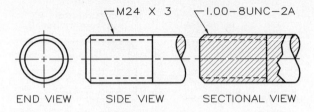

END VIEW | SIDE VIEW | SECTIONAL VIEW

17.32 These simplified thread symbols represent external threads in view and section.

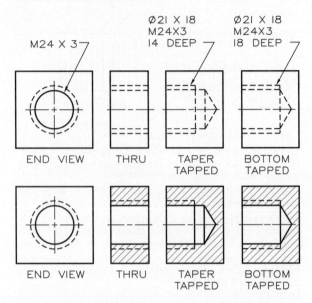

M24 X 3

Ø21 X 18
M24X3
14 DEEP

Ø21 X 18
M24X3
18 DEEP

END VIEW | THRU | TAPER TAPPED | BOTTOM TAPPED

END VIEW | THRU | TAPER TAPPED | BOTTOM TAPPED

17.33 These simplified thread symbols represent internal threads in view and section. Draw minor diameters at about 3/4 the major diameter.

17.9 Simplified Symbols

Examples of external threads drawn with simplified symbols and noted with both metric and English formats are shown in **Fig. 17.32**. Simplified symbols are the easiest to draw and are the best suited for drawing small threads where drawing using schematic and detailed symbols would be too crowded. Various techniques of applying simplified symbols to threads are shown in **Fig. 17.33**. The minor diameter is drawn as hidden lines spaced at about three-quarters the major diameter. **Figure 17.34** shows the steps of drawing simplified threads. With experience, you will be able to approximate the location of the minor diameter of simplified threads by eye.

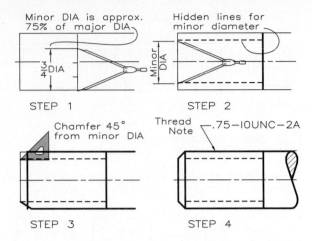

Minor DIA is approx. 75% of major DIA

Hidden lines for minor diameter

STEP 1 | STEP 2

Chamfer 45° from minor DIA

Thread Note   .75−10UNC−2A

STEP 3 | STEP 4

17.34 A simplified representation of threads.

Step 1 Lay out the major diameter. Locate the minor diameter (about three-quarters of the major diameter).

Step 2 Draw hidden lines to represent the minor diameter.

Step 3 Draw a 45° chamfer from the minor diameter to the major diameter.

Step 4 Darken the lines and add a thread note.

Drawing Small Threads

When drawing a thread, it is important to keep in mind that you are producing a pictorial symbol, and therefore you do not need to draw the true spacing of the threads. True spacing will be too close and too hard to read. Instead, select a wider spacing between root and crest lines. Your drawing will be easier to draw and to read (**Fig. 17.35**). This conventional practice of enlarging the thread's pitch to separate thread symbols is applied in drawing all three types of symbols (simplified, schematic, and detailed). Add a thread note to the drawing to give the necessary detailed specifications.

17.10 Nuts and Bolts

Nuts and bolts (**Fig. 17.36**) come in many forms and sizes for many different applications. Some common types of threaded fasteners are shown in **Fig. 17.37**. A **bolt** is a

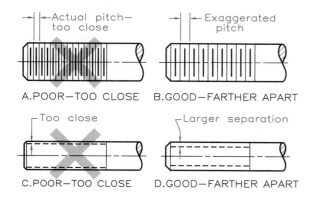

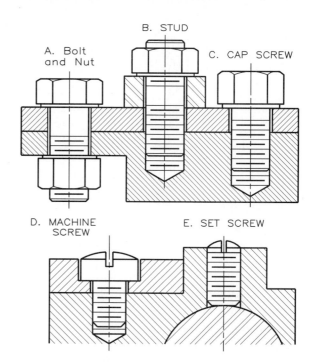

B. STUD

A. Bolt and Nut

C. CAP SCREW

D. MACHINE SCREW

E. SET SCREW

17.37 Some common types of threaded bolts.

A.POOR—TOO CLOSE

B.GOOD—FARTHER APART

C.POOR—TOO CLOSE

D.GOOD—FARTHER APART

17.35 Most threads must be drawn using exaggerated dimensions instead of actual measurements to prevent the drawing from having lines drawn too closely together.

17.36 Nuts, bolts, and washers can be used in combination. (Courtesy of Lamson and Sessions.)

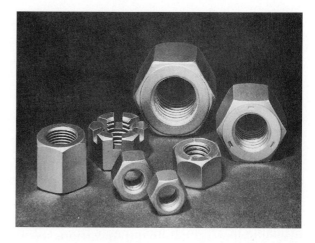

17.38 Some examples of nuts. (Courtesy of Russell, Burdsall and Ward Bolt and Nut Company.)

threaded cylinder with a head, and it is used with a **nut** to hold parts together. A **stud** is a headless bolt, threaded at both ends, that is screwed into one part with a nut attached to the other end.

A **cap screw** usually does not have a nut but passes through a hole in one part and screws into another threaded part. A hexagon-head **machine screw** is similar to but smaller than a cap screw. Machine screws also come with other types of heads. A **set screw** is used to hold one part fixed in place with another, usually to prevent rotation, as with a pulley on a shaft.

Types of heads used on regular and heavy bolts and nuts are shown in **Fig. 17.38** and **Fig. 17.39**. Heavy bolts have thicker heads than regular bolts for heavier usage. A **finished**

head (or nut) has a 1/64-inch-thick washer face (a circular boss) to provide a bearing surface for smooth contact. **Semifinished bolt heads** and **nuts** are the same as finished bolt heads and nuts. Unfinished bolt heads

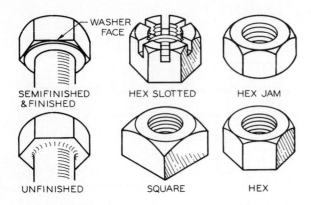

17.39 Finished and semifinished bolts and nuts with raised washer faces. Several types of nuts are shown here also.

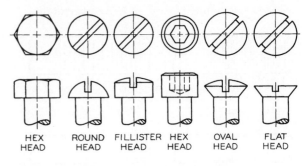

17.40 Standard types of bolt and screw heads.

and nuts have no bosses and no machined surfaces.

A **hexagon jam nut** does not have a washer face, but it is chamfered (beveled at its corners) on both sides. **Figure 17.40** shows other standard bolt and screw heads for cap screws and machine screws.

Dimensions

Figure 17.41 shows a properly dimensioned bolt. The ANSI tables in Appendixes 10–13 give nut and bolt dimensions, but you may use the following guides for hexagon-head and square-head bolts.

Overall Lengths Hexagon-head bolts are available in 1/4-inch increments up to 8 inches long, in 1/2-inch increments from 8 to 20 inches long, and in 1-inch increments from 20 to 30 inches long. Square-head bolts are available in 1/8-inch increments from 1/2 to 3/4 inch long, in 1/4-inch increments from 3/4 inch to 5 inches long, in 1/2-inch increments from 5 to 12 inches long, and in 1-inch increments from 12 to 30 inches long.

Thread-Lengths For both hexagon-head and square-head bolts up to 6 inches long,

$$\text{Thread length} = 2D + 1/4 \text{ inch.}$$

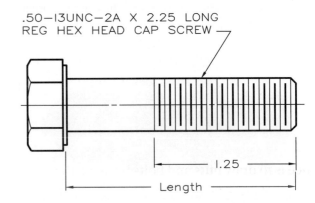

17.41 A properly dimensioned and noted hexagon-head bolt.

where D is the diameter of the bolt. For bolts more than 6 inches long,

$$\text{Thread length} = 2D + 1/2 \text{ inch.}$$

Threads for bolts can be coarse, fine, or 8-pitch threads. The class of fit for bolts and nuts is understood to be 2A and 2B if no class is specified in the note.

Dimension Notes

Designate standard square-head and hexagon-head bolts by notes in one of three forms:

3/8–16 × 3 1-1/2 SQUARE BOLT—STEEL;

1/2–13 × 3 HEX CAP SCREW—SAE GRADE 8—STEEL;

.75 × 3 5.00 UNC–2A HEX HD LAG SCREW.

The numbers (left to right) represent bolt diameter, threads per inch (omit for lag screws), bolt length, screw name, and material (material designation is optional). When not specified in a note, each bolt is assumed to have a class 2 fit. Three types of notes for designating nuts are:

1/2–13 SQUARE NUT—STEEL;

3/4–16 HEAVY HEX NUT;

1.00–8UNC–2B HEX HD THICK SLOTTED NUT—CORROSION-RESISTANT STEEL.

When nuts are not specified as heavy, they are assumed to be regular. When the class of fit is not specified in a note, it is assumed to be 2B for nuts.

17.11 Drawing Square Heads

Appendix 10 gives dimensions for square bolt heads and nuts. However, conventional practice is to draw nuts and bolts by using the general proportions shown in **Fig. 17.42.** Your first step in drawing a bolt head or nut is to determine whether the view is to be across corners or across flats—that is, whether the lines at either side of the view represent the square's corners or flats. Drawing across corners shows nuts and bolts best, but occasionally you must draw one across flats when the head or nut is truly in this orientation.

17.12 Drawing Hexagon Heads

Figure 17.43 shows the steps of drawing the head of a hexagon bolt across corners by using the bolt's major diameter, DIA, as the basis for all other proportions. Begin by drawing the top view of the head as a circle of a diameter of 1-1/2 DIA. For a regular head, the thickness is 2/3 DIA, and for a heavy head it is 7/8 DIA. Circumscribe a hexagon about the circle. Then draw outside arcs in the rectangular view and tangent chamfers (bevels) to complete the drawing.

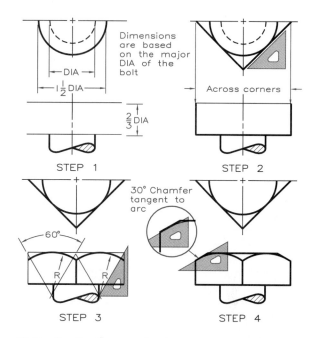

17.42 Drawing the square head.

Step 1 Draw the major diameter, DIA, of the bolt. Use 1.5 DIA to draw the hexagon-head's diameter and 2/3 DIA to establish its thickness.

Step 2 Draw the top view of the square head at a 45° angle to give an across-corners view.

Step 3 Show the chamfer in the front view by using a 30°–60° triangle to find the centers for the radii.

Step 4 Show a 30° chamfer tangent to the arcs in the front view. Darken the lines.

Computer Method You may draw a hexagon head for a bolt by computer, as shown in Step 1 of **Fig. 17.44.** Use a bolt diameter of 1 inch for easy scaling. You may then scale and rotate the block as desired when you use the INSERT command. Step 2 shows the drawing BLOCKed and inserted at scales of 50 percent and 75 percent. To insert a thread with a diameter of 0.50 inch, assign a size factor of 0.50 when prompted by the INSERT and BLOCK commands.

Drawing Nuts

Use the same techniques to draw a square and a hexagon nut (shown across corners in **Fig. 17.45**) that you did to draw bolt heads. The

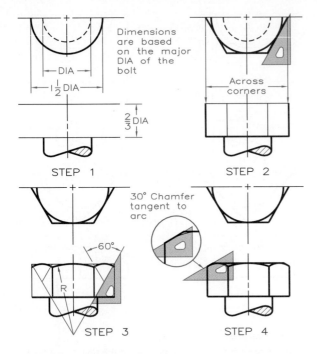

STEP 1 | STEP 2

Dimensions are based on the major DIA of the bolt

STEP 3 | STEP 4

30° Chamfer tangent to arc

17.43 Drawing the hexagon head.

Step 1 Draw the major diameter, DIA, of the bolt and use it to establish the head diameter (as 1.5 DIA) and thickness (as 2/3 DIA).

Step 2 Construct a hexagon head with a 30°–60° triangle to give an across-corners view.

Step 3 Find arcs in the front view to draw the chamfer of the head.

Step 4 Draw a 30° chamfer tangent to the arcs in the front view. Darken the lines.

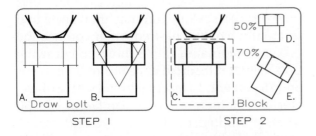

STEP 1 | STEP 2

17.44 A bolt head representation drawn by computer.

Step 1 Draw a hexagon head using the steps of geometric construction in Fig. 17.43. Base the size of the head on a bolt diameter of 1 inch to form a UNIT BLOCK for easy scaling.

Step 2 Convert the drawing into a BLOCK by using a window. INSERT the block at any size and position and scale with the desired factor as shown at D and E.

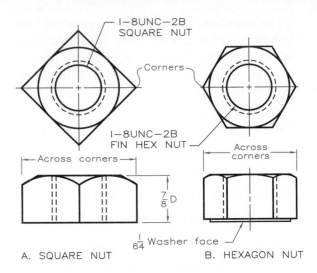

A. SQUARE NUT | B. HEXAGON NUT

17.45 Drawing square and hexagon nuts across corners involves the same steps used for drawing bolt heads. Add notes to give nut specifications.

difference is that nuts are thicker than bolt heads: The thickness of a regular nut is 7/8 D, and the thickness of a heavy nut is 1 D, where D is the bolt diameter. Hidden lines may be inserted in the front view to indicate threads, or omitted. Exaggerate the thickness of the 1/64-inch washer face on the finished and semifinished hexagon nuts to about 1/32 inch to make it more noticeable. Place thread notes on circular views with leaders when space permits. Square nuts that are not labeled heavy are assumed to be regular nuts.

Figure 17.46 shows how to construct square and hexagon nuts across flats. For regular nuts, the distance across flats is 1-1/2 × D (D = major diameter of the thread), and 1-5/8 D for heavy nuts. Draw the top views in the same way you did across-corner top views, but rotate them to give across-flat front views. **Figure 17.47** depicts dimensioned regular and heavy hexagon nuts drawn across corners, with notes added.

Drawing Nut and Bolt Combinations

Nuts and bolts in assembly are drawn in the same manner that they are drawn individually

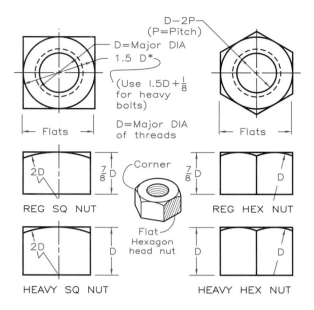

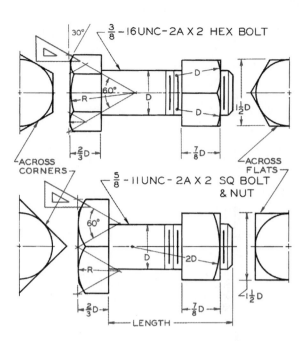

17.46 These square and hexagon nuts are drawn across flats with notes added to give their specifications. The square nuts are unfinished.

17.48 The proportions and the geometry for drawing square nuts and bolts and hexagon nuts and bolts are shown here.

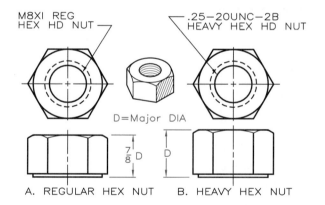

17.47 These regular and heavy hexagon nuts are drawn across corners, with notes added to give their specifications.

(**Fig. 17.48**). Use the major diameter, D, of the bolt as the basis for other dimensions. Here, the views of the bolt heads are across corners, and the views of the nuts are across flats, although both views could have been drawn across corners. The half end views are used to find the front views by projection. Add a note to give the specifications of the nut and bolt.

17.13 Types of Screws

Cap Screws
The cap screw passes through a hole in one part and screws into a threaded hole in the other part so that the two parts can be held together without a nut. Cap screws are usually larger than machine screws, and they may also be used with nuts. **Figure 17.49** shows the standard types of cap screw heads drawn on a grid that can be used as a guide for drawing cap screws of other sizes. Appendixes 14–18 gives cap screw dimensions, which can aid in drawing them.

Machine Screws
Smaller than most cap screws, **machine screws** usually are less than 1 inch in diameter. They screw into a threaded hole in a part or into a nut. Machine screws are fully threaded when their length is 2 inches or less. Longer screws have thread lengths of 2D + 1/4

17.13 TYPES OF SCREWS • 263

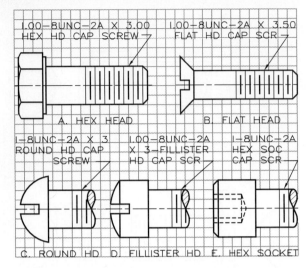

17.49 These cap screws are drawn on a grid to give the proportions for drawing them at different sizes. Notes give thread specifications, length, head type, and bolt name (cap screw).

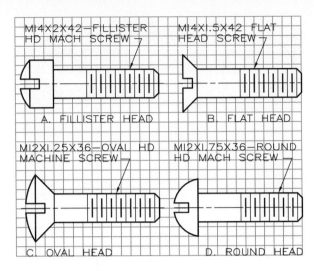

17.50 The four standard types of machine screws. The same proportions may be used to draw machine screws of all sizes.

inch (D = major diameter of the thread). **Figure 17.50** shows four types of machine screws, along with notes, drawn on a grid that may be used as an aid in drawing them without dimensions from a table. Machine screws range in diameter from No. 0 (0.060 inch) to 3/4 inch, as shown in Appendix 19, which gives the dimensions of round-head machine screws.

Set Screws

Set screws are used to hold parts together, such as pulleys and handles on a shaft, and to prevent rotation. **Figure 17.51** shows various types of set screws, with dimensions denoted by letters that correspond to the tables of dimensions in Appendix 20.

Set screws are available in combinations of points and heads. The shaft against which the set screw is tightened may have a machined flat surface to provide a good bearing surface for a **dog** or **flat-point** set screw end to press against. The **cup point** gives good gripping when pressed against round shafts. The **cone point** works best when inserted into holes

drilled in the part being held. The **headless set screw** has no head to protrude above a rotating part. An exterior square head is good for applications in which greater force must be applied with a wrench to hold larger set screws in position.

Wood Screws

A wood screw is a pointed screw with sharp coarse threads that will screw into wood while making its own internal threads in the process. **Figure 17.52** shows the three most common types of wood screws drawn on a grid to show their relative proportions.

Sizes of wood screws are specified by single numbers, such as 0, 6, or 16. From 0 to 10, each digit represents a different size. Beginning at 10, only even-numbered sizes are standard—that is, 10, 12, 14, 16, 18, 20, 22, and 24. Use the following formula to translate these numbers into the actual diameter sizes:

Actual DIA = 0.06 + (screw number × 0.013).

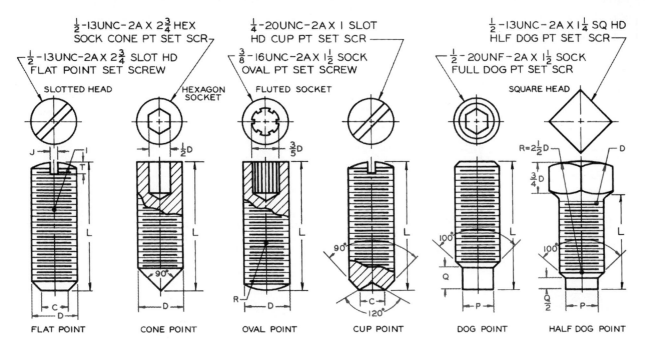

17.51 Set screws are available with various combinations of heads and points. Their measurements are given in notes. (See Appendix 20.)

For example, the diameter for the No. 7 wood screw shown in Fig. 17.52 is calculated as follows:

$$DIA = 0.06 + 7(0.013) = 0.151.$$

17.14 Other Threaded Fasteners

Only the more standard types of nuts and bolts are covered in this chapter. **Figure 17.53** illustrates a few of the many other types of threaded fasteners that have their own special applications. Three types of wing screws that are turned by hand are available in incremental lengths of 1/8 inch (**Fig. 17.54**). **Figure 17.55** shows two types of thumb screws, which serve the same purpose as wing screws, and **Fig. 17.56** shows wing nuts that can be screwed together by fingertip without wrenches or screwdrivers.

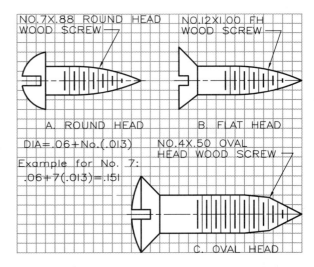

17.52 The three most common types of wood screws, drawn on a grid that gives the proportions for drawing them at other sizes.

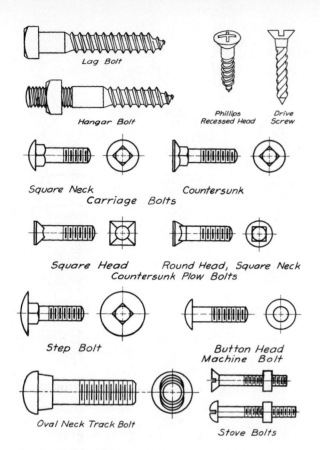

Lag Bolt

Hangar Bolt

Phillips Recessed Head   *Drive Screw*

Square Neck   *Countersunk*
Carriage Bolts

Square Head   *Round Head, Square Neck*
Countersunk Plow Bolts

Step Bolt   *Button Head Machine Bolt*

Oval Neck Track Bolt   *Stove Bolts*

17.53 A few of the other different types of bolts and screws, each with their special applications.

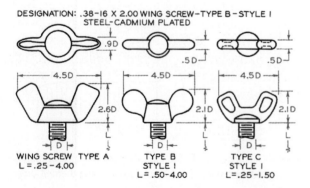

DESIGNATION: .38-16 X 2.00 WING SCREW-TYPE B-STYLE I
STEEL-CADMIUM PLATED

WING SCREW TYPE A
L = .25-4.00

TYPE B
STYLE I
L = .50-4.00

TYPE C
STYLE I
L=.25-1.50

17.54 These wing screw proportions are for screw diameters of about 5/16 inch. The same proportions may be used to draw wing screws of any diameter. Type A screws are available in diameters of 4, 6, 8, 10, 12, 0.25", 0.313", 0.375", 0.438", 0.50", and 0.625". Type B screws are available in diameters of 10 to 0.625". Type C screws are available in diameters of 6 to 0.375".

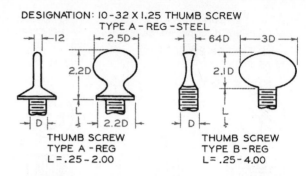

DESIGNATION: 10-32 X 1.25 THUMB SCREW
TYPE A - REG - STEEL

THUMB SCREW
TYPE A - REG
L = .25 - 2.00

THUMB SCREW
TYPE B - REG
L = .25 - 4.00

17.55 These thumb screw proportions are for screw diameters of about 1/4 inch. The same proportions may be used to draw thumb screws of any diameter. Type A screws are available in diameters of 6, 8, 10, 12, 0.25", 0.313", and 0.375". Type B thumb screws are available in diameters of 6 to 0.50".

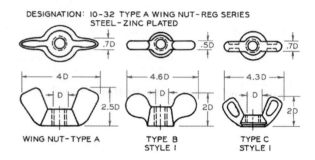

DESIGNATION: 10-32 TYPE A WING NUT-REG SERIES
STEEL - ZINC PLATED

WING NUT-TYPE A

TYPE B
STYLE I

TYPE C
STYLE I

17.56 These wing nut proportions are for screw diameters of 3/8 inch. The same proportions may be used to draw thumb screws of any size. Type A wing nuts are available in screw diameters of 3, 4, 5, 6, 8, 10, 12, 0.25", 0.313", 0.375", 0.438", 0.50", 0.583", 0.625", and 0.75". Type B nuts are available in sizes from 5 to 0.75". Type C nuts are available in sizes from 4 to 0.50".

17.15 Tapping a Hole

An internal thread is made by drilling a hole with a tap drill with a 120° point (**Fig. 17.57**). The depth of the drilled hole is measured to the shoulder of the conical point, not to the point. The diameter of the drilled hole is approximately equal to the root diameter, calculated as the major diameter of the screw thread minus its pitch (Appendixes 5 and 6). The hole is **tapped**, or threaded, with a **tool**.

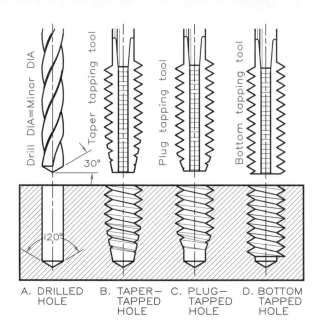

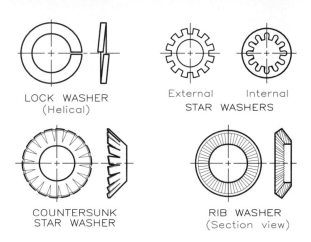

17.58 Lock washers are used to keep threaded parts from vibrating apart.

A. DRILLED HOLE   B. TAPER–TAPPED HOLE   C. PLUG–TAPPED HOLE   D. BOTTOM TAPPED HOLE

17.57 Three types of tapping tools are used to thread internal drilled holes: taper tap, plug tap, and bottom tap.

The **taper**, **plug**, and **bottom** hand taps have identical measurements, except for the chamfered portion of their ends. The taper tap has a long chamfer (8 to 10 threads), the plug tap has a shorter chamfer (3 to 5 threads), and the bottom tap has the shortest chamfer (1 to 1-1/2 threads).

When tapping is to be done by hand in open or "through" holes, the taper tap should be used for coarse threads and in harder metals because it ensures straighter alignment and starting. The plug tap may be used in soft metals and for fine-pitch threads. When a hole is tapped to its bottom, all three taps—taper, plug, and bottoming—are used in that sequence on the same internal threads.

Notes are added to specify the depth of a drilled hole and the depth of the threads within it. For example, a note reading 7/8 DIA-3 DEEP × 1/8 UNC-2A × 2 DEEP means that the hole is to be drilled deeper than it is threaded and that the last usable thread will be 2 inches deep in the hole.

17.16 Washers, Lock Washers, and Pins

Various types of washers are used with nuts and bolts to improve their assembly and increase their fastening strength.

Plain washers are noted on a drawing as

.938 × 1.750 × 0.134 TYPE A PLAIN WASHER,

where the numbers (left to right) represent the washer's inside diameter, outside diameter, and thickness (see Appendixes 26 and 27).

Lock washers reduce the likelihood that threaded parts will loosen because of vibration and movement. **Figure 17.58** shows several common types of lock washers. Appendix 28 contains a table of dimensions for regular and extra-heavy-duty helical-spring lock washers. Designate them with a note in the form:

HELICAL-SPRING LOCK WASHER 1/4 REGULAR—PHOSPHOR BRONZE,

where the 1/4 is the washer's inside diameter. Designate tooth lock washers with a note in one of two forms:

INTERNAL-TOOTH LOCK WASHER 1/4-TYPE A—STEEL;

EXTERNAL-TOOTH LOCK WASHER -562-TYPE B—STEEL.

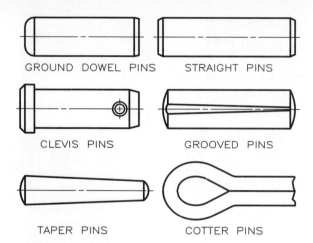

GROUND DOWEL PINS   STRAIGHT PINS

CLEVIS PINS   GROOVED PINS

TAPER PINS   COTTER PINS

17.59 Pins are used to hold parts together in assembly.

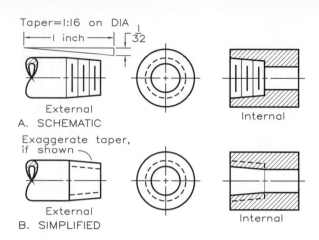

A. SCHEMATIC

B. SIMPLIFIED

17.60 Pipe threads are shown with schematic and simplified symbols here.

Pins (**Fig. 17.59**) are used to hold parts together in a fixed position. Appendix 23 gives dimensions for straight pins. The **cotter pin** is another locking device that everyone who has had a toy wagon is familiar with. Appendix 22 contains a table of dimensions for cotter pins.

17.17 Pipe Threads and Fittings

Pipe threads are used for connecting pipes, tubing, and various fittings including lubrication fittings. The most commonly used pipe thread is tapered at a ratio of 1 to 16 on its diameter, but straight pipe threads also are available (**Fig. 17.60**). Tapered pipe threads will engage only for an effective length of

$$L = (0.80D + 6.8)P,$$

where D is the outside diameter of the threaded pipe and P is the pitch of the thread.

The pipe threads shown in Fig. 17.60 have a taper exaggerated to 1:16 on radius (instead of on diameter) to emphasize it. Drawing them with no taper, obviously, is easier. You may use either schematic or simplified symbols to show the threaded features.

Use the following ANSI abbreviations in pipe thread notes. All begin with NP (for National Pipe thread).

NPT national pipe taper

NPTF national pipe thread (dryseal, for pressure-tight joints)

NPS straight pipe thread

NPSC straight pipe thread in couplings

NPSI national pipe straight internal thread

NPSF straight pipe thread (dryseal)

NPSM straight pipe thread for mechanical joints

NPSL straight pipe thread for locknuts and locknut pipe threads

NPSH straight pipe thread for hose couplings and nipples

NPTR taper pipe thread for railing fittings

To specify a pipe thread in note form, give the nominal pipe diameter (the common-fraction size of its internal diameter), the number of threads per inch, and the thread-type symbol:

1-1/4-11-1/2 NPT   or   3–8 NPTR

Appendix 9 gives a table of dimensions for pipe threads. **Figure 17.61** shows how to present specifications for external and internal threads in note form. Dryseal threads, either straight or tapered, provide a pressure-tight joint without the use of a lubricant or sealer.

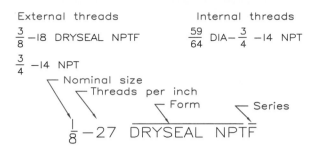

External threads

$\frac{3}{8}$–18 DRYSEAL NPTF

$\frac{3}{4}$–14 NPT

Internal threads

$\frac{59}{64}$ DIA–$\frac{3}{4}$–14 NPT

Nominal size
Threads per inch
Form
Series

$\frac{1}{8}$–27 DRYSEAL NPTF

17.61 Typical pipe thread notes used in specifications.

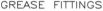

Thread size	$\frac{1}{8}$ 3mm		$\frac{1}{4}$ 6mm		$\frac{3}{8}$ 10mm	
Overall length	L=in.	mm	L=in.	mm	L=in.	mm
Straight	.625	16	1.000	25	1.200	30
90° Elbow	.800	20	1.250	32	1.400	36
45° Angle	1.000	25	1.500	38	1.600	41

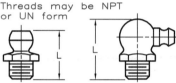

GREASE FITTINGS
Threads may be NPT
or UN form

A. STRAIGHT   B. 90° ANGLE   C. 45° ANGLE

17.62 Three standard types of grease fittings used to lubricate moving parts with a grease gun.

Grease Fittings

Grease fittings (**Fig. 17.62**) allow the application of lubricate to moving parts. Threads of grease fittings are available as tapered and straight pipe threads. The ends where grease is inserted with a grease gun are available straight or at 90° and 45° angles. A one-way valve, formed by a ball and spring, permits grease to enter the fitting (forced through by a grease gun) but prevents it from escaping.

17.18 Keys

Keys are used to attach pulleys, gears, or crank handles to shafts, allowing them to remain assembled while moving and transmitting power. The four types of keys shown in **Fig. 17.63** are the most commonly used. Appendix 25 contains tables of dimensions for keyways, keys, and keyseats.

17.19 Rivets

Rivets are fasteners that permanently join thin overlapping materials (**Fig. 17.64.**). The rivet is inserted in a hole slightly larger than the diameter of the rivet, and the application of pressure to the projecting end forms the

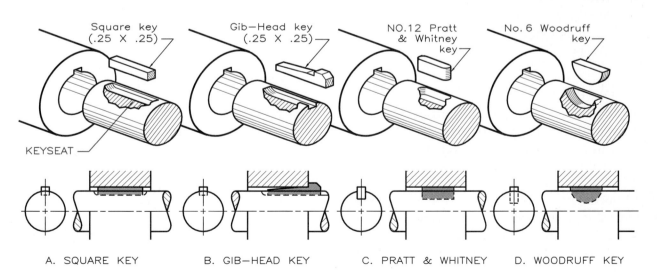

Square key
(.25 X .25)

Gib–Head key
(.25 X .25)

NO.12 Pratt
& Whitney
key

No. 6 Woodruff
key

KEYSEAT

A. SQUARE KEY   B. GIB–HEAD KEY   C. PRATT & WHITNEY   D. WOODRUFF KEY

17.63 Standard types of keys used to hold parts on a shaft.

17.64 Rivets are used to permanently fasten structural elements together. (Courtesy of Russell, Burdsall and Ward Bolt and Nut Company.)

Flat Head   Countersunk Head   Button Head

Pan Head   Truss or Wagon Box Head

17.65 The proportions of small rivets with shanks up to 1/2 inch are shown here.

headless end into shape. Forming may be done with either hot or cold rivets, depending on the application.

Figure 17.65 shows typical shapes and proportions of small rivets that vary in diameter from 1/16 to 1 3/4 inches. Rivets are used extensively in pressure-vessel fabrication, heavy construction (such as bridges and buildings), and sheet-metal construction.

Figure 17.66 shows some of the standard ANSI symbols for representing rivets. Rivets that are driven in the shop are called shop rivets; those assembled at the job site are called field rivets.

17.20 Springs

Springs are devices that absorb energy and react with an equal force (**Fig. 17.67**). Most springs are **helical**, as are bed springs, but they can also be **flat** (leaf), as in an automobile chassis. Some of the more common types of springs are **compression, torsion, extension, flat,** and **constant force.** In **Figure 17.68,** parts A–C show single-line conventional representations of the first three types. Parts D–F represent the types of ends used on compression springs.

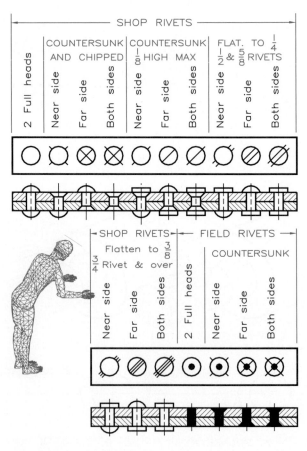

17.66 Rivets are represented in a drawing by these standard ANSI symbols.

17.67 Springs are available for numerous special applications.

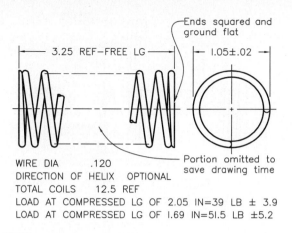

Ends squared and ground flat

3.25 REF—FREE LG

1.05±.02

WIRE DIA	.120
DIRECTION OF HELIX	OPTIONAL
TOTAL COILS	12.5 REF

LOAD AT COMPRESSED LG OF 2.05 IN=39 LB ± 3.9
LOAD AT COMPRESSED LG OF 1.69 IN=51.5 LB ±5.2

Portion omitted to save drawing time

17.69 A conventional double-line drawing showing a compression spring and its specifications.

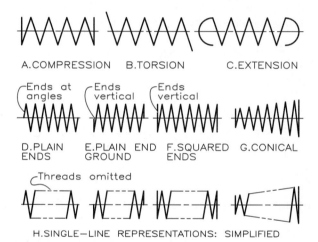

A. COMPRESSION   B. TORSION   C. EXTENSION

Ends at angles   Ends vertical   Ends vertical

D. PLAIN ENDS   E. PLAIN END GROUND   F. SQUARED ENDS   G. CONICAL

Threads omitted

H. SINGLE–LINE REPRESENTATIONS: SIMPLIFIED

17.68 These various types of springs are drawn as single-line representations as a means of saving drawing time.

Plain ends of springs simply end with no special modification of the coil. Ground plain ends are coils that have been machined by grinding to flatten the ends perpendicular to their axes. Squared ends are inactive coils that have been closed to form a circular flat coil at the end of a spring, which may also be ground.

In Fig. 17.68G a conical helical spring is shown in its simplified form. Figure 17.68H shows schematic single-line representations of the same types of springs depicted in Fig. 17.68D–G, with phantom outlines instead of all the coils. This is a conventional method of drawing springs that saves time and effort.

Working drawing specifications of a compression spring drawn as a double-line representation are shown in **Fig. 17.69**. Two coils are drawn at each end of the spring and phantom lines are drawn between them in order to save drawing time. A dimension is given with the diameter and free length of the spring on the drawing. The remaining specifications are given in a table placed near the drawing.

A working drawing of an extension spring (**Fig. 17.70**) is similar to that of a compression spring. An extending spring is designed to resist stretching, whereas a **compression spring** is designed to resist squeezing. In a drawing of a helical torsion spring, which resists and reacts to a twisting motion (**Fig. 17.71**), angular dimensions specify the initial and final positions of the spring as torsion is applied. Again, dimension the drawing and add specifications to describe details.

17.21 Drawing Springs

Springs may be represented with single-line drawings (**Fig. 17.68**) or as more realistic double-line drawings (**Fig. 17.72**). Draw each type shown by first laying out the diameters of

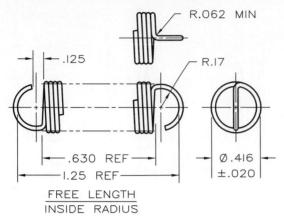

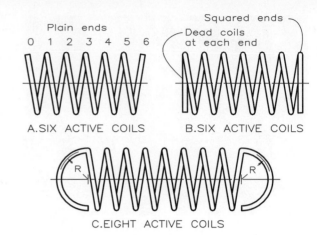

WIRE DIA   0.42
DIRECTION OF HELIX   OPTIONAL
TOTAL COILS   14 REF
RELATIVE POSITION OF ENDS   180° ±20°
EXTENDED LENGTH INSIDE ENDS
WITHOUT PERMANENT SET   2.45 IN (MAX)
INITIAL TENSION   1.00 LB ±.10 LB
LOAD   4.0 LB ±.4 LB AT 1.56 IN
EXTENDED LG INSIDE ENDS
LOAD   6.30 LB ±.63 LB AT 1.95

17.70 A conventional double-line drawing showing an extension spring and its specifications.

17.72 Double-line drawings of springs.

A A spring with six active coils.

B A spring with six coils and a "dead" (inactive) coil at each end.

C An extension spring with eight active coils.

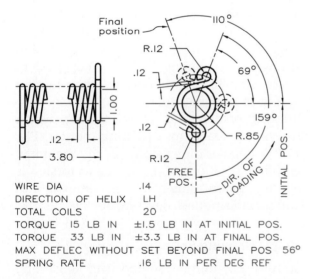

WIRE DIA   .14
DIRECTION OF HELIX   LH
TOTAL COILS   20
TORQUE   15 LB IN   ±1.5 LB IN AT INITIAL POS.
TORQUE   33 LB IN   ±3.3 LB IN AT FINAL POS.
MAX DEFLEC WITHOUT SET BEYOND FINAL POS 56°
SPRING RATE   .16 LB IN PER DEG REF

17.71 A conventional double-line drawing showing a helical torsion spring and its specifications.

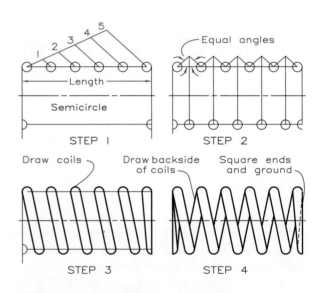

17.73 Drawing a compression spring in detail.

Step 1 Lay out the diameter and length of the spring and locate the five coils by the diagonal-line technique.

Step 2 Locate the coils on the lower side along the bisectors of the spaces between the coils on the upper side.

Step 3 Connect the coils on each side. This is a right-hand coil; a left-hand spring would slope in the opposite direction.

Step 4 Construct the back side of the spring and the end coils to complete the drawing. The spring has a square end that is to be ground.

272 • **CHAPTER 17 SCREWS, FASTENERS, AND SPRINGS**

the coils and lengths of the springs and then dividing the lengths into the number of active coils (Fig. 17.72A). In Fig. 17.72B, both end coils are "dead" (inactive) coils, and only six coils are active. Figure 17.72C depicts an extension spring with eight active coils.

The steps of drawing a double-line detailed representation of a compression spring are shown in **Fig. 17.73**. Springs can be right-hand or left-hand, but like threads, most are drawn as right-hand coils. The ends of the spring in this case are to be squared by grinding the ends to make them flat and perpendicular to the axis of the spring.

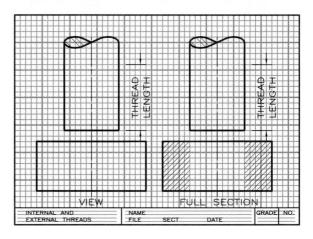

17.74 Problems 1–3.

Problems

Solve and draw these problems on size A sheets. Each grid space equals 0.20 inch, or 5 mm.

1. (**Fig. 17.74**) Draw detailed representations of Acme threads with major diameters of 2 in. Show both external and internal threads as views and sections. Provide a thread note by referring to Appendix 11.

2. Repeat Problem 1, but draw detailed internal and external representations of square threads.

3. Repeat Problem 1, but draw detailed internal and external representations of UN threads. Provide a thread note for a coarse thread with a class 2 fit.

4. Using the notes in **Fig. 17.75**, draw detailed representations of the internal threads and holes in section. Provide thread notes on each as specified.

5. Repeat Problem 4, but this time use schematic symbols.

6. Repeat Problem 4, but this time use simplified symbols.

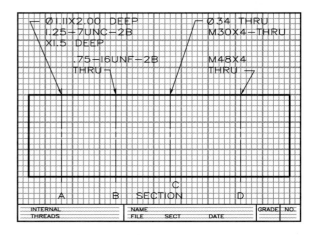

17.75 Problems 4–6.

7. Using the partial views in **Fig. 17.76** and detailed thread symbols, draw external, internal, and end views of the full-size threaded parts. Provide thread notes for UNC threads with a class 2 fit.

8. Repeat Problem 7, but this time use schematic symbols.

9. Repeat Problem 7, but this time use simplified symbols.

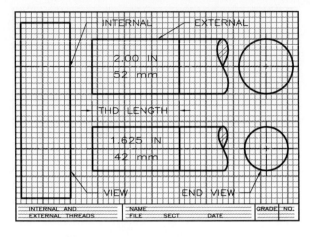

17.76 Problems 7–9.

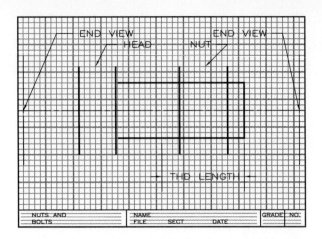

17.77 Problems 10–12.

10. (Fig. 17.77) Complete the drawing of the finished hexagon-head bolt and a heavy hexagon nut. Draw the bolt head and nut across corners using detailed thread symbols. Provide thread notes in either English or metric forms, as assigned.

11. Repeat Problem 10, but draw the nut and bolt as having unfinished square heads. Use schematic thread symbols.

12. Repeat Problem 10, but draw the bolt with a regular finished hexagon head across flats, using simplified thread symbols. Draw the nut across flats also and provide thread notes for both.

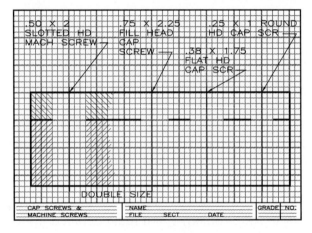

17.78 Problems 13–15.

13. Use the notes in **Fig. 17.78** to draw the screws in section and complete the sectional view showing all cross-hatching. Use detailed thread symbols and provide thread notes to the parts.

14. Repeat Problem 13, but use schematic thread symbols.

15. Repeat Problem 13, but use simplified thread symbols.

16. (Fig. 17.79) The pencil pointer has a 1/4-in. shaft that fits into a bracket designed to clamp onto a desk top. A set screw holds the shaft in position. Make a drawing of the

17.79 Problem 16.

bracket, estimating its dimensions. Show the details and the method of using the set screw to hold the shaft and provide a thread note.

17. (**Fig. 17.80**) On axes A and B, construct hexagon-head cap screws (across flats), with UNC threads and a class 2 fit. The cap screws should not reach the bottoms of the threaded holes. Convert the view to a half-section.

18. (**Fig. 17.81**) On axes A and B, draw studs having a hexagon-head nut (across flats) that hold the two parts together. The studs are to be fine series with a class 2 fit, and they should not reach the bottom of the threaded hole. Provide a thread note. Show the view as a half-section.

19. (**Fig. 17.82**) Draw a 2-in. (50-mm) diameter hexagon-head bolt, with its head across flats, using schematic symbols. Draw a plain washer and regular nut (across corners) at the right end. Design the size of the opening in the part at the left end to hold the bolt head so that it will not turn. Use a UNC thread with a series 2 fit, and provide a thread note.

20. (**Fig. 17.83**) Draw a 2-in. (50-mm) diameter hexagon-head cap screw that holds the two parts together. Determine the length of the bolt, show the threads with schematic thread symbols, and provide a thread note.

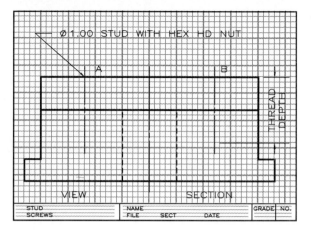

17.81 Problem 18.

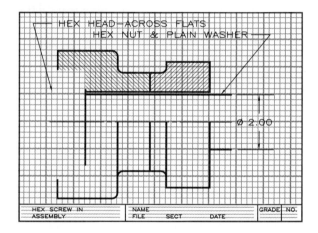

17.82 Problem 19.

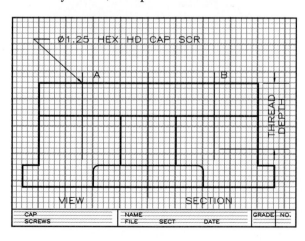

17.80 Problem 17.

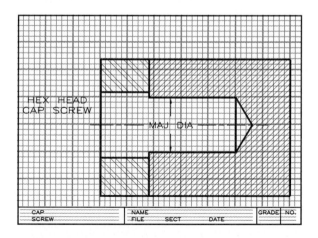

17.83 Problem 20.

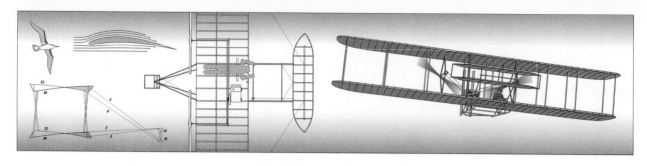

18
Gears and Cams

18.1 Introduction

Gears are toothed wheels whose circumferences mesh to transmit force and motion from one gear to the next. Multiple gears and cams in combination that mesh precisely can be seen in the multiple-spindle bar machine in **Fig. 18.1**. The three most common types of gears—spur, bevel, and worm—are illustrated in **Fig. 18.2**.

Cams are irregularly shaped plates and cylinders that control the motion of a follower as they revolve to produce a type of reciprocating action. For example, cams make the needle of a sewing machine move up and down.

18.2 Spur Gears

Terminology

The **spur gear** is a circular gear with teeth cut around its circumference. Two meshing spur gears transmit power from one shaft to a par-

18.1 Numerous gears and cams are shown in this detail of an Acme-Gridley multiple-spindle bar machine. (Courtesy of National Acme Company.)

allel shaft. When the two meshing gears are unequal in diameter, the smaller gear is called the **pinion** and the larger one the **spur**.

The following terms and corresponding formulas describe the parts of a spur gear, several of which are shown in **Fig. 18.3**.

18.2 The three basic types of gears: (A) spur gears, (B) bevel gears, and (C) worm gears. (Courtesy of the Process Gear Company.)

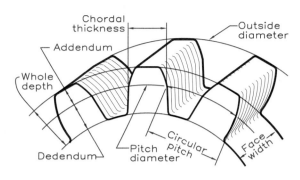

18.3 The terms describing the various parts of a spur gear.

Pitch circle (PC): the imaginary circle of a gear, as if it were a friction wheel without teeth that contacted another circular friction wheel.

Pitch diameter (PD): the diameter of the pitch circle; PD = N/DP, where N is the number of teeth and DP is the diametral pitch.

Diametral pitch (DP): the ratio between the number of teeth on a gear and its pitch diameter; DP = N/PD, where N is the number of teeth and is expressed as teeth per inch of diameter.

Circular pitch (CP): the circular measurement from one point on a tooth to the corresponding point on the next tooth measured along the pitch circle; CP = 3.14/DP.

Center distance (CD): the distance from the center of a gear to its mating gear's center; CD = (NP + NS)/(2DP), where NP and NS are the number of teeth in the pinion and spur, respectively.

Addendum (A): the height of a gear above its pitch circle; A = 1/DP.

Dedendum (D): the depth of a gear below the pitch circle; D = 1.157/DP.

Whole depth (WD): the total depth of a gear tooth; WD = A + D.

Working depth (WKD): the depth to which a tooth fits into a meshing gear; WKD = 2/DP, or WKD = 2A.

Circular thickness (CRT): the circular distance across a tooth measured along the pitch circle; CRT = 1.57/DP.

Chordal thickness (CT): the straight-line distance across a tooth at the pitch circle; CT = PD (sin 90°/N), where N is the number of teeth.

Face width (FW): the width across a gear tooth parallel to its axis; a variable dimension, but usually three to four times the circular pitch; FW = 3CP to 4CP.

Outside diameter (OD): the maximum diameter of a gear across its teeth; OD = PD + 2A.

Root diameter (RD): the diameter of a gear measured from the bottom of its gear teeth; RD = PD − 2D.

Pressure angle (PA): the angle between the line of action and a line perpendicular to the centerline of two meshing gears; angles of 14.5° and 20° are standard for involute gears.

Base circle (BC): the circle from which an involute tooth curve is generated or developed; BC = PD cos PA.

Tooth Forms

The most common gear tooth is an involute tooth with a 14.5° pressure angle. The 14.5°

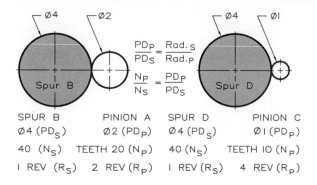

SPUR B	PINION A	SPUR D	PINION C
Ø4 (PD$_S$)	Ø2 (PD$_P$)	Ø4 (PD$_S$)	Ø1 (PD$_P$)
40 (N$_S$)	TEETH 20 (N$_P$)	40 (N$_S$)	TEETH 10 (N$_P$)
1 REV (R$_S$)	2 REV (R$_P$)	1 REV (R$_S$)	4 REV (R$_P$)

18.4 Examples of ratios between meshing spur gears and pinion gears.

angle is the angle of contact between two gears when the tangents of both gears are in contact. Gears with pressure angles of 20° and 25° are also used. Gear teeth with larger pressure angles are wider at the base and thus stronger than the standard 14.5° teeth.

18.3 Gear Ratios

The diameters of two meshing spur gears establish ratios that are important to their function (**Fig. 18.4**). If the diameter of a gear is twice that of its pinion (the small gear), the gear has twice as many teeth as the pinion. The pinion then must make twice as many turns as the spur; therefore, the revolutions per minute (RPM) of the pinion is twice that of the spur.

The relationship between two meshing gears may be determined by finding the velocity of a point on the pinion that is equal to $\pi PD \times RPM$. The velocity of a point on the large gear equals $\pi PD \times RPM$. The velocity of points on each gear must be equal, so

$$\pi PD_P (RPM_P) = \pi PD_S (RPM_S);$$

therefore

$$\frac{PD_P}{PD_S} = \frac{RPM_S}{RPM_P}.$$

If the radius of the pinion is 1 inch, the diameter of the spur is 4 inches, and the RPM of the pinion is 20, the RPM of the spur is

$$\frac{2(1)}{2(4)} = \frac{RPM_S}{20}.$$

or

$$RPM_S = \frac{2(20)}{2(4)} = 5 RPM$$

Thus, the RPM of the spur (5) is one-fourth that of the pinion (20).

The number of teeth on each gear is proportional to the diameters of a pair of meshing gears, or

$$\frac{N_P}{N_S} = \frac{PD_P}{PD_S}$$

where N_P and N_S are the number of teeth on the pinion and spur, respectively, and PD_P and PD_S are their pitch diameters.

Calculations

Before starting a working drawing of a gear, you have to calculate the gear's dimensions.

Problem 1   Calculate the dimensions for a spur that has a pitch diameter of 5 in., a diametral pitch of 4, and a pressure angle of 14.5°.

Solution

Number of teeth: PD(DP) = 5(4) = 20.
Addendum: 1/4 = 0.25″.
Dedendum: 1.157/4 = 0.2893″.
Circular thickness: 1.5708/4 = 0.3927″.
Outside diameter: (20 + 2)/4 = 5.50″.
Root diameter: 5 − 2(0.2893) = 4.421″.
Chordal thickness:
  5(sin 90°/20) = 5(0.079) = 0.392″.
Chordal addendum:
  0.25 + [0.3927^2/(4 × 5)] = 0.2577″.

Face width: 3.5(0.79) = 2.75.

Circular pitch: 3.14/4 = 0.785″.

Working depth: 0.6366(3.14/4) = 0.4997″.

Whole depth: 0.250 + 0.289 = 0.539″.

Use these dimensions to draw the spur and to provide specifications necessary for its manufacture.

Problem 2 shows the method of determining design information for two meshing gears when their working ratios are known.

Problem 2 Find the number of teeth and other specifications for a pair of meshing gears with a driving gear that turns at 100 RPM and a driven gear that turns at 60 RPM. The diametral pitch for each is 10, and the center-to-center distance between the gears is 6 in.

Solution

Step 1 Find the sum of the teeth on both gears:

Total teeth = 2(center-to-center distance)(DP)
= 2(6)(10) = 120 teeth.

Step 2 Find the number of teeth for the driving gear:

$$\frac{\text{Driver RPM}}{\text{Driven RPM}} + 1 = \frac{100}{60} + 1 = 2.667,$$

so

$$\frac{\text{Total Teeth}}{\frac{100}{60}+1} = \frac{120}{2.667} = 45 \text{ teeth.}$$

(The number of teeth must be a whole number since there cannot be fractional teeth on a gear.)

Step 3 Find the number of teeth for the driven gear:

Total teeth − teeth on driver =
teeth on driven gear

120 − 45 = 75 teeth.

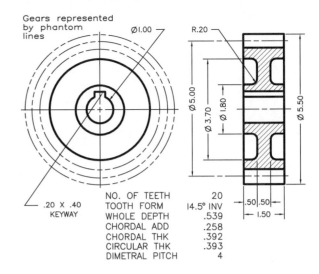

NO. OF TEETH	20
TOOTH FORM	14.5° INV
WHOLE DEPTH	.539
CHORDAL ADD	.258
CHORDAL THK	.392
CIRCULAR THK	.393
DIMETRAL PITCH	4

18.5 This detail drawing of a spur gear contains a table of values that supplements the dimensions shown on the view and section.

Step 4 Calculate the other dimensions for the gears as in Problem 1. It may be necessary to adjust the center distance to yield a whole number of teeth.

18.4 Drawing Spur Gears

Figure 18.5 shows a conventional drawing of a spur gear. Not having to draw the gear teeth in the circular view saves a lot of time. Showing only simplified circular and sectional views of the gear and providing a table of dimensions called **cutting data** is acceptable. Circular phantom lines represent the root circle, pitch circle, and outside circle of the gear in the circular view.

A table of dimensions is a necessary part of a gear drawing (**Fig. 18.6**). You may calculate these data or get them from tables of standards in gear handbooks such as *Machinery's Handbook*.

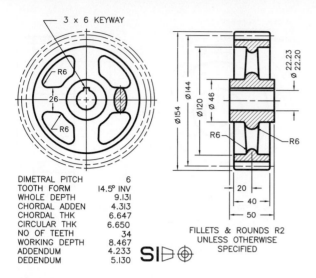

DIMETRAL PITCH	6
TOOTH FORM	14.5° INV
WHOLE DEPTH	9.131
CHORDAL ADDEN	4.313
CHORDAL THK	6.647
CIRCULAR THK	6.650
NO OF TEETH	34
WORKING DEPTH	8.467
ADDENDUM	4.233
DEDENDUM	5.130

18.6 A drawing of a spur gear that was produced on a computer.

18.5 Bevel Gears

Terminology

Bevel gears have axes that intersect at angles. The angle of intersection is usually 90°, but other angles also are used. The smaller of the two bevel gears is the **pinion**, as it is with spur gears; the larger is the **gear**.

Figure 18.7 illustrates the terminology of bevel gears. A further explanation of and the corresponding formula for each feature follow. You may also use gear handbooks to find these dimensions.

Pitch angle of pinion (PA_p): $\tan PA_p = N_p/N_g$, where N_g and N_p are the number of teeth on the gear and pinion, respectively.

Pitch angle of gear (PA_g): $\tan PA_g = N_g/N_p$.

Pitch diameter (PD): the number of teeth, N, divided by the diametral pitch, DP; $PD = N/P$.

Addendum (A): measured at the large end of the tooth; $A = 1/DP$.

Dedendum (D): is measured at the large end of the tooth; $D = 1.157/DP$.

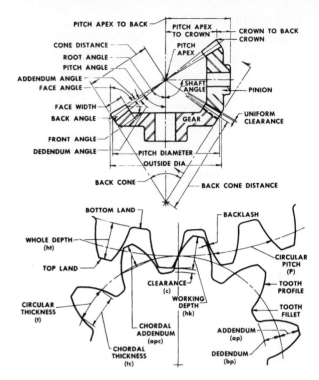

18.7 The terminology of bevel gears. (Courtesy of Philadelphia Gear Corporation.)

Whole tooth depth (WD): $WD = 2.157/DP$.

Thickness of tooth (TT): measured at the pitch circle; $TT = 1.571/DP$.

Diametral pitch: $DP = N/PD$, where N is the number of teeth.

Addendum angle (AA): the angle formed by the addendum and pitch cone distance; $\tan AA = A/PCD$.

Angular addendum: $AK = COS\ PA \times A$.

Pitch cone distance: (PCD): $PCD = PD/(2 \sin PA)$.

Dedendum angle (DA): the angle formed by the dedendum and the pitch cone distance; $\tan DA = D/PCD$.

Face angle (FA): the angle between the gear's centerline and the top of its teeth; FA = 90° − (PCD + AA).

Cutting angle (or root angle) (CA): the angle between the gear's axis and the roots of the teeth; CA = PCD − D.

Outside diameter (OD): the greatest diameter of a gear across its teeth; OD = PD + 2A.

Apex-to-crown distance (AC): the distance from the crown of the gear to the apex of the cone measured parallel to the axis of the gear; AC = OD/(2 tan FA).

Chordal addendum (CA): CA = A + [(TT² cos PA)/4PD].

Chordal thickness (CT): measured at the large end of the tooth; CT = PD (sin 90°/N).

Face width (FW): can vary, but should be approximately equal to the pitch cone distance divided by 3; FW = PCD/3.

Calculations

Problem 3 demonstrates use of the preceding formulas. Some of the formulas result in specifications that apply to both gear and pinion.

Problem 3 Two bevel gears intersect at right angles and have a diametral pitch of 3. The gear has 60 teeth and the pinion has 45 teeth. Find the dimensions of the gear.

Solution

Pitch cone angle of gear:
    tan PCA = 60/45 = 1.33; PCA = 53°7′
Pitch cone angle of pinion:
    tan PCA = 45/60; PCA = 36°52′
Pitch diameter of gear: 60/3 = 20.00″
Pitch diameter of pinion: 45/3 = 15.00″

The following calculations yield the same dimensions for both gear and pinion:

Addendum: 1/3 = 0.333″

Dedendum: 1.157/3 = 0.3857″

Whole depth: 2.157/3 = 0.719″

Tooth thickness on pitch circle:
    1.571/3 = 0.5237″

Pitch cone distance:
    20/(2 sin 53°7′) = 12.5015″

Addendum angle:
    tan AA = 0.333/12.5015 = 1°32′

Dedendum angle:
    DA = 0.3857/12.5015 = 0.0308 = 1°46′

Face width: PCD/3 = 4.00″

The following dimensions must be calculated separately for gears and pinions:

Chordal addendum of gear:
    0.333″ + [(0.5237² cos 53°7′)/(4 × 20)] = 0.336″

Chordal addendum of pinion:
    0.333″ + [(0.5237² cos 36°52′)/(4 × 15)] = 0.338″

Chordal thickness of gear:
    sin 90°/(60 × 20″) = 0.524″

Chordal thickness of pinion:
    sin 90°/(45 × 15″) = 0.523″

Face angle of gear:
    90° − (53°7′ + 1°32′) = 35°21′

Face angle of pinion:
    90° − (36°52′ + 1°32′) = 51°36′

Cutting angle of gear: 53°7′ − 1°46′ = 51°21′

Cutting angle of pinion: 36°52′ − 1°46′ = 35°6′

Angular addendum of gear:
    0.333″ cos 53°7′ = 0.1999″

Angular addendum of pinion:
    0.333″ cos 36°52′ = 0.2667″

Outside diameter of gear:
    20″ + 2(0.1999″) = 20.4000″

Outside diameter of pinion:
    15″ + 2(0.2667″) = 15.533″

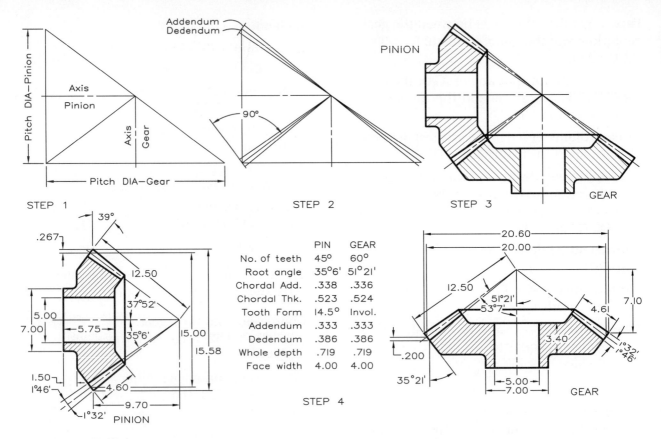

STEP 1

STEP 2

STEP 3

	PIN	GEAR
No. of teeth	45°	60°
Root angle	35°6'	51°21'
Chordal Add.	.338	.336
Chordal Thk.	.523	.524
Tooth Form	14.5°	Invol.
Addendum	.333	.333
Dedendum	.386	.386
Whole depth	.719	.719
Face width	4.00	4.00

STEP 4

18.8 Drawing bevel gears.

Step 1 Lay out the pitch diameters and axes of the two bevel gears with construction lines.

Step 2 Draw construction lines to establish the limits of the teeth by using the addendum and dedendum dimensions.

Step 3 Draw the pinion and gear using the specified or calculated dimensions.

Step 4 Complete the detail drawings of both gears and provide a table of cutting data.

Apex-to-crown distance of gear:
  $(20.400''/2)(\tan 35°7') = 7.173''$

Apex-to-crown distance of pinion:
  $(15.533''/2)(\tan 51°36') = 9.800''$

18.6 Drawing Bevel Gears

Use the calculated dimensions to lay out bevel gears in a detail drawing. Many of these dimen-

sions are difficult to measure with a high degree of accuracy on a drawing. Therefore, providing a table of cutting data for each gear is essential.

The steps involved in drawing bevel gears are shown in **Figure 18.8**. On the final drawing, notice that dimensions on the views are supplemented by a table of additional dimensions and data.

18.7 Worm Gears

A worm gear consists of a threaded shaft called a **worm** and a circular gear called a **spider** (Fig. 18.9). When the worm is revolved, it causes the spider to revolve about its axis. Figures 18.9 and **18.10** illustrate the terminology of worm gearing. The following lists further explain these terms and provide the formulas for calculating their dimensions.

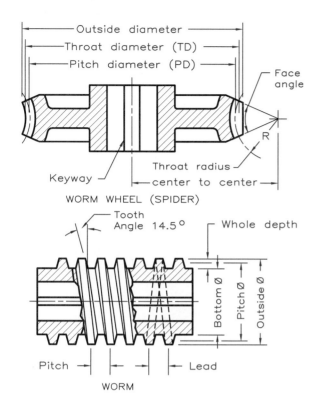

WORM WHEEL (SPIDER)

WORM

18.9 The terminology of worm gears.

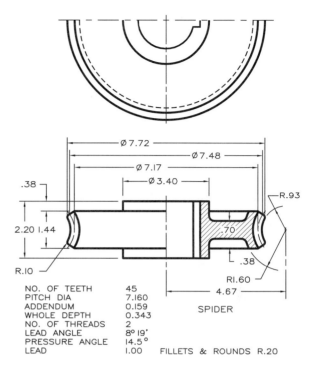

NO. OF TEETH	45
PITCH DIA	7.160
ADDENDUM	0.159
WHOLE DEPTH	0.343
NO. OF THREADS	2
LEAD ANGLE	8° 19'
PRESSURE ANGLE	14.5°
LEAD	1.00

SPIDER

FILLETS & ROUNDS R.20

18.10 A spider for a worm gear with a table of cutting data.

Worm Terminology

Linear pitch (P): the distance from one thread to the next, measured parallel to the worm's axis; $P = L/N$, where N is the number of threads (1 if a single thread, 2 if a double thread, and so on).

Lead (L): the distance a thread advances in a turn of 360°.

Addendum of tooth (AW): $AW = 0.3183P$.

Pitch diameter (PDW): $PDW = OD - 2AW$, where OD is the outside diameter.

Whole depth of tooth (WDT): $WDT = 0.6866P$.

Bottom diameter of worm (BD): $BD = OD - 2WDT$.

Width of thread at root (WT): $WT = 0.31P$.

Minimum length of worm (MLW):

$$MLW = \sqrt{8PDS\,(AW)},$$ where PDS is the pitch diameter of the spider.

Helix angle (HA): $\cot\beta = 3.14(PDW)/L$.
Outside diameter (OD): $OD = PD + 2A$.

Spider Terminology

Pitch diameter of spider (PDS):
  $PDS = N(P)/3.14$, where N is the number of teeth on the spider.
Throat diameter of spider (TD):
  $TD = PDS + 2A$.
Radius of spider throat (RST):
  $RST = OD$ of worm$/2 - 2A$.
Face angle (FA):
  may be selected between 60° and 80° for the average application.

Center-to-center distance (CD):
measured between the worm and spider;
CD = PDW + PDS/2.

Outside diameter of spider (ODS):
ODS = TD + 0.4775P.

Face width of gear (FW):
FW = 2.38P + 0.25.

Calculations

Problem 4 demonstrates use of the preceding formulas to find the dimensions for a worm gear.

Problem 4 Calculate the dimensions for a worm gear (worm and spider). The spider has 45 teeth, and the worm has an outside diameter of 2.50 in., a double thread, and a pitch of 0.5 in.

Solution

Lead: L = 0.5"(2) = 1".

Worm addendum: AW = 0.3183P = 0.1592".

Pitch diameter of worm:
PDW = 2.50" − 2(0.1592") = 2.1818".

Pitch diameter of spider:
PDS = (45" × 0.5)/3.14 = 7.166".

Center distance between worm and spider:
CD = (2.182" + 7.166")/2 = 4.674".

Whole depth of worm tooth:
WDT = 0.687(0.5") = 0.3433".

Bottom diameter of worm:
BD = 2.50" − 2(0.3433") = 1.813".

Helix angle of worm:
cotβ = 3.14(2.1816)/1 = 8°19'.

Width of thread at root:
WT = 0.31(1) = 0.155".

Minimum length of worm:
MLW = $\sqrt{8(0.1592)\,(7.1656)}$ = 3.02"

Throat diameter of spider:
TD = 7.1656" + 2(0.1592") = 7.484".

Radius of spider throat:
RST = (2.5/2) − (2 3 0.1592) = 0.9318".

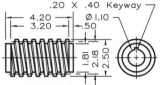

NO. OF THREADS	2
PITCH DIA	2.182
ADDENDUM	.159
WHOLE DEPTH	.343
LEAD ANGLE	8° 19'
PRESSURE ANGLE	14.5°
LEAD	1.00

18.11 This detail drawing of a worm is based on calculated dimensions.

18.12 Three types of machined cams. (Courtesy of Ferguson Machine Company.)

Face width: FW = 2.38(0.5) + 0.25 = 1.44".

Outside diameter of spider:
ODS = 7.484 + 0.4775 (0.5) = 7.723".

18.8 Drawing Worm Gears

Draw and dimension the worm and spider as shown in **Figs. 18.10** and **18.11**. The preceding calculations yield the dimensions needed for scaling and laying out the drawings and providing cutting data.

18.9 Cams

Plate cams are irregularly shaped machine elements that produce motion in a single plane, usually up and down (**Fig. 18.12**). As the cam revolves about its center, the cam's shape alternately raises and lowers the follower that is in contact with it. Cams utilize the principle of the inclined wedge, with the surface of the cam acting as the wedge, causing a change in

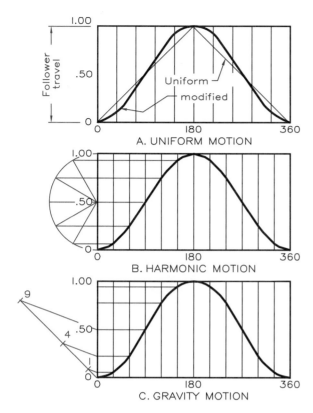

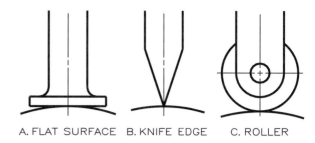

18.14 The three basic types of cam followers: (A) flat surface, (B) knife edge, and (C) roller.

18.13 Displacement diagrams showing three standard motions: uniform, harmonic, and gravity.

the slope of the plane and thereby producing the desired motion of the follower. Cams are designed primarily to produce (1) uniform or linear motion (2) harmonic motion, (3) gravity motion (uniform acceleration), or (4) combinations of these motions.

Uniform Motion

The uniform or linear motion depicted in **Fig. 18.13A** represents the motion of the cam follower as the cam rotates through 360°. This curve has sharp corners, indicating abrupt changes of velocity that cause the follower to bounce. Therefore, uniform motion is usually modified to smooth the changes of velocity. The radius of the modifying arc varies up to a

radius of one-half the total displacement, depending on the speed of operation.

Harmonic Motion

The harmonic motion plotted in **Fig. 18.13B** is a smooth, continuous motion based on the change of position of points on a circle. At moderate speeds, this displacement gives a smooth operation.

Gravity Motion

The gravity motion (uniform acceleration) illustrated in **Fig. 18.13C** is used for high-speed operation. The variation of displacement is analogous to the force of gravity, with the difference in displacement being 1, 3, 5, 5, 3, 1, based on the square of the number. For instance, $1^2 = 1$; $2^2 = 4$; $3^2 = 9$ give a uniform acceleration. This motion is repeated in reverse order for the remaining half of the follower's motion. Intermediate points are obtained by squaring fractional increments, such as $(2.5)^2$.

Cam Followers

The three basic types of cam followers are the flat surface, knife edge, and roller (**Fig. 18.14**). Use of flat-surface and knife-edge followers is limited to slow-moving cams, where minor force will be exerted during rotation. The roller follower is able to withstand higher speeds.

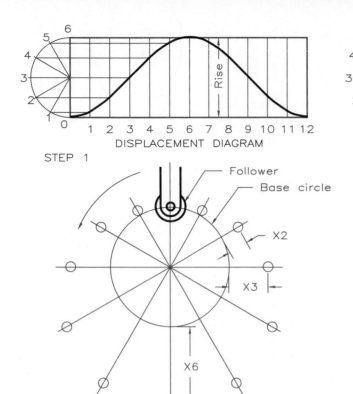

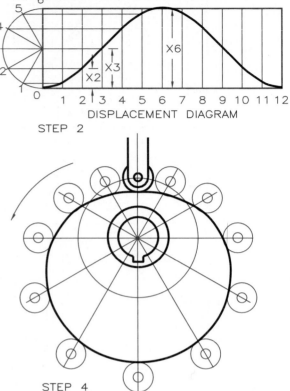

18.15 Drawing a plate cam for harmonic motion.

Step 1 Construct a semicircle on the vertical side of the displacement diagram whose diameter equals the rise of the follower. Divide the semicircle into the same number of segments as there are between 0° and 180° on the horizontal axis of the displacement diagram. Plot the displacement curve.

Step 2 Measure distances of rise and fall (X1, X2, X3, ... X6) at each interval from the base circle.

Step 3 Construct the base circle, draw the follower, and divide the circle into the same number of sectors as there are divisions on the displacement diagram. Transfer distances X1, X2, ... from the displacement diagram to their respective radial lines of the circle, measuring outward from it.

Step 4 Draw circles to represent the positions of the roller as the cam revolves counterclockwise. Draw the cam profile tangent to all the rollers to complete the drawing.

18.10 Designing Plate Cams

Harmonic Motion The steps involved in designing a plate cam for harmonic motion are shown in **Fig. 18.15**. Before laying out the drawing of a cam, you must know the motion and rise of the follower, the diameter of the base circle, and the direction of rotation. The displacement diagram shown in Step 1 of Fig. 18.15 gives the specifications graphically for the cam.

Gravity Motion The steps involved in designing a cam for gravity motion are shown in **Fig. 18.16**. The same steps used in designing a cam for harmonic motion apply, but the displacement diagram and knife-edge follower are different.

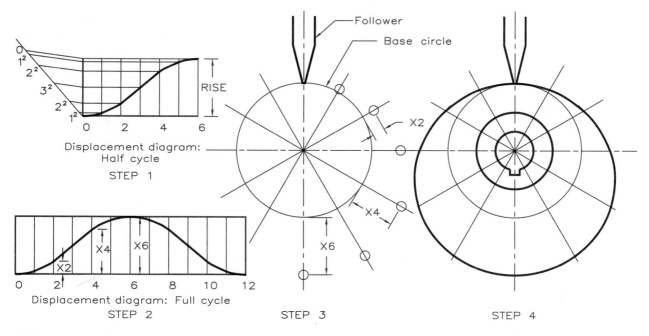

18.16 Drawing a plate cam for uniform acceleration.

Step 1 Construct a displacement diagram to represent the rise of the follower. Divide the horizontal axis into angular increments of 30°. Draw a construction line through point 0; locate the 1^2, 2^2, and 3^2 divisions and project them to the vertical axis to represent half the rise.

Step 2 Use the same construction to find the right half of the symmetrical curve.

Step 3 Construct the base circle and draw the knife-edge follower. Divide the circle into the same number of sectors as there are divisions in the displacement diagram. Transfer distances from the displacement diagram to their respective radial lines of the base circle and measure outward from the base circle.

Step 4 Connect the points found in Step 3 with a smooth curve to complete the cam profile. Show the cam hub and keyway.

Cam with an Offset Follower The cam shown in **Fig. 18.17** produces harmonic motion through 360°. In this case, plot the motion directly from the follower rather than from the usual displacement diagram.

Draw a semicircle with its diameter equal to the total motion of the follower beginning at the centerline of the follower roller. Draw the base circle to pass through the center of the roller of the follower. Extend the centerline of the follower downward and draw a circle tangent to the extension with its center at the center of the base circle. Divide the small circle into 30° intervals to establish points through which to draw construction lines tangent to the circle.

Lay out the distances from tangent points to the position points along the path of the follower along the tangent lines drawn at 30° intervals. Locate these points by measuring from the base circle, as shown. For example, point 3 is located distance X from the base circle. Draw the circular roller in all views, and then draw the profile of the cam tangent to the rollers at all positions.

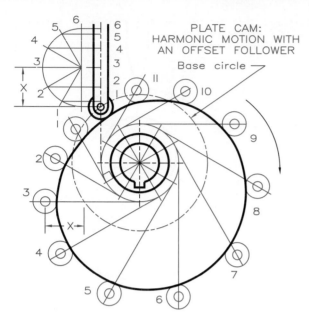

PLATE CAM:
HARMONIC MOTION WITH
AN OFFSET FOLLOWER

Base circle

18.17 A plate cam having an offset roller follower.

Problems

Gears

Use size A sheets for the following gear problems. Select appropriate scales so that the drawings will effectively use the available space.

1–5. Calculate the dimensions for the following spur gears, and make a detail drawing of each. Give the dimensions and cutting data for each gear. Provide any other dimensions needed.

Problem	Gear Teeth	Diametral Pitch	14.5° Involute
1	20	5	0
2	30	3	0
3	40	4	0
4	60	6	0
5	80	4	0

6–10. Calculate the gear sizes and number of teeth using the following ratios and data.

Problem	RPM Pinion	RPM Gear	Center to Center	Diametral Pitch
6	100 (driver)	60	6.0″	10
7	100 (driver)	50	8.0″	9
8	100 (driver)	40	10.0″	8
9	100 (driver)	35	12.0″	7
10	100 (driver)	25	14.0″	6

11–20. Make a detail drawing of each gear for which you made calculations in Problems 6–10. Provide a table of cutting data and other dimensions needed to complete the specifications.

21–25. Calculate the specifications for the bevel gears that intersect at 90°, and make detail drawings of each, including the necessary dimensions and cutting data.

Problem	Diametral Pitch	No. of Teeth on Pinion	No. of Teeth on Gear
21	3	60	15
22	4	100	40
23	5	100	60
24	6	100	50
25	7	100	30

26–30. Calculate the specifications for the worm gears and make a detail drawing of each, providing the necessary dimensions and cutting data.

Problem	No. of Teeth in Spider Gear	Outside DIA of Worm	Pitch of Worm	Thread of Worm
26	45	2.50	0.50	double
27	30	2.00	0.80	single
28	60	3.00	0.80	double
29	30	2.00	0.25	double
30	80	4.00	1.00	single

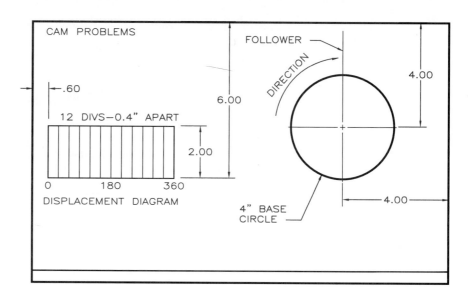

18.18 Layout for Problems 31–36 on size B sheets.

Cams

Use size B sheets for the following cam problems. The standard dimensions are base circle, 3.50 in.; roller follower, 0.60-in. diameter; shaft, 0.75-in. diameter; and hub, 1.25-in. diameter. The direction of rotation is clockwise. The follower is positioned vertically over the center of the base circle. Lay out the problems and displacement diagrams as shown in **Fig. 18.18.**

31. Draw a plate cam with a knife-edge follower for uniform motion and a rise of 1.00 in.

32. Draw a displacement diagram and a cam that will give a modified uniform motion to a knife-edge follower with a rise of 1.7 in. Modify the uniform motion with an arc of one-quarter the rise in the displacement diagram.

33. Draw a displacement diagram and a cam that will give a harmonic motion to a roller follower with a rise of 1.60 in.

34. Draw a displacement diagram and a cam that will give a harmonic motion to a knife-edge follower with a rise of 1.00 in.

35. Draw a displacement diagram and a cam that will give uniform acceleration to a knife-edge follower with a rise of 1.70 in.

36. Draw a displacement diagram and a cam that will give a uniform acceleration to a roller follower with a rise of 1.40 in.

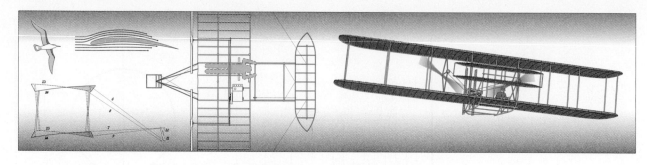

19

Materials and Processes

19.1 Introduction

Various materials and manufacturing processes are commonly used to make parts similar to those discussed in this textbook. A large proportion of parts designed by engineers are made of metal, but other materials (such as plastics, fibers, and ceramics) are available to the designer in increasingly useful applications.

Metallurgy, the study of metals, is a field that is constantly changing as new processes and alloys are developed (**Fig. 19.1**). These developments affect the designer's specification of metals and their proper application for various purposes. Three associations have standardized and continually update guidelines for designating various types of metals: the American Iron and Steel Institute (AISI), the Society of Automotive Engineers (SAE), and the American Society for Testing Materials (ASTM).

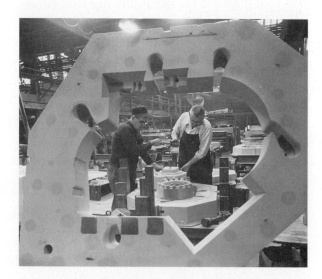

19.1 These workers are assembling a sand casting mold to produce a transmission housing weighing 186 pounds of aluminum alloy. The shape in the foreground is part of the mold assembly. (Courtesy of the Aluminum Company of America.)

DESIGNATION OF GRAY IRON (450 LBS/CF)

ATSM Grade (1000 psi)	SAE Grade	Typical Uses
ASTM 25 CI	G 2500 CI	Small engine blocks, pump bodies, clutch plates, transmission cases
ASTM 30 CI	G 3000 CI	Auto engine blocks, heavy castings, flywheels
ASTM 35 CI	G 3500 CI	Diesel engine blocks, tractor transmission cases, heavy and high—strength parts
ASTM 40 CI	G 4000 CI	Diesel cylinders, pistons, camshafts

19.2 The numerical designations for gray iron and its typical uses.

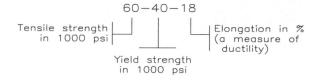

19.3 The numerical notes specifying ductile iron.

DESIGNATIONS OF DUCTILE IRON (490 LBS/CF)

Grade	Typical Uses
60—40—18 CI	Valves, steam fittings, chemical plant equipment, pump bodies
65—45—12 CI	Machine components that are shock loaded, disc brake calipers
80—55—6 CI	Auto crankshafts, gears, rollers
100—70—3 CI	High—strength gears and machine parts
120—90—2 CI	Very high—strength gears, rollers, and slides

19.4 The numerical designations for ductile iron and its typical uses.

19.2 Commonly Used Metals

Iron[*]

Metals that contain iron, even in small quantities, are called **ferrous metals**. Three common types of iron are **gray iron**, **white iron**, and **ductile iron**.

Gray iron contains flakes of graphite, which result in low strength and low ductility, making it easy to machine. Gray iron resists vibration better than other types of iron. **Figure 19.2** shows designations of and typical applications for gray iron.

White iron contains carbide particles that are extremely hard and brittle, enabling it to withstand wear and abrasion. Although the composition of white iron differs from one supplier to another, there are no designated grades of white iron. It is used for parts on grinding and crushing machines, digging teeth on earthmovers and mining equipment, and wear plates on reciprocating machinery used in textile mills.

[*]This section on iron was developed by Dr. Tom Pollack, a metallurgist at Texas A&M University.

Ductile iron (also called nodular or spheroidized iron) contains tiny spheres of graphite, making it stronger and tougher than most types of gray iron and more expensive to produce. Three sets of numbers (**Fig. 19.3**) describe the most important features of ductile iron. **Figure 19.4** shows the designations of and typical applications for the commonly used alloys of ductile iron.

Malleable iron is made from white iron by a heat-treatment process that converts carbides into carbon nodules (similar to ductile iron). The numbering system for designating grades of malleable iron is shown in **Fig. 19.5**. Some of the commonly used grades of malleable iron and their typical applications are shown in **Fig. 19.6**.

Irons are sometimes referred to as **cast iron** when they are melted and poured into a mold to form it by casting, a commonly used process of producing machine parts. Although cheaper and easier to machine than steel, iron does not have steel's ability to withstand shock and force.

```
              32510
                 ┐ ┌┐┌
Tensile strength ┘ └┘└  Elongation
        in MPa            in per cent
  A.
```

```
              M3210
                ┐┌┐┌┐
"Martensite,"   ┘└┘└┘  Elongation
which is not      │      in per cent
  essential     Tensile strength
  B.            in MPa X 0.1
```

19.5   The numerical designations for malleable iron.

DESIGNATIONS OF MALLEABLE IRON (490 LBS/CF)

ASTM Grade	Typical Uses
35018 CI	Marine and railroad valves and fittings, "black—iron" pipe fittings (similar to 60—40—18 ductile CI)
45006 CI	Machine parts (similar to 80—55—6 ductile CI)
M3210 CI	Low—stress components, brackets
M4504 CI	Crankshafts, hubs
M7002 CI	High—strength parts, connecting rods, universal joints
M8501 CI	Wear—resistant gears and sliding parts

19.6   The numerical designations of malleable iron and its typical uses.

Steel

Steel is an alloy of iron and carbon, which often contains other constituents such as manganese, chromium, or nickel. Carbon (usually between 0.20 percent and 1.50 percent) is the ingredient having the greatest effect on the grade of steel. The types of steel are **carbon steel** (free-cutting plain carbon—10XX and free-cutting resulphurized—11XX) and **alloy steel** (23XX—92XX) (Fig. 19.7). The types of steels and their SAE designation are given by four-digit numerals where the last two numbers (XX) give the percentage of carbon. The first digit indicates the type of steel: 1 is carbon steel, 2 is nickel steel, and so on. The second digit gives content (as a percentage) of

DESIGNATIONS OF STEEL (490 LBS/CF)

Type of steel	Number	Applications
Carbon steels		
Plain carbon	10XX	Tubing, wire, nails
Resulphurized	11XX	Nuts, bolts, screws
Manganese steel	13XX	Gears, shafts
Nickel steel	23XX	Keys, levers, bolts
	25XX	Carburized parts
	31XX	Axles, gears, pins
	32XX	Forgings
	33XX	Axles, gears
Molybdenum	40XX	Gears, springs
Chromium—moly.	41XX	Shafts, tubing
Nickel—chromium	43XX	Gears, pinions
Nickel—moly.	46XX	Cams, shafts
	48XX	Roller bearings, pins
Chromium steel	51XX	Springs, gears
	52XX	Ball bearings
Chrom. vanadium	61XX	Springs, forgings
Silicon manganese	92XX	Leaf springs

19.7   The SAE numerical designations of steel and its applications.

the material represented by the first digit. The last two or three digits give the percentage of carbon in the alloy: 100 equals 1 percent, and 50 equals 0.50 percent.

Steel weighs about 490 pounds per cubic foot. Some frequently used SAE steels are 1010, 1015, 1020, 1030, 1040, 1070, 1080, 1111, 1118, 1145, 1320, 2330, 2345, 2515, 3130, 3135, 3240, 3310, 4023, 4042, 4063, 4140, and 4320.

Copper

One of the first metals discovered, copper is easily formed and bent without breaking. Because it is highly resistant to corrosion and highly conductive, it is used for pipes, tubing, and electrical wiring. It is an excellent roofing and screening material because it withstands the weather well. Copper weighs about 555 pounds per cubic foot.

Copper has several alloys, including brasses, tin bronzes, nickel silvers, and copper nickels. Brass (about 530 pounds per cubic foot) is an alloy of copper and zinc, and

ALUMINUM DESIGNATIONS (169 LBS/CF)		
Composition	Alloy Number	Application
Aluminum (99% pure)	1XXX	Tubing, tank cars
Aluminum alloys		
Copper	2XXX	Aircraft parts, screws, rivets
Manganese	3XXX	Tanks, siding, gutters
Silicon	4XXX	Forging, wire
Magnesium	5XXX	Tubes, welded vessels
Magnesium and silicon	6XXX	Auto body, pipes
Zinc	7XXX	Aircraft structures
Other elements	8XXX	

19.8 The numerical designations of aluminum and aluminum alloys and their applications.

ALUMINUM CASTINGS AND INGOT DESIGNATIONS	
Composition	Alloy Number
Aluminum (99% pure)	1XX.X
Aluminum alloys	
Copper	2XX.X
Silicon with copper and/or magnesium	3XX.X
Silicon	4XX.X
Magnesium	5XX.X
Magnesium and silicon	6XX.X
Zinc	7XX.X
Tin	8XX.X
Other elements	9XX.X

19.9 The numerical designations of cast aluminum, ingots, and aluminum alloys.

bronze (about 548 pounds per cubic foot) is an alloy of copper and tin. Copper and copper alloys are easily finished by buffing or plating; joined by soldering, brazing, or welding; and machined.

Wrought copper has properties that permit it to be formed by hammering. A few of the numerical designations of wrought copper are C11000, C11100, C11300, C11400, C11500, C11600, C10200, C12000, and C12200.

Aluminum

Aluminum is a corrosion-resistant, light-weight metal (approximately 169 pounds per cubic foot) that has numerous applications. Most materials called aluminum actually are aluminum alloys, which are stronger than pure aluminum.

The types of wrought aluminum alloys are designated by four digits (**Fig. 19.8**). The first digit (2 through 9) indicates the alloying element that is combined with aluminum. The second digit indicates modifications of the original alloy or impurity limits. The last two digits identify other alloying materials or indicate the aluminum's purity.

Figure 19.9 shows a four-digit numbering system used to designate types of cast aluminum and alloys. The first digit indicates the alloy group, and the next two digits identify the aluminum alloy or aluminum purity. The number to the right of the decimal point represents the aluminum form: XX.0 indicates castings, XX.1 indicates ingots with a specified chemical composition, and XX.2 indicates ingots with a specified chemical composition other than the XX.1 ingot. Ingots are blocks of cast metal to be remelted, and billets are castings of aluminum to be formed by forging.

Magnesium

Magnesium is a light metal (109 pounds per cubic foot) available in an inexhaustible supply because it is extracted from seawater and natural brines. Magnesium is an excellent material for aircraft parts, clutch housings, crankcases for air-cooled engines, and applications where lightness is desirable.

Magnesium is used for die and sand castings, extruded tubing, sheet metal, and forging. Magnesium and its alloys may be joined by bolting, riveting, or welding. Some numerical designations of magnesium alloys are

M10100, M11630, M11810, M11910, M11912, M12390, M13320, M16410, and M16620.

19.3 Properties of Metals

All metals have properties that designers must utilize to the best advantage. The following terms describe these properties.

Ductility: a softness in some metals, such as copper and aluminum, which permits them to be formed by stretching (drawing) or hammering without breaking.

Brittleness: a characteristic that will not allow metals such as cast irons and hardened steels to stretch without breaking.

Malleability: the ability of a metal to be rolled or hammered without breaking.

Hardness: the ability of a metal to resist being dented when it receives a blow.

Toughness: the property of being resistant to cracking and breaking while remaining malleable.

Elasticity: the ability of a metal to return to its original shape after being bent or stretched.

Modifying Properties by Heat Treatment

The properties of metals can be changed by various types of heat treating. Although heat affects all metals, steels are affected to a greater extent than others.

Hardening: heating steel to a prescribed temperature and quenching it in oil or water.

Quenching: rapidly cooling heated metal by immersing it in liquids, gases, or solids (such as sand, limestone, or asbestos).

Tempering: reheating previously hardened steel and then cooling it, usually by air, to increase its toughness.

Annealing: heating and cooling metals to soften them, release their internal stresses, and make them easier to machine.

Normalizing: heating metals and letting them cool in air to relieve their internal stresses.

Case hardening: hardening a thin outside layer of a metal by placing the metal in contact with carbon or nitrogen compounds that it absorbs as it is heated; afterward, the metal is quenched.

Flame hardening: hardening by heating a metal to within a prescribed temperature range with a flame and then quenching the metal.

19.4 Forming Metal Shapes

The two major methods of forming shapes are casting and forging.

Casting

Casting involves preparing a mold in the shape of the part desired, pouring molten metal into it, and cooling the metal to form the part. The types of casting, which differ in the way the molds are made, are **sand casting**, **permanent-mold casting**, **die casting**, and **investment casting**.

Sand-Casting   In the first step of sand casting, a wood or metal form or pattern is made in the shape of the part to be cast. The pattern is placed in a metal box called a flask and molding sand is packed around the pattern. When the pattern is withdrawn from the sand, it leaves a void forming the mold. Molten metal is poured into the mold through sprues or gates. After cooling, the casting is removed and cleaned (**Fig. 19.10**).

Cores formed from sand may be placed in a mold to create holes or hollows within a

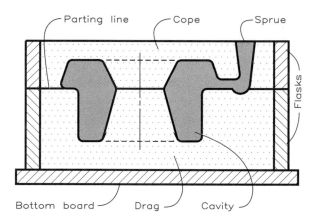

19.10 A two-section sand mold is used for casting a metal part.

19.11 This pattern is held in the bottom half (the drag) of a sand mold to form a mold for a casting.

casting. After the casting has been formed, the cores are broken apart and removed, leaving behind the desired void within the casting.

Because the patterns are placed in and removed from the sand before the metal is poured, the sides of the patterns follow the angle of taper, called draft, for ease of withdrawal from the sand. This angle depends on the depth of the pattern in the sand and varies from 2° to 8° in most applications. **Figure 19.11** shows a pattern held in the sand by a lower flask. Patterns are made oversize to compensate for shrinkage that occurs when the casting cools.

Because sand castings have rough surfaces, features that come into contact with other parts must be machined by drilling, grinding, finishing, or shaping. A casting of a tailstock base of a lathe shown in **Fig. 19.12** illustrates raised bosses that have been finished. The casting must be made larger than finished size where metal is to be removed by machining.

Fillets and rounds are used at the inside and outside corners of castings to increase their strength by relieving the stresses in the cast metal (**Fig. 19.13**). Fillets and rounds are also used because forming square corners by

19.12 The tailstock casting of a lathe has raised bosses and contact surfaces that were finished to improve the effectiveness of nuts and bolts. Fillets and rounds were added to the inside and outside corners. (Courtesy L. W. Chuck Company.)

the sand-casting process is difficult and because rounded edges make the finished product more attractive (**Fig. 19.13**).

Permanent Mold Casting Permanent molds are made for the mass production of parts. They are generally made of cast iron and coated to prevent fusing with the molten metal poured into them (**Fig. 19.14**).

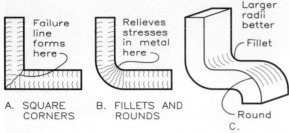

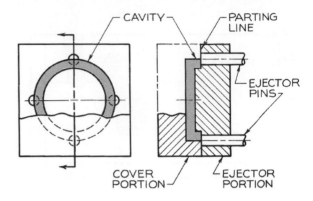

19.13

A Square corners cause a failure line to form, creating a weakness at this point.

B Fillets and rounds make the corners of a casting stronger and more attractive.

C The larger the radii of fillets and rounds the stronger the casting will be.

19.15 A die used for casting a simple part. The metal is forced into the die to form the casting.

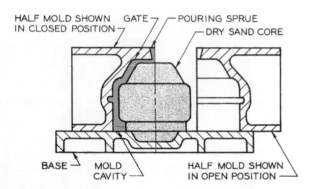

19.14 Permanent molds are made of metal for repetitive usage. Here, a sand core made from another mold is placed in the permanent mold to create a void within the casting.

Die Casting Die castings are used for the mass production of parts made of aluminum, magnesium, zinc alloys, copper, and other materials. Die castings are made by forcing molten metal into dies (or molds) under pressure. They are inexpensive, meet close tolerances, and have good surface qualities. The same general principles of sand castings—using fillets and rounds, allowing for shrinkage, and specifying draft angles—apply to die castings (**Fig. 19.15**).

Investment Casting Investment casting is used to produce complicated parts or artistic sculptures that would be difficult to form by other methods (**Fig. 19.16**). A new pattern must be used for each investment casting, so a mold or die is made for casting a wax master pattern. The wax pattern, identical to the casting, is placed inside a container and plaster or sand is poured (invested) around it. Once the investment has cured, the wax pattern is melted, leaving a hollow cavity to serve as the mold for the molten metal. After the casting has set, the plaster or sand is broken away from it.

Forging

The second major method of forming shapes is forging, which is the process of shaping or forming heated metal by hammering or forcing it into a die. Drop forges and press forges are used to hammer metal billets into forging dies. Forgings have the high strength and resistance to loads and impacts required for applications such as aircraft landing gears (**Fig. 19.17**).

Figure **19.18** shows three types of forging dies. A single-impression die gives an impression on one side of the parting line between the mating dies; a double-impression die

19.16 An investment casting (lost-wax process) used to produce complex metal pieces, decorative objects, and sculptures.

19.17 An aircraft landing-gear component formed by forging. (Courtesy of Cameron Iron Works.)

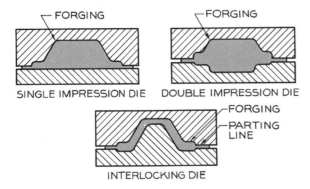

19.18 The three types of forging dies.

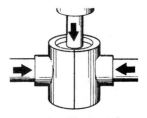

A. Side dies are closed on the billet, forming as required.

B. Vertical and horizontal rams enter the closed dies to form the part.

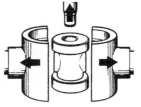

C. Rams are withdrawn, the dies open, and forging extracted

D. Result: A forging having multiple planes, no flash, and no draft.

19.19 The steps involved in forging a part with external dies and an internal ram. (Courtesy of Cameron Iron Works.)

gives an impression on both sides of the parting line; and the interlocking dies give an impression that may cross the parting line on either side. **Figure 19.19** shows how an object is forged with horizontal dies and a internal ram to hollow the object.

Figure 19.20 illustrates the sequence of forging a part from a billet by hammering it into different dies. It is then machined to its proper size within specified tolerances.

Figure 19.21 shows a working drawing for making a forged part. When preparing forging drawings, you must consider (1) draft angles and parting lines, (2) fillets and rounds, (3) forging tolerances, (4) extra material for machining, and (5) heat treatment of the finished forging.

Draft, the angle of taper, is crucial to the forging process. The minimum radii for inside corners (fillets) are determined by the height of the feature (**Fig. 19.22**). Similarly, the minimum radii for the outside corners (rounds)

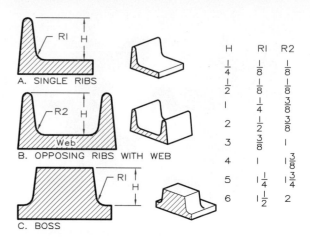

A. SINGLE RIBS

B. OPPOSING RIBS WITH WEB

C. BOSS

H	R1	R2
$\frac{1}{4}$	$\frac{1}{8}$	$\frac{1}{8}$
$\frac{1}{2}$	$\frac{1}{8}$	$\frac{1}{8}$
1	$\frac{1}{4}$	$\frac{3}{8}$
2	$\frac{1}{2}$	$\frac{3}{8}$
3	$\frac{3}{8}$	1
4	1	$1\frac{3}{4}$
5	$1\frac{1}{4}$	$1\frac{3}{4}$
6	$1\frac{1}{2}$	2

19.22 Guidelines for determining the minimum radii for fillets (inside corners) on forged parts.

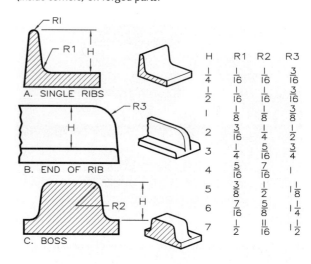

A. SINGLE RIBS

B. END OF RIB

C. BOSS

H	R1	R2	R3
$\frac{1}{4}$	$\frac{1}{16}$	$\frac{1}{16}$	$\frac{3}{16}$
$\frac{1}{2}$	$\frac{1}{16}$	$\frac{1}{16}$	$\frac{3}{16}$
1	$\frac{1}{8}$	$\frac{1}{8}$	$\frac{3}{8}$
2	$\frac{3}{16}$	$\frac{1}{4}$	$\frac{1}{2}$
3	$\frac{1}{4}$	$\frac{5}{16}$	$\frac{3}{4}$
4	$\frac{5}{16}$	$\frac{7}{16}$	1
5	$\frac{3}{8}$	$\frac{1}{2}$	$1\frac{1}{8}$
6	$\frac{7}{16}$	$\frac{5}{8}$	$1\frac{1}{4}$
7	$\frac{1}{2}$	$\frac{11}{16}$	$1\frac{1}{2}$

19.23 Guidelines for determining the minimum radii of rounds (outside corners) on forged parts.

are related to a feature's height (**Fig. 19.23**). The larger the radius of a fillet or round, the better it is for the forging process.

Some of the standard steels used for forging are designated by the SAE numbers 1015, 1020, 1025, 1045, 1137, 1151, 1335, 1340, 4620, 5120, and 5140. Iron, copper, and aluminum also can be forged.

Rolling Rolling is a type of forging in which the stock is rolled between two or more rollers to shape it. Rolling can be done at right angles or parallel to the axis of the part (**Fig. 19.24**). If

19.20 Steps A through G are required to forge a billet into a finished connecting rod. (Courtesy of the Drop Forging Association.)

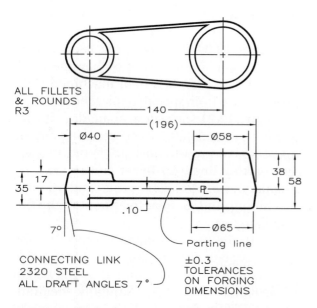

ALL FILLETS & ROUNDS R3

CONNECTING LINK
2320 STEEL
ALL DRAFT ANGLES 7°

Parting line

±0.3 TOLERANCES ON FORGING DIMENSIONS

19.21 A working drawing for a forging, showing draft angles and the parting line (PL) where the dies come together.

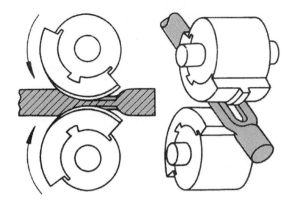

19.24 Features on parts may be formed by rolling. Here a part is being rolled parallel to its axes. (Courtesy of General Motors Corporation.)

19.25 A cylindrical rod being rolled to shape.

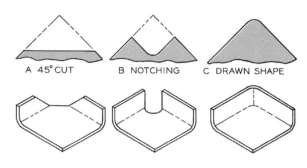

19.26 Box-shaped parts formed by stamping.

A A corner cut of 45° permits flanges to be folded with no further trimming.

B Notching has the same effect as the 45° cut and is often more attractive.

C A continuous corner flange requires that the blank be developed so that it can be drawn into shape.

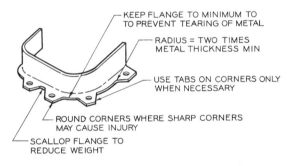

19.27 A sheet metal flange design with notes that explain design details.

a high degree of shaping is required, the stock is usually heated before rolling. If the forming requires only a slight change in shape, rolling can be done without heating the metal, which is called **cold rolling** (**CR**); CRS means cold-rolled steel. **Figure 19.25** shows a cylindrical rod being rolled.

Stamping

Stamping is a method of forming flat metal stock into three-dimensional shapes. The first step of stamping is to cut out the shapes, called **blanks**, which are formed by bending and pressing them against forms. **Figure 19.26** shows three types of box-shaped parts formed by stamping, and **Fig. 19.27** shows a design for a flange to be formed by stamping. Holes in stampings are made by punching, extruding, or piercing (**Fig. 19.28**).

19.5 Machining Operations

After metal parts have been formed, machining operations must be performed to complete them. The machines used most often are

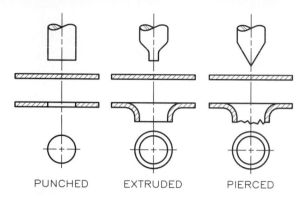

PUNCHED　　　EXTRUDED　　　PIERCED

19.28　The three methods used to form holes in sheet metal by punching.

19.29　This typical metal lathe holds and rotates the work piece between its centers for machining. (Courtesy of the Clausing Corporation.)

the **lathe, drill press, broaching machine, milling machine, shaper,** and **planer.** Some of these machines require manual operation; others are computer programmed to run at high speeds automatically, and require minimal or no operator attention.

Lathe

The **lathe** shapes cylindrical parts while rotating the work piece between its centers (**Fig. 19.29**). The fundamental operations performed on the lathe are **turning, facing, drilling, boring, reaming, threading,** and **undercutting** (**Fig. 19.30**).

Turning forms a cylinder with a tool that advances against and moves parallel to the cylinder being turned between the centers of the lathe (**Fig. 19.31**). **Facing** forms flat surfaces perpendicular to the axis of rotation of the part being rotated.

Drilling is performed by mounting a drill in the tail stock of the lathe and rotating the work while the bit is advanced into the part (**Fig. 19.32A–C**). **Boring** makes large holes by enlarging smaller drilled holes with a tool mounted on a boring bar (**Fig. 19.33A**). **Undercutting** is a groove cut inside a cylindrical hole with a tool mounted on a boring bar. The groove is cut as the tool advances from the center of the axis of revolution into the

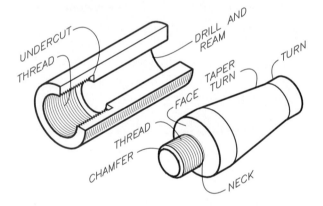

19.30　The basic operations that are performed on a lathe.

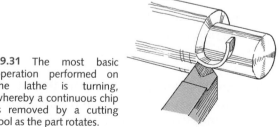

19.31　The most basic operation performed on the lathe is turning, whereby a continuous chip is removed by a cutting tool as the part rotates.

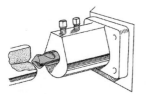

A. Start drilling

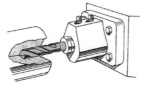

B. Twist drilling

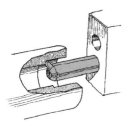

C. Core drilling

19.32 The three steps in drilling a hole in the end of a cylinder: (A) start drilling, (B) twist drilling, and (C) core drilling.

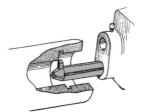

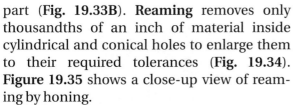

A. Boring

B. Undercutting

19.33 (A) This hole is being bored with a cutting tool attached to a boring bar of a lathe. (B) An undercut is being done by the tool and boring bar.

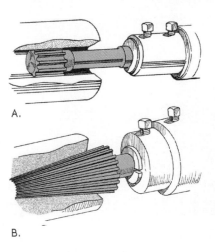

A.

B.

19.34 Fluted reamers can be used to finish inside (A) cylindrical and (B) conical holes within a few thousandths of an inch.

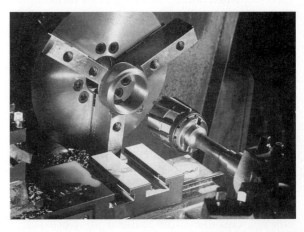

19.35 A hole being reamed by honing. (Courtesy Barber-Coleman Company.)

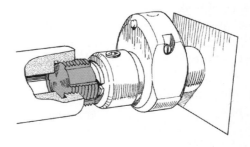

19.36 Internal threads can be cut on a lathe with a die called a *tap*. A recess, called a *thread relief*, was formed at the end of the threaded hole.

part (**Fig. 19.33B**). **Reaming** removes only thousandths of an inch of material inside cylindrical and conical holes to enlarge them to their required tolerances (**Fig. 19.34**). **Figure 19.35** shows a close-up view of reaming by honing.

Threading of external shafts and internal holes can be done on the lathe. The die used for cutting internal holes is called a **tap** (**Fig. 19.36**). A tapping die being used to thread a hole held in the chuck of a lathe is illustrated in **Fig. 19.37**.

The **turret lathe** is a programmable lathe that can perform sequential operations, such as drilling a series of holes, boring them, and

19.37 A die being used on a lathe to cut internal threads in a part. (Courtesy the Landis Machine Company.)

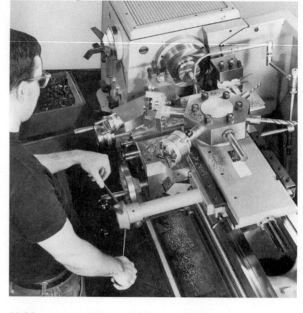

19.38 A turret lathe performs a sequence of operations by revolving the turret on which various tools are mounted.

then reaming them. The turret is a multisided tool holder that sequentially rotates each tool into position for its particular operation (**Fig. 19.38**).

Drill Press

The **drill press** is used to drill small and medium-sized holes (**Fig. 19.39**). The stock being drilled is held securely by fixtures or clamps. The drill press can be used for counterdrilling, reaming, boring, counterboring, spotfacing, countersinking, and tapping or threading (**Fig. 19.40**). Multiple-head drill presses can be programmed to perform a series of drilling operations for mass production applications (**Fig. 19.41**).

Measuring Cylinders The diameters of cylindrical features of parts made on a drill press or a lathe are measured, not their radii, to determine their sizes. Internal and external micrometer calipers are used for this purpose to measure to within one ten-thousandth of an inch.

Broaching Machine

Cylindrical holes can be converted into square, rectangular, or hexagonal holes with a

19.39 This small drill press is used to make holes in parts. (Courtesy of Clausing Corporation.)

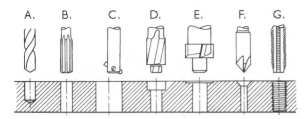

19.40 The basic operations performed on the drill press are (A) drilling, (B) reaming, (C) boring, (D) counterboring, (E) spotfacing, (F) countersinking, and (G) tapping (threading).

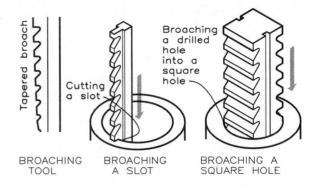

19.42 A broaching tool can be used to cut slots and holes with square corners on the interior and exterior of parts. Other shapes may also be broached.

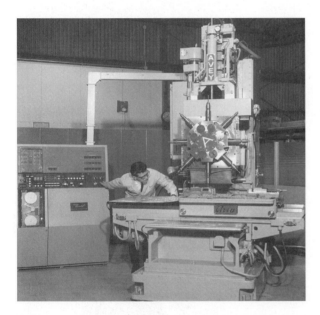

19.41 A multiple-head drill press, which can be programmed to perform a series of operations sequentially.

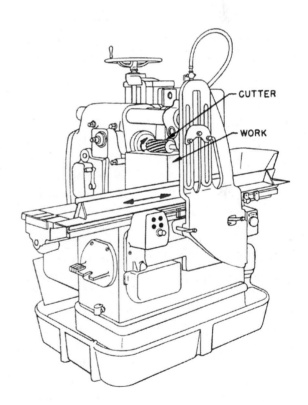

19.43 The milling machine operates by mounting the work on a bed that moves beneath revolving cutters. (Courtesy of the General Motors Corporation.)

broach mounted on a special machine (**Fig. 19.42**). A broach has a series of teeth graduated in size along its axis, beginning with teeth that are nearly the size of the hole to be broached and tapering to the final size of the hole. The broach is forced through the hole by pushing or pulling in a single pass, with each tooth cutting more from the hole as it passes through. Broaches can be used to cut external grooves, such as keyways or slots, in a part.

Milling Machine

The **milling machine** uses a variety of cutting tools, rotated about a shaft (**Fig. 19.43**), to form different grooved slots, threads, and gear teeth. The milling machine can cut irregular grooves in cams and finish surfaces on a part within a high degree of tolerance. The cutters revolve about a stationary axis while the work is passed beneath them.

Shaper

The shaper is a machine that holds a work piece stationary while the cutter passes back and forth across it to shape the surface or to cut a groove one stroke at a time (**Fig. 19.44**). With each stroke of the cutting tool, the material is shifted slightly to align the part for the next overlapping stroke (**Fig. 19.45.**).

Planer

Unlike the shaper, which holds the work piece stationary, the **planer** passes the piece under the cutters to machine large flat surfaces (**Fig. 19.46**). Like the shaper, the planer can cut grooves or slots and finish surfaces that must meet close tolerances.

19.6 Surface Finishing

Surface finishing produces a smooth, uniform surface. It may be accomplished by grinding, polishing, lapping, buffing, and honing.

    Grinding involves holding a flat surface against a rotating abrasive wheel (**Fig. 19.47**). Grinding is used to smooth surfaces, both cylindrical and flat, and to sharpen edges used for cutting, such as drill bits (**Fig. 19.48**).

    Polishing is done in the same way as grinding, except that the polishing wheel is flexible because it is made of felt, leather, canvas, or fabric.

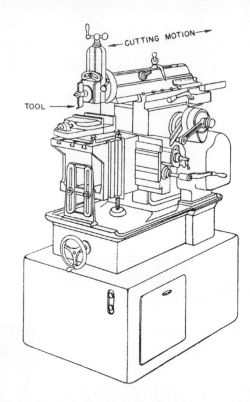

19.44 The shaper holds the work stationary while the machine's cutting tool makes strokes back and forth across the part to cut the desired shape in the work piece. (Courtesy of the General Motors Corporation.)

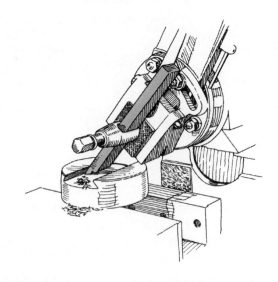

19.45 The shaper moves back and forth across the part, removing metal as it advances, to shape surfaces, cut slots, and perform other operations.

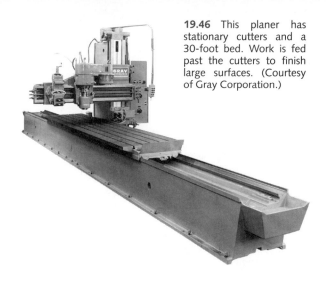

19.46 This planer has stationary cutters and a 30-foot bed. Work is fed past the cutters to finish large surfaces. (Courtesy of Gray Corporation.)

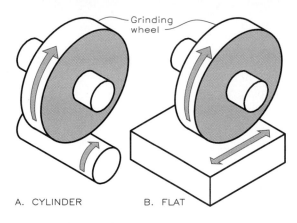

A. CYLINDER   B. FLAT

19.48 Grinding may be used to finish (A) cylindrical and (B) flat surfaces.

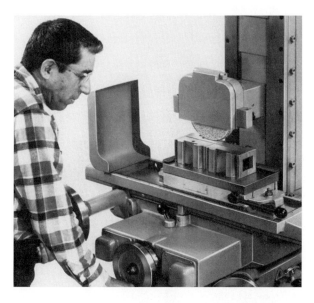

19.47 The operator is grinding the upper surface of this part to a smooth finish with a grinding wheel. (Courtesy of the Clausing Corporation.)

Lapping produces very smooth surfaces. The surface to be finished is held against a lap, which is a large, flat surface coated with a fine abrasive powder that finishes a surface as the lap rotates. Lapping is done only after the surface has been previously finished by a less accurate technique, such as grinding or pol-

ishing. Cylindrical parts can be lapped by using a lathe with the lap.

Buffing removes scratches from a surface with a belt or rotating buffer wheel made of wool, cotton, felt, or other fabric. To enhance the buffing, an abrasive mixture is applied to the buffed surface during the process.

Honing finishes the outside or inside of holes within a high degree of tolerance (see again Fig. 19.35). The honing tool is rotated as it is passed through the holes to produce the types of finishes found in gun barrels, engine cylinders, and other products requiring a high degree of smoothness.

19.7 Plastics and Other Materials*

Plastics (polymers) are widely used in numerous applications ranging from clothing, containers, and electronics to automobile bodies and components. Plastics are easily formed into irregular shapes, have a high resistance to weather and chemicals, and are available in limitless colors. The three basic types of plastics are thermoplastics, thermosetting plastics, and elastomers.

*Parts of this section are based on Serope Kalpakjian, *Manufacturing Engineering and Technology,* 2nd ed. (Reading, MA: Addison-Wesley, 1992).

	MACHINABILITY	FORMABILITY	CASTABILITY	WELDABILITY	CORROSION RES.	ABRASION RES.	LB/CU FT	YIELD: 1000 PSI	Typical Applications
THERMOPLASTICS									
ACRYLIC	G	G	E	A	E	F	74	9	Aircraft windows, TV parts, Lenses, skylights
ABS	G	G	G	A	E	G	66	66	Luggage, boat hulls, tool handles, pipe fittings
POLYMIDES (NYLON)	E	G	G	—	G	E	73	15	Helmets, gears, drawer slides, hinges, bearings
POLYETHYLENE	G	F	G	A	F	F	58	2	Chemical tubing, containers, ice trays, bottles
POLYPROPYLENE	G	G	G	A	E	G	56	5.3	Card files, cosmetic cases, auto pedals, luggage
POLYSTYRENE	G	E	G	A	P	G	67	7	Jugs, containers, furniture, lighted signs
POLYVINYL CHLORIDE	E	E	G	A	G	G	78	4.8	Rigid pipe and tubing, house siding, packaging
THERMOSETS									
EPOXY	F	G	G	—	E	G	69	17	Circuit boards, boat bodies, coatings for tanks
SILICONE	F	G	G	—	G	G	109	28	Flexible hoses, heart valves, gaskets
ELASTOMERS									
POLYURETHANE	G	G	G	A	G	E	74	6	Rigid: Solid tires, bumpers; Flexible: Foam, sponges
SBR RUBBER	—	—	E	—	F	E	39	3	Belts, handles, hoses, cable coverings
GLASSES									
GLASS	F	G	—	—	F	F	160	10+	Bottles, windows, tumblers, containers
FIBERGLASS	G	—	E	A	G	G	109	20+	Boats, shower stalls, auto bodies, chairs, signs

E=Excellent
G=Good
F=Fair
P=Poor
A=Adhesives

19.49 These are the characteristics of and typical applications for commonly used plastics and other materials.

Thermoplastics may be softened by heating and formed to the desired shape. If a polymer returns to its original hardness and strength after being heated, it is classified as a **thermoplastic**. In contrast, thermosetting plastics cannot be changed in shape by reheating after they have permanently set. **Elastomers** are rubberlike polymers that are soft, expandable, and elastic, which permits them to be deformed greatly and then to return to their original size.

Figure 19.49 shows commonly used plastics and other materials, including **glass** and **fiberglass**. The weights and yields of the materials are given, along with examples of their applications.

The motorized wheelchair shown in **Fig. 19.50** is made of plastic. It has fewer parts and weighs less than motorized wheelchairs made of metal. Its rounded corners make it safe, eliminate joints, and make it easy to fabricate and clean.

19.50 The use of Dow plastic in this motorized wheelchair (made by Amigo, Inc.) reduced the number of parts by 97 percent and weight by 10 percent. It is also safe and easy to clean. (Courtesy of Dow Chemical Corporation.)

Review Questions

If you generally know the answers to these questions, you have read the chapter pretty well. If not, scan it again.

1. What is a ferrous metal? List the types of ferrous metals. Is steel ferrous or nonferrous?

2. What type of iron contains spheres of graphite that make it stronger than most other types of iron?

3. What type of iron is used for engine blocks because it resists vibration better than other types of iron?

4. What type of iron is known for withstanding wear and abrasion? What does it contain that gives it this characteristic?

5. For a ductile iron that is noted as 80-55-6, what is its tensile strength? What is its yield strength, and what is its percent of elongation?

6. For a malleable iron that is noted as 45006, what is its tensile strength and what is its percent of elongation.

7. What is the main ingredient that makes steel stronger than iron? Explain what the number designation, 1015 steel, means.

8. In general, what is the comparison of the weight of steel and iron?

9. When the number designation of aluminum is 5232, what is the main alloying element used with the aluminum?

10. How much would a 9 in. $\times$ 6 in. $\times$ 8 in. block of aluminum weigh?

11. How much would the block in Question 10 weigh if it were magnesium?

12. If a part were to be made by forging (hammering into shape), what metal characteristic would be most desirable?

13. What are the main differences in brittleness and hardness of metals?

14. In a sand casting mold, what are the top and bottom parts of the two-piece mold called?

15. Fillets and rounds are used to remove sharp corners from a metal part and to make it look more attractive. Are there other reasons for fillets and rounds? Explain.

16. What is the name of the casting process for molding an irregular-shaped piece of an art sculpture?

17. How much would the connecting link in **Fig. 19.21** weigh if it were made of steel, brass, or magnesium?

18. What are the major methods of forming sheet metal?

19. What is the difference between boring and drilling?

20. What is a thread relief and why is it important?

21. What is the process for making a square hole?

22. What machining process is used to make interior holes smoother and accurate to a higher level of tolerance?

23. What is the difference between a shaper and a planer? Explain.

24. What are the methods of surface finishing? What are the characteristics of each?

25. What is a polymer? What are the three types of plastics and what are their respective applications?

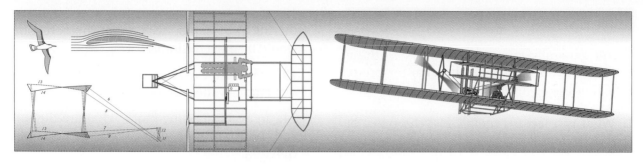

20

Dimensioning

20.1 Introduction

Working drawings show dimensions and notes that convey sizes, specifications, and other information necessary to build a project. With their full dimensions and specifications drawings serve as construction documents that become legal contracts.

The techniques of dimensioning presented here are based primarily on the standards of the American National Standards Institute (ANSI), especially Y14.5M, *Dimensioning and Tolerancing for Engineering Drawings*. The standards issued by companies such as the General Motors Corporation are also used.

20.2 Terminology

The strap shown in **Fig. 20.1** is described in **Fig. 20.2** with orthographic views to which dimensions were added. Refer to this drawing

20.1 This tapered strap is a part of a clamping device that is dimensioned in Fig. 20.2.

as various dimensioning terms are introduced.

Dimension lines: Thin lines (2H–4H pencil) with arrows at each end and numbers placed near their midpoints to specify size.

Extension lines: Thin lines (2H–4H pencil) extending from the part and between which dimension lines are placed.

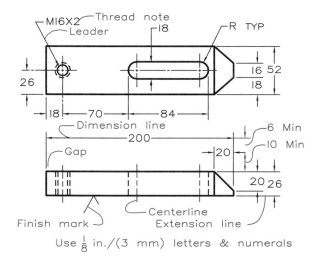

Use $\frac{1}{8}$ in./(3 mm) letters & numerals

20.2 This typical dimensioned drawing of the tapered strap shown in Fig. 20.1 introduces the terminology of dimensioning.

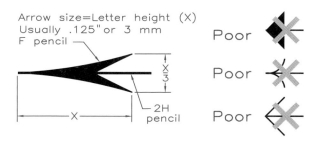

20.3 Draw arrowheads as long as the height of the letters used on the drawing and one-third as wide as they are long.

Centerlines: Thin lines (2H–4H pencil) used to locate the centers of cylindrical parts such as holes.

Leaders: Thin lines (2H–4H pencil) drawn from a note to the feature to which it applies.

Arrowheads: Symbols drawn at the ends of dimension lines and leaders; the same length as the height of the letters or numerals, usually 1/8 inch, as shown in **Fig. 20.3**.

Dimension numbers: Numerals placed near the middle of the dimension line and usually 1/8-inch high, with no units of measurement (″, in., or mm) shown.

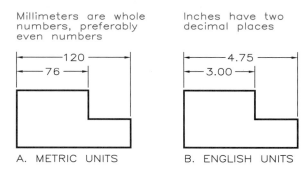

20.4 For the metric system, round millimeters to the nearest whole number. For the English system, show inches with two decimal places, even for whole numbers such as 3.00.

20.3 Units of Measurement

The two commonly used units of measurement are the decimal inch in the English (imperial) system and the millimeter in the metric (SI) system (**Fig. 20.4**). Giving fractional inches as decimals rather than common fractions makes arithmetic easier.

Figure 20.5 demonstrates the proper and improper dimensioning techniques with millimeters, decimal inches, and fractional inches. In general, round off dimensions in millimeters to whole numbers without fractions. However, when you must show a metric dimension of less than a millimeter, use a zero before the decimal point. Do not use a zero before the decimal point when inches are the unit.

Show decimal inch dimensions with two-place decimal fractions, even if the last numbers are zeros. Omit units of measurement from the dimension because they are understood to be in millimeters or inches. For example, use 112 (not 112 mm) and 67 (not 67″ or 5′-7″).

Architects use combinations of feet and inches in dimensioning and show foot marks, but usually omit inch marks—for example, 7′-2. Engineers use feet and decimal fractions of feet to dimension large-scale projects such as road designs—for example 252.7′.

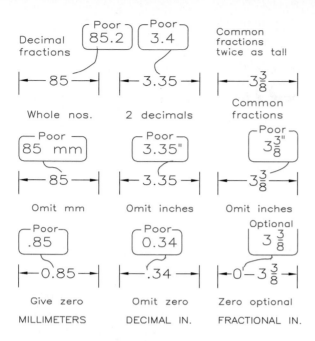

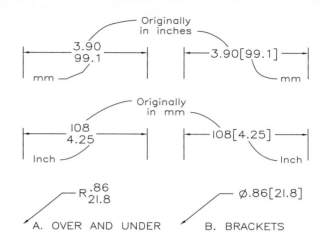

20.6 In dual dimensioning, place size equivalents in millimeters under or to the right of the inches (in brackets). Place the equivalent measurement in inches under or to the right of millimeters (in brackets). Show millimeters converted from inches as decimal fractions.

20.5 Basic principles of specifying measurements in SI and English units on a drawing.

20.4 English/Metric Conversions

To convert dimensions in inches to millimeters, multiply by 25.4. Similarly, to convert dimensions in millimeters to inches, divide by 25.4.

When millimeter fractions are required as a result of conversion from inches, one-place fractions usually are sufficient, but two-place fractions are used in some cases. Find the decimal digit by applying the following rules.

- Retain the last digit unchanged if it is followed by a number less than 5; for example, round 34.43 to 34.4.

- Increase the last digit retained by 1 if it is followed by a number greater than 5; for example, round 34.46 to 34.5.

- Retain the last digit unchanged if it is even and is followed by the digit 5; for example, round 34.45 to 34.4.

- Increase the last digit retained by 1 if it is odd and is followed by the digit 5; for example, round 34.75 to 34.8.

20.5 Dual Dimensioning

Some drawings may have dual dimensioning to give both metric and English units (**Fig. 20.6**). Place the millimeter equivalent either over or under the inch units, or place the converted dimension in brackets to the right of the original dimension. Be consistent in the arrangement you use on any set of drawings.

Computer Method As **Fig. 20.7** shows, the dimensioning variable DIMALT must be set to ON to obtain alternative (dual) dimensions in brackets following the units originally used. Set the variable DIMALTF (scale factor) to the value of the multiplier to be used to change the first dimension. Use DIMALTD to assign the desired number of decimal places for the second dimension. Then select dimensions by using the DIM command the same way you do to find single-value dimensions (Fig. 20.34).

20.6 Metric Units

Recall that in the metric system (SI), the first angle of projection positions the front view over the top view and the right-side view to

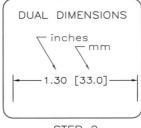

DIM VARS	DUAL DIMENSIONS
DIMALT=ON 　Alternate units DIMALTF=25.4 　Scale factor DIMALTD=1 　Decimal places	inches 　mm ├─ 1.30 [33.0] ─┤
STEP 1	STEP 2

20.7 Alternate (dual) dimensions by computer.

Step 1 Set the `Dim Vars` to `DIMALT` to set dual dimensions to ON, assign the scale factor (`DIMALTF`), and specify the number of decimal places (`DIMALTD`).

Step 2 Find the linear dimensions by using the same steps as later shown in Fig. 20.34. The dimension in brackets is the metric equivalent of the inch dimensions.

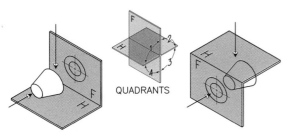

QUADRANTS

A. 1ST ANGLE PROJECTION   B. 3RD ANGLE PROJECTION

20.8 Projection systems.

A The SI system uses the first angle of orthographic projection, which places the top view under the front view.

B The American system uses the third angle of projection, which places the top view over the front view.

the left of the front view (**Fig. 20.8**). You should label metric drawings with one of the symbols shown in **Fig. 20.9** to designate the angle of projection. Display either the letters SI or the word METRIC prominently in or near the title block to indicate that the measurements are metric.

20.7 Numerals and Symbols

Vertical Dimensions

Vertical numeric dimensions on a drawing may be *aligned* or *unidirectional*. In the unidirectional method, all dimensions appear in

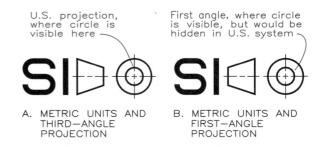

U.S. projection, where circle is visible here	First angle, where circle is visible, but would be hidden in U.S. system

A. METRIC UNITS AND THIRD—ANGLE PROJECTION

B. METRIC UNITS AND FIRST—ANGLE PROJECTION

20.9 The SI symbol.

A The SI symbol indicates that the millimeter is the unit of measurement, and the truncated cone specifies that third-angle projection was used to position the orthographic views.

B Again, the SI symbol denotes use of the millimeter, but the truncated cone designates that the first-angle projection was used.

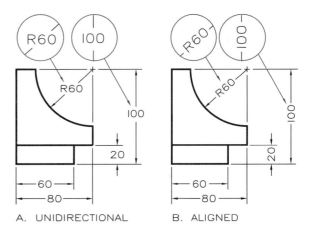

A. UNIDIRECTIONAL   B. ALIGNED

20.10 Unidirectional and aligned dimensions.

A Dimensions are unidirectional when they are horizontal in both vertical and horizontal dimension lines.

B Dimensions are aligned when they are lettered parallel to angular and vertical dimension lines to read from the right-hand side of the drawing (not from the left).

the standard horizontal position (**Fig. 20.10A**). In the aligned method numerals are parallel with vertical and angular dimension lines and read from the right-hand side of the drawing, never from the left-hand side (**Fig. 20.10B**). Aligned dimensions are used almost entirely in architectural drawings where dimensions (such as 22'-10 1/2) composed of feet, inches, and fractions are too long to fit well unidirectionally.

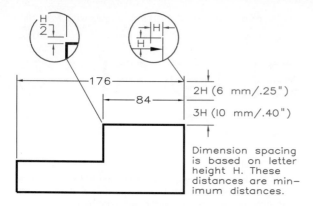

2H (6 mm/.25")

3H (10 mm/.40")

Dimension spacing is based on letter height H. These distances are min- imum distances.

20.11   Place dimensions on a view as shown here where all dimensioning geometry is based on the letter height (H) used.

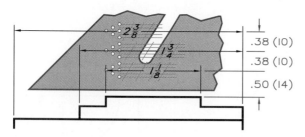

.38 (10)

.38 (10)

.50 (14)

20.12   Draw guidelines for common fractions in dimensions by aligning the center holes in the Braddock-Rowe triangle with the dimension line.

Computer Method   The variable `DIMTIH` (text inside dimension lines is horizontal), a vari- able of the `Dim` command of AutoCAD, must be set to `OFF` for aligned dimensions and to `ON` for unidirectional dimensions. The `DIMTOH` mode controls the position of text lying out- side dimension lines where the numerals do not fit within a short dimension line. When `DIMTOH` is `ON`, numerals will be horizontal; when `OFF`, numerals will align with the dimension line.

Placement

Dimensions should be placed on the most descriptive views of the part being dimen- sioned. The first row of dimensions should be at least three times the letter height (3H) from the object (**Fig. 20.11**). Successive rows of dimensions should be spaced equally at least two times the letter height apart (0.25 inch, or

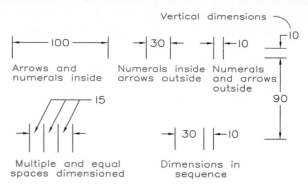

Vertical dimensions

Arrows and numerals inside

Numerals inside arrows outside

Numerals and arrows outside

Multiple and equal spaces dimensioned

Dimensions in sequence

20.13   When space permits, place numerals and arrows inside extension lines. For smaller spaces use other placements, as shown.

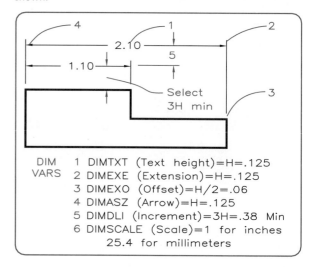

Select 3H min

DIM VARS		
	1	DIMTXT (Text height)=H=.125
	2	DIMEXE (Extension)=H=.125
	3	DIMEXO (Offset)=H/2=.06
	4	DIMASZ (Arrow)=H=.125
	5	DIMDLI (Increment)=3H=.38 Min
	6	DIMSCALE (Scale)=1 for inches
		25.4 for millimeters

20.14   The assignment of dimensioning variables to be applied by AutoCAD is shown here. Once set, these variables remain active with the file in use. `DIMSCALE` can be used to enlarge or reduce all of these variables.

6 mm, when 1/8-inch letters are used). Use the Braddock-Rowe lettering guide triangle to space the dimension lines (**Fig. 20.12**).

Figure 20.13 illustrates how to place dimen- sions in limited spaces. Regardless of space lim- itations, do not make numerals smaller than they appear elsewhere on the drawing.

Computer   Method   Dimensioning variables and their minimum settings are shown in **Fig. 20.14**. You may change variables set at these proportions at the same time by using `DIM-SCALE`. For example, `DIMSCALE` = 25.4 would

Dim Vars	Default	Description
DIMADEC	-1	Decimal places for ang. dims.
DIMALT	OFF	Alternate units selected
DIMALTD	2	Alternate unit decimal places
DIMALTF	25.4	Alternate unit scale factor
DIMALTTD	2	Alternate tolerance dec. places
DIMALTTZ	0	Alternate tolerance zero suppress.
DIMALTU	2	Alternate units
DIMALTZ	0	Alternate unit zero suppression
DIMAPOST	—	Default suffix for alternate text
DIMASO	ON	Create associative dimensions
DIMASZ	.125	Arrow length
DIMAUNIT	0	Angular unit format
DIMBLK	—	Arrow block name
DIMBLK1	—	First arrow block name
DIMBLK2	—	Second arrow block name
DIMCEN	.09	Center mark size
DIMCLRD	BYLAYER	Dimension line color
DIMCLRE	BYLAYER	Extension line & leader color
DIMCLRT	BYLAYER	Dimension & extension color
DIMDEC	4	Decimal places for dimensions
DIMDLE	0	Dimension line extension
DIMDLI	.38	Dim. increment for cintinuation
DIMEXE	.125	Extension beyond dimension line
DIMEXO	.06	Extension line offset
DIMFIT	3	Fit text
DIMGAP	.06	Justification of text on dim. line
DIMJUST	0	Gap from dimension line to text
DIMLFAC	1	Length factor
DIMLIM	OFF	Gives tolerances in limit form
DIMPOST	—	Character suffix after dimensions
DIMRND	0	Rounding value for distances
DIMSAH	OFF	Separate arrowheads at each end
DIMSCALE	1	Scale factor for all dim. vars.
DIMSD1	OFF	Suppress first dimension line
DIMSD2	OFF	Suppress second dimension line
DIMSE1	OFF	Suppress first extension line
DIMSE2	OFF	Suppress second extension line
DIMSHO	ON	Changes dimens. while dragging
DIMSOXD	OFF	Suppress outside dimension lines
DIMSTYLE	STANDARD	Current dimensioning style
DIMTAD	0	Text placed above dimension line
DIMTDEC	4	Tolerance decimal places
DIMTFAC	1	Tolerance text scale factor
DIMTIH	ON	Text inside extension lines horiz.
DIMTIX	OFF	Text forced inside extension lines
DIMTM	0	Minus tolerance value
DIMTOFL	OFF	Forces dim. line inside, text out
DIMTOH	ON	Text outside ext. lines is horiz
DIMTOL	OFF	Applies tolerances to dimensions
DIMTOLJ	1	Tolerance vertical justification
DIMTP	0	Plus tolerance value
DIMTSZ	0	Tick size
DIMTVP	0	Text over or under dimen. line
DIMTXSTY	STANDARD	Text style
DIMTXT	.125	Text height
DIMTZIN	0	Tolerance zero suppression
DIMUNIT	2	Unit format
DIMUPT	OFF	User positioned text
DIMZIN		Zero suppression

20.15 AutoCAD's 59 dimensioning variables. Chapter 37 covers more details of their application.

Symbols

Figure 20.16 shows standard dimensioning symbols and their sizes based on the letter height, usually 1/8 inch. Using these symbols instead of lettering full notes saves drawing time.

20.8 Dimensioning by Computer

The AutoCAD program offers many features that are helpful in dimensioning. For exam-

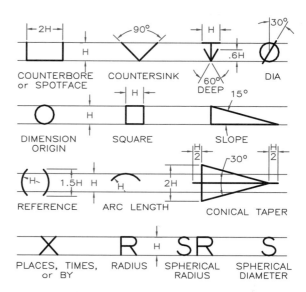

20.16 Standard dimensioning symbols can be used instead of words to indicate parts. Their proportions are based on the letter height, H, which usually is 1/8 inch.

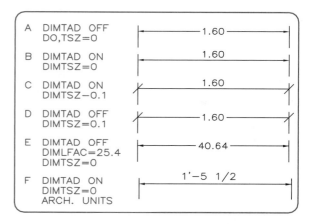

A	DIMTAD OFF DO,TSZ=0	1.60
B	DIMTAD ON DIMTSZ=0	1.60
C	DIMTAD ON DIMTSZ—0.1	1.60
D	DIMTAD OFF DIMTSZ=0.1	1.60
E	DIMTAD OFF DIMLFAC=25.4 DIMTSZ=0	40.64
F	DIMTAD ON DIMTSZ=0 ARCH. UNITS	1'-5 1/2

20.17 These dimension lines illustrate the effects of using the different dimensioning variables in AutoCAD.

ple, **Fig. 20.17** shows several combinations of dimensioning variables that are available with the DIM command.

You can place text inside the dimension line (DIMTAD: OFF) or above it (DIMTAD: ON). You may place arrowheads at the ends of dimension lines DIMASZ>0 or use tick marks (slashes) instead when DIMTSZ is set to a value greater than 0, usually about half the letter height. You may select UNITS as architectural (feet and inches), metric (no decimal

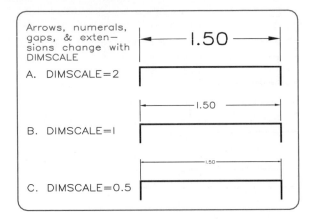

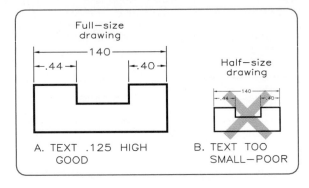

20.18 DIMSCALE, a subcommand under DIM:, permits changing the sizes of all dimensioning variables by typing a single factor. It changes the text size, arrows, offsets, and extensions at the same time.

fractions), decimal inches (two or more decimal fractions), or engineering units (feet and decimal inches).

You may use DIMSCALE (an option with the DIM command) to change all variables of dimensioning by typing a new DIMSCALE value (**Fig. 20.18**). When invoked, the command changes the lengths of arrows, text size, and extension line offsets, without affecting other previously applied dimensions.

The current text STYLE is the dimensioning text. Had you set text to a specified height under the STYLE command, you could not change it with a different DIMSCALE setting, but it would remain at its constant specified height. For text height to be changeable, its height set by the STYLE command, must be 0 (zero). Then, DIMSCALE can be used to enlarge or reduce text height along with all the other dimensioning variables (**Fig. 20.19**).

20.9 Dimensioning Rules

There are many rules of dimensioning you should become familiar with in order to place dimensions and notes on drawings most effectively. Each geometrical shape—prisms, angular surfaces, cylindrical features, pyramids, cones, spheres, and arcs—has its own set of rules.

20.19 When dimensioning a drawing, be aware of its final plotted size so that you can size the dimensioning variables for reduction or enlargement. DIMSCALE is the most efficient command for assigning the scale to dimensioning variables.

RULE I: First row is placed 3 times letter height from part, minimum. (X=Letter height)

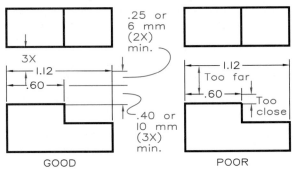

20.20 Place the first row of dimensions at least three times the letter height from the object. Successive rows should be at least two times the letter height apart.

Dimensioning rules should be looked upon as guidelines more than rules since in many drawings it is difficult to apply rules rigidly. Quite often rules of dimensioning must be violated or applied in different manner due to the complexity of the part or the lack of space that is available.

Prisms
Figures 20.20–20.32 illustrate the fundamental rules of dimensioning prisms. The rules are presented in the simplest of examples in order to focus on the specific rules one point at a time. These generally accepted dimensioning rules were not arbitrarily arrived at; they are based on a logical approach to aid in their application and interpretation.

RULE 2: Place dimensions between the views.

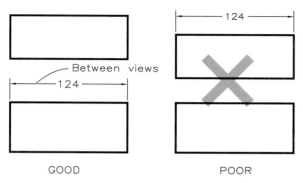

20.21 Place dimensions between the views sharing these dimensions.

RULE 3: Dimension the most descriptive views.

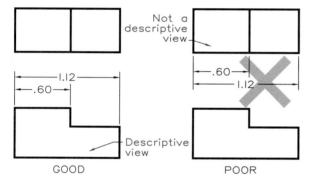

20.22 Place dimensions on the most descriptive views of an object.

RULE 4: Dimension from visible lines, not hidden lines.

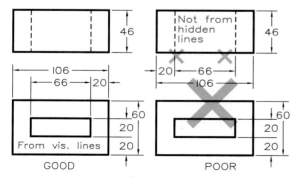

20.23 Dimension from visible features, not hidden features.

RULE 5: Give an overall dimension and omit one of the chain dimensions.

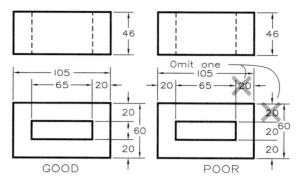

20.24 Leave the last dimension blank in a chain of dimensions and give an overall dimension.

RULE 5 (DEVIATION): When one chain dimension is not omitted, mark one dimension as a reference dimension.

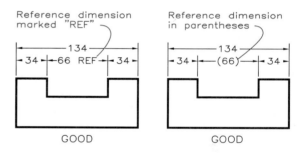

20.25 If you give all dimensions in a chain, mark the reference dimension (the one that would be omitted) with REF or place it in parentheses. Giving a reference dimension is a way of eliminating mathematical calculations in the shop.

RULE 6: Organize and align dimensions for ease of reading.

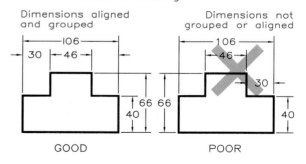

20.26 Place dimensions in well-organized lines for uncluttered drawings.

20.9 DIMENSIONING RULES • 315

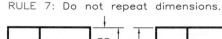

RULE 7: Do not repeat dimensions.

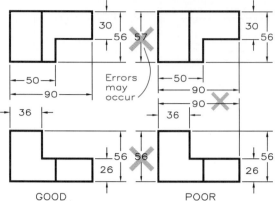

GOOD   POOR

20.27 To avoid errors or confusion, do not duplicate dimensions on a drawing.

RULE 8: Dimension lines should not cross other lines.

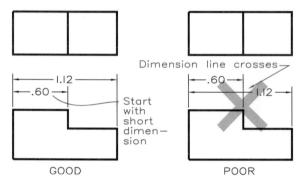

GOOD   POOR

20.28 Dimension lines should not cross any other lines unless absolutely necessary.

RULE 9: Extension lines may cross other lines if they must.

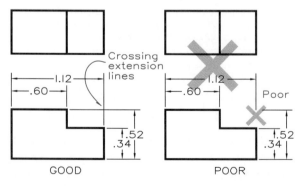

GOOD   POOR

20.29 Extension lines may cross other extension lines or object lines if necessary.

316 • **CHAPTER 20   DIMENSIONING**

RULE 9 (Cont.): Extension lines may cross other lines.

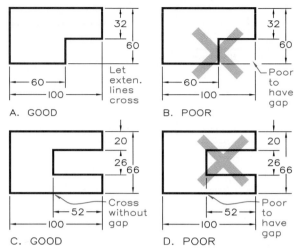

A. GOOD   B. POOR

C. GOOD   D. POOR

20.30 Leave a small gap from the edges of an object to extension lines that extend from them. Do not leave gaps where extension lines cross object lines or other extension lines.

RULE 10: Do not place dimensions within the views unless necessary.

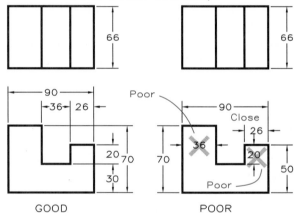

GOOD   POOR

20.31 Whenever possible place dimensions outside objects rather than inside their outlines.

Computer Method Dimension the part shown in **Fig. 20.33** by entering the DIM command and selecting the LINEAR and HORI-ZONTAL options. Obtain extension lines automatically by responding to the First extension line origin prompt with a (Enter) to get the prompt Select line, arc, or circle. Select the line to be measured and locate the dimension line when prompted.

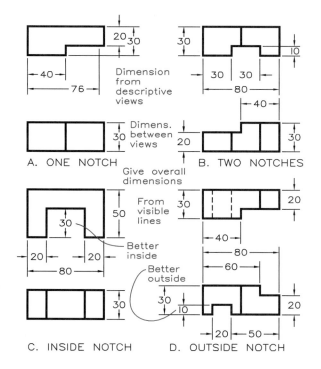

20.32 Dimensioning prisms.

A–B Dimension prisms from descriptive views and between views.

C You may dimension a notch inside the object if doing so improves clarity.

D Dimension visible lines, not hidden lines.

If you set the dimensioning variable DIMASO to ON, dimensions will be associative dimensions. That is, use of the STRETCH command updates the dimensioning numerals as you increase the part's size (**Fig. 20.34**). Associative dimensions will erase as a single unit (extension lines, dimension lines, arrows, and number). When DIMSHO is ON, you will be able to see the dimensioning numerals changing dynamically on the screen as you STRETCH the part to a new size.

Angles

You may dimension angles either by using coordinates that locate the ends of sloping surfaces or by angular measurements in degrees (**Fig. 20.35**). Fractional angles can be specified in decimal units or in degrees, minutes, and seconds. Recall that there are 60

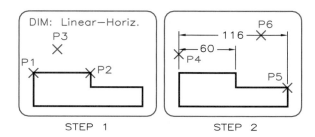

20.33 Linear dimensioning by computer.

Step 1 Command: DIM (Enter)
Dim: HORIZONTAL (Enter)
First extension line origin or press ENTER to select: (Select P1.)
Second extension line origin: (Select P2.)
Dimension line location (Text/Angle): (Select P3.)
Dimension text [current]: ((Enter) to select default value or input type value.)

Step 2 Dim: HORIZONTAL
First extension line origin or press ENTER to select: (Select P4.)
Second extension line origin: (Select P5.)
Dimension line location: (Select P6.)
Dimension text [current]: ((Enter) to select default or input correct value from the keyboard.)

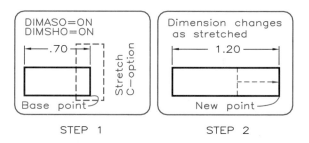

20.34 Associative dimensions.

Step 1 Set variables, DIMASO, and DIMSHO to ON to associate the dimensions with the size of the part to which they apply. Use the C option of the STRETCH command to window the ends of an extension line and dimension lines.

Step 2 elect a base point and a new point. Dragging the size of the part recalculates the dimensions dynamically.

minutes in a degree and 60 seconds in a minute. It is seldom that you will you need to measure angles to the nearest second. **Figures 20.36** and **20.37** illustrate basic rules for dimensioning angles.

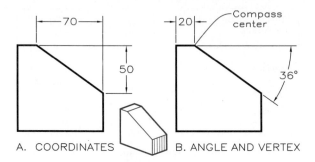

20.35 Dimensioning angles.

A Dimension angular planes by using coordinates.

B Measure angles by locating the vertex and measuring the angle in degrees. When accuracy is essential, specify angles in degrees, minutes, and seconds.

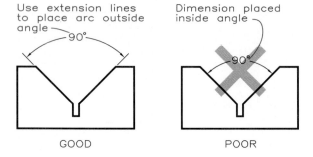

20.36 Place angular dimensions outside the object by using extension lines.

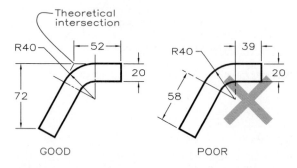

20.37 Dimension a bent surface rounded corner by locating its theoretical point of intersection with extension lines.

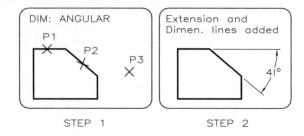

20.38 Producing angular dimensions by computer.

Step 1 Command: `DIM` (Enter)
Dim: `ANGULAR` (Enter)
Select first line: (Select P1.)
Second line: (Select P2.)
Enter dimension line arc location: (Select P3.)
Dimension text [41]: ((Enter) to accept default value 41.)

Step 2 Enter text location: (Enter)
Dim: ((Enter) to continue angular dimensioning or Esc to return to the Command prompt.)

Computer Method Select the ANGULAR option under the DIM command to dimension angles (**Fig. 20.38**). If room is not available for the arrows between the extension lines, AutoCAD will generate them outside the extension lines.

Cylindrical Parts and Holes

The diameters of cylinders are measured with either an internal micrometer (**Fig. 20.39**) or an external micrometer (**20.40**). Therefore, dimension cylinders in their rectangular views with a diameter instead of with a radius (**Figs. 20.41** and **20.42**). Place diameter symbols in front of the diametral dimensions to indicate that the dimension is a diameter. You will recall that the diametric symbol is a circle with a slash through it. In the English system the abbreviation DIA placed after the diametral dimension is still sometimes used, but the metric diameter symbol is preferred.

Space is almost always a problem in dimensioning, and all means of conserving space must be used. Stagger dimensions for concentric cylinders to avoid crowding, as

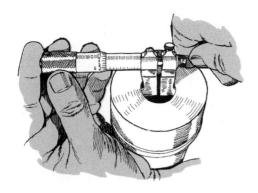

20.39 The internal micrometer caliper measures internal cylindrical diameters (radii cannot be measured).

20.40 The external micrometer caliper measures the diameter of a cylinder.

RULE 13: Dimension cylinders in their rectangular views with diameters.

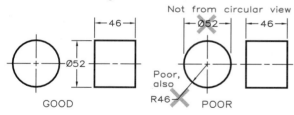

Not from circular view

GOOD   POOR

20.41 Dimension the diameter (not the radius) of a cylinder in the rectangular view.

shown in **Fig. 20.43**. Dimension cylindrical holes in their circular view with leaders (**Fig. 20.44**). The circular view is what would be used when the hole is located and drilled. Draw leaders specifying hole sizes as shown in **Fig. 20.45**. When you must place diameter

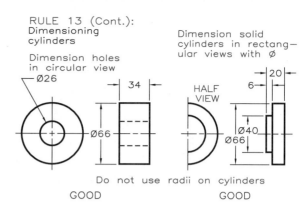

RULE 13 (Cont.):
Dimensioning cylinders

Dimension holes in circular view

Dimension solid cylinders in rectangular views with ⌀

Do not use radii on cylinders
GOOD   GOOD

20.42 Dimension holes in their circular views with leaders and concentric cylinders with a series of diameters.

RULE 14: Stagger dimension numerals to prevent crowding.

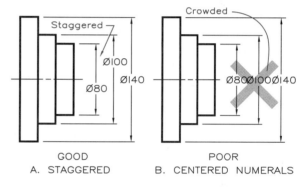

GOOD   POOR
A. STAGGERED   B. CENTERED NUMERALS

20.43 Dimensions on concentric cylinders are easier to read if they are staggered within their dimension lines.

RULE 15: Hole sizes are best given as diameters with leaders in circular views.

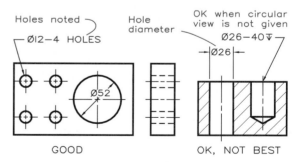

GOOD   OK, NOT BEST

20.44 Dimension holes in their circular view with leaders whenever possible, but dimension them in their rectangular views if necessary.

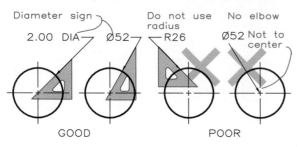

RULE 16: Leaders should have horizontal elbows and point toward the hole centers.

GOOD   POOR

20.45 Draw leaders pointing toward the centers of holes.

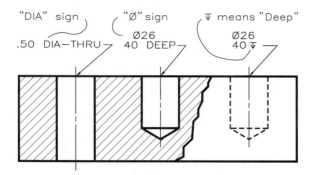

20.46 Diameter notes for holes in their rectangular views.

notes for holes in the rectangular view instead of the circular view, draw them as shown in **Fig. 20.46**. Examples of correctly dimensioned parts with cylindrical features are shown in **Fig. 20.47** and **20.48**.

Computer Method Dimension circles in the circular view as shown in **Fig. 20.49**. Dimension lines begin with the point selected on the circle and pass through the circle's center. The diameter symbol appears in front of the dimension numerals. Small circles are dimensioned with the arrows inside the circle and the dimension numerals outside, connected by a leader. Circles that are smaller have both the arrows and dimension outside the circle.

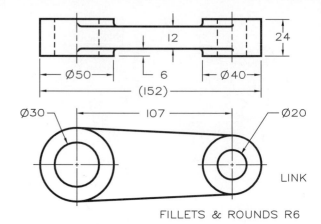

LINK

FILLETS & ROUNDS R6

20.47 The application of dimensions to cylindrical features. (F&R R6 means that fillets and rounds have a 6 mm radius.)

20.48 A cylindrical part, a collar (A) is shown drawn and dimensioned (B).

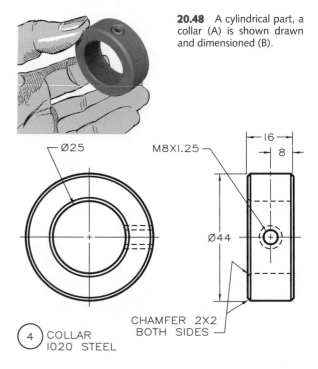

CHAMFER 2X2 BOTH SIDES

④ COLLAR
1020 STEEL

Pyramids, Cones, and Spheres

In **Figure 20.50** parts A–C show three methods of dimensioning pyramids and parts D and E show two acceptable methods of dimensioning cones. Dimension a complete sphere by giving its diameter as shown in part F. If the

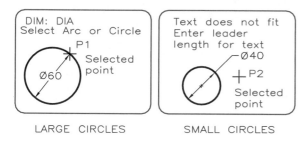

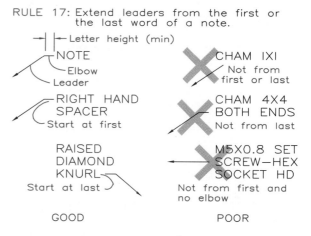

GOOD   POOR

20.49 Dimensioning circles by computer.

Step 1 Select the diameter option (DIA) under the DIM command to dimension a circle. Select P1 on the arc (an endpoint of the dimension) to produce the dimension. You have the option of replacing the dimension measured by the computer with a different value.

Step 2 When text does not fit, you get the prompt Dimension line location (Text/Angle): Select a point, P2, and the leader and dimension are drawn.

20.51 Extend leaders from the first or last word of a note with a horizontal elbow.

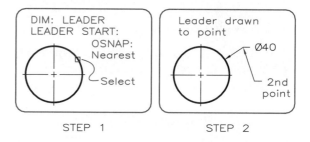

STEP 1   STEP 2

20.52 Producing leaders by computer.

Step 1 Command: DIM (Enter)
Dim: LEADER (Enter)
Leader start: (Pick P1 on the circle.)

Step 2 To point: (Select next point.) (Enter)
Dimension text [0.20]: %%C40 (Enter) (Type the desired dimension; this command does not measure.)
Dim: ((Enter) to continue leader dimensioning or Esc to return to a command prompt.)

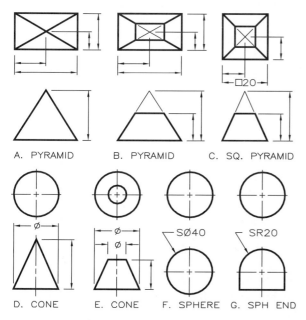

A. PYRAMID   B. PYRAMID   C. SQ. PYRAMID

D. CONE   E. CONE   F. SPHERE   G. SPH END

20.50 The proper way to dimension pyramids, cones, and spheres.

spherical shape is less than a hemisphere (part G), use a spherical radius (SR). Only one view is needed to describe a sphere.

Leaders

Leaders—thin lines used to reference notes and dimensions to features—are most often drawn at standard angles of triangles (see

again Fig. 20.45). Leaders should begin at either the first or last word of a note, with a short horizontal line (elbow) from the note, and extend to the feature being described, as shown in **Fig. 20.51**.

Computer Method To produce a leader that begins with an arrow, select the LEADER option under the DIM command (**Fig. 20.52**). To ensure that the arrow touches the circle, use OSNAP and NEAREST to snap the point of the arrow to the circumference. The program will prompt you To point until you press (Enter).

RULE 18: Dimension arcs (less than 180°)
with radii.

Arrow and dim. inside   Arrow inside dim. outside   Arrow and dim. outside

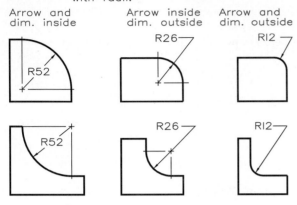

20.53 When space permits, place dimensions and arrows between the center and the arc. When space is not available for the number, place the arrow between the center and the arc number outside. If there is no space for the arrow inside, place both the dimension and arrow outside the arc with a leader.

The diameter of the circle will then appear on the screen. You must type in the desired diameter because this command does not measure. It applies the last number it used.

If the previous option was DIAMETR, the diameter symbol will precede the dimension. If the previous option was RADIUS, an R, the radius symbol, will precede the dimension.

Arcs and Radii

Full circles are dimensioned with diameters by arcs with radii (**Fig. 20.53**). Current standards specify that radii be dimensioned with an R preceding the dimension (for example, R10). The previous standard was for R to follow the dimension (10R, for example). Thus, both methods are seen on drawings.

You may dimension large arcs with a false radius (**Fig. 20.54**) by drawing a zigzag to indicate that the line is not the true radius. Where space is not available for radii, dimension small arcs with leaders.

Computer Method Use the RADIUS option of the DIM command to dimension arcs by selecting a point on the arc as the starting point of the arrow (**Fig. 20.55**). If space permits, the dimension will appear between the

FILLETS AND ROUNDS RIO UNLESS OTHER-WISE SPECIFIED

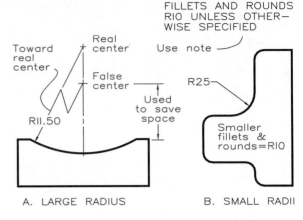

A. LARGE RADIUS   B. SMALL RADII

20.54 Dimensioning radii.

A Show a long radius with a false radius (a line with a zigzag) to indicate that it is not true length. Show its false center on the centerline of the true center.

B Specify fillets and rounds with a note to reduce repetitive dimensions of small arcs.

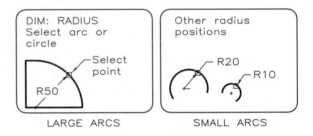

LARGE ARCS   SMALL ARCS

20.55 Use the RADIUS option of the DIM command to select a point on the arc. The program generates the dimension with its arrow at this point and precedes the dimension with an R. The program dimensions smaller arcs by positioning the dimension outside the arc or, because of even more limited space, by placing both the arrow and dimension outside the arc.

center and the arrow. For smaller arcs, the arrows will appear inside and the dimensions outside the arc.

When space is not available for arrows inside, both the arrow and dimension will appear outside the arc. Decimal values are preceded with a zero unless you override them by typing in R and the value without a preceding zero. Leading zeros will be omitted if the variable DIMZIN = 4.

Fillets and Rounds

When all fillets and rounds are equal in size, you may place a note on the drawing stating

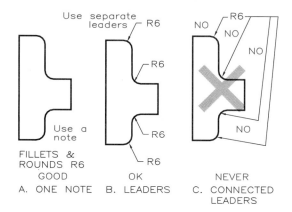

FILLETS &
ROUNDS R6
GOOD
A. ONE NOTE

OK
B. LEADERS

NEVER
C. CONNECTED
LEADERS

20.56 Indicate fillets and rounds by (A) notes or (B) separate leaders and dimensions. Never use confusing leaders (C).

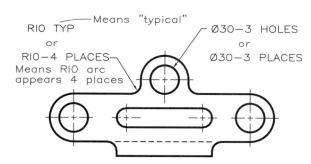

20.57 Use notes to indicate that identical features and dimensions are repeated to simplify dimensioning.

that condition or use separate notes (**Fig. 20.56**). If most, but not all, of the fillets and rounds have equal radii, the note may read ALL FILLETS AND ROUNDS R6 UNLESS OTHERWISE SPECIFIED (or abbreviated as F&R R6), with the fillets and rounds of different radii dimensioned separately.

You may note repetitive features, as shown in **Fig. 20.57** by using the notes TYPICAL, or TYP, which means that the dimensioned feature is typical of those not dimensioned. You may use the note PLACES, or PL, to specify the number of places that identical features appear, although only one is dimensioned.

Figure 20.58 shows a pulley that is drawn and dimensioned. The drawing demonstrates proper application of many of the rules discussed in this section.

20.58 This pulley, composed of cylindrical features, is dimensioned below.

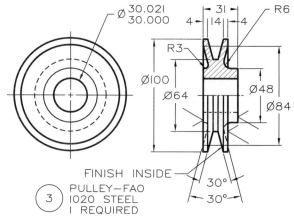

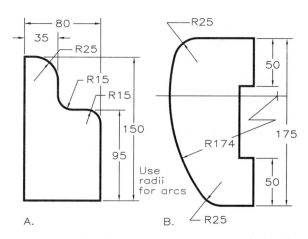

20.59 Use radii to dimension parts composed of arcs and partial circles.

20.10 Curved and Symmetrical Parts

You may dimension an irregular shape comprised of tangent arcs of varying sizes by using a series of radii (**Fig. 20.59**). Irregular curves

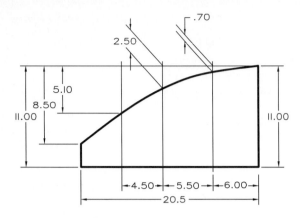

20.60 Use coordinates to dimension points along an irregular curve on a part.

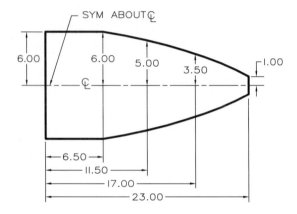

20.61 Use coordinates to dimension points along the curves of a symmetrical part.

may be dimensioned by using coordinates to locate a series of points along the curve (**Fig. 20.60**). You must use your judgement in determining how many located points are necessary to define the curve. Placing extension lines at an angle provides additional space for showing dimensions.

Symmetrical Parts

Dimension an irregular symmetrical curve with coordinates (**Fig. 20.61**). Note the use of dimension lines as extension lines, a permissible violation of dimensioning rules in this case.

Dimension symmetrical objects by using coordinates to imply that the dimensions are

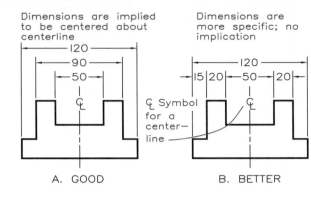

20.62 Symmetrical parts.

A You may dimension symmetrical parts implicitly about their centerlines.

B The better way to dimension symmetrical parts is explicitly about their centerlines.

symmetrical about the centerline (abbreviated CL), as shown in **Fig. 20.62A. Figure 20.62B** shows a better method of dimensioning this type of object, where symmetry is dimensioned with no interpretation required by the reader.

20.11 Finished Surfaces

Parts formed in molds, called **castings**, have rough exterior surfaces. If these parts are to assemble with and move against other parts, they will not function well unless their contact surfaces are machined to a smooth finish by a process such as grinding, shaping, or lapping.

Finish marks are drawn on the edge views of surfaces to indicate that they are to be finished (**Fig. 20.63**). Finish marks should be shown in every view where finished surfaces appear as edges, even if they are hidden lines. The preferred finish mark for the general cases is the uneven mark shown in **Fig. 20.63B**. When an object is finished on all surfaces, the note FINISHED ALL OVER (abbreviated FAO) is placed on the drawing.

20.12 Location Dimensions

Location dimensions give the positions, not the sizes, of geometric shapes (**Fig. 20.64**).

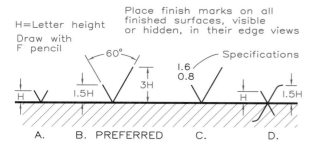

H=Letter height
Draw with
F pencil

Place finish marks on all finished surfaces, visible or hidden, in their edge views

60°

Specifications

1.6
0.8

3H

1.5H

H

H

1.5H

A.   B. PREFERRED   C.   D.

20.63 Finish marks indicate that a surface is to be machined to a smooth surface.

A The traditional V can be used for general applications.

B The unequal finish mark is the best for general applications.

C Where surface texture must be specified, this finish mark is used with texture values.

D The f-mark is the oldest and least-used symbol.

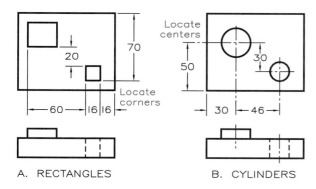

Locate centers

70

20

Locate corners

60   16 16

50   30

30   46

A. RECTANGLES   B. CYLINDERS

20.64 Location dimensions give the positions but not the sizes of geometric features with respect to other geometric.

Locate rectangular shapes by using coordinates of their corners and cylindrical shapes by using coordinates of their centerlines. In each case, dimension the view that shows both measurements. Always extend coordinates from any finished surfaces (even if a finished surface is a hidden line) because smooth machined surfaces allow the most accurate measurements. Locate and dimension single holes as shown in **Fig. 20.65** and multiple holes as shown in **Fig. 20.66**. **Figure 20.67** shows the application of location dimensions to a typical part with size dimensions omitted.

RULE 19: Locate holes in circular views and dimension diameters with a leader.

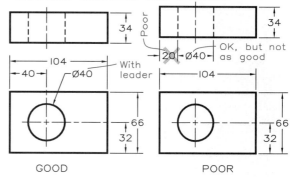

34

Poor

34

104

With leader

20 Ø40

OK, but not as good

40

Ø40

104

66

32

66

32

GOOD   POOR

20.65 Locate cylindrical holes in their circular views by coordinates to their centers.

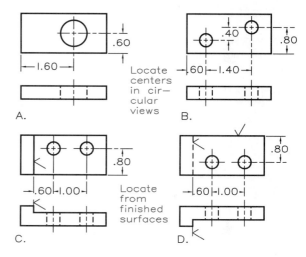

.60

1.60

Locate centers in cir—cular views

.40

.80

60 1.40

A.

B.

.80

.60 1.00

Locate from finished surfaces

.80

.60 1.00

C.

D.

20.66 Rules in summary.

A Locate cylindrical holes in their circular views from two surfaces of the object.

B Locate multiple holes from center to center.

C Locate holes from finished surfaces, even if the finished surfaces are hidden as in (D).

Baseline dimensions extend from two baselines in a single view (**Fig. 20.68**). The use of baselines eliminates the possible accumulation of errors in size that can occur from chain dimensioning.

Holes through circular plates can be located by using coordinates or a note, as shown in **Fig. 20.69**. Dimension the diameter of the imaginary circle passing through the centers of the holes in the circular view as a reference dimension and locate the holes by using coordinates

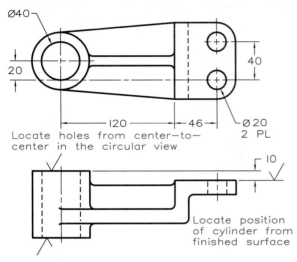

Ø40

20

120 — 46 — Ø20
2 PL

Locate holes from center—to—
center in the circular view

10

Locate position
of cylinder from
finished surface

20.67 A dimensioned shaft arm showing the application of location dimensions, with sizes omitted.

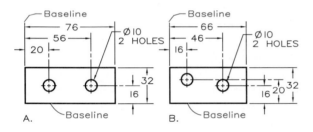

A.

Baseline
76
56
20
Ø10
2 HOLES
32
16
Baseline

B.

Baseline
66
46
16
Ø10
2 HOLES
32
20
16
Baseline

20.68 Measuring holes from two datum planes is the most accurate way to locate them and reduces the accumulation of errors possible in chain dimensioning.

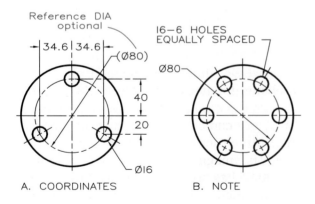

Reference DIA optional
34.6 | 34.6
(Ø80)
40
20
Ø16

16—6 HOLES EQUALLY SPACED
Ø80

A. COORDINATES   B. NOTE

20.69 Locate holes on a bolt circle by using (A) coordinates or (B) notes.

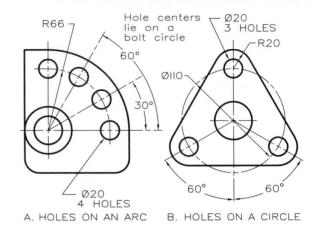

R66

Hole centers lie on a bolt circle
60°
Ø110
30°

Ø20
3 HOLES
R20

Ø20
4 HOLES

60°   60°

A. HOLES ON AN ARC   B. HOLES ON A CIRCLE

20.70 Locate centers of holes by using (A) a combination of radii and degrees or (B) a bolt circle.

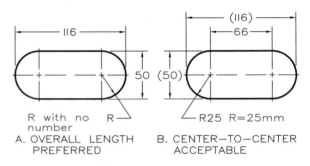

116

50 (50)

R with no number

A. OVERALL LENGTH PREFERRED

(116)
66

R25 R=25mm

B. CENTER—TO—CENTER ACCEPTABLE

20.71 Rounded ends.
A Dimension objects having rounded ends from end-to-end and give the height.
B A less desirable choice is to dimension the rounded ends from center-to-center and give the radius.

(A) or a note (B). This imaginary circle is called the **bolt circle** or **circle of centers**.

You may also locate holes with radial dimensions and their angular positions in degrees (**Fig. 20.70**). Holes may be located on their bolt circle even if the shape of the object is not circular.

Objects with Rounded Ends

Dimension objects with rounded ends from one rounded end to the other (**Fig. 20.71A**) and show their radii as R without dimensions to specify that the ends are arcs. Obviously, the end radius is half the height of the part. If you dimension the object from center to center (**Fig. 20.71B**), you must give the radius size.

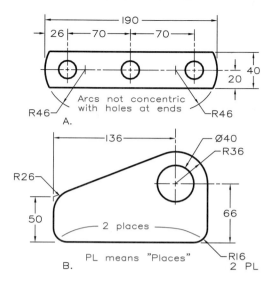

Arcs not concentric
with holes at ends

R46

A.

PL means "Places"

B.

20.72 These drawings show how to dimension parts having (A) rounded ends that are not concentric with the holes and (B) rounds and cylinders.

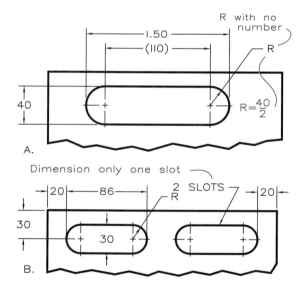

A.

Dimension only one slot

B.

20.73 The methods of dimensioning parts that have (A) one slot and (B) more than one slot.

You may specify the overall width as a reference dimension (116) to eliminate the need for calculations.

Dimension parts with partially rounded ends as shown in **Fig. 20.72A**. Dimension objects with rounded ends that are smaller than a semicircle with a radius and locate the arc's center (**Fig. 20.72B**).

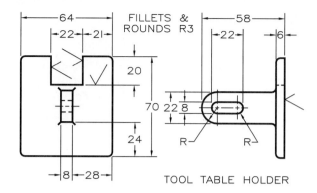

FILLETS &
ROUNDS R3

TOOL TABLE HOLDER

20.74 A dimensioned part having both slots and arcs.

Dimension a single slot with its overall width and height (**Fig. 20.73A**). When there are two or more slots, dimension one slot and use a note to indicate that there are other identical slots (**Fig. 20.73B**).

The tool table holder shown in **Fig. 20.74** illustrates dimensioning of arcs and slots. To prevent dimension lines from crossing, several dimensions are placed on a less descriptive view. Notice that the diameters of the semicircular features are given; therefore, notes of R are given to indicate radii but the radius size is unnecessary. Radii are half the diameters.

20.13 Outline Dimensioning

Now that you are familiar with most of the rules of dimensioning, you can better understand outline dimensioning, which is a way of applying dimensions to the outline (silhouette) of a part. By taking this approach, you have little choice but to place dimensions in the most descriptive views of the part. For example, imagine that the T-block shown in **Fig. 20.75** has no lines inside its outlines. It is dimensioned beginning with its location dimensions. When the inside lines are considered, additional dimensions are seldom needed.

Outline dimensioning is the placement of dimensions on views as if they had no internal lines. Practice in applying this concept will help you place dimensions on their most descriptive views.

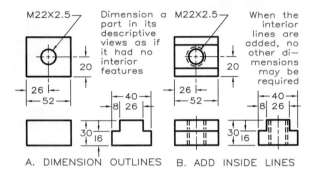

A. DIMENSION OUTLINES   B. ADD INSIDE LINES

M22X2.5

Dimension a part in its descriptive views as if it had no interior features

M22X2.5

When the interior lines are added, no other dimensions may be required

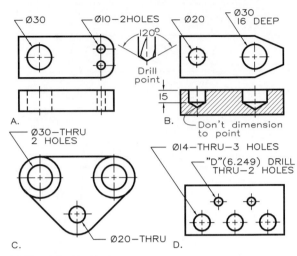

A.

B. Don't dimension to point

Drill point

C. Ø30—THRU 2 HOLES   Ø20—THRU

D. Ø14—THRU—3 HOLES "D"(6.249) DRILL THRU—2 HOLES

20.77 Cylindrical holes.

A and **B** Dimension cylindrical holes by either of these methods.

C and **D** When you use only one view, you have to note the THRU holes or specify their depths.

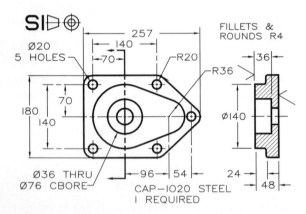

20.76 The use of the outline method to dimension a cap in its most descriptive views.

Figure 20.76 shows an example of outline dimensioning. Note the extension of all dimensions from the outlines of well-defined features.

20.14 Machined Holes

Machined holes are formed by operations such as drilling, boring, or reaming (**Fig. 20.77**). Occasionally, a machining operation is specified in the note, such as 32 DRILL, but it is preferred to omit the specific machining operation. Give the diameter of the hole with its symbol in front of its dimension (for example, O32) with a leader extending from the circular view. You may also note hole diameters with DIA after their size (for example, 2.00 DIA).

Drilling is the basic method of making holes. Dimension the size of a drilled hole with a leader extending from its circular view. You may give its depth in the note from the circular view or dimension it in the rectangular view (**Fig. 20.77B**). Dimension the depth of a drilled hole to the usable part of the hole, not to the conical point.

Counterdrilling involves drilling a large hole inside a smaller hole to enlarge it (**Fig. 20.78**). The drill point leaves a 120° conical shoulder as a byproduct of counterdrilling.

Countersinking is the process of forming conical holes for receiving screw heads (**Fig. 20.79**). Give the diameter of a countersunk hole (the maximum diameter on the surface)

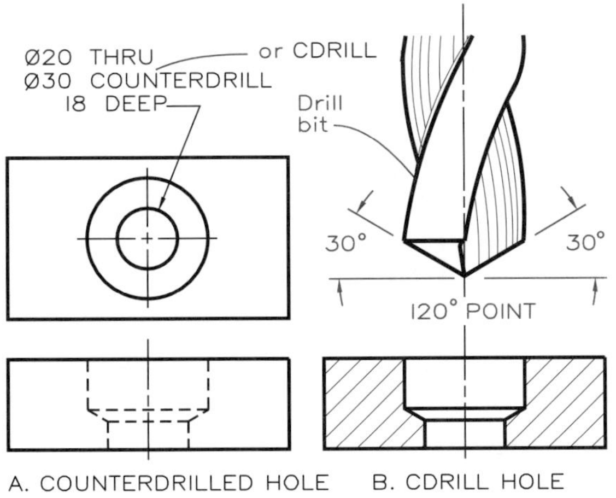
Ø20 THRU
Ø30 COUNTERDRILL
18 DEEP
or CDRILL
Drill bit
30°
30°
120° POINT
A. COUNTERDRILLED HOLE
B. CDRILL HOLE

20.78 Counterdrilling notes give the specifications for drilling a larger hole inside a smaller hole. Do not dimension the 120° angle because it is a byproduct of the drill point. Noting the counterdrill with a leader from the circular view is preferable.

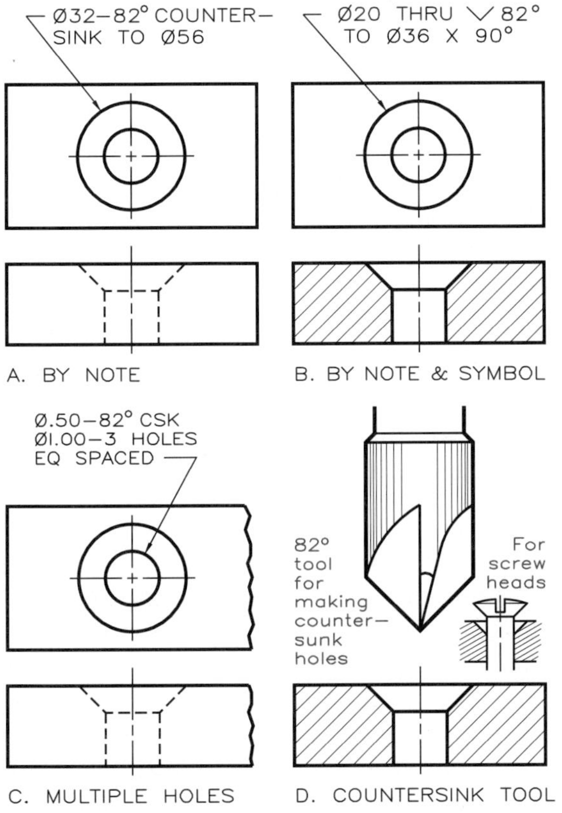
Ø32–82° COUNTER–
SINK TO Ø56
Ø20 THRU ∨82°
TO Ø36 X 90°
A. BY NOTE
B. BY NOTE & SYMBOL
Ø.50–82° CSK
Ø1.00–3 HOLES
EQ SPACED
82° tool for making counter–sunk holes
For screw heads
C. MULTIPLE HOLES
D. COUNTERSINK TOOL

20.79 The methods of noting and specifying countersunk holes for receiving screw heads.

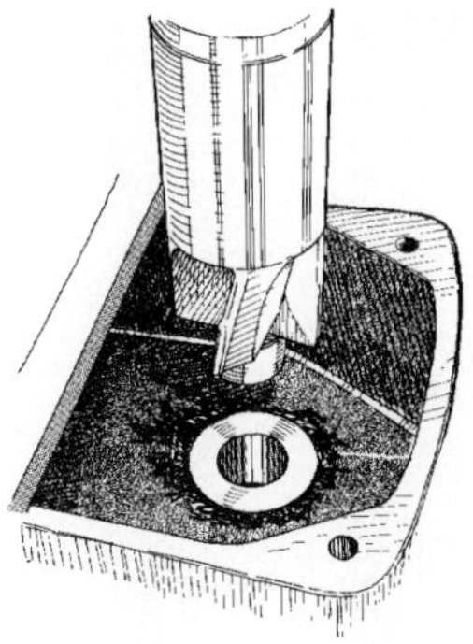

20.80 A spotfacing tool used to finish the cylindrical boss to provide a smooth seat for a bolt head. Spotfacing is the process of smoothing the surface where it will contact a washer, nut, or bolt. It is dimensioned as shown below.

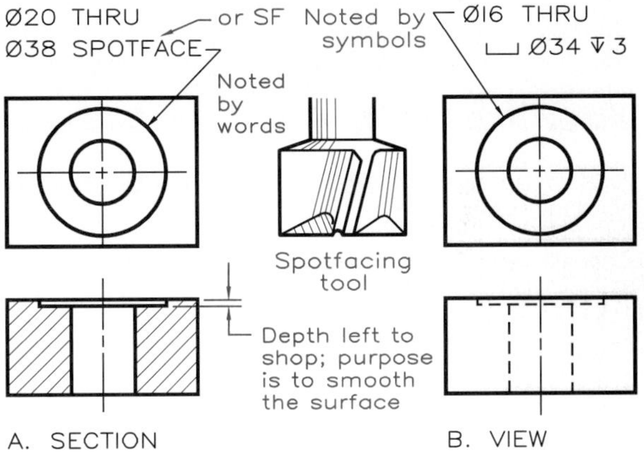
Ø20 THRU
Ø38 SPOTFACE
or SF
Noted by words
Noted by symbols
Ø16 THRU
⌴ Ø34 ↧3
Spotfacing tool
Depth left to shop; purpose is to smooth the surface
A. SECTION
B. VIEW

and the angle of the countersink in a note. Countersunk holes are also used as guides in shafts, spindles, and other cylindrical parts held between the centers of a lathe.

Spotfacing is the process of finishing the surface around holes to provide bearing surfaces for washers, nuts, or bolts (**Fig. 20.80A**). **Figure 20.80B** shows the method of spotfacing a boss (a raised cylindrical element).

Boring is the process of making large holes, usually on a lathe with a bore or a boring bar (**Fig. 20.81**).

Counterboring is the process of enlarging the diameter of a drilled hole (**Fig. 20.82**) to give a flat hole bottom without the tapers as in counterdrilled holes.

Reaming is the operation of finishing or slightly enlarging drilled or bored holes within their prescribed tolerances. A ream is similar to a drill bit.

20.81 A lathe boring a large hole with a boring bar. (Courtesy of Clausing Corporation.)

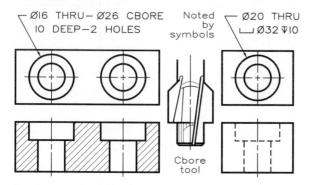

20.82 Counterbored holes are similar to counterdrilled holes but have flat bottoms instead of tapered sides. Dimension them as shown.

20.15 Chamfers

Chamfers are beveled edges cut on cylindrical parts, such as shafts and threaded fasteners, to eliminate sharp edges and to make them easier to assemble. When the chamfer angle is 45°, use a note in either of the forms shown in **Fig. 20.83A**. Dimension chamfers of other angles as shown in **Fig. 20.83B**. When inside openings of holes are chamfered, dimension them as shown in **Fig. 20.84**.

20.16 Keyseats

A keyseat is a slot cut into a shaft for aligning and holding a pulley or a collar on a shaft.

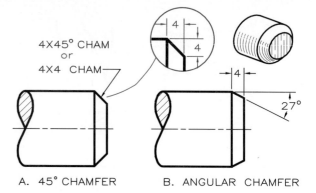

20.83 Chamfers.
A Dimension 45° chamfers by one of the methods shown.
B Dimension chamfers of all other angles as shown.

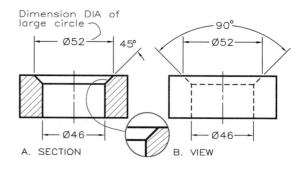

20.84 Dimension chamfers on the insides of cylinders as shown.

Figure 20.85 shows how to dimension keyways and keyseats with dimensions taken from the tables in the appendix. The double dimensions on the diameter are tolerances (discussed in Chapter 21).

20.17 Knurling

Knurling is the operation of cutting diamond-shaped or parallel patterns on cylindrical surfaces for gripping, decoration, or press fits between mating parts that are permanently assembled. Draw and dimension diamond knurls and straight knurls as shown in **Fig. 20.86**, with notes that specify type, pitch, and diameter.

The diametral pitch (DP) is the ratio of the number of grooves on the circumference (N)

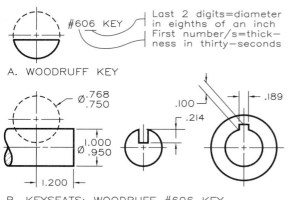

A. WOODRUFF KEY

#606 KEY

Last 2 digits=diameter in eighths of an inch. First number/s=thickness in thirty-seconds

B. KEYSEATS: WOODRUFF #606 KEY

20.85 Methods of dimensioning (A) Woodruff keys and (B) keyways used to hold a part on a shaft. Appendix 24 gives their tables of sizes.

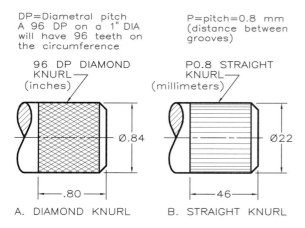

DP=Diametral pitch
A 96 DP on a 1" DIA will have 96 teeth on the circumference

P=pitch=0.8 mm (distance between grooves)

96 DP DIAMOND KNURL (inches)

P0.8 STRAIGHT KNURL (millimeters)

A. DIAMOND KNURL

B. STRAIGHT KNURL

20.86 The diamond knurl has a diametral pitch (DP) of 96 and the straight knurl has a linear pitch (P) of 0.8 mm. Pitch is the distance between the grooves on the circumference.

to the diameter (D), expressed as DP = N/D. The preferred diametral pitches for knurling are 64 DP, 96 DP, 128 DP, and 160 DP.

For diameters of 1 inch, knurling of 64 DP, 96 DP, 128 DP, and 160 DP will have 64, 96, 128, and 160 teeth, respectively, on the circumference. The note P0.8 means that the knurling grooves are 0.8 mm apart. Make knurling calculations in inches and then convert them to millimeters. Specify knurls for press fits with the diameter size before knurling and with the minimum diameter size after knurling.

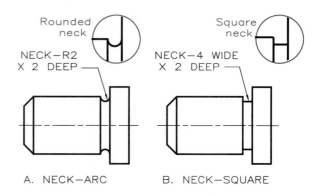

A. NECK—ARC

B. NECK—SQUARE

20.87 Necks are recesses cut in cylinders, with rounded or square bottoms, usually at the intersections of concentric cylinders. Dimension necks as shown.

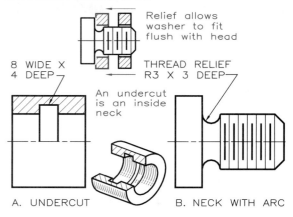

A. UNDERCUT

B. NECK WITH ARC

20.88 Undercuts and necks.

A An undercut is a groove cut inside a cylinder.

B A thread relief is a groove cut at the end of a thread to improve the screw's assembly. Dimension both types of necks as shown.

20.18 Necks and Undercuts

A **neck** is a recess cut around the circumference of a cylindrical part. If cut where cylinders of different diameters join (**Fig. 20.87**), a neck ensures that the assembled parts fit flush at the shoulder of the larger cylinder and allows trash that would cause binding to drop out of the way.

An **undercut** is a groove inside a cylindrical hole (**Fig. 20.88A**). A thread relief is a groove that has been cut at the end of a thread to ensure that the head of the threaded part will fit flush against the part it screws into (**Fig. 20.88B**).

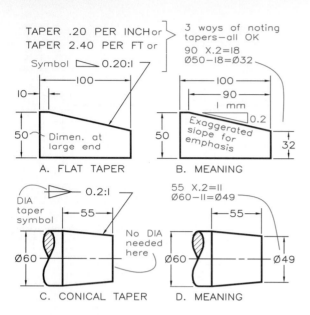

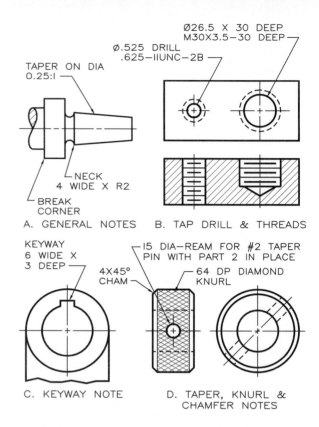

20.89 Tapers may be specified for either flat or conical surfaces: dimensioning and interpretation for (A and B) a flat taper and (C and D) for a conical taper.

20.19 Tapers

Tapers for both flat planes and conical surfaces may be specified with either notes or symbols. Flat taper is the ratio of the difference in the heights at each end of a surface to its length (**Fig. 20.89A** and **B**). Tapers on flat surfaces may be expressed as inches per inch (.20 per inch), inches per foot (2.40 per foot), or millimeters per millimeter (0.20:1).

Conical taper is the ratio of the difference in the diameters at each end of a cone to its length (**Fig. 20.89C** and **D**). Tapers on conical surfaces may be expressed as inches per inch (.25 per inch), inches per foot (3.00 per foot), or millimeters per millimeter (0.25:1).

20.20 Miscellaneous Notes

Notes on detail drawings provide information and specifications that would be difficult to represent by drawings alone (**Figs. 20.90–20.92**). Place notes horizontally on the sheet whenever possible, because they are easier to letter and read in that position.

20.90 Miscellaneous dimensioning.

A The notes for this part indicate a neck, a taper, and a break corner, which is a slight round to remove corner sharpness.

B Threaded holes are sometimes dimensioned by giving the tap drill size in addition to the thread specifications, but selection of the tap drill size is usually left to the shop.

C This note is for dimensioning a keyway.

D The notes for this collar call for knurling, chamfering, and drilling for a no. 2 taper pin.

Several notes in sequence on the same line should be separated with short dashes between them (for example, 15 DIA–30 DIA SPOTFACE). Use standard abbreviations in notes to save space and time.

Problems

1-36. (**Figs. 20.93** and **20.94**) Solve these problems on size A paper, one per sheet, if you draw them full size. If you draw them double

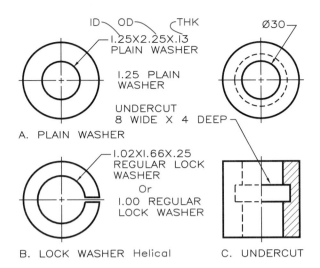

A. PLAIN WASHER

ID — OD — THK
1.25X2.25X.13
PLAIN WASHER

1.25 PLAIN WASHER

Ø30

UNDERCUT
8 WIDE X 4 DEEP

1.02X1.66X.25
REGULAR LOCK WASHER

Or

1.00 REGULAR LOCK WASHER

B. LOCK WASHER Helical

C. UNDERCUT

20.91 Washers and undercuts.

A and **B** Dimension washers and lock washers as shown by taking sizes from the tables in the appendix.

C Dimension an undercut with a note.

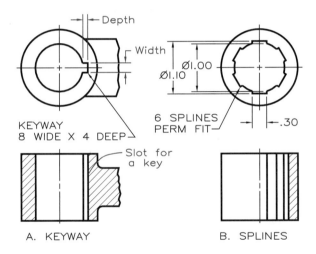

Depth

Width

Ø1.00

Ø1.10

6 SPLINES PERM FIT

.30

KEYWAY
8 WIDE X 4 DEEP

Slot for a key

A. KEYWAY

B. SPLINES

20.92 These drawings illustrate how to dimension (A) keyways and (B) splines.

size, use size B paper. The views are drawn on a 0.20-in. (5 mm) grid. You will need to vary the spacing between views to provide adequate room for the dimensions. Sketching the views and dimensions to determine the required spacing before laying out the solutions with instruments would be helpful. Supply lines that may be missing in all views.

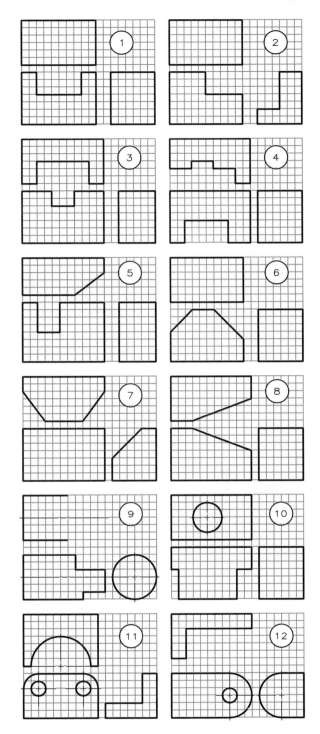

20.93 Problems 1–12.

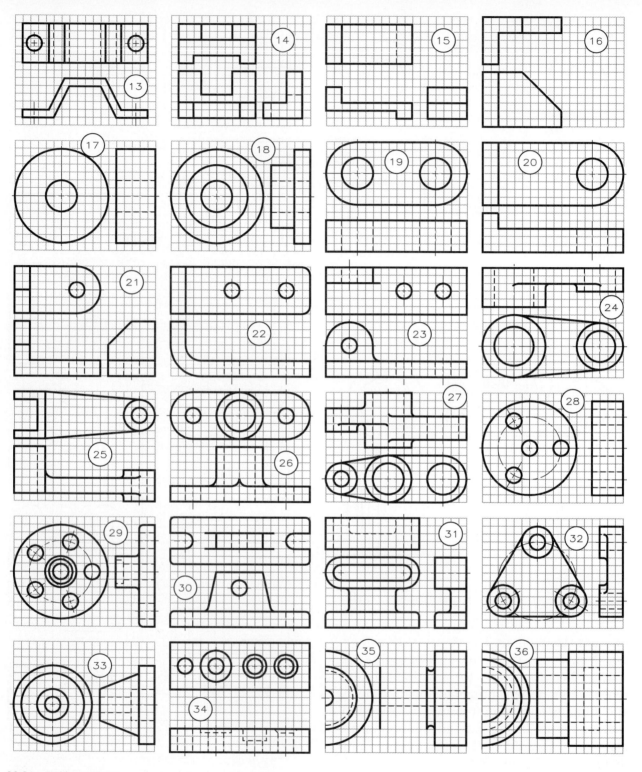

20.94 Problems 13–36.

37–39. (Figs. 20–37 thru 20–39) Lay out the assigned problems on an size-A sheet. The views are drawn on 0.20-in. (5 mm) grid. Dimension the parts using the specifications below as assigned:

  A. Dimension the parts with decimal dimensions and 1/8-in. numerals and letters.

  B. Dimension the parts with metric dimensions (mm) and 3 mm numerals and letters.

  C. Add finish marks to the part where you think they would be appropriate. This could modify the present design of the parts to some degree.

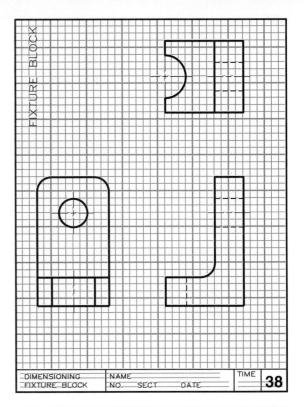

DIMENSIONING	NAME		TIME	
FIXTURE BLOCK	NO. SECT DATE			**38**

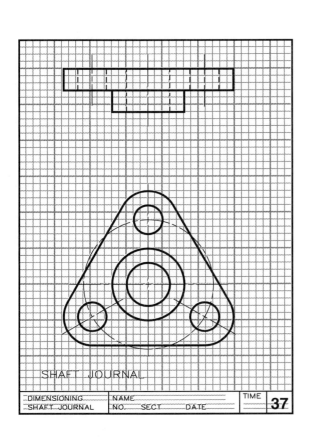

SHAFT JOURNAL

DIMENSIONING	NAME		TIME	
SHAFT JOURNAL	NO. SECT DATE			**37**

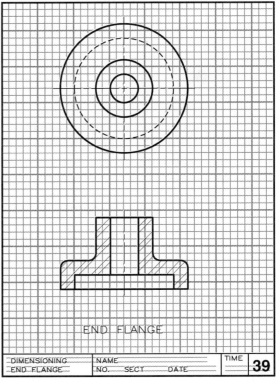

END FLANGE

DIMENSIONING	NAME		TIME	
END FLANGE	NO. SECT DATE			**39**

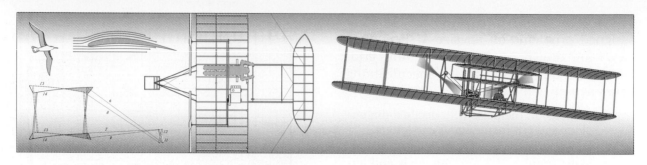

21

Tolerances

21.1 Introduction

Today's technology requires that parts be specified with increasingly exact dimensions. Many parts made by different companies at widely separated locations must be interchangeable, which requires precise size specifications and production.

The technique of dimensioning parts within a required range of variation to ensure interchangeability is called **tolerancing**. Each dimension is allowed a certain degree of variation within a specified zone, or tolerance. For example, a part's dimension might be expressed as 20 ± 0.50, which allows a tolerance (variation in size) of 1.00 mm.

A tolerance should be as **large as possible** without interfering with the function of the part to minimize production costs. Manufacturing costs increase as tolerances become smaller.

The cutting-tool holder in **Fig. 21.1** illustrates a number of parts that must fit within a

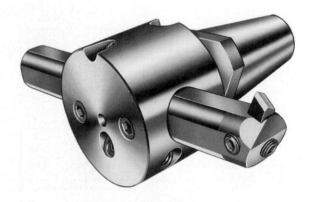

21.1 This cutting-tool holder will function precisely while being used to make cuts on a lathe because all of its parts have been finished within close tolerances.

high degree of precision: The holder must hold the tool exactly if it is to work properly with a lathe.

21.2 Tolerance Dimensions

Three methods of specifying tolerances on dimensions—**unilateral**, **bilateral**, and **limit**

336

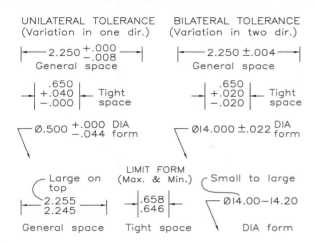

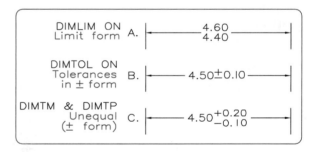

21.2 The three methods of specifying tolerances: unilateral, bilateral, and limit forms for general and tight spaces and dimensions on leaders.

21.3 Place upper limits either above or to the right of lower limits. In plus-and-minus tolerancing, place the plus limits above the minus limits.

21.4 The spacing and proportions of numerals used to specify tolerances on dimensions.

21.5 AutoCAD gives toleranced dimensions in these forms. When DIMLIM is ON, tolerances will be applied in limit form; when DIMTOL is ON, tolerances will be in plus-or-minus form.

forms—are shown in **Fig. 21.2**. When plus-or-minus tolerancing is used, it is applied to a theoretical dimension called the **basic dimension**. When dimensions can vary in only one direction from the basic dimension (either larger or smaller), tolerancing is **unilateral**. Tolerancing that permits variation in both directions from the basic dimension (larger and smaller) is **bilateral**.

Tolerances may be given in **limit form**, with dimensions representing the largest and smallest sizes for a feature. When tolerances are shown in limit form, the basic dimension will be unknown.

The customary methods of applying tolerance values on dimension lines are shown in **Fig. 21.3**. The spacing and proportions of the numerals used to specify tolerance dimensions are shown in **Fig. 21.4**.

Tolerances by Computer

AutoCAD provides an automatic means of showing toleranced dimensions in limit form or plus-and-minus form (**Fig. 21.5**). DIMLIM must be turned ON to obtain dimensions in limit form. Assign tolerances to the DIMTM and DIMTP modes under the DIM: command.

In addition to linear dimensions, diametral and radial dimensions are automatically given with either a circle symbol or an R preceding the dimensions (**Figs. 21.6** and **21.7**). Angular measurements can be toleranced in the plus-and-minus form or the limit form

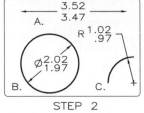

SET Dim. variables
DIMLIM to ON
DIMTP to 0.02
DIMTM to 0.03
(Dimensions will be in LIMIT form)

STEP 1

STEP 2

21.6 Limit tolerances by computer.

Step 1 Set the dimensioning variables as shown for tolerances to be given in LIMIT form.

Step 2 The program gives linear dimensions in limit form, diametral dimensions as limits preceded by O, and radial dimensions preceded by R.

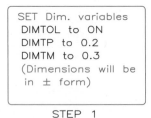

SET Dim. variables
DIMTOL to ON
DIMTP to 0.2
DIMTM to 0.3
(Dimensions will be in ± form)

STEP 1

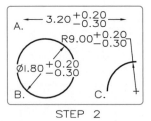

STEP 2

21.7 Plus-or-minus tolerances by computer.

Step 1 Set the dimensioning variables as shown and set DIMTOL to ON.

Step 2 The program will give linear dimensions as a basic diameter followed by plus-and-minus tolerances, diametral dimensions as linear dimensions preceded by O, and radial dimensions preceded by R.

(**Fig. 21.8**). The program measures the angle and automatically computes the upper and lower limits.

21.3 Mating Parts

Mating parts must be toleranced to fit within a prescribed degree of accuracy (**Fig. 21.9**). The upper part is dimensioned with limits indicating its maximum and minimum sizes. The slot in the lower part is toleranced to be slightly larger, allowing the parts to assemble with a clearance fit that allows freedom of movement.

Mating parts may be cylindrical forms, such as a pulley, bushing, and shaft (**Fig. 21.10**). The bushing should force fit inside the pulley to provide a good bearing surface for the rotating shaft. At the same time, the shaft and the bushing should mate so that the pul-

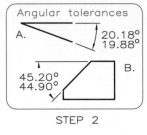

SET Dim. variables
DIMLIM to ON
DIMTP to 0.2
DIMTM to 0.1
DIMDEC to 2
(Dimensions will be in LIMIT form)

STEP 1

STEP 2

21.8 Angular tolerances by computer.

Step 1 Set these dimensioning variables. Set DIMDEC to 2 to have two decimal places.

Step 2 Select the lines forming the angles and the location of the arc, and the toleranced measurements will be given in limit form. Using the UNITS command, you may obtain angular measurements as decimal degrees, minutes and seconds, grads, radians, or surveyor's units.

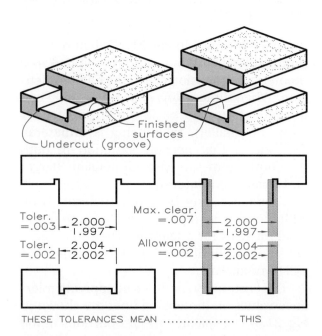

21.9 These mating parts have tolerances (variations in size) of 0.003″ and 0.002″, respectively. The allowance (tightest fit) between the assembled parts is 0.002″.

ley and bushing will rotate on the shaft with a free running fit.

ANSI tables (see Appendixes 29–33) prescribe cylindrical-fit tolerances for different applications. Familiarity with the terminology of cylindrical tolerancing is essential to apply the data in these tables.

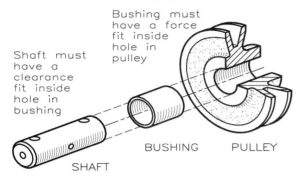

21.10 These parts must be assembled with cylindrical fits that give a clearance and an interference fit.

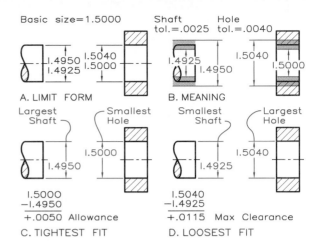

21.11 The allowance (tightest fit) between these mating cylindrical parts is +0.005″. The maximum clearance is +0.0115″.

21.4 Tolerancing Terms: English Units

Figure 21.11 illustrates the following terminology and definitions of tolerancing:

Tolerance: The difference between the limits of size prescribed for a single feature, or 0.0025 in. for the shaft and 0.0040 for the hole in Fig. 21.11A.

Limits of tolerance: The maximum and minimum sizes of a feature, or 1.4925 and 1.4950 for the shaft and 1.5000 and 1.5040 for the hole in Fig. 21.11B.

Allowance: The tightest fit between two mating parts, or +0.0050 in Fig. 21.11C. Allowance is negative for an interference fit.

Nominal size: A general size of a shaft or hole, usually expressed with common fractions, 1 1/2 in. or 1.50 in., as in Fig. 21.11.

Basic size: The size to which plus-and-minus tolerances are applied to obtain the limits of size, 1.5000 in Fig. 21.11. The basic diameter cannot be determined from tolerances that are expressed in limit form.

Actual size: The measured size of the finished part.

Fit: The degree of tightness or looseness between two assembled parts, which can be one of the following: **clearance, interference, transition**, and **line**.

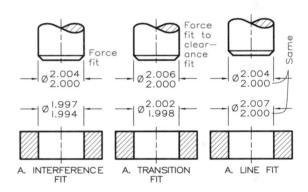

21.12 Three types of fits between mating parts. Clearance fit is shown in Fig. 21.11.

Clearance fit: The clearance between two assembled mating parts. The shaft and the hole in Fig. 21.11C and D have a minimum clearance of 0.0050″ and a maximum clearance of 0.0115″.

Interference fit: A binding fit that requires parts to be forced together, much as if they are welded (**Fig. 21.12A**).

Transition fit: A fit that ranges from an interference to a clearance between the assembled parts. The shaft may be either smaller or larger than the hole and still be within the prescribed tolerances shown in **Fig. 21.12B**.

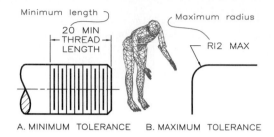

Minimum length · 20 MIN THREAD LENGTH · Maximum radius · R12 MAX

A. MINIMUM TOLERANCE   B. MAXIMUM TOLERANCE

21.13 Single tolerances in maximum (MAX) or minimum (MIN) form can be given in applications of this type.

Line fit: A fit that results in surface contact or clearance when the limits are reached (**Fig. 21.12C**).

Selective assembly: A method of selecting and assembling parts by hand and by trial and error that allows parts to be made with larger tolerances at less cost as a compromise between a high manufacturing accuracy and ease of assembly.

Single limits: Dimensions designated by either minimum (MIN) or maximum (MAX), as shown in **Fig. 21.13**. Depths of holes, lengths, threads, corner radii, and chamfers are sometimes dimensioned in this manner.

21.5 Basic Hole System

The **basic hole system** uses the smallest hole size as the **basic diameter** for calculating tolerances and allowances. The basic hole system is best when drills, reamers, and machine tools are used to give precise hole sizes.

The smallest hole size is the basic diameter because a hole can be enlarged by machining but not reduced in size. Looking again at Fig. 21.11, the smallest diameter of the hole is 1.500″. Subtract the allowance, 0.0050″, from it to find the diameter of the largest shaft, 1.4950″. To find the smallest limit for the shaft diameter, subtract the tolerance from 1.4950 in.

21.6 Basic Shaft System

The **basic shaft system** uses the largest diameter as the **basic diameter** to which the tolerances are applied. This system is applicable when shafts are available in uniform standard sizes.

The largest shaft size is used as the basic diameter because shafts can be machined to smaller size but not enlarged. For example, if the largest permissible shaft size is 1.500″, add the allowance to this dimension to obtain the smallest hole diameter into which the shaft fits. If the parts are to have an allowance of 0.0040″, the smallest hole would have a diameter of 1.5040″.

21.7 Cylindrical Fits

The ANSI B4.1 standard gives a series of fits between cylindrical features in inches for the basic hole system. The following types of fit are covered in this standard:

RC: running or sliding clearance fits

LC: clearance locational fits

LT: transition locational fits

LN: interference locational fits

FN: force and shrink fits

Appendixes 29–33 list these five types of fit, each of which has several classes.

Running or sliding clearance fits (RC) provide a similar running performance, with suitable lubrication allowance. The clearance for the first two classes (RC 1 and RC 2), which are used chiefly as slide fits, increases more slowly with diameter size than other classes to maintain an accurate location, even at the expense of free relative motion.

Locational fits (LC, LT, LN) determine only the location of mating parts; they may provide nonmoving rigid locations (interference fits) or permit some freedom of location (clear-

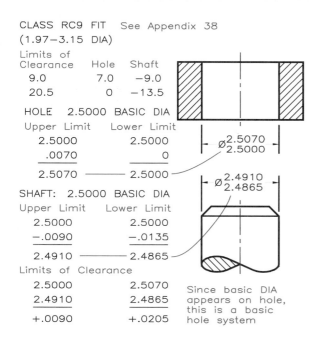

CLASS RC9 FIT See Appendix 38
(1.97–3.15 DIA)

Limits of Clearance	Hole	Shaft
9.0	7.0	−9.0
20.5	0	−13.5

HOLE 2.5000 BASIC DIA

Upper Limit	Lower Limit
2.5000	2.5000
.0070	0
2.5070 ——	2.5000

SHAFT: 2.5000 BASIC DIA

Upper Limit	Lower Limit
2.5000	2.5000
−.0090	−.0135
2.4910 ——	2.4865

Limits of Clearance

2.5000	2.5070
2.4910	2.4865
+.0090	+.0205

ø 2.5070 / 2.5000
ø 2.4910 / 2.4865

Since basic DIA appears on hole, this is a basic hole system

21.14 This example shows how to calculate limits and allowances for an RC9 fit between a shaft and hole with a basic diameter of 2.5000″. Values are taken from Appendix 29.

ance fits). The three locational fits are **clearance fits** (LC), **transition fits** (LT), and **interference fits** (LN).

Force fits (FN) are interference fits characterized by a constant bore pressure throughout the range of sizes. There are five types of force fits: FN1 through FN5 varying from light drive to heavier drives, respectively.

The method of applying tolerance values from the tables in Appendix 29 for an RC9 fit is shown in **Figure 21.14.** The basic diameter of 2.5000″ falls between 1.97″ and 3.15″ in the *Size* column of the table. Limits are in thousandths, which requires that the decimal point be moved three places to the left. For example, +7 is +0.0070″.

Add the limits to the basic diameter when a *plus sign* precedes the values and *subtract* when a *minus sign* is given. Add the limits (+0.0070″ and 0.0000″) to the basic diameter to find the upper and lower limits of the hole (2.5070″ and 2.5000″). Subtract the limits

(−0.0090″ and −.0135″) from the basic diameter to find the limits of the shaft (2.4910″ and 2.4865″).

The tightest fit (the allowance) between the assembled parts (+0.0090″) is the difference between the largest shaft and the smallest hole. The loosest fit (+0.0205″) is the difference between the smallest shaft and the largest hole. These values appear in the *Limit* column of the table (Appendix 29).

This method of using tables of fits is applied to other types of fits by using their respective tables: force fit, interference fit, transition fit, and locational fit. Subtract negative limits from the basic diameter and add positive limits to it. A *minus sign* preceding limits of clearance in the tables indicates an *interference fit* between the assembled features, and a *positive limit* of clearance indicates a *clearance fit*.

21.8 Tolerancing: Metric Units

The system recommended by the International Standards Organization (ISO) in ANSI B4.2 for metric measurements are fits that usually apply to cylinders—holes and shafts—but these tables can be used to specify fits between parallel contact surfaces such as a key in a slot.

Basic size: The size, usually a diameter from which limits or deviations are calculated (**Fig. 21.15**). Select basic sizes from the *First Choice* column in the table in **Fig. 21.16.**

Deviation: The difference between the hole or shaft size and the basic size.

Upper deviation: The difference between the maximum permissible size of a part and its basic size (Fig. 21.15).

Lower deviation: The difference between the minimum permissible size of a part and its basic size (Fig. 21.15).

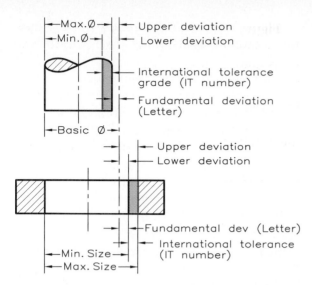

21.15 The terminology and definitions of the metric system of cylindrical fits.

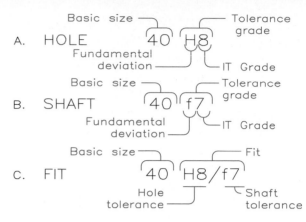

21.17 These tolerance symbols and their definitions apply to holes and shafts.

First Choice	Second Choice	First Choice	Second Choice	First Choice	Second Choice
I	1.1	10	11	100	110
1.2	1.4	12	14	120	140
1.6	1.8	16	18	160	180
2	2.2	20	22	200	220
2.5	2.8	25	28	250	280
3	3.5	30	35	300	350
4	4.5	40	45	400	450
5	5.5	50	55	500	550
6	7	60	70	600	700
8	9	80	90	800	900
				1000	

21.16 Basic sizes for metric fits selected first from the first-choice column are preferred over those in the second-choice column.

Fundamental deviation: The deviation closest to the basic size (Fig. 21.15). In note 40 H8 in **Fig. 21.17**, H represents the fundamental deviation for a hole, and in note 40 f7, the f represents the fundamental deviation for a shaft.

Tolerance: The difference between the maximum and minimum allowable sizes of a part.

International tolerance (IT) grade: A series of tolerances that vary with basic size to provide a uniform level of accuracy within a given grade (Fig. 21.15). In note 40 H8 in Fig. 21.17,

the 8 represents the IT grade. There are 18 IT grades: IT01, IT0, IT1, . . . , IT16.

Tolerance zone: A combination of the fundamental deviation and the tolerance grade. The H8 portion of note 40 H8 in Fig. 21.17 is the tolerance zone.

Hole basis: A system of fits based on the minimum hole size as the basic diameter. The fundamental deviation for a hole-basis system is "H." Appendixes 40 and 41 give hole-basis data for tolerances.

Shaft basis: A system of fits based on the maximum shaft size as the basic diameter. The fundamental deviation for a shaft-basis system is "h." Appendixes 42 and 43 give shaft-basis data for tolerances.

Clearance fit: A fit resulting in a clearance between two assembled parts under all tolerance conditions.

Interference fit: A force fit between two parts, requiring that they be driven together.

Transition fit: A fit that may result in either a clearance or an interference between assembled parts.

Tolerance symbols: Notes giving the specifications of tolerances and fits (Fig. 21.17). The basic size is a number, followed by the fundamental deviation letter and the IT number, which give the tolerance zone. **Uppercase letters** (H) indicate the fundamental deviations for **holes**, and **lowercase letters** (f) indicate fundamental deviations for **shafts**.

Preferred Sizes and Fits

The table in Fig. 21.16 shows the preferred basic sizes for computing tolerances. Under the *First Choice* heading, each number increases by about 25 percent from the preceding value. Each number in the *Second Choice* column increases by about 12 percent. To minimize cost, select basic diameters from the first column because they correspond to standard stock sizes for round, square, and hexagonal metal products.

Hole Basis		Shaft Basis	Description
H11/c11	Clearance Fits	C11/h11	Loose Running Fit for wide commerical tolerances on external members
H9/d9		D9/h9	Free Running Fit for large temperature variations, high running speeds, or high journal pressures
H8/f7		F8/h7	Close Running Fit for accurate location and moderate speeds and journal pressures
H7/g6		G7/h6	Sliding Fit for accurate fit and location and free moving and turning, not free running
H7/h6	Transition Fits	H7/h6	Locational Clearance for snug fits for parts that can be freely assembled
H7/k6		K7/h6	Locational Transition Fit for accurate locations
H7/n6		N7/h6	Locational Transition Fit for more accurate locations and greater interference
H7/p6	Interference Fits	P7/h6	Locational Interference Fit for rigidity and alignment without special bore pressures
H7/s6		S7/h6	Medium Drive Fit for shrink fits on light sections; tightest fit usable for cast iron
H7/u6		U7/h6	Force Fit for parts that can be highly stressed and for shrink fits.

21.18 The preferred hole-basis and shaft-basis fits for the metric system.

Figure 21.18 shows preferred clearance, transition, and interference fits for the hole-basis and shaft-basis systems. Appendixes 40–43 contain the complete tables.

Preferred Fits: Hole-Basis System The preferred fits for the hole-basis system, in which the smallest hole is the basic diameter, are shown in **Fig. 21.19**. **Clearance, transition**, and **interference fits** are options of the hole-basis system. **Figure 21.20** illustrates the preferred fits for a hole-basis system where the lower deviation of the hole is zero which means that the smallest hole is the basic size. Variations in fit between parts range from a clearance fit of H11/c11 to an interference fit of H7/u6 (see Fig. 21.18).

Preferred Fits: Shaft-Basis System The preferred fits of the shaft-basis system, in which the largest shaft is the basic diameter, are shown in **Fig. 21.21**. Variations in fit range from a clearance fit of C11/h11 to an interference fit of U7/h6 (see Fig. 21.18).

Standard Cylindrical Fits

The following examples demonstrate how to calculate and apply tolerances to cylindrical parts. You must use Fig. 21.16, Fig. 21.18, and data from Appendixes 35–38.

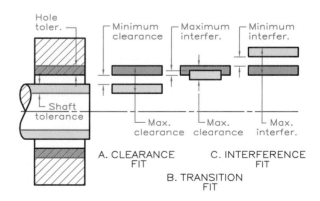

21.19 Types of fits: (A) clearance fit, where space is between the parts, (B) transition fit, where there can be either interference or clearance, and (C) interference fit, where the parts must be forced together.

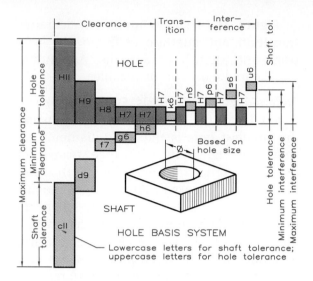

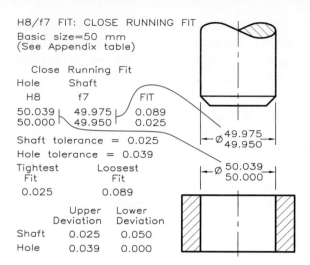

21.20 The preferred fits for the hole-basis system listed in Fig. 21.18. Appendixes 35 and 36 give values for these fits.

21.22 The method for calculating and applying metric limits and fits to a shaft and hole (Appendix 35).

Hole: Find the upper and lower limits of the hole in Appendix 35 under H8 and across from 50 mm. These limits are 50.000 and 50.039 mm.

Shaft: Find the upper and lower limits of the shaft under f7 and across from 50 mm in Appendix 35. These limits are 49.950 and 49.975 mm.

Symbols: **Figure 21.22** shows how to apply toleranced dimensions to the hole and shaft.

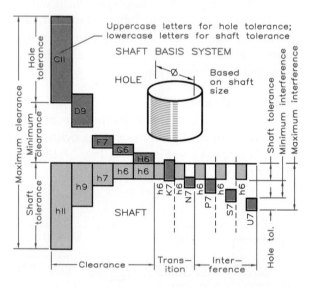

21.21 The preferred fits for a shaft-basis system listed in Fig. 21.18. Appendixes 37 and 38 give values for these fits.

Example 1 (Fig. 21.24)

Required: Use the hole-basis system, a close running fit, and a basic diameter of 49 mm.

Solution: Use a preferred basic diameter of 50 mm (Fig. 21.16) and fit of H8/f7 (Fig. 21.18).

Example 2 (Fig. 21.23)

Required: Use the hole-basis system, a location transition fit, and a basic diameter of 57 mm.

Solution: Use a preferred basic diameter of 60 mm (Fig. 21.16) and a fit of H7/k6 (Fig. 21.18).

Hole: Find the upper and lower limits of the hole in Appendix 36 under H7 and across from 60 mm. These limits are 60.000 and 60.030 mm.

Shaft: Find the upper and lower limits of the shaft under k6 and across from 60 mm in Appendix 36. These limits are 60.021 and 60.002 mm.

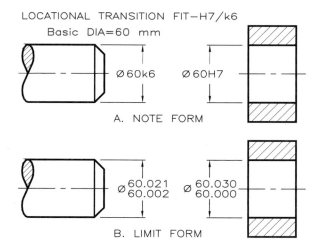

LOCATIONAL TRANSITION FIT—H7/k6
Basic DIA=60 mm

A. NOTE FORM

Ø 60k6   Ø 60H7

B. LIMIT FORM

Ø 60.021
  60.002   Ø 60.030
            60.000

21.23 The method for applying metric tolerances to a hole and a shaft with a transition fit (Appendix 36).

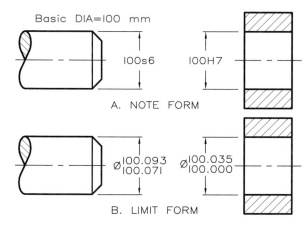

Basic DIA=100 mm

A. NOTE FORM

100s6   100H7

B. LIMIT FORM

Ø 100.093
  100.071   Ø 100.035
            100.000

21.24 Either of these formats can be used to apply metric tolerances to a hole and shaft that have an interference fit (Appendix 36).

Symbols: **Figure 21.23** shows two methods of applying the tolerance symbols to a drawing.

Example 3 (Fig. 21.24)

Required: Use the hole-basis system, a medium drive fit, and a basic diameter of 96 mm.

Solution: Use a preferred basic diameter of 100 mm (Fig. 21.16) and a fit of H7/s6 (Fig. 21.18).

Hole: Find the upper and lower limits of the hole in Appendix 36 under H7 and across from 100 mm. These limits are 100.035 and 100.000 mm.

Shaft: Find the upper and lower limits of the shaft under s6 and across from 100 mm in Appendix 36. These limits are 100.093 and 100.071 mm. Appendix 36 gives the tightest fit as an interference of −0.093 mm, and the loosest fit as an interference of −0.036 mm. Minus signs in front of these numbers indicate an interference fit.

Symbols: **Figure 21.24** shows how to apply toleranced dimensions to the hole and shaft.

Nonstandard Fits: Nonpreferred Sizes

Limits of tolerances for nonstandard sizes can be calculated for any of the preferred fits shown in Fig. 21.18 for nonstandard sizes that do not appear in Appendixes 35–38. Limits of tolerances for nonstandard hole sizes are in Appendix 39, and limits of tolerances for nonstandard shaft sizes are in Appendix 40.

Figure 21.25 shows the hole and shaft limits for an H8/f7 fit and a 45-mm DIA. The tolerance limits of 0.000 and 0.039 mm for an H8 hole are from Appendix 39, across from the size range of 40–50 mm. The tolerance limits of −0.025 and −0.050 mm for the shaft are from Appendix 40. Calculate the hole limits by adding the positive tolerances to the 45-mm basic diameter and the shaft limits by subtracting the negative tolerances from the 45-mm basic diameter.

21.9 Chain Versus Datum Dimensions

When parts are dimensioned to locate surfaces or geometric features by a chain of dimensions laid end-to-end (**Fig. 21.26A**), variations may accumulate in excess of the specified tolerance. For example, the tolerance between surfaces A and B is 0.02, between A and C it is 0.04, and between A and D it is 0.06.

21.9 CHAIN VERSUS DATUM DIMENSIONS • 345

CALCULATION OF NONSTANDARD LIMITS

FIT: H8/f7   Ø45 BASIC DIA

From Appendix		Hole Limits	45.039
Hole	Shaft		45.000
H8	f7		
0.039	−0.025	Shaft Limits	44.975
0.000	−0.050		44.950

21.25 This calculation is for an H8/f7 nonstandard diameter of 45 mm (Appendixes 39 and 40).

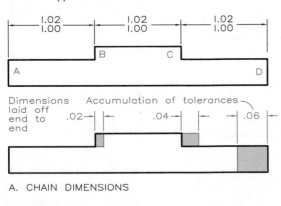

A. CHAIN DIMENSIONS

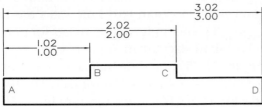

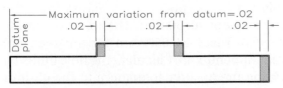

B. DATUM PLANE (BASELINE) DIMENSIONS

21.26 Chain versus datum dimensioning.

A Dimensions given end-to-end in a chain fashion may result in an accumulation of tolerances of up to 0.06″ at D instead of the specified 0.02″.

B When dimensioned from a single datum, the variations of B, C, and D cannot deviate more than the specified 0.02″ from the datum.

Tolerance accumulation can be eliminated by measuring from a single plane called a **datum plane** or **baseline.** A datum plane is usually on the object, but it can also be on the machine used to make the part. Because each

346 • **CHAPTER 21   TOLERANCES**

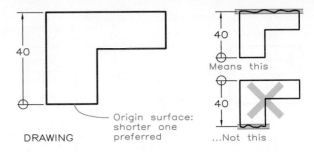

DRAWING   Origin surface: shorter one preferred

Means this

...Not this

21.27 Selection of the shorter surface as the origin surface for locating a longer parallel surface gives the greatest accuracy.

plane in **Fig. 21.26B** is located with respect to a datum plane, the tolerances between the intermediate planes do not exceed the maximum tolerance of 0.02. Always base the application of tolerances on the function of a part in relationship to its mating parts.

Origin Selection

When the shorter of two surfaces is chosen to be the mounting surface, it should be specified as the origin for locating a parallel surface. Had the longer surface been chosen as the origin surface, the shorter surface would have had a greater angular variation from the same tolerance zone (**Fig. 21.27**).

21.10 Tolerance Notes

You should tolerance all dimensions on a drawing either by using the rules previously discussed or by placing a note in or near the title block. For example, the note

TOLERANCE ±1/64

might be given on a drawing for less critical dimensions.

Some industries give dimensions in inches with two, three, and four decimal place fractions. A note for dimensions with two and three decimal places might be given on the drawing as

TOLERANCES XX.XX ±0.10

XX.XXX ±0.005.

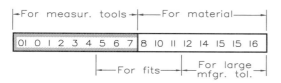

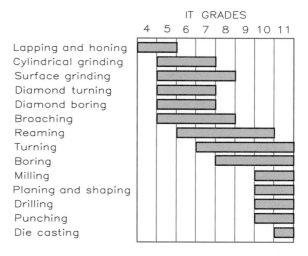

21.28 Tolerances on angles can be specified by one of these methods.

21.29 International tolerance (IT) grades and their applications. Appendix 34 lists IT grade values.

21.30 International tolerance (IT) values may be selected from this table, which is based on the general capabilities of various machining processes.

Tolerances of four places would be given directly on the dimension lines.

The most common method of noting tolerances is to give as large a tolerance as feasible in a note, such as

TOLERANCES ±0.05,

and to give tolerances on the dimension lines for dimensions requiring closer tolerances. Give angular tolerances in a general note in or near the title block:

ANGULAR TOLERANCES ±0.50° or 30′.

Use one of the formats shown in **Fig. 21.28** to give specific angular tolerances directly on angular dimensions.

21.11 General Tolerances: Metric Units

All dimensions on a drawing must be specified within certain tolerance ranges when they are not shown on dimension lines. Tolerances not shown on dimension lines should be specified by a general tolerance note on the drawing.

Linear-Dimensions Tolerance linear dimensions by indicating plus and minus (±) one half of an international tolerance (IT) grade as given in Appendix 34. You may select the IT grade from the chart in **Fig. 21.29**, where IT

grades for mass-produced items range from IT12 through IT16. IT grades can be selected from **Fig. 21.30** for a particular machining process.

General tolerances using IT grades may be expressed in a note as follows:

UNLESS OTHERWISE SPECIFIED

ALL UNTOLERANCED

DIMENSIONS ARE IT14.

This note means that a tolerance of ± 0.700 mm is allowed for a dimension between 315 and 400 mm. The value of the tolerance, 1.400 mm, is extracted from Appendix 34.

Figure 21.31 shows recommended tolerances for fine, medium, and coarse series for ranges of size. A medium tolerance, for example, can be specified by the following note:

GENERAL TOLERANCES SPECIFIED

IN ANSI B4.3 MEDIUM SERIES APPLY.

Equivalent tolerances may be given in table form (**Fig. 21.32**) on the drawing, the grade (medium in this example) selected from Fig. 21.31. General tolerances may be given in a table for dimensions expressed with one or no decimal places (**Fig. 21.33**).

GENERAL TOLERANCES: LINEAR DIMENSIONS (mm)

Basic Dimensions	Fine Series	Medium Series	Coarse Series
0.5 to 3	± 0.05	± 0.1	– –
Over 3 to 6	± 0.05	± 0.1	± 0.2
Over 6 to 30	± 0.1	± 0.2	± 0.5
Over 30 to 120	± 0.15	± 0.3	± 0.8
Over 120 to 315	± 0.2	± 0.5	± 1.2
Over 315 to 1000	± 0.3	± 0.8	± 2
Over 1000 to 2000	± 0.5	± 1.2	± 3

21.31 General tolerance values for fine, medium, and coarse series. Tolerances vary with dimensions.

Get values from previous table

Specifies a medium series

GENERAL TOLERANCES (mm)
UNLESS OTHERWISE SPECIFIED, THE FOLLOWING TOLERANCES ARE APPLICABLE

LINEAR	Over to	0.5 6	6 30	30 120	120 315	315 1000	1000 2000
TOL.	±	0.1	0.2	0.3	0.5	0.8	1.2

21.32 This table for a medium series of values was extracted from Fig. 21.31 for insertion on a working drawing to provide the tolerances for a medium series of sizes.

Medium series for numbers with one decimal place

Coarse series for numbers with no decimal places

GENERAL TOLERANCES (mm)
UNLESS OTHERWISE SPECIFIED, THE FOLLOWING TOLERANCES ARE APPLICABLE

LINEAR	OVER TO	– 120	120 315	315 1000	1000 –
TOL. ONE DECIMAL ±		0.3	0.5	0.8	1.2
TOL. NO DECIMALS ±		0.8	1.2	2	3

21.33 Placed on a drawing, this table of tolerances would indicate the tolerances for dimensions having one or no decimal places, such as 24.0 and 24, denoting medium and coarse series.

General tolerances may also be notated in the following form:

UNLESS OTHERWISE SPECIFIED

ALL UNTOLERANCED DIMENSIONS

ARE ±0.8 mm.

Use this method only when the dimensions on a drawing are similar in size.

Length of shorter leg (mm)	Up to 10	Over 10 to 50	Over 50 to 120	Over 120 to 400
Degrees	± 1°	± 0° 30'	± 0°20'	± 0°10'
mm per 100	± 1.8	± 0.9	± 0.6	± 0.3

21.34 General tolerances for angular and taper dimensions may be taken from this table of values.

ANGULAR TOLERANCES				
LENGTH OF SHORTER LEG (mm)	UP TO 10	OVER 10 TO 50	OVER 50 TO 120	OVER 120 TO 400
TOLERANCE	±1°	± 0°30'	± 0°20'	±0°10'

Values in degrees and minutes taken from previous table

21.35 This table, extracted from Fig. 21.34, is placed on the drawing to indicate the general tolerance for angles in degrees and minutes.

Angular Tolerances Express angular tolerances as (1) an angle in decimal degrees or in degrees and minutes, (2) a taper expressed in percentage (mm per 100 mm), or (3) milliradians. (To find milliradian, multiply the degrees of an angle by 17.45.) **Figure 21.34** shows the suggested tolerances for decimal degrees and taper, based on the length of the shorter leg of the angle. General angular tolerances may be notated on the drawing as follows:

UNLESS OTHERWISE SPECIFIED

THE GENERAL TOLERANCES

IN ANSI B4.3 APPLY.

A second method involves showing a portion of the table from **Fig. 21.34** as a table of tolerances on the drawing (**Fig. 21.35**). A third method is a note with a single tolerance such as:

UNLESS OTHERWISE SPECIFIED

ANGULAR TOLERANCES ARE ±0°30'.

21.12 Geometric Tolerances

Geometric tolerancing is a system that specifies tolerances that control **location form, profile, orientation, location,** and **runout** on

GEOMETRIC SYMBOLS

Tolerance		Characteristic	Symbol
INDIVIDUAL FEATURES	Form	Straightness	—
		Flatness	▱
		Circularity	○
		Cylindricity	⌭
BOTH	Profile	Profile: Line	⌒
		Profile: Surface	⌓
RELATED FEATURES	Orientation	Angularity	∠
		Perpendicularity	⊥
		Parallelism	∥
	Location	Position	⊕
		Concentricity	◎
		Symmetry	≡
	Runout	Runout: Circular	↗
		Runout: Total	↗↗

21.36 These symbols specify the geometric characteristics of a part's features.

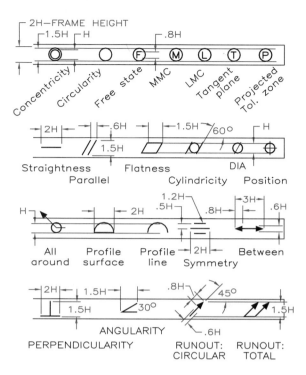

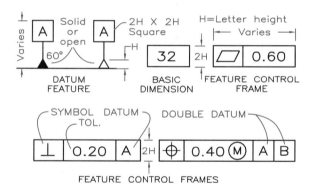

21.37 The proportions of these feature control symbols are based on letter height, usually 1/8 inch.

21.38 Examples of geometric tolerancing feature control frames and feature control symbols used to indicate geometric tolerances. H is the letter height.

a dimensioned part as covered by the *ANSI Y14.5M-1994 Standards* and the *Military Standards* (Mil-Std) of the U.S. Department of Defense. Before discussing those types of tolerancing, however, we need to introduce symbols, size limits, rules, three-datum-plane concepts, and applications.

Symbols

The most commonly used symbols for representing geometric characteristics of dimensioned drawings are shown in **Fig. 21.36**. The proportions of feature control symbols in relation to their feature control frames, based on letter height, are shown in **Fig. 21.37**. On most drawings, a 1/8-in. or 3-mm letter height is recommended. Examples of feature control frames and their proportions are shown in **Fig. 21.38**.

Size Limits

Three conditions of size are used when geometric tolerances are applied: maximum material condition, least material condition, and regardless of feature size.

Maximum material condition (MMC) indicates that a feature contains the maximum amount of material. For example, the

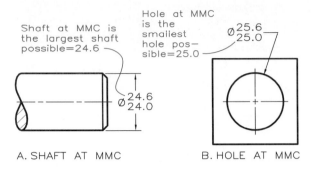

A. SHAFT AT MMC   B. HOLE AT MMC

21.39 A shaft is at maximum material condition (MMC) when it is at the largest diameter permitted by its tolerance. A hole is at MMC when it is at its smallest size.

shaft shown in **Fig. 21.39** is at MMC when it has the largest permitted diameter of 24.6 mm. The hole is at MMC when it has the most material, or the smallest diameter of 25.0 mm.

Least material condition (LMC) indicates that a feature contains the least amount of material. The shaft in Fig. 21.39 is at LMC when it has the smallest diameter of 24.0 mm. The hole is at LMC when it has the least material, or the largest diameter of 25.6 mm.

Regardless of feature size (RFS) indicates that tolerances apply to a geometric feature regardless of its size, ranging from MMC to LMC.

21.13 Rules of Tolerancing

Two general rules of tolerancing geometric features should be followed:

Rule 1 (individual feature of size): When only a tolerance of size is specified on a feature, the limits of size control the variation in its geometric form. The forms of the shaft and hole shown in **Fig. 21.40** are permitted to vary within the tolerance ranges of the dimensions.

Rule 2 (all applicable geometric tolerances): If no modifying symbol is specified, RFS applies with respect to the individual tolerance or datum reference, or both. The modifiers MMC or LMC must be specified where required on a drawing.

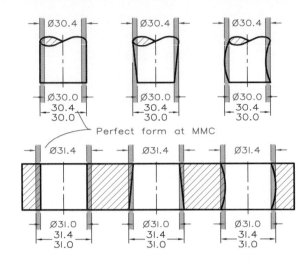

21.40 When only a tolerance of size is specified on a feature, the limits prescribe the form of the features, as shown for these shafts and holes having identical limits.

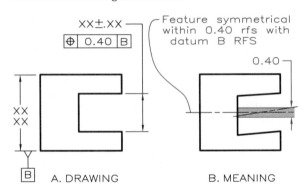

A. DRAWING   B. MEANING

21.41 Tolerances of position should include the note of M or L to indicate maximum material condition or least material condition, if either is required.

Alternate Practice: For a tolerance of position, RFS may be specified on the drawing with respect to the individual tolerance or datum reference, or both.

The specification of symmetry for the part in **Fig. 21.41** is based on a tolerance at RFS from a datum at RFS since LMC or MMC was not specified.

Three-Datum-Plane Concept

A datum plane is used as the origin of a part's features that have been toleranced. Datum planes usually relate to manufacturing equipment, such as machine tables or locating pins.

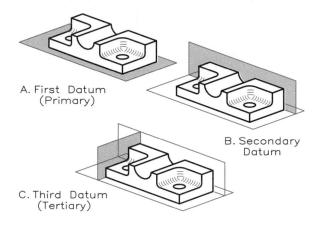

A. First Datum (Primary)

B. Secondary Datum

C. Third Datum (Tertiary)

21.42 When an object is referenced to a primary datum plane, it contacts the datum at three or more points. The vertical surface contacts the secondary datum plane at two or more points. The third surface contacts the third datum at one or more points. Datum planes are listed in order of priority in the feature control frame.

Three mutually perpendicular datum planes are required to dimension a part accurately. For example, the part shown in **Fig. 21.42** sits on the primary datum plane, with at least three points of its base in contact with the datum. The part is related to the secondary plane by at least two contact points. The third (tertiary) datum is in contact with at least one point on the object.

The priority of datum planes is presented in sequence in feature control frames. For example, the primary datum in **Fig. 21.43** is surface A, the secondary datum is surface B, and the tertiary datum is surface C. **Figure 21.44** lists the order of priority of datum planes A–C sequentially in the feature control frames.

21.14 Cylindrical Datum Features

A part with a cylindrical datum feature that is the axis of a cylinder is illustrated in **Fig. 21.45**. Datum K is the primary datum. Datum M is associated with two theoretical planes—the second and third in a three-plane relationship. The two theoretical planes are represented in the circular view by perpendicular centerlines that intersect at the point view of

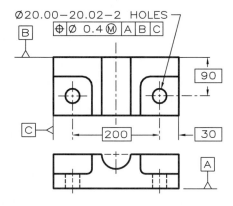

21.43 The three planes of the reference system are noted where they appear as edges. The primary datum plane (A) is given first in the feature control frame; the secondary plane (B), and the tertiary plane (C). Single numbers in frames are basic dimensions.

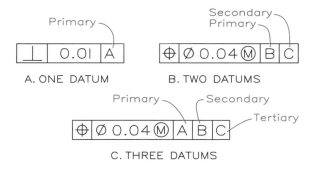

21.44 Use feature control frames to indicate from one to three datum planes in order of priority.

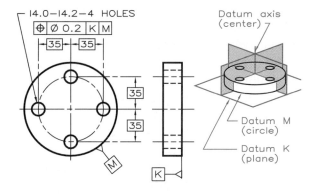

21.45 These true-position holes are located with respect to primary datum K and secondary datum M. Because datum M is a circle, the holes are located about two intersecting datum planes at the crossing centerlines in the circular view, satisfying the three-plane concept.

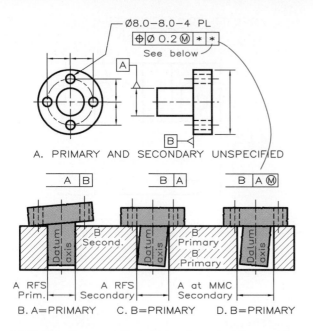

A. PRIMARY AND SECONDARY UNSPECIFIED

B. A=PRIMARY   C. B=PRIMARY   D. B=PRIMARY

21.46 For the unspecified datum planes in (A), examples (B)–(D) illustrate the effects of selecting the datum planes in order of priority and of RFS and MMC.

the datum axis. All dimensions originate from the datum axis perpendicular to datum K; the other two intersecting datum planes are used for measurements in the *x* and *y* directions.

The priority of the datum planes in the feature control frame is significant in the manufacturing and inspection processes. The part shown in **Fig. 21.46** is dimensioned in three ways to show the effects of datum-plane selection and material condition on the location of the hole pattern.

The effect of specifying diameter A at RFS as the primary datum plane and surface B as the secondary datum plane is illustrated in Fig. 21.46B. During production, the part is centered on cylinder A, then mounted in a chuck, mandrel, or centering device on the processing equipment, which centers the part at RFS. Any variation from perpendicular in surfaces A and B will affect the degree of contact of surface B with its datum plane.

If surface B were specified as the primary datum feature, it would contact datum plane B at no fewer than three points (Fig. 21.46C).

The axis of datum cylinder A will be gauged by the smallest cylinder that is perpendicular to the first datum that will contact cylinder A at RFS. This cylinder identifies variation from perpendicular between planes A and B and size variations.

In Fig. 21.46D, plane B is specified as the primary datum feature and cylinder A as the secondary datum feature at MMC. The part is mounted on the processing equipment so that at least three points on feature B come into contact with datum B. The datum axis is the axis of a circumscribed cylinder of a fixed size that is perpendicular to datum B. Using the modifier to specify MMC gives a more liberal tolerance zone than when RFS is specified.

Datum Features at RFS

When size dimensions are applied to a feature at RFS, the processing equipment that comes into contact with surfaces of the part establishes the datum. Variable machine elements, such as chucks or center devices, are adjusted to fit the external or internal features and establish datums.

Primary-Diameter Datums For an external cylinder (shaft) at RFS, the datum axis is the axis of the smallest circumscribed cylinder that contacts the cylindrical feature (**Fig. 21.47A**). That is, the largest diameter of the part making contact with the smallest cylinder of the machine element holding the part is the datum axis.

For an internal cylinder (hole) at RFS, the datum axis is the axis of the largest inscribed cylinder making contact with the hole. That is, the smallest diameter of the hole making contact with the largest cylinder of the machine element inserted in the hole is the datum axis (**Fig. 21.47B**).

Primary External Parallel Datums The datum for external features at RFS is the center plane between two parallel planes—at minimum

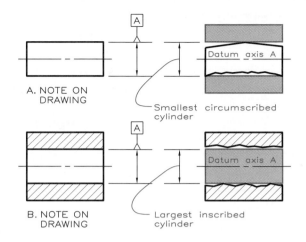

21.47 The datum axis of a shaft is the smallest circumscribed cylinder in contact with the shaft. The datum axis of a hole is the centerline of the largest inscribed cylinder in contact with the hole.

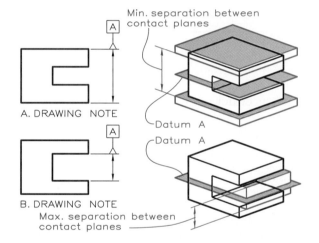

21.48 The datum plane for external parallel surfaces is the center plane between two contact parallel planes at their minimum separation. The datum plane for internal parallel surfaces is the center plane between two contact parallel surfaces at their maximum separation.

separation—that contact the planes of the object (**Fig. 21.48A**). These are planes of a vise-like device at minimum separation that holds the part.

Primary Internal Parallel Datums The datum for internal features is the center plane between two parallel planes—at their maximum separation—that contact the inside planes of the object (**Fig. 21.48B**).

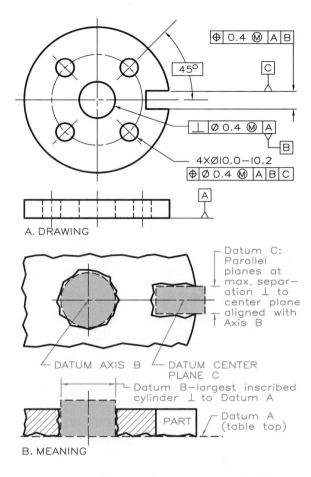

21.49 The features of this part have been dimensioned with respect to primary, secondary, and tertiary datum planes.

Secondary Datums The secondary datum (axis or center plane) for both external and internal diameters (or distances between parallel planes) has the additional requirement that the cylinder in contact with the parallel elements of the hole be perpendicular to the primary datum (**Fig. 21.49**). Datum axis B is the axis of cylinder B.

Tertiary Datums The third datum (axis or center plane) for both external and internal features has the further requirement that either the cylinder or parallel planes be oriented angularly to the secondary datum. Datum C in Fig. 21.49 is the tertiary datum plane.

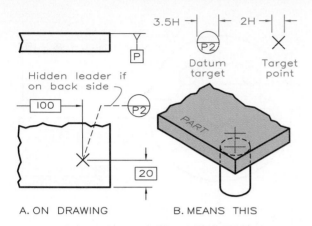

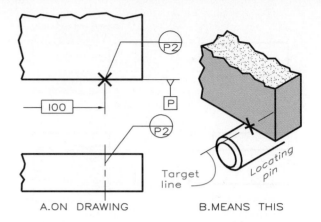

21.50 Target points from which a datum point is established are located with an X and a target symbol. The leader to the target is a hidden line since the target is on the back side.

21.51 An X and a leader from a datum symbol locate the position of a target line.

21.15 Datum Targets

Instead of using a plane surface as a datum, special datum targets can be indicated on the surface of a part where the part is supported by spherical or pointed locating pins. The symbol X is used to indicate target points, which are the points of support for the pins (**Fig. 21.50**). Datum target symbols are placed outside the outline of the part with a leader from the target. When the target is on the near (visible) surface, the leaders are solid lines. When the target is on the far side (invisible) surface, the leaders are drawn as hidden lines, as shown in Fig. 21.50.

Three target points are required to establish the primary datum plane, two for the secondary, and one for the tertiary (third). Notice that the target symbol in Fig. 21.50 is labeled P2 to match the designation of the primary datum. The other two points (not shown) would be labeled P2 and P3 to establish the primary datum.

A datum target line is specified in **Fig. 21.51** for a part that is supported on a datum line instead of a datum point. An X and a centerline is used to locate the line of support.

Target areas are specified for cases where spherical or pointed locating pins are inadequate to support a part. The diameters of the

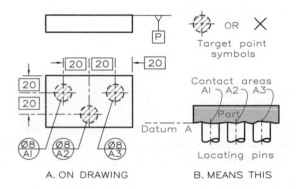

21.52 Target points with areas are located with basic dimensions and target symbols that give the diameters of the targets. Hidden leaders indicate that the targets are on the hidden side of the plane.

targets are specified with crosshatched circles surrounded by dashed lines (**Fig. 21.52**). Target symbols give both the diameter of the targets and their number designations. The X symbol could be used as an alternative method of indicating targets with areas.

The part is located on its datum plane by placing it on the three locating pins with 8-mm diameters, as shown in **Fig. 21.52**. The leaders from the target areas to the target symbols are solid lines to indicate that the targets are on the visible side of the part.

21.16 Location Tolerancing

Tolerances of location deal with **position**, **concentricity**, and **symmetry**.

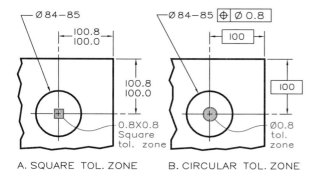

A. SQUARE TOL. ZONE   B. CIRCULAR TOL. ZONE

21.53 Toleranced and untoleranced dimensions.

A Toleranced location dimensions give a square tolerance zone for the axis of the hole.

B Untoleranced basic dimensions (in frames) locate the true position about which a circular tolerance zone of 0.8 mm is specified.

Position Location dimensions that are toleranced result in a square (or rectangular) tolerance zone for locating the center of a hole (**Fig. 21.53A**). In contrast, untoleranced location dimensions, called **basic dimensions**, locates the **true position** of a hole's center, about which a circular tolerance zone is specified (**Fig. 21.53B**).

In both methods, the size of the hole's diameter is toleranced by identical notes. In the true-position method, a feature control frame specifies the diameter of the circular tolerance zone inside which the hole's center must lie. A circular position zone gives a more precise tolerance of the hole's true position than a square.

Figure 21.54 shows an enlargement of the square tolerance zone resulting from the use of toleranced location dimension to locate a hole's center. The diagonal across the square zone is greater than the specified tolerance by a factor of 1.4. Therefore, the true-position method, shown enlarged in **Fig. 21.55**, can have a larger circular tolerance zone by a factor of 1.4 and still have the same degree of accuracy specified by the 0.1 square zone. If a variation of 0.14 across the diagonal of the square tolerance zone is acceptable in the coordinate

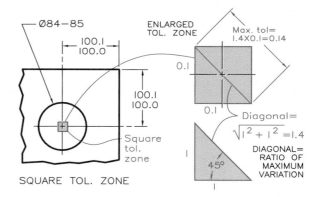

SQUARE TOL. ZONE

21.54 Toleranced coordinates give a square tolerance zone with a diagonal that exceeds the specified tolerance by a factor of 1.4.

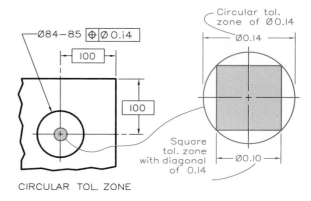

CIRCULAR TOL. ZONE

21.55 The true-position method of locating holes results in a circular tolerance zone, which can be 1.4 times greater than the square tolerance zone and still have the same degree of accuracy.

method, a circular tolerance zone of 0.14, which is greater than the 0.1 tolerance permitted by the square zone, should be acceptable in the true-position tolerance method.

The circular tolerance zone specified in the circular view of a hole extends the full depth of the hole. Therefore, the tolerance zone for the centerline of the hole is a cylindrical zone inside which the axis must lie. Because both the size of the hole and its position are toleranced, these two tolerances establish the diameter of a gauge cylinder for checking conformance of hole sizes and their locations against specifications (**Fig. 21.56**).

Subtracting the true-position tolerance from the hole at MMC (the smallest permissible

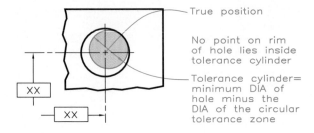

True position

No point on rim of hole lies inside tolerance cylinder

Tolerance cylinder= minimum DIA of hole minus the DIA of the circular tolerance zone

21.56 When a hole at MMC is located at true position, no element of the hole will be inside the imaginary cylinder obtained by subtracting the circular tolerance zone from the minimum diameter of the hole.

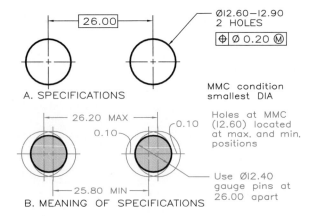

Ø12.60–12.90
2 HOLES

⊕ | Ø 0.20 Ⓜ

A. SPECIFICATIONS

MMC condition smallest DIA

26.20 MAX   0.10
0.10

Holes at MMC (12.60) located at max. and min. positions

25.80 MIN

Use Ø12.40 gauge pins at 26.00 apart

B. MEANING OF SPECIFICATIONS

21.57 Holes at MMC: True position.

A The two holes at MMC are to be located at true position as specified.

B The two holes may be gauged with pins 12.52 mm in diameter located 26.00 mm apart.

hole) yields the circle that represents the least favorable condition when the part is gauged or assembled with a mating part. When the hole is not at MMC, it is larger and permits greater tolerance and easier assembly.

Gauging a Two-Hole Pattern Gauging is a technique that checks dimensions to determine whether they meet specified tolerances (**Fig. 21.57**). The two holes with diametral size limits of 12.60–12.90 are located at true position 26.00 mm apart within a diameter of 0.20 at MMC. The gauge pin diameter is calculated to be 12.40 mm (the smallest hole's size, 12.60, minus the true-position tolerance, 0.20), as Fig. 21.57B shows. Thus, two pins with diameters of

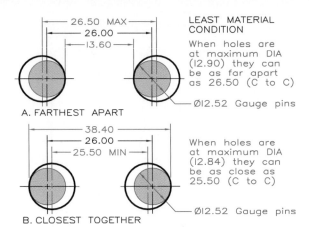

26.50 MAX
26.00
13.60

LEAST MATERIAL CONDITION

When holes are at maximum DIA (12.90) they can be as far apart as 26.50 (C to C)

A. FARTHEST APART

Ø12.52 Gauge pins

38.40
26.00
25.50 MIN

When holes are at maximum DIA (12.84) they can be as close as 25.50 (C to C)

B. CLOSEST TOGETHER

Ø12.52 Gauge pins

21.58 Holes at MMC: Maximum and minimum spacing.

A These holes at MMC can have their centers spaced as far apart as 26.50 mm and still be acceptable.

B The holes may be placed as close as 25.50 mm when they are at maximum size.

12.40 mm spaced exactly 26.00 mm apart could be used to check the diameters and positions of the holes at MMC, the most critical size. If the pins can be inserted into the holes, the holes are properly sized and located.

When the holes are not at MMC, or larger than the minimum size, the gauge pins permit a greater range of variation (**Fig. 21.58**). When the holes are at their maximum size of 12.90 mm, they can be located as close as 25.50 mm from center to center or as far apart as 26.50 mm.

Concentricity Concentricity (closely related to a new term, coaxiality) is a feature of location because it specifies the relationship of two cylinders that share the same axis. In **Fig. 21.59**, the large cylinder is labeled as datum A to be used as the datum for locating the small cylinder's axis.

Feature control frames of the type shown in **Fig. 21.60** are used to specify concentricity and other geometric characteristics throughout the remainder of this chapter.

Symmetry Symmetry is also a characteristic of location in which a feature is symmetrical

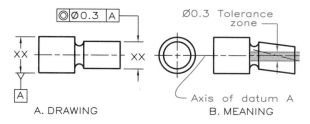

A. DRAWING B. MEANING

21.59 Concentricity (related to coaxiality) is a tolerance of location. The feature control frame specifies that the axis of the small cylinder be concentric to datum cylinder A, within a tolerance of 0.3-mm diameter.

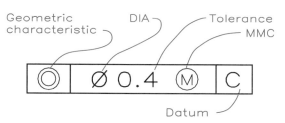

21.60 This typical feature control frame indicates that a surface is concentric to datum C within a cylindrical diameter of 0.4 mm at MMC.

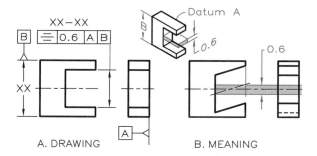

A. DRAWING B. MEANING

21.61 Symmetry is a tolerance of location specifying that a part's features be symmetrical about the center plane between parallel surfaces of the part.

with the same contour and size on opposite sides of a central plane. **Figure 21.61A** shows how to apply a symmetry feature symbol to the notch that is symmetrical about the part's central datum plane B for a zone of 0.6 mm (**Fig. 21.61B**).

21.17 Form Tolerancing

Flatness A surface is flat when all its elements are in one plane. A feature control frame specifies flatness within a 0.40-mm tol-

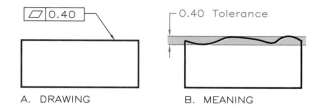

A. DRAWING B. MEANING

21.62 Flatness is a tolerance of form that specifies a tolerance zone within which a surface must lie.

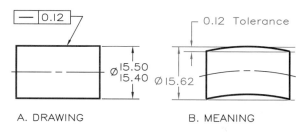

A. DRAWING B. MEANING

21.63 Straightness is a tolerance of form that indicates that elements of a surface are straight lines. The tolerance frame is applied to the views in which elements appear as straight lines, not points.

erance zone in **Fig. 21.62**, where no point on the surface may vary more than 0.40 from the highest to the lowest point.

Straightness A surface is straight if all its elements are straight lines within a specified tolerance zone. The feature control frame shown in **Fig. 21.63** specifies that the elements of a cylinder must be straight within 0.12 mm. On flat surfaces, straightness is measured in a plane passing through control-line elements, and it may be specified in two directions (usually perpendicular) if desired.

Circularity (Roundness) A surface of revolution (a cylinder, cone, or sphere) is circular when all points on the surface intersected by a plane perpendicular to its axis are equidistant from the axis. In **Fig. 21.64**, the feature control frame specifies circularity of a cone and cylinder, permitting a tolerance of 0.54 mm on the radius. **Figure 21.65** specifies a 0.3-mm tolerance zone for the roundness of a sphere.

Cylindricity A surface of revolution is cylindrical when all its elements lie within a cylindrical

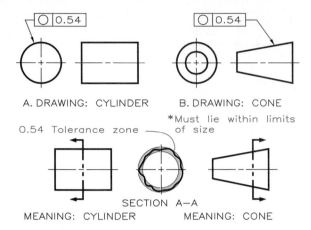

A. DRAWING: CYLINDER   B. DRAWING: CONE

*Must lie within limits of size

0.54 Tolerance zone

SECTION A—A

MEANING: CYLINDER   MEANING: CONE

21.64 Circularity (roundness) is a tolerance of form. It indicates that a cross-section through a surface of revolution is round and lies within two concentric circles.

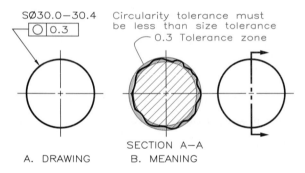

A. DRAWING   B. MEANING

SECTION A—A

21.65 Circularity of a sphere means that any cross-section through it is round within the specified tolerance.

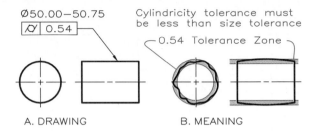

A. DRAWING   B. MEANING

21.66 Cylindricity is a tolerance of form that is a combination of roundness and straightness. It indicates that the surface of a cylinder lies within a tolerance zone formed by two concentric cylinders.

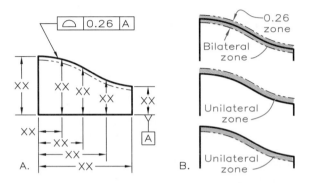

21.67 Profile is a tolerance of form for irregular curving planes. (A) A curving plane, located by coordinates and toleranced unidirectionally. (B) Some methods for applying tolerance.

tolerance zone, which is a combination of tolerances of roundness and straightness (**Fig. 21.66**). Here, a cylindricity tolerance zone of 0.54 mm on the radius of the cylinder is specified.

21.18 Profile Tolerancing

Profile tolerancing involves specifying tolerances for a contoured shape formed by arcs or irregular curves; it can apply to a **surface** or a **single line**. The surface with the unidirectional profile tolerance shown in **Fig. 21.67A** is defined by coordinates. **Figure 21.67B** shows how to specify bilateral and unilateral tolerance zones.

A profile tolerance for a single line is specified as shown in **Fig. 21.68.** The curve is

formed by tangent arcs whose radii are given as basic dimensions. The radii are permitted to vary ±0.10 mm from the basic radii.

21.19 Orientation Tolerancing

Tolerances of orientation include **parallelism**, **perpendicularity**, and **angularity**.

Parallelism A surface or line is parallel when all its points are equidistant from a datum plane or axis. There are two types of parallelism tolerance zones:

1. A planar tolerance zone parallel to a datum plane within which the axis or surface of the feature must lie (**Fig. 21.69**). This tolerance also controls flatness.

2. A cylindrical tolerance zone parallel to a datum feature within which the axis of a feature must lie (**Fig. 21.70**).

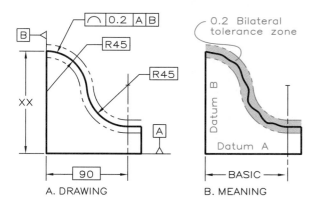

A. DRAWING   B. MEANING

21.68 The profile of a line is a tolerance of form that specifies the variation allowed from the path of a line. Here, the line is formed by tangent arcs. The tolerance zone may be either bilateral or unilateral, as shown in Fig. 21.67.

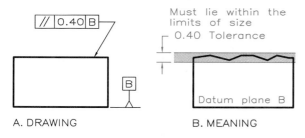

A. DRAWING   B. MEANING

21.69 Parallelism is a tolerance of form. It indicates that a plane is parallel to a datum plane within specified limits. Here, plane B is the datum plane.

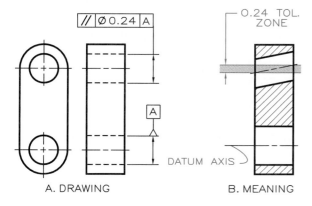

A. DRAWING   B. MEANING

21.70 You may specify parallelism of one centerline to another by using the diameter of one of the holes as the datum.

Figure 21.71 shows the effect of specifying parallelism at MMC, where the modifier M is given in the feature control frame. Tolerances of form apply at RFS when not specified. Specifying parallelism at MMC means that the

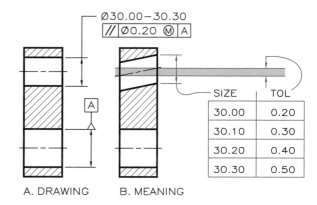

A. DRAWING   B. MEANING

SIZE	TOL
30.00	0.20
30.10	0.30
30.20	0.40
30.30	0.50

21.71 The critical tolerance exists when features are at MMC. (A) The upper hole must be parallel to the hole used as datum A within a 0.20-mm DIA. (B) As the hole approaches its maximum size of 30.30 mm, the tolerance zone approaches 0.50 mm.

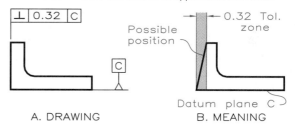

A. DRAWING   B. MEANING

21.72 Perpendicularity is a tolerance of form that gives a tolerance zone of 0.32 for a plane perpendicular to a specified datum plane.

axis of the cylindrical hole must vary no more than 0.20 mm when the holes are at their smallest permissible size.

As the hole approaches its upper limit of 30.30, the tolerance zone increases to a maximum of 0.50 DIA. Therefore, a greater variation is given at MMC than at RFS.

Perpendicularity The specifications for the perpendicularity of a plane to a datum are shown in **Fig. 21.72**. The feature control frame shows that the surface perpendicular to datum plane C has a tolerance of 0.32 in. In **Fig. 21.73**, a hole is specified as perpendicular to datum plane A.

Angularity A surface or line is angular when it is at an angle (other than 90°) from a datum or an axis. The angularity of the surface shown in **Fig. 21.74** is dimensioned with a basic angle (exact angle) of 30° and an angularity toler-

21.19 ORIENTATION TOLERANCING • 359

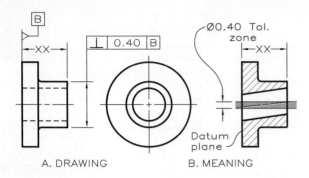

A. DRAWING   B. MEANING

21.73 Perpendicularity can apply to the axis of a feature, such as the centerline of a cylinder.

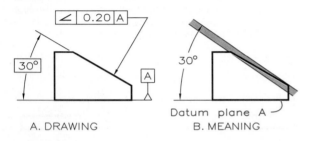

A. DRAWING   B. MEANING

21.74 Angularity is a tolerance of form specifying a tolerance zone for an angular surface with respect to a datum plane. Here, the 30° angle is an exact, or basic, angle to which a tolerance of 0.20 mm is applied.

ance zone of 0.20 mm inside of which the plane must lie.

21.20 Runout Tolerancing

Runout tolerancing is a way of controlling multiple features by relating them to a common datum axis. Features so controlled are surfaces of revolution about an axis and surfaces perpendicular to the axis.

The datum axis, such as diameter B in **Fig. 21.75**, is established by a circular feature that rotates about the axis. When the part is rotated about this axis, the features of rotation must fall within the prescribed tolerance at full indicator movement (FIM).

There are two types of runout: **circular** and **total**. One arrow in the feature control frame indicates circular runout; two arrows indicate total runout.

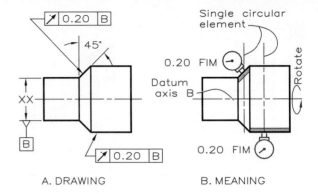

A. DRAWING   B. MEANING

21.75 Circular runout provides control of circular elements. The tolerance is applied independently to each circular measuring position as the part is rotated 360°.

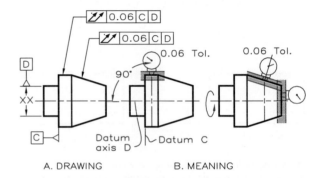

A. DRAWING   B. MEANING

21.76 Runout tolerance is measured here by mounting the object on the primary datum plane C and the secondary datum cylinder D. The cylinder and conical surface are gauged to check their conformity to a tolerance zone of 0.06 mm. The runout at the end of the cone could have been noted.

Circular Runout Rotating an object about its axis 360° determines whether a circular cross section exceeds the permissible runout tolerance at any point (**Fig. 21.76**). This same technique is used to measure the amount of wobble in surfaces perpendicular to the axis of rotation.

Total Runout Used to specify cumulative variations of circularity, straightness, concentricity, angularity, taper, and profile of a surface (Fig. 21.76), total runout tolerances are measured for all circular and profile positions as the part is rotated 360°. When applied to surfaces perpendicular to the axis, total

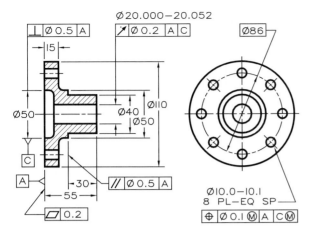

21.77 A combination of notes and symbols describe this part's geometric features.

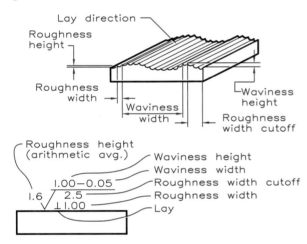

21.78 The definitions of surface texture for a finished surface.

runout tolerances control variations in perpendicularity and flatness.

The dimensioned part shown in **Fig. 21.77** illustrates several of the techniques of geometric tolerancing described in this and previous sections.

21.21 Surface Texture

Because the surface texture of a part affects its function, it must be precisely specified instead of giving an unspecified finished mark such as a V. **Figure 21.78** illustrates most of the terms that apply to surface texture (surface control):

Surface texture: The variation in a surface, including roughness, waviness, lay, and flaws.

Roughness: The finest of the irregularities in the surface caused by the manufacturing process used to smooth the surface.

Roughness height: The average deviation from the mean plane of the surface measured in microinches ([mu]in.) or micrometers ([mu]m), or millionths of an inch or a meter, respectively.

Roughness width: The width between successive peaks and valleys forming the roughness measured in microinches or micrometers.

Roughness width cutoff: The largest spacing of repetitive irregularities that includes average roughness height (measured in inches or millimeters). When not specified, a value of 0.8 mm (0.030 in.) is assumed.

Waviness: A widely spaced variation that exceeds the roughness width cutoff measured in inches or millimeters. Roughness may be regarded as a surface variation superimposed on a wavy surface.

Waviness height: The peak-to-valley distance between waves measured in inches or millimeters.

Waviness width: The spacing between wave peaks or wave valleys measured in inches or millimeters.

Lay: The direction of the surface pattern caused by the production method used.

Flaws: Irregularities or defects occurring infrequently or at widely varying intervals on a surface, including cracks, blow holes, checks, ridges, scratches, and the like. The effect of flaws is usually omitted in roughness height measurements.

Contact area: The surface that will make contact with a mating surface.

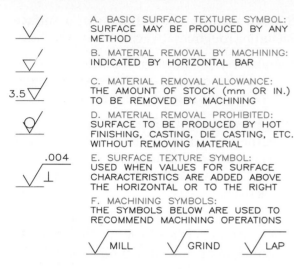

A. BASIC SURFACE TEXTURE SYMBOL: SURFACE MAY BE PRODUCED BY ANY METHOD

B. MATERIAL REMOVAL BY MACHINING: INDICATED BY HORIZONTAL BAR

C. MATERIAL REMOVAL ALLOWANCE: THE AMOUNT OF STOCK (mm OR IN.) TO BE REMOVED BY MACHINING

D. MATERIAL REMOVAL PROHIBITED: SURFACE TO BE PRODUCED BY HOT FINISHING, CASTING, DIE CASTING, ETC. WITHOUT REMOVING MATERIAL

E. SURFACE TEXTURE SYMBOL: USED WHEN VALUES FOR SURFACE CHARACTERISTICS ARE ADDED ABOVE THE HORIZONTAL OR TO THE RIGHT

F. MACHINING SYMBOLS: THE SYMBOLS BELOW ARE USED TO RECOMMEND MACHINING OPERATIONS

MILL   GRIND   LAP

21.79 The surface texture symbols that are used to specify surface finish on the edge views of finished surfaces.

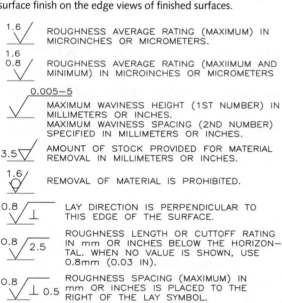

ROUGHNESS AVERAGE RATING (MAXIMUM) IN MICROINCHES OR MICROMETERS.

ROUGHNESS AVERAGE RATING (MAXIIMUM AND MINIMUM) IN MICROINCHES OR MICROMETERS

MAXIMUM WAVINESS HEIGHT (1ST NUMBER) IN MILLIMETERS OR INCHES.
MAXIMUM WAVINESS SPACING (2ND NUMBER) SPECIFIED IN MILLIMETERS OR INCHES.

AMOUNT OF STOCK PROVIDED FOR MATERIAL REMOVAL IN MILLIMETERS OR INCHES.

REMOVAL OF MATERIAL IS PROHIBITED.

LAY DIRECTION IS PERPENDICULAR TO THIS EDGE OF THE SURFACE.

ROUGHNESS LENGTH OR CUTOFF RATING IN mm OR INCHES BELOW THE HORIZONTAL. WHEN NO VALUE IS SHOWN, USE 0.8mm (0.03 IN.).

ROUGHNESS SPACING (MAXIMUM) IN mm OR INCHES IS PLACED TO THE RIGHT OF THE LAY SYMBOL.

21.80 Values may be added to surface control symbols for more precise specifications.

Symbols for specifying surface texture are shown in **Fig. 21.79**. The point of the V must touch the edge view of the surface, an extension line from it, or a leader pointing to the surface. **Figure 21.80** shows how to specify values as a part of surface texture symbols.

Roughness height values are related to the processes used to finish surfaces and may be taken from the chart in **Fig. 21.81**. The pre-

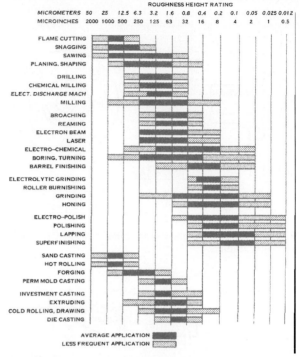

21.81 Various types of production methods result in the surface roughness heights shown in micrometers and microinches (millionths of a meter or an inch).

PREFERRED ROUGHNESS AVERAGE VALUES

Micrometers μm	Microinches μin.		Micrometers μm	Microinches μin.
0.025	1		1.6	63
0.050	2		3.2	125
0.10	4		6.3	250
0.20	8		12.5	500
0.40	16		25	1000
0.80	32		Micrometers=0.001 mm	

21.82 The roughness heights recommended in the *ANSI Y14.36* standards.

MILLIMETERS	0.08	0.25	0.80	2.5	8.0	25
INCHES	.003	.010	.030	.1	.3	1

21.83 The roughness width cutoff values recommended in the *ANSI Y14.36* standards. When unspecified, assume a value of 0.8.

ferred values of roughness height are listed in **Fig. 21.82**.

The preferred roughness width cutoff values in **Fig. 21.83** are for specifying the sampling width used to measure roughness height. A value of 0.80 mm is assumed if no

MAXIMUM WAVINESS HEIGHT VALUES

mm	in.	mm	in.
0.0005	.00002	0.025	.001
0.0008	.00003	0.05	.002
0.0012	.00005	0.08	.003
0.0020	.00008	0.12	.005
0.0025	.0001	0.20	.008
0.005	.0002	0.25	.010
0.008	.0003	0.38	.015
0.012	.0005	0.50	.020
0.020	.0008	0.80	.030

21.84 The maximum waviness height values recommended in the *ANSI Y14.36* standards.

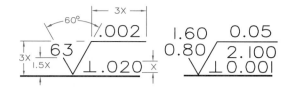

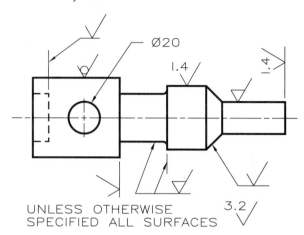

21.86 Examples and proportions of typical, fully specified surface texture symbols.

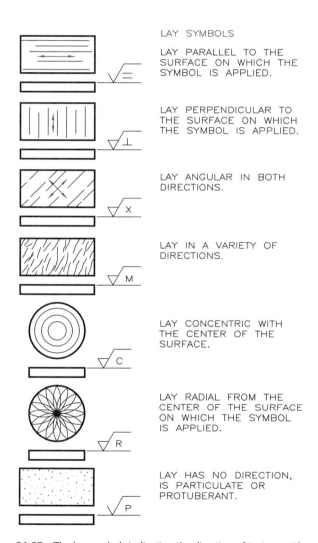

LAY SYMBOLS

LAY PARALLEL TO THE SURFACE ON WHICH THE SYMBOL IS APPLIED.

LAY PERPENDICULAR TO THE SURFACE ON WHICH THE SYMBOL IS APPLIED.

LAY ANGULAR IN BOTH DIRECTIONS.

LAY IN A VARIETY OF DIRECTIONS.

LAY CONCENTRIC WITH THE CENTER OF THE SURFACE.

LAY RADIAL FROM THE CENTER OF THE SURFACE ON WHICH THE SYMBOL IS APPLIED.

LAY HAS NO DIRECTION, IS PARTICULATE OR PROTUBERANT.

21.85 The lay symbols indicating the direction of texture with respect to the surface where the control symbol is placed.

UNLESS OTHERWISE SPECIFIED ALL SURFACES

21.87 Various techniques of applying surface texture symbols to a part.

value is given. When required, maximum waviness height values may be selected from the recommended values shown in **Fig. 21.84**.

Lay symbols indicating the direction of texture (markings made by the machining operation) on a surface (**Fig. 21.85**) may be added to surface texture symbols as shown in **Fig. 21.86**. The perpendicular sign indicates that lay is perpendicular to the edge view of the surface in this view (where the surface control symbol appears). **Figure 21.87** illustrates how to apply a variety of surface texture symbols to a part.

Problems

Solve the following problems on size A sheets laid out on a grid of 0.20 in. or 5 mm.

Cylindrical Fits
1. (**Fig. 21.88**) Draw the shaft and hole shown (it need not be to scale), give the limits for

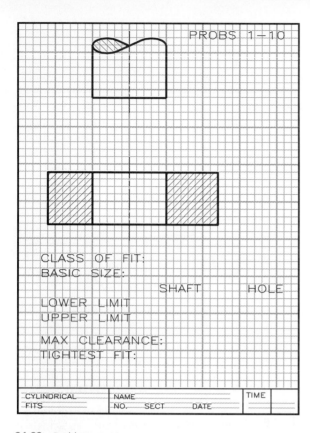

21.88 Problems 1–10.

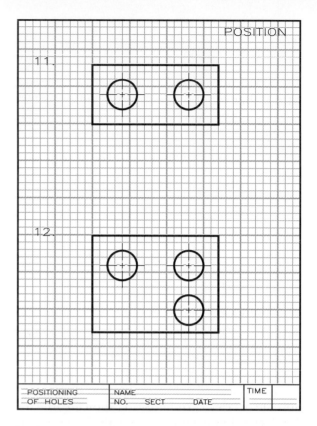

21.89 Problems 11–13.

each diameter, and complete the table of values. Use a basic diameter of 1.00 in. (25 mm) and a class RC 1 fit or a metric fit of H8/f7.

2. Repeat Problem 1, but use a basic diameter of 1.75 in. (45 mm) and a class RC 9 fit or a metric fit of H11/c11.

3. Repeat Problem 1, but use a basic diameter of 2.00 in. (51 mm) and a class RC 5 fit or a metric fit of H9/d9.

4. Repeat Problem 1, but use a basic diameter of 12.00 in. (305 mm) and a class LC 11 fit or a metric fit of H7/h6.

5. Repeat Problem 1, but use a basic diameter of 3.00 in. (76 mm) and a class LC 1 fit or a metric fit of H7/h6.

6. Repeat Problem 1, but use a basic diameter of 8.00 in. (203 mm) and a class LC 1 fit or a metric fit of H7/k6.

7. Repeat Problem 1, but use a basic diameter of 102 in. (2591 mm) and a class LN 3 fit or a metric fit of H7/n6.

8. Repeat Problem 1, but use a basic diameter of 11.00 in. (279 mm) and a class LN 2 fit or a metric fit of H7/p6.

9. Repeat Problem 1, but use a basic diameter of 6.00 in. (152 mm) and a class FN 5 fit or a metric fit of H7/s6.

10. Repeat Problem 1, but use a basic diameter of 2.60 in. (66 mm) and a class FN 1 fit or a metric fit of H7/u6.

Position Tolerancing

11. (**Fig. 21.89**) Make an instrument drawing of the part shown. Locate the two holes with a size tolerance of 1.00 mm and a position tolerance of 0.50 DIA. Insert the proper symbols and dimensions.

12. Repeat Problem 11, but locate three holes using the same tolerances for size and position.

13. Give the specifications for a two-pin gauge that can be used to measure the correctness of the two holes specified in Problem 11. Make a sketch of the gauge and show the proper dimensions on it.

14. (**Sheet 1**) Using positioning tolerances, locate the holes and properly note them to provide a size tolerance of 1.50 mm and a locational tolerance of 0.60 DIA.

15. Repeat Problem 14, but locate six equally spaced, equally sized holes using the same tolerances of position.

16. (**Sheet 2**) Using a feature control symbol and the necessary dimensions, indicate that the notch is symmetrical to the left-hand end of the part within 0.60 mm.

17. (**Sheet 2**) Using a feature control symbol and the necessary dimensions, indicate that the small cylinder is concentric with the large one (the datum cylinder) within a tolerance of 0.80.

18. (**Sheet 3**) Using a feature control symbol and the necessary dimensions, indicate that the elements of the cylinder are straight within a tolerance of 0.20 mm.

19. (**Sheet 3**) Using a feature control symbol and the necessary dimensions, indicate that surface A of the object is flat within a tolerance of 0.08 mm.

20–22. (**Sheet 4**) Using feature control symbols and the necessary dimensions, indicate that the cross sections of the cylinder, cone, and sphere are round within a tolerance of 0.40 mm.

23. (**Sheet 5**) Using a feature control symbol and the necessary dimensions, indicate that the profile of the irregular surface of the object lies within a bilateral or unilateral tolerance zone of 0.40 mm.

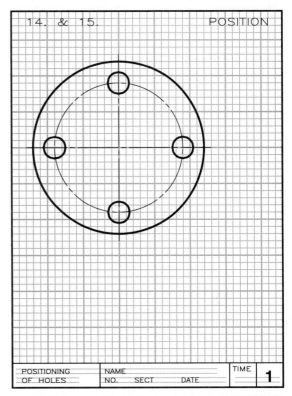

14. & 15.   POSITION

| POSITIONING OF HOLES | NAME NO.   SECT   DATE | TIME | 1 |

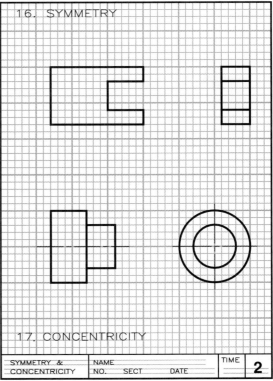

16. SYMMETRY

17. CONCENTRICITY

| SYMMETRY & CONCENTRICITY | NAME NO.   SECT   DATE | TIME | 2 |

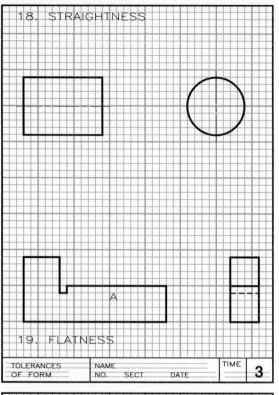

18. STRAIGHTNESS

A

19. FLATNESS

TOLERANCES OF FORM	NAME			TIME	**3**
	NO.	SECT	DATE		

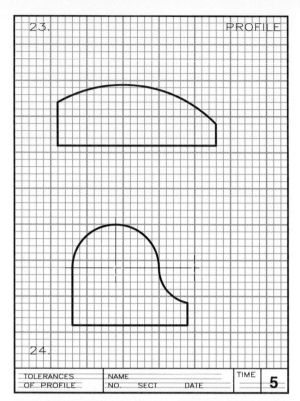

23.   PROFILE

24.

TOLERANCES OF PROFILE	NAME			TIME	**5**
	NO.	SECT	DATE		

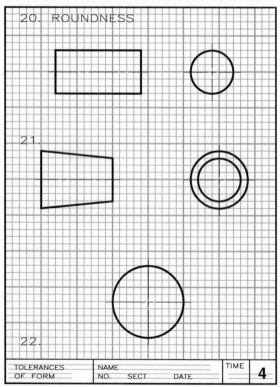

20. ROUNDNESS

21.

22.

TOLERANCES OF FORM	NAME			TIME	**4**
	NO.	SECT	DATE		

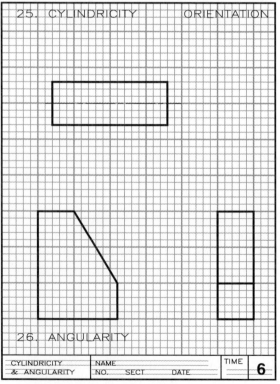

25. CYLINDRICITY   ORIENTATION

26. ANGULARITY

CYLINDRICITY & ANGULARITY	NAME			TIME	**6**
	NO.	SECT	DATE		

24. (**Sheet 5**) Using a feature control symbol and the necessary dimensions, indicate that the profile of the line formed by tangent arcs lies within a bilateral or unilateral tolerance zone of 0.40 mm.

25. (**Sheet 6**) Using a feature control symbol and the necessary dimensions, indicate that the cylindricity of the cylinder is 0.90 mm.

26. (**Sheet 6**) Using a feature control symbol and the necessary dimensions, indicate that the angularity tolerance of the inclined plane is 0.7 mm from the bottom of the object, the datum plane.

27. (**Sheet 7**) Using a feature control symbol and the necessary dimensions, indicate that surface A of the object is parallel to datum B within 0.30 mm.

28. (**Sheet 7**) Using a feature control symbol and the necessary dimensions, indicate that the small hole is parallel to the large hole, the datum, within a tolerance of 0.80 mm.

29. (**Sheet 8**) Using a feature control symbol and the necessary dimensions, indicate that the vertical surface B is perpendicular to the bottom of the object, the datum C, within a tolerance of 0.20 mm.

30. (**Sheet 8**) Using a feature control symbol and the necessary dimensions, indicate that the hole is perpendicular to datum A within a tolerance of 0.08 mm.

31. (**Sheet 9**) Using the appropriate geometric tolerancing symbols and cylinder A as the datum, indicate that the conical feature has a runout of 0.80 mm.

32. (**Sheet 9**) Using a feature control symbol with cylinder B as the primary datum and surface C as the secondary datum, indicate that surfaces D, E, and F have a runout of 0.60 mm.

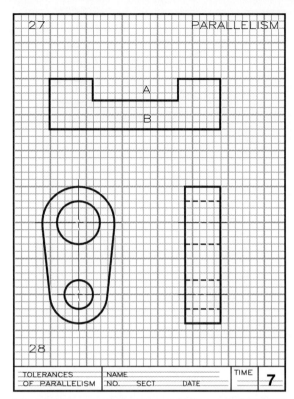

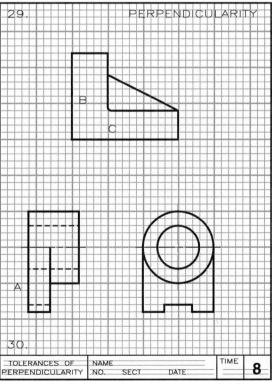

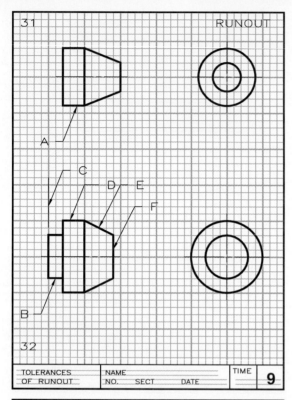

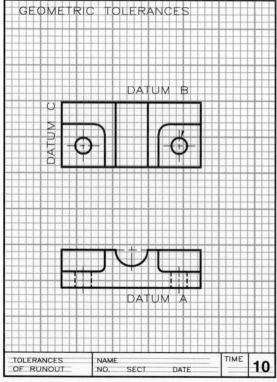

DATUM C
(TERTIARY)

DATUM B
(SECONDARY)

DATUM A
(PRIMARY)

JOURNAL BASE
1020 STEEL

21.100 Problems 33 and 34.

33. (**Sheet 10**) Draw the journal base as it is shown, which is a half-size, two-view drawing. Using metric units and the three datum planes, locate the two holes to lie within a tolerance zone of 1.6 DIA. Indicate that datum A is flat within 0.8 and that datum B is perpendicular to datum A within 1.2. (Refer to **Fig. 21.100.**)

34. Same as Problem 33, but in addition to these specifications, give the complete dimensions necessary to fully describe the part. Also indicate that the upper surface in the front view is parallel to the datum A within 1.4.m.

35. (**Sheet 11**) Draw a feature control symbol on datum A. Complete the tolerance table to show how the tolerance increases as the hold departs from MMC. Refer to Fig. 21.71.

36. (**Sheet 12**) Draw a feature control symbol on datum A. Complete the tolerance table to show the tolerances, regardless of feature size (RFS). Refer to Fig. 21.71.

368 • **CHAPTER 21 TOLERANCES**

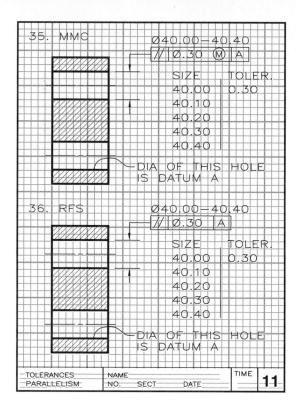

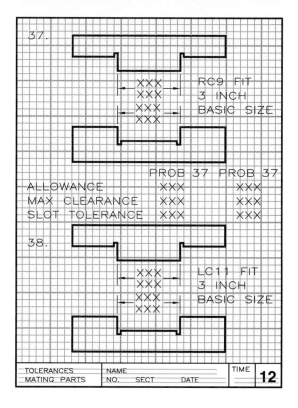

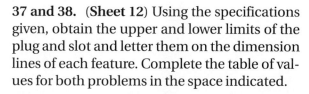

37 and 38. (**Sheet 12**) Using the specifications given, obtain the upper and lower limits of the plug and slot and letter them on the dimension lines of each feature. Complete the table of values for both problems in the space indicated.

39 and 40. (**Sheet 13**) By referring to Figs. 21.35 and 21.35 as guides, draw tables of general tolerances for linear dimensions (coarse series) and angular dimensions (in decimal degrees). Size the table to be adequate for 1/8-in. lettering.

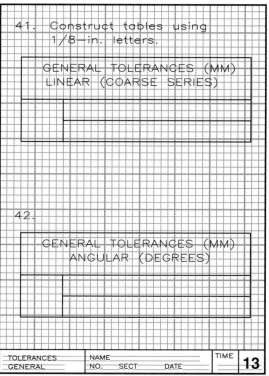

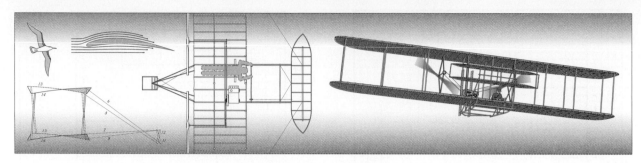

22

Welding

22.1 Introduction

Welding is the process of permanently joining metal by heating a joint to a suitable temperature with or without applying pressure and with or without using filler material. The welding practices described in this chapter comply with the standards developed by the American Welding Society and the American National Standards Institute (ANSI).

Welding is done in shops, on assembly lines, or in the field as shown in **Fig. 22.1** where a welder is joining pipes. Welding is a widely used method of fabrication with its own language of notes, specifications, and symbology. You must become familiar with this system of notations in order to make and read drawings containing welding specifications.

The advantages of welding over other methods of fastening include (1) simplified fabrication, (2) economy, (3) increased strength and rigidity, (4) ease of repair, (5) cre-

22.1 A welder joining two pipes in accordance with specifications on a set of drawings. (Courtesy of Texas Eastern; *TE Today*; photo by Bob Thigpen.)

ation of gas- and liquid-tight joints, and (6) reduction in weight and size.

22.2 Welding Processes

Figure 22.2 shows various types of welding processes. The three main types are gas weld-

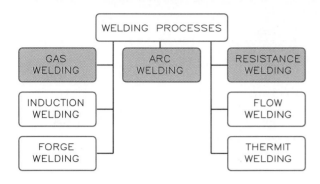

22.2 The three main types of welding processes are gas welding, arc welding, and resistance welding.

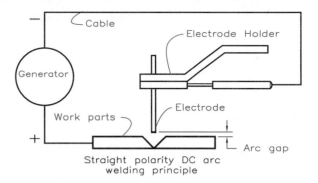

22.4 In arc welding, either AC or DC current is passed through an electrode to heat the joint.

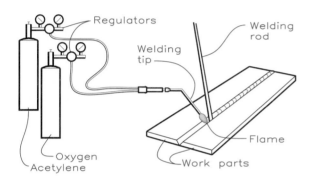

22.3 The gas welding process burns gases such as oxygen and acetylene in a torch to apply heat to a joint. The welding rod supplies the filler material.

ing, arc welding, and resistance welding.

Gas welding involves the use of gas flames to melt and fuse metal joints. Gases such as acetylene or hydrogen are mixed in a welding torch and burned with air or oxygen (**Fig. 22.3**). The oxyacetylene method is widely used for repair work and field construction.

Most oxyacetylene welding is done manually with a minimum of equipment. Filler material in the form of welding rods is used to deposit metal at the joint as it is heated. Most metals, except for low- and medium-carbon steels, require fluxes to aid the process of melting and fusing the metals.

Arc welding involves the use of an electric arc to heat and fuse joints, with pressure

sometimes required in addition to heat (**Fig. 22.4**). The filler material is supplied by a consumable or nonconsumable electrode through which the electric arc is transmitted. Metals well-suited to arc welding are wrought iron, low- and medium-carbon steels, stainless steel, copper, brass, bronze, aluminum, and some nickel alloys. In electric-arc welding, the flux is a material coated on the electrodes that forms a coating on the metal being welded. This coating protects the metal from oxidation so that the joint will not be weakened by overheating.

Flash welding is a form of arc welding, but it is similar to resistance welding because both pressure and electric current (**Fig. 22.5**) are applied. The pieces to be welded are brought together, and an electric current is passed through them, causing heat to build up between them. As the metal burns, the current is turned off, and the pressure between the pieces is increased in order to fuse them.

Resistance welding comprises several processes by which metals are fused both by the heat produced from the resistance of the parts to an electric current and by pressure. Fluxes and filler materials normally are not used. All resistance welds are either lap- or butt-type welds.

Resistance spot welding is performed by pressing the parts together. An electric cur-

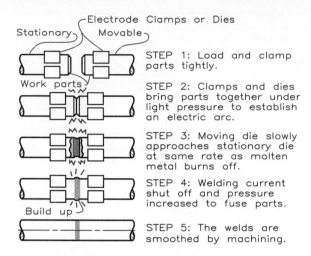

STEP 1: Load and clamp parts tightly.

STEP 2: Clamps and dies bring parts together under light pressure to establish an electric arc.

STEP 3: Moving die slowly approaches stationary die at same rate as molten metal burns off.

STEP 4: Welding current shut off and pressure increased to fuse parts.

STEP 5: The welds are smoothed by machining.

22.5 Flash welding, a type of arc welding, uses a combination of electric current and pressure to fuse two parts.

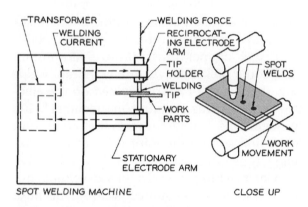

22.6 Resistance spot welding may be used to join lap and butt joints.

Material	Spot Welding	Flash Welding
Low—carbon mild steel		
SAE 1010	Rec.	Rec.
SAE 1020	Rec.	Rec.
Medium—carbon steel		
SAE 1030	Rec.	Rec.
SAE 1050	Rec.	Rec.
Wrought alloy steel		
SAE 4130	Rec.	Rec.
SAE 4340	Rec.	Rec.
High—alloy austenitic stainless steel		
SAE 30301–30302	Rec.	Rec.
SAE 30309–30316	Rec.	Rec.
Ferritic and martensistic stainless steel		
SAE 51410–51430	Satis.	Satis.
Wrought heat—resisting alloys		
19–9–DL	Satis.	Satis.
16–25–6	Satis.	Satis.
Cast iron	NA	Not Rec.
Gray iron	NA	Not Rec.
Aluminum & alum. alloys	Rec.	Satis.
Nickel & nickel alloys	Rec.	Satis.

Rec.—Recommended   Satis.—Satisfactory
Not Rec.— Not recommended   NA—Not applicable

22.7 Resistance welding processes for various materials.

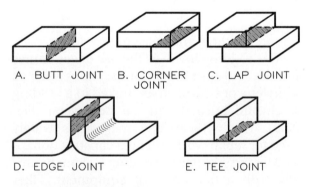

22.8 The five standard weld joints.

rent then fuses them, as illustrated in the lap joint weld in **Fig. 22.6**. A series of small welds spaced at intervals, called **spot welds**, secure the parts. **Figure 22.7** lists the recommended materials and processes to be used for resistance welding.

22.3 Weld Joints and Welds

Figure 22.8 shows the five standard weld joints: (1) **butt**, (2) **corner**, (3) **lap**, (4) **edge**, and (5) **tee**. The **butt joint** can be joined with the square groove, V-groove, bevel groove,

U-groove, and J-groove welds. The **corner joint** can be joined with these welds and with the fillet weld. The **lap joint** can be joined with the bevel groove, J-groove, fillet, slot, plug, spot, projection, and seam welds. The **edge**

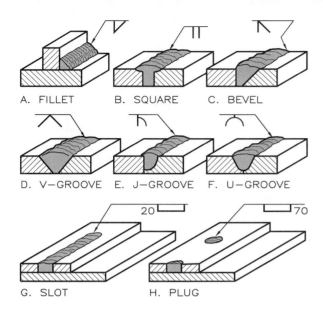

A. FILLET B. SQUARE C. BEVEL

D. V-GROOVE E. J-GROOVE F. U-GROOVE

20 70

G. SLOT H. PLUG

22.9 Standard welds and their corresponding ideographs.

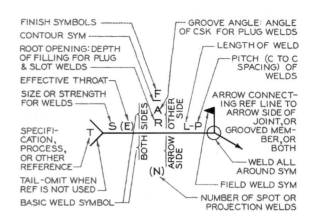

FINISH SYMBOLS
CONTOUR SYM
ROOT OPENING: DEPTH OF FILLING FOR PLUG & SLOT WELDS
EFFECTIVE THROAT
SIZE OR STRENGTH FOR WELDS
SPECIFICATION, PROCESS, OR OTHER REFERENCE
TAIL-OMIT WHEN REF IS NOT USED
BASIC WELD SYMBOL

GROOVE ANGLE: ANGLE OF CSK FOR PLUG WELDS
LENGTH OF WELD
PITCH (C TO C SPACING) OF WELDS
ARROW CONNECTING REF LINE TO ARROW SIDE OF JOINT, OR GROOVED MEMBER, OR BOTH
WELD ALL AROUND SYM
FIELD WELD SYM
NUMBER OF SPOT OR PROJECTION WELDS

22.10 The welding symbol. Usually, it is modified to a simpler form for use on drawings.

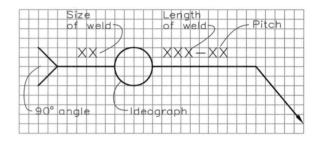

22.11 Welding symbol proportions are based on the letter height used on a drawing, usually 1/8 in. or 3 mm. This grid is equal to the letter height.

joint uses the same welds as the lap joint along with the square groove, V-groove, U-groove, and seam welds. The **tee joint** can be joined by the bevel groove, J-groove, and fillet welds.

Figure 22.9 depicts commonly used welds and their corresponding ideographs (symbols). The fillet weld is a built-up weld at the intersection (usually 90°) of two surfaces. The square, bevel, V-groove, J-groove, and U-groove welds all have grooves, and the weld is made in these grooves. Slot and plug welds have intermittent holes or openings where the parts are welded. Holes are unnecessary when resistance welding is used.

22.4 Welding Symbols

If a drawing has a general welding note—ALL JOINTS ARE WELDED THROUGHOUT—the designer has transferred responsibility to the welder. Welding is too important to be left to chance and should be specified more precisely.

Symbols are used to convey welding specifications on a drawing. The complete welding symbol is shown in **Figure 22.10**, but it usually

appears on a drawing in modified, more general form with less detail. The scale of the welding symbol is based on the letter height used on the drawing, which is the size of the grid on which the symbol is drawn in **Fig. 22.11**. The standard height of lettering on a drawing is usually 1/8 in. or 3 mm.

The **ideograph** is the symbol that denotes the type of weld desired, and it generally depicts the cross-section representation of the weld. **Figure 22.12** shows the ideographs used most often. They are drawn to scale on the 1/8-in. (3-mm) grid (equal to the letter height), which represents their full size when added to the welding symbol.

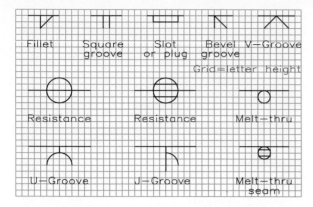

22.12 The sizes of the ideographs shown on the 1/8-in. (3-mm) grid (the letter height) are proportional to the size of the welding symbol (Fig. 22.11).

22.5 Application of Symbols

Fillet Welds

In **Fig. 22.13A**, placement of the fillet weld ideograph below the horizontal line of the symbol indicates that the weld is at the joint on the arrow side—the right side, in this case. The vertical leg of the ideograph is always on the left side.

A numeral (either a common fraction or a decimal value) to the right of the ideograph indicates the size of the weld. You may omit this number from the symbol if you insert a general note elsewhere on the drawing to specify the fillet size, such as:

ALL FILLET WELDS 1/4 IN.

UNLESS OTHERWISE NOTED.

Placing the ideograph above the horizontal line, as in **Fig. 22.13B**, indicates that the weld is to be on the other side—that is, the joint on the other side of the part away from the arrow. When the part is to be welded on both sides, use the ideograph shown in **Fig. 22.13C**. You may omit the tail and other specifications from the symbol when you provide detailed specifications elsewhere.

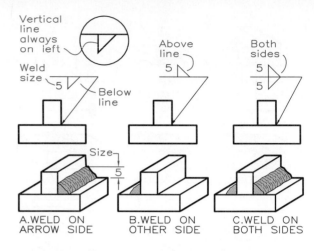

22.13 Fillet welds with abbreviated symbols.

A When the ideograph appears below the horizontal line, it specifies a weld on the arrow side.

B When it is above the line, it specifies a weld on the opposite side.

C When it is on both sides of the line, it specifies a weld on each side.

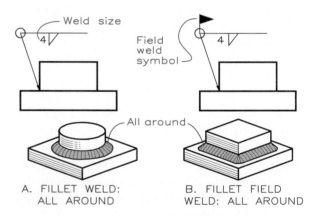

22.14 These symbols indicate fillet welds all around two types of parts.

A single arrow is often used to specify a weld that is to be made all around two joining parts (**Fig. 22.14A**). A circle of 6 mm (twice the letter height) in diameter, drawn at the bend in the leader of the symbol denotes this type of weld. If the welding is to be done in the field

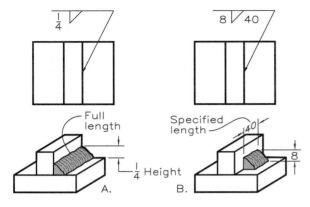

22.15 Fillet weld symbols.

A A symbol indicating full-length fillet welds.

B A symbol indicating fillet welds of a specified but less than full length.

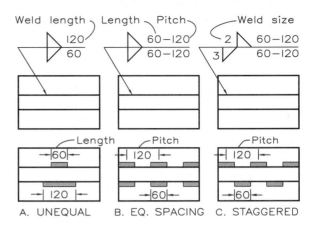

A. UNEQUAL B. EQ. SPACING C. STAGGERED

22.16 These symbols specify intermittent welds of varying lengths and alignments.

rather than in the shop, a solid black triangular "flag" is also added (**Fig. 22.14B**).

You may specify a fillet weld that is to run the full length of the two parts, as in **Fig. 22.15A**. The ideograph is on the lower side of the horizontal line, so the weld is on the arrow side. You may specify a fillet weld that is to run shorter than full length, as in **Fig. 22.15B**, where 40 represents the weld's length in millimeters.

You may specify fillet welds to run different lengths and to be positioned on both sides of a part, as in **Fig. 22.16A**. The dimension on the lower side of the horizontal gives the length of the weld on the arrow side, and the dimension on the upper side of the horizontal gives the length on the opposite side.

Intermittent welds have a specified length and are spaced uniformly, center to center, at an interval called the pitch. In **Fig. 22.16B**, the welds are equally spaced on both sides, are 60 mm long, and have pitches of 120 mm, as indicated by the symbol shown. In **Fig. 22.16C** the symbol shown specifies intermittent welds that are staggered in alternate positions on opposite sides.

Groove Welds

Figure 22.17 illustrates the standard types of groove welds: **V-groove**, **bevel groove**, **double V-groove**, **U-groove** (Fig. 22.17), and **J-groove** shown in **Fig. 22.18**. When you do not give the depth of the grooves, the angle of the chamfer, and the root openings on a symbol, you must specify them elsewhere on the drawing or in supporting documents. In Fig. 22.17A and B, the angles of the V-joints are labeled 60° and 90° under the ideographs. In Fig. 22.17B, the depths of the weld (6) and the root opening (2)—the gap between the two parts—are given.

In a bevel groove weld, only one of the parts is beveled. The symbol's leader is bent and pointed toward the beveled part to call attention to it (**Figs. 22.17C** and **22.18B**). This practice also applies to J-groove welds, where one side is grooved and the other is not (**Fig. 22.18A**). Notate double V-groove welds by weld size, bevel angle, and root opening (Fig. 22.17D and E). Omit root opening sizes or show a zero on the symbol when parts fit flush. Give the angle and depth of the groove in the symbol for a U-groove weld (Fig. 22.17F).

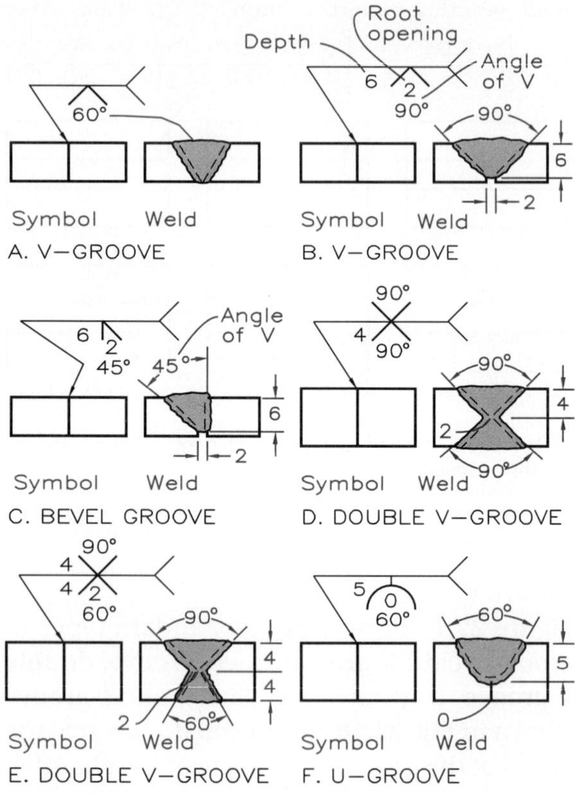
Root
opening
Depth
Angle
of V
Symbol
Weld
A. V—GROOVE
B. V—GROOVE
C. BEVEL GROOVE
D. DOUBLE V—GROOVE
E. DOUBLE V—GROOVE
F. U—GROOVE

22.17 The standard types of groove welds and their general specifications.

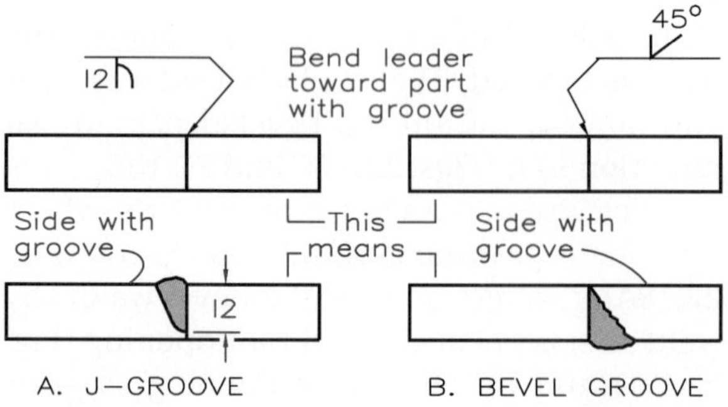
Bend leader
toward part
with groove
Side with
groove
This
means
A. J—GROOVE
B. BEVEL GROOVE

22.18 J-groove welds and bevel welds are specified by bent arrows pointing to the side of the joint to be grooved or beveled.

WELDING ABBREVIATIONS

CAW	Carbon—arc w.	IB	Induction brazing
CW	Cold welding	IRB	Infrared brazing
DB	Dip brazing	OAW	Oxyacetylene w.
DFW	Diffusion welding	OHW	Oxyhydrogen w.
EBW	Electric beam w.	PGW	Pressure gas w.
ESW	Electroslag welding	RB	Resist. brazing
EXW	Explosion welding	RPW	Projection weld.
FB	Furnace brazing	RSEW	Resist. seam w.
FOW	Forge welding	RSW	Resist. spot w.
FRW	Friction welding	RW	Resist. welding
FW	Flash welding	TB	Torch brazing
GMAW	Gas metal arc w.	UW	Upset welding
GTAW	Gas tungsten w.		*w.=welding

22.19 These abbreviations represent the various types of welding processes and are used in the tail of the welding symbol.

Seam Welds

A seam weld joins two lapping parts with either a continuous weld or a series of closely spaced spot welds. The seam weld process to be used is identified by abbreviations in the tail of the welding symbol (**Fig. 22.19**). The circular ideograph for a resistance weld is about 12 mm (four times the letter height) in diameter and is centered over the horizontal line of the symbol (**Fig. 22.20A**). The weld's width, length, and pitch are given.

When the seam weld is to be made by carbon arc welding (CAW), the diameter of the ideograph is about 6 mm (twice the letter height) and goes on the upper or lower side of the symbol's horizontal line to indicate whether the seam is to be applied to the arrow side or opposite side (**Fig. 22.20B**). When the length of the weld is not shown, the seam weld is understood to extend between abrupt changes in the direction of the seam.

Spot welds are similarly specified with ideographs and specifications by diameter, number of welds, and pitch between the welds. The process of resistance spot welding (RSW) is noted in the tail of the symbol (**Fig. 22.21A**). For arc welding, the arrow side or

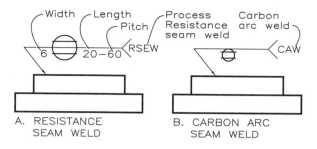

A. RESISTANCE
   SEAM WELD

B. CARBON ARC
   SEAM WELD

22.20 The process used for (A) resistance seam welds and (B) arc-seam welds is indicated in the tail of the symbol. For the arc weld, the symbol must specify the arrow side or the other side of the piece.

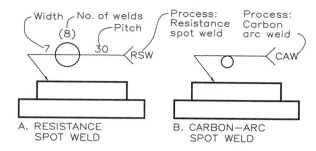

A. RESISTANCE
   SPOT WELD

B. CARBON–ARC
   SPOT WELD

22.21 The process to be used for (A) resistance spot welding and (B) arc spot welding is indicated in the tail of the symbol. For arc welding, the symbol must specify the arrow side or the other side of the piece.

other side must be indicated by a symbol (**Fig. 22.21B**).

Built-Up Welds

When the surface of a part is to be enlarged, or built-up, by welding, indicate this process with the symbol shown in **Fig. 22.22**. Dimension the width of the built-up weld in the view. Specify the height of the weld above the surface in the symbol to the left of the ideograph. The radius of the circular segment is 6 mm (twice the letter height).

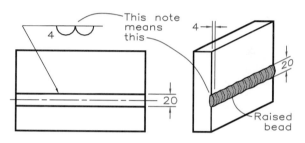

22.22 The symbol indicating a built-up weld on a surface.

22.6 Surface Contouring

Contour symbols are used to indicate which of the three types of contours—**flush, concave**, or **convex**—is desired on the surface of the weld (**Fig. 22.23**). Flush contours are smooth with the surface or flat across the hypotenuse of a fillet weld. Concave contours bulge inward with a curve, and convex contours bulge outward with a curve.

Finishing the weld by an additional process to obtain the desired contour often is necessary. These processes, which may be indicated by their abbreviations, are **chipping** (C), **grinding** (G), **hammering** (H), **machining** (M), **rolling** (R), and **peening** (P), as shown in **Fig. 22.24**.

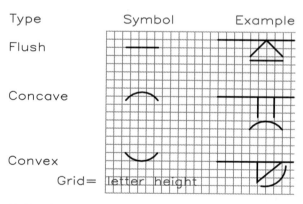

Type	Symbol	Example
Flush		
Concave		
Convex		

Grid = letter height

22.23 These contour symbols specify the desired surface finish of a weld.

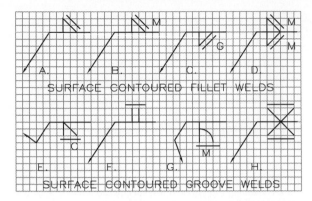

22.24 These are a few examples of contoured welding symbols with letters added indicate the type of finishing to be applied to the weld (M, machining; G, grinding; C, chipping).

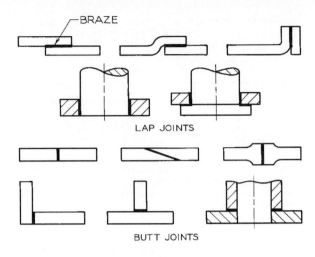

22.25 The two basic types of brazing joints.

22.7 Brazing

Brazing is a method much like welding for joining pieces of metal. Brazing entails heating joints to more than 800°F and distributing by capillary action a nonferrous filler material, with a melting point below that of the base materials, between the closely fitting parts.

Before brazing, the parts must be cleaned and the joints fluxed. The brazing filler is added before or just as the joints are heated beyond the filler's melting point. After the filler material has melted, it is allowed to flow between the parts to form the joint. As **Fig. 22.25** shows, there are two basic brazing joints: lap joints and butt joints.

Brazing is used to join parts, to provide gas- and liquid-tight joints, to ensure electrical conductivity, and to aid in repair and salvage. Brazed joints withstand more stress, higher temperature, and more vibration than soft-soldered joints.

22.8 Soldering

Soldering is the process of joining two metal parts with a third metal that melts below the temperature of the metals being joined. Solders are alloys of nonferrous metals that melt below 800°F. Widely used in the automotive and electrical industries, soldering is one of the basic techniques of welding and often is done by hand with a soldering iron like the one depicted in **Fig. 22.26**. The iron is placed on the joint to heat it and to melt the solder. Basic soldering is notated on drawing with a leader simply as SOLDER, as shown in Fig. 22.26.

By necessity, the coverage in this chapter is introductory in nature, but adequate for a basic understanding of how to specify welding on an engineering drawing. More detailed information on welding is available from the American Welding Society, 2501 N.W. 7th Street, Miami, Florida 33125. This society maintains and publishes guidelines and standards for the technology of welding.

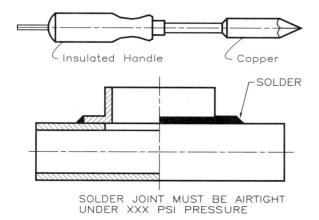

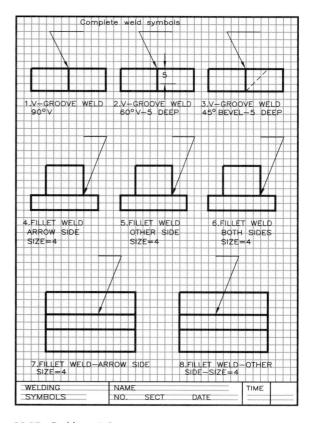

Insulated Handle   Copper

SOLDER

SOLDER JOINT MUST BE AIRTIGHT
UNDER XXX PSI PRESSURE

22.26 This typical hand-held soldering iron is used to soft-solder two parts together. The method of notating a drawing for soldering is also shown.

Complete weld symbols

5

1. V—GROOVE WELD
90°V

2. V—GROOVE WELD
60°V—5 DEEP

3. V—GROOVE WELD
45° BEVEL—5 DEEP

4. FILLET WELD
ARROW SIDE
SIZE=4

5. FILLET WELD
OTHER SIDE
SIZE=4

6. FILLET WELD
BOTH SIDES
SIZE=4

7. FILLET WELD—ARROW SIDE
SIZE=4

8. FILLET WELD—OTHER
SIDE—SIZE=4

WELDING SYMBOLS	NAME		TIME	
	NO. SECT DATE			

22.27 Problems 1–8.

Problems

Solve these problems on size A sheets laid out on a grid of 0.20 inches (5 mm).

1–8. (Fig. 22.27) Give welding notes to include the information specified for each problem. Omit instructional information from the solution.

9. (Fig. 22.28) The shaft socket has a base of 4 in. × 4 in. Draw a top and front view of it; approximate its dimensions and show the appropriate welding notes for its fabrication.

22.28 Problem 9.

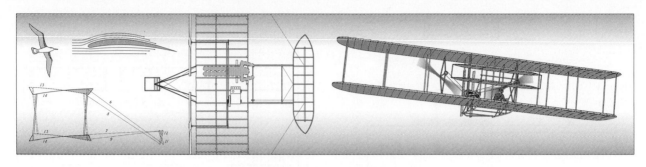

23

Working Drawings

23.1 Introduction

Working drawings provide the visual base from which a design is implemented. All principles of orthographic projection and techniques of graphics can be used to communicate the details of a project in working drawings. A **detail drawing** is a working drawing of a single part (or detail) within the set of working drawings.

Specifications are the written instructions that accompany working drawings. When the design can be represented on a few sheets, the specifications are usually written on the drawings to consolidate the information into a single format.

All parts must interact with each other to some degree to yield the desired function from a design. Before detail drawings of individual parts are made, the designer must thoroughly analyze the working drawing to ensure that the parts fit properly with mating parts, that the correct tolerances are applied, that

the contact surfaces are properly finished, and that the proper motion is possible between the parts. Much of the work in preparing these drawings is done by the drafter, but the designer, who is usually an engineer, is responsible for their correctness. It is working drawings that bring products and systems into being.

23.2 Working Drawings as Legal Documents

Working drawings are legal contracts that document the design details and specifications as directed by the engineer. Therefore, they must be as clear, precise, and thorough as possible. Revisions and modifications of a project at the time of production or construction are much more expensive than when done in the preliminary design stages.

Poorly executed working drawings result in wasted time and resources and increase implementation costs. To be economically competitive, drawings must be as error-free as possible.

Working drawings specify all aspects of the design, reflecting the soundness of engineering and function of the finished product and economy of fabrication. The working drawing is the instrument that is most likely to establish the responsibility for any failure to meet specifications during implementation.

23.3 Dimensions and Units

English System

The inch is the basic unit of the English system, and virtually all shop drawings made with English units are dimensioned in inches. This practice is followed even when dimensions are several feet in length.

The base flange shown in **Fig. 23.1** is an example of a relatively simple working drawing of a single part—one of the many drawings that must be designed, developed, and detailed in order for the overall project to come into being.

The base flange is dimensioned with two-place decimal inches except where four-place decimal inches are used for toleranced dimensions. Inch marks (″) are omitted from dimensions on working drawings because the units are understood to be in inches, and their

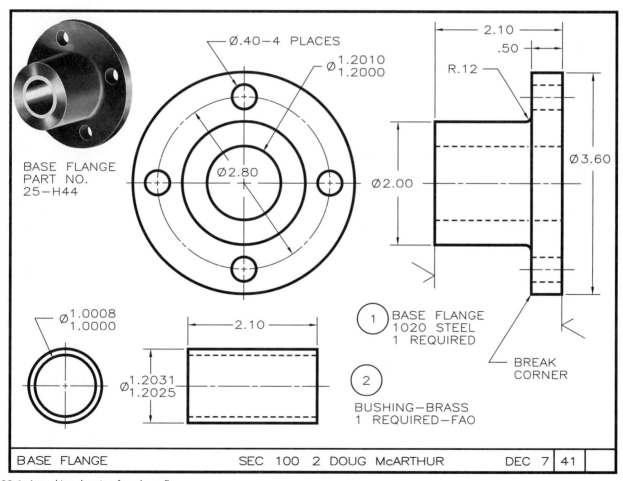

23.1 A working drawing for a base flange.

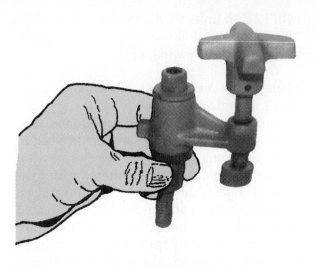

23.2 This revolving clamp assembly holds parts while they are being machined. (Courtesy of Jergens, Incorporated.)

omission saves drafting time. Finish marks are applied to the surfaces that must be machined smooth. Notice that the dimensions are spaced and applied in accordance with the principles covered in Chapters 20 and 21. A photograph showing a three-dimensional view of the flange has been inserted in the corner of the drawing as a raster image by using AutoCAD.

The revolving clamp assembly illustrated in **Fig. 23.2** is detailed in three sheets of a working drawing (**Figs. 23.3–23.5**), which are also dimensioned in inches. Decimal fractions are preferable to common fractions, but common fractions are still used (although mostly by architects). Arithmetic can be done with greater ease with decimal fractions than with common fractions.

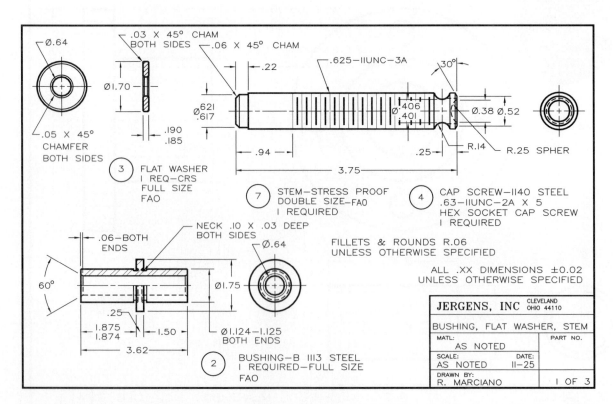

23.3 Sheet of 1 of 3: A computer-generated working drawing of the parts of the clamp assembly shown in Fig. 23.2. (Figures 23.3–23.5 courtesy of Jergens, Incorporated.)

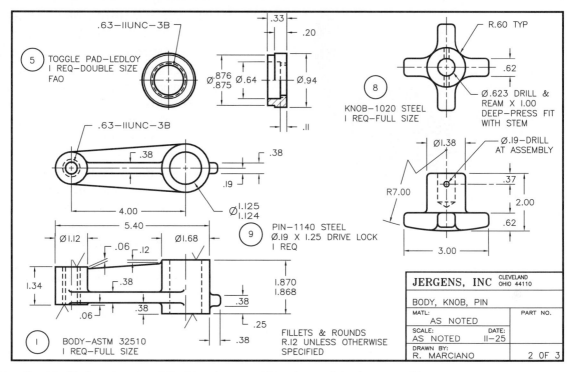

23.4 Sheet 2 of 3: A continuation of Fig. 23.2, showing additional parts of the clamp assembly.

23.5 Sheet 3 of 3: The pad assembly, an orthographic assembly drawing of the clamp, and a parts list are shown.

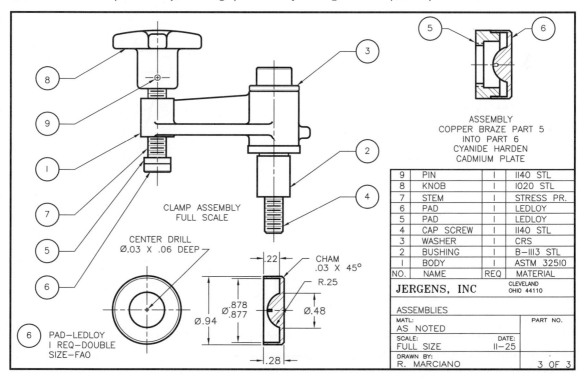

23.6 A left-end handcrank, detailed in the working drawing in Figs. 23.7 and 23.8.

Usually, several dimensioned orthographic views of parts may be shown on each sheet. However, some companies have policies that views of only one part be drawn on a sheet, even if the part is extremely simple, such as a threaded fastener or the base flange shown in Fig. 23.1.

The arrangement of views of parts on the sheet need not attempt to show the relationship of the parts when assembled; the views are simply positioned to best fit the available space on the sheet. The views of each part are labeled with a part number, a name for identification, the material it is made of, the number of the parts required, and any other notes necessary to explain manufacturing procedures.

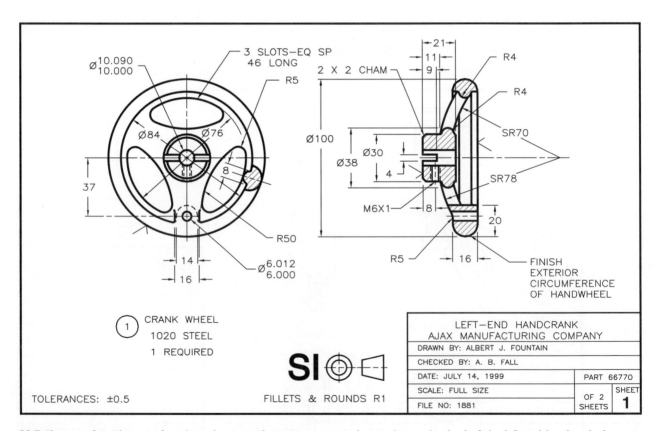

23.7 Sheet 1 of 2: This set of working drawings (dimensions in mm) depicts the crank wheel of the left-end handcrank shown in Fig. 23.6.

The orthographic assembly drawing shown on Sheet 3 (Fig. 23.5) illustrates how the parts are to fit together. Each part is numbered and cross-referenced with the part numbers in the parts list, which serves as a bill of materials.

Metric System

The **millimeter** (**mm**) is the basic unit of the metric system, and dimensions are usually given to the nearest whole millimeter without decimal fractions (except to specify tolerances that may require three-place decimals). Metric abbreviations after the numerals are omitted from dimensions because the SI symbol near the title block indicates that all units are metric. If you have trouble relating to the length of a millimeter, recall that the fingernail of your index finger is about 10-mm wide.

The left-end handcrank (**Fig. 23.6**) is depicted in the working drawings shown in **Fig. 23.7** and **Fig. 23.8** on two size B sheets. The dimensions (in millimeters), notes, and descriptive views give all the information needed to construct the four pieces. The orthographic, sectioned assembly drawing of the left-end handcrank shown in Fig. 23.8 illustrates how the parts are to be put together. The part numbers in the balloons are cross-referenced to the parts list placed just above the title block.

Dual Dimensions

Some working drawings carry both inch and millimeter dimensions, as shown in **Fig. 23.9** where the dimensions in brackets are millimeters. The units may also appear in millimeters first, and with inches indicated in

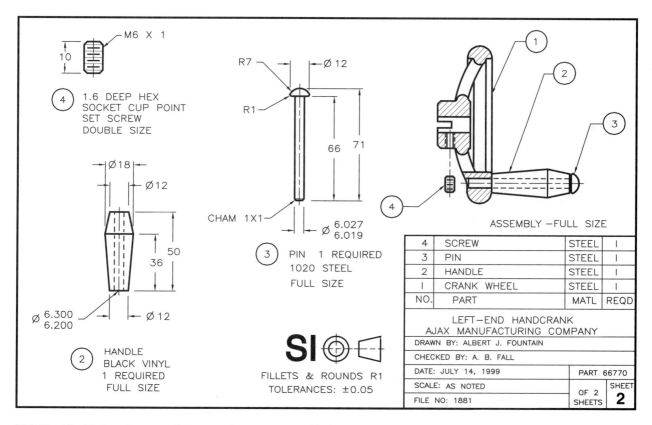

23.8 Sheet 2 of 2: A continuation of Fig. 23.7, showing an assembly drawing and parts list.

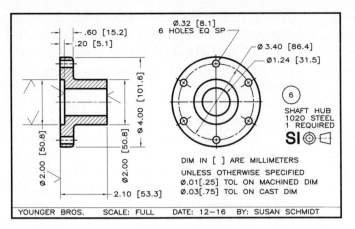

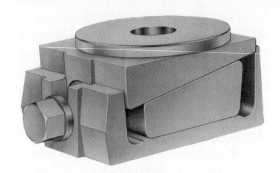

23.9 In this dual-dimensioned drawing, dimensions are shown in millimeters with their equivalents in inches given in brackets.

23.10 A Lev-L-Line lifting device, used to level heavy machinery. (Courtesy of Unisorb Machinery Installation Systems.)

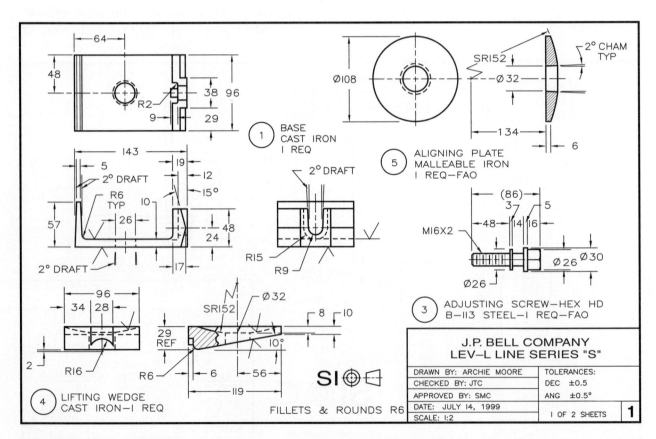

23.11 Sheet 1 of 2: A working drawing (dimensioned in millimeters) of the lifting device shown in Fig. 23.10. (Courtesy of Unisorb Machinery Installation Systems.)

brackets. Converting from one unit to the other results in fractional round-off errors. An explanation of the primary unit system for each drawing should be noted next to the title block.

Metric Working Drawing Example

Figure 23.10 illustrates a lifting device used to level heavy equipment such as lathes and milling machines. The device raises or lowers the machinery when the screw is rotated and the two wedges slide together. A two-sheet working drawing that gives the details of the parts of the lifting device is shown in **Figs. 23.11** and **23.12**. The SI symbol indicates that the dimensions are in millimeters and the

truncated cone shows that the orthographic views are drawn using third-angle projection. The assembly drawing in **Fig. 23.12** illustrates how the parts are to be assembled after they have been made.

23.4 Laying Out a Detail Drawing

By Computer

The process of laying out a simple working drawing of a single part by computer should begin with the border (**Fig. 23.13**). Although this example is elementary, it is useful in describing the procedures for laying out more complex drawings. Allow a margin of at least 0.25 in. (7 mm) between the edge of the sheet and the borders. Outline space for the title

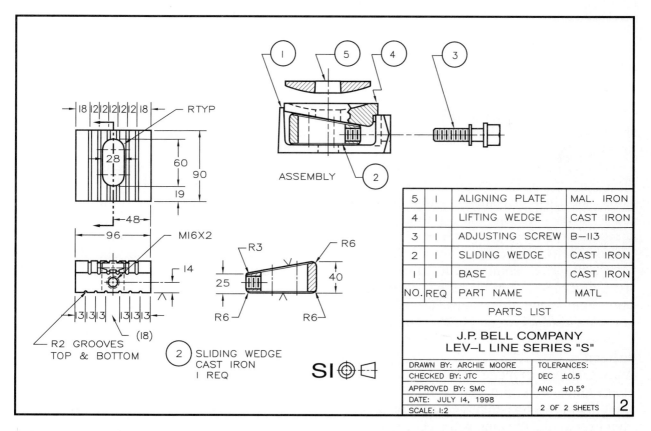

23.12 Sheet 2 of 2: A continuation of the working and assembly drawings of the lifting device shown in Fig. 23.10. (Courtesy of Unisorb Machinery Installation Systems.)

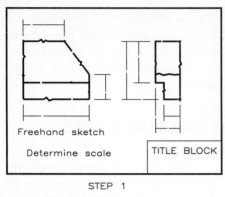

Freehand sketch

Determine scale

TITLE BLOCK

STEP 1

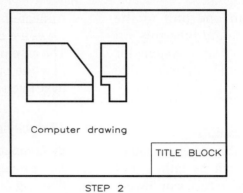

Computer drawing

TITLE BLOCK

STEP 2

Move views apart
with the MOVE command

STEP 3

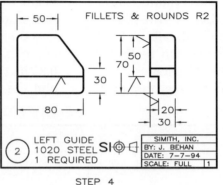

| 50 |

FILLETS & ROUNDS R2

50
70
30
80
20
30

② LEFT GUIDE
1020 STEEL SI⊕⊡
1 REQUIRED

SIMITH, INC.
BY: J. BEHAN
DATE: 7-7-94
SCALE: FULL   1

STEP 4

23.13 Laying out a working drawing by computer.

Step 1 On a sheet of paper, sketch the views and dimensions, determine their placement, and select the scale for the drawing.

Step 2 On the computer screen, insert the border and title block; draw the orthographic views close together to make projection easier.

Step 3 Separate the views with the MOVE command to make room for dimensions.

Step 4 Apply dimensions, notes, SI symbol, and title block information.

block in the lower right-hand corner of the sheet to ensure that the drawing or notes do not occupy this area.

First, make a freehand sketch to determine the necessary views, the number of dimensions, and their placement so that you can select the proper scale. You can then use the computer to arrange the views close together to make projection from view to view easier.

Use the MOVE command to separate the views and make room for notes and dimensions. Finally, add dimensions, notes, SI symbol, and title block to complete the drawing.

By Hand
When making a drawing by hand with instruments on paper or film, first lay out the views and dimensions on a different sheet of paper, then overlay the drawing with vellum or film and trace it to obtain the final drawing. You must use guidelines for lettering for each

dimension and note. Lightly draw the guidelines or underlay the drawing with a sheet containing guidelines.

Figure 23.14 shows the standard sheet sizes for working drawings. Paper, film, cloth, and reproduction materials are available in these modular sizes; good practice requires that you make drawings in one of these standard sizes. Drawings in modular sizes can be folded to fit standard-sized envelopes and fil-

ENGLISH SIZES			METRIC SIZES		
A	11 × 8.5		A4	297 × 210	
B	17 × 11		A3	420 × 297	
C	22 × 17		A2	594 × 420	
D	34 × 22		A1	841 × 594	
E	44 × 34		A0	1189 × 841	

23.14 The standard sheet sizes for working drawings, dimensioned in English sizes (inches) and metric sizes (millimeters).

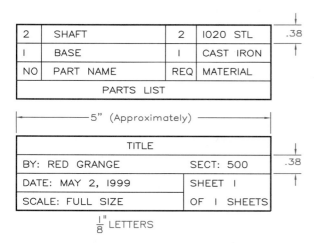

2	SHAFT	2	1020 STL
I	BASE	I	CAST IRON
NO	PART NAME	REQ	MATERIAL
	PARTS LIST		

.38

|← 5" (Approximately) →|

TITLE	
BY: RED GRANGE	SECT: 500
DATE: MAY 2, 1999	SHEET I
SCALE: FULL SIZE	OF I SHEETS

.38

$\frac{1}{8}$" LETTERS

23.15 A typical title block and parts list, suitable for most student assignments.

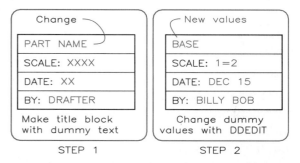

REVISIONS	COMPANY NAME COMPANY ADDRESS	
CHG. HEIGHT	TITLE: LEFT—END BEARING	
FAO	DRAWN BY: JOHNNY RINGO	
	CHECKED BY: FRED J. DODGE	
	DATE: JULY 14, 1999	
	SCALE HALF SIZE	SHEET 1 OF 3 SHEETS

23.16 A title block and a revision block, typical of those used in industry.

Change	New values
PART NAME	BASE
SCALE: XXXX	SCALE: 1=2
DATE: XX	DATE: DEC 15
BY: DRAFTER	BY: BILLY BOB
Make title block with dummy text	Change dummy values with DDEDIT
STEP 1	STEP 2

23.17 Producing a title block by computer.

Step 1 Draw the title block and add dummy text values, using the desired text style and size. Make a BLOCK of the title block.

Step 2 Position the title block against the lower bottom and right borders. EXPLODE the BLOCK and use the DDEDIT command to convert the dummy values to actual values.

ing cabinets and to match the sizes of print paper.

23.5 Notes and Other Information

Title Blocks and Parts Lists

Figure 23.15 shows a title block and parts list suitable for most student assignments. Title blocks usually are placed in the lower right-hand corner of the drawing sheet against the border. The parts list should be placed directly over the title block (see also Figs. 23.8 and 23.12).

Title Blocks In practice, title blocks usually contain the title or part name, drafter, date, scale, company, and sheet number. Other information, such as tolerances, checkers, and materials, may also be given. **Figure 23.16** shows another example of a title block, which is typical of those used by various industries. Any modifications or changes added after the first version to improve the design is shown in the revision blocks.

Depending on the complexity of the project, a set of working drawings may contain from one to more than a hundred sheets. Therefore, giving the number of each sheet and the total number of sheets in the set on each sheet is important (for example, sheet 2 of 6, sheet 3 of 6, and so on).

Computer Method A computer shortcut for filling in a title block that will be used on several sheets is shown in **Fig. 23.17**. You draw the title block only once, filling it in with dummy values to establish the positions of the text. By using the DDEDIT command, you can update the dummy entries with applicable values after the block is inserted or copied into a new drawing.

A similar shortcut in the design of a title block involves the use of ATTRIBUTES. In that case, the program will prompt you to INSERT entries into the title block one at a time.

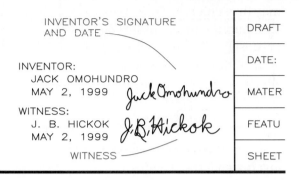

23.18 A note next to the title block names the inventor and is witnessed by an associate to establish ownership of a design for patent purposes.

Parts List The numbers and names in the parts list correspond to those given to each part depicted on the working drawings. In addition, the number of identical parts required are given along with the material used to make each part. Because the exact material (for example, 1020 STEEL) is designated for each part on the drawing, the material in the parts list may be shortened to STEEL, which requires less space.

Patent Rights Note

In **Fig. 23.18**, a note near the title block that names Jack Omohundro as the inventor of the part or process is used to establish ownership of the design. An associate, J. B. Hickok, signs and dates the drawing as a witness to the designer's work. This type of note establishes ownership of the ideas and dates of their development to help the inventor obtain a patent. An even better case for design ownership is made if a second witness signs and dates the drawing. As modifications to the design are made, those drawings should receive the same documentation.

Scale Specification

If all working drawings in a set are the same scale, you need to indicate it only once in the title block on each sheet. If several detail drawings on a working drawing are different

```
SCALE:  1=2    (implies  inches)
SCALE:  1:2    (implies  millimeters)
    SI OR
METRIC  }      (implies  SI  or  metric)

0        50      100
▰▱▰▱▰▱          (graphical  scale)
```

23.19 The methods used to specify scales in English and SI or metric units on working drawings.

scales, indicate them on the drawing under each set of views. In this case, indicate "as shown" in the title block opposite *scale*. When a drawing is not to scale, place the abbreviation NTS (not to scale) in the title block.

Figure 23.19 shows several methods of indicating scales. Use of the colon (for example, 1:2) implies the metric system; use of the equal sign (for example, 1=2) implies the English system—but these are not absolute rules. The SI symbol or metric designation on a drawing specifies that millimeters are the units of measurement.

In some cases, you may want to show a graphical scale with calibrations on a drawing to permit the interpretation of linear measurements by transferring them with dividers from the drawing to the scale.

Tolerances

Recall from Chapter 21 that you may use general notes on working drawings to specify the dimension tolerances. **Figure 23.20** shows a table of values with boxes in which you can indicate with a check mark whether the units are in inches or millimeters. Position plus-and-minus tolerances under each common or decimal fraction. In this example, the table specifies that each dimension with two-place decimals will have a tolerance of ± 0.10 in. You may also give angular tolerances in general notes ($\pm 0.5°$, for example).

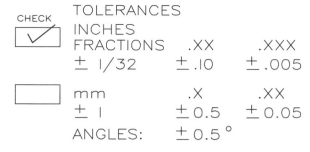

23.20 General tolerance notes on working drawing specify the dimension tolerances permitted.

23.21 Name and number each part on a working drawing for use in the parts list, and indicate the number of parts of this particular part that are needed.

Part Labeling

Give each part a name and number, using letters and numbers 1/8-in. (.125 in. or 3 mm) high (**Fig. 23.21**). Place part numbers inside circles, called balloons, having diameters approximately four times the height of the numerals.

Place part numbers near the views to which they apply, so their association will be clear. On assembly drawings, balloons are especially important because the same part numbers are used in the parts list. Show the number of parts required near the part name.

23.6 Checking a Drawing

The process of checking a drawing must be done by those who are experienced in drafting, design, and manufacturing methods. Their ability to identify errors and to suggest modifications that will improve the product's function and ease of manufacture will result in a better product at a lower cost. A checker

may be a chief drafter experienced in drafting and manufacturing processes, or the engineer or designer who originated the project. In large companies, personnel in the various shops involved in production review the drawings to ensure that the most efficient production methods are specified for each part.

Checkers never work on the original drawing; instead, they mark corrections with a colored pencil on a diazo (blue-line) print. They return the marked-up print to the drafter, who revises the original and makes another print for final approval.

In **Fig. 23.22**, the various modifications made by checkers are labeled with letters that are circled and placed near the revisions. The drafter lists and dates changes in the revision record, which lists the revisions made.

Checkers inspect a working or detail drawing for correctness and soundness of design. In addition, they are responsible for the drawing's completeness, quality, readability, and clarity, which are related to the lettering and drafting techniques used. However, it is the project engineer who is ultimately responsible for correctness of the drawings and documents. Lettering and text quality is especially important because the shop person is guided by lettered notes and dimensions.

Student Working Drawings

The best way for you to check your drawings for adequate dimensions is to rapidly make a scale drawing of the part using the dimensions from the working drawings. You can identify missing dimensions more easily in this way than by reading a drawing by eye. **Figure 23.23** shows a grading scale for checking working drawings prepared by students, with hypothetical grades shown. Use this list as an outline for reviewing working drawings to ensure that you have met the main requirements.

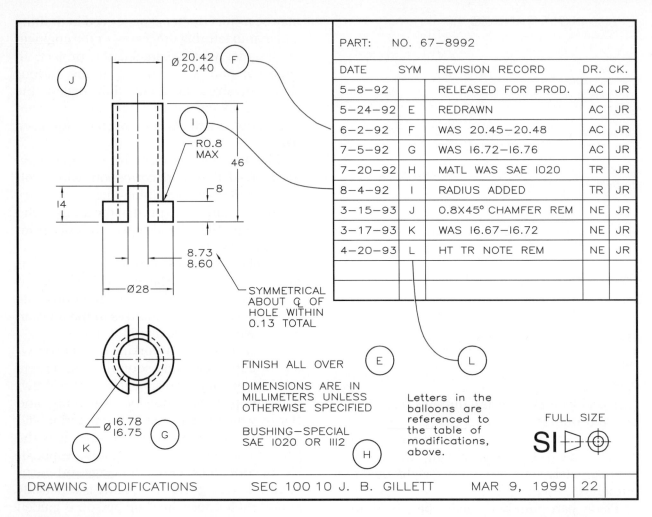

PART:	NO. 67−8992				
DATE	SYM		REVISION RECORD	DR.	CK.
5−8−92			RELEASED FOR PROD.	AC	JR
5−24−92	E		REDRAWN	AC	JR
6−2−92	F		WAS 20.45−20.48	AC	JR
7−5−92	G		WAS 16.72−16.76	AC	JR
7−20−92	H		MATL WAS SAE 1020	TR	JR
8−4−92	I		RADIUS ADDED	TR	JR
3−15−93	J		0.8X45° CHAMFER REM	NE	JR
3−17−93	K		WAS 16.67−16.72	NE	JR
4−20−93	L		HT TR NOTE REM	NE	JR

FINISH ALL OVER

DIMENSIONS ARE IN MILLIMETERS UNLESS OTHERWISE SPECIFIED

BUSHING—SPECIAL SAE 1020 OR 1112

Letters in the balloons are referenced to the table of modifications, above.

FULL SIZE

SYMMETRICAL ABOUT ℄ OF HOLE WITHIN 0.13 TOTAL

| DRAWING MODIFICATIONS | SEC 100 10 J. B. GILLETT | MAR 9, 1999 | 22 | |

23.22 The modifications to this working drawing are noted near the details revised. The letters in balloons correspond to those in the revision table.

23.7 Drafter's Log

In addition to the individual revision records, drafters should keep a log of all changes made during a project. As the project progresses, the drafter should record the changes, dates, and people involved. Such a log allows anyone reviewing the project in the future to understand easily and clearly the process used in arriving at the final design.

Calculations are often made during a drawing's preparation. If they are lost or poorly done, they may have to be redone dur-

ing a later revision; therefore, they should be a permanent part of the log in order to preserve previously expended work.

23.8 Assembly Drawings

After parts have been made according to the specifications of the working drawings, they will be assembled in accordance with the directions of an assembly drawing (**Fig. 23.24**). There are two general types of assembly drawings: **orthographic assemblies** and **pictorial assemblies**. Dimensions usually are omitted from assembly drawings.

The lifting device that was shown in Fig. 23.10 is depicted in an isometric assembly in

23.23 This checklist may be used to evaluate a student's working-drawing assignment.

23.24 An assembly drawing will explain how the parts of a product, such as this Ford tractor, are to be assembled. (Courtesy of Ford Motor Company.)

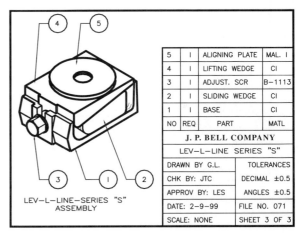

5	1	ALIGNING PLATE	MAL. 1
4	1	LIFTING WEDGE	CI
3	1	ADJUST. SCR	B-1113
2	1	SLIDING WEDGE	CI
1	1	BASE	CI
NO	REQ	PART	MATL

J. P. BELL COMPANY

LEV-L-LINE SERIES "S"

DRAWN BY G.L.	TOLERANCES
CHK BY: JTC	DECIMAL ±0.5
APPROV BY: LES	ANGLES ±0.5
DATE: 2-9-99	FILE NO. 071
SCALE: NONE	SHEET 3 OF 3

LEV-L-LINE-SERIES "S" ASSEMBLY

23.25 An isometric assembly drawing depicting the parts of the lifting device shown in Fig. 23.10 fully assembled. Dimensions are usually omitted from assembly drawings; a parts list is given.

Fig. 23.25. All parts are numbered with balloons and leaders to cross-reference them to the parts list, where more information about each part is given.

Figure 23.26 shows an exploded orthographic assembly drawing. In many applications, the arrangement of parts may be easier to understand when the parts are shown exploded along their centerlines. These views are shown as regular orthographic views, with some lines shown as hidden lines and others omitted.

Assembly of the same part is shown in **Fig. 23.27** in an orthographic assembly drawing, in which the parts are depicted in their assembled positions. The views are sectioned to make them easier to understand.

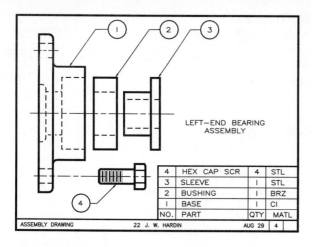

23.26 An exploded orthographic assembly drawing illustrating how the parts shown are to be put together.

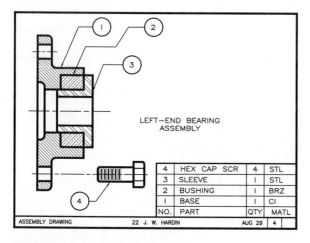

23.27 A sectioned orthographic assembly showing the left-end bearing assembly from Fig. 23.27. All parts except for the exploded bolt are in their assembled positions.

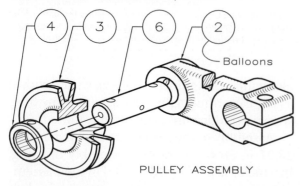

PULLEY ASSEMBLY

23.28 An exploded pictorial assembly drawing of a pulley assembly.

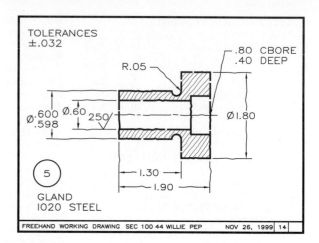

23.29 For this simple part, a freehand working drawing with the essential dimensions is as adequate as an instrument-drawn detail drawing.

Figure 23.28 shows an exploded pictorial assembly drawing of a pulley assembly illustrating how the parts fit together. Part numbers are given in balloons that match the numbers of the parts on the detail drawings and in the parts list.

23.9 Freehand Working Drawings

A freehand sketch can serve the same purpose as an instrument drawing, provided that the part is sufficiently simple and that the essential dimensions are shown (**Fig. 23.29**). The same principles can be followed when making working drawings with instruments and freehand. Freehand sketches have certain advantages. They can be executed quickly, and they can be made in the field, fabrication shop, or other locations where drafting-room instruments are not readily available.

23.10 Working Drawings for Forged Parts and Castings

The two versions of the part shown in **Fig. 23.30** illustrate the difference between a part that has been forged into a shape (sometimes called a **blank**) and its final state after the forg-

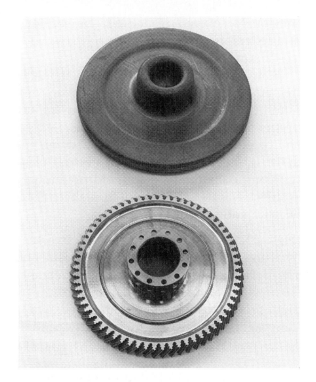

23.30 The part on the left, a blank that has been forged, will look like the part on the right after it has been machined.

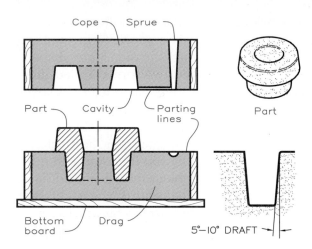

23.31 A two-part sand mold is used to produce a casting. A draft of from 5° to 10° is needed to permit withdrawal of the pattern from the sand. Some machining is usually required to finish various features of the casting within specified tolerances.

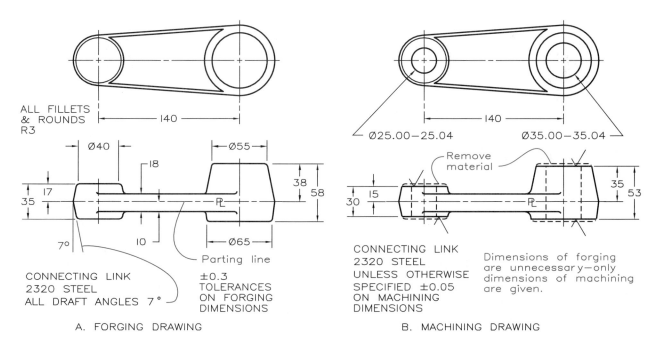

ALL FILLETS
& ROUNDS
R3

Ø40
18
Ø55
38
58
17
35
PL
7°
10
Ø65
Parting line

CONNECTING LINK
2320 STEEL
ALL DRAFT ANGLES 7°

±0.3
TOLERANCES
ON FORGING
DIMENSIONS

A. FORGING DRAWING

140

Ø25.00−25.04   Ø35.00−35.04

Remove material

30   15   PL   35   53

CONNECTING LINK
2320 STEEL
UNLESS OTHERWISE
SPECIFIED ±0.05
ON MACHINING
DIMENSIONS

Dimensions of forging
are unnecessary—only
dimensions of machining
are given.

B. MACHINING DRAWING

23.32 These separate working drawings, (A) a forging drawing and (B) a machining drawing, give details of the same part. This information is often combined into a single drawing.

ing has been machined. Recall from Chapter 19 that a forging is a rough form made by hammering (forging) the metal into shape or pressing it between two forms (called **dies**). The forged part is then machined to its specified finished dimensions and tolerances so it will function as intended.

A casting, like a forging, must be machined so that it too will fit and function with other parts when assembled; therefore, additional material is added to the areas where metal will be removed by the machining processes. As covered in Chapter 19, a casting is formed by pouring molten metal into a mold formed by a pattern that is slightly larger than the finished part to compensate for metal shrinkage (**Fig. 23.31**). For the pattern to be removable from the sand that forms the mold, its sides must have a taper—called a **draft**—of about 5° to 10°.

Some industries that work extensively with cast and forged parts require that separate working drawing be made for forged and cast parts (**Fig. 23.32**). More often, however, the parts are detailed on the regular working drawing with the understanding that the features that are to be machined by operations such as grinding or shaping are made oversize by the fabrication shop.

Problems

Dimensions given on many of the pictorial three-dimensional problems do not always represent good dimensioning practices because of space limitations and the nature of pictorial drawings, but they are adequate to provide dimensions from which working drawings can be made. In cases where dimensions may be missing, use your own judgment to approximate them. Provide all the necessary information, notes, and dimensions to describe the views completely. Use any of the previously covered principles, conventions, and techniques to present the views with the maximum clarity and simplicity.

Your working drawings can be made as freehand sketches or as instrument or computer drawings. The determination of the proper scale, selection of sheet sizes, and the choice and positioning of the views on the drawing sheet will be a major portion of all problem assignments. Freehand preliminary sketches will be very helpful in making these decisions and saving layout time.

(Figs. 23.33–23.40). These example working drawings have been layed out in size-A and size-B sheets with notes and dimensions that are needed to make the parts. By drawing these problems, you will obtain a worthwhile exercise in the use of equipment (by hand or by computer), the drawing of views, and the applications of notes and dimensions.

You may wish to add different types of sections, convert sectional views into plain views, or make other conventional changes. Dimension units can be changed from inches to millimeters or from millimeters to inches, if you like. Complete the title block with the appropriate information.

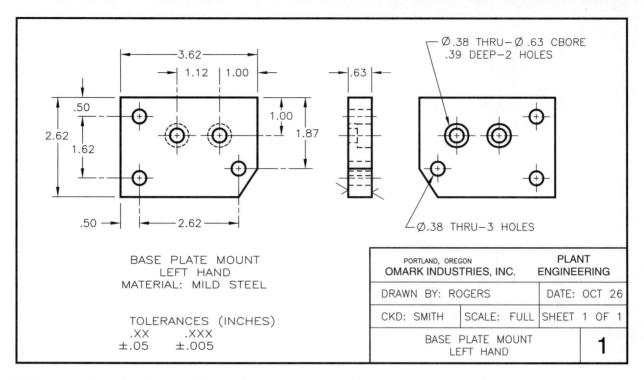

Ø.38 THRU–Ø .63 CBORE
.39 DEEP–2 HOLES

3.62

1.12 | 1.00

.63

1.00

.50

2.62

1.62

1.87

.50

2.62

Ø.38 THRU–3 HOLES

BASE PLATE MOUNT
LEFT HAND
MATERIAL: MILD STEEL

TOLERANCES (INCHES)
.XX   .XXX
±.05   ±.005

PORTLAND, OREGON OMARK INDUSTRIES, INC.	PLANT ENGINEERING	
DRAWN BY: ROGERS	DATE: OCT 26	
CKD: SMITH	SCALE: FULL	SHEET 1 OF 1
BASE PLATE MOUNT LEFT HAND		1

23.33 Duplicate the working drawing of the base plate mount on a size B sheet. (Courtesy of Omark Industries, Incorporated.)

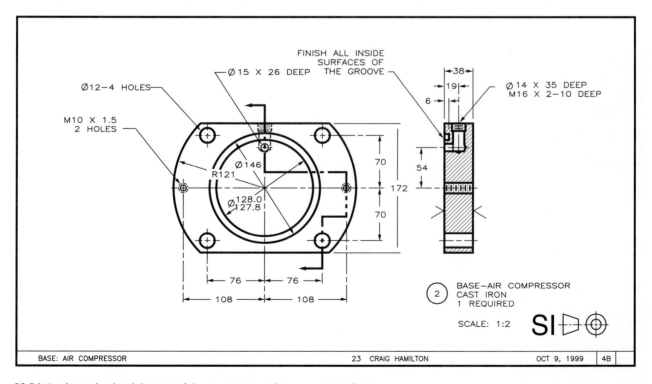

FINISH ALL INSIDE
SURFACES OF
Ø15 X 26 DEEP   THE GROOVE

38

19

6

Ø14 X 35 DEEP
M16 X 2–10 DEEP

Ø12–4 HOLES

M10 X 1.5
2 HOLES

Ø146

R121

Ø128.0
127.8

70

172

70

54

76   76

108   108

BASE–AIR COMPRESSOR
CAST IRON
1 REQUIRED

2

SCALE: 1:2

SI

BASE: AIR COMPRESSOR		23 CRAIG HAMILTON	OCT 9, 1999	4B

23.34 Duplicate the detail drawing of the air compressor base on a size B sheet.

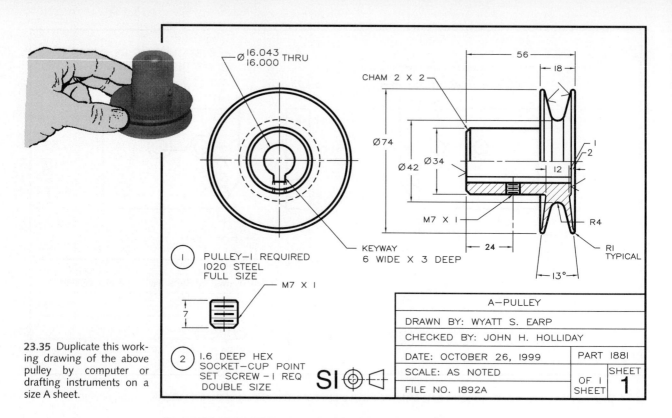

Ø 16.043 / 16.000 THRU

CHAM 2 X 2

Ø74

Ø42   Ø34

M7 X 1

56

18

12

24

13°

R4

R1
TYPICAL

1

2

KEYWAY
6 WIDE X 3 DEEP

① PULLEY—1 REQUIRED
1020 STEEL
FULL SIZE

M7 X 1

7

② 1.6 DEEP HEX
SOCKET—CUP POINT
SET SCREW—1 REQ
DOUBLE SIZE

SI

23.35 Duplicate this working drawing of the above pulley by computer or drafting instruments on a size A sheet.

A—PULLEY		
DRAWN BY: WYATT S. EARP		
CHECKED BY: JOHN H. HOLLIDAY		
DATE: OCTOBER 26, 1999	PART 1881	
SCALE: AS NOTED	OF 1 SHEET	SHEET 1
FILE NO. 1892A		

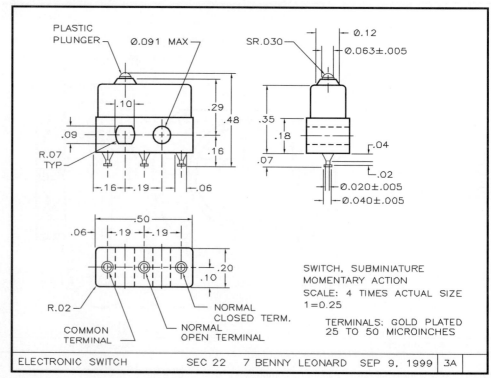

PLASTIC
PLUNGER

Ø.091 MAX

SR.030

Ø.12

Ø.063±.005

.10

.29

.48

.09

.16

R.07
TYP

.16   .19   .06

.35

.18

.07

.04

.02

Ø.020±.005

Ø.040±.005

.50

.06   .19   .19

R.02

COMMON
TERMINAL

NORMAL
OPEN TERMINAL

NORMAL
CLOSED TERM.

.20

.10

SWITCH, SUBMINIATURE
MOMENTARY ACTION

SCALE: 4 TIMES ACTUAL SIZE
1=0.25

TERMINALS: GOLD PLATED
25 TO 50 MICROINCHES

23.36 Duplicate this working drawing of a switch by computer or drafting instruments on a size A sheet. Note that the views are enlarged by a factor of 4.

ELECTRONIC SWITCH	SEC 22 7 BENNY LEONARD SEP 9, 1999	3A	

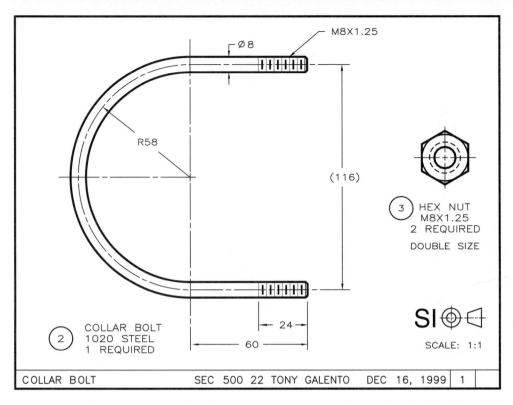

M8X1.25

Ø 8

R58

(116)

3 HEX NUT
M8X1.25
2 REQUIRED
DOUBLE SIZE

SI⊕◁

SCALE: 1:1

24

60

2 COLLAR BOLT
1020 STEEL
1 REQUIRED

COLLAR BOLT   SEC 500 22 TONY GALENTO   DEC 16, 1999 | 1

23.37 Sheet 1 of 3: Duplicate this full-size working drawing and assembly (in mm) of a pipe hanger on size A sheets.

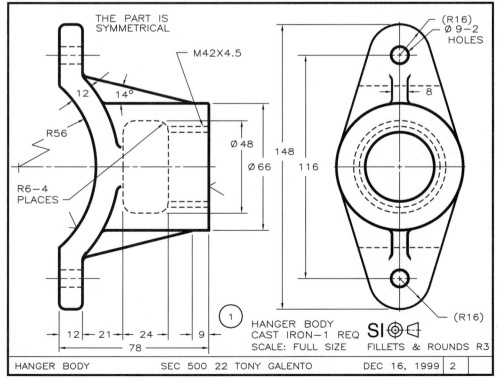

THE PART IS
SYMMETRICAL

M42X4.5

(R16)
Ø 9−2
HOLES

12   14°

8

R56

Ø 48
Ø 66

148

116

R6−4
PLACES

1

12 | 21 | 24 | 9

78

(R16)

HANGER BODY
CAST IRON−1 REQ
SCALE: FULL SIZE

SI⊕◁

FILLETS & ROUNDS R3

HANGER BODY   SEC 500 22 TONY GALENTO   DEC 16, 1999 | 2

23.38 Sheet 2 of 3: Duplicate this second sheet of the pipe hanger working drawing that shows the collar bolt.

23.39 Sheet 3 of 3: Duplicate this third sheet of the pipe hanger, which shows its assembly.

23.40 The ball crank (above) has been detailed in the working drawing below. Duplicate the working drawing on a size B sheet. On a second size B sheet make an assembly drawing of the parts.

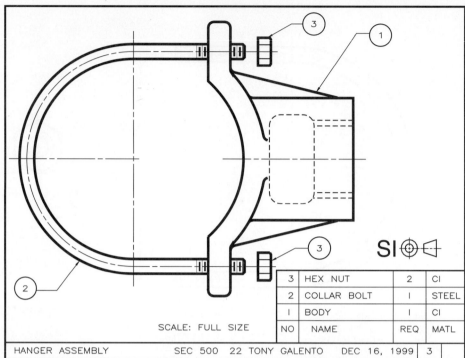

SCALE: FULL SIZE

3	HEX NUT	2	CI
2	COLLAR BOLT	I	STEEL
I	BODY	I	CI
NO	NAME	REQ	MATL

HANGER ASSEMBLY   SEC 500   22 TONY GALENTO   DEC 16, 1999 | 3

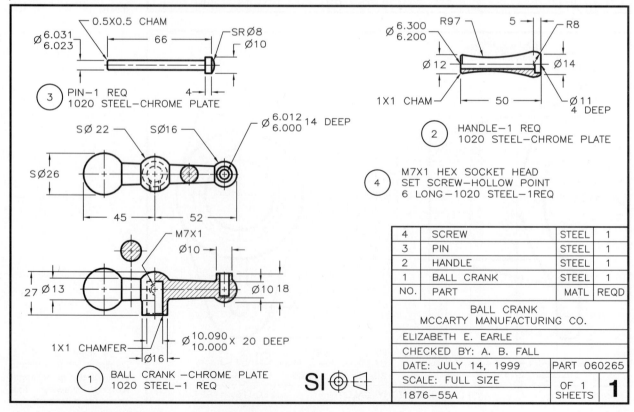

③ PIN—1 REQ
1020 STEEL—CHROME PLATE

0.5X0.5 CHAM
Ø 6.031 / 6.023
66
SR Ø8
Ø10
4

② HANDLE—1 REQ
1020 STEEL—CHROME PLATE

Ø 6.300 / 6.200
R97
5
R8
Ø12
Ø14
1X1 CHAM
50
Ø11 4 DEEP

④ M7X1 HEX SOCKET HEAD
SET SCREW—HOLLOW POINT
6 LONG—1020 STEEL—1REQ

SØ 22
SØ16
Ø 6.012 / 6.000 14 DEEP
SØ26
45
52

M7X1
Ø10
27 Ø13
Ø10 18
1X1 CHAMFER
Ø 10.090 / 10.000 X 20 DEEP
Ø16

① BALL CRANK —CHROME PLATE
1020 STEEL—1 REQ

4	SCREW	STEEL	1
3	PIN	STEEL	1
2	HANDLE	STEEL	1
1	BALL CRANK	STEEL	1
NO.	PART	MATL	REQD

BALL CRANK
MCCARTY MANUFACTURING CO.

ELIZABETH E. EARLE	
CHECKED BY: A. B. FALL	
DATE: JULY 14, 1999	PART 060265
SCALE: FULL SIZE	OF 1
1876—55A	SHEETS 1

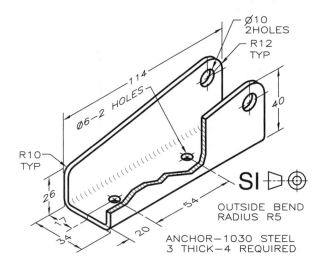

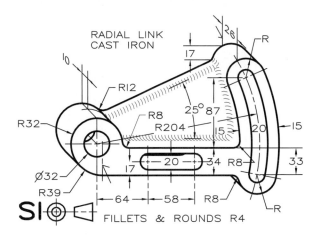

23.41 Make a detail drawing of this anchor on a size A sheet.

23.43 Make a detail drawing of this radial link on a size B sheet.

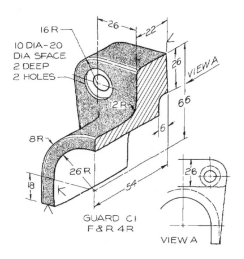

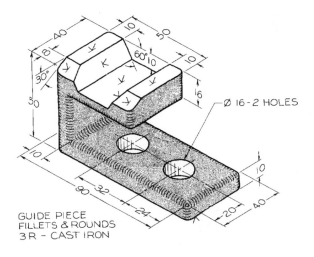

23.42 Make a detail drawing of this guard on a size B sheet.

23.44 Make a detail drawing of this guide piece on a size B sheet.

Working Drawing Practice

Reproduce the drawings shown in **Figs. 23.41–23.53**, as directed. Some have only one sheet and others have more. You may use the dimensions given or convert them to the other system (from millimeters to inches, for example). The purpose of these assignments is to give you experience in laying out a working drawing and improving your draftsmanship on the board or at the computer.

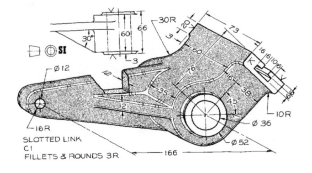

23.45 Make a detail drawing of this slotted link on a size B sheet.

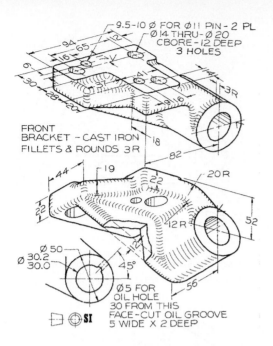

23.46 Make a detail drawing of this front bracket on a size B sheet.

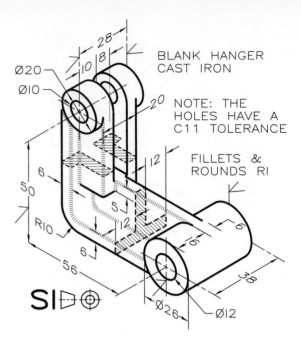

23.48 Make a detail drawing of this blank hanger on a size B sheet.

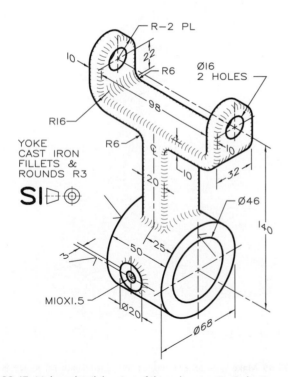

23.47 Make a detail drawing of this yoke on a size B sheet.

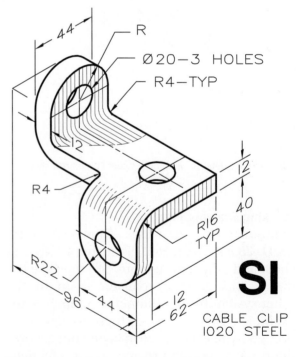

23.49 Make a detail drawing of this cable clip on a size B sheet.

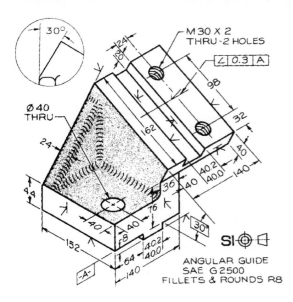

23.50 Make a detail drawing of this angular guide on a size B sheet.

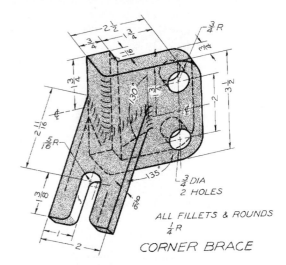

CORNER BRACE

23.52 Make a detail drawing of this bracket brace on a size B sheet.

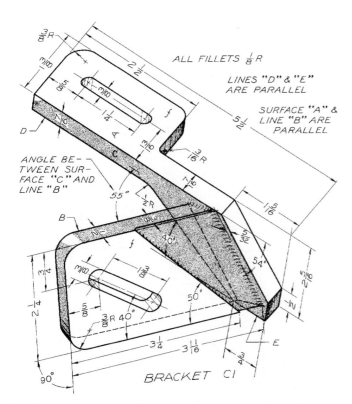

BRACKET C1

ALL FILLETS ⅛ R

LINES "D" & "E" ARE PARALLEL

SURFACE "A" & LINE "B" ARE PARALLEL

ANGLE BE-TWEEN SUR-FACE "C" AND LINE "B"

23.51 Make a detail drawing of this corner brace on a size B sheet.

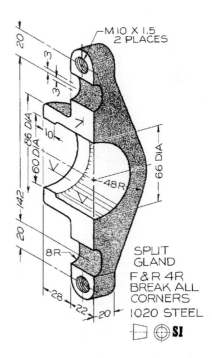

SPLIT GLAND
F & R 4R
BREAK ALL CORNERS
1020 STEEL

23.53 Make a detail drawing of this split gland on a size B sheet.

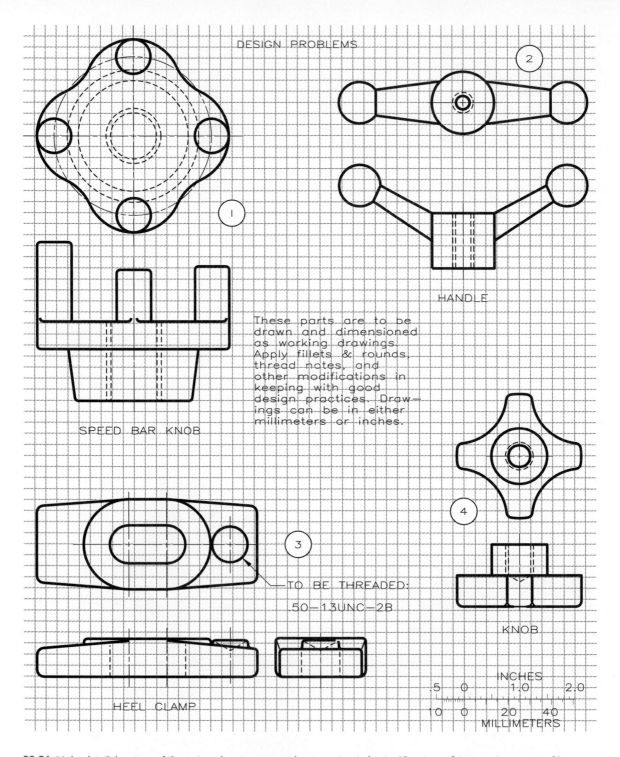

DESIGN PROBLEMS

HANDLE

SPEED BAR KNOB

These parts are to be drawn and dimensioned as working drawings Apply fillets & rounds, thread notes, and other modifications in keeping with good design practices. Drawings can be in either millimeters or inches.

KNOB

3

TO BE THREADED:
.50—13UNC—2B

HEEL CLAMP

INCHES
.5   0   1.0   2.0
10   0   20   40
MILLIMETERS

23.54 Make detail drawings of the assigned parts, one per sheet, on size A sheets. (Courtesy of Jergens, Incorporated.)

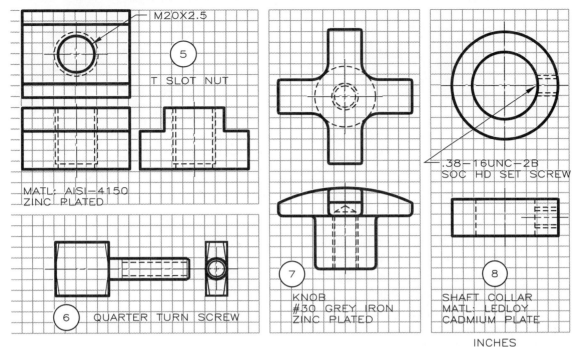

M20X2.5

(5)

T SLOT NUT

MATL: AISI-4150
ZINC PLATED

(6) QUARTER TURN SCREW

(7) KNOB
#30 GREY IRON
ZINC PLATED

.38-16UNC-2B
SOC HD SET SCREW

(8) SHAFT COLLAR
MATL: LEDLOY
CADMIUM PLATE

INCHES
.5   0   1.0   2.0
10   0   20   40
MILLIMETERS

23.55 Make detail drawings of the assigned parts, one per sheet, on size A sheets.

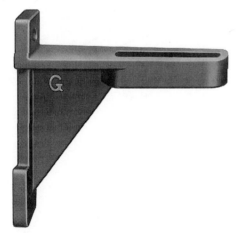

23.56 Make detail drawings of this hanger bracket that extends 8 in. from the wall and supports a 0.50-in. diameter bolt. Use size A sheet.

Single Parts Involving Design Features

Make working drawings of the single parts shown in **Figs. 23.54–23.59** on size A sheets. These problems require that you use your judgment and creativity to depict design features. Include a title block and the dimensions and notes necessary for making the part.

23.57 Make a detail drawing of this 3-in.-inside diameter shaft socket on a size A sheet.

23.58 Make a detail drawing of this 4-in. high angle on a size A sheet.

23.59 Make a detail drawing of this crank arm on a size B sheet.

Working Drawings of Products with Multiple Parts

Make working drawings by hand or by computer, as assigned, of the products consisting of multiple parts shown in **Figs. 23.60–23.82** on the sheet sizes specified. Include a title block and the dimensions and notes necessary for manufacturing the part. Draw an assembly that shows how the parts fit together. More than one sheet may be required for a solution.

23.60 Make working drawings, with an assembly drawing, of this fixture guide on size A sheets.

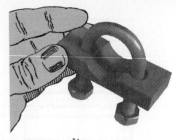

23.61 Make working drawings of this U-bolt pipe strap on size A sheets.

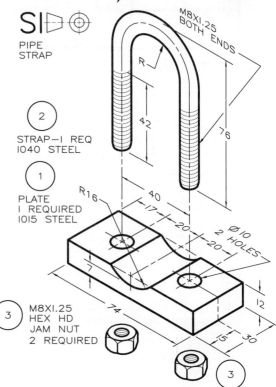

SI ▷◁ ⊕

PIPE STRAP

②
STRAP—I REQ
1040 STEEL

①
PLATE
I REQUIRED
1015 STEEL

③
M8X1.25
HEX HD
JAM NUT
2 REQUIRED

M8X1.25
BOTH ENDS

R

42

76

R16

40

17

20

20

Ø10
2 HOLES

7

74

12

15

30

3

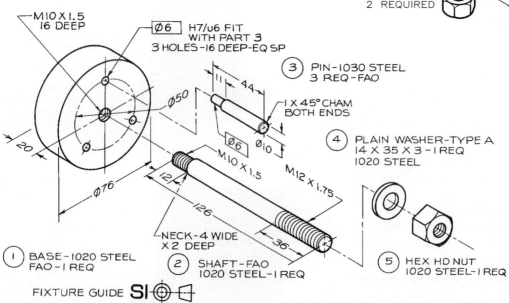

M10 X 1.5
16 DEEP

Ø6 H7/u6 FIT
WITH PART 3
3 HOLES–16 DEEP–EQ SP

③ PIN–1030 STEEL
3 REQ–FAO

I X 45° CHAM
BOTH ENDS

④ PLAIN WASHER–TYPE A
14 X 35 X 3 – I REQ
1020 STEEL

⑤ HEX HD NUT
1020 STEEL–I REQ

Ø50

Ø76

20

Ø6

Ø10

M10 X 1.5

M12 X 1.75

11 44

12

126

36

NECK–4 WIDE
X 2 DEEP

① BASE–1020 STEEL
FAO–I REQ

② SHAFT–FAO
1020 STEEL–I REQ

FIXTURE GUIDE SI ⊕ ◁

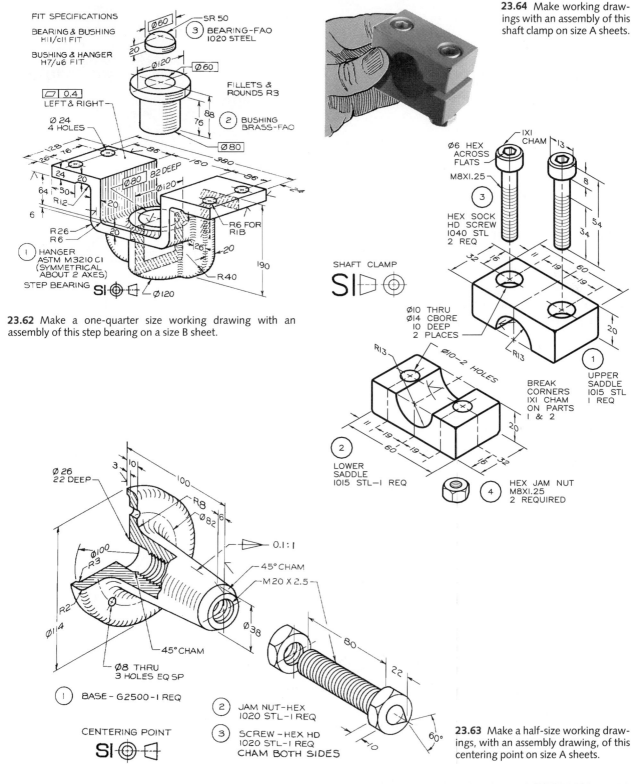

FIT SPECIFICATIONS

BEARING & BUSHING
H11/c11 FIT

BUSHING & HANGER
H7/u6 FIT

SR 50
③ BEARING-FAO
1020 STEEL

Ø60

20

Ø120

Ø60

FILLETS &
ROUNDS R3

88

76

② BUSHING
BRASS-FAO

Ø80

□ 0.4
LEFT & RIGHT

Ø 24
4 HOLES

128

76

26

24   20

R12

50

64

6

R26
R6

20

Ø80   82 DEEP

86

160

380

Ø120

96

24

R6 FOR
RIB

20

26   20

R40

190

① HANGER
ASTM M3210 CI
(SYMMETRICAL
ABOUT 2 AXES)

STEP BEARING SI ⊕ ⊟ Ø120

23.62 Make a one-quarter size working drawing with an assembly of this step bearing on a size B sheet.

23.64 Make working drawings with an assembly of this shaft clamp on size A sheets.

IXI
CHAM   13

Ø6 HEX
ACROSS
FLATS

M8X1.25

8

③

HEX SOCK
HD SCREW
1040 STL
2 REQ

54

34

SHAFT CLAMP

SI ▷ ⊕

32   16   11   19   60   19

Ø10 THRU
Ø14 CBORE
10 DEEP
2 PLACES

20

R13

① UPPER
SADDLE
1015 STL
1 REQ

R13

Ø10-2 HOLES

BREAK
CORNERS
IXI CHAM
ON PARTS
1 & 2

11   19   60   19

20

② LOWER
SADDLE
1015 STL-1 REQ

16   32

④ HEX JAM NUT
M8X1.25
2 REQUIRED

Ø 26
22 DEEP

3   10

100

R 8

Ø 82   6

0.1:1

Ø100

R3

45° CHAM

M 20 X 2.5

R2

Ø114

Ø 38

45° CHAM

Ø8 THRU
3 HOLES EQ SP

80

22

① BASE - G2500-1 REQ

② JAM NUT-HEX
1020 STL-1 REQ

CENTERING POINT

SI ⊕ ⊟

③ SCREW -HEX HD
1020 STL-1 REQ
CHAM BOTH SIDES

60°

10

23.63 Make a half-size working drawings, with an assembly drawing, of this centering point on size A sheets.

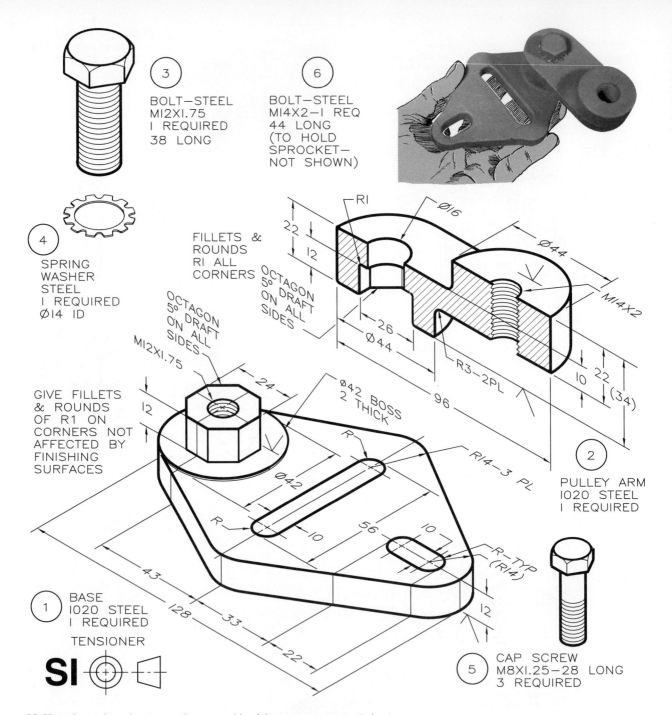

③ BOLT—STEEL
M12X1.75
1 REQUIRED
38 LONG

⑥ BOLT—STEEL
M14X2—1 REQ
44 LONG
(TO HOLD
SPROCKET—
NOT SHOWN)

④ SPRING
WASHER
STEEL
1 REQUIRED
Ø14 ID

FILLETS &
ROUNDS
R1 ALL
CORNERS

OCTAGON
5° DRAFT
ON ALL
SIDES

R1

Ø16

Ø44

M14X2

22

12

26

Ø44

R3—2PL

10

22

(34)

96

② PULLEY ARM
1020 STEEL
1 REQUIRED

OCTAGON
5° DRAFT
ON ALL
SIDES

M12X1.75

GIVE FILLETS
& ROUNDS
OF R1 ON
CORNERS NOT
AFFECTED BY
FINISHING
SURFACES

12

24

Ø42

R

Ø42 BOSS
2 THICK

R

10

56

10

R14—3 PL

R—TYP
(R14)

12

43

128

33

22

① BASE
1020 STEEL
1 REQUIRED

TENSIONER

SI ⊕ ⊡

⑤ CAP SCREW
M8X1.25—28 LONG
3 REQUIRED

23.65 Make working drawings with an assembly of this tensioner on size B sheets.

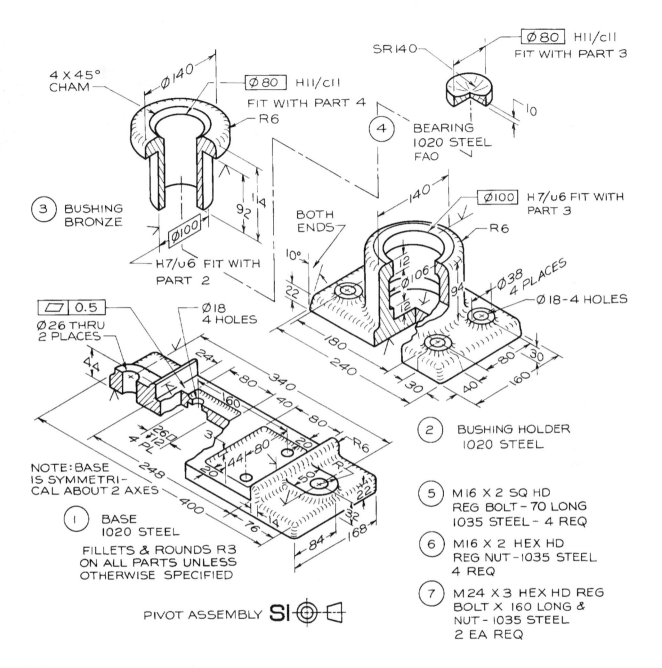

4 X 45° CHAM

Ø140

Ø80 H11/c11
FIT WITH PART 4

R6

③ BUSHING
BRONZE

Ø100

92

14

H7/u6 FIT WITH
PART 2

SR140

Ø80 H11/c11
FIT WITH PART 3

10

④ BEARING
1020 STEEL
FAO

140

Ø100 H7/u6 FIT WITH
PART 3

R6

12

Ø106

12

94

Ø38
4 PLACES

Ø18-4 HOLES

BOTH
ENDS

10°

22

180

240

30

30

40

80

160

30

② BUSHING HOLDER
1020 STEEL

0.5

Ø26 THRU
2 PLACES

Ø18
4 HOLES

24

80

340

60

40

80

26□
4 PL

12

3

248

44

80

20

20

50

R6

R

400

44

76

14

84

168

32

22

NOTE: BASE
IS SYMMETRI-
CAL ABOUT 2 AXES

① BASE
1020 STEEL

FILLETS & ROUNDS R3
ON ALL PARTS UNLESS
OTHERWISE SPECIFIED

PIVOT ASSEMBLY SI

⑤ M16 X 2 SQ HD
REG BOLT – 70 LONG
1035 STEEL – 4 REQ

⑥ M16 X 2 HEX HD
REG NUT –1035 STEEL
4 REQ

⑦ M24 X 3 HEX HD REG
BOLT X 160 LONG &
NUT – 1035 STEEL
2 EA REQ

23.66 Make a working drawing, with an assembly, of this pivot assembly on size B sheets.

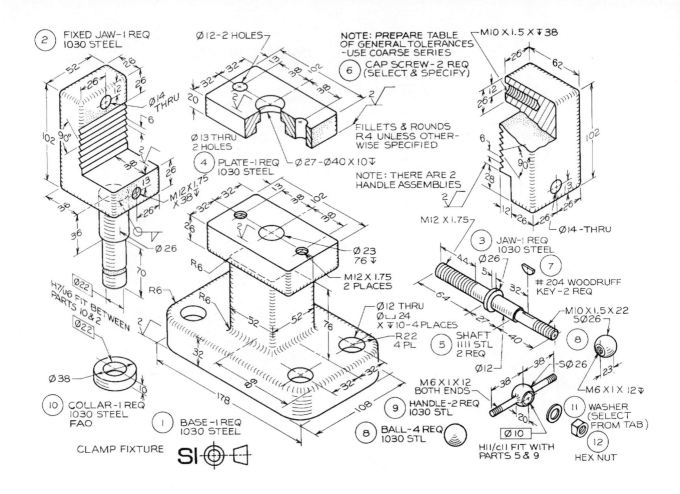

(2) FIXED JAW–1 REQ
1030 STEEL

Ø12–2 HOLES

NOTE: PREPARE TABLE
OF GENERAL TOLERANCES
–USE COARSE SERIES

(6) CAP SCREW–2 REQ
(SELECT & SPECIFY)

M10 X 1.5 X ↧ 38

Ø14 THRU

Ø13 THRU
2 HOLES

(4) PLATE–1 REQ
1030 STEEL

Ø 27–Ø40 X 10 ↧

FILLETS & ROUNDS
R4 UNLESS OTHER-
WISE SPECIFIED

NOTE: THERE ARE 2
HANDLE ASSEMBLIES

M12 X 1.75

M12 X 1.75
X 38 ↧

Ø 26

M12 X 1.75
2 PLACES

Ø 23
76 ↧

Ø12 THRU
Ø↧ 24
X ↧ 10–4 PLACES

R22
4 PL

(3) JAW–1 REQ
1030 STEEL

(7) # 204 WOODRUFF
KEY–2 REQ

Ø26

M10 X 1.5 X 22
S Ø26

(8)

Ø14 –THRU

(5) SHAFT
1111 STL
2 REQ

Ø12

M6 X 1 X 12
BOTH ENDS

S Ø 26

M6 X 1 X 12 ↧

H7/U6 FIT BETWEEN
PARTS 10 & 2

Ø22

Ø22

Ø 26

(10) COLLAR–1 REQ
1030 STEEL
FAO

Ø 38

(1) BASE–1 REQ
1030 STEEL

178

108

(9) HANDLE–2 REQ
1030 STL

(8) BALL–4 REQ
1030 STL

(11) WASHER
(SELECT
FROM TAB)

(12) HEX NUT

H11/c11 FIT WITH
PARTS 5 & 9

Ø 10

CLAMP FIXTURE

GENERAL TOLERANCES—MILLIMETERS
UNLESS OTHERWISE SPECIFIED

LINEAR TOLERANCE	OVER TO	— 120	120 315	315 1000	1000 —
		±0.3	±0.5	±0.8	±1.2

23.67 Make a half-size working drawing, with an assembly drawing, of this clamp fixture on size B sheets.

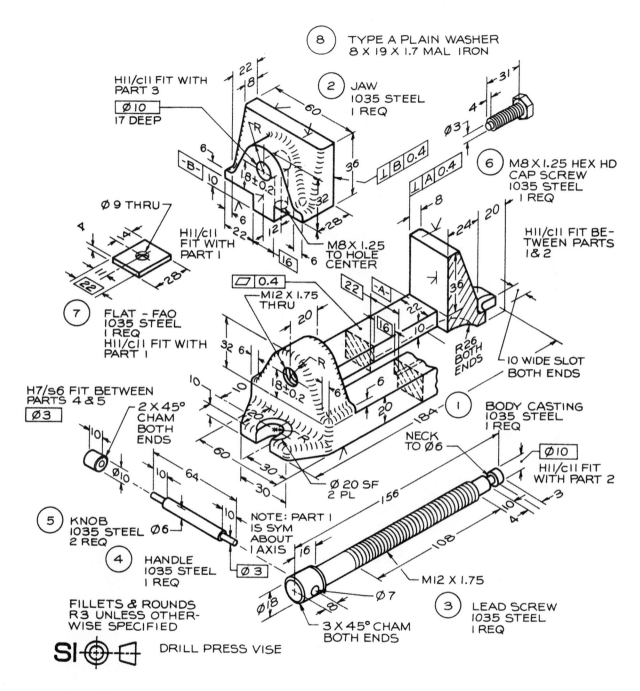

8 TYPE A PLAIN WASHER
8 X 19 X 1.7 MAL IRON

HII/cII FIT WITH
PART 3
Ø10
17 DEEP

2 JAW
1035 STEEL
I REQ

22
8

60

R

18±0.2

36

32

28

6

10

6

22

12

16

-B-

6 M8 X 1.25 HEX HD
CAP SCREW
1035 STEEL
I REQ

Ø3

31

4

⊥ B 0.4

⊥ A 0.4

HII/cII FIT BE-
TWEEN PARTS
1 & 2

8

24

20

36

R26
BOTH
ENDS

10 WIDE SLOT
BOTH ENDS

Ø 9 THRU

HII/cII
FIT WITH
PART 1

22

28

7 FLAT – FAO
1035 STEEL
I REQ
HII/cII FIT WITH
PART 1

M8 X 1.25
TO HOLE
CENTER

0.4

M12 X 1.75
THRU

20

22

16

10

-A-

6

20

32 6

10

18±0.2

6

R

H7/s6 FIT BETWEEN
PARTS 4 & 5
Ø3

2 X 45°
CHAM
BOTH
ENDS

10

Ø10

10

64

10

20

20

R

1 BODY CASTING
1035 STEEL
I REQ

184

NECK
TO Ø6

Ø10

HII/cII FIT
WITH PART 2

3

10

4

5 KNOB
1035 STEEL
2 REQ

Ø6

60

30

30

Ø 20 SF
2 PL

156

4 HANDLE
1035 STEEL
I REQ

NOTE: PART I
IS SYM
ABOUT
I AXIS

16

Ø3

10

108

M12 X 1.75

FILLETS & ROUNDS
R3 UNLESS OTHER-
WISE SPECIFIED

Ø18

8

Ø7

3 X 45° CHAM
BOTH ENDS

3 LEAD SCREW
1035 STEEL
I REQ

SI ⊕ ⊏⊐ DRILL PRESS VISE

23.68 Make working drawings, with an assembly drawing, of this drill press vise on size B sheets.

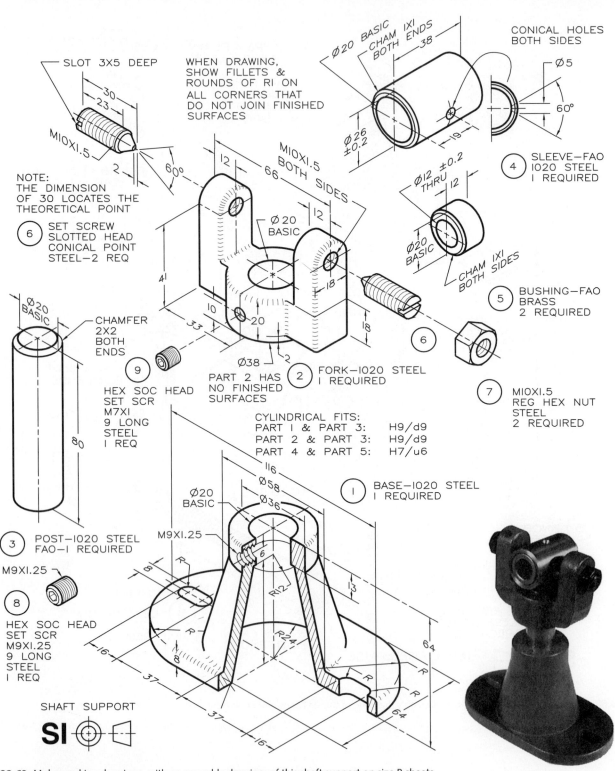

SLOT 3X5 DEEP

30
23

MIOXI.5

2

60°

NOTE:
THE DIMENSION
OF 30 LOCATES THE
THEORETICAL POINT

⑥ SET SCREW
SLOTTED HEAD
CONICAL POINT
STEEL—2 REQ

WHEN DRAWING,
SHOW FILLETS &
ROUNDS OF RI ON
ALL CORNERS THAT
DO NOT JOIN FINISHED
SURFACES

Ø20 BASIC
CHAM IXI
BOTH ENDS
38

CONICAL HOLES
BOTH SIDES

Ø5

60°

Ø26
±0.2

19

④ SLEEVE—FAO
1020 STEEL
I REQUIRED

MIOXI.5
BOTH SIDES

12
66

12

Ø 20
BASIC

Ø12 ±0.2
THRU

12

Ø20
BASIC

CHAM IXI
BOTH SIDES

⑤ BUSHING—FAO
BRASS
2 REQUIRED

Ø20
BASIC

CHAMFER
2X2
BOTH
ENDS

41

10

18

18

⑥

Ø38

2

② FORK—1020 STEEL
I REQUIRED

PART 2 HAS
NO FINISHED
SURFACES

20

33

⑨

HEX SOC HEAD
SET SCR
M7XI
9 LONG
STEEL
I REQ

⑦ MIOXI.5
REG HEX NUT
STEEL
2 REQUIRED

80

CYLINDRICAL FITS:
PART I & PART 3:   H9/d9
PART 2 & PART 3:   H9/d9
PART 4 & PART 5:   H7/u6

① BASE—1020 STEEL
I REQUIRED

116
Ø58

Ø20
BASIC

Ø36

M9XI.25

6

③ POST—1020 STEEL
FAO—I REQUIRED

M9XI.25

⑧

HEX SOC HEAD
SET SCR
M9XI.25
9 LONG
STEEL
I REQ

8

R

R12

13

R

R24

64

R

R

16

8

37

37

16

64

SHAFT SUPPORT

SI ⊕ ⊏

23.69 Make working drawings, with an assembly drawing, of this shaft support on size B sheets.

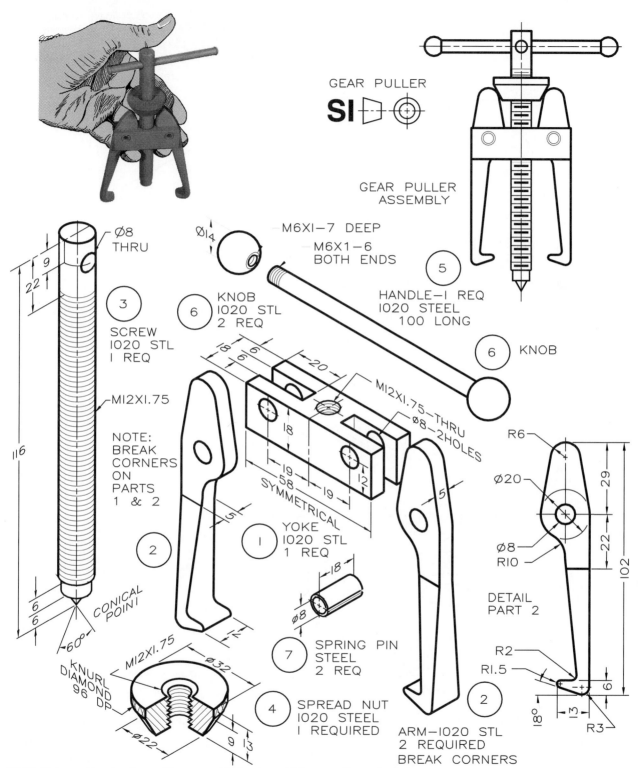

GEAR PULLER

SI ▭▷⊕

GEAR PULLER
ASSEMBLY

Ø8
THRU

Ø14

M6X1–7 DEEP
M6X1–6
BOTH ENDS

③

⑤

SCREW
1020 STL
1 REQ

⑥ KNOB
1020 STL
2 REQ

HANDLE–1 REQ
1020 STEEL
100 LONG

M12X1.75

⑥ KNOB

9
22

116

6
6

CONICAL
POINT

60°

M12X1.75
Ø32

KNURL
DIAMOND
96 DP

Ø22

9 13

M12X1.75–THRU
Ø8–2HOLES

18
6
6
20

18

19
58 19
SYMMETRICAL

12

②

5

YOKE
1020 STL
1 REQ

①

5

2

18
Ø8

⑦ SPRING PIN
STEEL
2 REQ

④ SPREAD NUT
1020 STEEL
1 REQUIRED

R6

Ø20

Ø8
R10

DETAIL
PART 2

R2
R1.5

18°

13

6

R3

②

ARM–1020 STL
2 REQUIRED
BREAK CORNERS

NOTE:
BREAK
CORNERS
ON
PARTS
1 & 2

29

22

102

23.70 Make working drawings, with an assembly drawing, of this gear puller on size B sheets.

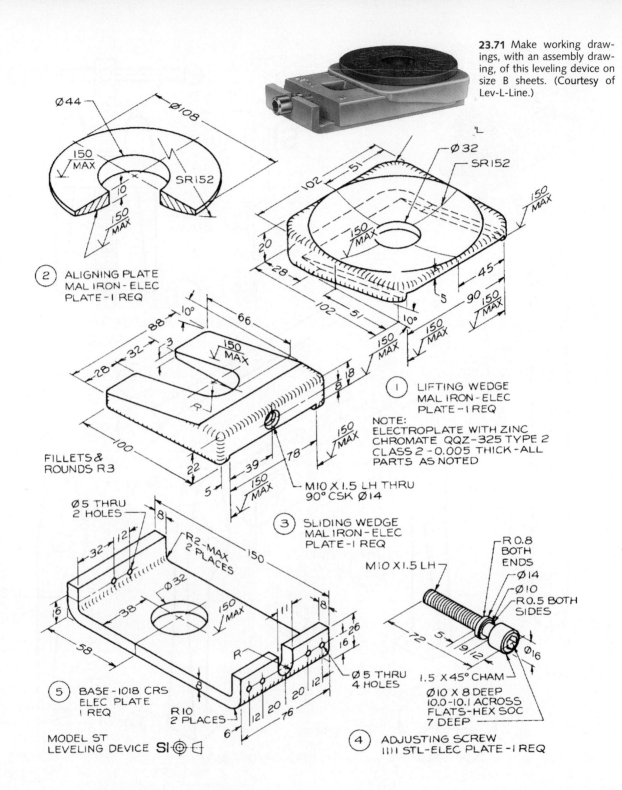

23.71 Make working drawings, with an assembly drawing, of this leveling device on size B sheets. (Courtesy of Lev-L-Line.)

Ø44

Ø108

150 MAX

10

150 MAX

SR152

(2) ALIGNING PLATE
MAL IRON - ELEC
PLATE - 1 REQ

'L

Ø32

SR152

102

51

20

150 MAX

28

102

51

45

5

90

150 MAX

10°

150 MAX

150 MAX

150 MAX

(1) LIFTING WEDGE
MAL IRON - ELEC
PLATE - 1 REQ

88

10°

66

32

3

150 MAX

28

R

18

8

150 MAX

100

22

39

5

78

150 MAX

FILLETS &
ROUNDS R3

NOTE:
ELECTROPLATE WITH ZINC
CHROMATE QQZ–325 TYPE 2
CLASS 2 – 0.005 THICK – ALL
PARTS AS NOTED

M10 X 1.5 LH THRU
90° CSK Ø14

(3) SLIDING WEDGE
MAL IRON - ELEC
PLATE - 1 REQ

Ø5 THRU
2 HOLES

8

R2 - MAX
2 PLACES

150

32

12

Ø32

150 MAX

11

8

16

38

R

26

16

58

R10
2 PLACES

8

Ø5 THRU
4 HOLES

12 20 20 12

6

76

M10 X 1.5 LH

R 0.8
BOTH
ENDS

Ø14

Ø10

R 0.5 BOTH
SIDES

72

5

9 12

Ø16

1.5 X 45° CHAM

Ø10 X 8 DEEP
10.0-10.1 ACROSS
FLATS-HEX SOC
7 DEEP

(5) BASE - 1018 CRS
ELEC PLATE
1 REQ

MODEL ST
LEVELING DEVICE SI ⦿ ⊟

(4) ADJUSTING SCREW
|||| STL-ELEC PLATE - 1 REQ

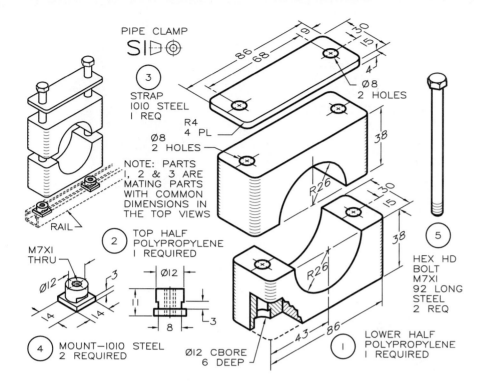

PIPE CLAMP

SI⊕

③ STRAP
1010 STEEL
1 REQ

Ø8
2 HOLES

R4
4 PL

Ø8
2 HOLES

86
68
9
30
15
4

NOTE: PARTS
1, 2 & 3 ARE
MATING PARTS
WITH COMMON
DIMENSIONS IN
THE TOP VIEWS

RAIL

② TOP HALF
POLYPROPYLENE
1 REQUIRED

38

R26

30
15

38

R26

⑤ HEX HD
BOLT
M7X1
92 LONG
STEEL
2 REQ

M7X1
THRU

Ø12

3

14   14

11

8   3

④ MOUNT—1010 STEEL
2 REQUIRED

Ø12 CBORE
6 DEEP

43   86

① LOWER HALF
POLYPROPYLENE
1 REQUIRED

23.72 Make working drawings,
with an assembly drawing, of this
pipe clamp on size A sheets.

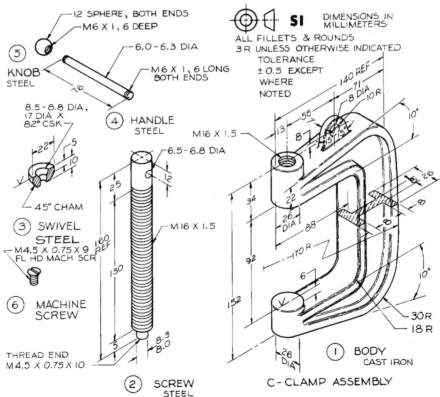

12 SPHERE, BOTH ENDS
M6 X 1, 6 DEEP

⑤ KNOB
STEEL

6.0 - 6.3 DIA

M 6 X 1, 6 LONG
BOTH ENDS

④ HANDLE
STEEL

⊕⊕◁ SI   DIMENSIONS IN
MILLIMETERS

ALL FILLETS & ROUNDS
3 R UNLESS OTHERWISE INDICATED
TOLERANCE
± 0.5 EXCEPT
WHERE
NOTED

8.5 - 8.8 DIA,
17 DIA X
82° CSK

22

5
10

45° CHAM

③ SWIVEL
STEEL

6.5 - 6.8 DIA

25

12

M 16 X 1.5

M16 X 1.5

140 REF
71
8 DIA
10 R

56
13
8

10°

22
34
26
DIA
88

170 R

92

6

10°

30 R
18 R

M4.5 X 0.75 X 9
FL HD MACH SCR

⑥ MACHINE
SCREW

160
REF

130

5   8.3
8.0

THREAD END
M4.5 X 0.75 X 10

② SCREW
STEEL

152

26
DIA

① BODY
CAST IRON

C- CLAMP ASSEMBLY

23.73 Make working drawings,
with an assembly drawing, of
this C-clamp on size B sheets.

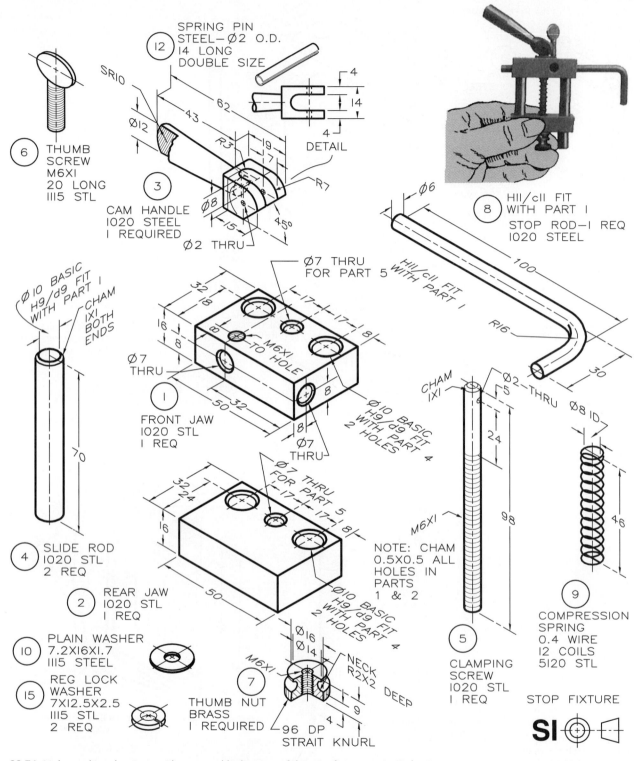

② SPRING PIN
STEEL—Ø2 O.D.
14 LONG
DOUBLE SIZE

DETAIL
4
14
4

⑥ THUMB
SCREW
M6X1
20 LONG
1115 STL

SR10
Ø12
62
43
R3
19
7
③ CAM HANDLE
1020 STEEL
1 REQUIRED
Ø8
15
45°
Ø2 THRU
R7

⑧ H11/c11 FIT
WITH PART 1
STOP ROD—1 REQ
1020 STEEL

Ø6
H11/c11 FIT
WITH PART 1
100
R16
30
Ø2 THRU

Ø10 BASIC
H9/d9 FIT
WITH PART 1
CHAM
1X1
BOTH
ENDS

④ SLIDE ROD
1020 STL
2 REQ
70

Ø7 THRU
FOR PART 5
32
18
16
8
8
Ø7
THRU
① FRONT JAW
1020 STL
1 REQ
M6X1
TO HOLE
50
32
8
8
Ø7
THRU
Ø10 BASIC
H9/d9 FIT
WITH PART 4
2 HOLES

② REAR JAW
1020 STL
1 REQ
32
24
16
Ø7 THRU
FOR PART 5
17
17
8
50
Ø10 BASIC
H9/d9 FIT
WITH PART 4
2 HOLES

NOTE: CHAM
0.5X0.5 ALL
HOLES IN
PARTS
1 & 2

CHAM
1X1
Ø2 THRU
5
24
M6X1
98
⑤ CLAMPING
SCREW
1020 STL
1 REQ

Ø8 ID
⑨ COMPRESSION
SPRING
0.4 WIRE
12 COILS
5120 STL
46

⑩ PLAIN WASHER
7.2X16X1.7
1115 STEEL

⑮ REG LOCK
WASHER
7X12.5X2.5
1115 STL
2 REQ

⑦ THUMB NUT
BRASS
1 REQUIRED
M6X1
Ø16
Ø14
NECK
R2X2
DEEP
9
4
96 DP
STRAIT KNURL

STOP FIXTURE

SI

23.74 Make working drawings, with an assembly drawing, of this stop fixture on size B sheets.

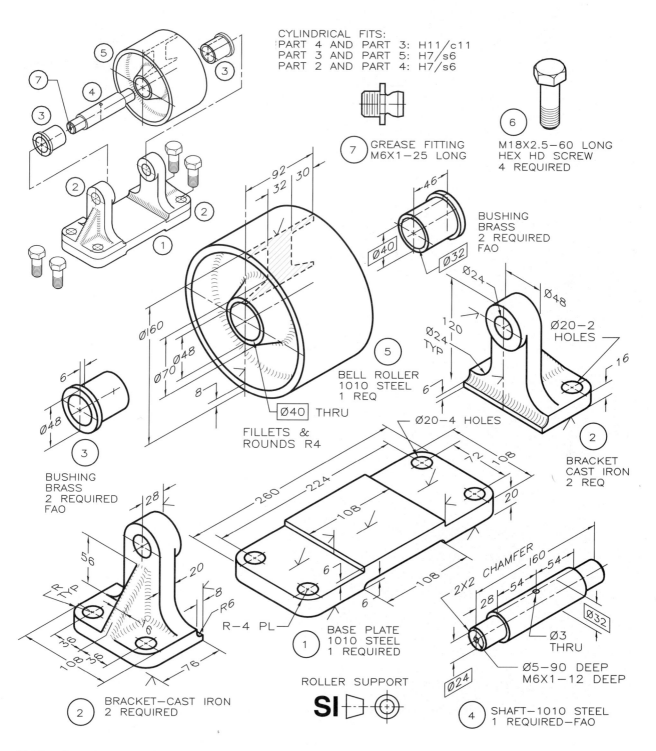

CYLINDRICAL FITS:
PART 4 AND PART 3: H11/c11
PART 3 AND PART 5: H7/s6
PART 2 AND PART 4: H7/s6

7 GREASE FITTING
M6X1–25 LONG

6 M18X2.5–60 LONG
HEX HD SCREW
4 REQUIRED

5 BELL ROLLER
1010 STEEL
1 REQ

Ø40 THRU

FILLETS &
ROUNDS R4

BUSHING
BRASS
2 REQUIRED
FAO

3 BUSHING
BRASS
2 REQUIRED
FAO

Ø20–2
HOLES

2 BRACKET
CAST IRON
2 REQ

Ø20–4 HOLES

1 BASE PLATE
1010 STEEL
1 REQUIRED

ROLLER SUPPORT

SI

2 BRACKET–CAST IRON
2 REQUIRED

2X2 CHAMFER

Ø3
THRU

Ø5–90 DEEP
M6X1–12 DEEP

4 SHAFT–1010 STEEL
1 REQUIRED–FAO

23.75 Make working drawings, with an assembly drawing, of this pulley support on size B sheets.

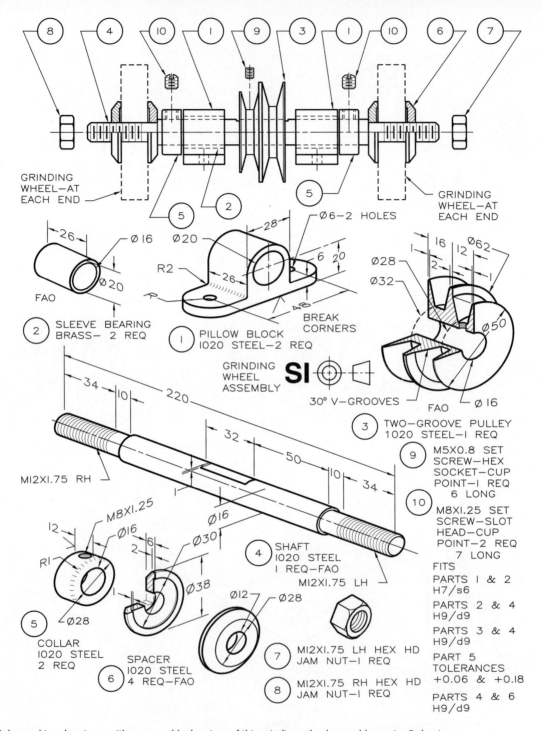

GRINDING WHEEL—AT EACH END

GRINDING WHEEL—AT EACH END

26
Ø16
Ø20

FAO
Ø20

② SLEEVE BEARING BRASS— 2 REQ

28
Ø6–2 HOLES

R2
26
6 20

① PILLOW BLOCK 1020 STEEL–2 REQ

48
BREAK CORNERS

GRINDING WHEEL ASSEMBLY

SI ⊕ ◁

30° V–GROOVES

16 12 Ø62
Ø28
Ø32

Ø50

FAO
Ø16

③ TWO–GROOVE PULLEY 1020 STEEL–1 REQ

34 10 220

32

50

Ø16

10 34

MI2X1.75 RH

M8X1.25
12
Ø16
2 6
R1
Ø28

⑤ COLLAR 1020 STEEL 2 REQ

Ø30
Ø38

④ SHAFT 1020 STEEL 1 REQ–FAO
MI2X1.75 LH

⑥ SPACER 1020 STEEL 4 REQ–FAO

Ø12 Ø28

⑦ MI2X1.75 LH HEX HD JAM NUT–1 REQ

⑧ MI2X1.75 RH HEX HD JAM NUT–1 REQ

⑨ M5X0.8 SET SCREW–HEX SOCKET–CUP POINT–1 REQ 6 LONG

⑩ M8X1.25 SET SCREW–SLOT HEAD–CUP POINT–2 REQ 7 LONG

FITS
PARTS 1 & 2 H7/s6

PARTS 2 & 4 H9/d9

PARTS 3 & 4 H9/d9

PART 5 TOLERANCES +0.06 & +0.18

PARTS 4 & 6 H9/d9

23.76 Make working drawings, with an assembly drawing, of this grinding wheel assembly on size B sheets.

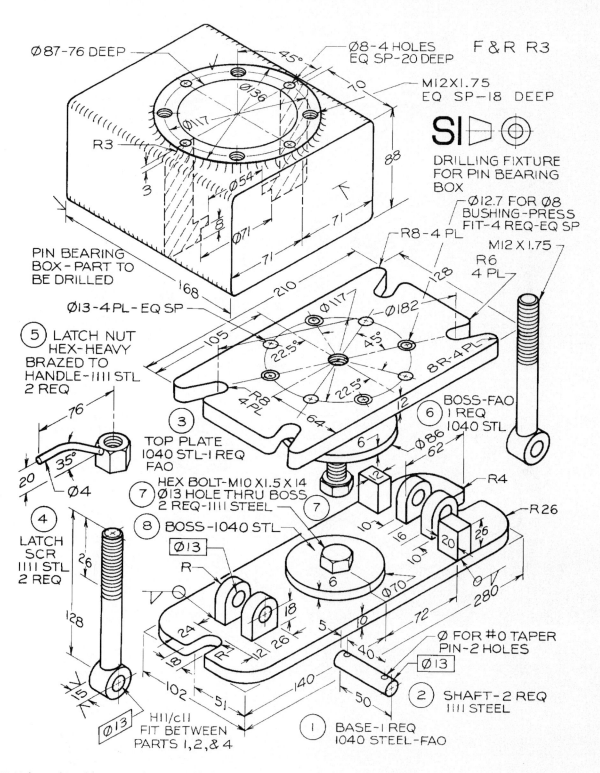

Ø87-76 DEEP

45°

Ø8-4 HOLES
EQ SP-20 DEEP

F & R R3

Ø136

70

M12X1.75
EQ SP-18 DEEP

SI ▷ ◉

DRILLING FIXTURE
FOR PIN BEARING
BOX

Ø117

R3

3

Ø54

88

Ø71

8

71

Ø12.7 FOR Ø8
BUSHING-PRESS
FIT-4 REQ-EQ SP

M12 X 1.75

R8-4 PL

R6
4 PL

PIN BEARING
BOX - PART TO
BE DRILLED

71

71

168

210

128

Ø13-4 PL-EQ SP

Ø117

Ø182

⑤ LATCH NUT
HEX-HEAVY
BRAZED TO
HANDLE-1111 STL
2 REQ

105

22.5°

45°

8R-4 PL

76

③

22.5°

12

BOSS-FAO
⑥ 1 REQ
1040 STL

35°

R8
4 PL

64

6

Ø86
62

Ø4

TOP PLATE
1040 STL-1 REQ
FAO

2

R4

20

HEX BOLT-M10 X1.5 X14
⑦ Ø13 HOLE THRU BOSS
2 REQ-1111 STEEL

⑦

R26

④

⑧ BOSS-1040 STL

10

16

26

LATCH
SCR
1111 STL
2 REQ

26

Ø13

R

10

20

6

Ø70

280

18

128

24

6

5

10

72

R

12 26

Ø FOR #0 TAPER
PIN-2 HOLES

15

18

40

Ø13

Ø13

102

51

140

50

H11/c11
FIT BETWEEN
PARTS 1, 2, & 4

① BASE-1 REQ
1040 STEEL-FAO

② SHAFT-2 REQ
1111 STEEL

23.77 Make working drawings, with an assembly drawing, of this drilling fixture on size B sheets.

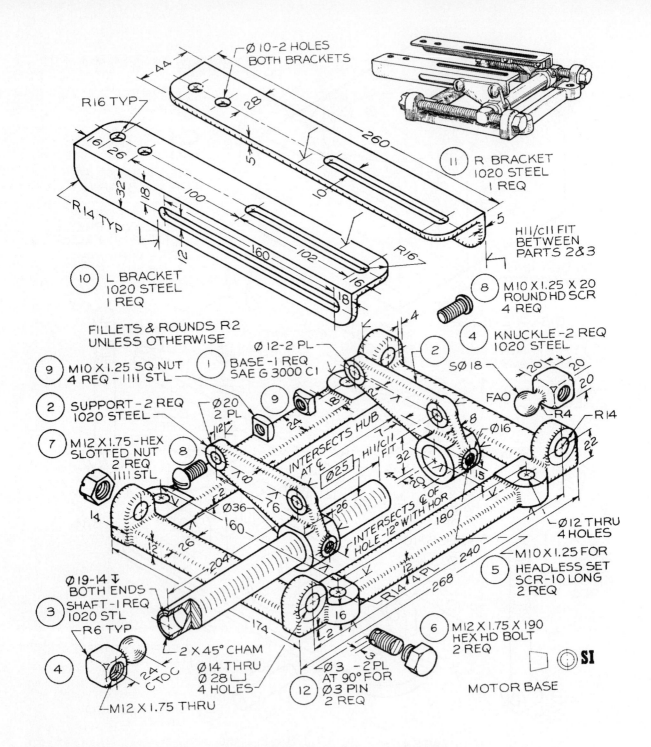

Ø 10–2 HOLES
BOTH BRACKETS

44

28

260

R16 TYP

16 | 26

32

18

100

R14 TYP

12

160

102

16

18

R16

5

⑪ R BRACKET
1020 STEEL
I REQ

H11/C11 FIT
BETWEEN
PARTS 2&3

⑩ L BRACKET
1020 STEEL
I REQ

FILLETS & ROUNDS R2
UNLESS OTHERWISE

⑧ M10 X 1.25 X 20
ROUND HD SCR
4 REQ

④ KNUCKLE – 2 REQ
1020 STEEL

S Ø 18

FAO

20

20

20

R4

R14

Ø 12–2 PL

②

⑨ M10 X 1.25 SQ NUT
4 REQ – 1111 STL

① BASE – I REQ
SAE G 3000 C1

⑨

Ø 20
2 PL

12

② SUPPORT – 2 REQ
1020 STEEL

⑦ M12 X 1.75 – HEX
SLOTTED NUT
2 REQ
1111 STL

⑧

24

18

INTERSECTS HUB
AT ℄

Ø 25

H11/C11
FIT

32

8

Ø 16

15

22

14

Ø 36

76

160

26

180

Ø 12 THRU
4 HOLES

Ø 19–14 ⌄
BOTH ENDS

③ SHAFT – I REQ
1020 STL

R6 TYP

④

12

26

204

174

2 X 45° CHAM

Ø 14 THRU
Ø 28 ⌴
4 HOLES

24
C TO C

M12 X 1.75 THRU

INTERSECTS ℄ OF
HOLE – 12° WITH HOR

26

20

12

R14 4 PL

16

268

240

⑤ M10 X 1.25 FOR
HEADLESS SET
SCR – 10 LONG
2 REQ

2

3

Ø 3 – 2 PL
AT 90° FOR

⑫ Ø 3 PIN
2 REQ

⑥ M12 X 1.75 X 190
HEX HD BOLT
2 REQ

SI

MOTOR BASE

23.78 Make working drawings, with an assembly drawing, of this motor base on size B sheets.

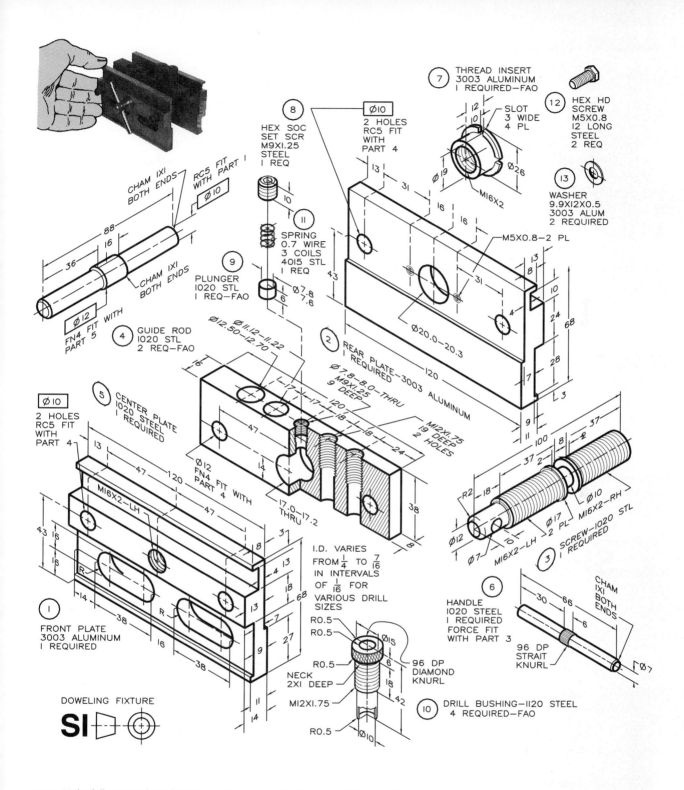

CHAM IXI BOTH ENDS
RC5 FIT WITH PART I
Ø10
88
16
36
Ø12 FN4 FIT WITH PART 5
CHAM IXI BOTH ENDS

8
HEX SOC SET SCR M9X1.25 STEEL I REQ
10

11
SPRING 0.7 WIRE 3 COILS 4015 STL I REQ

9
PLUNGER 1020 STL I REQ—FAO
Ø7.8 7.6
6

Ø10
2 HOLES RC5 FIT WITH PART 4

7
THREAD INSERT 3003 ALUMINUM I REQUIRED—FAO
12
10
SLOT 3 WIDE 4 PL
Ø19
Ø26
M16X2

12
HEX HD SCREW M5X0.8 12 LONG STEEL 2 REQ

13
WASHER 9.9X12X0.5 3003 ALUM 2 REQUIRED

13
31
16
16
43
M5X0.8—2 PL
13
8
31
Ø20.0—20.3
10
24
4
68
7
28
9
11
3

2
REAR PLATE—3003 ALUMINUM I REQUIRED
120

4
GUIDE ROD 1020 STL 2 REQ—FAO

5
CENTER PLATE 1020 STEEL I REQUIRED

Ø10
2 HOLES RC5 FIT WITH PART 4

Ø11.12—11.22
Ø12.50—12.70
6
17
17
120
18
18
24
Ø7.8—8.0—THRU
M9X1.25 9 DEEP
M12X1.75 19 DEEP 2 HOLES

Ø12 FN4 FIT WITH PART 4 WITH
47
14
17.0—17.2 THRU
38
8

13
47
120
MI6X2—LH
47
3
8
13
4
18
68
7
9
27
11
14

43 16
16
R
14
R
38
16
38

1
FRONT PLATE 3003 ALUMINUM I REQUIRED

I.D. VARIES FROM $\frac{1}{4}$ TO $\frac{7}{16}$ IN INTERVALS OF $\frac{1}{16}$ FOR VARIOUS DRILL SIZES

R0.5
R0.5
Ø15
R0.5
6
96 DP DIAMOND KNURL
NECK 2X1 DEEP
18
M12X1.75
42
R0.5
Ø10

10
DRILL BUSHING—1120 STEEL 4 REQUIRED—FAO

100
8
2
37
37
2
R2
18
Ø10
Ø17 2 PL M16X2—RH
Ø12
Ø7
M16X2—LH
SCREW—1020 STL I REQUIRED

3

6
HANDLE 1020 STEEL I REQUIRED FORCE FIT WITH PART 3
30
66
6
CHAM IXI BOTH ENDS
96 DP STRAIT KNURL
Ø7

DOWELING FIXTURE
SI

23.79 Make full-size working drawings, with an assembly drawing, of this doweling fixture on size B sheets.

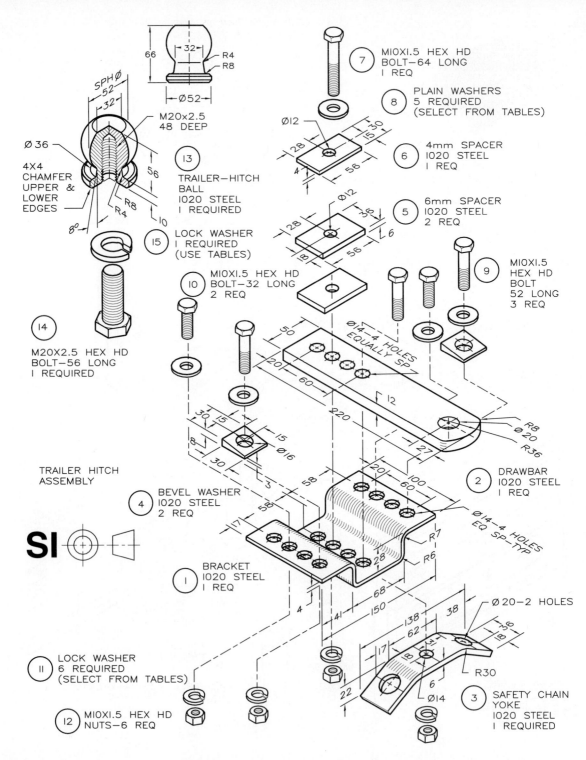

66
─32─
R4
R8
─Ø52─

SPHØ
52
─32─

M20x2.5
48 DEEP

Ø 36

4X4
CHAMFER
UPPER &
LOWER
EDGES

56

R8
R4

8°

10

⑬ TRAILER—HITCH
BALL
1020 STEEL
1 REQUIRED

⑮ LOCK WASHER
1 REQUIRED
(USE TABLES)

⑭

M20X2.5 HEX HD
BOLT—56 LONG
1 REQUIRED

⑦ M10X1.5 HEX HD
BOLT—64 LONG
1 REQ

⑧ PLAIN WASHERS
5 REQUIRED
(SELECT FROM TABLES)

Ø12
30
15
28
56
4

⑥ 4mm SPACER
1020 STEEL
1 REQ

Ø12
30
28
56
18
6

⑤ 6mm SPACER
1020 STEEL
2 REQ

⑨ M10X1.5
HEX HD
BOLT
52 LONG
3 REQ

⑩ M10X1.5 HEX HD
BOLT—32 LONG
2 REQ

50
Ø14—4 HOLES
EQUALLY SP
20
60
220
12
R8
Ø 20
R36
27

② DRAWBAR
1020 STEEL
1 REQ

Ø14—4 HOLES
EQ SP—TYP

TRAILER HITCH
ASSEMBLY

④ BEVEL WASHER
1020 STEEL
2 REQ

30
15
8
15
Ø16
30
3

58

58
17

20
100
60

58

R7
R6
28

① BRACKET
1020 STEEL
1 REQ

SI

Ø 20—2 HOLES
16
18

⑪ LOCK WASHER
6 REQUIRED
(SELECT FROM TABLES)

4
41
150
138
62
17
18
18
6
Ø14
22

R30

6
Ø14

③ SAFETY CHAIN
YOKE
1020 STEEL
1 REQUIRED

68
38

⑫ M10X1.5 HEX HD
NUTS—6 REQ

23.80 Make full-size working drawings, with an assembly drawing, of this trailer hitch on size B sheets.

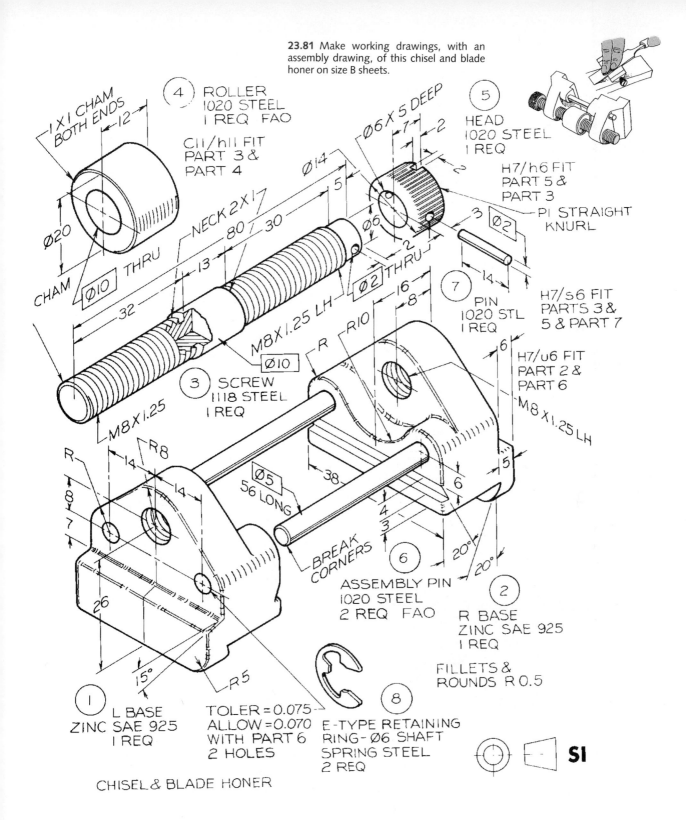

23.81 Make working drawings, with an assembly drawing, of this chisel and blade honer on size B sheets.

1×1 CHAM BOTH ENDS
12
④ ROLLER
1020 STEEL
1 REQ FAO

Ø6 X 5 DEEP
7
2
2

⑤ HEAD
1020 STEEL
1 REQ

C11/h11 FIT
PART 3 &
PART 4

Ø14
5
Ø20

CHAM Ø10 THRU

H7/h6 FIT
PART 5 &
PART 3

P1 STRAIGHT
KNURL

3 Ø2

NECK 2×1
80 30
Ø6
2

13
32
Ø2 THRU

14

⑦ PIN
1020 STL
1 REQ

H7/s6 FIT
PARTS 3 &
5 & PART 7

M8×1.25 LH

Ø10

③ SCREW
1118 STEEL
1 REQ

16
8

6

H7/u6 FIT
PART 2 &
PART 6

M8 X1.25 LH

M8×1.25

R R10

R8
14
14

Ø5
56 LONG

38

5

R
8
7

BREAK CORNERS

⑥ ASSEMBLY PIN
1020 STEEL
2 REQ FAO

6
4
3

20°
20°

② R BASE
ZINC SAE 925
1 REQ

26

15°
R5

FILLETS & ROUNDS R 0.5

① L BASE
ZINC SAE 925
1 REQ

TOLER = 0.075
ALLOW = 0.070
WITH PART 6
2 HOLES

⑧ E-TYPE RETAINING
RING-Ø6 SHAFT
SPRING STEEL
2 REQ

SI

CHISEL & BLADE HONER

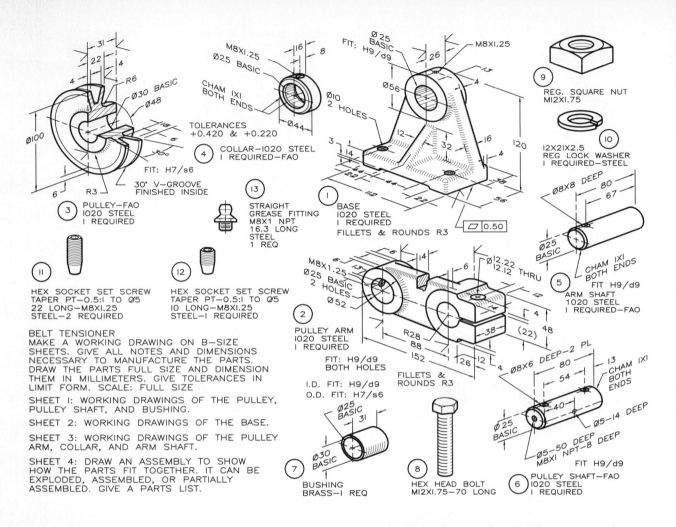

BELT TENSIONER
MAKE A WORKING DRAWING ON B-SIZE
SHEETS. GIVE ALL NOTES AND DIMENSIONS
NECESSARY TO MANUFACTURE THE PARTS.
DRAW THE PARTS FULL SIZE AND DIMENSION
THEM IN MILLIMETERS. GIVE TOLERANCES IN
LIMIT FORM. SCALE: FULL SIZE

SHEET 1: WORKING DRAWINGS OF THE PULLEY,
PULLEY SHAFT, AND BUSHING.

SHEET 2: WORKING DRAWINGS OF THE BASE.

SHEET 3: WORKING DRAWINGS OF THE PULLEY
ARM, COLLAR, AND ARM SHAFT.

SHEET 4: DRAW AN ASSEMBLY TO SHOW
HOW THE PARTS FIT TOGETHER. IT CAN BE
EXPLODED, ASSEMBLED, OR PARTIALLY
ASSEMBLED. GIVE A PARTS LIST.

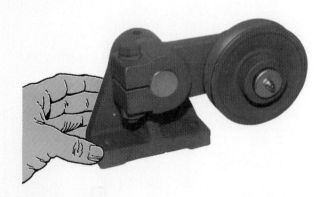

23.82 Make working drawings, with an assembly drawing, of
this belt tensioner on size B sheets.

Working Drawings of Multiple Parts Involving Design Features

Make dimensioned working drawings of the
multiple parts shown in **Figs. 23.83–23.93** on a
sheet size of your choice with the necessary
dimensions and notes to fabricate the parts.
Each part is given in a general format, which
requires some design on your part. You must
consider the addition of fillets and rounds, the
application of finish marks, and the modification of features of the parts to make them
functional and practical. Apply the tolerance
to the parts in limit form by using the tables of
cylindrical fits in the Appendix. Make an
assembly drawing and parts list to show how
the parts are to be put together.

23.83 Design: Make working drawings, with an assembly drawing, of this turn buckle vise on size B sheets.

23.84 Design: Make a working drawing of this base; add four ribs plus fillets and rounds where appropriate. Refer to Fig. 20.1 for dimensions.

23.85 Design: Make working drawings of this 6-in. high angle.

23.86 Design: Make working drawings, with an assembly, of this roller chain puller (for stretching a chain for assembly) on size B sheets. The prongs must join when closed.

23.87 Design: Make working drawings, with an assembly, of this lathe dog with a 3/4-in. diameter screw on size B sheets.

23.88 Design: Make working drawings with an assembly drawing of this I-beam clamp with a 3/4-in. diameter screw on size B sheets. (Courtesy of Grinnel Corporation.)

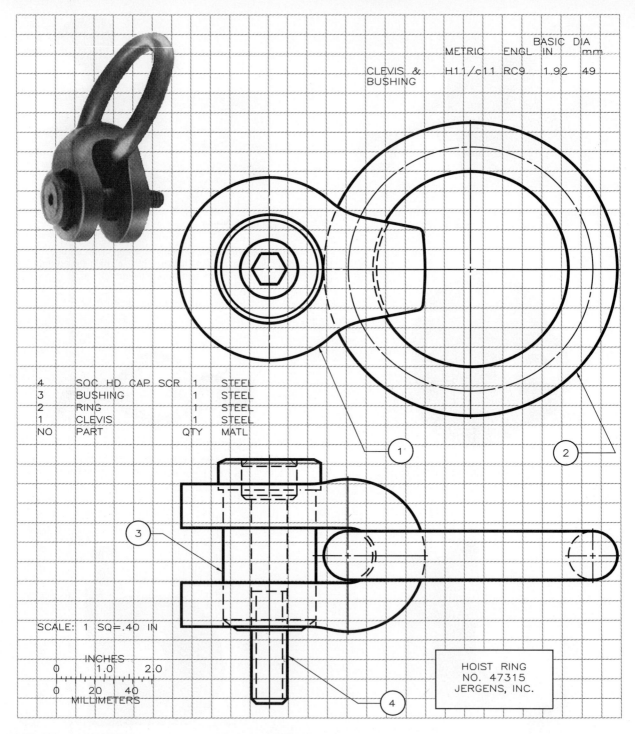

METRIC   ENGL IN   mm

CLEVIS &   H11/c11 RC9   1.92   49
BUSHING

4	SOC HD CAP SCR	1	STEEL
3	BUSHING	1	STEEL
2	RING	1	STEEL
1	CLEVIS	1	STEEL
NO	PART	QTY	MATL

SCALE: 1 SQ=.40 IN

INCHES
0   1.0   2.0

0   20   40
MILLIMETERS

HOIST RING
NO. 47315
JERGENS, INC.

23.89 Make working drawings, with an assembly drawing, of this hoist ring on size B sheets.

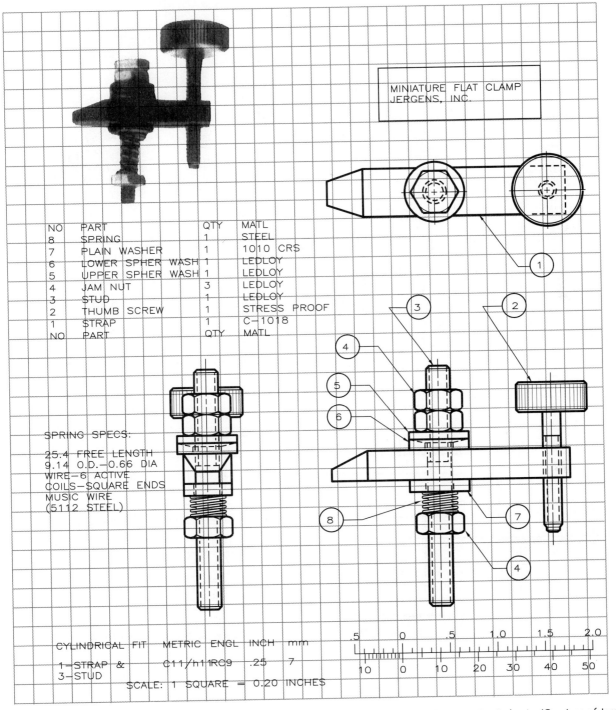

NO	PART	QTY	MATL
8	SPRING	1	STEEL
7	PLAIN WASHER	1	1010 CRS
6	LOWER SPHER WASH	1	LEDLOY
5	UPPER SPHER WASH	1	LEDLOY
4	JAM NUT	3	LEDLOY
3	STUD	1	LEDLOY
2	THUMB SCREW	1	STRESS PROOF
1	STRAP	1	C-1018
NO	PART	QTY	MATL

MINIATURE FLAT CLAMP
JERGENS, INC.

SPRING SPECS:

25.4 FREE LENGTH
9.14 O.D.-0.66 DIA
WIRE-6 ACTIVE
COILS-SQUARE ENDS
MUSIC WIRE
(5112 STEEL)

CYLINDRICAL FIT	METRIC	ENGL INCH	mm
1-STRAP &	C11/h11RC9	.25	7
3-STUD			

SCALE: 1 SQUARE = 0.20 INCHES

23.90 Design: Make working drawings, with an assembly drawing, of this miniature flat clamp on size B sheets. (Courtesy of Jergens, Incorporated.)

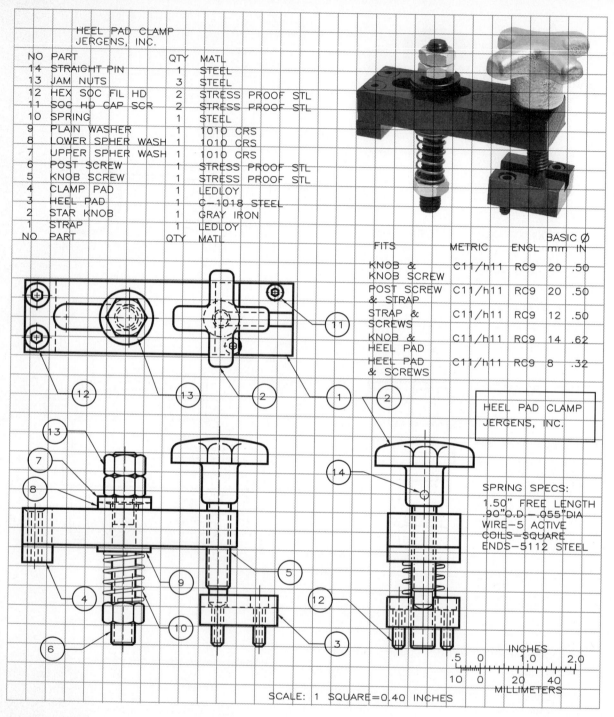

HEEL PAD CLAMP
JERGENS, INC.

NO	PART	QTY	MATL
14	STRAIGHT PIN	1	STEEL
13	JAM NUTS	3	STEEL
12	HEX SOC FIL HD	2	STRESS PROOF STL
11	SOC HD CAP SCR	2	STRESS PROOF STL
10	SPRING	1	STEEL
9	PLAIN WASHER	1	1010 CRS
8	LOWER SPHER WASH	1	1010 CRS
7	UPPER SPHER WASH	1	1010 CRS
6	POST SCREW	1	STRESS PROOF STL
5	KNOB SCREW	1	STRESS PROOF STL
4	CLAMP PAD	1	LEDLOY
3	HEEL PAD	1	C-1018 STEEL
2	STAR KNOB	1	GRAY IRON
1	STRAP	1	LEDLOY
NO	PART	QTY	MATL

FITS	METRIC	ENGL	BASIC Ø mm	IN
KNOB & KNOB SCREW	C11/h11	RC9	20	.50
POST SCREW & STRAP	C11/h11	RC9	20	.50
STRAP & SCREWS	C11/h11	RC9	12	.50
KNOB & HEEL PAD	C11/h11	RC9	14	.62
HEEL PAD & SCREWS	C11/h11	RC9	8	.32

HEEL PAD CLAMP
JERGENS, INC.

SPRING SPECS:
1.50" FREE LENGTH
.90"O.D.—.055"DIA
WIRE—5 ACTIVE
COILS—SQUARE
ENDS—5112 STEEL

INCHES
.5   0   1.0   2.0
10 0   20   40
MILLIMETERS

SCALE: 1 SQUARE=0.40 INCHES

23.91 Design: Make working drawings, with an assembly drawing, of this heel pad clamp on size B sheets. (Courtesy of Jergens, Incorporated.)

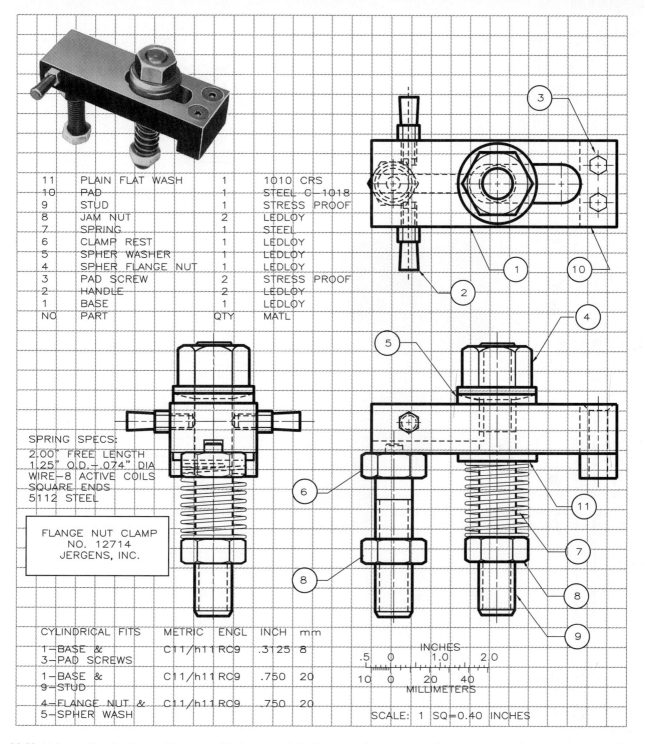

NO	PART	QTY	MATL
11	PLAIN FLAT WASH	1	1010 CRS
10	PAD	1	STEEL C 1018
9	STUD	1	STRESS PROOF
8	JAM NUT	2	LEDLOY
7	SPRING	1	STEEL
6	CLAMP REST	1	LEDLOY
5	SPHER WASHER	1	LEDLOY
4	SPHER FLANGE NUT	1	LEDLOY
3	PAD SCREW	2	STRESS PROOF
2	HANDLE	2	LEDLOY
1	BASE	1	LEDLOY

SPRING SPECS:
2.00" FREE LENGTH
1.25" O.D. – .074" DIA
WIRE–8 ACTIVE COILS
SQUARE ENDS
5112 STEEL

FLANGE NUT CLAMP
NO. 12714
JERGENS, INC.

CYLINDRICAL FITS	METRIC	ENGL	INCH	mm
1–BASE & 3–PAD SCREWS	C11/h11	RC9	.3125	8
1–BASE & 9=STUD	C11/h11	RC9	.750	20
4=FLANGE NUT & 5–SPHER WASH	C11/h11	RC9	.750	20

INCHES
.5 0   1.0   2.0
10 0   20   40
MILLIMETERS

SCALE: 1 SQ=0.40 INCHES

23.92 Make working drawings, with an assembly drawing, of this flange nut clamp on size B sheets. (Courtesy of Jergens, Incorporated.)

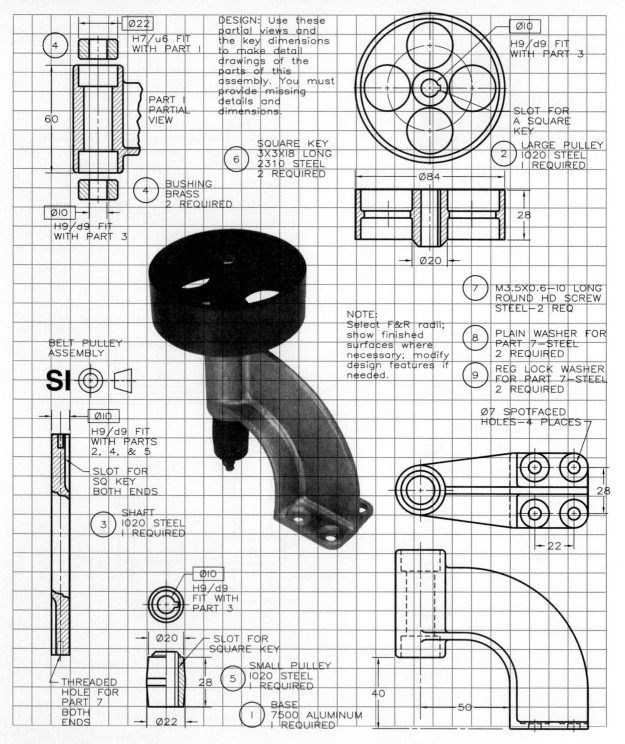

Ø22
H7/u6 FIT
WITH PART 1

④

PART 1
PARTIAL
VIEW

60

④ BUSHING
BRASS
2 REQUIRED

Ø10
H9/d9 FIT
WITH PART 3

DESIGN: Use these
partial views and
the key dimensions
to make detail
drawings of the
parts of this
assembly. You must
provide missing
details and
dimensions.

SQUARE KEY
⑥ 3X3X18 LONG
2310 STEEL
2 REQUIRED

Ø10
H9/d9 FIT
WITH PART 3

SLOT FOR
A SQUARE
KEY

② LARGE PULLEY
1020 STEEL
1 REQUIRED

Ø84

28

Ø20

⑦ M3.5X0.6—10 LONG
ROUND HD SCREW
STEEL—2 REQ

NOTE:
Select F&R radii;
show finished
surfaces where
necessary; modify
design features if
needed.

⑧ PLAIN WASHER FOR
PART 7—STEEL
2 REQUIRED

⑨ REG LOCK WASHER
FOR PART 7—STEEL
2 REQUIRED

BELT PULLEY
ASSEMBLY

SI

Ø10
H9/d9 FIT
WITH PARTS
2, 4, & 5

SLOT FOR
SQ KEY
BOTH ENDS

SHAFT
③ 1020 STEEL
1 REQUIRED

Ø7 SPOTFACED
HOLES—4 PLACES

28

22

Ø10
H9/d9
FIT WITH
PART 3

Ø20 — SLOT FOR
SQUARE KEY

SMALL PULLEY
⑤ 1020 STEEL
1 REQUIRED

28

THREADED
HOLE FOR
PART 7
BOTH
ENDS

Ø22

BASE
① 7500 ALUMINUM
1 REQUIRED

40

50

23.93 Design: Make working drawings, with an assembly, of this belt pulley assembly on size B sheets. Use your design ability to supply missing details, specifications, and notes.

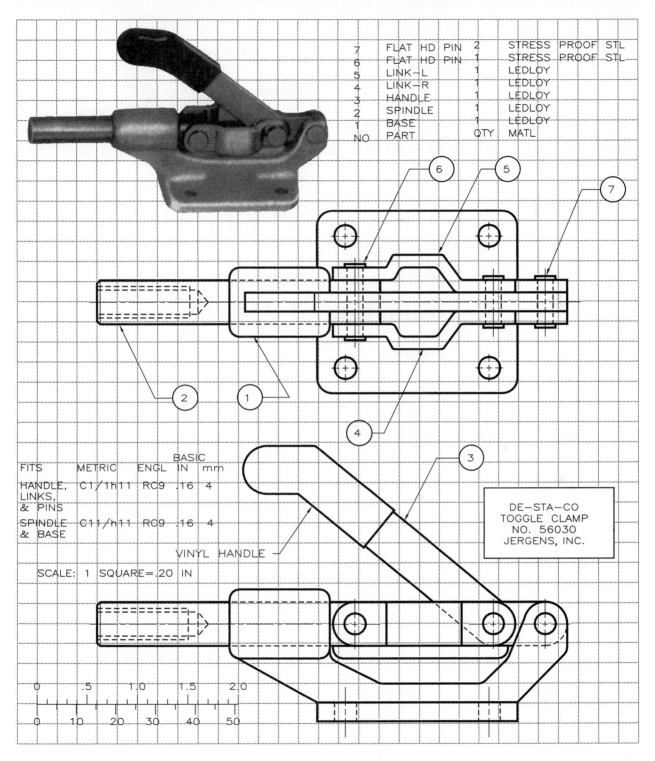

NO	PART	QTY	MATL		
7	FLAT HD PIN	2	STRESS PROOF STL		
6	FLAT HD PIN	1	STRESS PROOF STL		
5	LINK—L	1	LEDLOY		
4	LINK—R	1	LEDLOY		
3	HANDLE	1	LEDLOY		
2	SPINDLE	1	LEDLOY		
1	BASE	1	LEDLOY		

FITS	METRIC	ENGL	BASIC IN	mm
HANDLE, LINKS, & PINS	C1/1h11	RC9	.16	4
SPINDLE & BASE	C11/h11	RC9	.16	4

VINYL HANDLE —

SCALE: 1 SQUARE=.20 IN

DE—STA—CO
TOGGLE CLAMP
NO. 56030
JERGENS, INC.

0   .5   1.0   1.5   2.0

0   10   20   30   40   50

23.94 Design: Make working drawings, with an assembly, of this belt pulley assembly on size B sheets. Use your design ability to supply missing details, specifications, and notes.

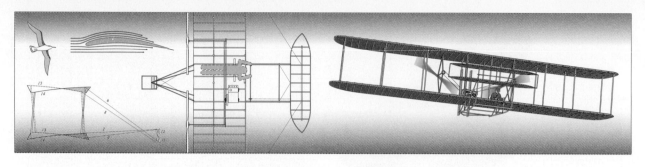

24

Reproduction of Drawings

24.1 Introduction

So far, we have discussed the preparation of drawings and specifications through the working-drawing stage where detailed drawings are completed on tracing film or paper. Now the drawings must be reproduced, folded, and prepared for transmittal to those who will use them to prepare bids or to fabricate the parts.

Several methods of reproduction are available to engineers and technologists for making copies of their drawings. However, most reproduction methods require strong, well-executed line work on the originals in order to produce good copies.

24.2 Computer Drawing Types

Three major types of computer reproduction are (1) pen plotting, (2) ink jet printing, and (3) laser printing.

Pen plotting is done by plotter with a single or a multiple ink pen holder with a fiber point that "draws" on the paper or film by

moving the pen in x and y directions. Multiple strokes of the pen will give various thicknesses of lines.

Ink jet printing is the process of spraying ink from tiny holes in a flat, disposable printhead onto the drawing surface as it passes through the printer. Prints can be obtained in color or black and white. Ink jet printers vary in size from 8 1/2 × 11 output (**Fig. 24.1**) to large engineering print sizes (**Fig. 24.2**).

Laser printing is an electrophotographic process that uses a laser beam to draw an image on a photosensitive drum where it is electrostatically charged to attract the toner. The electrostatically charged paper is rolled against the drum, the image is transferred, and toner fused to the paper by heat (**Fig. 24.3**). Laser printers make sharp drawings of the highest quality in color or black and white.

24.3 Types of Reproduction

Drawings made by a drafter are of little use in their original form. If original drawings were

24.1 The DesignJet 200 printer provides quiet high-speed operation and high print quality. Its large-format monochrome plotter prints five times faster than pen plotters and at about the same cost. (Courtesy of Hewlett-Packard Company.)

24.2 The DesignJet 350C large-format ink jet printer is excellent for printing full color architectural and engineering drawings with photographic quality. (Courtesy of Hewlett-Packard Company.)

24.3 The HP LaserJet 5Si MX printer has a speed of 24 pages per minute to accommodate more users and higher print volumes. (Courtesy of Hewlett-Packard Company.)

handled by checkers and by workers in the field or shop, they would quickly be soiled and damaged, and no copy would be available as a permanent record of the job. Therefore, the reproduction of drawings is necessary for making inexpensive, expendable copies for use by the people who need to use them.

The most often used processes of reproducing engineering drawings are (1) diazo printing, (2) microfilming, (3) xerography, and (4) photostating.

Diazo Printing

The **diazo print** more correctly is called a **whiteprint** or **blue-line print**, rather than a blueprint, because it has a white background and blue lines. Other colors of lines are avail-

24.4 A typical whiteprinter, which operates on the diazo process. (Courtesy of Blu-Ray, Incorporated, Essex, CT.)

able, depending on the type of diazo paper used. (Blueprinting, which creates a print with white lines and a blue background, is a wet process that is almost obsolete at the present.) **Figure 24.4** shows a typical diazo printer.

Diazo printing requires that original drawings be made on semitransparent tracing paper, cloth, or film that light can pass through, except where lines have been drawn. The diazo paper on which the blue-line print is copied is chemically treated, giving it a yellow tint on one side. Diazo paper must be stored away from heat and light to prevent spoilage.

The sequential steps of making a diazo print are shown in **Fig. 24.5**. The drawing is placed face up on the yellow side of the diazo paper and then fed through the diazo-process machine, which exposes the drawing to a built-in light. Light rays pass through the tracing paper and burn away the yellow tint on the diazo paper except where the drawing lines have shielded the paper from the light, a process similar to developing a photographic negative. (In order to make a good print, it is important that your lines be adequately dense to shield the diazo paper.) The exposed diazo paper becomes a duplicate of the original drawing, except that the lines are light yellow and are not permanent.

When the diazo paper is passed through the developing unit of the diazo machine, ammonia fumes develop the yellow lines on it

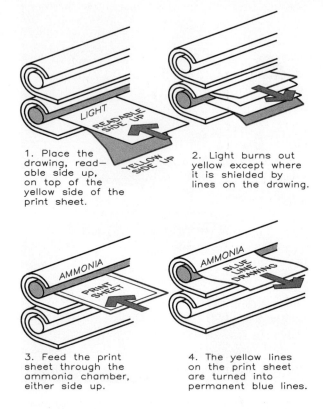

1. Place the drawing, readable side up, on top of the yellow side of the print sheet.

2. Light burns out yellow except where it is shielded by lines on the drawing.

3. Feed the print sheet through the ammonia chamber, either side up.

4. The yellow lines on the print sheet are turned into permanent blue lines.

24.5 Diazo (blue-line) prints are made by placing the original readable side up and on top of the yellow side of the diazo paper and feeding them under the light, as shown in the steps above.

into permanent blue lines. The speed at which the drawing passes under the light determines the darkness of the blue-line copy; the faster the speed, the darker the print is. A slow speed burns out more of the yellow and produces a clear white background, but some of the lighter lines of the drawing may be lost. Most diazo copies are made at a speed fast enough to give a light tint of blue in the background to obtain the darkest lines on the copy. Ink drawings, whether made by hand or by computer, give the best reproductions.

Diazo printing has been enhanced by the advent of computer drawings, which are made in ink and produce a much better print quality than pencil drawings. Also, computer drawings are more uniform in line weight, lettering, and technique than drawings made by hand by different drafters.

Microfilming

Microfilming is a photographic process that converts large drawings into film copies—either aperture cards or roll film. Drawings are placed on a copy table and photographed on either 16-mm or 35-mm film.

The roll film or aperture cards are placed in a microfilm enlarger-printer, where the individual drawings can be viewed on a built-in screen. The selected drawings can be printed from the film in standard sizes. Microfilm copies are usually made smaller than the original drawings to save paper and make the drawings easier to use.

Microfilming eliminates the need for large, bulky files of drawings because hundreds of drawings can be stored in permanent archives in miniature on a small amount of film. This is the same process used to preserve newspapers and other large materials by libraries and archives.

Xerography

Xerography is an electrostatic process of duplicating drawings on ordinary, unsensitized paper. Originally developed for business and clerical uses, xerography is currently used for the reproduction of engineering drawings. The xerographic process can also reduce the sizes of the drawings being copied to more convenient and easier to use sizes. The Xerox 2080 can reduce a 24 × 36 inch drawing to 8 × 10 inches.

Photostating

Photostating is a method of enlarging or reducing drawings photographically. The drawing is placed under the glass of the exposure table, which is lit by built-in lamps. The image appears on a glass plate inside the darkroom where it is exposed on photographically sensitive paper. The exposed negative

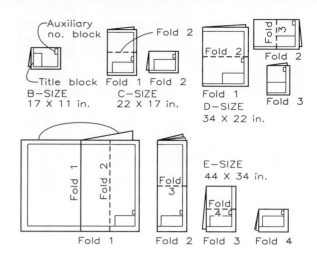

24.6 All standard drawing sheets can be folded to 8 1/2 × 11 inches for filing and storage.

paper is placed in contact with receiver paper, and the two are fed through the developing solution to obtain a photostatic copy. Photostating also can be used to make reproductions on transparent films and for reproducing halftones (photographs with tones of gray).

24.4 Assembling Drawing Sets

After the original drawings have been copied, they should be stored flat and unfolded in a flat file for future use and updating. Prints made from the originals, however, usually are folded or rolled for ease of transmittal from office to office. The methods of folding size B, C, D, and E sheets so that the image will appear on the outside of the fold are shown in **Fig. 24.6**. Drawings should be folded to show the title block always on the outside at the right, usually in the lower right-hand corner of the page (**Fig. 24.7**). The final size after folding is 8 1/2 × 11 inches (or 9 × 12 inches).

An alternative method of folding and stapling size B sheets is often used for student assignments so that they can be kept in a three-ring binder (**Fig. 24.8**). The basic rules of assembling drawings are listed in **Fig. 24.9**.

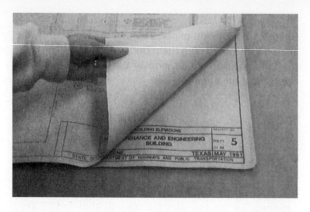

24.7 The title block should appear on the right, usually in the lower right-hand corner of the sheet.

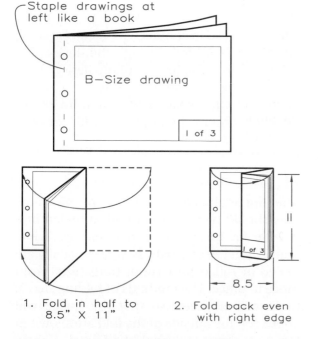

24.8 A set of size B drawings can be assembled by stapling, punching, and folding, as shown here, for safekeeping in a three-ring binder with the title block visible on top.

1. Staple along left edge, like a book. Use several staples, never just one.

2. Fold with drawing on outside.

3. Fold drawings as a set, not one at a time separately.

4. Fold to an 8.5"X 11" modular size.

5. The title block must be visible after folding.

6. Sheets of a set should be uniform in size.

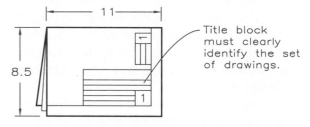

24.9 The basic rules for assembling sets of working drawing prints.

24.5 Transmittal of Drawings

Prints of drawings are delivered to contractors, manufacturers, fabricators, and others who must use the drawings for implementing the project. Prints are usually placed in standard 9 × 12-inch envelopes for delivery by hand or mail. Sets of large drawings, which may be 30 × 40 inches in size and contain four or more sheets, usually are rolled and sent in a mailing tube when folding becomes impractical. It is not uncommon for a set of drawings to have 40 or 50 sheets.

An advanced method of transmitting drawings is by use of large fax machines. Within minutes, large documents can be scanned and transmitted to their destination sites.

Computer drawings can be transmitted on disk by mailing them to their destination, where hard copies can be plotted and reproduced. This procedure offers substantial savings in shipping charges.

24.10 Hewlett Packard's OmniShare conferencer enables people in two locations to "meet" and collaborate on the same document, at the same time, over a single phone line. (Courtesy of Hewlett-Packard Company.)

24.11 The fast, high-quality output and paper-handling flexibility required of today's business user can be found in the HP LaserJet 6L printer (right). In addition, users can add copy, fax, file, and read capabilities through the LaserJet Companion printer accessory. (Courtesy of Hewlett-Packard Company.)

Computer drawings can also be transmitted electronically in the form of data that are downloaded at their destination. The downloaded data are then printed in the form of a drawing and maintained in the database of the computer. In the future, more drawings, documents, and photographs will be sent electronically as data and as scanned images over telephone wires, making them available instantaneously at the desired location.

Only a few years ago, transmission of information and data across the state or nation was time-consuming with the risk of loss. Today, any document can be transmitted overnight with certainty of delivery, and most can be transmitted to the receiver within minutes. The OmniShare conferencer (**Fig. 24.10**) lets people in two locations collaborate on the same document at the same time over a single phone line.

Hewlett-Packard's LaserJet printers have accessories available for fax, copy, file, and read capabilities. Today, the communication of engineering data can be done instantaneously and easily, contributing to an increased productivity (**Fig. 24.11**).

Numerically controlled manufacturing systems can be actuated directly from engineering data once the designs have been digitized. Such systems can be controlled from remote sites to produce products that previously required a high intensity of work hours by individuals. The future will hold many unique innovations in the manner in which business, manufacturing, and construction is done.

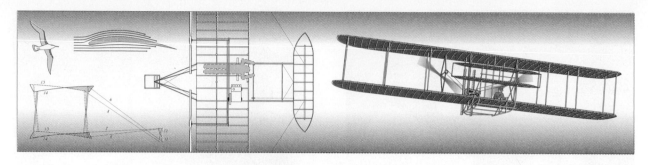

25

Three-Dimensional Pictorials

25.1 Introduction

A **three-dimensional pictorial** is a drawing that shows an object's three principal planes, much as they would be captured by a camera. This type of pictorial is an effective means of illustrating a part that is difficult to visualize when only orthographic views are given. Pictorials are especially helpful when a design is complex and when the reader of the drawings is unfamiliar with orthographic drawings.

Sometimes called **technical illustrations**, pictorials are widely used to describe products in catalogs, parts manuals, and maintenance publications (**Fig. 25.1**). The ability to sketch pictorials rapidly to explain a detail to an associate in the field is an important communication skill.

The commonly used types of pictorials are (1) obliques, (2) isometrics, (3) axono-metrics, and (4) perspectives (**Fig. 25.2**).

25.1 Many objects cannot be seen as well in real life as they can in a drawing, as shown in this pen set. (Courtesy of Keuffel & Esser Co.)

Oblique pictorials: Three-dimensional drawings made by projecting from the object with parallel projectors that are oblique to the picture plane (Fig. 25.2B).

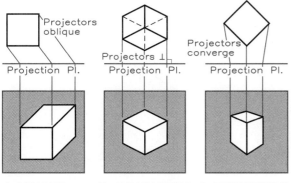

A. OBLIQUE   B. AXONOMETRIC   C. PERSPECTIVE

25.2 The three pictorial projection systems are: (A) oblique pictorials, with parallel projectors oblique to the projection plane; (B) axonometric (including isometric) pictorials, with parallel projectors perpendicular to the projection plane; and (C) perspectives, with converging projectors that make varying angles with the projection plane.

Isometric and axonometric pictorials: Three-dimensional drawings made by projecting from the object with parallel projectors that are perpendicular to the picture plane (Fig. 25.2A).

Perspective pictorials: Three-dimensional drawings made with projectors that converge at the viewer's eye and make varying angles with the picture plane (Fig. 25.2C).

25.2 Oblique Drawings

The pulley arm shown in **Fig. 25.3** is illustrated by orthographic views and an oblique pictorial. Because most parts are drawn before they are made, photographs cannot be taken; therefore the next best option is to draw a three-dimensional pictorial of the part. Details can usually be drawn with more clarity than can be shown in a photograph.

Oblique pictorials are easy to draw. If you can draw an orthographic view of a part, you are but one step away from drawing an oblique. For example, **Fig. 25.4**, shows that drawing a front view of a box twice and connecting its corners yields an oblique drawing.

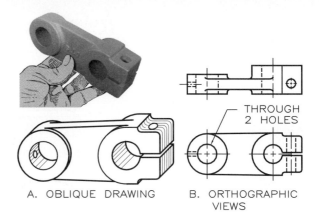

A. OBLIQUE DRAWING   B. ORTHOGRAPHIC VIEWS

25.3 The oblique drawing of this part makes it easier to visualize it than do its orthographic views.

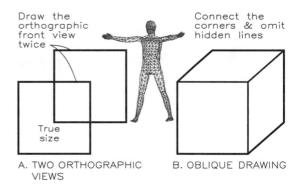

A. TWO ORTHOGRAPHIC VIEWS   B. OBLIQUE DRAWING

25.4 Draw two true-size surfaces of the box, connect them at the corners, to create an oblique drawing.

Thus, an oblique is no more than an orthographic view with a receding axis, drawn at an angle to show the depth of the object. An oblique is a pictorial that does not exist in reality (a camera cannot give an oblique). This type of pictorial is called an oblique because its parallel projectors from the object are oblique to the picture plane. The underlying principles of oblique projection are covered in Section 25.3.

Types of Obliques

The three basic types of oblique drawings are: (1) **cavalier**, (2) **cabinet**, and (3) **general** (**Fig. 25.5**). For each type, the angle of the receding axis with the horizontal can be at any angle

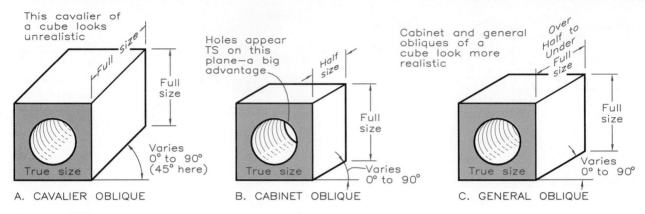

A. CAVALIER OBLIQUE

B. CABINET OBLIQUE

C. GENERAL OBLIQUE

25.5 The three types of obliques:

A The cavalier oblique has a receding axis at any angle and true-length measurements on the receding axis.

B The cabinet oblique has a receding axis at any angle and half-size measurements along the receding axis.

C The general oblique has a receding axis at any angle and measurements along the receding axis larger than half size and less than full size.

between 0° and 90°. Measurements along the receding axes of the cavalier oblique are laid off true length, and measurements along the receding axes of the cabinet oblique are laid off half size. The general oblique has measurements along the receding axes that are greater than half size and less than full size.

Figure 25.6 shows three examples of cavalier obliques of a cube. The receding axes for each is drawn at a different angle, but the receding axes are drawn true length. **Figure 25.7** compares cavalier with cabinet obliques.

Constructing Obliques

You can easily begin a cavalier oblique by drawing a box using the overall dimensions of height, width, and depth with light construction lines. As demonstrated in **Fig. 25.8**, first draw the front view as a true-size orthographic view. True measurements must be made parallel to the three axes and transferred from the orthographic views with your dividers. Then remove the notch from the blocked-in construction box to complete the oblique.

Angles

Angular measurements can be made on the true-size plane of an oblique, but not on the

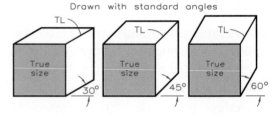

25.6 A cavalier oblique usually has its receding axis at one of the standard angles of drafting triangles. Each gives a different view of a cube.

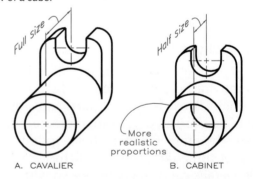

A. CAVALIER

B. CABINET

25.7 Measurements along the receding axis of a cavalier oblique are full size and those in a cabinet oblique are half size.

other two planes. Note in **Fig. 25.9** that a true angle can be measured on a true-size surface, but in **Fig. 25.10** it cannot because angles along receding planes are either smaller or larger than their true sizes. A better, easier-to-draw oblique is obtained when angles are drawn to appear true size.

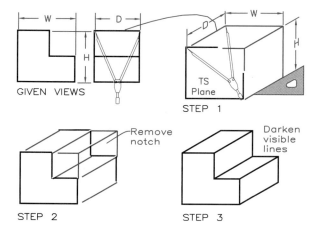

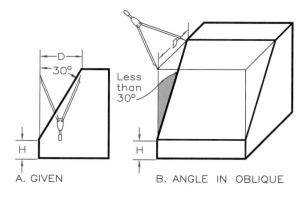

A. GIVEN   B. ANGLE IN OBLIQUE

25.10 Angles that do not lie in a true-size plane of an oblique must be located with coordinates.

25.8 Constructing a cavalier oblique.

Step 1 Draw the front surface of the object as a true-size plane. Draw the receding axis at a convenient angle and transfer the true distance D from the side view to it with your dividers.

Step 2 Draw the notch on the front plane and project it to the rear plane.

Step 3 Darken the lines to complete the drawing.

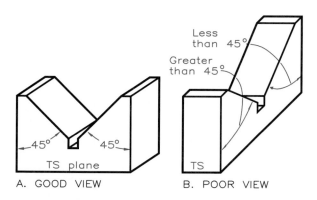

A. GOOD VIEW   B. POOR VIEW

25.9 Objects with angular features should be drawn in oblique so that the angles appear true size. This results in a better pictorial and one that is easier to draw.

To construct an angle in an oblique on one of the receding planes, you must use coordinates, as shown in Fig. 25.10. To find the surface that slopes 30° from the front surface, locate the vertex of the angle, H distance from the bottom. To find the upper end of the sloping plane, measure the distance D along the receding axis. Transfer H and D to the oblique with your dividers. The angle in the oblique is not equal to the 30° angle in the orthographic view.

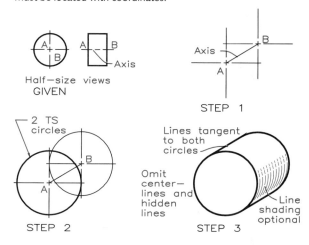

25.11 Drawing a cylinder in oblique.

Step 1 Draw axis AB and locate the centers of the circular ends of the cylinder at A and B. Because the axis is true length, this will be a cavalier oblique.

Step 2 Draw a true-size circle with its center at A by using a compass or computer-graphics techniques.

Step 3 Draw the other circular end with its center at B and connect the circles with tangent lines parallel to axis AB.

Cylinders

The **major advantage** of an oblique is that circular features can be drawn as true circles on its frontal plane (**Fig. 25.11**). Draw the centerlines of the circular end at A and construct the receding axis at the desired angle. Locate the end at B by measuring along the axis, draw true-size circles at each end at centers A and B, and draw tangents to both circles.

These same principles apply to construction of the object having semicircular features

25.2 OBLIQUE DRAWINGS • 441

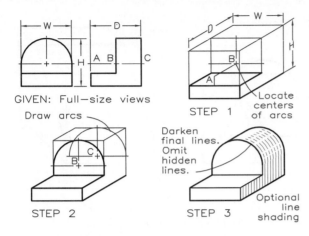

GIVEN: Full—size views

STEP 1 — Locate centers of arcs

STEP 2

Darken final lines. Omit hidden lines.

STEP 3 — Optional line shading

25.12 Drawing semicircular features in oblique.

Step 1 Block in the overall dimensions of the cavalier oblique with light construction lines, ignoring the semicircular feature.

Step 2 Locate centers B and C and draw arcs with a compass or by computer tangent to the sides of the construction boxes.

Step 3 Connect the arcs with lines tangent to each arc and parallel to axis BC; darken the lines.

shown in **Fig. 25.12**. Position the oblique so that the semicircular features are true size. Locate centers A, B, and C and the two semicircles. Then complete the cavalier oblique.

Circles

Circular features drawn as true circles on a true-size plane of an oblique pictorial appear on the receding planes as ellipses.

The four-center ellipse method is a technique of constructing an approximate ellipse with a compass and four centers (**Fig. 25.13**). The ellipse is tangent to the inside of a rhombus drawn with sides equal to the circle's diameter. Drawing the four arcs produces the ellipse.

The four-center ellipse method will not work for the cabinet or general oblique, but coordinates must be used. **Figure 25.14** illustrates the method of locating coordinates on the planes of cavalier and cabinet obliques. For the cabinet oblique, the coordinates along the receding axis are half size, and the coordinates along the horizontal axis (true-size axis)

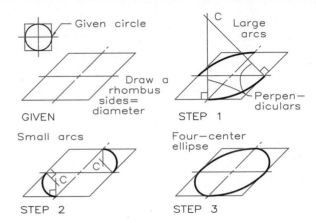

25.13 Constructing a four-center ellipse in oblique.

Given Block in the circle to be drawn in oblique with a square tangent to the circle. This square becomes a rhombus on the oblique plane.

Step 1 Draw construction lines perpendicular to the points of tangency to locate the centers for drawing two segments of the ellipse.

Step 2 Locate the centers for the two remaining arcs with perpendiculars drawn from adjacent tangent points.

Step 3 Draw the four arcs, which yield an approximate ellipse.

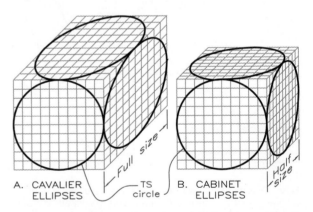

A. CAVALIER ELLIPSES — TS circle — Full size

B. CABINET ELLIPSES — Half size

25.14 Circular features on the faces of cavalier and cabinet obliques are compared here. Ellipses on the receding planes of cabinet obliques must be plotted by coordinates. The spacing of the coordinates along the receding axis of cabinet obliques is half size.

are full size. Draw the ellipse with an irregular curve or an ellipse template that approximates the plotted points.

Whenever possible, oblique drawings of objects with circular features should be positioned so that circles can be drawn as true circles instead of ellipses. The circular view in

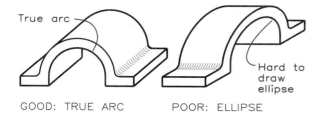

GOOD: TRUE ARC   POOR: ELLIPSE

25.15 An oblique should be positioned so that circular and curving features can be drawn most easily.

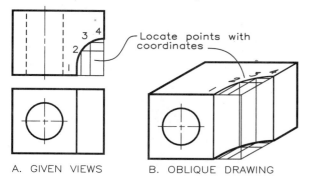

Locate points with coordinates

A. GIVEN VIEWS   B. OBLIQUE DRAWING

25.16 Coordinates are used to find points along irregular curves in oblique pictorials. Projecting the points downward a distance equal to the height of the object yields the lower curve.

Fig. 25.15 is better than the elliptical view in Fig. 25.15 because it gives a more descriptive view of the part and is easier to draw.

Curves

Irregular curves in oblique pictorials must be plotted point by point with coordinates (**Fig. 25.16**). Transfer the coordinates from the orthographic to the oblique view and draw the curve through the plotted points with an irregular curve. If the object has a uniform thickness, project the points for the lower curve vertically downward from the upper points a distance equal to the object's height.

To obtain the elliptical feature on the inclined surface shown in **Fig. 25.17**, use a series of coordinates to locate points along its curve. Connect the plotted points by using an irregular curve or ellipse template.

Sketching

Understanding the principles of oblique construction is essential for sketching obliques

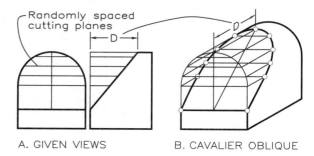

Randomly spaced cutting planes

A. GIVEN VIEWS   B. CAVALIER OBLIQUE

25.17 Construction of an elliptical feature on an inclined surface in oblique requires the use of three-dimensional coordinates to locate points on the curve.

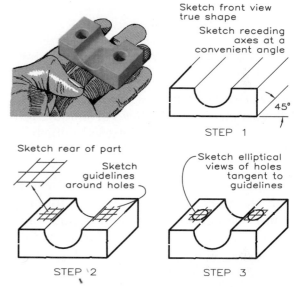

Sketch front view true shape

Sketch receding axes at a convenient angle

STEP 1

Sketch rear of part

Sketch guidelines around holes

Sketch elliptical views of holes tangent to guidelines

STEP 2   STEP 3

25.18 Sketching obliques.

Step 1 Sketch the front of the object as true-size surface and draw a receding axis from each corner.

Step 2 Lay off the depth, D, along the receding axes to locate the rear of the part. Lightly sketch pictorial boxes as guidelines for drawing the holes.

Step 3 Sketch the holes inside the boxes and darken all lines.

freehand. The sketch of the part shown in **Fig. 25.18** is based on the principles discussed, but its proportions were determined by eye instead of with scales and dividers.

Lightly drawn guidelines need not be erased when you darken the final lines. When sketching on tracing vellum, you can place a printed grid under the sheet to provide guidelines. Refer to Chapter 13 to review sketching techniques if needed.

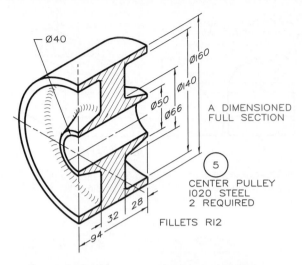

25.19 Oblique pictorials can be drawn as sections and dimensioned to serve as working drawings.

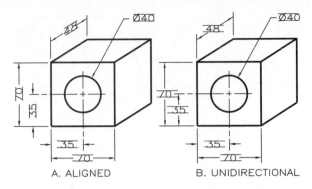

25.20 The two methods of lettering that are used to dimension obliques.

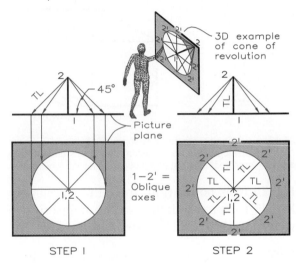

25.21 This drawing demonstrates the underlying principle of the cavalier oblique projection by using a series of projectors to form a cone.

Step 1 Each element from point 2 makes a 45° angle with the picture plane.

Step 2 The projected lengths of 1–2′ are equal in length to line 1–2, which is perpendicular to the picture plane. Thus, the receding axis of a cavalier oblique is true length and can be drawn at any angle.

Dimensioned Obliques

Dimensioned sectional views of obliques provide excellent, easily understood depictions of objects (**Fig. 25.19**). Apply numerals and lettering in oblique pictorials by using either the **aligned** method (with numerals aligned with the dimension lines) or the **unidirectional** method (with numerals positioned horizontally, regardless of the direction of the dimension lines), as shown in **Fig. 25.20**. Notes connected with leaders are positioned horizontally in both methods.

25.3 Oblique Projection Theory

Now that you have a general understanding of oblique pictorials, you should know the theory on which this system is based. The oblique projection shown in **Fig. 25.21** is the basis of oblique drawings. Receding axis 1–2 is perpendicular to the frontal projection plane. Projectors drawn from point 2 at 45° to the projection plane yield lengths on the front surface that are the same length as 1–2 (true length, in other words). Infinitely many 45° projectors form a cone of projectors with its apex at 2.

The true-length projections of lines 1–2′ represent receding axes that can be used for cavalier obliques, which by definition, have true-length dimensions along their receding axes. Do not use a vertical or a horizontal receding axis, but one between those limits.

To distinguish oblique projection from oblique drawing, as described in this chapter so far, observe the top and side views of a part

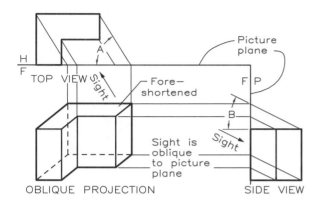

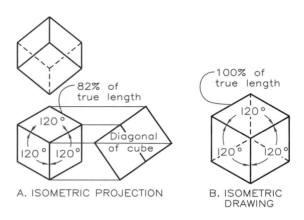

25.22 An oblique projection may be drawn at varying angles of sight. However, a line of sight making an angle of less than 45° with the picture plane would result in a receding axis longer than its true length, thereby distorting the pictorial.

25.24 Projection versus drawing.

A A true isometric projection is found by constructing a view in which the diagonal of a cube appears as a point and the axes are foreshortened.

B An isometric drawing is not a true projection because the dimensions are true size rather than foreshortened.

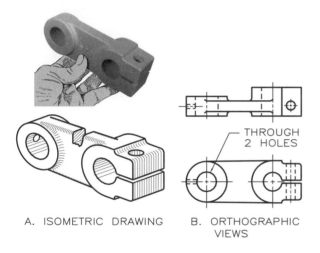

A. ISOMETRIC DRAWING   B. ORTHOGRAPHIC VIEWS

25.23 An isometric drawing gives a more realistic view of a part than an oblique drawing.

and the picture planes shown in **Fig. 25.22**. In an oblique projection, projectors from the top and side views are oblique to the edge views of the projection planes—hence, the name oblique.

Your line of sight can yield obliques with receding axes longer than true length (which should be avoided). Because of this shortcoming and the complexity of construction, oblique pictorials usually are **oblique drawings** rather than **oblique projections**.

25.4 Isometric Pictorials

In **Fig. 25.23**, the pulley arm is drawn in orthographic views and as a three-dimensional pictorial drawing. The pictorial is an isometric drawing in which the three planes of the object are equally foreshortened, representing the object more realistically than an oblique drawing can. With more realism comes more difficulty of construction. In particular, circles and curves do not appear true shape on any of the three isometric planes.

Isometric Projection Versus Drawing

In **isometric projection**, parallel projectors are perpendicular to the imaginary projection (picture) plane in which the diagonal of a cube appears as a point (**Fig. 25.24**). An isometric pictorial constructed by projection is called an isometric projection, with the three axes foreshortened to 82 percent of their true lengths and 120° apart. The name isometric, which means equal measurement, aptly describes this type of projection because the planes are equally foreshortened.

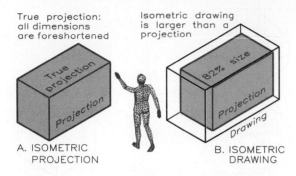

True projection: all dimensions are foreshortened

Isometric drawing is larger than a projection

True projection

Projection

82% size

Projection

Drawing

A. ISOMETRIC PROJECTION

B. ISOMETRIC DRAWING

25.25 The true isometric projection is foreshortened to 82 percent of full size. The isometric drawing is drawn full size for convenience.

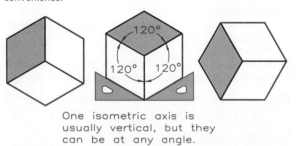

120°

120° 120°

One isometric axis is usually vertical, but they can be at any angle.

25.26 Isometric axes are spaced 120° apart, but they can be revolved into any position. Usually, one axis is vertical, but it can be at any angle with axis spacing remaining the same.

An **isometric drawing** is a convenient approximate isometric pictorial in which the measurements are shown full size along the three axes rather than at 82 percent, as in isometric projection (**Fig. 25.25**). Thus, the isometric drawing method allows you to measure true dimensions with standard scales and lay them off with dividers along the three axes. The only difference between the two is the larger size of the drawing. Consequently, isometric drawings are used much more often than isometric projections.

The axes of isometric drawings are separated by 120° (**Fig. 25.26**), but more often than not, one of the axes selected is vertical, since most objects have vertical lines. However, isometrics without a vertical axis are still isometrics.

25.5 Isometric Drawings

An isometric drawing is begun by drawing three axes 120° apart. Lines parallel to these

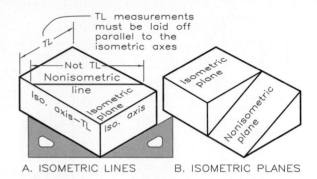

TL measurements must be laid off parallel to the isometric axes

TL

Not TL Nonisometric line

Iso. axis—TL

Isometric plane

Iso. axis

Isometric plane

Nonisometric plane

A. ISOMETRIC LINES

B. ISOMETRIC PLANES

25.27 Isometric lines and planes.

A Isometric lines (parallel to the three axes) give true measurements, but nonisometric lines do not.

B Here, the three isometric planes are equally foreshortened, and the nonisometric plane is inclined at an angle to one of the isometric planes.

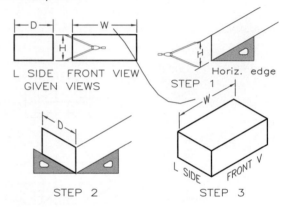

D

W

H

L SIDE FRONT VIEW
GIVEN VIEWS

H

Horiz. edge

STEP 1

D

W

L SIDE

FRONT V

STEP 2

STEP 3

25.28 Drawing an isometric pictorial of a box.

Step 1 Use a 30°–60° triangle and a horizontal straight-edge to construct a vertical line equal to the height, H, and draw two isometric lines through each end.

Step 2 Draw two 30° lines and locate the depth, D, by transferring depth from the given views with dividers.

Step 3 Locate the width, W, of the object, complete the surfaces of the isometric box, and darken the lines.

axes are called **isometric lines** (**Fig. 25.27A**). You can make true measurements along isometric lines but not along nonisometric lines. The three surfaces of a cube in an isometric drawing are called **isometric planes** (**Fig. 25.27B**). Planes parallel to those planes also are isometric planes.

To draw an isometric pictorial, you need a scale, dividers, and a 30°–60° triangle (**Fig. 25.28**). Begin by selecting the three axes and then constructing a plane of the isometric

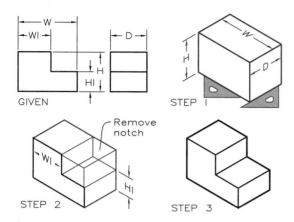

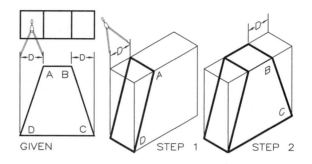

25.31 Use coordinates measured along the isometric axes to obtain inclined surfaces. Angular lines are not true length in isometric.

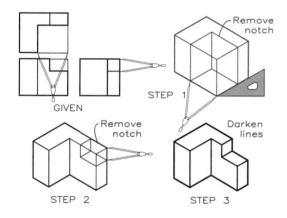

25.29 Constructing an isometric of a simple part.

Step 1 Construct an isometric drawing of a box with the overall dimensions W, D, and H from the given views.

Step 2 Locate the notch by transferring dimensions W1 and H1 from the given views with your dividers.

Step 3 Darken the lines to complete the drawing.

25.30 Laying out an isometric drawing.

Step 1 Use the overall dimensions given to block in the object with light lines and remove the large notch.

Step 2 Remove the small notch.

Step 3 Darken the lines to complete the drawing.

from the dimensions of height, H, and depth, D. Add the third dimension of width, W, and complete the isometric drawing.

Use light construction lines to block in all isometric drawings (**Fig. 25.29**) and the overall dimensions W, D, and H. Take other dimensions from the given views with dividers and measure along their isometric lines to locate notches in the blocked-in drawing.

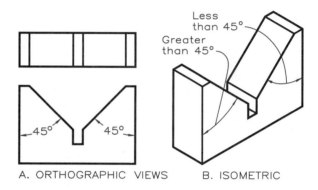

25.32 Angles in isometric may appear larger or smaller than they actually are.

Figure 25.30 shows an isometric drawing of a slightly more complex object, with two notches. The object was blocked in by using the H, W, and D dimensions. The notches in the block are removed to complete the drawing.

Angles

You cannot measure an angle's true size in an isometric drawing because the surfaces of an isometric are not true size. Instead, you must locate angles with isometric coordinates measured parallel to the axes (**Fig. 25.31**). Lines AD and BC are equal in length in the orthographic view, but they are shorter and longer than true length in the isometric drawing. **Figure 25.32** shows a similar situation, where two angles drawn in isometric are less than and greater than their true dimensions

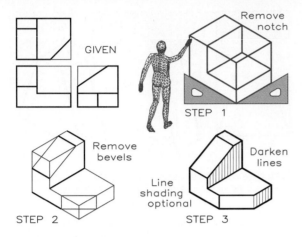

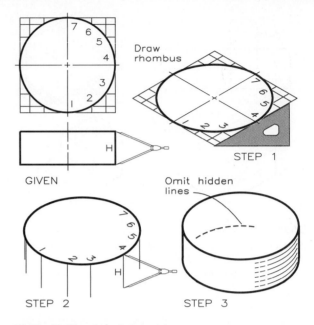

25.33 Drawing inclined planes in isometric.

Step 1 Block in the object with light lines, using the overall dimensions, and remove the notch.

Step 2 Locate the ends of the inclined planes by using measurements parallel to the isometric axes.

Step 3 Darken the lines to complete the drawing.

25.34 Plotting circles in isometric.

Step 1 Block in the circle by using its overall dimensions. Transfer the coordinates that locate points on the circle to the isometric plane and connect them with a smooth curve.

Step 2 Project each point a distance equal to the height of the cylinder to obtain the lower ellipse.

Step 3 Connect the two ellipses with tangent lines and darken all lines.

in the orthographic view. **Figure 25.33** shows how to construct an isometric drawing of an object with inclined surfaces. Blocking in the object with its overall dimensions with light construction lines is followed by removal of the inclined portions.

Circles

Three methods of constructing circles in isometric drawings are (1) point plotting, (2) four-center ellipse construction, and (3) ellipse template usage.

Point plotting is a method of using a series of coordinates to locate points on a circle in the given orthographic views. The coordinates are then transferred with dividers to the isometric drawing to locate the points on the ellipse one at a time (**Fig. 25.34**).

Block in the cylinder with light construction lines and show the centerlines as an isometric rhombus. Draw coordinates on the upper plane and use the height dimension to locate the points on the lower plane. Draw the ellipses with an irregular curve or an ellipse template.

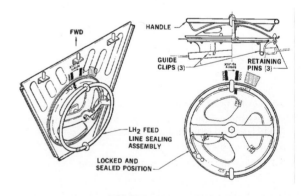

25.35 This handwheel assembly proposed for use in an orbital workshop is an example of parts with circular features drawn as ellipses in isometric. (Courtesy of NASA.)

A plotted ellipse is a true ellipse and is equivalent to a 35° ellipse drawn on an isometric plane. An example of a design composed of circular features drawn in isometric is the handwheel shown in **Fig. 25.35**.

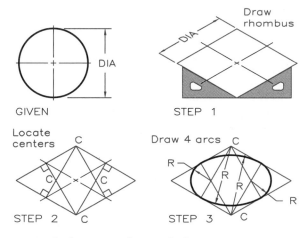

25.36 The four-center ellipse method.

Step 1 Use the diameter of the given circle to draw an isometric rhombus and the centerlines.

Step 2 Draw light construction lines perpendicularly from the midpoints of each side to locate four centers.

Step 3 Draw four arcs from the centers to represent an ellipse tangent to the rhombus.

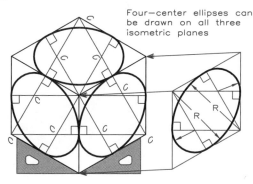

25.37 Four-center ellipses may be drawn on all three surfaces of an isometric drawing.

Four-center ellipse-construction is the method of producing an approximate ellipse (**Fig. 25.36**) by using four arcs drawn with a compass. Draw an isometric rhombus with its sides equal to the diameter of the circle to be represented. Find the four centers by constructing perpendiculars to the sides of the rhombus at the midpoints of each side, and draw the four arcs to complete the ellipse. You may draw four-center ellipses on all three isometric planes because each plane is equally foreshortened (**Fig. 25.37**). Although it is only

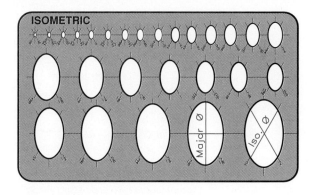

25.38 The isometric template (a 35° ellipse angle) is designed for drawing elliptical features in isometric. The isometric diameters of the ellipses are not major diameters of the ellipses but are diameters that are parallel to the isometric axes.

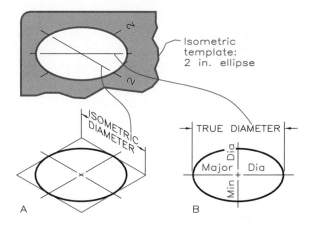

25.39 Ellipse terminology.

A Measure the diameter of a circle in isometric along the isometric axes. The major diameter of an isometric ellipse thus is larger than the measured diameter.

B The minor diameter is perpendicular to the true major diameter.

an approximate ellipse, the four-center ellipse technique is acceptable for drawing large ellipses and as a way to draw ellipses when an ellipse template is unavailable.

Isometric ellipse templates are specially designed for drawing ellipses in isometric (**Fig. 25.38**). The numerals on the templates represent the isometric diameters of the ellipses because diameters are measured parallel to the isometric axes of an isometric drawing (**Fig. 25.39**). Recall that the maximum diameter

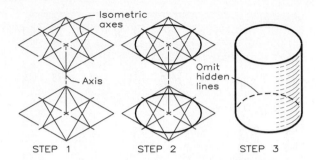

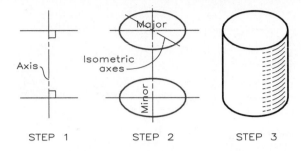

25.40 A cylinder drawn with the four-center method.

Step 1 Draw an isometric rhombus at each end of the cylinder's axis.

Step 2 Draw a four-center ellipse within each rhombus.

Step 3 Draw lines tangent to each rhombus to complete the drawing.

25.41 A cylinder drawn with the ellipse template method.

Step 1 Establish the length of the axis of the cylinder and draw perpendiculars at each end.

Step 2 Draw the elliptical ends by aligning the major diameter of the ellipse template with the perpendiculars at the ends of the axis. The isometric diameters of the isometric ellipse template will align with two isometric axes.

Step 3 Connect the ellipses with tangent lines to complete the drawing; omit hidden lines.

across the ellipse is its major diameter, which is a true diameter. Thus the size of the diameter marked on the template is less than the ellipse's major diameter. You may use the isometric ellipse template to draw an ellipse by constructing centerlines of the ellipse in isometric and aligning the ellipse template with those isometric lines (see Fig. 25.39).

Cylinders

A cylinder may be drawn in isometric by using the four-center ellipse method (**Fig. 25.40**). Use the isometric axes and centerline axis to construct a rhombus at each end of the cylinder. Then draw the ellipses at each end, connect them with tangent lines, and darken the lines to complete the drawing.

An easier way to draw a cylinder is to use an isometric ellipse template (**Fig. 25.41**). Draw the axis of the cylinder and construct perpendiculars at each end. Because the axis of a right cylinder is perpendicular to the major diameter of its elliptical ends, position the ellipse template with its major diameter perpendicular to the axis. Draw the ellipses at each end, connect them with tangent lines, and darken the visible lines to complete the drawing.

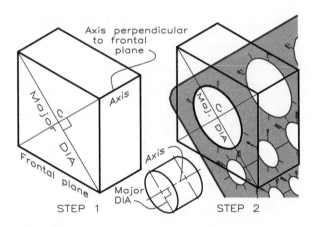

25.42 Constructing a cylindrical hole in a block.

Step 1 Locate the center of the hole on a face of the isometric drawing. Draw the axis of the cylinder from the center parallel to the isometric axis perpendicular to the plane of the circle. The major diameter is perpendicular to this axis.

Step 2 Use the 2-in. ellipse template to draw the ellipse by aligning guidelines on the template with the major and minor diameters drawn on the front surface.

To construct a cylindrical hole in a block (**Fig. 25.42**), begin by locating the center of the hole on the isometric plane. Draw the axis of the cylinder parallel to the isometric axis that is perpendicular to the plane of the ellipse through its center. Align the ellipse template

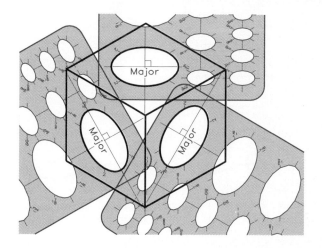

25.43 Position the isometric ellipse template as shown for drawing ellipses of various sizes on the three isometric planes.

with the major diameter, which makes a 90° angle with the cylindrical axis and complete the elliptical view of the cylindrical hole.

The isometric ellipse template can be used to draw ellipses on all three planes of an isometric drawing. On each plane, the major diameter is perpendicular to the isometric axis of the adjacent perpendicular plane. The isometric diameters marked on the template align with the isometric axes. All ellipses drawn on isometric planes must align in the directions shown in **Fig. 25.43**.

Rounded Corners

The rounded corners of an object can be drawn with an ellipse template (**Fig. 25.44**). Block in each corner with light construction lines, draw centerlines, draw the major diameter, and construct ellipses at each corner by positioning the template as shown. The rounded corners may also be constructed by using the four-center ellipse method (see Fig. 25.40) or by plotting points with coordinates (see Fig. 25.34).

A similar drawing involving the construction of ellipses is the conical shape shown in **Fig. 25.45**. Block in the ellipses on the upper and lower surfaces. Then draw the circular

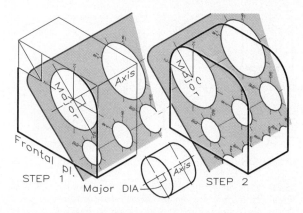

25.44 Drawing rounded corners.

Step 1 Draw the centerlines and isometric axes at the corners. Align the ellipse template with these guidelines and draw one-quarter of the ellipse.

Step 2 Draw the other elliptical corner in the same manner with the same size ellipse.

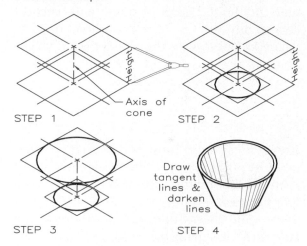

25.45 Constructing a cone in isometric.

Step 1 Draw the axis of the cone and block in the ellipses on the upper and lower surfaces.

Step 2 Block in the smaller end of the cone.

Step 3 Block in the larger end of the cone.

Step 4 Connect the ellipses with tangents, draw the cone's wall thickness, and darken the lines to complete the drawing.

features by using a template or the four-center ellipse method, and draw lines tangent to each ellipse.

Inclined Planes

Inclined planes in isometric may be located by coordinates, but they cannot be measured

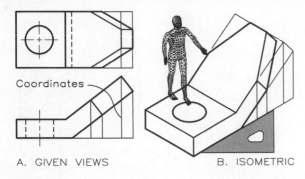

A. GIVEN VIEWS **B. ISOMETRIC**

25.46 Inclined surfaces in isometric must be located with three-dimensional coordinates parallel to the isometric axes. True angles cannot be measured in isometric drawings.

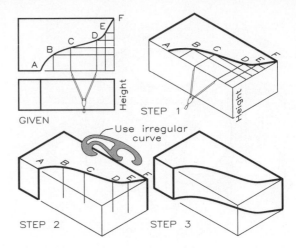

GIVEN **STEP 1**

Use irregular curve

STEP 2 **STEP 3**

25.47 Plotting irregular curves.

Step 1 Block in the shape by using the overall dimensions. Locate points on the irregular curve with coordinates transferred from the orthographic views.

Step 2 Project these points downward the height distance from the upper points to obtain the lower curve.

Step 3 Connect the points and darken the lines.

with a protractor because they do not appear true size. **Figure 25.46** illustrates the coordinate method. Use horizontal and vertical coordinates (in the x and y directions) to locate key points on the orthographic views. Transfer these coordinates to the isometric drawing with dividers to show the features of the inclined surface.

Curves

Irregular curves in isometric must be plotted point by point, with coordinates locating each point. Locate points A through F in the orthographic view with coordinates of width and depth (**Fig. 25.47**), then transfer them to the isometric view of the blocked-in part and connect them with an irregular curve.

Project points on the upper curve downward a distance of height, the height of the part, to locate points on the lower ellipse. Connect these points with an irregular curve and darken the lines to complete the isometric.

Ellipses on Nonisometric Planes

Ellipses on nonisometric planes in an isometric drawing, such as the one shown in **Fig. 25.48**, must be found by locating a series of points on the curve. Locate three-dimen-

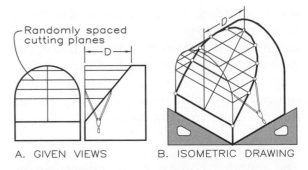

A. GIVEN VIEWS **B. ISOMETRIC DRAWING**

25.48 To construct ellipses on inclined planes, draw coordinates to locate points in the orthographic views, then transfer the three-dimensional coordinates to the isometric drawing and connect them with a smooth, irregular curve.

sional coordinates in the orthographic views and then transfer them to the isometric with your dividers. Connect the plotted points with an irregular curve or an ellipse template selected to approximate the plotted points. The more points you select, the more accurate will be the final ellipse. It will not be an isometric ellipse template, but one that fits the plotted points.

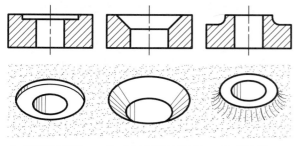

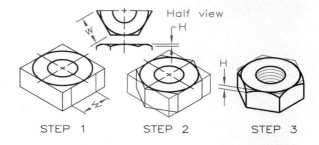

A. SPOTFACE   B. COUNTERSINK   C. BOSS

25.49 These examples of circular features in isometric may be drawn by using ellipse templates.

STEP 1   STEP 2   STEP 3

25.51 Drawing a hexagon-head nut.

Step 1 Use the overall dimensions of the nut to block in the nut.

Step 2 Construct the hexagonal sides at the top and bottom.

Step 3 Draw the chamfer with an irregular curve. Draw the threads to complete the drawing.

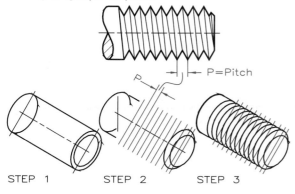

STEP 1   STEP 2   STEP 3

25.50 Threads in isometric.

Step 1 Using an ellipse template, draw the cylinder to be threaded.

Step 2 Lay off perpendiculars, spacing them apart at a distance equal to the pitch of the thread, P.

Step 3 Draw a series of ellipses to represent the threads. Draw the chamfered end by using an ellipse whose major diameter is equal to the root diameter of the threads.

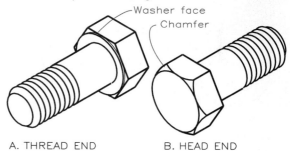

A. THREAD END   B. HEAD END

25.52 Isometric drawings of the lower and upper sides of a hexagon-head bolt.

25.6 Technical Illustration

Orthographic and isometric views of a spotface, countersink, and boss are shown in **Fig. 25.49**. These features may be drawn in isometric by point-by-point plotting of the circular features or by the four-center, or ellipse template method (the easiest method of the three).

A threaded shaft may be drawn in isometric, as shown in **Fig. 25.50**. First draw the cylinder in isometric. Draw the major diameters of the crest lines equally separated by distance P, the pitch of the thread. Then draw ellipses by aligning the major diameter of the

ellipse template with the perpendiculars to the cylinder's axis. Use a smaller ellipse at the end for the 45° chamfered end.

Figure 25.51 shows how to draw a hexagon-head nut with an ellipse template. Block in the nut and draw an ellipse tangent to the rhombus. Construct the hexagon by locating distance W across a flat parallel to the isometric axes. To find the other sides of the hexagon, draw lines tangent to the ellipse. Lay off distance H at each corner to establish the chamfers.

Figure 25.52 depicts a hexagon-head bolt in two positions. The washer face is on the lower side of the head, and the chamfer is on the upper side.

A portion of a sphere is drawn to represent a round-head screw in **Fig. 25.53**. Construct a hemisphere and locate the centerline of the

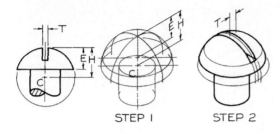

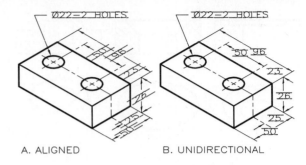

25.53 Drawing spherical features.

Step 1 Use an isometric ellipse template to draw the elliptical features of a round-head screw.

Step 2 Draw the slot in the head and darken the lines to complete the drawing.

25.55 Either of the techniques shown—aligned or unidirectional—is acceptable for placing dimensions on isometric drawings. Guidelines should always be used for lettering.

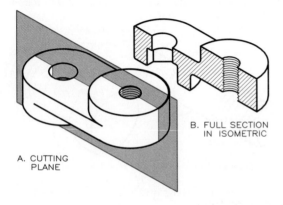

25.54 Isometric sections can be used to clarify the internal features of a part.

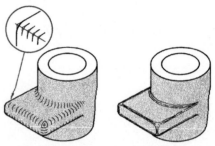

25.56 Two methods that can be used to representing fillets and rounds on the pictorial view of a part.

slot along one of the isometric planes. Measure the head's thickness, E, from the highest point on the sphere.

Sections

A full section drawn in isometric can clarify internal details that might otherwise be overlooked (**Fig. 25.54**). Half sections may also be used advantageously.

Dimensioned Isometrics

When you dimension isometric drawings, place numerals on the dimension lines, using either **aligned** or **unidirectional** numerals (**Fig. 25.55**). In both cases, notes connected with leaders usually are positioned horizontally, but drawing them to lie in an isometric plane is permissible. Always use guidelines for your lettering and numerals.

Fillets and Rounds

Fillets and rounds in isometric may be represented by either of the techniques shown in **Fig. 25.56** for added realism. The enlarged detail in the balloon shows how to draw intersecting guidelines equal in length to the radii of the fillets and rounds with arcs drawn tangent to them. These arcs may be drawn either freehand or with an ellipse template. The stipple shading was applied by using an adhesive overlay film.

When fillets and rounds of a dimensional part are shown in three-dimensional drawings, it is much easier to understand its features than it is when the part is represented by orthographic views (**Fig. 25.57**).

Assemblies

Assembly drawings illustrate how to put parts together. **Figure 25.58A** shows common mistakes in applying leaders and balloons to an

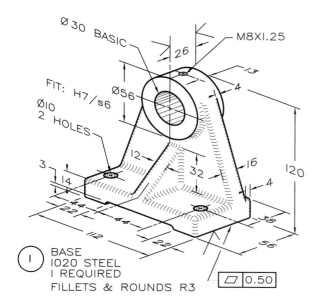

FIT: H7/s6
Ø30 BASIC
M8X1.25
26
13
Ø56
4
Ø10
2 HOLES
12
3
14
32
16
120
44
22
44
4
112
22
38
56

① BASE
1020 STEEL
1 REQUIRED
FILLETS & ROUNDS R3

▱ 0.50

25.57 This three-dimensional drawing has been drawn to show fillet and rounds, dimensions, and notes in order for it to be used as a working drawing.

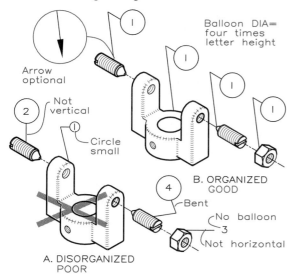

Balloon DIA= four times letter height

Arrow optional

Not vertical

Circle small

B. ORGANIZED GOOD

Bent

No balloon

Not horizontal

A. DISORGANIZED POOR

25.58 Common mistakes (A) in applying leaders and part numbers in balloons to an assembly, and acceptable techniques (B) of applying leaders and part numbers to an assembly.

assembly, and **Fig. 25.58B** shows the correct method of applying them. The numbers in the balloons correspond to the part numbers in the parts list. The exploded isometric assembly in **Fig. 25.59** illustrates the relationship of

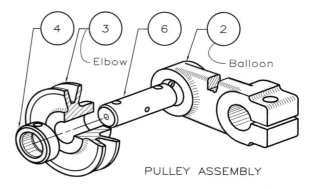

Elbow

Balloon

PULLEY ASSEMBLY

25.59 An exploded isometric assembly shows how parts are to be put together.

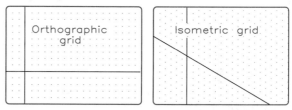

Orthographic grid

Isometric grid

A. ORTHOGRAPHIC GRID   B. ISOMETRIC GRID

25.60 The SNAP command permits you to use the orthographic grid (Standard) or the isometric grid option (I) for drawing isometric pictorials.

four mating parts. Illustrations of this type are excellent for inclusion in parts catalogs and maintenance manuals.

25.7 Isometrics by Computer

AutoCAD provides an ISOMETRIC grid for drawing isometrics. The STYLE option of the SNAP command allows changing the rectangular GRID, called STANDARD (S), to ISOMETRIC (I) with dots shown vertically and at 30° to the horizontal (**Fig. 25.60**). In this mode, you can make the cursor's cross hairs SNAP to the grid points and align with the axes of isometric drawings.

Isometric drawings drawn with this technique (**Fig. 25.61**) are not a true three-dimensional drawings system. Instead, they are two-dimensional isometrics that cannot be rotated to show other views.

The ISOCIRCLE option of the ELLIPSE command draws isometric ellipses automatically. When using this command, the cursor is

25.7 ISOMETRICS BY COMPUTER • 455

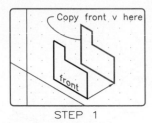 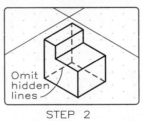

| STEP 1 | STEP 2 |

25.61 Producing isometrics by computer.

Step 1 Set the isometric grid on the screen (SNAP and I), and set SNAP to the grid. Draw the front view as an isometric and copy it to the backside with the COPY command.

Step 2 Connect the visible corner points and ERASE hidden lines to complete the drawing.

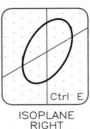 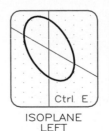 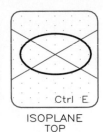

| ISOPLANE LEFT | ISOPLANE RIGHT | ISOPLANE TOP |

25.62 Use the ELLIPSE command and the ISOCIRCLE option to draw circles in isometric. By pressing Ctrl-E, you may alternatively rotate the isometric ellipses 120° to fit the three isometric planes.

aligned with each of the three isometric planes by pressing Ctrl-E on the keyboard (**Fig. 25.62**). When the cursor is aligned with the proper axes of an isometric plane, you may select the center of the isometric ellipse or the endpoints of its diameter (**Fig. 25.63**).

The ISOPLANE command changes the position of the cursor in the same way Ctrl-E does. ISOPLANE will prompt you to select from Left/Top/Right/[Toggle]: options. To use the Toggle option, press (Enter) to successively move the cursor position from plane to plane.

25.8 Axonometric Projection

An axonometric projection is a type of orthographic projection in which the pictorial view is projected perpendicularly onto the picture plane with parallel projectors. The object is positioned at an angle to the picture plane so that its pictorial projection will be a three-dimensional view. The three types of axonometric projections are (1) isometric, (2) dimetric, and (3) trimetric (**Fig. 25.64**).

Recall that the **isometric projection** is a pictorial in which the diagonal of a cube is seen as a point, the three axes and planes of the cube are equally foreshortened, and the axes are equally spaced 120° apart. Measure-

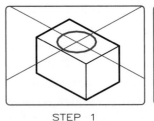 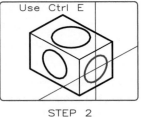

| STEP 1 | STEP 2 |

25.63 Isometric ellipses by computer.

Step 1 Use SNAP's Isometric-grid mode to draw isometric ellipses.
Command: ELLIPSE (Enter)
[Axis endpoint 1]:/Center/Isocircle: I (Enter)
Center of Circle: (Select with cursor.)
[Circle radius]/Diameter: (Select radius with cursor.)

Step 2 Change the orientation of the cursor for drawing isometric ellipses on the other two planes by pressing Ctrl-E. Repeat the process in Step 1.

ments along the three axes will be equal but less than true length because the isometric projection is true projection.

A **dimetric projection** is a pictorial in which two planes are equally foreshortened and two of the axes are separated by equal angles. Measurements along two axes of the cube are equal.

A **trimetric projection** is a pictorial in which all three planes are unequally foreshortened. The lengths of the axes are unequal, and the angles between them are different.

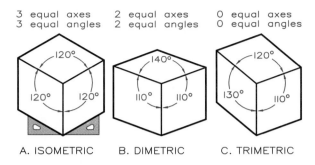

25.64   The three types of axonometric projection.

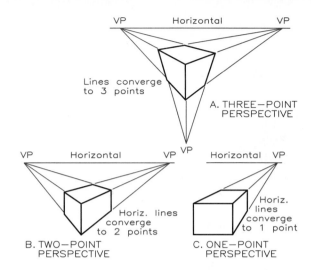

25.65   The three basic types of perspectives.

25.9 Perspective Pictorials

A perspective pictorial most closely resembles the view seen by the eye or camera and is the most realistic form of pictorial. In a perspective, parallel lines converge at vanishing points (VPs) as the lines recede from the observer. The three basic types of perspectives are (1) one point, (2) two point, and (3) three point, depending on the number of vanishing points used in their construction (**Fig. 25.65**).

One-point perspectives have one surface of the objective that is parallel to the picture plane, making it a true shape. The other sides vanish to a single vanishing point on the horizon.

Two-point perspectives are positioned with two sides at an angle to the picture plane, requiring two vanishing points. All horizontal lines converge at the vanishing points on the horizon, but vertical lines remain vertical and have no vanishing point.

Three-point perspectives have three vanishing points because the object is positioned so that all of its sides make an angle with the picture plane. Three-point perspectives are used for drawing large objects, such as buildings. They are the most realistic perspectives and the most complex to draw. Because of their complexity, we do not show how to construct them in this section.

Constructing One-Point Perspectives

Figure 25.66 shows how to draw a one-point perspective. It shows the top and side views of the object, picture plane, station point, horizon, and ground line. The picture plane (PP) appears as an edge in the top view and is the plane onto which the perspective is projected.

The station point (SP) is the location of the observer's eye in the top view and lies on the horizon in the front view. The horizon is a horizontal line in the front view that represents an infinite horizontal, such as the surface of the ocean, and it is aligned with the viewer's eye. The ground line (GL) is an infinite horizontal line parallel to the horizon from which vertical measurements are made.

Constructing Two-Point Perspectives

If two surfaces of an object are positioned at angles to the picture plane, two vanishing points are required to draw it as a perspective. Placing the horizon above the ground line and the height of the object in the front view yields an aerial view (**Fig. 25.67**). Placing the ground line and horizon on top of each other in the front view gives a ground-level view (worm's-eye view). Placing the horizon above the

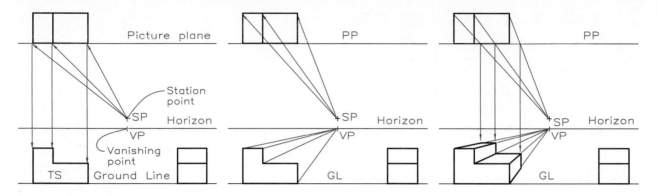

25.66 One-point perspective.

Step 1 Since the object is parallel to the picture plane, there will be only one vanishing point, located on the horizon below the station point. Projections from the top and side views establish the true-size front plane, which lies in the picture plane.

Step 2 Draw projectors from the station point to the rear points of the object in the top view and from the front view to the vanishing point on the horizon. In a one-point perspective, the vanishing point is the front view of the station point on the horizon.

Step 3 Construct vertical projectors from the top view to the front view from the points where the projectors cross the picture plane. These projectors intersect the lines that are drawn to the single vanishing point (VP), which is located the back side of the perspective.

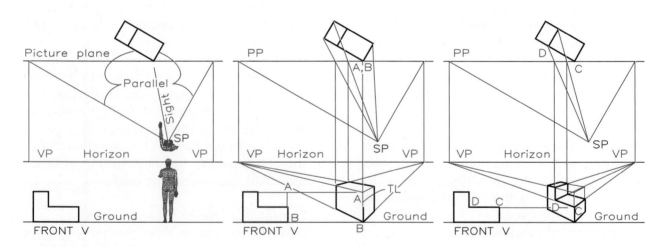

25.67 Two-point perspective.

Step 1 Extend projectors from the top view of the station point to the picture plane parallel to the forward edges of the object. Project these points vertically to the horizon in the front view to locate vanishing points. Draw the ground line below the horizon and construct the side view on the ground line.

Step 2 Lines in the picture plane are true length, so AB is true length. Project AB from the side view to determine its height. Project each end of AB to the vanishing points. Draw projectors from the station point to the exterior edges of the top view and project the intersections of these projectors with the picture plane to the front view.

Step 3 Find point C in the front view by projecting from the side view to AB. Draw a projector from point C to the left vanishing point. Point D lies on this projector beneath the point where a projector from the station point to the top view of point D crosses the picture plane. Draw the notch by projecting to the respective vanishing points.

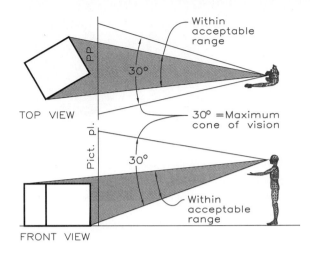

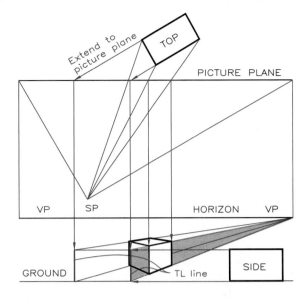

25.68 The station point should be far enough away from the object to permit the cone of vision to be less than 30° to reduce distortion.

25.69 This two-point perspective is of an object that does not come into contact with the picture plane.

ground line and through the object, usually at a person's height for large objects such as buildings, results in a general view.

Figure 25.67 shows how to construct a two-point perspective. Because line AB lies in the picture plane, it will be true length in the perspective. All height dimensions originate at this vertical line because it is the only true-length line of the object.

When drawing any perspective, you should position the station point far enough away from the object that the perspective can be contained in a cone of vision of 30° or less (**Fig. 25.68**). A larger cone of vision will distort the perspective.

The object shown in **Fig. 25.69** does not come into contact with the picture plane in the top view as it does in Fig. 25.67. To draw a perspective of this object, the planes of the object are extended to the picture plane. Measure the height on this line and draw an infinite plane to the right vanishing point. Locate the corner of the object on this infinite plane by projecting the object's right corner to the picture plane in the top view with a projector from the station point and then projecting this point downward to the infinite plane.

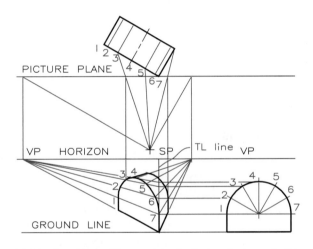

25.70 An object with a semicircular feature is drawn in this two-point perspective.

Arcs in Perspective

Draw arcs in perspective by using coordinates to locate points along the curves (**Fig. 25.70**). Transfer points 1–7 from the semicircular arc in the orthographic view to the perspective by projecting coordinates from the top and side

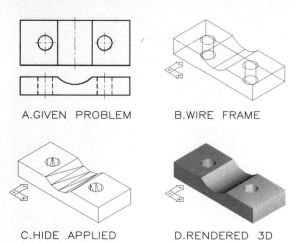

A.GIVEN PROBLEM   B.WIRE FRAME

C.HIDE APPLIED   D.RENDERED 3D

25.71 Modeling a simple part.

A Two orthographic views of the part are given.

B A wire-frame drawing of the part is made .

C The hidden lines are suppressed by HIDE to give a three-dimensional model.

D The model is rendered to give it a realistic look.

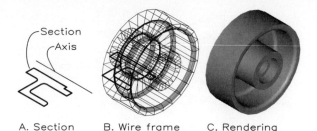

A. Section   B. Wire frame   C. Rendering

25.72 A model by revolution.

A A typical section of the pulley and its axis is drawn.

B The section is rotated about the axis to obtain a wire-frame drawing.

C The wire frame is rendered to obtain a realistic view of the pulley.

views. All heights are projected to the TL line from the orthographic view. These points form an egg-shaped oval, not a true ellipse. Connect the points with an irregular curve.

25.10 Three-Dimensional Modeling

You can draw objects with AutoCAD as true three-dimensional solids that can be rotated and viewed from any angle as if they were held in your hand. **Figure 27.71** illustrates a simple object represented by two orthographic views, a wire-frame drawing, a hidden-line wire-frame drawing, and a rendered solid. The capability to depict objects as rendered solids is a powerful design and communications tool.

Another example of a three-dimensional part that would be difficult to draw by hand is the pulley shown in **Fig. 25.72**. A typical section through the pulley and its axis are drawn, the section is revolved about the axis, and the wire-frame diagram is rendered. In addition to being able to select various views of the

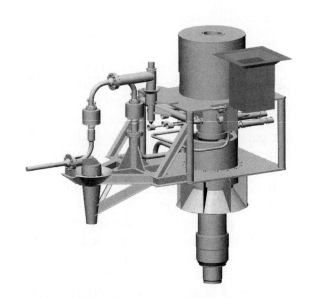

25.73 A rendered three-dimensional model of a moderately complex apparatus. (Courtesy of Cameron.)

pulley, different lighting combinations and materials can be applied to it in infinite combinations of effects.

An example of an industrial application is given in **Fig. 25.73**, which shows an apparatus of a higher degree of complexity that would be a rigorous assignment if drawn by hand. Although it is no easy chore to draw it as a series of solids by computer, the computer drawing makes it possible to obtain many dif-

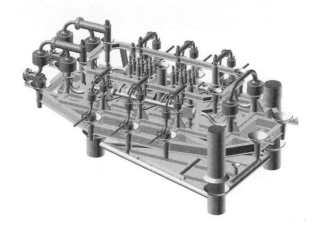

25.74 The apparatus in Fig. 25.73 is replicated a number of times in this equipment assembly used in subsea production. (Courtesy of Cameron.)

25.75 Human figures drawn by MANNEQUIN can be made to move and walk in orthographic, isometric, and perspective. (Courtesy of HUMANCAD.)

ferent views of the parts and to replicate drawings in combination. For example, the apparatus in **Fig. 25.73** is applied repetitively in the subsea production-equipment assembly in **Fig. 25.74**. The savings in time and effort becomes highly significant, and the final rendering greatly improves the understanding of the unit as a whole.

An introduction to three-dimensional modeling by several methods is given in Chapter 38. You will find that solid modeling begins with an understanding of the underlying fundamentals covered in this chapter. The ability to sketch three-dimensional drawings is an invaluable skill that you will use to develop and communicate design applications.

25.11 The Human Figure

An ultimate aspiration of the illustrator has always been the ability to represent the human form in a realistic manner. Several software packages have been developed that can be used with AutoCAD and other programs to do just that.

The package produced by MANNEQUIN provides a database of males and females—large and small, young and old—that can be placed on a drawing to determine how well a design fits the users of the product. This software enables the selected human figure to be walked across the screen in orthographic, isometric, or perspective (**Fig. 25.75**). Figures can be manipulated one finger at a time or one limb at a time. Also, the scope of vision of the mannequin can be shown on the screen to simulate what a person in that position would see.

The software Poser by Fractal Design offers many options for representing the human body—from stick figures to formally dressed figures (**Fig. 25.76**). Choices of body styles can be made from many categories, a few of which are age, sex, weight, and pose. Bodies can be positioned and controlled to fit almost any application. Figures can be rotated to obtain orthographic, axonometric, or perspective views of them (**Fig. 25.77**).

Clothing options range from casual to a formal business suit or dress, in the case of a

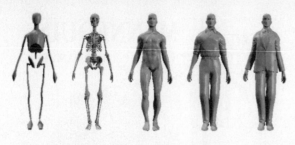

A.STICK B.SKELETON C.NUDE D.CASUAL E.FORMAL

25.76 Several of the rendering options that are available as part of POSER 2.

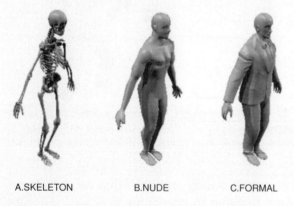

A.SKELETON   B.NUDE   C.FORMAL

25.77 With POSER 2, the human figure can be viewed as orthographic, axonometric, and perspective views.

female. Since all designs and projects are to fulfill the needs of people, it is important that design concepts reflect the human body in all stages of its development.

The Future
The future of three-dimensional graphics is truly exciting. What is available today for the microcomputer was not possible even on much larger and more expensive computers just a few years ago. The capabilities of 3-dimensional programs will continue to become more powerful and easier to use. Future graphics will include more solid modeling, animation, and sound effects. Get ready for an exciting trip!

Problems

Draw your solutions to the following problems (**Fig. 25.78**) on size A or B sheets, as assigned. Select an appropriate scale to take advantage of the space available on each sheet. By letting each square represent 0.20 in. (5 mm), you can draw two solutions on each size A sheet. By setting each square to 0.40 inch (10 mm), you can draw one solution on each size B sheet.

Oblique Pictorials
1–24. Construct cavalier, cabinet, or general obliques of the parts assigned.

Isometric Pictorials
1–24. Construct isometrics of the parts assigned.

Perspective Pictorials
1–24. On size B sheets, lay out perspective views of the parts assigned.

Summary
Graphic illustrations are important tools of the communication design process without which the transfer of ideas would be significantly hampered. Three-dimensional pictorials are the most powerful of all illustrations since they show objects in a real-world format that we are accustomed to seeing, much as shown by photographs. Three-dimensional pictorials do not require the viewer to mentally convert two-dimensional views into 3D as required by orthographic views.

Your ability to make 3D drawings, by hand or by computer, will make you more productive in engineering or any similar field where creativity and communication must be applied.

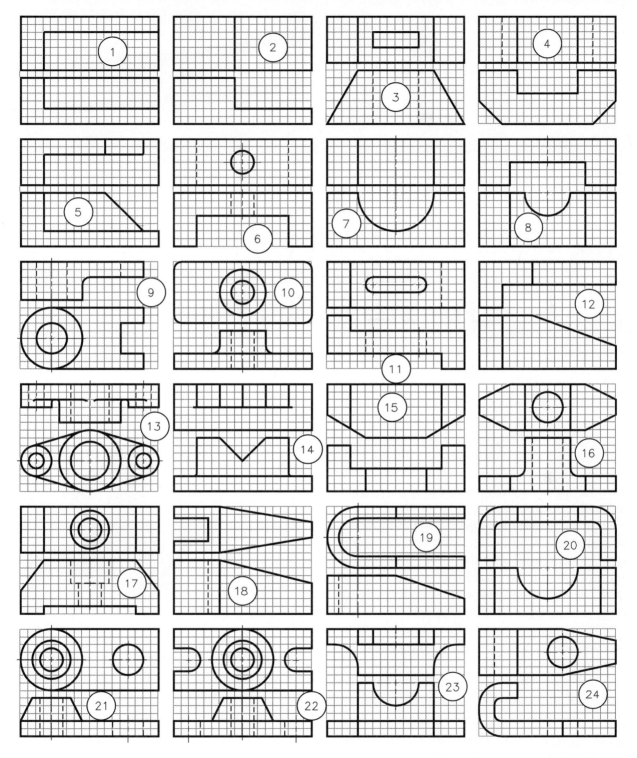

25.78 Problems 1–24.

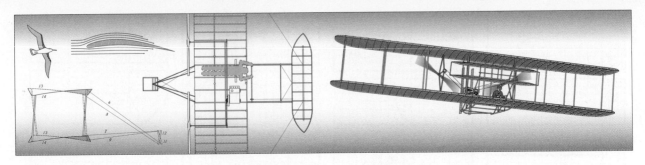

26

Points, Lines, and Planes

26.1 Introduction

Points, lines, and planes are the basic geometric elements used in three-dimensional (3D) spatial—also called descriptive—geometry. You need to understand how to locate and manipulate these elements in their simplest form because they will be applied to 3D spatial problems in Chapters 26 through 31.

The huge antenna in **Fig. 26.1** is composed of many points, lines, and planes that represent its structural members and shapes. Its geometry had to be established one point at a time with great precision in order for it to function and to be properly supported.

The labeling of points, lines, and planes is an essential part of 3D projection because it is the primary means of analyzing all spatial relationships. **Figure 26.2** illustrates the fundamental requirements for properly labeling these elements in a drawing:

Lettering: Use 1/8-inch letters with guidelines for labels; label lines at each end and planes at each corner with either letters or numbers.

26.1 It is easy to see the numerous applications of points, lines, and planes that were encountered by the team of designers who created this massive antenna.

Points: Mark with two short perpendicular dashes forming a cross, not a dot; each dash should be approximately 1/8 inch long.

Points on lines: Mark with a short perpendicular dash crossing the line, not a dot.

Reference lines: Label these thin, dark lines as described in Chapter 14.

464

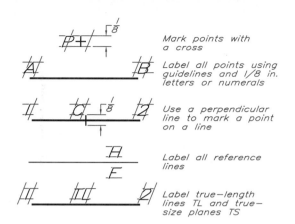

26.2 Standard practices for labeling points, lines, and planes.

Object lines: Draw these lines used to represent points, lines, and planes twice as thick as hidden lines with an F or HB pencil; draw hidden lines twice as thick as reference lines.

True-length lines: Label true length or TL.

True-size planes: Label true size or TS.

Projection lines: Draw precisely with a 2H or 4H pencil as thin lines, just dark enough to be visible so they need not be erased.

26.2 Projection of Points

A point is a theoretical location in space having no dimensions other than its location. However, a series of points establishes lengths, areas, and volumes of complex shapes.

A point must be located in at least two adjacent orthographic views to establish its position in 3D space (**Fig. 26.3**). When the planes of the projection box (Fig. 26.3A) are opened onto the plane of the drawing surface (Fig. 26.3C), the projectors from each view of point 2 are perpendicular to the reference lines between the views. Letters **H**, **F**, and **P** represent the **horizontal**, **frontal**, and **profile planes**, the three principal projection planes.

A point may be located from verbal descriptions with respect to the principal planes. For example, point 2 in Fig. 26.3 may be described as being (1) 5 units left of the

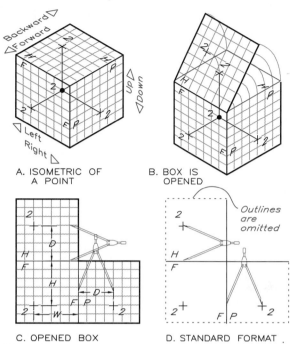

A. ISOMETRIC OF A POINT B. BOX IS OPENED

C. OPENED BOX D. STANDARD FORMAT

26.3 Three views of a point.

A The point is projected to three projection planes.

B The projection planes are opened into a single plane.

C In the opened box, point 2 is 5 units to the left of the profile, 5 units below the horizontal, and 4 units behind the frontal plane.

D The outlines of the projection planes are omitted in orthographic projection.

profile plane, (2) 5 units below the horizontal plane, and (3) 4 units behind the frontal plane.

When you look at the front view of the box, the horizontal and profile planes appear as edges. In the top view, the frontal and profile planes appear as edges. In the side view, the frontal and horizontal planes appear as edges.

26.3 Lines

A line is the straight path between two points in 3D space. A line may appear as foreshortened or true-length, or as a point (**Fig. 26.4**). Oblique lines are neither parallel nor perpendicular to a principal projection plane (**Fig. 26.5**). When line 1–2 is projected onto the

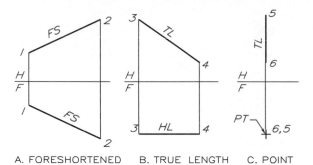

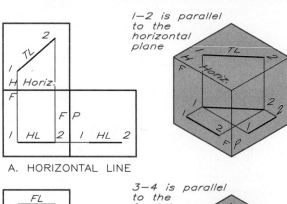

A. FORESHORTENED   B. TRUE LENGTH   C. POINT

26.4 A line in orthographic projection can appear as foreshortened (FS), or true length (TL), or as a point (PT).

A. HORIZONTAL LINE

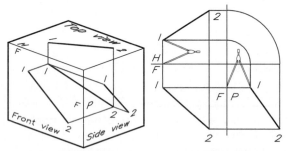

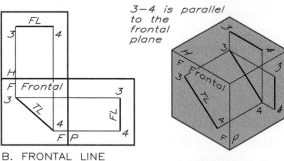

A. THREE—DIMENSIONAL VIEW   B. ORTHOGRAPHIC VIEWS

26.5 A line in space.

A Three views of a line, projected onto the three principal planes.

B Three standard orthographic views of a line.

horizontal, frontal, and profile planes, it appears foreshortened in each view.

Principal lines are parallel to at least one of the principal projection planes. A principal line is true length in the view where the principal plane to which it is parallel appears true size. The three types of principal lines are **horizontal**, **frontal**, and **profile lines**.

Figure 26.6A shows a horizontal line (HL) which appears true length in the horizontal (top) view. Any line shown in the top view will appear true length as long as it is parallel to the horizontal plane.

When looking at the top view, you cannot tell whether the line is horizontal. You must look at the front or side views to do so. In those views, an HL will be parallel to the edge view of the horizontal, the HF fold line (**Fig. 26.7**). A line

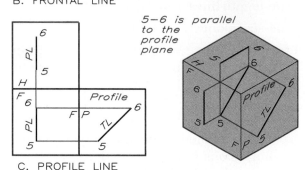

B. FRONTAL LINE

C. PROFILE LINE

26.6 Principal lines.

A A horizontal line is true length in the horizontal (top) view. It is parallel to the edge view of the horizontal plane in the front and side views.

B The frontal line is true length in the front view. It is parallel to the edge view of the frontal plane in the top and side views.

C The profile line is true length in the profile (side) view. It is parallel to the edge view of the profile plane in the top and front views.

that projects as a point in the front view is a combination horizontal and profile line.

A frontal line (FL) is parallel to the frontal projection plane. It appears true length in the front view because your line of sight is perpendicular to it in this view. In **Fig. 26.6B**, line 3–4 is an FL because it is parallel to the edge of the frontal plane in the top and side views.

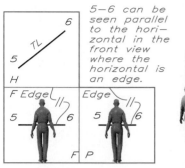
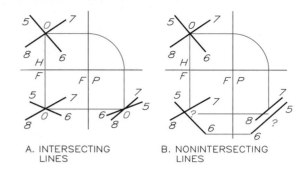

A. ORTHOGRAPHIC VIEWS   B. PICTORIAL VIEW

26.7 In order to determine that a line is horizontal, you must look at the front or side views in which the horizontal projection plane is an edge. Line 5–6 is seen parallel to the horizontal and is a horizontal line, too.

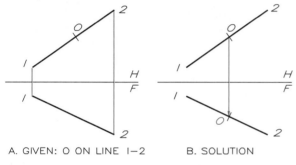

A. GIVEN: O ON LINE 1–2   B. SOLUTION

26.8 Point 0 on the top view of line 1–2 can be found in the front view by projection. The projector is perpendicular to the HF reference line between the views.

A profile line (PL) is parallel to the profile projection planes and appears true length in the side (profile) views. To tell whether a line is a PL, you must look at a view adjacent to the profile view, or the top or front view. In **Fig. 26.6C**, line 5–6 is parallel to the edge view of the profile plane in both the top and side views.

Locating a Point on a Line

Figure 26.8 shows the top and front views of a line 1–2 with point 0 located at its midpoint. To find the front view of the point, recall that, in orthographic projection, the projector between the views is perpendicular to the HF fold line. Use that projector to transfer point 0 to line 1–2 in the front view. A point located at a line's midpoint will be at the line's midpoint in all orthographic views of the line.

A. INTERSECTING   B. NONINTERSECTING
   LINES   LINES

26.9 Crossing lines.

A These lines intersect because 0, the point of intersection, projects as a common point of intersection in all views.

B The lines cross in the top and front views, but they do not intersect because there is no common point of intersection in all views.

Intersecting and Nonintersecting Lines

Lines that intersect have a common point of intersection lying on both lines. Point 0 in **Fig. 26.9A** is a point of intersection because it projects to a common crossing point in all three views. However, the crossing point of the lines in **Fig. 26.9B** in the top and front views is not a point of intersection. Point 0 does not project to a common crossing point in the top and front views, so the lines do not intersect; they simply cross, as shown in the profile view.

26.4 Visibility

Crossing Lines

In **Fig. 26.10**, nonintersecting lines AB and CD cross in certain views. Therefore, portions of the lines are visible or hidden at the crossing points (here, line thickness is exaggerated for purposes of illustration). Determining which line is above or in front of the other is referred to as finding a line's visibility, a requirement of many 3D problems.

You have to determine line visibility by analysis. For example, select a crossing point in the front view and project it to the top view to determine which line is in front of the other. Because the projector contacts line AB first, you know that line AB is in front of CD and is visible in the front view.

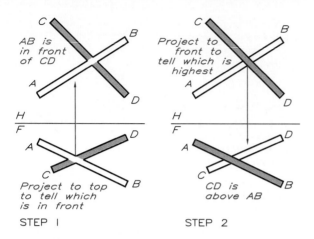

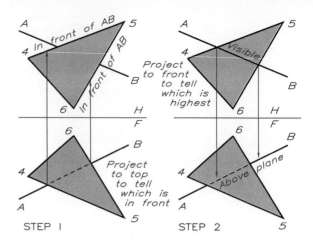

26.10 Determining visibility of lines.
Step 1 Project the crossing point from the front to the top view. This projector strikes line AB before it strikes line CD, indicating that line AB is in front and thus is visible in the front view.
Step 2 Project the crossing point from the top view to the front view. This projector strikes line CD before it strikes line AB, indicating that line CD is above line AB and thus is visible in the top view.

26.11 Determining visibility of a line and a plane.
Step 1 Project the points where line AB crosses the plane from the front view to the top view. These projectors intersect lines 4–6 and 5–6 of the plane first, indicating that the plane is in front of the line and making line AB hidden in the front view.
Step 2 Project the points where line AB crosses the plane in the top view to the front view. These projectors encounter line AB first, indicating that line AB is higher than the plane; thus, the line is visible in the top view.

Repeat this process by projecting downward from the intersection in the top view to find that line CD is above line AB and is visible in the top view. If only one view were available, visibility would be impossible to determine.

Lines and Planes

The principles of visibility analysis also apply to determining the visibility of a line and a plane (**Fig. 26.11**). First, project the intersections of line AB with lines 4–5 and 5–6 to the top view to determine that the lines of the plane (4–5 and 5–6) lie in front of line AB in the front view. Therefore, line AB is a hidden line in the front view.

Similarly, project the two intersections of line AB in the top view to the front view, where line AB is found to lie above lines 4–5 and 5–6 of the plane. Because line AB is above the plane, it is a visible line in the top view.

26.5 Planes

A plane may be represented in orthographic projection by any of the four combinations shown in **Fig. 26.12**. In orthographic projec-

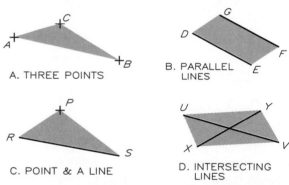

26.12 A plane can be represented as (A) three points not on a straight line, (B) two parallel lines, (C) a line and a point not on the line or its extension, and (D) two intersecting lines.

tion, a plane may appear as (1) an edge, (2) a true-size plane, or (3) a foreshortened plane (**Fig. 26.13**).

Oblique planes (the general case) are not parallel to principal projection planes in any view (**Fig. 26.14**). Principal planes are parallel to principal projection planes (**Fig. 26.15**). The three types of principal planes are horizontal, frontal, and profile planes.

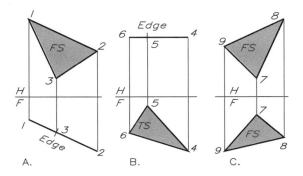

A. B. C.

26.13 A plane in orthographic projection can appear as (A) an edge, (B) a true-size (TS) plane, or (C) a foreshortened (FS) plane. A plane that is foreshortened in all principal views is an oblique plane.

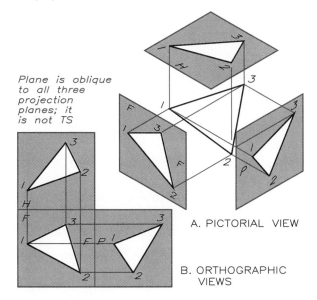

Plane is oblique to all three projection planes; it is not TS

A. PICTORIAL VIEW

B. ORTHOGRAPHIC VIEWS

26.14 An oblique plane is neither parallel nor perpendicular to a projection plane. It is the general-case plane.

A horizontal plane is parallel to the horizontal projection plane and is true size in the top view (Fig. 26.15A). To determine that the plane is horizontal, you must observe the front or profile views, where you can see its parallelism to the edge view of the horizontal plane.

A frontal plane is parallel to the frontal projection plane and appears true size in the front view (Fig. 26.15B). To determine that the plane is frontal, you must look at the top or profile views, where you can see its parallelism to the edge view of the frontal plane.

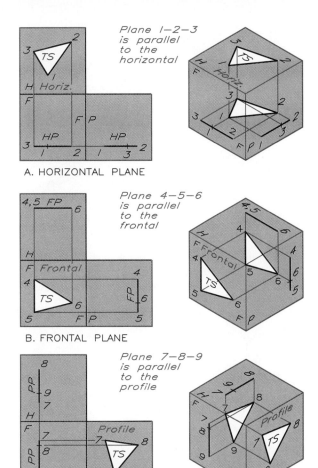

Plane 1-2-3 is parallel to the horizontal

A. HORIZONTAL PLANE

Plane 4-5-6 is parallel to the frontal

B. FRONTAL PLANE

Plane 7-8-9 is parallel to the profile

C. PROFILE PLANE

26.15 Principal planes.

A The horizontal plane is true size in the horizontal (top) view. It is parallel to the edge view of the horizontal plane in the front and profile views.

B The frontal plane is true size in the front view. It is parallel to the edge view of the frontal plane in the top and profile views.

C The profile plane is true size in the profile view. It is parallel to the edge view of the profile plane in the top and front views.

A profile plane is parallel to the profile projection plane and is true size in the profile view (Fig. 26.15C). To determine that the plane is profile, you must observe the top or front views, where you can see its parallelism to the edge view of the profile plane.

Point on a Plane

Point 0 on the front view of plane 4–5–6 in **Fig. 26.16** is to be located on the plane in the top

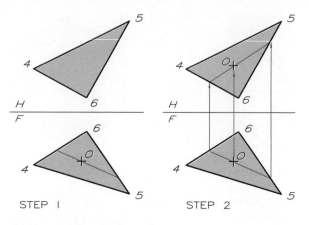

26.16 Locating a point on a plane.

Step 1 In the front view, draw a line through point 0 in any convenient direction except vertical.

Step 2 Project the ends of the line to the top view and draw the line. Project point 0 to this line.

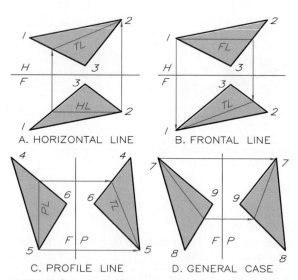

26.17 Finding principal lines on a plane.

A Draw a horizontal line in the front view parallel to the edge view of the horizontal plane, then project it to the top view, where it is true length.

B Draw a frontal line in the top view parallel to the edge view of the frontal plane, then project it to the front view, where it is true length.

C Draw a profile line in the front view parallel to the edge view of the profile plane, then project it to the profile view, where it is true length.

D A general-case line is not parallel to the frontal, horizontal, or profile planes and is not true length in any principal view.

view. First, draw a line in any direction (except vertical) through the point to establish a line on the plane, then project this line to the top view. Project point 0 from the front view to the top view of the line.

Principal Lines on a Plane

Principal lines may be found in any view of a plane when at least two orthographic views of the plane are given. Any number of principal lines can be drawn on any plane.

Figure 26.17A shows a horizontal line parallel to the edge view of the horizontal projection plane in the front view. When projected to the top view, this line is true length.

Figure 26.17B shows a frontal line parallel to the edge view of the frontal projection plane in the top view. When projected to the front view, this line is true length.

Figure 26.17C shows a profile line parallel to the edge view of the profile projection plane in the front. When projected to the profile view, this line is true length.

In the general case (oblique plane), a line is not parallel to the edge view of any principal projection plane (**Fig. 26.17D**). Therefore, it is not true length in any principal view.

26.6 Parallelism

Lines

Two parallel lines appear parallel in all views, except in views where both appear as points. Parallelism of lines in 3D space cannot be determined without at least two adjacent orthographic views. In **Fig. 26.18**, line AB was drawn parallel to the horizontal view of line 3–4 and through point 0, which is the midpoint. Projecting points A and B to the front view with projectors perpendicular to the HF reference plane yields the length of line AB, which is parallel to line 3–4 through point 0.

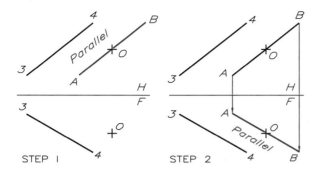

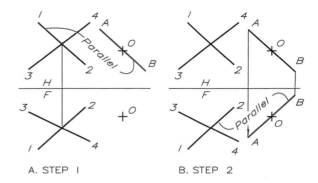

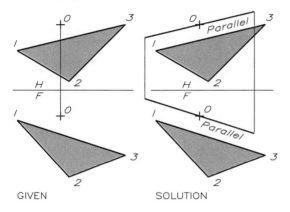

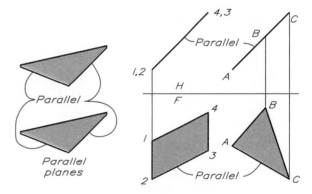

26.18 Constructing a line parallel to a line through 0.

Step 1 Draw line AB parallel to the top view of line 3–4, with its midpoint at 0.

Step 2 Draw the front view of line AB parallel to the front view of 3–4 through point 0.

26.20 Constructing a line parallel to plane.

Step 1 Draw line AB parallel to line 1–2 through point 0.

Step 2 Draw line AB parallel to the same line, line 1–2, in the front view, which makes line AB parallel to the plane.

26.19 A line may be drawn through point 0 parallel to plane 1–2–3 if the line is parallel to any line in the plane. Draw line AB parallel to line 1–3 of the plane in the front and top views, making it parallel to the plane.

26.21 Two planes are parallel when intersecting lines in one are parallel to intersecting lines in the other. When parallel planes appear as edges, their edges are parallel.

Line and Plane

A line is parallel to a plane when it is parallel to any line in the plane. In **Fig. 26.19**, a line with its midpoint at point 0 is to be drawn parallel to plane 1–2–3. In this case, line AB was drawn parallel to a line in the plane, or line 1–3, in the top and front views. The line could have been drawn parallel to any line in the plane, making infinite solutions possible.

Figure 26.20 shows a similar example. Here, a line parallel to the plane, with its midpoint at 0, was drawn. In this case, the plane is represented by two intersecting lines instead of an outlined area.

Planes

Two planes are parallel when intersecting lines in one plane are parallel to intersecting lines in the other (**Fig. 26.21**). Determining whether planes are parallel is easy when both appear as edges in a view.

In **Fig. 26.22**, a plane is to be drawn through point 0 parallel to plane 1–2–3. First, draw line EF through point 0 parallel to line 1–2 in the top and front views, then draw a second line through point 0 parallel to line 2–3 of the plane in the front and top views. These two intersecting lines form a plane parallel to plane 1–2–3, as intersecting lines on one plane are parallel to intersecting lines on the other.

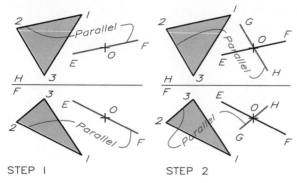

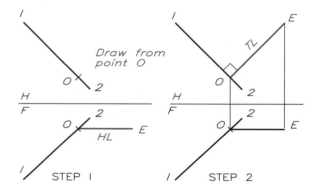

STEP 1   STEP 2

26.22 Constructing a plane through a point parallel to a plane.

Step 1 Draw line EF parallel to any line in the plane (line 1–2 in this case). Show the line in both views.

Step 2 Draw a second line parallel to line 2–3 in the top and front views. These intersecting lines passing through 0 represent a plane parallel to 1–2–3.

26.24 A line perpendicular to a principal line through 0.

Step 1 Line 5–6 is a frontal line and is true length in the front view, so a perpendicular from point 0 makes a true 90° angle with it in the front view.

Step 2 Project point P to the top view and connect it to point 0. As neither line is true length in the top view, they do not intersect at 90°.

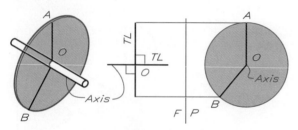

26.23 Perpendicular lines have a true angle of 90° between them in a view where one or both of them appear true length.

26.7 Perpendicularity

Lines

When two lines are perpendicular, draw them with a true 90° angle of intersection in views where one or both of them appears true length (**Fig. 26.23**). In a view where neither of two perpendicular lines is true length, the angle between is not a true 90° angle.

In Fig. 26.23, the axis is true length in the front view; therefore, any spoke of the circular wheel is perpendicular to the axis in the front view. Spokes OA and OB are examples of true length foreshortened axes, respectively, in the front view.

Line Perpendicular to a Principal Line In **Fig. 26.24**, a line is to be constructed through point 0 perpendicular to frontal line 5–6,

26.25 A line perpendicular to an oblique line through 0.

Step 1 Draw a horizontal line (OE) from 0 in the front view.

Step 2 Horizontal line OE is true length in the top view, so draw it perpendicular to line 1–2 in this view.

which is true length in the front view. First, draw OP perpendicular to line 5–6 because it is true length, then project point P to the top view of line 5–6. In the top view, line OP is not perpendicular to line 5–6 because neither of the lines is true length in this view.

Line Perpendicular to an Oblique Line In **Fig. 26.25**, a line is to be constructed from point 0 perpendicular to oblique line 1–2. First, draw a horizontal line from 0 to some convenient length in the front view, say, to E. Locate point 0 in the top view by projection and draw line OE to make a 90° angle with the top view of

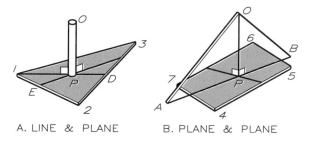

A. LINE & PLANE   B. PLANE & PLANE

26.26 Perpendicularity: lines and planes.

A A line is perpendicular to a plane when it is perpendicular to two intersecting lines on the plane.

B A plane is perpendicular to another plane if it contains a line that is perpendicular to the other plane.

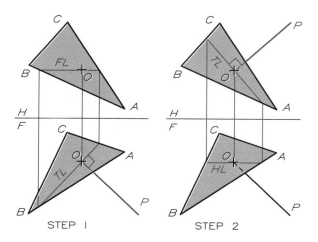

STEP 1   STEP 2

26.27 Constructing a line perpendicular to a plane.

Step 1 Construct a frontal line on the plane through 0 in the top view. This line is true length in the front view, so draw line OP perpendicular to this true-length line.

Step 2 Construct a horizontal line through point 0 in the front view. This line is true length in the top view; draw line OP perpendicular to it.

1–2. Line OE is true length in the top view, so it makes a true 90° angle with line 1–2.

Planes

A line is perpendicular to a plane when it is perpendicular to any two intersecting lines in the plane (**Fig. 26.26A**). A plane is perpendicular to another plane when a line in one plane is perpendicular to the other plane (**Fig. 26.26B**).

Line Perpendicular to a Plane

In **Fig. 26.27**, a line is to be drawn perpendicular to the plane from point 0 on the plane. First, draw a frontal line on the plane in the top view through 0. Project the line to the front view, where it is true length. Draw line OP at a convenient length perpendicular to the true-length line. Draw a horizontal line through point 0 in the front view and project it to the top view of the plane, perpendicular to the true-length line. This construction results in a line perpendicular to the plane because the line is perpendicular to two intersecting lines, a horizontal and a frontal line, in the plane.

Problems

Use size A sheets for the following problems and lay out your solutions with instruments. Each square on the grid is equal to 0.20 in. (5 mm). Use either grid paper or plain paper. Label all reference planes and points in each problem with 1/8-in. letters and numbers using guidelines.

1. (Sheet 1)

(A–D) Draw three views (top, front, and right-side views) of the given point.

(E–F) Draw the three views of the points and connect them to form lines.

2. (Sheet 2)

(A–C) Draw three views (top, front, and right-side views) of the partially drawn lines.

(D–F) Draw the missing views of the lines so that 7–8 is a horizontal line, 1–2 is a frontal line, and 3–4 is a profile line.

3. (Sheet 3)

(A–B) Draw the right-side view of line 1–2 and plane 3–4–5.

(C–E) Draw the missing views of the planes so that 6–7–8 is a frontal plane, 1–2–3 is a horizontal plane, and 4–5–6 is a profile plane.

PROBLEMS • 473

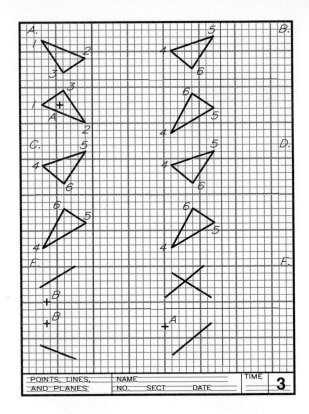

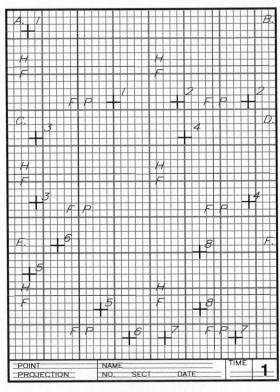

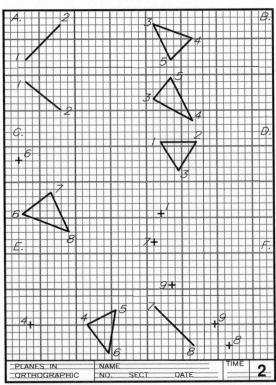

(F) Complete the top and side views of the plane that appears as an edge in the front view.

Problems

4. (Sheet 4)

(A–B) Draw 1.50-in. lines that pass through point O and are parallel to their respective planes.

(C–D) Through point O draw the top and front views of lines that are perpendicular to their respective planes.

5. (Sheet 5)

(A–E) The missing views of the planes are to be drawn as three-view projections with top, front, and right-side views. The missing views are to be drawn in the areas of the question marks.

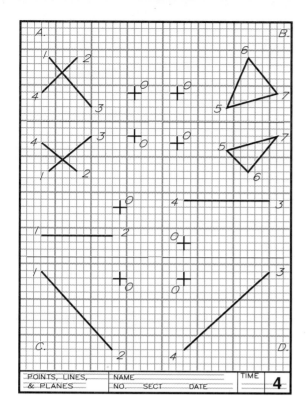

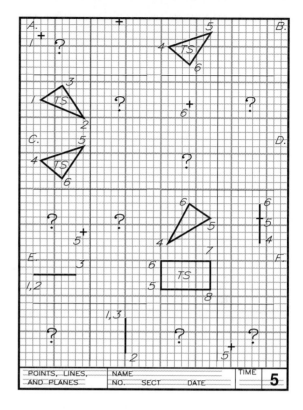

Review Questions

1. How many principal lines are there, and what are their names?

2. What are the ways in which a line can appear on a drawing?

3. The three orthographic dimensions are height, width, and depth. Which of these dimensions are necessary to locate a line in the side view? The front view? The top view?

4. What are the three ways in which a plane can appear on a drawing?

5. When is a line perpendicular to a plane?

6. When is a line parallel to a plane?

7. When can two perpendicular lines be drawn as perpendicular on a drawing?

8. If a plane appears as an edge in the front view, what can be said about its top view?

9. A line that is vertical appears how in the top view? The side view?

10. A line that is true length in the top view is what type of principal line? How will it appear in the front view?

11. A line that is perpendicular to a plane in the top view makes what angle with a frontal line on the plane?

12. How many principal lines can be drawn on any view of a plane?

13. A plane that is vertical is true size in which views?

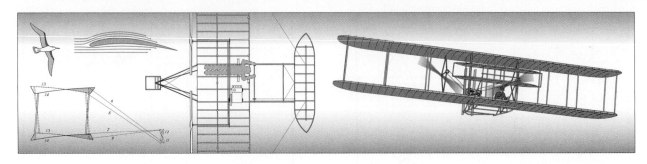

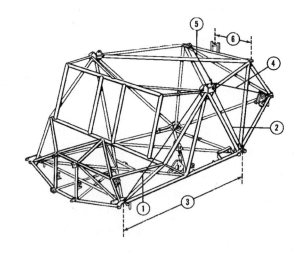

27

Primary Auxiliary Views in Descriptive Geometry

27.1 Introduction

Descriptive geometry is the projection of three-dimensional (3D) orthographic views onto a two-dimensional (2D) plane of paper to allow graphical determination of lengths, angles, shapes, and other geometric information. Orthographic projection is the basis for laying out and solving problems by descriptive geometry.

The primary auxiliary view, which permits analysis of 3D geometry, is essential to descriptive geometry. For example, the design of the helicopter frame shown in **Fig. 27.1** contains many complex geometric elements (lines, angles, and surfaces) that were analyzed by descriptive geometry prior to its fabrication.

27.2 Geometry by Computer

Five useful computer routines for solving descriptive geometry problems are covered in

27.1 The designers of this helicopter frame used the principles of descriptive geometry to determine lengths, angles, and areas. (Courtesy of Bell Helicopter.)

this section and documented in Appendix 41. The commands—PERPLINE, PARALLEL, TRANSFER, COPYDIST, and BISECT*—are from a LISP program called ACAD, and they must be typed and saved in the SUPPORT

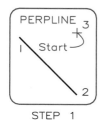 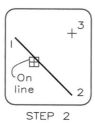 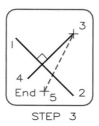

| STEP 1 | STEP 2 | STEP 3 |

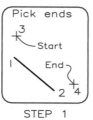 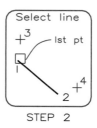 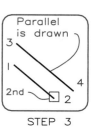

| STEP 1 | STEP 2 | STEP 3 |

27.2 The PERPLINE command.

Step 1 Command: <u>PERPLINE</u> (Enter)
Select START point of perpendicular line: **(Select pt. 3.)**

Step 2 Select ANY point on line to which perp'lr: **(Select point on line.)**

Step 3 Select END point of desired perpendicular (for length only): **(Select pt. 5.) (3–4 is drawn.)**

27.3 The PARALLEL command.

Step 1 Command: <u>PARALLEL</u> (Enter)
Select START point of parallel line: **(Select pt. 3.)**
Select END point of parallel line: **(Select pt. 4.)**

Step 2 Select 1st point on line for parallelism: **(Select pt. 1.)**

Step 3 Select 2nd point on line for parallelism: **(Select pt. 2.) (3–4 is drawn.)**

directory of AutoCAD. To open ACAD, type (Load "ACAD"), being sure to include the parentheses and quotation marks. Access the individual commands by typing their names one at a time—PERPLINE, for example.

Figure 27.2 shows how to draw line 3–4 perpendicular to line 1–2 with PERPLINE. Locate the starting point, select the line to which the constructed line is to be perpendicular, and pick a third point in the general area of the line's endpoint.

In **Fig. 27.3**, line 3–4 is drawn parallel to line 1–2 with PARALLEL. Locate the starting point of the parallel (point 3) and the general location of its endpoint (point 4). Select the endpoints of line 1–2 in the same order (from 1 to 2), then draw the line.

The TRANSFER command (**Fig. 27.4**) transfers distances from reference lines in the same way in which you use your dividers. Select endpoint 2 in the front view and then the reference line. Select endpoint 2 in the top view, then the auxiliary reference line. A circle will locate the endpoint at its center in the auxiliary view.

*These LISP commands were written by Professor Leendert Kersten of the University of Nebraska at Lincoln.

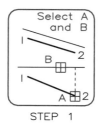 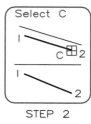 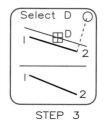

| STEP 1 | STEP 2 | STEP 3 |

27.4 The TRANSFER command.

Step 1 Command: <u>TRANSFER</u> (Enter)
Select start of transfer distance: **(Select pt. A.)**
Select the reference plane: **(Select pt. B.)**

Step 2 Select point to be projected: **(Select pt. C.)**

Step 3 Select other reference plane: **(Select pt. D.)** (Point is located at the center of the circle.)

A distance may be copied from one position to another with COPYDIST, as shown in **Fig. 27.5**. Select the endpoints of the line to be copied; locate its beginning point in the new position and locate the direction of the line to be copied. A circle will locate the endpoint of the line at its center the exact distance being copied.

The BISECT command bisects angles. Because it is sufficiently self-explanatory in its screen prompts, it is not necessary to provide coverage here.

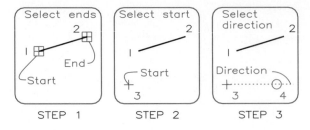

STEP 1 STEP 2 STEP 3

27.5 The COPYDIST command.

Step 1 Command: <u>COPYDIST</u> (Enter)
Select start point of line distance to be copied: (Select end 1.)
End point?: (Select end 2.)

Step 2 Start point of new distance location: (Select pt. 3.)

Step 3 Which direction?: (Select with cursor, and end-point 4 is located at the center of the circle.)

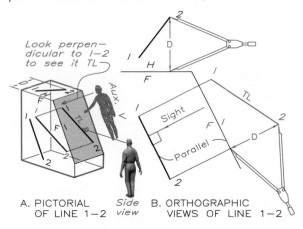

A. PICTORIAL OF LINE 1—2 *Side view* B. ORTHOGRAPHIC VIEWS OF LINE 1—2

27.6 True-length line by auxiliary view.

A A pictorial of line 1–2 is shown inside a projection box where an auxiliary plane is parallel to the line and perpendicular to the frontal plane.

B The auxiliary view is projected from the front orthographic view to find 1–2 true length.

27.3 True-Length Lines

Primary Auxiliary View

Figure 27.6 shows the top and front views of line 1–2 pictorially and orthographically. Line 1–2 is not a principal line, so it is not true length in a principal view. Therefore, a primary auxiliary view is required to find its true-length view. In Fig. 27.6A, the line of sight is perpendicular to the front view of the line and reference line F1 is parallel to the line's frontal

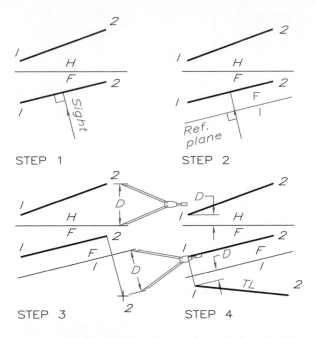

STEP 1 STEP 2

STEP 3 STEP 4

27.7 The true length of a line:

Step 1 To find the true length of line 1—2, the line of sight must be perpendicular to one of its views, the front view here.

Step 2 Draw the F1 reference line parallel to the line and perpendicular to the line of sight.

Step 3 Project point 2 perpendicularly from the front view. Transfer distance D from the top view to locate point 2 in the auxiliary view.

Step 4 Locate point 1 in the same manner to find line 1—2 true length in the auxiliary view.

view. The auxiliary plane is parallel to the line and perpendicular to the frontal plane, accounting for its label, F1, where F and 1 are abbreviations for frontal and primary planes, respectively.

Projecting parallel to the line of sight and perpendicular to the F1 reference line yields the auxiliary view (Fig. 27.6B). Transferring distance D with dividers to the auxiliary view locates point 2 because the frontal plane appears as an edge in both the top and auxiliary views. Point 1 is located in the same manner, and the points are connected to find the true-length view of the line.

Figure 27.7 summarizes the steps of find-

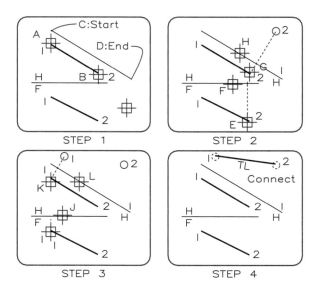

STEP 1   STEP 2

STEP 3   STEP 4

27.8 True length by computer.

Step 1 Command: <u>PARALLEL</u> (Enter) (Follow the prompts in the caption of Fig. 27.3 to produce the reference line parallel to line 1–2 with PARALLEL.)

Step 2 Command: <u>TRANSFER</u> (Enter) (Follow the prompts in the caption of Fig. 27.4 to locate point 2 in the auxiliary view with TRANSFER.)

Step 3 Locate point 1 in the auxiliary view by using TRANS-FER.

Step 4 Draw a line from the centers of the circles obtained in the preceding steps by using the CENTER option of the OSNAP command. ERASE the circles after this step.

ing the true-length view of an oblique line. Letter all reference planes using the notation suggested in Chapter 26 and as shown in the examples throughout this chapter, with the exception of noted dimensions such as D. Use your dividers to transfer dimensions.

Computer Method The method of using the PARALLEL and TRANSFER commands to find a line's true length by an auxiliary view are illustrated in **Fig. 27.8**. Draw a line parallel to the top view of line 1–2, and TRANSFER the endpoints of the line. Use the CENTER option of the OSNAP command to draw line 1–2 from the centers of the circles. Erase the circles afterward.

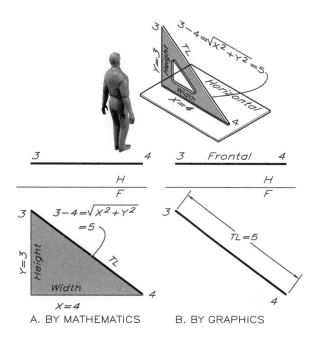

A. BY MATHEMATICS   B. BY GRAPHICS

27.9 Apply the Pythagorean theorem to calculate the length of a line that appears true length in a view, the front view here. Because line 3–4 is true length in the front view, it can be measured to find its length.

True Length by Analytical Geometry

The method of finding the true length of frontal line 3–4 mathematically is shown in **Fig. 27.9**. The Pythagorean theorem states that the hypotenuse of a right triangle is equal to the square root of the sum of the squares of the other two sides. Because the line is true length in the front view, measuring that length provides a check on the mathematical solution.

The true length of a line shown pictorially in **Fig. 27.10A** (line 1-2) is determined by analytical geometry from its length in the front view where the X and Y distances form a right triangle. **Figure 27.10B** shows a second right triangle, 1–0–2, whose hypotenuse is the true length of line 1–2. Thus, the true length of an oblique line is the square root of the sum of the squares of the X, Y, and Z distances that correspond to the width, height, and depth of the triangles.

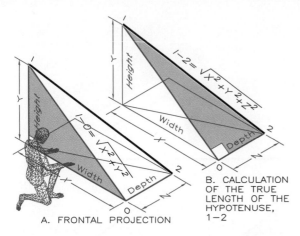

A. FRONTAL PROJECTION

B. CALCULATION OF THE TRUE LENGTH OF THE HYPOTENUSE, 1–2

27.10 To calculate the true length of a 3D line that is not true length in principal view, find (A) the frontal projection, line 1–0, by using the X and Y distances, and (B) the hypotenuse of the right triangle 1–0–2 by using the length of line 1—0 and the Z distance. Then apply the Pythagorean theorem to find its length of 5.

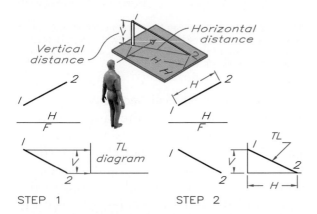

STEP 1   STEP 2

27.11 Using a true-length diagram.

Step 1 Transfer the vertical distance between the ends of line 1–2 to the vertical leg of the TL diagram.

Step 2 Transfer the horizontal length of the line in the top view to the horizontal leg of the TL diagram. The diagonal is the true length of line 1–2.

True-Length Diagram

A true-length diagram is two perpendicular lines used to find a line of true length (**Fig. 27.11**). The two measurements laid out on the true-length diagram may be transferred from any two adjacent orthographic views. One measurement is the distance between the endpoints in one of the views. The other mea-

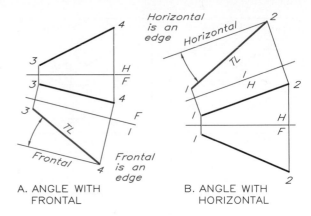

A. ANGLE WITH FRONTAL

B. ANGLE WITH HORIZONTAL

27.12 Angles between lines and principal planes.

A An auxiliary view projected from the front view that shows the line true length will show the frontal plane as an edge, where its angle with the frontal plane can be measured.

B An auxiliary view projected from the top that shows the line true length will show the horizontal plane as an edge, where its angle with the horizontal plane can be measured.

surement, from the adjacent view, is the distance between the endpoints perpendicular to the reference line between the two views. Here, these dimensions are vertical, V, and horizontal, H, between points 1 and 2. This method does not give the line's direction, only its true length.

27.4 Angles Between Lines and Principal Planes

To measure the angle between a line and a plane, the line must appear true length and the plane as an edge in the same view (Fig. 27.12). A principal plane appears as an edge in a primary auxiliary view projected from it, so the angle that a line makes with this principal plane can be measured if the line is true length in this auxiliary view.

27.5 Sloping Lines

Slope is the angle that a line makes with the horizontal plane when the line is true length and the plane is an edge. **Figure 27.13** shows

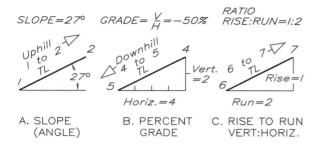

A. SLOPE (ANGLE)
B. PERCENT GRADE
C. RISE TO RUN VERT:HORIZ.

27.13 The three methods for specifying slope are: (A) slope angle, (B) percent grade, and (C) slope ratio (rise to run).

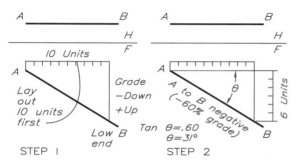

27.14 The percent grade of a line.

Step 1 The percent grade of a line can be measured in the view where the horizontal appears as an edge and the line is true length (here, the front view). Lay off 10 units parallel to the horizontal from the end of the line.

Step 2 A vertical distance from the end of the 10 units to the line measures 6 units. The percent grade is 6 divided by 10, or 60 percent. This is a negative grade from A to B because the line slopes downward from A. The tangent of this slope angle is 6/10, or 0.60, which can be used to verify the slope of 31° from trigonometric tables.

the three methods for specifying slope: slope angle, percent grade, and slope ratio.

The slope angle of line AB in **Fig. 27.14** is 31°. It can be measured in the front view where the line is true length.

Percent grade is the ratio of the vertical (rise) divided by the horizontal (run) between the ends of a line, expressed as a percentage. The percent grade of line AB is determined in the front view of Fig. 27.14 where the line is true length and the horizontal plane is an edge. Line AB has a −60 percent grade from A to B because the line slopes downward; it would be positive (slope upward) from B to A.

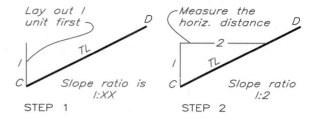

27.15 Slope ratio.

Step 1 Because the slope ratio always begins with 1, lay out a vertical distance of 1 from end C.

Step 2 Lay out a horizontal distance from the end of the vertical line and measure it. It is 2, so the slope ratio (always expressed as 1:XX) of this line is 1:2.

Trigonometric tables verify that an angle whose tangent is 0.60 (6/10) is 31°.

The **slope ratio** is the ratio of a rise of 1 to the run. The rise is always written as 1, followed by a colon and the run (for example, 1:10, 1:200). **Figure 27.15** illustrates the graphical method of finding the slope ratio. The rise of 1 unit is laid out on the true-length view of CD. The corresponding run measures 2 units, for a slope ratio of 1:2.

The slope of oblique lines is found true length in an auxiliary view projected from the top view so that the horizontal reference plane will appear as an edge (**Fig. 27.16A**). The slope is expressed as an angle (26°).

To find the percent grade of an oblique line (**Fig. 27.16B**), lay off 10 units horizontally, parallel to the H1 reference line. The corresponding vertical distance measures 4.5 units, for −45 percent grade from point 3 to point 4.

The principles of true-line length and angles between lines and planes are useful in applications such as the design of aggregate conveyors (**Fig. 27.17**), where slope is crucial to optimal operation of the equipment.

27.6 Bearings and Azimuths of Lines

Two types of bearings of a line's direction are **compass bearings** and **azimuths**. Compass bearings are angular measurements from

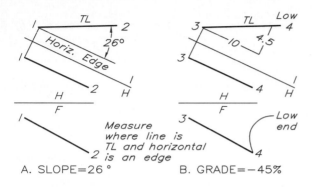

A. SLOPE=26° B. GRADE=−45%

27.16 Slope of an oblique line.

A Find the slope angle of an oblique line (26°, in this case) in a view where the horizontal appears as an edge and the line is true length.

B Find the percent grade in an auxiliary view projected from the top view where line 3–4 is true length (−45 percent from 3 to 4, the low end, in this case).

27.17 The design of these aggregate conveyors requires the application of sloping-line principles in order to obtain their optimal slopes. (Courtesy of Link-Belt.)

north or south. The line in **Fig. 27.18A** that makes a 30° angle with north has a bearing of N 30° W. The line making a 60° angle with south toward the east has a bearing of south 60° east, or S 60° E. Because a compass can be read only when held level, bearings of a line must be found in the top, or horizontal, view.

Azimuths are measured clockwise from north through 360° (**Fig. 27.18B**). Azimuth bearings are written N 120°, N 210°, and so on, indicating that they are measured from north.

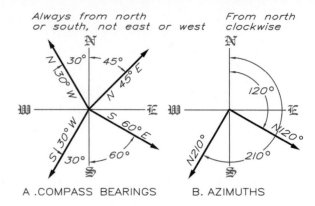

A. COMPASS BEARINGS B. AZIMUTHS

27.18 Compass directions.

A Compass bearings are measured with respect to north and south.

B Azimuths are measured clockwise from north up to 360°.

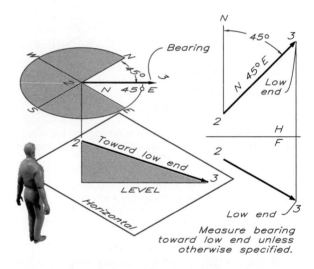

27.19 Measure the compass bearing of a line in the top view toward its low end (unless otherwise specified). Line 2–3 has a bearing of N 45° E from 2 to 3 toward the low end at point 3.

The bearing of a line is toward the low end of the line unless otherwise specified. For example, line 2–3 in **Fig. 27.19** has a bearing of N 45° E because the line's low end is point 3 in the front view.

Figure 27.20 shows how to find the bearing and slope of a line. This information may be used verbally to describe the line as having a bearing of S 60° E and a slope of 26° from point 5 to point 6. This information and the location

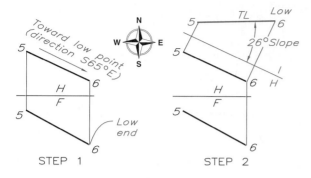

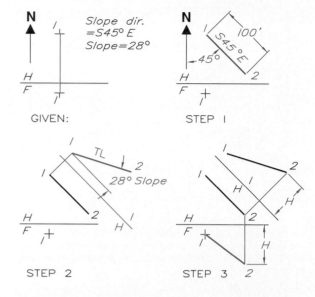

STEP 1   STEP 2

27.20 The bearing and slop of a line.

Step 1 Measure the bearing in the top view toward its low end, or S 65° E, in this case.

Step 2 Measure the slope angle of 26° from the H1 reference line in an auxiliary view projected from the top view where the line is true length.

of one point in the top and front views is sufficient to complete a 3D drawing of a line as illustrated in **Fig. 27.21**.

27.7 Application: Plot Plans

A typical plot plan for a tract of land is shown in **Fig. 27.22**. The boundary lines of the tract, their bearings, and the interior angles are used to legally define the property. AutoCAD provides an option called Surveyor's units that is an excellent means of drawing and labeling a plot plan.

Computer Method To obtain the Surveyor's Units option, type `UNITS (Enter)` to get the Text Window on the screen. You will be given the following prompts:

```
System of angle measure: (Example)
1. Decimal degrees 45.0000
2. Degree/minutes/seconds 45d0'0"
3. Grads 50.0000g
4. Radians 0.7854r
5. Surveyor's units N45d0'0"E
Enter choice, 1 to 5 <default>: 5
(Enter)
```

By entering 5, you obtain surveyor's units that

27.21 A line from slope specifications.

Required Draw a line through point 1 that bears S 45° E for 100 ft horizontally and slopes 28°.

Step 1 Draw the bearing and the horizontal distance in the top view.

Step 2 Project an auxiliary view from the top view and draw the line at a slope of 28°.

Step 3 Find the front view of line 1–2 by locating point 2 in the front view.

give directional angles with respect to north or south in east or west directions, such as N30d45'10", where d = degrees, ' = minutes, and " = seconds. Interior angles measured with the `DIM` command give the angles in the same form, but without reference to compass directions, such as 152d34'17".

Next, the `UNITS` command gives a prompt for setting the direction of east to establish the relationship of the drawing to the directional north arrow:

```
Direction for angle 0:
East    3 o'clock  =      0
North  12 o'clock  =     90
West    9 o'clock  =    180
South   6 o'clock  =    270
Enter direction for angle E <cur-
rent>:
```

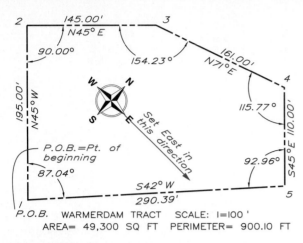

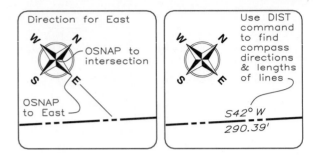

27.22 A typical plot plan showing the lengths and bearings of each side of a tract of land, the interior angles, and the north arrow. AutoCAD's Surveyor's Units option has been used to create this.

27.23 Compass direction by computer.

Step 1 Insert the compass arrow on the plot plan. Then use the UNITS command and select 5, Surveyor's units. It will prompt you to give the direction for east. Select a point on the drawing with the cursor and select two points (west to east) to indicate the direction of east.

Step 2 Using the DIST command, select the endpoints of each side clockwise about the plot from the point of beginning (P.O.B.). Label the sides with direction inside and lengths outside the plot.

With the cursor, select an area on the drawing; the Text Window will be suppressed. Select two west-to-east points to indicate the direction of east in your drawing (**Fig. 27.23**). By picking these two points, you establish the direction of north. The last prompt reads:

Do you want angles measured clockwise? <N>: N (Enter)

By selecting NO (N), you obtain angles measured in the standard direction (Fig. 27.22). Use the DIST command to find the lengths and directions of lines by selecting endpoints of the lines clockwise about the plot from the point of beginning (P.O.B.).

27.8 Contour Maps and Profiles

A contour map depicts variations in elevation of the earth in two dimensions (**Fig. 27.24**). Three-dimensional representations involve (1) conventional orthographic views of the contour map combined with profiles and (2) contoured surface views (often in the form of models).

Contour lines are horizontal (level) lines that represent constant elevations from a

horizontal datum such as sea level. The vertical interval of spacing between the contours shown in Fig. 27.24 is 10 ft. Contour lines may be thought of as the intersection of horizontal planes with the surface of the earth.

Contour maps contain contour lines that connect points of equal elevation on the earth's surface and therefore are continuous (Fig. 27.24). The closer the contour lines are to each other, the steeper the terrain is.

Profiles are vertical sections through a contour map that show the earth's surface at any desired location (Fig. 27.24). Contour lines represent edge views of equally spaced horizontal planes in profiles. True representation of a profile involves use of a vertical scale equal to the scale of the contour map; however, the vertical scale is usually drawn larger to emphasize changes in elevation that often are slight compared to horizontal dimensions.

Contoured surfaces are also depicted in drawings with contour lines (Fig. 27.24) or on models. When applied to objects other than the earth's surface—such as airfoils, automobile bodies, ship hulls, and household appli-

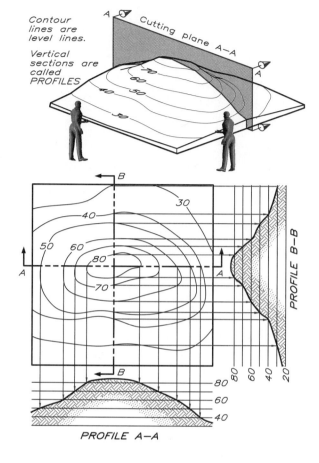

27.24 A contour map shows variations in elevation on a surface. A profile is a vertical section through the contour map. To construct a profile, draw elevation lines parallel to the cutting planes, spacing them equally to show the difference in elevations of the contours (10 ft, in this case). Project crossing points of contours and the cutting plane to their respective elevations in the profile and connect them.

ances—this technique of showing contours is called **lofting**.

Station numbers identify distances on a contour map. Because surveyors use a chain (metal tape) 100 ft long, primary stations are located 100 ft apart (**Fig. 27.25**). For example, station 7 is 700 ft from the beginning point, station 0; a point 32 feet beyond station 7 is station 7 + 32.

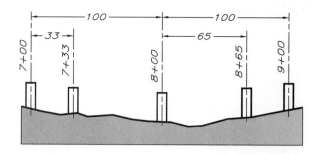

27.25 Primary station points are located 100 ft apart; for example, station 7 is 700 ft from station 0 (not shown). A point 32 ft beyond station 7 is labeled station 7 + 32. A point 870 ft from the origin is labeled station 8 + 70.

A plan-profile combines a section of a contour map (a plan view) and a vertical section (a profile view). Engineers use plan-profile drawings extensively for construction projects such as pipelines, roadways, and waterways.

Application: Vertical Sections

In **Fig. 27.26**, a vertical section passed through the top view of an underground pipe gives a profile view. The pipe is known to have elevations of 90 ft at point 1 and 60 ft at point 2. Project an auxiliary view perpendicularly from the top view, locate contour lines, and draw the top of the earth over the pipe in profile. To measure the true lengths and angles of slope in the profile, use the same scale for both the contour map and the profile.

Pipeline installation (**Fig. 27.27**) requires major outlays for engineering design and construction. The use of profiles, found graphically, is the best way to make cost estimations for constructing ditches for laying underground pipe.

27.9 Plan-Profiles

A plan-profile drawing shows an underground drainage system from manhole 1 to manhole 3 in **Figs. 27.28** and **27.29**. The profile has a larger vertical scale to emphasize

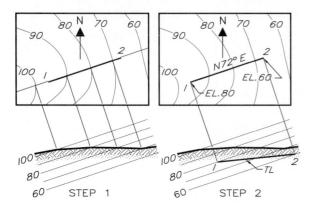

27.26 Drawing profiles (vertical sections).

Step 1 An underground pipe has elevations of 90 ft and 60 ft at its ends. Project an auxiliary view perpendicularly from the top view and draw contours at 10-ft intervals corresponding to their elevations in the plan view. Locate the ground surface by projecting from the contour lines in the plan view.

Step 2 Locate points 1 and 2 at elevations of 90 ft and 60 ft in the profile. Line 1–2 is TL in the section, so measure its slope (percent grade) here and label its bearing and slope in the top view.

27.27 Pipeline construction applies the principles of descriptive geometry, true-length lines, and slopes of 3D lines. (Courtesy of Consumers Power Company.)

variations in the earth's surface and the grade of the pipe, although the vertical scale may be drawn at the same scale as the plan if desired.

The location of manhole 1 is projected to the profile orthographically, but the remaining points are not (Fig. 27.28). Instead, the distances where the contour lines cross the top view of the pipe are transferred to their respective elevations in the profile with dividers to show the surface of the ground over the pipe.

Figure. 27.29 shows the manholes, their elevations, and the bottom line of the pipe. Find the drop from manhole 1 to manhole 2 (4.40 ft) by multiplying the horizontal distance of 220.00 ft by a −2.00 percent grade. The pipes intersect at manhole 2 at an angle, so the flow of the drainage is disrupted at the turn. A drop of 0.20 ft (2.4 in.) across the bottom of the manhole compensates for the loss of pressure (head) through the manhole.

You cannot measure the true lengths of the pipes in the profile view when the vertical scale is different from the horizontal scale. Instead, you must use trigonometry to calculate them.

27.10 Edge Views of Planes

The edge view of a plane appears in a view where any line on the plane appears as a point. Recall that you can find a line as a point by projecting from its true-length view (**Fig. 27.30**). You may obtain a true-length line on any plane by drawing a line parallel to one of the principal planes and projecting it to the adjacent view (**Fig. 27.31**). You then get the edge view of the plane in an auxiliary view by finding the point view of line 3–4 on its surface.

Dihedral Angles

The angle between two planes, the **dihedral angle**, is found in a view where the line of intersection between two planes appears as a point. In such a view, both planes appear as edges and the angle between them is true size. The line of intersection, line 1–2, between the

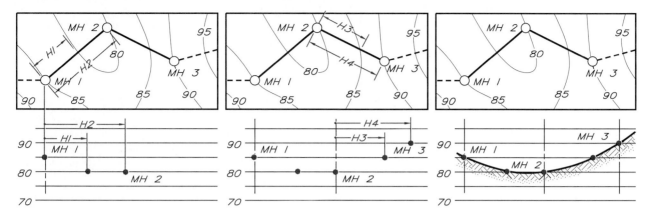

27.28 Plan-profile: vertical section.

Required: Find the profile of the earth=s surface over the pipeline.

Step 1 Transfer distances H1 and H2 from MH 1 in the plan to their respective elevations in the profile view.

Step 2 Measure distances H3 and H4 from manhole 2 in plan and transfer them to their respective elevations in profile. These points represent elevations of points on the earth above the pipe.

Step 3 Connect these points with a free-hand line and crosshatch the drawing to represent the earth's surface. Draw center-lines to show the locations of the three manholes.

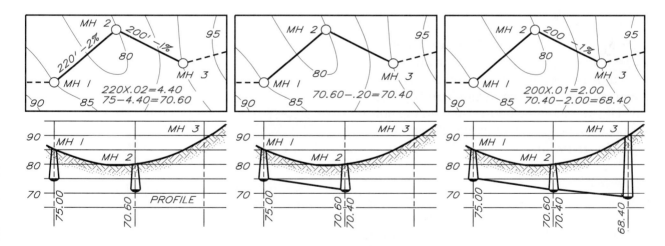

27.29 Plan-profile: manhole location.

Step 1 Multiply the horizontal distance from MH1 to MH2 by −2 percent. Find the elevation of the bottom of MH2 by subtracting the amount of fall from the elevation of MH1 (70.60').

Step 2 The lower side of MH2 is 0.20 ft lower than the inlet side to compensate for loss of head (pressure) because of the turn in the pipeline. Find the elevation on the lower side (70.40') and label it.

Step 3 Calculate the elevation of MH3 (200 ft) at a −1 percent grade from MH2 (68.40'). Draw the flow line of the pipeline from manhole to manhole and label the elevations at each manhole.

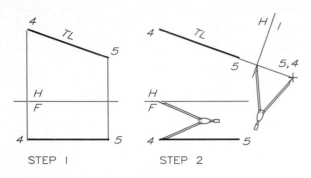

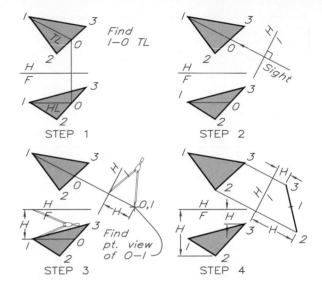

27.30 The point view of a line.

Step 1 Line 4–5 is horizontal in the front view and therefore is true length in the top view.

Step 2 Find the point view of line 4–5 by projecting parallel to its true length to the auxiliary view.

27.31 The edge view of a plane.

Step 1 To find the edge view of plane 1–2–3, draw horizontal line 1–0 on its front view of the plane and project it to the top view, where it is true length.

Step 2 Draw a line of sight parallel to the true-length line 1–0. Draw H1 perpendicular to the line of sight.

Step 3 Find the point view of 1–0 in the auxiliary view by transferring height (H) from the front view.

Step 4 Locate points 2 and 3 in the same manner to find the edge view of the plane.

two planes shown in **Fig. 27.32** is true length in the top view. Project an auxiliary view from the top view to find the point view of line 1–2 and the edge views of both planes.

27.11 Planes and Lines

Piercing Points

By Projection The piercing point of line 1–2 passing through the plane by projection is shown in **Fig. 27.33**. Pass cutting planes through the line and plane in the top view. Then project the trace of this cutting plane, line DE, to the front view to find piercing point P. Locate the top view of P and determine the visibility of the line.

By Auxiliary View You may also find the piercing point of a line and a plane by auxiliary view in which the plane is an edge (**Fig. 27.34**). The location of piercing point P in step 2 is where line AB crosses the edge view of the plane. Project point P to AB in the top view from the auxiliary view, and then to the front view. To verify the location of point P in the front view, transfer dimension H from the auxiliary view with dividers.

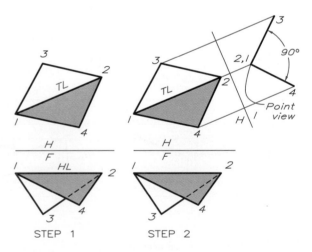

27.32 A dihedral angle.

Step 1 The line of intersection between two planes, line 1–2, is true length in the top view.

Step 2 The angle between the planes (the dihedral angle) is found in the auxiliary view where the line of intersection appears as a point and both planes are edges.

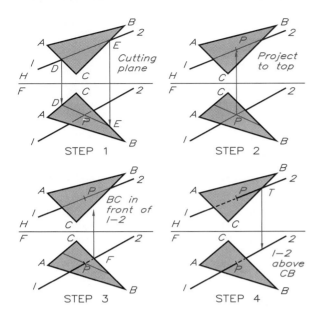

27.33 A piercing point by projection.

Step 1 Pass a vertical cutting plane through the top view of line 1–2, which cuts the plane along line DE. Project line DE to the front view to locate piercing point P.

Step 2 Project point P to the top view of line 1–2.

Step 3 Determine visibility in the front view by projecting the crossing point of lines CB and 1–2 to the top view. Because CB is encountered first, it is in front of 1–2, making segment PF hidden in the front view.

Step 4 Determine visibility in the top view by projecting the crossing point of lines CB and 1–2 to the front view. Because line 1–2 is encountered first, it is above line CB, making TP visible in the top view.

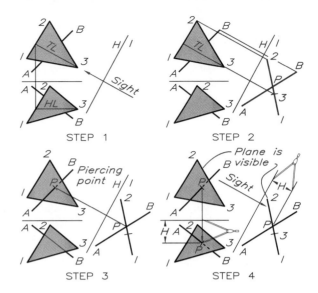

27.34 A piercing point by auxiliary view.

Step 1 Draw a horizontal line on the plane in the front view and project it to the top view where it is true length on the plane.

Step 2 Find the edge view of the plane in an auxiliary view and project AB to this view. P is the piercing point.

Step 3 Project point P to line AB in the top view. Line AP is nearest the H1 reference line, so it is the highest end of the line and is visible in the top view.

Step 4 Project P to line AB in the front view. AP is visible in the front view because line AP is in front of 1–2.

You can easily determine visibility for the top view because you see in the auxiliary view that line AP is higher than the plane and, therefore, is visible in the top view. Similarly, the top view shows that endpoint A is the forward-most point and line AP, therefore, is visible in the front view.

Perpendicular to a Plane

A perpendicular line appears true length and perpendicular to a plane where the plane appears as an edge. In **Fig. 27.35**, a line is to be drawn from point 0 perpendicular to the plane. Obtain an edge view of the plane and draw the true-length perpendicular to locate piercing point P. Locate point P in the top view by drawing line OP parallel to the H1 reference line. (This principle is reviewed in **Fig. 27.36**.) Line OP is also perpendicular to a true-length line in the top view of the plane. Obtain the front view of point P, along with its visibility, by projection.

Intersection Between Planes

To find the intersection between planes, find the edge view of one of the planes (**Fig. 27.37**), then project piercing points L and M from the auxiliary view to their respective lines, 5–6 and 4–6, in the top view. Plane 4–5–L–M is visible in the top view because sight line 1 has an

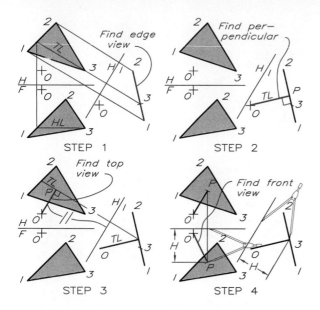

STEP 1

STEP 2

STEP 3

STEP 4

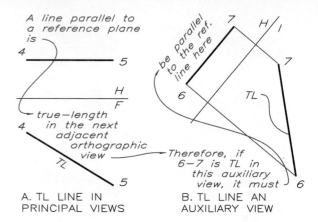

A. TL LINE IN PRINCIPAL VIEWS

B. TL LINE AN AUXILIARY VIEW

27.36 In the top view of Fig. 27.35, line OP is parallel to the H1 reference line because it is true length in the auxiliary view. Here, lines 4–5 and 6–7 are examples of this principle: Both are true length in one view and parallel to the reference line in the preceding view.

27.35 A line perpendicular to a plane.

Step 1 Find the edge view of the plane by finding the point view of a line on it in an auxiliary view. Project point 0 to this view also.

Step 2 Draw line OP perpendicular to the edge view of the plane, which is true length in this view.

Step 3 Because line OP is true length in the auxiliary view, it must be parallel to the H1 reference line in the preceding view. Line OP is visible in the top view because it appears above the plane in the auxiliary view.

Step 4 Project point P to the front view and locate it by transferring height H from the auxiliary view with dividers.

unobstructed view of the 4–5–L–M portion of the plane in the auxiliary view. Plane 4–5–L–M is visible in the front view because sight line 2 has an unobstructed view of the top view of this portion of the plane.

27.12 Sloping Planes

Slope and Direction of Slope

The slope of a plane is described using the following definitions:

Angle of Slope: The angle that the plane's edge view makes with the edge of the horizontal plane.

Direction of Slope: The compass bearing of a line perpendicular to a true-length line in the top view of a plane toward its low side (the direction in which a ball would roll on the plane).

As **Fig. 27.38** shows, a ball would roll perpendicular to all horizontal lines on the roof toward the low side. This direction is the slope direction, and it can be measured in the top view as a compass bearing.

The steps of determining the slope and direction of slope of a plane are shown in **Figure 27.39**. Conversely, a plane can be drawn in three-dimensional space by working from slope and direction specifications, as shown in **Fig. 27.40**. Draw the direction of slope in the top view to locate a perpendicular true-length line on the plane. Find the edge view of the plane by locating point 1 and constructing a slope of 30° through it in an auxiliary view. Transfer points 3 and 2 to the front view from the auxiliary view.

Application: Cut and Fill

A level roadway through irregular terrain and the embankment for an earthen dam (**Fig. 27.41**) involve the principles of cut and fill.

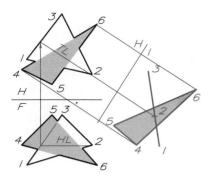
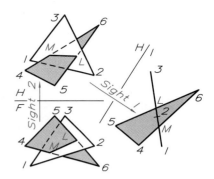

27.37 The intersection of planes by auxiliary view.

Step 1 Locate the edge view of one of the planes in an auxiliary view and project the other plane to this view.

Step 2 Piercing points L and M are found on the edge view of the plane in the auxiliary view. Project the line of intersection, LM, back to the top and front views.

Step 3 The line of sight from the top view strikes L–5 first in the auxiliary view, indicating that L–5 is visible in the top view, where 4–5 is farthest forward and is visible in the front.

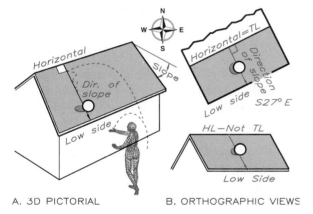

A. 3D PICTORIAL   B. ORTHOGRAPHIC VIEWS

27.38 Slope definition.

A The direction of slope of a plane is the compass bearing of the direction in which a ball on the plane will roll.

B Slope direction is measured in the top view toward the low side of the plane and perpendicular to a horizontal line on the plane.

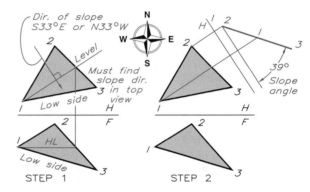

27.39 The slope and bearing of a plane.

Step 1 Slope direction is perpendicular to a true-length, level line in the top view toward the low side of the plane, or S 33° E in this case.

Step 2 Find the slope in an auxiliary view where the horizontal is an edge and the plane is an edge, or 39° in this case.

Cut and fill is the process of cutting away high ground and filling low areas, generally of equal volumes.

Figure 27.42 illustrates how a level roadway at an elevation of 60 ft is to be constructed along a specified centerline with specified angles of cut and fill. First, draw the roadway in the top view. Use contour intervals in the profile view of 10 ft to match those in the top view. Next, measure and draw the cut angles on both sides of the roadway. Project the points where the cut angles cross each elevation line to the respective contour lines in the plan view to find the limits of cut. Then, measure and draw the fill angles in the profile view. Project the points where the fill angles cross each elevation line to the respective contour lines in the plan view to find the limits of fill. Finally, draw new contour lines inside the areas of cut parallel to the centerline.

Some of the terms associated with the design of a dam are (1) **crest** (the top of the dam); (2) **water level**; and (3) **freeboard** (the height of the crest above the water level). These terms are illustrated in **Fig. 27.43**.

27.12 SLOPING PLANES • 491

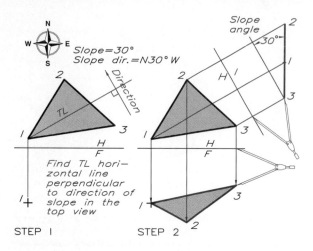

27.40 A plane from slope specifications.

Step 1 If the top view of a plane, the front view of point 1, and slope specifications are given, you can complete the front view. Draw the direction of slope in the top view and a true-length horizontal line on the plane perpendicular to the slope direction.

Step 2 Find a point view of the TL line in the auxiliary view to locate point 1. Draw the edge view of the plane through point 1 at a slope of 30°, according to the specifications. Find the front view by transferring height dimensions from the auxiliary view to the front view.

27.41 A dam built by applying the principles of cut and fill. (Courtesy of the Bureau of Reclamation, U.S. Department of the Interior.)

Strike and Dip

Strike and dip are terms used in geology and mining engineering to describe the location of strata of ore under the surface of the earth:

Strike: The compass bearings (two are possible) of a level line in the top view of a plane.

Dip: The angle that the edge view of a plane makes with the horizontal and its general compass direction, such as NW or SW.

The dip angle lies in the primary auxiliary view projected from the top view. The dip direction is perpendicular to the strike and toward its low side.

Figure 27.44 demonstrates how to find the strike and dip of a plane. Here, the true-length line in the top view of the plane has a strike of N 66° W or S 66° E. The dip angle appears in an auxiliary view projected from the top view that shows the horizontal (H1) and the sloping plane as edges.

You can construct a plane from the strike and dip specifications shown in **Fig. 27.45**. First, draw the strike as a true-length horizontal line on the plane and the dip direction perpendicular to the strike. Then find the edge view of the plane in the auxiliary view through point 1 at a dip of 30°. Locate points 2 and 3 in the front view by transferring them from the auxiliary view.

27.13 Ore-Vein Applications

The principles of descriptive geometry can be applied to find the distance from a point to a plane. Techniques for finding such distances are often used to solve mining and geological problems. For example, test wells are drilled into coal seams to learn more about them (**Fig. 27.46**).

Application: Underground Ore Veins Geologists and mining engineers usually assume that strata of ore veins have upper and lower planes that are parallel. In **Fig. 27.47**, point 0 is

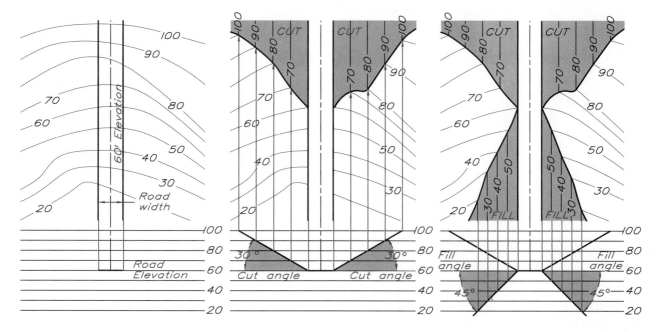

27.42 Cut and fill for a level roadway.

Step 1 Draw and label a series of elevation planes in the front view at the same scale as the contour map. Draw the width and elevation (60 ft in this case) of the roadway in the top and front views.

Step 2 Draw the cut angles on the higher sides of the road in the front view. Project the points of intersection between the cut angles and the contour planes in the front view to their respective contour lines in the top view to determine the limits of cut.

Step 3 Draw the fill angles on the lower sides of the road in the front. Project the points in the front where the fill angles cross the contour lines to these respective contour lines in the top view. Draw new contours parallel to the centerline in the cut-and-fill areas.

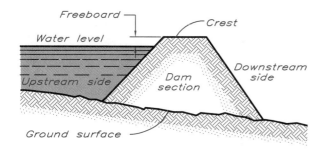

27.43 These terms and symbols are used in the design of a dam.

on the upper surface of the earth, and plane 1–2–3 is an underground ore vein. Point 4 is on the lower plane of the vein.

Find the edge view of plane 1–2–3 by projecting from the top view, and then draw the lower plane through point 4 parallel to the upper plane. Draw the horizontal distance from point 0 to the plane parallel to the H1 ref-

erence line and the vertical distance perpendicular to line H1. The shortest distance is perpendicular to the ore vein. These three lines from point 0 are true length in the auxiliary view where the ore vein appears as an edge.

Application: Ore-Vein Outcrop The same assumption regarding parallel planes is made in **Fig. 27.48** in analyzing the orientation of underground ore veins that are inclined to the earth's surface and outcrop on it. When ore veins outcrop, open-pit mining can be used to reduce costs. To find the outcrop of an ore vein, the locations of sample drillings are given on a contour map. Points A, B, and C are located on the upper plane of the ore vein, and point D is located on the lower plane of the vein. Draw them in the front view at their determined elevations.

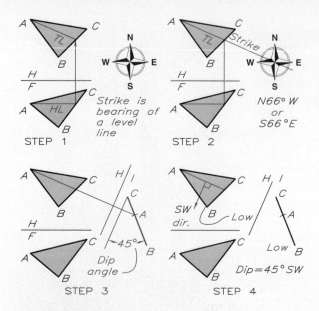

STEP 1

STEP 2

STEP 3

STEP 4

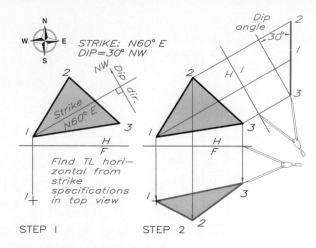

STEP 1

STEP 2

27.45 Strike and dip specifications.

Step 1 Draw the strike in the top view of the plane as a true-length horizontal line. Draw the direction of dip perpendicular to the strike toward the NW as specified.

Step 2 Find the point view of strike in the auxiliary view to locate point 1, where the edge view of the plane passes through it at a 30° dip, as specified. Complete the front view by transferring height (H) dimensions from the auxiliary to the front view.

27.44 The strike and dip of a plane.

Step 1 Draw a horizontal line on the plane in the front view and project it to the top view, where it is true length.

Step 2 Determine strike, the compass direction of a level line on the plane in the top view, either N 66° W or S 66° E.

Step 3 Find the edge view of the plane in the auxiliary view. The dip angle of 45° is the angle between the H1 reference line and the edge view of the plane.

Step 4 The general compass direction of dip is toward the low side, and perpendicular to a strike in the top view, or SW in this case. Dip direction is 45°SW.

Find the edge view of the ore vein in an auxiliary view projected from the top view, then project points on the upper surface where the vein crosses elevation lines back to their respective contour lines in the top view. Also, project points on the lower surface of the vein (through point D) to the top. If the ore vein extends uniformly at its angle of inclination to the earth's surface, the area between these two lines will be the outcrop of the vein.

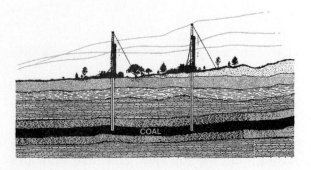

27.46 Test wells are drilled into coal seams to determine the elevations of coal seams that may contribute to the exploration for gas. (Courtesy of Texas Eastern *News*.)

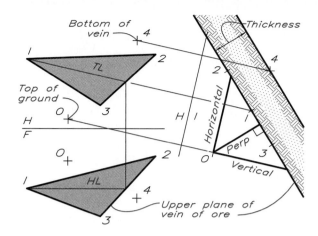

27.47 To find the vertical, horizontal, and perpendicular distances from a point to an ore vein, project an auxiliary view from the top view, where the vein appears as an edge. The thickness of an ore vein is perpendicular to the upper and lower planes of the vein.

27.14 Intersections Between Planes

Strike and Dip Method The method of locating the intersection of two planes with strike and dip specifications is shown in **Fig. 27.49**. The given strike lines are true-length level lines in the top view, so the edge view of the planes appears in the auxiliary views where the strikes appear as points. Draw the edge views using the given dip angles and directions.

Use the additional horizontal datum plane HRP1 to find lines on each plane at equal elevations that intersect when projected to the top view from their auxiliary views. Connect points A and B as the line of intersection between the two planes in the top view and project that line to the front view to establish line AB in three dimensions.

Cutting-Plane-Method In **Fig. 27.50**, top and front views of two planes are given and the line of intersection between them, if they were extended, is to be determined. Draw cutting

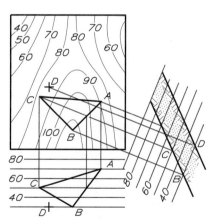

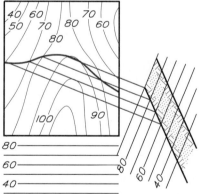

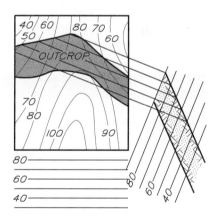

27.48 Locating an ore vein outcrop.

Step 1 Use points A, B, and C on the upper surface of the ore vein to find its edge view by projecting an auxiliary from the top view. Draw the lower surface of the vein parallel to the upper plane through point D.

Step 2 Project points of intersection between the upper plane of the vein and the contour lines in the auxiliary view to their respective contours in the top view to find a line of the outcrop.

Step 3 Project points from the lower plane in the auxiliary view to their respective contours in the top view to find the second line of outcrop. Crosshatch the area between the lines to depict the outcrop of the vein.

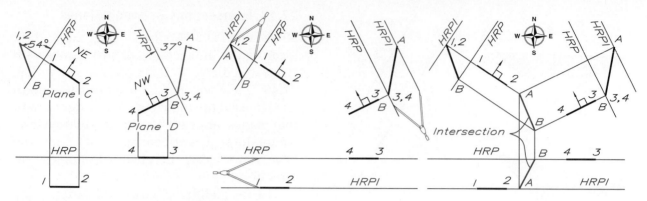

27.49 Intersection of planes: Strike and dip method.

Step 1 Lines 1–2 and 3–4 are strike lines and are true length in the top view. Use a common reference plane, HRP, to find the point view of each strike line by auxiliary views. Find the edge views by drawing the dip angles with the HRP line through the point views.

Step 2 Draw a supplementary horizontal plane, HRP1, at a convenient location in the front view. This plane, shown in both auxiliary views, is located H distance from HRP. The HRP1 cuts through each edge in both auxiliary views, locating A and B.

Step 3 Project points A from both auxiliary views of HRP1 to the top view to their intersection at A. Project points B and HRP1 to their intersection in the top view. Project points A and B to their respective planes in the front view. AB is the line of intersection.

planes through both planes in either view at any angle and project them to the top view. Find points L and M in the top view to establish the line of intersection. Find the front view of line LM, the line of intersection, by projecting its endpoints from the top view to their respective planes in the front view.

Problems

Use size A sheets for the following problems; lay out the solutions using instruments. Each square on the grid is equal to 0.20 inch (5 mm). Use either grid or plain paper. Label all reference planes and points with 1/8-in. (3 mm) letters or numbers with guidelines.

1. (Sheet 1) True-length lines.

(A–D) Find the true-length views of the lines by auxiliary view as indicated by the given lines of sight. *Alternative method:* Find the true-length of the lines by the Pythagorean theorem.

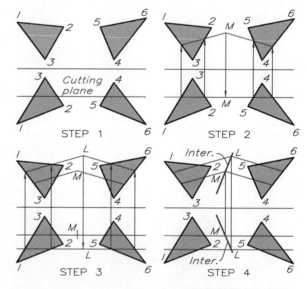

27.50 Intersection of planes by the cutting-plane method.

Step 1 Draw a cutting plane that passes through both planes in the front view in any convenient direction.

Step 2 Project the intersections of the cutting plane to the top views of the planes. Find intersection point M in the top view and project it to the front view.

Step 3 Draw a second cutting plane through the front view of the planes and project it to the top view. Find point L in the top view and project it to the front view.

Step 4 Connect L and M in the top and front views to represent the line of intersection of the extended planes.

2. (Sheet 2) True-length lines.

(**A–D**) Find the true-length views of the lines by auxiliary view as indicated by the lines of sight. *Alternative method:* Find the true length of the lines by the Pythagorean theorem.

3. (Sheet 3) True-length diagram.

(**A–B**) Find the true length of the lines by using a true-length diagram.

(**C–D**) Find the point views of the lines.

4. (Sheet 4) Sloping lines.

(**A–D**) Find the slope angle, tangent of the slope angle, and the percent grade of the four lines.

5. (Sheet 5) Edge views of planes.

(**A–B**) Find the edge views of the planes.

6. (Sheet 6) Intersections: lines and planes.

(**A**) Find the angle between the planes.

(**B**) By projection, find the point of intersection between the line and plane, and show visibility.

(**C**) By the auxiliary view method, find the point of intersection between the line and plane and show visibility.

7. (Sheet 7) Perpendiculars to planes.

(**A**) Construct a 1-in. line perpendicular from point 0 on the plane and show it in all views.

(**B**) Draw a line perpendicular to the plane from 0, find the piercing point, and show visibility.

8. (Sheet 8) Dihedral angles. By using auxiliary views, find the following:

(**A**) The line of intersection between the planes.

(**B**) The angle between the planes.

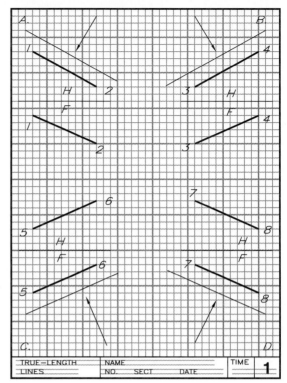

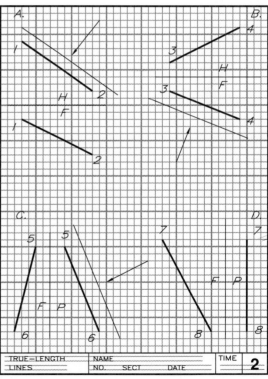

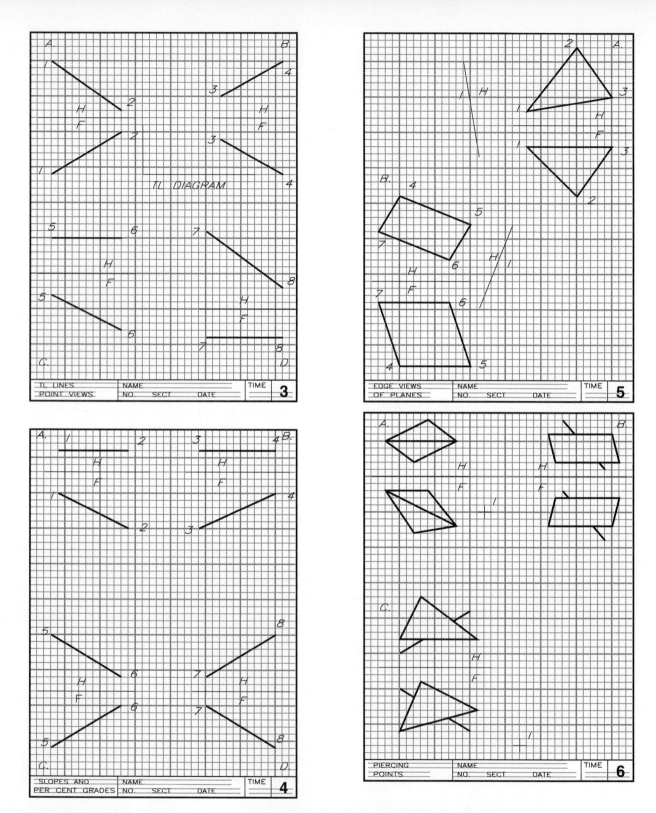

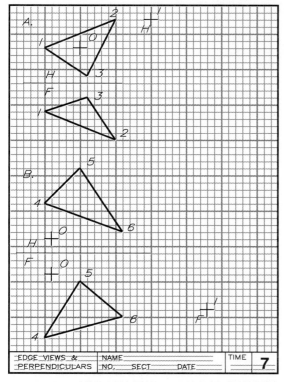

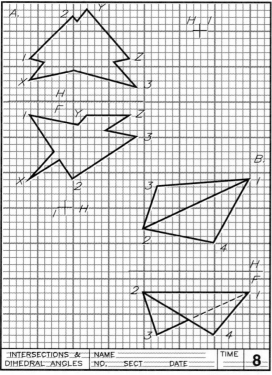

9. (**Sheet 9**) Slope and direction of slope.

(**A–B**) Find the direction of slope and the slope angle of the planes. *Alternative method:* Find the strike and dip of the planes.

10. (**Sheet 10**) Distances to a plane.

Find the shortest distance, the horizontal distance, and the vertical distance from point 0 to the underground ore vein represented by plane 1–2–3. Point B is on the lower plane of the vein. Find the thickness of the vein. Label your solutions.

11. (**Sheet 11**) Intersecting planes.

(**A**) Find the line of intersection between the two planes by the cutting plane method.

(**B**) Find the line of intersection between the two planes indicated by strike lines 1–2 and 3–4. The plane with strike line 1–2 has a dip of 30°, and the plane with strike line 3–4 has a dip of 55°.

12. (**Sheet 12**) Cut and fill.

Find the limits of cut and fill in the plan view of the roadway. Use a cut angle of 35° and fill angle of 40°.

13. (**Sheet 13**) Outcrop.

Find the outcrop of the ore vein represented by plane 1–2–3 on its upper surface. Point B is on the lower surface.

14. (**Sheet 14**) Plan-profile.

Complete the plan-profile drawing of the drainage system from manhole 1 through manhole 2 to manhole 3, using the grades indicated. Allow a drop of 0.20 ft across each manhole to compensate for loss of pressure.

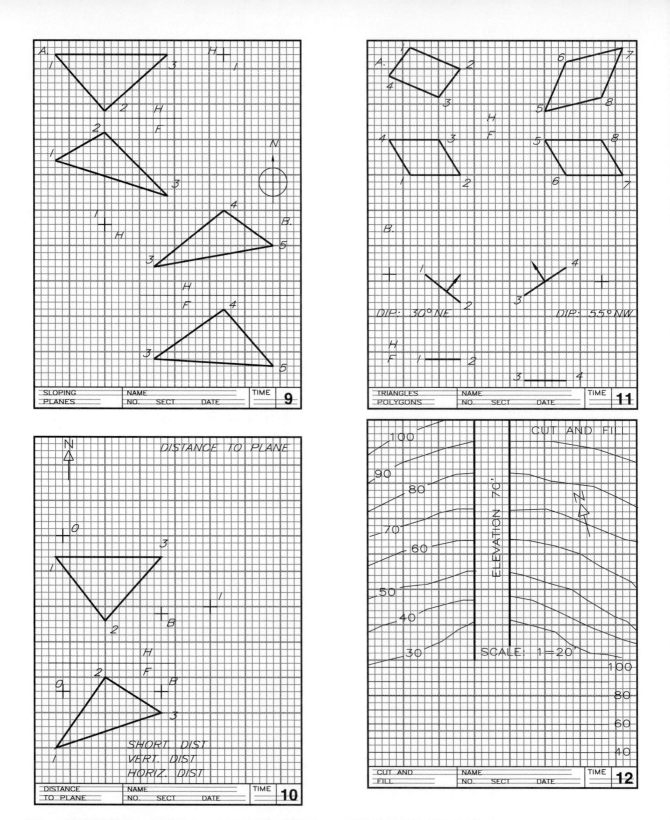

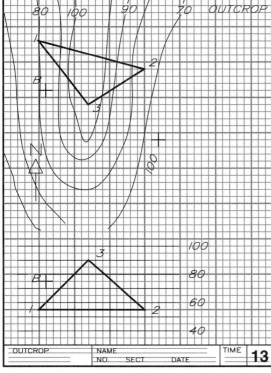

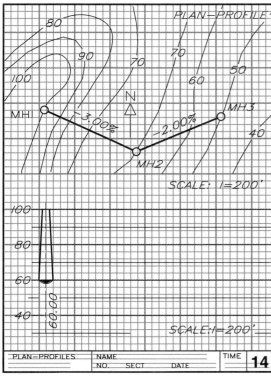

15. (**Sheet 15**) Contour map.

Draw the contour map with its contour lines. Give the lengths of the sides, their compass directions, interior angles, scale, and north arrow.

16. (**Sheet 16**) Vertical section.

Draw the contour map and construct the profile (vertical section), as indicated by the cutting plane line in the plan view. Note that the profile scale is different from the plan scale.

Review Questions

1. What conditions must exist in order for you to measure the angle between a line and plane?

2. In which view can you give the compass bearing of a sloping line?

3. In what direction is a sloping line assumed to slope if not specified?

4. What is the percent grade of a line that is 100 ft long in the top view, and has a 5 ft vertical distance between its end in the front view? What is the rise-to-run ratio of this line? What is its angle of slope?

5. Which view must an auxiliary view be projected from in order to obtain the slope of a line?

6. What is the azimuth bearing of a line that has a compass bearing of S30°W?

7. When given top and front views of a line, which view must you project from in order to find a true-length view of the line?

8. Where can you measure the slope of a line?

9. What is the difference between a level line and a contour line?

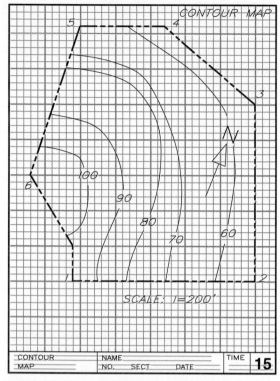

CONTOUR MAP

SCALE: 1=200'

| CONTOUR | NAME | | | TIME | **15** |
| MAP | NO. SECT DATE | | | |

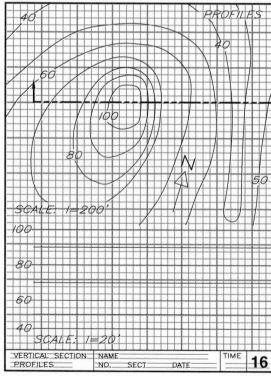

PROFILES

SCALE: 1=200'

SCALE: 1=20'

| VERTICAL SECTION | NAME | | | TIME | **16** |
| PROFILES | NO. SECT DATE | | | |

10. What can be said about the terrain where contour lines are closer together? Farther apart?

11. What is the distance between two points when one has a station number of 8+20 and the other has a station number of 9+10?

12. Why does a manhole usually have a drop of about 0.2 ft between pipes coming into and leaving it? How many inches is 0.2 ft?

13. Describe the view in which a dihedral angle can be found.

14. In what type of view can the piercing point between a line and a plane be easily seen?

15. In what type of view can a line that is perpendicular to a plane be drawn at a true right angle to it?

16. When will a plane appear as an edge?

17. How many frontal lines can be drawn on a given plane? How many horizontal lines?

18. What is the slope of a plane?

19. In what view can the direction of slope be measured and specified? What is the definition of slope direction?

20. If a line in the top view is parallel to the HF reference line, what do you know about its front view?

21. Define the strike of a plane.

22. In what view must strike be measured?

23. Define the dip of a plane.

24. What is the difference between a level line and a horizontal line?

25. What is the freeboard of a dam?

26. If a level line of a plane has an azimuth bearing of N60° E, what is the strike of the plane?

27. What is the area called where an underground ore vein extends to the surface of the earth?

28. What is the slope of a level line?

29. Can you find the point view of a line by projecting from the front view of a horizontal line? Explain.

30. How many orthographic views of a line are required in order to find its true length by the Pythagorean theorem? Explain.

Thinking with a Pencil:

31. Sketch a TL vertical line in the front view about 2 in. tall. Sketch its top and side views. How will it appear in the top? In the side? What type of line is it?

32. Sketch line AB that is 2 in. long in the top view and has a direction of N45° E. Draw its front view with B 0.5 in. lower than A. What is its percent grade? In which view is it TL?

33. Sketch line AB (Problem 32) in a TL diagram. Approximately how long is it?

34. Sketch a right cone with a horizontal base approximately 3 in. in diameter with a height of 4 in. Sketch contour lines in both views at 0.5-in. intervals to describe the cone.

35. Sketch a plot plan (4 in. × 4 in.) using contour lines spaced 10 ft apart vertically to represent a hilltop located in the center of the plot. The highest point is 80 ft, and the lowest point is 20 ft. Multiple solutions are possible.

36. Sketch a profile of your plot made in Problem 35.

37. Same as Problem 35, but modify the hilltop to have a vertical face on one side from an elevation of 30 ft to 80 ft.

38. Using the plot plan sketched in Problem 35, sketch the path of a ball rolling from its peak in four directions: north, south, east, and west.

39. Sketch top and front views of a horizontal plane that is triangular in shape. Draw top and front views of a line that is higher than the line and show visibility.

40. Sketch top and front views of a three-legged stool with a 12-in. circular seat. The top of the seat is to be 12 in. from the floor; each leg makes a 10° angle with the vertical. Sketch a view that will show one of the legs TL. What angle must the drilled holes make with the seat and in what direction?

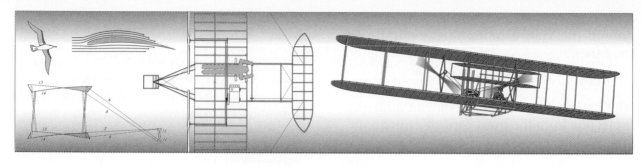

28

Successive Auxiliary Views

28.1 Introduction

A detailed drawing and specifications for a design cannot be completed without determining its geometry, a process that usually requires the application of descriptive geometry. The structural supports for the roof system of the shopping mall shown in **Fig. 28.1** are examples of complex spatial geometry problems in which lengths must be determined, angles between lines and planes calculated, and three-dimensional connectors designed.

The process of determining the 3D geometry of a design requires the use of secondary and successive auxiliary views. **Secondary auxiliary views are views projected from primary auxiliary views, and successive auxiliary views are views projected from secondary auxiliary views**.

28.2 Point View of a Line

Recall that, when a line appears true length, you can find its point view in a primary auxil-

28.1 The structural support of the roof system of a shopping mall, designed and fabricated through the application of descriptive geometry. (Courtesy of Lorna Stuckgold, Kaiser Engineers.)

504

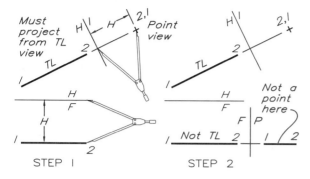

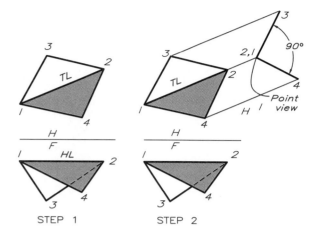

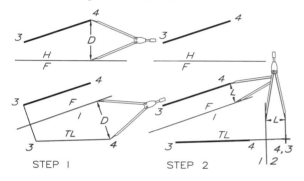

28.2 To find the point view of a line, project an auxiliary view from the true-length view of the line.

28.3 A point view of an oblique line.

Step 1 Draw a line of sight perpendicular to one of the views—the front view, in this case. Line 3–4 is found true length in an auxiliary view projected perpendicularly from the front view.

Step 2 Draw a secondary reference line, 1–2, perpendicular to the true-length view of line 3–4. Find the point view by transferring dimension L from the front view to the secondary auxiliary view.

iary view projected parallel from it. In **Fig. 28.2**, line 1–2 is true length in the top view because it is horizontal in the front view. To find its point view in the primary auxiliary view, first construct reference line H1 perpendicular to the true-length line. Transfer the height dimension, H, to the auxiliary view in order to locate the point view of 1–2.

Line 3–4 in **Fig. 28.3** is not true length in either view. Finding the line's true length by a primary auxiliary view enables you to find its point view. To obtain a true-length view of line 3–4, project an auxiliary view from the front view (or from the top view). Projecting parallel from the true-length view to a secondary auxiliary view gives the point view of line 3–4.

28.4 The angle between planes (the dihedral angle) appears in the view where their line of intersection projects as a point. The line of intersection, 1–2, is true length in the top view, so it can be found as a point in a view projected from the top view.

Label the line 4–3 because you see point 4 first in the secondary auxiliary view. Label the reference line between the primary and secondary planes 1–2 to represent the primary (1) and secondary (2) planes.

28.3 Dihedral Angles

Recall that the angle between two planes is called **a dihedral angle and can be found in a view where the line of intersection appears as a point**. The line of intersection lies on both planes, so both appear as edges when the intersection is a point view.

The planes shown in **Fig. 28.4** represent a special case because their line of intersection, line 1–2, is true length in the top view. This condition permits you to find the line's point view in a primary auxiliary view and measure the true angle between the planes.

Figure 28.5 presents a more typical case. Here, the line of intersection between the two planes is not true length in either view. The line of intersection, line 1–2, is true length in a primary auxiliary view, and the point view of the line appears in the secondary auxiliary view, where you measure the dihedral angle.

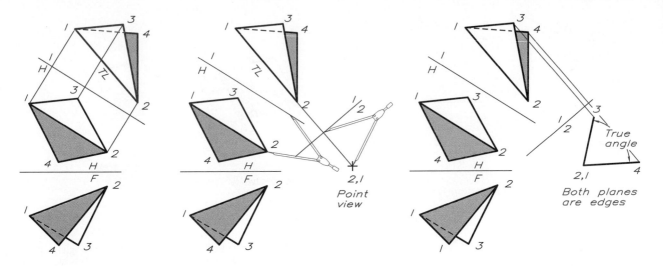

28.5 Angle between two planes.

Step 1 The angle between two planes is found in a view where the line of intersection (1–2) appears as a point. Find the TL view of the intersection in an auxiliary view.

Step 2 Obtain the point view of the line of intersection in the secondary auxiliary view by projecting parallel to the true-length view of line 1–2 in the primary auxiliary view.

Step 3 Complete the edge views of the planes in the secondary auxiliary view by locating points 3 and 4. Measure the angle between the planes (the dihedral angle) in this view.

This principle was applied to determine the angles between the wall panels of the control tower shown in **Fig. 28.6**. That allowed the corner braces to be designed and the structure to be assembled correctly.

28.4 True Size of a Plane

A plane can be found true size in a view projected perpendicularly from an edge view of a plane. The front view of plane 1–2–3 in **Fig. 28.7** appears as an edge in the front view as a special case. The plane's true size is in a primary auxiliary view projected perpendicularly from the edge view.

Figure 28.8 depicts a general case in which you can find the true-size view of plane 1–2–3 by finding the edge view of the plane and constructing a secondary auxiliary view projected perpendicularly from the edge view to find the plane's true size.

This principle can be used to find the angle between lines—for example, the angles of bend in an automobile exhaust pipe (**Fig. 28.9**). The method of solving a problem of this

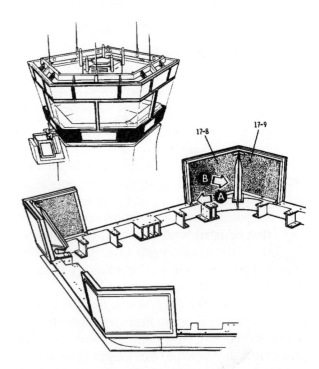

28.6 The wall panels of this control tower, intersecting at compound angles, illustrate the need to determine and measure dihedral angles. The designer had to calculate the angles between the wall panels in order to design connectors for securing the panels at their joints.

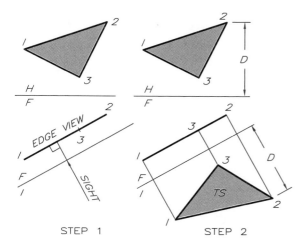

STEP 1 STEP 2

28.7 The true size of a plane (special case).

Step 1 Because plane 1–2–3 appears as an edge in the front view, it is a special case. Draw the sight line perpendicular to its edge and the F1 parallel to the edge.

Step 2 Find the true size of plane 1–2–3 in the primary auxiliary view by locating the vertex points with the depth D dimension.

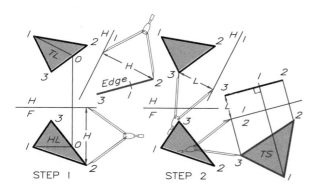

STEP I STEP 2

28.8 The true size of a plane (general case).

Step 1 Find the edge view of plane 1–2–3 by obtaining the point view of true-length line 1-0 in an auxiliary view.

Step 2 Find a true-size view by projecting a secondary auxiliary view perpendicularly from the edge view of the plane found in the primary auxiliary view.

type is shown in **Fig. 28.10**. The top and front views of intersecting centerlines are given; the angles of bend and the radii of curvature must be found. Angle 1–2–3 is an edge in the primary auxiliary view and true size in the secondary view, where it can be measured and

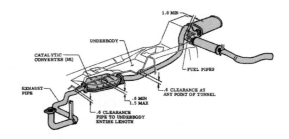

28.9 The designer determined the angles of bend in this automobile exhaust pipe by applying the principles of the angle between two lines. (Courtesy of General Motors Corporation.)

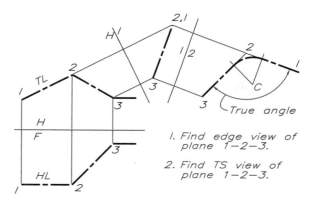

1. Find edge view of plane 1–2–3.

2. Find TS view of plane 1–2–3.

28.10 The angle between two lines is obtained by finding a true-size view of the plane formed by lines 1–2 and 2–3.

28.11 The support frame of this materials conveyor is an example of the application of determining lengths and angles during the design stages. It is used on construction sites to rapidly move cement, aggregate, and sand. (Courtesy of Speed King Manufacturing Co.)

the radius of curvature drawn. The support frame of the materials conveyor (**Fig. 28.11**) is an example of a design involving planes and lines in geometric combinations that was designed with applications of auxiliary views.

28.5 Shortest Distance from a Point to a Line: Line Method

The shortest distance from a point to a line can be measured in the view where the line appears as a point. The shortest distance from point 3 to line 1–2 that appears in a primary auxiliary view in **Fig. 28.12** (Step 1) is a special case. The distance from point 3 to the line is true length in the auxiliary view where the line is a point, so it is parallel to reference line F1 in the front view.

Figure 28.13 shows how to solve a general-case problem of this type, where neither line appears true length in the given views. Line 1–2 is true length in the primary auxiliary projected perpendicularly from the front view. The point view of line 1–2 lies in the secondary auxiliary view, where the distance from point 3 is true length. Because line 0–3 is true length in this view, it will be parallel to reference line 1–2 in the preceding view, the primary auxiliary view. It is also perpendicular to the true-length view of line 1–2 in the primary auxiliary view.

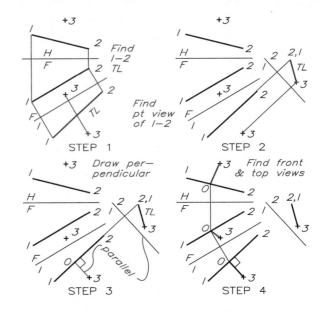

28.13 The shortest distance from a point to a line.

Step 1 The shortest distance from a point to a line is found in the view where the line appears as a point. Find the true length of line 1–2 by projecting from the front view.

Step 2 Line 1–2 is a point in a secondary auxiliary view projected from the true-length view of line 1–2. The shortest distance to it is true length in this view.

Step 3 Since 3–0 is true length in the secondary auxiliary view, it is parallel to the 1–2 reference line in the primary auxiliary view and perpendicular to the line.

Step 4 Find the front and top views of 3–0 by projecting from the primary auxiliary view in sequence.

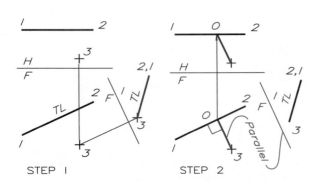

28.12 The shortest distance from a point to a line.

Step 1 The shortest distance from a point to a line is the true length where the line (1–2) appears as a point. The true-length view of the connecting line appears in the primary auxiliary view.

Step 2 When the connecting line is projected back to the front view, it must be parallel to the F1 reference line in the front view. Project line 3–0 back to the top view.

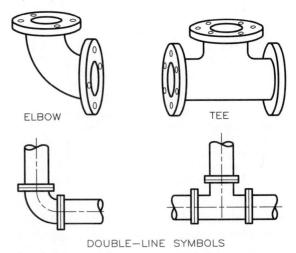

28.14 The shortest distance between two lines, the perpendicular distance, is the most economical connector between them. This also permits the use of standard fittings, 90° tees and elbows.

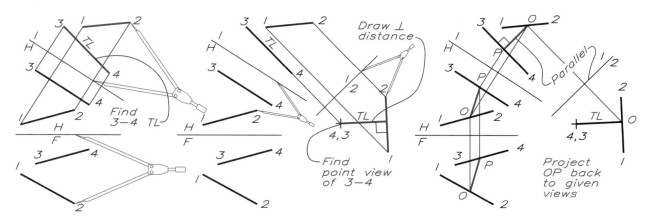

28.15 Shortest distance between skewed lines (line method).

Step 1 The shortest distance between skewed lines appears where one of the lines is a point. Find the 3–4 TL by projecting from the top view.

Step 2 Find the point view of line 3–4 in a secondary auxiliary view projected from the true-length view of line 3–4. The shortest distance between the lines is perpendicular to line 1–2.

Step 3 The shortest distance is true length in the secondary auxiliary view, so it must be parallel to the 1–2 reference line in the preceding view. Project line OP back to the given views.

28.6 Shortest Distance Between Skewed Lines: Line Method

Randomly positioned (nonparallel) lines are called **skewed lines**. The shortest distance between two skewed lines is found in the view where one of the lines appears as a point.

The shortest distance between two lines is a line perpendicular to both lines. The location of the shortest distance between lines is both functional and economical. **Figure 28.14** shows standard 90° pipe connectors (tees and elbows) that are used to make the shortest connections between skewed pipes.

Figure 28.15 illustrates how to use the line method to find the shortest distance between skewed lines. Find the true length of line 3–4 and then its point view in the secondary auxiliary view, where the shortest distance is perpendicular to line 1–2. Because the distance between the lines is true length in the secondary auxiliary view, it is parallel to reference line 1–2 in the primary auxiliary view. Find point 0 by projection and draw OP perpendicular to line 3–4. Project the line back to the given principal views.

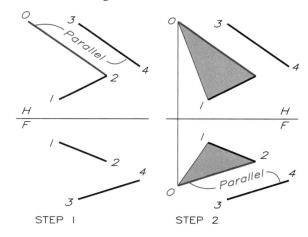

28.16 A plane through a line parallel to another line.

Step 1 Draw line 0–2 parallel to line 3–4 to a convenient length.

Step 2 Draw the front view of line 0–2 parallel to the front view of line 3–4. Find the length of line 0–2 in the front view by projecting from the top view of 0. Plane 1–2–0 is parallel to line 3–4.

28.7 Shortest Distance Between Skewed Lines: Plane Method

You may also determine the shortest distance between skewed lines by the plane method, which requires construction of a plane through a line parallel to another line (**Fig. 28.16**). The top and front views of line 0–2 are parallel to their respective views of line 3–4.

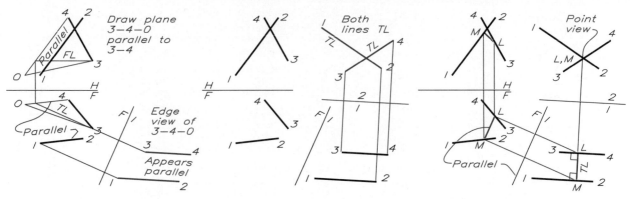

28.17 Shortest distance between skewed line (plane method).

Step 1 Construct a plane through 3–4 parallel to 1–2. Find plane 3–4–0 as an edge by projecting it from the front view. The lines will appear parallel.

Step 2 The shortest distance is true length in the primary auxiliary view and perpendicular to both lines. Project the secondary auxiliary view perpendicularly from the lines in the primary auxiliary to find both lines true length.

Step 3 The crossing point of the two lines is the point view of the perpendicular distance (LM) between them. Project LM to the primary auxiliary view, where it is true length, and back to the given views.

Therefore, plane 1–2–0 is parallel to line 3–4. Both lines will appear parallel in an auxiliary view, where plane 1–2–0 appears as an edge.

Figure 28.17 demonstrates this principle. First, construct plane 3–4–0. When its edge view is found in a primary auxiliary view, the lines appear parallel. To find the secondary auxiliary view where both lines are true length and cross, project a secondary auxiliary view perpendicularly from these parallel lines. The crossing point is the point view of the shortest distance between the lines. That distance is true length and perpendicular to both lines when projected to the primary auxiliary view as line LM. Project line LM back to the given top and front views to complete the solution.

28.8 Shortest Level Distance Between Skewed Lines

The shortest level (horizontal) distance between two skewed lines can be found by the plane method but not by the line method. In **Fig. 28.18**, plane 3–4–0 is constructed parallel to line 1–2, and its edge view is found in the primary auxiliary view. Lines 1–2 and 3–4 appear parallel in this view, and the horizontal reference plane H1 appears as an edge.

A line of sight parallel to H1 is used, and the secondary reference line, 1–2, is drawn perpendicular to H1. The crossing point of the lines in the secondary auxiliary view locates the point view of the shortest horizontal distance between the lines. This line, LM, is true length in the primary auxiliary view and parallel to the H1 plane. Line LM is projected back to the given views. As a check on construction, LM must be parallel to the HF line in the front view, verifying that it is a level or horizontal line.

28.9 Shortest Grade Distance Between Skewed Lines

Features of many applications (such as highways, power lines, or conveyors) are connected to other features at specified grades other than horizontal or perpendicular. For example, the design of the refinery installation shown in **Fig. 28.19** involved the application of slopes and grades of conveyor chutes that were critical to their optimum operation.

If you need to find a 40-percent grade connector between two lines (**Fig. 28.20**), use the plane method. To obtain an edge view of the horizontal plane from which the 40-percent

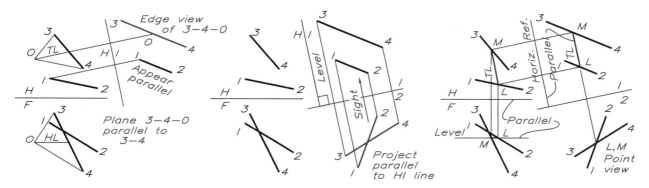

28.18 Shortest level distance between skewed lines (plane method).

Step 1 Construct plane 0–3–4 parallel to 1–2 by drawing 0–4 parallel to 1–2. Find the edge view of plane 0–3–4 by projecting off the top view; the lines appear parallel. Project the auxiliary view from the top view to find the horizontal plane as an edge.

Step 2 Infinitely many horizontal (level) lines may be drawn parallel to reference line H1 in the auxiliary view, but the shortest one appears true length. Construct the secondary auxiliary view by projecting parallel to H1 to find the point view of the shortest level line.

Step 3 The crossing point of the lines in the secondary auxiliary view is the point view of the level connector, LM. Project LM back to the given views. LM is parallel to the horizontal reference plane in the front view, verifying that it is a level line.

28.19 This massive refinery installation with numerous conveyors involved many applications of skewed lines at specified percent grades.

grade is constructed, you must project the primary auxiliary view from the top view. Construct a view in which the lines appear parallel and draw a 40-percent grade line from the edge view of the horizontal (H1) by laying off rise and run units of 4 and 10, respectively. The grade line may be constructed in two directions from the H1 reference line, but the shortest distance is the direction most nearly perpendicular to both lines.

Project the secondary auxiliary view parallel to this 40-percent grade line to find the crossing point of the lines to locate the shortest connector, LM. Project line LM back to all views; it is true length in the primary auxiliary view, where the given lines appear parallel.

The shortest distances between skewed lines—perpendicular, horizontal, and perpendicular—are true length in the view where the lines appear parallel. The plane method is the general-case method that can be used to find the shortest connector between any two lines.

28.10 Angular Distance to a Line

Standard connectors used to connect pipes and structural members are available in standard angles of 90° and 45° (see again Fig. 28.14). Specifying these standard connectors in a design is far more economical than calling for fabrication of specially made connectors.

In **Fig. 28.21**, a line from point 0 that makes an angle of 45° with line 1–2 is to be found. Connect point 0 with the line's endpoints, 1 and 2, to find plane 1–2–0 in the top and front

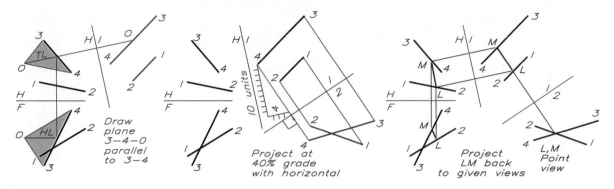

28.20 Grade distance between skewed lines.

Step 1 To find a level line or a line on a grade between two skewed lines, the primary auxiliary must be projected from the top view. Construct plane 3–4–0 parallel to 1–2. Find the edge view of the plane; the lines appear parallel.

Step 2 Construct a 40-percent grade line from the edge view of the H1 reference line in the primary auxiliary view that is most nearly perpendicular to the lines. Project the secondary auxiliary view parallel to the grade line. The shortest grade distance appears true length in the primary auxiliary.

Step 3 The point of crossing of the two lines in the secondary auxiliary view establishes the point view of the 40-percent grade line, LM. Project LM back to the primary auxiliary view to find its true length. Project LM back to the top and front views to complete the problem.

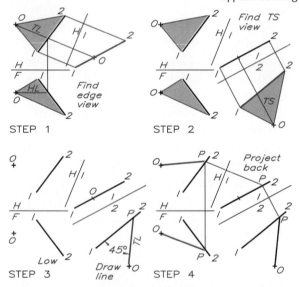

28.21 A line through a point with a given angle to a line.

Step 1 Connect 0 to each end of the line to form plane 1–2–0 in both views. Draw a horizontal line in the front view of the plane and project it to the top view where it is TL. Find the point view of AO and the edge view of the plane.

Step 2 Find the true size of plane 1–2–0 in the primary auxiliary view by projecting perpendicularly from its edge view. Omit the outline of the plane in this view and show only line 1–2 and point 0.

Step 3 Construct line OP at an angle of 45° with line 1–2. Draw the angle toward point 2 (the low end) if it slopes downward or toward point 1 if it slopes upward.

Step 4 Project line OP back to the previous views in sequence to complete the problem.

views, and find its edge view in a primary auxiliary view. Find the true-size view of plane 1–2–0 by projecting perpendicularly from its edge view. Measure the angle of the line from point 0 in this view, where the plane of the line and point is true size.

Draw the 45° connector from point 0 toward point 2 (the low point) if it slopes downward or toward point 1 if it slopes upward. Determine the upper and lower ends of line 1–2 by referring to the front view, where the height is easily seen. Project the 45° line, OP, back to the given views.

28.11 Angle Between a Line and a Plane: Plane Method

The angle between a line and a plane can be measured in the view where the plane appears as an edge and the line appears true length. In **Fig. 28.22**, the edge view of plane 1–2–3 lies in a primary auxiliary view projected from the top view and is true size (Step 2) where the line appears foreshortened. Line AB is true length in a third successive auxiliary view projected perpendicularly from the secondary auxiliary view of line AB. The line

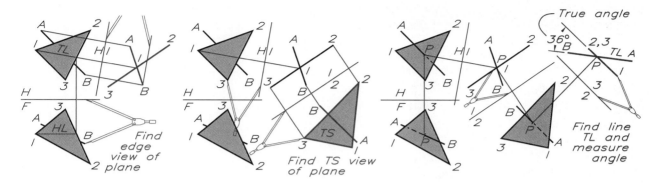

28.22 Angle between a line and a plane (plane method).

Step 1 The angle between a line and a plane is found in the view where the plane is an edge and the line is TL. Find the plane as an edge by projecting it from the top view.

Step 2 To find the plane's true size, project a secondary auxiliary view perpendicularly from the edge view of the plane. A view projected in any direction from a true-size plane will show the plane as an edge.

Step 3 Project a third successive auxiliary view perpendicular from AB. The line is TL, the plane is an edge, and the angle is TS. Project AB back in sequence to the given views; find the piercing points and the visibility of the views.

appears true length and the plane appears as an edge in the third successive auxiliary view. Therefore, the angle between the line and the plane can be measured here. A tracking antenna that sends signals into outer space (**Fig. 28.23**) provides a vivid example of the need for finding the angles between lines and planes as well as other applications covered in this chapter.

28.12 Angle Between a Line and a Plane: Line Method

An alternative method (not illustrated here) of finding the angle between a line and a plane is the line method, where the line, rather than the plane, is the primary geometric element that is projected. The line is found as true length in a primary auxiliary view; it is found as a point in the secondary auxiliary view; and the plane is found as an edge in the third successive auxiliary view.

Because this last view was projected from a point view of the line, the line appears true length in this view where the plane appears as an edge. Project the piercing point back to the secondary, primary, top, and front views in sequence to complete the problem.

28.23 The design of this giant tracking antenna involved the application of theories about the intersections of planes and lines. (Courtesy of Ryan Aeronautics.)

Problems

Use size A sheets for the following problems and lay out your solutions with instruments on grid or plain paper. Each square on the grid is equal to 0.20 in. (5 mm). Label all reference planes and points in each problem with 1/8-in. letters or numbers, using guidelines.

Use the crosses marked "1" and "2" for positioning the primary and secondary reference lines. Primary reference lines should pass through "1" and secondary reference lines through "2."

1. (Sheet 1) Lines and planes.

  (A–B) Find the point views of the lines.

  (C–D) Find the angles between the planes.

2. (Sheet 2) True-size planes.

  (A–B) Find the true-size views of the planes.

3. (Sheet 3) Angles between lines.

  (A–B) Find the angles between the lines.

4. (Sheet 4) Distance to lines.

  (A–B) Find the shortest distances from the points to the lines. Show this distance in all views.

5. (Sheet 5) Distances between lines.

  (A–B) Find the shortest distances between the lines by the line method. Show this distance in all views.

6. (Sheet 6) Skewed line problems.

Find the shortest distance between the lines by the plane method. Show the line in all views. Alternative problem: Find the shortest horizontal distance between the two lines and show the distance in all views.

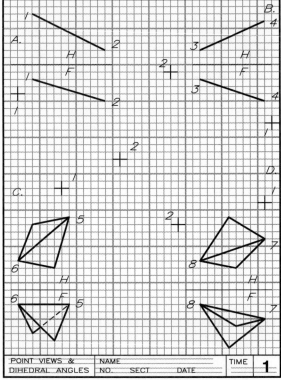

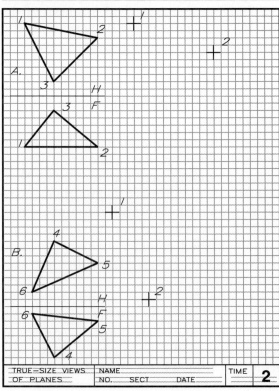

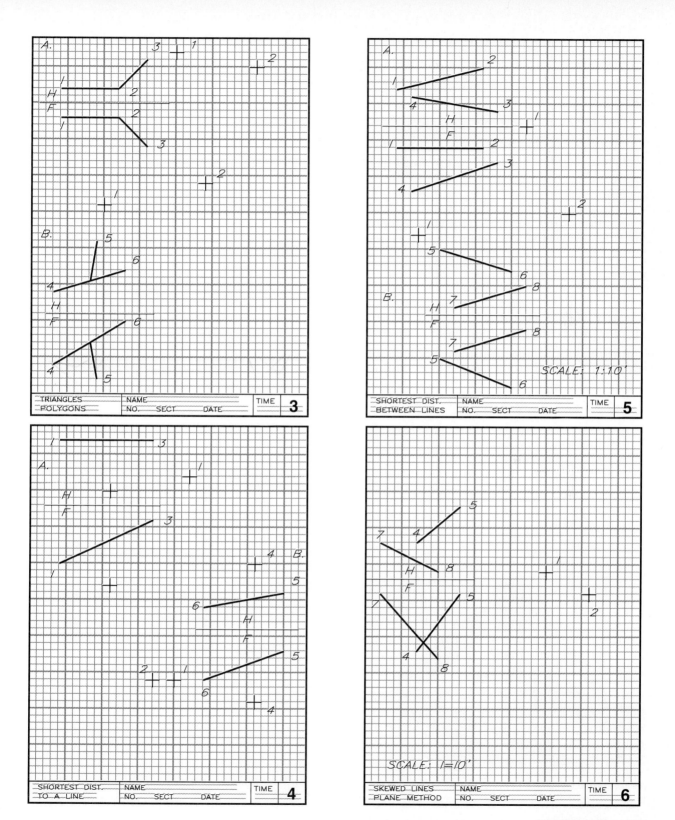

TRIANGLES POLYGONS | NAME | NO. SECT DATE | TIME **3**

SHORTEST DIST. BETWEEN LINES | NAME | NO. SECT DATE | TIME **5**

SCALE: 1:10'

SHORTEST DIST. TO A LINE | NAME | NO. SECT DATE | TIME **4**

SKEWED LINES PLANE METHOD | NAME | NO. SECT DATE | TIME **6**

SCALE: 1=10'

PROBLEMS • 515

7. (Sheet 7) Grade distance between lines.

Find the shortest 20-percent grade distance between the two lines. Show this distance in all views.

8. (Sheet 8) Angular distance to a line.

Find the connector from point 0 that intersects intersect line 1–2 at 60°. Show this line in all views. Project from the top view. Scale: full size.

9. (Sheet 9) Angle between a line and plane. Find the angle between the line and the plane by the plane method. Show visibility in all views.

Design Problems

10. Railroad bumper (Fig 28.24): Approximate the dimensions to construct the needed orthographic views on a size-B sheet for finding the following:

    A. The lengths of all members.

    B. The angles the members make with the tracks.

    C. Angles between the members.

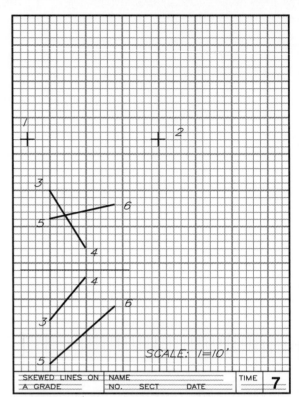

SCALE: 1=10'

| SKEWED LINES ON A GRADE | NAME NO.   SECT   DATE | TIME | 7 |

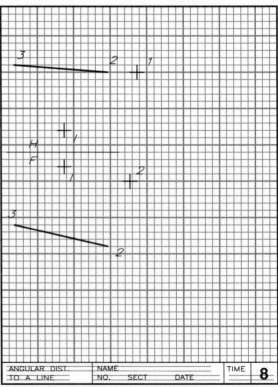

| ANGULAR DIST. TO A LINE | NAME NO.   SECT   DATE | TIME | 8 |

28.24 A railroad bumper for stopping cars at the end of the track. (Courtesy of Hays Track Appliance Company.)

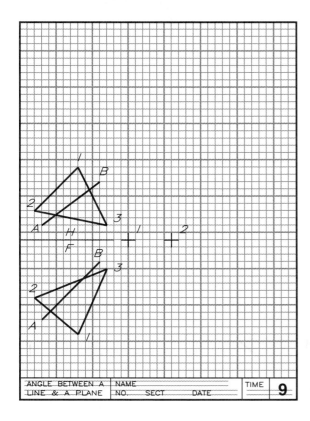

ANGLE BETWEEN A LINE & A PLANE | NAME | | TIME | **9**
NO. SECT DATE

12. Bulkhead clearance (Fig. 28.26): Lay out the orthographic views on a size-B sheet that are necessary to find the clearance between the bulkhead and the electrical connector box. If necessary, relocate the connector so it will have a .50 clearance with the bulkhead.

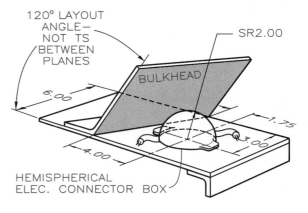

28.26 Problem 12

11. Member connector (Fig. 28.25): A connector for attaching a support member for a bumper brace similar to that in **Fig. 28.24** is shown in **Fig. 28.25.** Using the information determined in Problem 10, make a detail drawing of the connector on a size-B sheet.

13. Web clearance (Fig. 28.27): Layout the orthographic views on a size-B sheet that are necessary to find the clearance between the tube and web in an aircraft. If necessary, modify the web to permit a .25 clearance with the tube.

28.25 A connector for installing a railroad bumper to a track. (See Fig. 28.24.) (Courtesy of Hayes Track Appliance Company.)

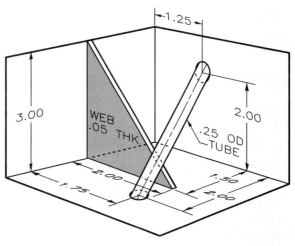

28.27 Problem 13

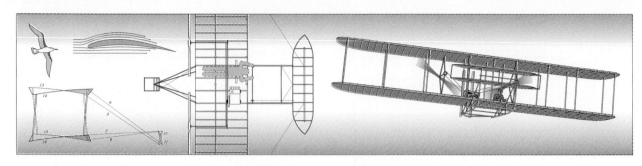

29

Revolution

29.1 Introduction

Many products and systems involve the principles of revolution. An automobile's front suspension, for example, must revolve about several axes at each wheel, a design that is just one of many based on the principles of revolution. Revolution is an alternative technique of solving countless spatial problems in orthographic views—from the true-size view of a surface to the angle between planes or lines. Revolution was used to solve descriptive geometry problems before the introduction of the auxiliary-view method.

29.2 True-Length Lines: Front View

Auxiliary view and revolved view methods of obtaining the true size of inclined surfaces are compared in **Fig. 29.1**. In the auxiliary view method, the observer changes position to an auxiliary vantage point and looks perpendicularly at the object's inclined surface. In the revolved view method, the top view of the

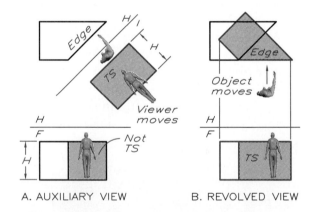

29.1 Auxiliary views vs. revolved views.

A The surface is found true size in an auxiliary view.

B The surface is revolved to be seen true size in the front view.

object is revolved about the axis until the edge view of the inclined plane is parallel to the frontal plane and perpendicular to the standard line of sight from the front view. In other words, the observer's line of sight does not change, but the object is revolved until the

518

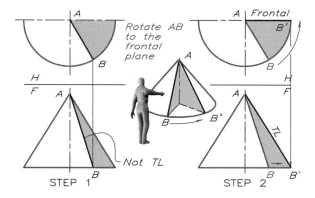

29.2 Determining a true-length line in the front view.

Step 1 Use the top view of line AB as a radius to draw the base of a cone with point A as the apex. Draw the front view of the cone with a horizontal base through point B.

Step 2 Revolve the top view of line AB to be parallel to the frontal plane. When projected to the front view, frontal line AB′, the outside element of the cone, is TL in the front view.

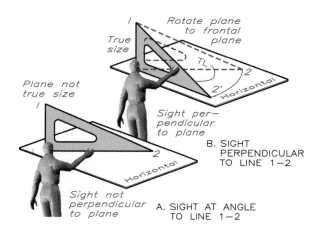

29.3 Determining a true-length line by revolution.

A Line 1–2 does not appear true length in the front view because the observer's line of sight is not perpendicular to it.

B When the triangle is revolved into the frontal plane, the observer's line of sight is perpendicular to it and line 1–2′ is seen true length.

plane appears true size in the observer's normal line of sight.

To find a true-length line in the front view by revolution (**Fig. 29.2**), revolve line AB into the frontal plane. The top view represents the circular base of a right cone, and the front view is the triangular view of a cone. Line AB′ is the outside element of the cone's frontal line and is true length in the front view.

Figure 29.3 illustrates the technique of finding line 1–2 true length in the front view. The observer's line of sight is not perpendicular to the triangle containing line 1–2 in its first position, and line 1–2 is not seen true length. When the triangle is revolved into the frontal plane, the observer's line of sight is perpendicular to the plane, and line 1–2 is seen true length.

In the Top View

A surface that appears as an edge in the front view may be found true size in the top view by a primary auxiliary view or by a single revolution (**Fig. 29.4**). The axis of revolution is a point in the front view and true length in the top view. Revolving the edge view of the plane into the horizontal in the front view and projecting it to the top view yields the surface's true size. As in the auxiliary view method, the depth dimension, D, does not change.

In **Fig. 29.5**, revolving line CD into the horizontal gives its true length in the top view. The arc of revolution in the front view represents the base of the cone of revolution. Line CD′ is true length in the top view because it is horizontal and an outside element of the cone. Note that the depth in the top view does not change.

In the Profile View

In **Fig. 29.6**, revolving the front view of line EF into the profile plane gives a true-length view of it. Projecting the circular view of the cone to the side view gives a triangular view of the cone. Because EF′ is a profile line, it is true length in the side view where it is the outside element of the cone.

Alternative Points of Revolution

In the preceding examples, each line is revolved about one of its ends. However, a line

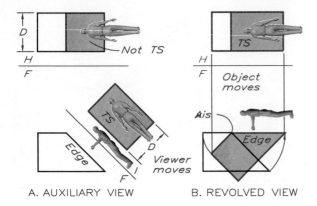

A. AUXILIARY VIEW   B. REVOLVED VIEW

29.4 Auxiliary views vs. revolved view.

A When viewed from a perpendicular, the edge view of a plane appears true size in the primary auxiliary view.

B When the edge view of the plane is revolved to become horizontal, it appears true size in the top view.

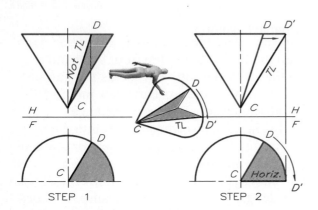

STEP 1   STEP 2

29.5 Finding the true length of a line in the top view.

Step 1 Use the front view of line CD as a radius to draw the base of a cone with C as the apex. Draw the top view of the cone with the base as a frontal plane.

Step 2 Revolve the front view of line CD into a horizontal position, CD'. When projected to the top view, CD' is the outside element of the cone and is true length.

may be revolved about any point on its length. **Figure 29.7** shows how to find line 5–6 true length by revolving it about point 0.

29.3 True Size of a Plane

When a plane appears as an edge in a principal view (the top view in **Fig. 29.8**), it can be revolved parallel to the frontal reference

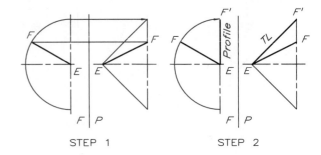

STEP 1   STEP 2

29.6 Finding the true length of a line in the side view.

Step 1 Use the front view of line EF as a radius to draw the circular view of the base of a cone. Draw the side view of the cone with its base through F.

Step 2 Revolve EF in the frontal view to position EF' where it is a profile line, the outside element of the cone, and true length in the side view.

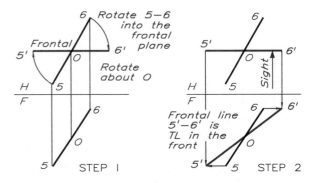

STEP 1   STEP 2

29.7 True-length views of lines may be obtained by revolving them about any point on the lines, not just their endpoints. Line 5–6 is revolved about its midpoint in the top view until it's parallel to the frontal plane and is true length in the front view.

plane. The new front view is true size when projected horizontally across from its original front view.

The combination of an auxiliary view and a single revolution finds the plane in **Fig. 29.9** true size. After finding the plane as an edge by determining the point view of a true-length line in the plane, revolve the edge view to be parallel to the F1 reference line. To find the true size of the plane, project the original points (1, 2, and 3) in the front view parallel to the F1 line to intersect the projectors from 1' and 2'. The true size of the plane may also be found by projecting from the top view to find the edge view.

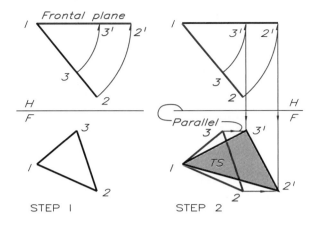

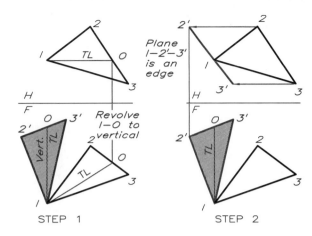

29.8 Determining the true size of a plane.

Step 1 Revolve the edge view of the plane until it is parallel to the frontal plane.

Step 2 Project points 2′ and 3′ to the horizontal projectors from points 2 and 3 in the front view.

29.10 Finding the edge view of a plane.

Step 1 Draw a true-length frontal line on the plane. Revolve the front view until the TL line is vertical.

Step 2 The true-length line, 1–A′ is vertical and appears as a point in the top view, and the plane appears as edge 1–2′–3′.

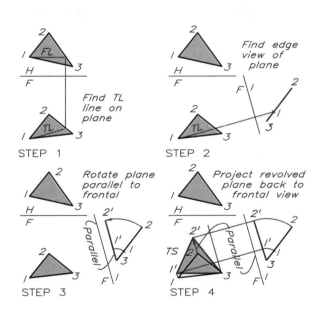

29.9 Determining the true size of a plane by revolution.

Step 1 To find the edge view of the plane by revolution, draw a frontal line on the plane that is true length in the front view.

Step 2 Find the edge view of the plane by finding the point view of the frontal line.

Step 3 Revolve the edge view of the plane until it is parallel to the F1 reference line.

Step 4 Project the revolved points 1′ and 2′ to the front view to the projectors from points 1 and 2 that are parallel to the F1 reference line.

Double Revolution

The edge view of a plane can be found by revolution without using auxiliary views (**Fig. 29.10**). Draw a frontal line on plane 1–2–3, and project it to the front view, where it is true length. Revolve the plane until the true-length line is vertical in the front view. The true-length line projects as a point in the top view; therefore, the plane appears as an edge in this view. Projectors from points 2 and 3 from the top view are parallel to the HF reference line.

A second revolution, called a double revolution, positions this edge view of the plane parallel to the frontal plane, as shown in step 1 of **Fig. 29.11**. Projecting the top views of points 1″ and 2″ to the front view gives a true-size plane 1″–2″–3″ (step 2). We could have shown this second revolution in Fig. 29.10, but it would have resulted in overlapping views, making observation of the separate steps difficult.

Figure 29.12 shows how to use double revolution to find the true size of the oblique plane (1–2–3) of an object. Revolve the true-length line 1–2 on the plane in the top view until it is perpendicular to the frontal plane.

29.3 TRUE SIZE OF A PLANE • 521

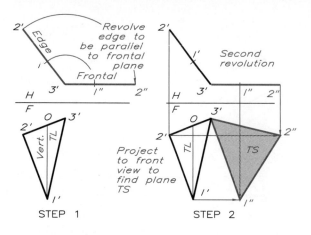

STEP 1   STEP 2

29.11 Finding the true size of a plane.

Step 1 The plane in Fig. 29.10 is revolved to a position parallel to the frontal line.

Step 2 Project points 1′ and 2′ to the front view to intersect with the horizontal projectors from the original points 1′ and 2′. Plane 1′–2′–3′ is true size in this view.

Line 1–2 appears as a point in the front view, and the plane appears as an edge. This revolution changes the width and depth but not the height. Then, revolve the edge view of the plane into a vertical position parallel to the profile plane. To find the plane in true size, project to the profile view, where the depth remains unchanged but the height is greater.

29.4 Angle Between Planes

The angle between the planes and members of the Saturn S-IVB were found by using principles of revolution (**Fig. 29.13**). This application of this geometry was an essential principle in designing the spacecraft.

In **Fig. 29.14**, finding the dihedral angle involves drawing its edge view perpendicular to the line of intersection and projecting the plane of the angle to the front view. When the edge view of the angle is revolved

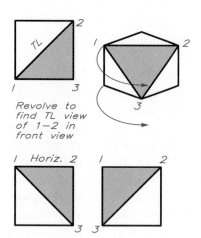

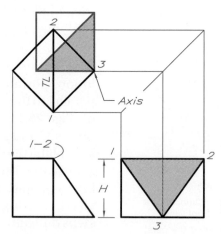

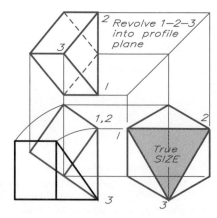

29.12 True size of a plane by double revolution.

Required Find the true size of the plane by revolution.

Step 1 Line 1–2 is horizontal in the frontal view and true length in the top view. Revolve the top view so that line 1–2 appears as a point in the front view.

Step 2 Plane 1–2–3 is an edge; revolve it into a vertical position in the front view to find its true size in the side view. The depth does not change.

29.13 The principles of revolution were used to determine the angles between the planes of the Saturn S-IVB during its design. (Courtesy of National Aeronautics and Space Administration.)

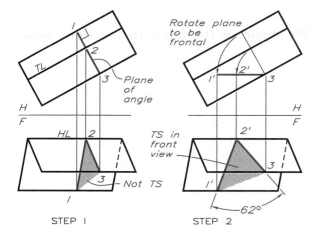

29.14 Finding the dihedral angle between planes.

Step 1 Draw the edge of the angle perpendicular to the TL line of intersection between the planes in the top view; project it to the front view.

Step 2 Revolve the edge view of the plane of the angle to position angle 1′–2′–3 in the top view parallel to the frontal plane. Project this angle to the frontal view, where it is true size and can be measured.

into a frontal plane and projected to the front view, the angle appears true size and can be measured.

Figure 29.15 shows how to solve a similar problem. Here, the line of intersection does not appear true length in the given views;

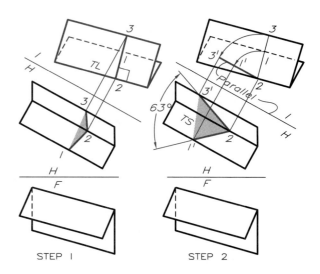

29.15 Determining the angle between oblique planes.

Step 1 Find the true-length view of the line of intersection by projecting perpendicularly from its top view. Draw the edge view of the angle perpendicular to the true length line of intersection and project it to the top view.

Step 2 Revolve the edge view of the plane of the angle (1–2–3) parallel to the H1 reference line and project 1′ and 3′ to the top view. The angle (1′–2–3′) is true size in the top view.

therefore, an auxiliary view is needed to find its true length. Draw the plane of the dihedral angle as an edge perpendicular to the true-length line of intersection. Project the foreshortened view of plane 1–2–3 to the top view, then revolve the edge view of that plane in the primary auxiliary view until it is parallel to the H1 reference line. Project the revolved edge view of the angle back to the top view, where it is true size.

29.5 Determining Direction

To solve more advanced problems of revolution, you must be able to locate the basic directions of up, down, forward, and backward in any given view. In **Fig. 29.16A**, directional arrows in the top and front views identify the directions of backward and up. Pointing backward in the top view, line 4–5 appears as a point in the front view. Projecting

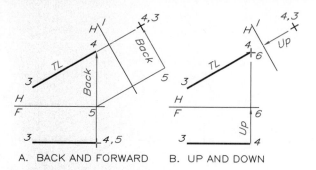

A. BACK AND FORWARD   B. UP AND DOWN

29.16 The directions of backward, forward, up, and down can be identified in the given views with arrows pointing in these directions. Directional arrows can be projected to successive auxiliary views. This drawing shows the directions of (A) backward and (B) up.

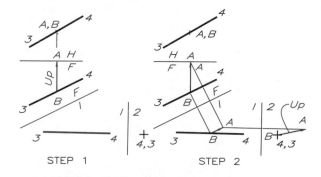

STEP 1   STEP 2

29.17 Direction in a secondary auxiliary view.

Step 1 To find the direction of up in the secondary auxiliary view, draw arrow AB pointing up in the front view. It appears as a point in the top view.

Step 2 Project arrow AB to the primary and secondary auxiliary views to show the direction of up.

arrow 4–5 to the auxiliary view as you would any other line determines the direction of backward. By drawing the arrow on the other end of the line, you would find the direction of forward.

Locate the direction of up in **Fig. 29.16B** by drawing line 4–6 in the direction of up in the front view and as a point in the top view. Then find the arrow in the primary auxiliary by the usual projection method. The direction of down is in the opposite direction.

The location of directions in secondary auxiliary views are found in the same way. To determine the direction of up in **Fig. 29.17**, begin with an arrow that points up in the front view and appears as a point in the top view. Project the arrow AB from the front view to the primary auxiliary view and then to a secondary auxiliary view to show the direction of up. Identify the other directions in the same way by beginning with the two principal views of a known direction.

29.6 Revolution: Point About an Axis

In **Fig. 29.18**, point 0 is to be revolved about axis 3–4 to its most forward position. Find the axis as a point in the primary auxiliary view and draw the circular path of revolution. Draw the direction of forward and find the new

location of point 0 at 0′. Project back through the successive views to find point 0′ in each view. Note that point 0′ lies on the line in the front view, verifying that point 0′ is in its most forward position.

In **Fig. 29.19**, an additional auxiliary view is needed in order to revolve a point about an axis because axis 3–4 is not true length in the given views. You must find the true length of the axis before you can find it as a point in the secondary auxiliary view, where the path of revolution appears as a circle. Revolve point 0 into its highest position, 0′, and locate the up arrow, 3–5, in the secondary auxiliary view. Project back to the given views to locate 0 in each view. Its position in the top view is over the axis, which verifies that the point is at its highest position.

The paths of revolution appear as edges when their axes are true length and as ellipses when their axes are not true length. The angle of the ellipse template for drawing the ellipse in the front view is the angle the projectors from the front view make with the edge view of the revolution in the primary auxiliary view. To find the ellipse in the top view, project an auxiliary view from the top view to obtain the

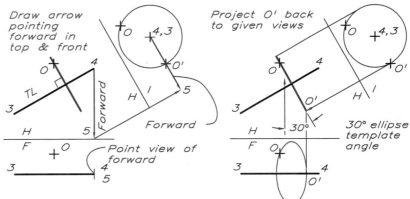

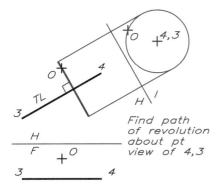

29.18 Revolving a point about an axis.

Step 1 To rotate O about axis 3–4 to its most forward position, find the point view of 3–4. The path of revolution is a circle in the auxiliary view and an edge perpendicular to 3–4 in the top view.

Step 2 Locate the most forward position of point O by drawing an arrow pointing forward in the top view that appears as a point in the front view. Find the arrow, 4–5, in the auxiliary view to locate point O' on the circular path of revolution.

Step 3 Project O' back to the given views. The path of revolution appears as an ellipse in the front view because the axis is not true length. Draw a 30° ellipse; this is the angle of your line of sight with the circular path in the front view.

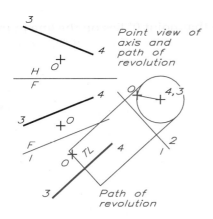

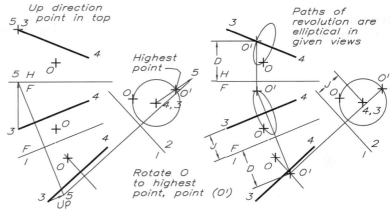

29.19 Revolution about an oblique axis.

Step 1 To rotate O about axis 3–4 to its highest position, find the point view of 3–4 and draw the circular path. The path of revolution is perpendicular to the 3–4 in the primary auxiliary views.

Step 2 To locate the highest position on the path of revolution, draw arrow 3–5 pointing up in the top view and project it to the secondary auxiliary view to find O'.

Step 3 Project point O' back to the given views by transferring the dimensions J and D with your dividers. The highest point lies over the line in the top view. The path of revolution is elliptical wherever the axis is not true length.

path of revolution as an edge perpendicular to the true-length axis.

The handcrank of a casement window (**Fig. 29.20**) is an example of the application of revolution techniques. The designer must determine the clearances between the sill and the window frame when designing the crank in order for it to operate properly.

A Right Prism

The coal chute shown in **Fig. 29.21** conveys coal continuously between two buildings. The sides of the enclosed chute must be vertical and the bottom of the chute's right section must be horizontal. The design of this chute required the application of the technique of revolving a prism about its axis.

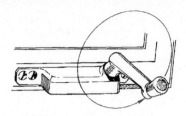

29.20 The handcrank on a casement window is an example of a problem solved by applying revolution principles. The handle must be properly positioned so as not to interfere with the windowsill or wall.

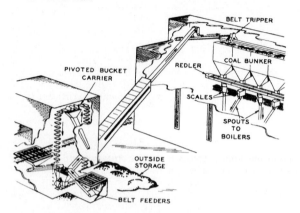

29.21 A conveyor chute must be installed so that two edges of its right section are vertical for the conveyors to function properly. (Courtesy of Stephens-Adamson Manufacturing Company.)

In **Fig. 29.22**, the right section is to be positioned about centerline AB so that two of its sides will be vertical. To do so, find the point view of the axis and project the direction of up to this view. Draw the right section about the axis so that two of its sides are parallel to the up arrow. Find the right section in the other views, then construct the sides of the chute parallel to the axis. The bottom of the chute's right section will be horizontal and properly positioned for conveying coal.

29.7 A Line at Specified Angles

In **Fig. 29.23**, a line is to be drawn through point 0 that makes angles of 35° with the frontal plane and 44° with the horizontal plane and slopes forward and down. First, draw the cone containing elements making 35° with the frontal plane and then the cone with elements making 44° with the horizontal plane. The length of the elements of both cones must be equal so that the cones will intersect with equal elements. Finally, find lines 0–1 and 0–2, which are elements that lie on each cone and make the specified angles with the principal planes.

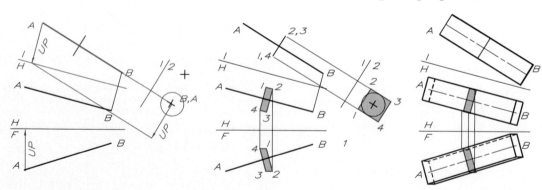

29.22 Revolving a prism about its axis.

Step 1 To draw a square chute with two of its sides vertical, find the point view of centerline AB in the secondary auxiliary view. Draw a circle about the axis with a diameter equal to the square section. Draw a vertical arrow in the front and top views; project it to the secondary auxiliary view to show the direction of vertical.

Step 2 Draw the right section, 1–2–3–4, in the secondary auxiliary view with two sides parallel to the vertical directional arrow. Project this section back to the previous views by transferring measurements with dividers. Locate the edge view of the section anywhere along AB in the primary auxiliary view.

Step 3 Draw the lateral edges of the prism through the corners of the right section parallel to AB in all views. Terminate the ends of the prism in the primary auxiliary view where they appear as edges perpendicular to the center line. Project the corner points of the ends to the top and front views to establish the ends in these views.

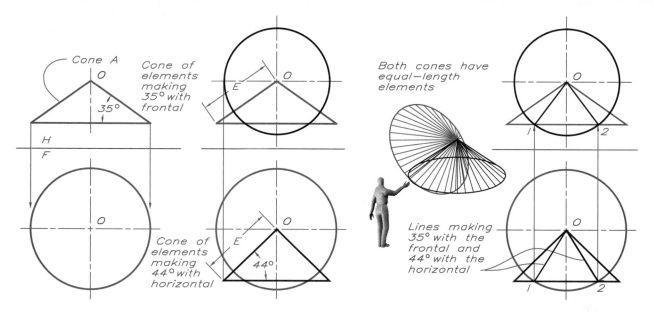

29.23 A line at specified angles.

Step 1 To draw lines at angles of 35° and 44° with the frontal and horizontal, respectively, draw a cone in the top view with outside elements at 35° with the frontal plane. Draw the circular view of the cone in the front with 0 as the apex. All elements of the cone are 35° with the frontal plane.

Step 2 Draw a second cone in the front view with outside elements that make an angle of 44° with the horizontal plane. Draw the elements of this cone equal in length to element E of cone A. All elements of cone B are 44° with the horizontal plane.

Step 3 Because elements A and B are equal in length, two elements lie on the surface of each cone: lines 0–1 and 0–2. Locate points 1 and 2 at the point where the bases of the cones intersect in both views. These lines slope forward and down from this point at the specified angles.

Problems

Use size A sheets for the following problems and lay out your solutions using instruments on grid or plain paper. Each square on the grid is equal to 0.20 in. (5mm). Label all reference planes and points in each problem with 1/8-in. letters or numbers, using guidelines.

Use the crosses marked "1" and "2" for positioning primary and secondary reference lines. The primary reference line should pass through "1" and the secondary reference line through "2."

1. (Sheet 1) True-length lines.

(A–B) Find the true-length views of the lines in their front views by revolution.

(C–D) Find the true-length views of the lines in their top views by revolution.

2. (Sheet 2) True-size planes.

(A–B) By revolution, find the true-size views of plane 1–2–3 in the front view and plane 4–5–6 in the top view.

(C) By using an auxiliary view projected from the top view and one revolution, find the true-size view of plane 7–8–9.

3. (Sheet 3) Angles between planes.

(A–B) Find the angles between the planes by revolution. Show construction.

4. (Sheet 4) Revolution of a point.

(A) Perform the construction necessary to show point 0 revolved into its highest position.

(B) Repeat (A) but show point 0 in its most forward position.

5. (Sheet 5) Chute design.

Construct a chute from A to B that has the cross-section shown. The longer sides are to be vertical sides.

6. (Sheet 6) Lines at specified angles.

Draw the views of the line that is 3.2 in. long that makes a 30° angle with the frontal plane and a 52° angle with the horizontal plane.

Thought Questions

7. In order to find a line TL in the front, the axis of revolution will appear as a point in which view? What will happen if a line appears TL in the horizontal view? In the profile view?

8. When a plane is revolved until a line on it appears as a point, how will the plane appear?

9. What is the difference between finding a line TL by revolution and by auxiliary view?

10. What is the largest view of a plane that can be found?

11. Can a vertical line in the front view be found TL in the top view? Sketch a solution.

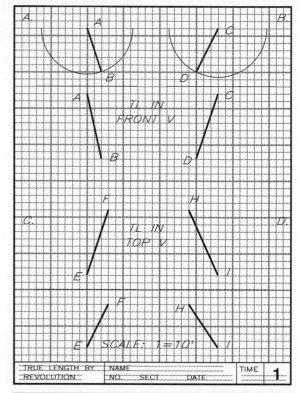

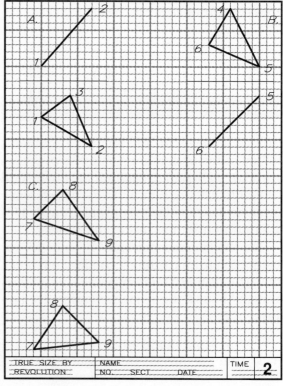

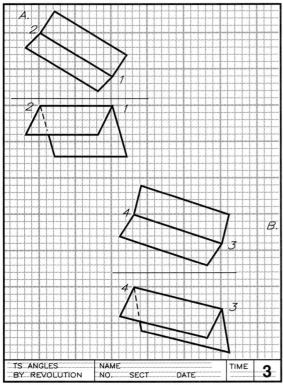

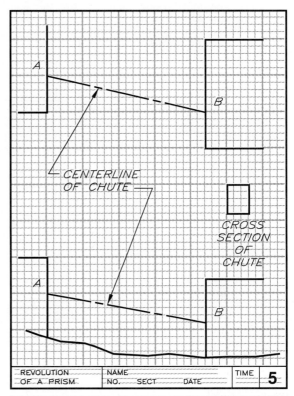

A.

2

1

2   1

4

3

B.

4   3

| TS ANGLES | NAME | | TIME | **3** |
| BY REVOLUTION | NO. SECT DATE | | | |

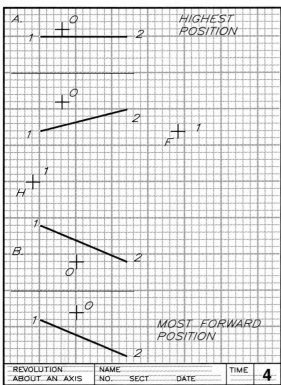

A.   0
1   +   2   HIGHEST
                     POSITION

        +0

1   2   + 1
                      F

H + 1

        1

B.   2

     0+

          + 0

1
                     MOST FORWARD
        2   POSITION

| REVOLUTION | NAME | | TIME | **4** |
| ABOUT AN AXIS | NO. SECT DATE | | | |

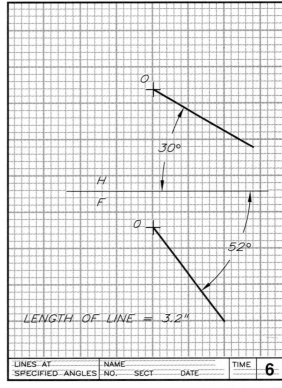

A

B

CENTERLINE
OF CHUTE

CROSS
SECTION
OF
CHUTE

A

B

| REVOLUTION | NAME | | TIME | **5** |
| OF A PRISM | NO. SECT DATE | | | |

0

30°

H

F

0

52°

LENGTH OF LINE = 3.2"

| LINES AT | NAME | | TIME | **6** |
| SPECIFIED ANGLES | NO. SECT DATE | | | |

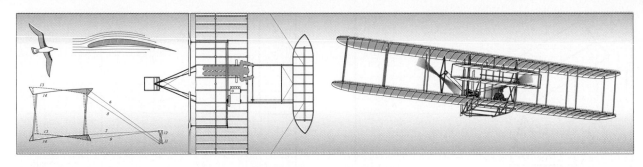

30

Vector Graphics

30.1 Introduction

The design of a structural system requires analysis of each member to determine the loads it must support and whether those loads are in tension or compression. Forces may be represented graphically by vectors and their magnitudes and directions determined in three-dimensional space. Graphical methods are useful in the solution of vector problems as alternatives to conventional trigonometric and algebraic methods. Quantities such as distance, velocity, and electrical properties may also be represented as vectors for graphical solution.

30.2 Definitions

The following definitions will help you more easily understand the discussion of vectors in this chapter:

Force: A push or pull tending to produce motion. All forces have (1) magnitude, (2) direction, and (3) a point of application. The person shown pulling the rope in **Fig. 30.1A** is applying a force to the weight (W).

Vector: A graphical representation of a force drawn to scale and depicting magnitude, direction, and point of application. The vector in **Fig. 30.1B** represents the force applied through the rope to pull the weight (W).

Magnitude: The amount of push or pull represented by the length of the vector line, usually measured in pounds or kilograms.

Direction: The inclination of a force (with respect to a reference coordinate system) indicated by a line with an arrow at one end.

Point of application: The point through which the force is applied on the object or member (point A in Fig. 30.1A).

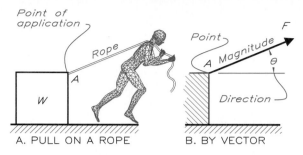

30.1 A force applied to an object (A) may be represented by vectors depicting the magnitude and direction of the force (B).

Compression: The state created in a member by forces that tend to shorten it. Compression is represented by the letter C or a plus sign (+).

Tension: The state created in a member by pulling forces that tend to stretch it. Tension is represented by the letter T or a minus sign (−).

System of forces: The combination of all forces acting on an object as shown (forces A, B, and C in **Fig. 30.2**).

Resultant: A single force that can replace all the forces of a force system and have the same effect (force R1 in Fig. 30.2).

Equilibrant: The opposite of a resultant; the single force that can be used to counterbalance all forces of a force system.

Components: Separate forces that, if combined, would result in a single force; forces A and B are components of resultant R1 in Fig. 30.2.

Space diagram: A visual depiction of the physical relationship between structural members, as given in Fig. 30.2.

Vector diagram: A visual depiction of vectors representing the forces in a system and used to solve for unknown vectors in the system.

Metric units: Standard units of weights and measures. The kilogram (kg) is the unit of mass (load), and one kilogram is approximately 2.2 pounds.

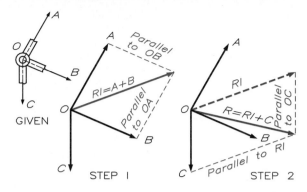

30.2 Determining the resultant by the parallelogram method.

Step 1 Draw a parallelogram with its sides parallel to vectors A and B. The diagonal R1 is the resultant of forces A and B.

Step 2 Draw a parallelogram using vectors R1 and C to find diagonal R, or the overall resultant that can replace forces A, B, and C.

30.3 Coplanar, Concurrent Forces

When several forces, represented by vectors, act through a common point of application, the system is **concurrent**. In Fig. 30.2, vectors A, B, and C act through a single point; therefore, this system is concurrent. When all vectors lie in the same plane, the system is **coplanar** and only one view is necessary to show them true length.

The **resultant** is the single vector that can replace all forces acting on the point of application. Resultants may be found graphically by (1) the parallelogram method and (2) the polygon method.

An **equilibrant** has the same magnitude, orientation, and point of application as the resultant in a system of forces, but in the opposite direction. The resultant of the system of forces shown in **Fig. 30.3** is balanced by the equilibrant applied at point O, thereby causing the system to be in equilibrium.

Resultant: Parallelogram Method
In **Fig. 30.2**, the vectors lie in the same plane, act through a common point, and are scaled

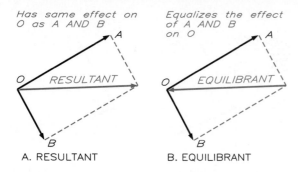

A. RESULTANT

B. EQUILIBRANT

30.3 The (A) resultant and (B) equilibrant are equal in all respects except in direction (shown by arrowhead).

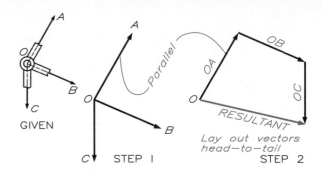

GIVEN   STEP 1

Lay out vectors head—to—tail   STEP 2

30.4 The resultant of a coplanar, concurrent system may be determined by the polygon method, in which the vectors are drawn head-to-tail. The vector that closes the polygon is the resultant.

to their known magnitudes. Use of the parallelogram method to determine resultants requires that the vectors be drawn to scale. Vectors A and B form two sides of a parallelogram. Constructing parallels to these vectors completes the parallelogram. Its diagonal, R1, is the resultant of forces A and B; that is, resultant R1 is the vector sum of vectors A and B.

Replaced by R1, vectors A and B now may be disregarded. Resultant R1 and vector C are two sides of a second parallelogram. Its diagonal, R, is the vector sum of R1 and C and the resultant of the entire system. Resultant R may be thought of as the only force acting on the point, thereby simplifying further analysis.

Resultant: Polygon Method
Figure 30.4 shows the same system of forces, but here the resultant is determined by the polygon method. Again, the vectors are drawn to scale but in this case head-to-tail, in their true directions to form the polygon. The vectors are laid out in a clockwise sequence beginning with vector A. The polygon does not close, so the system is not in equilibrium but tends to be in motion. The resultant R (from the tail of vector A to the head of vector C) closes the polygon.

30.4 Noncoplanar, Concurrent Forces
When vectors lie in more than one plane of projection, they are noncoplanar, requiring three-dimensional views for analysis of their spatial relationships. The resultant of a system of noncoplanar forces may be obtained by the parallelogram method if their projections are given in two adjacent orthographic views. Otherwise, it must be determined by the polygon method.

Resultant: Parallelogram Method
In **Fig. 30.5**, vectors 1 and 2 were used to construct the top and front views of a parallelogram and its diagonal R1 in both views. The front view of R1 must be an orthographic projection of its top view.

Then, resultant R1 and vector 3 are resolved to form the overall resultant in both views. The top and front views of the resultant must project orthographically. The overall resultant replaces vectors 1, 2, and 3. However, it is an oblique line, so an auxiliary view (**Fig. 30.6**) or revolution must be used to obtain its true length.

Resultant: Polygon Method
Figure 30.6 shows the solution of the same system of forces for the resultant by the poly-

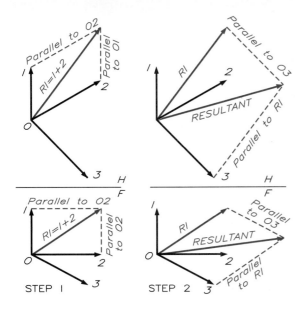

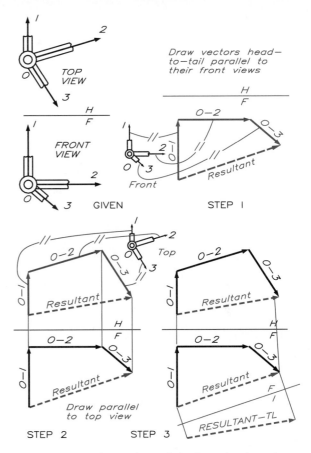

30.5 Finding the resultant by the parallelogram method.

Step 1 Use vectors 1 and 2 to construct a parallelogram in the top and front views. Diagonal R1 is the resultant of vectors 1 and 2.

Step 2 Use vectors 3 and R1 to construct a second parallelogram to find the overall resultant, R.

30.6 Obtaining the resultant of the forces by the polygon method.

Step 1 Lay off each vector head-to-tail parallel to the given view and find the front view of the resultant.

Step 2 Draw the same vectors head-to-tail in the top view, a 3D polygon projected above the front view.

Step 3 The resultant is found true length in an auxiliary view projected from the front view.

gon method. **Each vector is laid head-to-tail clockwise, beginning with vector 1 in the front view**. The vectors are then projected orthographically from the front view to the top view of the vector polygon. The vector polygon does not close, so the system is not in equilibrium. In both views, the resultant (from the tail of vector 1 to the head of vector 3) closes the polygon. However, the resultant is an oblique line, requiring an auxiliary view to obtain its true length.

30.5 Forces in Equilibrium

The manufacturing hoist shown in **Fig. 30.7** can be analyzed graphically to determine the loads carried by each member and cable since it is a coplanar, concurrent structure in equilibrium. **A structure in equilibrium is static with no motion taking place; the members balance each other.**

The coplanar, concurrent structure depicted in **Fig. 30.8** is designed to support a load of W = 165 kg. The maximum loading of each structural member determines the material and size of the members to be used in the design.

A single view of a vector polygon in equilibrium allows you to find only two unknown values. (Later, we show how to solve for three

30.7 The loads in the members of this manufacturing hoist can be determined by vector graphics as a coplanar system in equilibrium. (Courtesy of Pacific Hoist Company.)

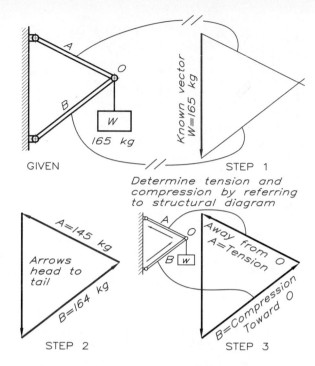

30.8 Coplanar forces in equilibrium: Finding the forces in the structural members supporting the 165-kg load.

Step 1 Draw the load of 165 kg as a vector. Draw vectors A and B parallel to their directions from the ends of the load and head-to-tail.

Step 2 The direction of the 165 kg vector is known. Draw the arrows on the polygon head-to-tail to find the directions of A and B.

Step 3 Vector A points away from point O when transferred to the structural diagram and thus is in tension. Vector B points toward point O and is in compression.

unknowns by using descriptive geometry.) Lay off the only known force, W = 165 kg, parallel to its given direction (here pointing vertically downward). Then draw the unknown forces A and B parallel to the supports to form the force polygon and scale (or calculate) the magnitude of these forces.

Analyze vectors A and B to determine whether they are in tension or compression and thus find their direction. Vector B points upward to the left, which is toward point O when transferred to the structural diagram shown in the small drawing. Vectors that act toward their point of application are in compression. Vector A points away from point O when transferred to the structural diagram and is in tension.

Figure 30.9 is a similar example involving determination of the loads in the structural members caused by the 110-lb weight acting through a pulley. The only difference between this solution and the previous one is the construction of two equal vectors at the outset to represent the cable loads on both sides of the pulley.

30.6 Coplanar Truss Analysis

Designers use vector polygons to determine the loads in each member of a truss by two graphical methods: (1) joint-by-joint analysis and (2) Maxwell diagrams.

Joint-by-Joint Analysis

In the **Fink truss** shown in **Fig. 30.10**, 3000-lb loads are applied at its joints. This method of designating forces is called **Bow's notation**. The exterior forces on the truss are labeled

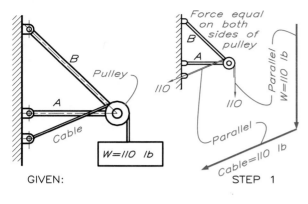

GIVEN:

STEP 1

with letters placed between them, and numerals are placed between the interior members. Each vector is referred to by the number on each of its sides clockwise about its joint. For example, the vertical load at the left is denoted AB, with A at the tail and B at the head of the vector.

First analyze the joint at the left end with a reaction of 4500 lb. When you read clockwise about the joint, the force is EA, where E is the tail and A is the head of the vector. Continuing

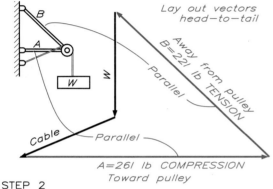

STEP 2

30.9 Forces in equilibrium (pulley application): Finding the forces in the members.

Step 1 The force in the cable is equal to 110 lb on both sides of the pulley. Draw these two forces as vectors head-to-tail and parallel to their directions in the space diagram.

Step 2 Draw A and B head-to-tail to close the polygon. Vector A points toward the point of application and thus is in compression. Vector B points away from the point and is in tension.

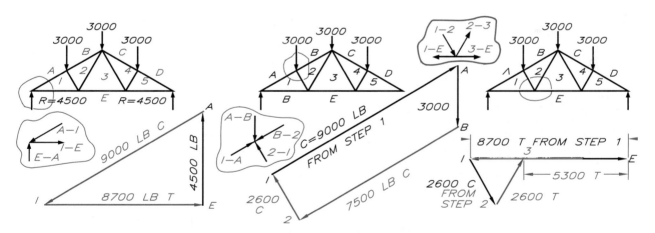

30.10 Truss analysis: joint-by-joint.

Step 1 Label the truss, with letters between the exterior loads and numbers between interior members. Analyze the left joint with only two unknowns, A–1 and 1–E. Find vectors A–1 and 1–E, by drawing them parallel to their directions from both ends of EA in a head-to-tail sequence.

Step 2 Use vector 1–A and load AB from step 1 and find B–2 and 2–1. Draw 1–A first, then AB, and draw B–2 and 2–1 to close the polygon, moving clockwise about the joint. A vector pointing toward the point of application is in compression; a vector pointing away from the point of application is in tension.

Step 3 Lay out vectors E–1 and 1–2 from the preceding steps. Vectors 2–3 and 3–E close the polygon and are parallel to their directions in the space diagram. Vectors 2–3 and 3–E point away from the point of application and thus are in tension.

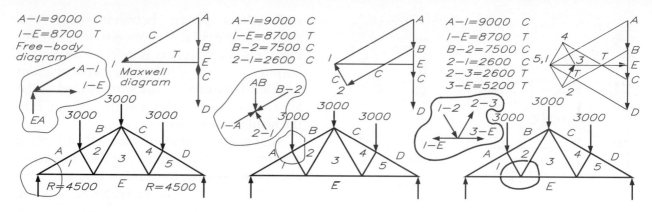

30.11 Truss analysis: Maxwell diagram method.

Step 1 Label the outer spaces with letters and the internal spaces with numbers. Draw the loads head-to-tail in a Maxwell diagram; sketch a free-body diagram of the first joint. Use EA, A–1, and 1–E (head-to-tail) to draw a vector diagram. A–1 is in compression because it points toward the joint; 1–E is in tension because it points away from it.

Step 2 Sketch the next joint to be analyzed. Because AB and A–1 are known, only 2–1 and B–2 are unknown. Draw them parallel to their direction (head-to-tail) in the Maxwell diagram using the previously found vector. Vectors B–2 and 2–1 are in compression, as each points toward the joint. Vector A–1 becomes vector 1–A when read in a clockwise direction.

Step 3 Sketch a free-body diagram of the next joint to be analyzed, where the unknowns are 2–3 and 3–E. Draw their vectors in the Maxwell diagram parallel to their given members to find point 3. Vectors 2–3 and 3–E are in tension because they point away from the joint. Repeat this process to find the vectors on the opposite side.

clockwise, the next forces are A–1 and 1–E, which close the polygon at E, the beginning letter. Place arrowheads in a head-to-tail sequence beginning with the known vector EA.

Determine tension and compression by relating the directions of each vector to the original joint. For example, A–1 points toward the joint and is in compression, whereas 1–E points away and is in tension. The truss is symmetrical and equally loaded, so the loads in the members on the right will be equal to those on the left.

Analyze the other joints in the same way. The directions of the vectors are opposite at each end. For example, vector A–1 is toward the left in step 1 and toward the right in step 2.

Maxwell Diagrams

The **Maxwell diagram** is virtually the same as the joint-by-joint analysis, with the exception that the polygons overlap, with some vectors common to more than one polygon. In **Fig. 30.11** (step 1), the exterior loads are laid out

head-to-tail in clockwise sequence—AB, BC, CD, DE, and EA—with a letter placed at each end of each vector. The forces are parallel so this force diagram is a vertical line.

Vector analysis begins at the left end where the force EA of 4500 lb is known. A free-body diagram is sketched to isolate this joint. The two unknowns, A–1 and 1–E, are drawn parallel to their directions in the truss, with A–1 beginning at point A, 1–E beginning at point E, and both extended to point 1.

Because resultant EA points upward, A–1 must have its tail at A and its direction toward point 1. The free-body diagram shows that the direction is toward the point of application, which means that A–1 is in compression. Vector 1–E points away from the joint, which means that it is in tension. The vectors are coplanar and may be scaled to determine their magnitudes.

In step 2, where vectors 1–A and AB are known, the unknown vectors, B–2 and 2–1 may be determined. Vector B–2 is drawn par-

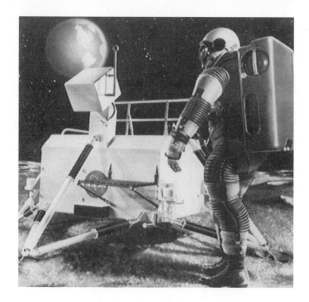

30.12 The structural members of this tripod support for a moon vehicle may be analyzed graphically to determine design load requirement. (Courtesy of NASA.)

allel to its structural member through point B in the Maxwell diagram, and the line of vector 2–1 is extended from point 1 to intersect with B–2 at point 2. The arrows of each vector are drawn head-to-tail. Vectors B–2 and 2–1 point toward the joint in the free-body diagram and therefore are in compression.

In step 3, the next joint is analyzed to find the forces in 2–3 and 3–E. The truss and its loading are symmetrical, so the Maxwell diagram will be symmetrical when completed.

If the last force polygon in the series does not close perfectly, an error in construction has occurred. A slight error may be disregarded, as a rounding error may be disregarded in mathematics. Arrowheads are unnecessary and are usually omitted on Maxwell diagrams because each vector will have the opposite direction when applied to a different joint.

30.7 Noncoplanar Vector Analysis

Special Case

The solution of 3D vector systems requires the use of descriptive geometry because the sys-

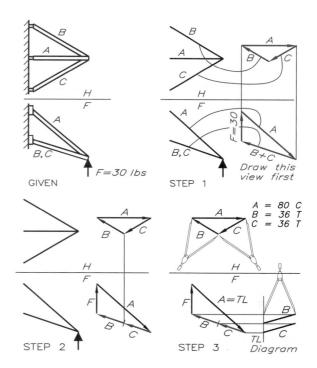

30.13 Noncoplanar structural analysis (special case).

Step 1 Forces B and C coincide in the front view, resulting in only two unknowns. Draw vector F (30 lb) and the two unknown forces parallel to their front view in the front view of the vector polygon. Find the top view of A by projecting from the front. Draw vectors B and C parallel to their top views.

Step 2 Project the point of intersection of vectors B and C in the top view to the front view to separate the head-to-tail vectors.

Step 3 Vectors B and C are in tension because they point away from the point of application in the space diagram. Vector A is in compression because it points toward the point of application.

tem must be analyzed in 3D space. An example is the manned flying system (MFS) shown in **Fig. 30.12**, which was analyzed to determine the loads on its support members. Weight on the moon is 0.165 of earth weight. Thus, a tripod that must support 182 lb on earth needs to support only 30 lb on the moon.

In general, only two unknown vectors can be determined in a single view of a vector polygon that is in equilibrium. However, the system shown in **Fig. 30.13** is a special case

30.14 Noncoplanar structural analysis (general case): Finding the loads in the members.

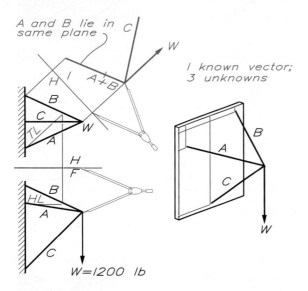

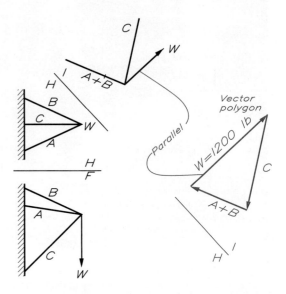

Step 1 Draw an auxiliary view where vectors A and B lie in the edge view of a plane. Draw a vector polygon parallel to the members in the auxiliary view in which W = 1200 lb is the only known vector.

Step 2 Construct an orthographic projection of the vector polygon from step 1, with its vectors parallel to the members in the top view. The reference plane between the two views is parallel to the H1 plane. (This portion of the solution is closely related to that in Fig 30.13.)

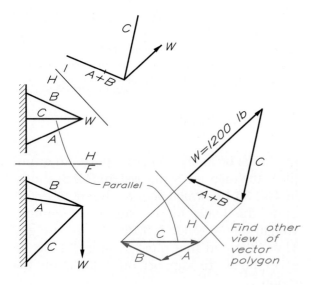

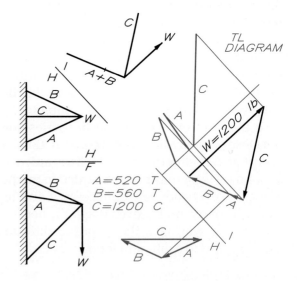

Step 3 Project the intersection of A and B in the top view of the vector polygon to the auxiliary view polygon. Find the magnitudes of vectors A, B, and C in a true-length diagram and analyze them for tension or compression by referring to the top and auxiliary views.

Step 4 The lengths of vectors B and C are not true length in their top and front views. Construct a true-length diagram and scale the lengths of these lines to obtain their magnitudes.

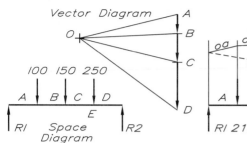
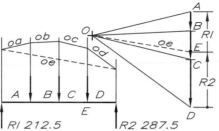
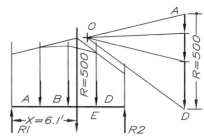

30.15 Parallel, nonconcurrent loads.

Step 1 Letter the spaces between the loads using Bow's notation. Find the sum of the vertical loads by drawing them head-to-tail in a vector diagram. Locate pole point O at a convenient location and draw strings from O to each end of the vectors.

Step 2 Extend the lines of vertical loads and draw a funicular diagram with string oa in the A space, ob in the B space, oc in the C space, and so on. The last string, oe, closes the diagram. Transfer oe to the vector polygon to locate E, thus establishing R1 and R2, which are EA and DE, respectively.

Step 3 The resultant of the three downward forces equals their graphical summation, line AD. Locate the resultant by extending strings oa and od in the funicular diagram to a point of intersection. The resultant, R = 500 lb, acts through this point in a downward direction at distance X from the left end.

because members B and C lie in the same edge view of the plane in the front view. Therefore, solving for three unknowns is possible in this case.

Construct a vector polygon in the front view by drawing force F as a vector and using the other vectors as the sides of the polygon. Draw the top view using vectors B and C to form the polygon that closes at each end of vector A. Then find the front view of vectors B and C.

A true-length diagram gives the lengths of the vectors; measure them to determine their magnitudes. Vector A is in compression because it points toward the point of application. Vectors B and C are in tension because they point away from the point.

General Case

The structural frame shown in **Fig. 30.14** is attached to a vertical wall to support a load of W = 1200 lb. There are three unknowns in each of the views, so begin by projecting an auxiliary view from the top view to obtain the edge view of a plane containing vectors A and B, thereby reducing the number of unknowns to two. You no longer need refer to the front view.

Draw a vector polygon with vectors parallel to their members in the auxiliary view. Then draw an adjacent orthographic view of the vector polygon with vectors parallel to their members in the top view. Use a true-length diagram to find the true length of the vectors and measure their magnitudes.

30.8 Resultant of Parallel, Nonconcurrent Forces

The beam in **Fig. 30.15** supports the three loads shown. It is necessary to determine the magnitude of supports R1 and R2, the magnitude of the resultant of the loads, and the resultant's location. Begin by labeling the spaces between all vectors clockwise with Bow's notation and draw a vector diagram.

Extend the lines of force in the space diagram and draw the strings from the vector diagram in their respective spaces, parallel to their original directions. For example, string oa is parallel to string oA in space A between forces EA and AB, and string ob is in space B, beginning at the intersection of oa with vector AB. The last string, oe, closes the diagram, called a **funicular diagram**.

Transfer the direction of string oe to the force diagram, and lay it off through point O to intersect the load line at E. Vector DE represents R2 (refer to Bow's notation as it was applied in step 1), and vector EA represents R1. It is easy to see that DE and EA are equal to the sum of the downward loads represented by AD.

To find the location of the resultant from the R1, extend the outside strings of the funicular diagram, oa and od, to their intersection. The resultant will pass through this point of intersection. The resultant has a magnitude of 500 lb, a vertical downward direction, and a point of application at X = 6.1 ft. Notice that two scales are used in this problem: one for the vectors in pounds and one in feet for the space diagram.

Problems

Draw your solutions to these problems with instruments on size A grid or plain sheets. Each grid square represents 0.20 in. (5mm). Be neat when executing all notes, sketches, drawings, and graphical work and use good design practices. Letter written matter legibly, using 1/8-in. guidelines.

1. (**Sheet 1**) Resultants.
(**A–B**) Find the resultants of the force systems by the parallelogram and polygon methods. Scale: 1" = 100 lb.

2. (**Sheet 2**) Resultants.
(**A–B**) Find the resultants of the force systems by the parallelogram and polygon methods. Scale: 1" = 100 lb.

3. (**Sheet 3**) Concurrent, coplanar.
(**A–B**) Find the forces in the coplanar force systems. Label the members, assign the forces in each of them, and indicate whether the forces are compression or tension.

4. (**Sheet 4**) Truss analysis.
Find the forces in each member of the truss by using a Maxwell diagram. Make a table of forces and indicate compression and tension.

5. (**Sheet 5**) Noncoplanar, special case.
Find the forces in the members of the concurrent noncoplanar force system. Make a table of forces and indicate compression and tension.

6. (**Sheet 6**) Noncoplanar, general case.
Find the forces in the members of the concurrent noncoplanar force system. Make a table of forces and indicate compression and tension.

7. (**Sheet 7**) Beam analysis.
(**A**) Find the forces in reactions R1 and R2 necessary to equalize the loads applied to the beam.
(**B**) Find the value and location of the single support that could replace both R1 and R2.

8. (**Sheet 8**) Beam analysis.
Repeat Problem 7 for this configuration.

9. (**Sheet 9**) Concurrent, noncoplanar forces.
Find the forces in the support members. Make a table of forces and indicate compression and tension.

10. (**Sheet 10**) Concurrent, coplanar forces.
Find the forces in the coplanar force system. Label the members, make a table of forces, and indicate compression and tension.

11. Repeat Problem 4 but use the joint-by-joint analysis instead of the Maxwell diagram method.

12. Find the loads in the structural members of the truss, list them in a table, and show whether they are in tension (T) or compression (C).

13. (**Sheet 11**) Determine the forces in the three members of the tripod used to lift concrete slabs into a vertical position when a pull of 480 lbs is applied at O.

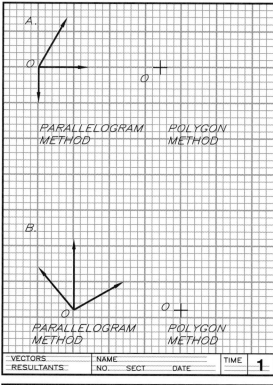

A.

PARALLELOGRAM
METHOD

POLYGON
METHOD

B.

PARALLELOGRAM
METHOD

POLYGON
METHOD

VECTORS	NAME		TIME	**1**
RESULTANTS	NO. SECT DATE			

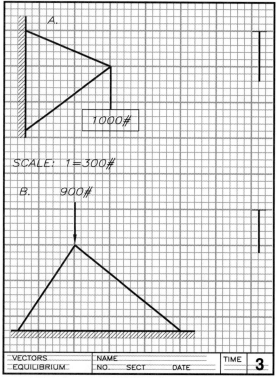

A.

1000#

SCALE: 1=300#

B.   900#

VECTORS	NAME		TIME	**3**
EQUILIBRIUM	NO. SECT DATE			

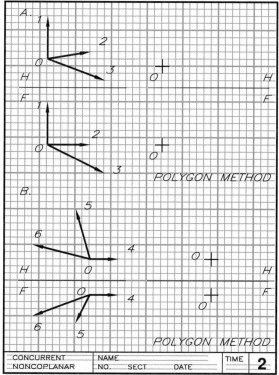

A.   1
        2
    O   3
H   H
F
    1
        2
    O
            3
                POLYGON   METHOD

B.
        5
    6
            4
H   O   H
F   O   4   F

    6
        5
                POLYGON   METHOD

CONCURRENT	NAME		TIME	**2**
NONCOPLANAR	NO. SECT DATE			

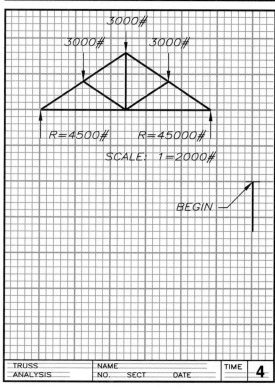

3000#

3000#   3000#

_R=4500#_   _R=45000#_

SCALE: 1=2000#

BEGIN

TRUSS	NAME		TIME	**4**
ANALYSIS	NO. SECT DATE			

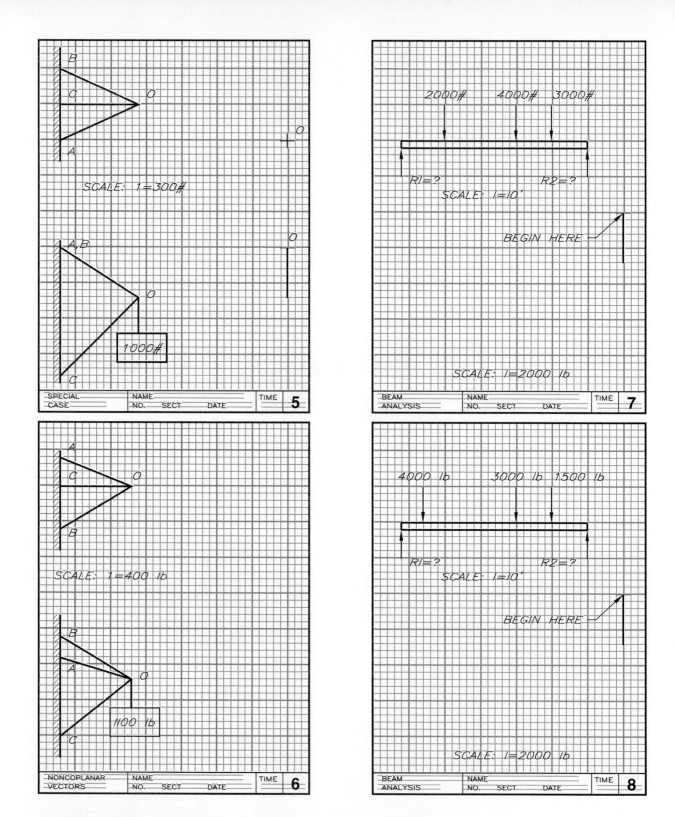

SCALE: 1=300#

1000#

SCALE: 1=400 lb

1100 lb

2000#   4000#   3000#

R1=?   R2=?

SCALE: 1=10'

BEGIN HERE

SCALE: 1=2000 lb

4000 lb   3000 lb 1500 lb

R1=?   R2=?

SCALE: 1=10'

BEGIN HERE

SCALE: 1=2000 lb

| SPECIAL | NAME | | TIME | 5 |
| CASE | NO. SECT DATE | | | |

| BEAM | NAME | | TIME | 7 |
| ANALYSIS | NO. SECT DATE | | | |

| NONCOPLANAR | NAME | | TIME | 6 |
| VECTORS | NO. SECT DATE | | | |

| BEAM | NAME | | TIME | 8 |
| ANALYSIS | NO. SECT DATE | | | |

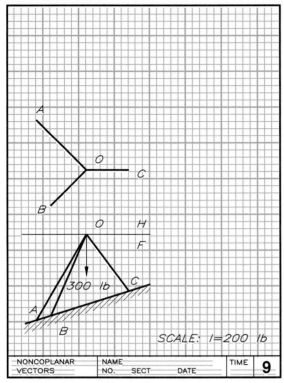

A
O
C
B

O   H
         F

A 300 lb C

B

SCALE: 1=200 lb

NONCOPLANAR VECTORS	NAME		TIME	**9**
	NO.	SECT DATE		

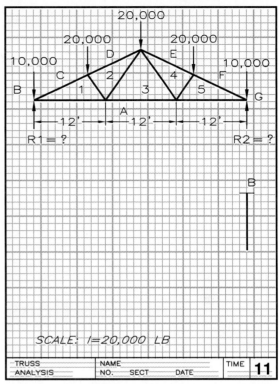

20,000

20,000   20,000

10,000   D   E   10,000

C   2   4   F

B   1   3   5   G

|←—12'—→|←—12'—→|←—12'—→|

R1 = ?   R2 = ?

B

SCALE: 1=20,000 LB

TRUSS ANALYSIS	NAME		TIME	**11**
	NO.	SECT DATE		

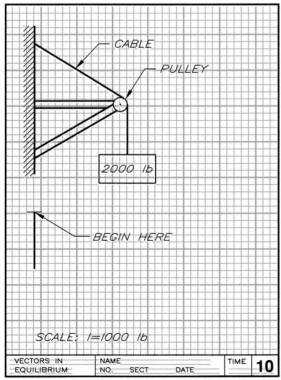

CABLE

PULLEY

2000 lb

BEGIN HERE

SCALE: 1=1000 lb

VECTORS IN EQUILIBRIUM	NAME		TIME	**10**
	NO.	SECT DATE		

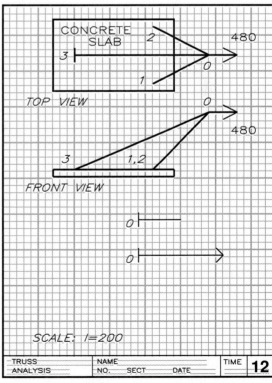

CONCRETE SLAB   2   480

3   1   O

TOP VIEW

O

480

3   1,2

FRONT VIEW

O

O

SCALE: 1=200

TRUSS ANALYSIS	NAME		TIME	**12**
	NO.	SECT DATE		

PROBLEMS • 543

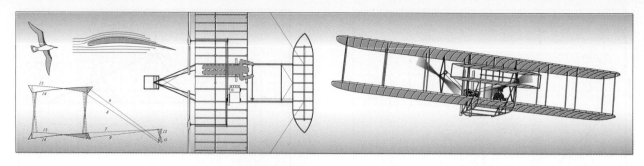

31

Intersections and Development

31.1 Introduction

Several methods may be used to find lines of intersection between parts that join. Usually such parts are made of sheet metal, or of plywood if used as forms for concrete. After **intersections** are found, **developments**, or flat patterns, can be laid out on sheet metal and cut to the desired shape. Common examples of intersections and developments range from air-conditioning ducts to massive refineries.

31.2 Intersections of Lines and Planes

Figure 31.1 illustrates the fundamental principle of finding the intersection between a line and a plane. This example is a special case in which the point of intersection clearly shows in the view where the plane appears as an edge. Projecting the piercing point P to the front view completes the visibility of the line.

This same principle is applied to finding the line of intersection between two planes

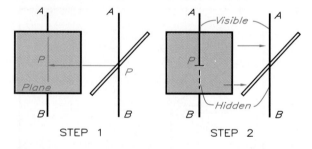

31.1 Intersection of a line and a plane.

Step 1 Find the point of intersection in the view where the plane appears as an edge (the side view in this case) and project it to the front view.

Step 2 Determine visibility in the front view by looking from the front view to the right-side view.

(**Fig. 31.2**). By locating the piercing points of lines AB and DC and connecting these points, the line of intersection is found.

The angular intersection of two planes at a corner gives a line of intersection that bends around the corner (**Fig. 31.3**). First, find piercing points 2' and 1', then project corner point 3 from the side view where the vertical corner

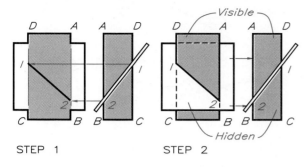

STEP 1   STEP 2

31.2   Intersection of planes.

Step 1 Find the piercing points of lines AB and DC with the plane where the plane appears as an edge and project them to the front view.

Step 2 Line 1–2 is the line of intersection. Determine visibility by looking from the front view to the right-side view.

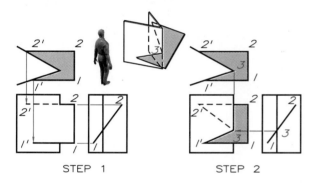

STEP 1   STEP 2

31.3   The intersection of a plane at a corner.

Step 1 The intersecting plane appears as an edge in the side view. Project intersection points 1′ and 2′ from the top and side views to the front view.

Step 2 The line of intersection from 1′ to 2′ must bend around the vertical corner at 3′ in the top and side views. Project point 3′ to the front view to locate line 1′–3′–2′.

pierces the plane to the front view of the corner. Point 2′ is hidden in the front view because it is on the back side.

Figure 31.4 shows how to find the intersection between a plane and prism where the plane appears as an edge. Obtain the piercing points for each corner line and connect them to form the line of intersection. Show visibility to complete the intersection.

Figure 31.5 depicts a more general case of an intersection between a plane and prism.

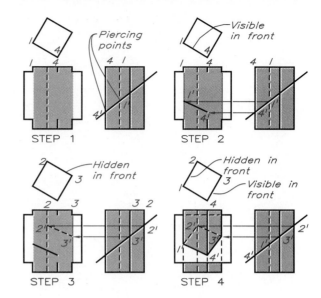

STEP 1   STEP 2

STEP 3   STEP 4

31.4   The intersection of a plane and a prism.

Step 1 Vertical corners 1 and 4 intersect the edge view of the plane in the side view at points 1′ and 4′.

Step 2 Project points 1′ and 4′ from the side view to lines 1 and 4 in the front view. Connect them to form a visible line of intersection.

Step 3 Vertical corners 2 and 3 intersect the edge view of the plane at points 2′ and 3′ in the side view. Project points 2′ and 3′ to the front view to form a hidden line of intersection.

Step 4 Connect points 1′, 2′, 3′, and 4′ and determine visibility by analyzing the top and side views.

Passing vertical cutting planes through the planes of the prism in the top view yields traces (cut lines) on the front view of the oblique plane on which the piercing points of the vertical corner lines lie. Connect the points and determine visibility to complete the solution.

In **Fig. 31.6**, finding the intersection between a foreshortened plane and an oblique prism involves finding an auxiliary view to obtain the edge view of the plane and simplify the problem. The piercing points of the corner lines of the prism lie in the auxiliary view and project back to the given views. Points 1, 2, and 3, projected from the auxiliary view to the given views, are shown as exam-

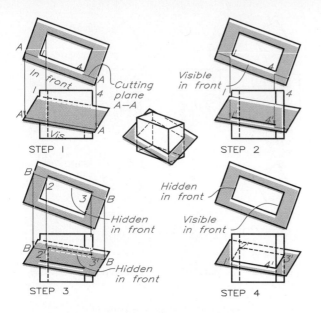

STEP 1

STEP 2

STEP 3

STEP 4

31.5 The intersection of an oblique plane and a prism.

Step 1 Pass vertical cutting plane A–A through corners 1 and 4 in the top view and project endpoints to the front view.

Step 2 Locate piercing points 1′ and 4′ in the front view where line A–A crosses lines 1 and 4.

Step 3 Pass vertical cutting plane B–B through corners 2 and 3 in the top view and project them to the front view to locate piercing points 2′ and 3′.

Step 4 Connect the four piercing points and determine visibility by analysis of the top view.

ples. Analysis of crossing lines determines visibility to complete the line of intersection in the top and front views.

31.3 Intersections Between Prisms

The techniques used to find the intersection between planes and lines also apply to finding the intersection between two prisms (**Fig. 31.7**). Project piercing points 1, 2, and 3 from the side and top views to the front view. Point X lies in the side view where line of intersection 1–2 bends around the vertical corner of the vertical prism. Connect points 1, X, and 2 and determine visibility.

Figure 31.8 illustrates how to find the line of intersection between an inclined and a vertical

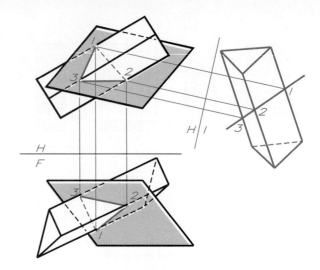

31.6 To find the intersection between a foreshortened plane and an oblique prism, construct a view in which the plane appears as an edge. Project piercing points 1, 2, and 3 back to the top and front views.

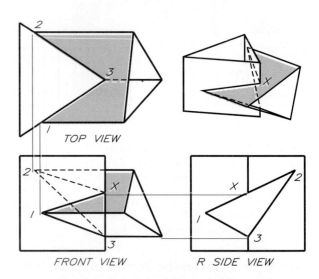

TOP VIEW

FRONT VIEW

R SIDE VIEW

31.7 Three views of intersecting prisms. The points of intersection are best found where intersecting planes appear as edges.

prism. An auxiliary view reveals the end view of the inclined prism where its planes appear as edges. In the auxiliary view, plane 1–2 bends around corner AB at point P. Project points of intersection 1′ and 2′ from the top and auxiliary to their intersections in the front view. Then draw the line of intersection 1′–P–2′ for

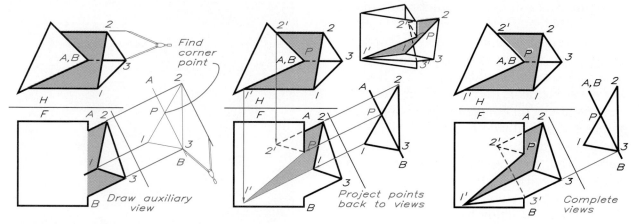

31.8 The intersection of an inclined and a vertical prism by auxiliary view.

Step 1 Draw the end view of the inclined prism in an auxiliary view from the front view. Show line AB of the vertical prism in the auxiliary view.

Step 2 Locate piercing points 1' and 2' in the top and front views. Intersection line 1'–2' bends around corner AB at P (projected from the auxiliary view).

Step 3 The intersection lines from 2' and 1' to 3' do not bend around the corner but are straight lines. Line 1'–3' is visible, and line 2'–3' is invisible.

this portion of the line of intersection. Connect the remaining lines, 1'–3' and 2'–3' to complete the solution.

Figure 31.9 shows an alternative method of solving this type of problem. Piercing points 1' and 2' appear in the front view as projections from the top view. Point 5 is the point where line 1'–5–2' bends around vertical corner AB. To find point 5 in the front view, pass a cutting plane through corner AB in the top view and project its trace to the front view. Draw the lines of intersection, 1'–5–2'.

Some applications of intersections and developments are massive in size, as illustrated by the blast furnace under construction in **Fig. 31.10**. The workers are barely visible on various levels of the enormous structure.

31.4 Intersections Between Planes and Cylinders

The standard sheet-metal vent pipe that is common to all homes is an example of a cylinder intersecting a plane. **Figure 31.11** shows how to find the intersection between a plane and a cylinder. Cutting planes passed verti-

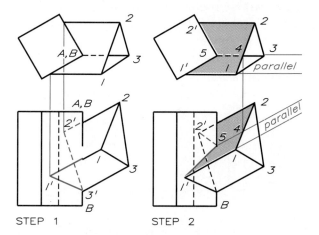

31.9 The intersection of prisms by projection.

Step 1 Project the piercing points of lines 1, 2, and 3 from the top view to the front view to locate piercing points 1', 2', and 3'.

Step 2 Pass a cutting plane through corner AB in the top view to locate point 5, where intersection line 1'–2' bends around the vertical prism. Find point 5 in the front view and draw line 1'–5–2'.

cally through the top view of the cylinder establish pairs of elements on the cylinder and their piercing points. Space the cutting planes conveniently apart by eye. Then project the piercing points to each view and draw the elliptical line of intersection.

Figure 31.12 shows the solution of a more general problem. Here, the cylinder is vertical and the plane is oblique and does not appear as an edge. Passing vertical cutting planes through the cylinder and the plane in the top view gives elements on the cylinder and their piercing points on the plane. Projecting these points to the front view completes the elliptical line of intersection. The more cutting planes used, the more accurate the line of intersection will be.

Figure 31.13 demonstrates the general case of the intersection between a plane and cylinder, where both the plane and cylinder are oblique in the given views. An auxiliary view is used to show the edge view of the plane. Cutting planes passed through the cylinder parallel to its axis in the auxiliary view establish elements on the cylinder and their piercing points. The points are projected back to the given views and connected to give an elliptical line of intersection in the front and side views.

31.10 The design of this huge blast furnace used applications of large-scale intersections and developments. (Courtesy of the Jones & Laughlin Steel Corporation.)

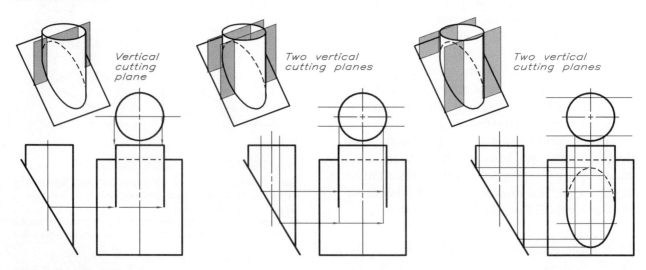

31.11 The intersection of a plane and a cylinder.

Step 1 Pass a vertical cutting plane through the cylinder parallel to its axis to find two points of intersection.

Step 2 Use two more cutting planes to find four additional points in the top and left-side views. Project these points to the front view.

Step 3 Use additional cutting planes to find more points. Connect these points to give an elliptical line of intersection.

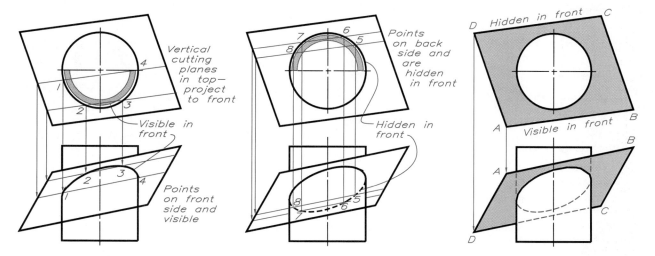

31.12 Intersection of a vertical cylinder and an oblique plane.

Step 1 Pass vertical cutting planes through the cylinder in the top view to find elements on it and the plane. Project points 1, 2, 3, and 4 to the front view of their respective lines and connect them with a visible line.

Step 2 Use additional cutting planes to find other piercing points—5, 6, 7, and 8—and project them to the front view of their respective lines on the oblique plane. Since the points are on the back side, they are hidden. Connect these points with a hidden line.

Step 3 Determine visibility of the plane and cylinder in the front view. Line AB is visible by inspection of the top view, since it is the farthest out in front; line CD is the farthest back in the top view and is therefore hidden in the front view.

31.5 Intersections Between Cylinders and Prisms

An inclined prism intersects a vertical cylinder in **Fig. 31.14**. A primary auxiliary view is drawn to show the end view of the inclined prism where its planes appear as edges. A series of vertical cutting planes in the top view establish lines lying on the surfaces of the cylinder and prism. The cutting planes, also shown in the auxiliary view, are the same distance apart as in the top view.

Projecting the line of intersection from 1 to 3 from the auxiliary view to the front view yields an elliptical line of intersection. The visibility of this line changes from visible to hidden at point X, which appears in the auxiliary view and is projected to the front view. Continuing this process gives the lines of intersection of the other two planes of the prism.

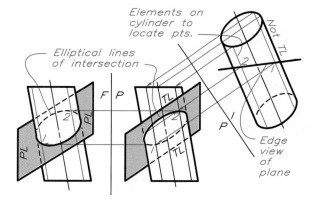

31.13 To find the intersection between an oblique cylinder and an oblique plane, construct a view that shows the plane as an edge. Cutting planes passed through the cylinder locate points on the line of intersection.

31.6 Intersections Between Cylinders

To find the line of intersection between two perpendicular cylinders, pass cutting planes through them parallel to their centerlines (**Fig. 31.15**). Each cutting plane locates a pair of elements on both cylinders that intersect at

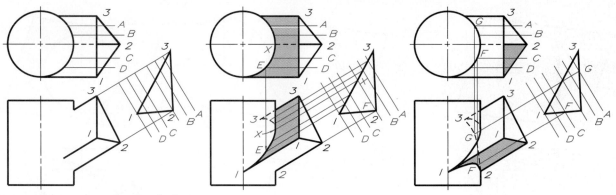

31.14 The intersection of a vertical cylinder and an inclined prism.

Step 1 Find the edge views of the planes of the triangular prism in an auxiliary view. Project from the front view. Draw frontal cutting planes through the top view and locate them in the auxiliary view with dividers.

Step 2 Locate points along intersection line 1–3 in the top view and project them to the front view. For example, find point E on cutting plane D in the top and auxiliary views and project it to the front view where the projectors intersect. Visibility changes in the front view at point X.

Step 3 Determine the remaining points of intersection by using the other cutting planes. Project point F, shown in the top and auxiliary views, to the front view of line 1–2. Connect the points and determine visibility.

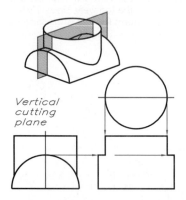
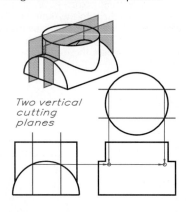
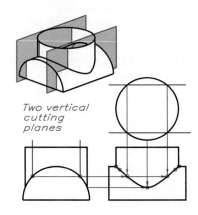

31.15 Intersection of perpendicular cylinders.

Step 1 Pass a cutting plane through the cylinders parallel to their axes, locating two points of intersection.

Step 2 Use two more cutting planes to find four more additional points of intersection.

Step 3 Use two more cutting planes to locate four more points. Connect the points with a smooth curve to complete the line of intersection.

a piercing point. Connecting the points and determining visibility completes the solution. An example of a transmission pipe fabricated with intersecting cylinders is shown in **Fig. 31.16**. Each cylinder had to be precisely cut to form accurate intersections for tight joints before welding them together.

Figure 31.17 illustrates how to find the intersection between nonperpendicular cylinders. This method involves passing a series of vertical

cutting planes through the cylinders parallel to their centerlines. Points 1 and 2, labeled on cutting plane D, are typical of points on the line of intersection. Other points may be found in the same manner. Although the auxiliary view is not essential to the solution, it is an aid in visualizing the problem. Projecting points 1 and 2 on cutting plane D in the auxiliary view to the front view provides a check on the projections from the top view.

31.16 Intersections between cylinders have been welded in shop fabrication of a transmission pipe as shown in these two views.

31.7 Intersections Between Planes and Cones

To find points of intersection on a cone, use cutting planes that are (1) perpendicular to the cone's axis or (2) parallel to the cone's axis. The vertical planes in the top view of **Fig. 31.18A** cut radial lines on the cone and establish elements on its surface. The horizontal planes in **Fig. 31.18B** cut circular sections that appear true size in the top view of a right cone.

A series of radial cutting planes define elements on a cone (**Fig. 31.19**). These elements cross the edge view of the plane in the front view to locate piercing points of each element that, when projected to the top view of the same elements, lie on the line of intersection.

A series of horizontal cutting planes may be used to determine the line of intersection between a cone and an oblique plane (**Fig. 31.20**). The sections cut by these imaginary planes are circles in the top view. The cutting planes also locate lines on the oblique plane that intersect the circular sections cut by each

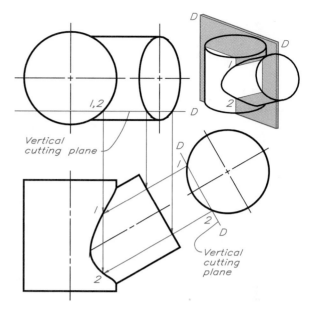

31.17 To find the intersection between nonperpendicular cylinders, find the end view of the inclined cylinder in an auxiliary view. Use vertical cutting planes to find the piercing points of the cylindrical elements and the line of intersection.

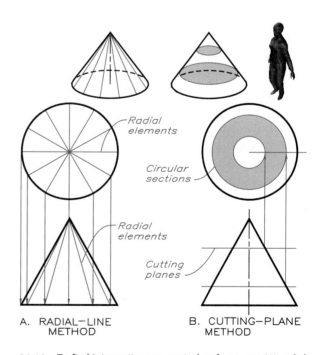

31.18 To find intersections on conical surfaces, use (A) radial cutting planes that pass through the cone's centerline and are perpendicular to its base, or (B) cutting planes that are parallel to the cone's base.

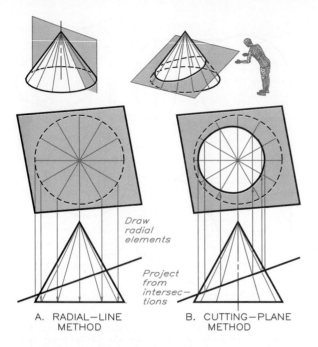

A. RADIAL−LINE METHOD

B. CUTTING−PLANE METHOD

Draw radial elements

Project from intersections

31.19 The intersection of a plane and a cone.

Step 1 Divide the base evenly in the top view and connect these points with the apex to establish elements on the cone. Project these elements to the front view.

Step 2 Project the piercing point of each element on the edge view of the plane to the top view of the same elements, and connect them to form the line of intersection.

respective cutting plane. The points of intersection found in the top view project to the front view. We could have used the horizontal cutting-plane method in Fig. 31.19 to obtain the same results.

31.8 Intersections Between Cones and Prisms

A primary auxiliary view gives the end view of the inclined prism that intersects the cone in **Fig. 31.21**. Cutting planes that radiate from the apex of the cone in the top view locate elements on the cone's surface that intersect the prism in the auxiliary view. These elements project to the front view.

Wherever the edge view of plane 1–3 intersects an element in the auxiliary view, the piercing points project to the same element in the front and top views. Passing an extra cutting plane through point 3 in the auxiliary view locates an element that projects to the front and top views. Piercing point 3 projects to this element in sequence from the auxiliary view to the top view. This same procedure yields the piercing points of the other two planes of the prism. All projections of points of intersection originate in the auxiliary view, where the planes of the prism appear as edges.

In **Fig. 31.22**, horizontal cutting planes passed through the front view of the cone and cylinder give a series of circular sections in the top view. Points 1 and 2, shown on cutting plane C in the top view, are typical and project to the front view. The same method produces other points.

This method is feasible only when the centerline of the cylinder is perpendicular to the axis of the cone, producing circular sections in the top view (rather than elliptical sections, which would be difficult to draw). A series of intersecting cylinders can be seen in this gas-powered turbine in **Fig. 31.23**.

31.9 Intersections Between Prisms and Pyramids

Figure 31.24 shows how to find the intersection of an inclined prism with a pyramid. An auxiliary view shows the end view of the inclined prism and the pyramid. The radial lines OB and OA drawn through corners 1 and 3 in the auxiliary view project back to the front and top views. Projection locates intersecting points 1 and 3 on lines OB and OA in each view. Point P is the point where line 1–3 bends around corner OC. Finding lines of intersection 1–4 and 4–3 and determining visibility completes the solution.

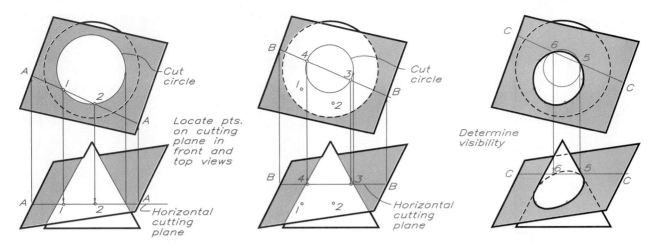

31.20 The intersection of an oblique plane and a cone.

Step 1 Pass a horizontal cutting plane through the front view to find a circular section on the cone and a line on the plane in the top view. The piercing points of this line are on the circle. Project points 1 and 2 to the front view.

Step 2 Pass horizontal cutting plane B–B through the front view in the same manner to locate piercing points 3 and 4 in the top view. Project these points to the horizontal plane in the front view from the top view.

Step 3 Use additional horizontal planes to find a sufficient number of points to complete the line of intersection in the same manner as covered in the previous steps. Draw the intersection and determine visibility.

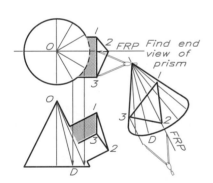
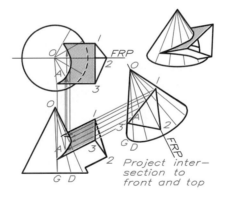
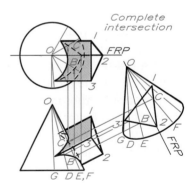

31.21 Intersection of a cone and an inclined prism.

Step 1 Draw an auxiliary view to obtain the edge views of the planes of the prism. In the top view, pass cutting planes through the cone through the apex. Project these elements to the front and auxiliary views.

Step 2 Find the piercing points of the cone's elements with the edge view of plane 1–3 in the auxiliary view and project them to the front and top views. For example, point A lies on element OD in the auxiliary view, so project it to the front and top views of OD.

Step 3 Locate the piercing points where the conical elements intersect the edge views of the planes of the prism in the auxiliary view. For example, find point B on O in the primary auxiliary view and project it to the front and top views of OE.

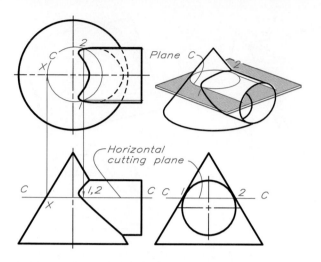

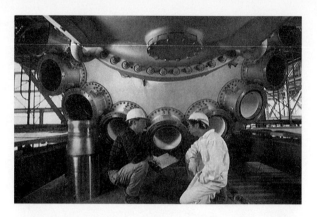

31.23 Examples of intersecting cylinders are shown in this gas turbine power plant in Hong Kong. (Courtesy of General Electric Company.)

31.22 Horizontal cutting planes are used to find the intersection between the cone and the cylinder. The cutting planes cut circles in the top view. Only one cutting plane is shown here as an example.

Figure 31.25 shows a horizontal prism that intersects a pyramid. An auxiliary view depicts the end view of the horizontal prism with its planes as edges. Passing a series of horizontal cutting planes through the corner points of the horizontal prism and the pyramid in the auxiliary view gives the lines of intersection, which form triangular sections in the top view.

The cutting plane through corner point P in the auxiliary view is an example of a typical cutting plane. At point P the line of intersection of this plane bends around the corner of the pyramid. Other cutting planes are passed through the corner lines of the prism in the auxiliary and front views. Each corner line extends in the top view to intersect the triangular section formed by the cutting plane, as shown at P.

31.10 Principles of Developments

The bodies of aircraft (**Fig. 31.26**) are designed to be fabricated from sheet metal stock formed into shapes. Although aircraft design

is one of the most advanced applications of **developments**, the same principles can be used to design commonplace objects such as a garbage can.

Figure 31.27 illustrates some of the standard edges and joints for sheet metal. The application determines the type of seam that is used.

The development of patterns for four typical shapes is shown in **Fig. 31.28**. The sides of a box are unfolded into a common plane. The cylinder is rolled out along a stretch-out line equal in length to its circumference. The pattern of a right cone and right pyramid are developed with the length of an element serving as a radius for drawing the base arc.

The construction of patterns for geometric shapes with parallel elements, such as the prisms and cylinders shown in **Fig. 31.29A** and **B**, begins with drawing stretch-out lines parallel to the edge views of the shapes' right sections. The distance around the right section becomes the length of the stretch-out line. The prism and cylinder in **Fig. 31.29C** and **D** are inclined, so their right sections are perpendicular to their sides, not parallel to their bases.

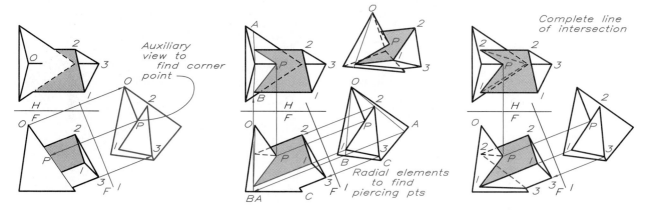

31.24 Intersection of an inclined prism and a pyramid.

Step 1 Find the edge views of the planes of the prism in an auxiliary view. Project the pyramid into this view also, showing only the visible surfaces.

Step 2 Pass planes A and B through 0 and points 1 and 3 in the auxiliary view. Project OA and OB to the front and top views; project 1 and 3 to them. Point 2 lies on OC. Connect 1, 2, and 3 for the intersection of this plane.

Step 3 Point 3 lies on OC in the auxiliary view. Project this point to the principal views. Connect 3 to points 1 and 2 to complete the intersections; show visibility. Assume that these shapes are constructed of sheet metal.

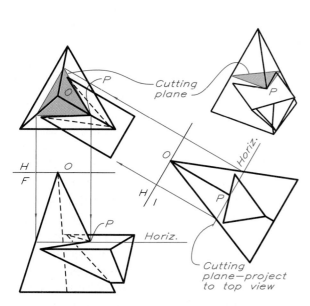

31.25 The intersection of this pyramid and prism is found by obtaining the end view of the prism in an auxiliary view. Horizontal cutting planes are passed through the fold lines of the prism to find the piercing points on the line of intersection. One cutting plane used to find corner point P is shown.

31.26 Essentially the entire body of an aircraft, such as the Super Hornet shown in the foreground, is a series of applications of intersections and developments. (Courtesy of Northrup Grumman Corporation.)

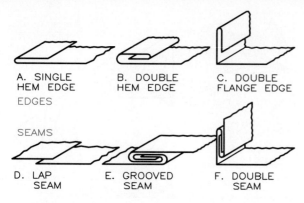

A. SINGLE HEM EDGE   B. DOUBLE HEM EDGE   C. DOUBLE FLANGE EDGE

EDGES

SEAMS

D. LAP SEAM   E. GROOVED SEAM   F. DOUBLE SEAM

31.27 Six types of edges and seams used to join sheet-metal developments. Other seams are joined by riveting and welding.

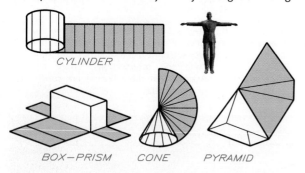

CYLINDER

BOX–PRISM   CONE   PYRAMID

31.28 Four standard types of developments are the box, cylinder, cone, and pyramid.

In development, an inside pattern is preferable to an outside pattern for two reasons: (1) most bending machines are designed to fold metal inward, and (2) markings and scribings will be hidden. The designer labels patterns with a series of lettered or numbered points on the layouts. All lines on developments must be true length. Patterns should be laid out so that the seam line (a line where the pattern is joined) is the shortest line in order to reduce the expense of riveting or welding the seams.

31.11 Development of Rectangular Prisms

The development of a flat inside pattern for a rectangular prism is illustrated in **Fig. 31.30**. The edges of the prism are vertical and true length in the front view. The right section is

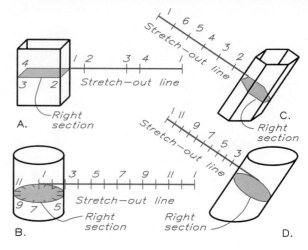

A.   B.   C.   D.

31.29 Stretch-out lines.

A—B To obtain the developments of right prisms and right cylinders, roll out the right sections along a stretch-out line.

C—D Draw stretch-out lines parallel to the edge views of the right section of cylinders and prisms, or perpendicular to their true-length elements.

perpendicular to these sides and the right section is true size in the top view. The stretch-out line begins with point 1 and is drawn parallel to the edge view of the right section.

If an inside pattern is to be laid out to the right, you must determine which point is to the right of the beginning point, point 1. Let's assume that you are standing inside the top view and are looking at point 1: You will see point 2 to the right of point 1.

To locate the fold lines of the pattern, transfer lines 2–3, 3–4, and 4–1 with your dividers from the right section in the top view to the stretch-out line. The length of each fold line is its projected true length from the front view. Connect the ends of the fold lines to form the boundary of the developed surface. Draw the fold lines as thin dark lines and the outside lines as thicker, visible object lines.

The fuselage of the Premier aircraft in **Fig. 31.31** was designed using the principles of intersections and development. Development of the prism depicted in **Fig. 31.32** is similar to

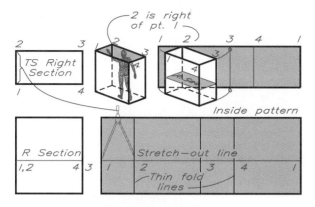

31.30 To develop a rectangular prism for an inside pattern, draw the stretch-out line parallel to the edge view of the right section. Transfer the distances between the fold lines from the true-size right section to the stretch-out line.

31.31 The assembly of the fuselage of the Premier aircraft is based on the design principles of intersections and developments. (Courtesy of Raytheon Aircraft.)

that shown in Fig. 31.30. Here, though, one of its ends is beveled (truncated) rather than square. The stretch-out line is parallel to the edge view of the right section in the front view. Lay off the true-length distances around the right section along the stretch-out line (beginning with the shortest one) and locate the fold lines. Find the lengths of the fold lines by projecting from the front view of these lines.

31.12 Development of Oblique Prisms

The prism shown in **Fig. 31.33** is inclined to the horizontal plane, but its fold lines are true length in the front view. The right section is an edge perpendicular to these fold lines, and the stretch-out line is parallel to the edge of the right section. A true-size view of the right section is found in the auxiliary view.

Transfer the distances between the fold lines from the true-size right section to the stretch-out line. Find the lengths of the fold lines by projecting from the front view. Determine the ends of the prism and attach them to the pattern so that they can be folded into position.

In **Fig. 31.34**, the fold lines of the prism are true length in the top view, and the edge view

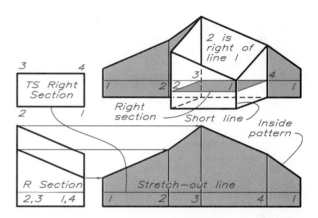

31.32 To develop an inside pattern of a rectangular prism with a beveled end, draw the stretch-out line parallel to the right section. You can then find the fold lines by transferring distances between the fold lines from the true-size right section to the stretch-out line.

of the right section is perpendicular to them. The stretch-out line is parallel to the edge view of the right section, and the true size of the right section appears in an auxiliary view projected from the top view. Transfer the distances about the right section to the stretch-out line to locate the fold lines, beginning with the shortest line. Find the lengths of the fold lines by projecting from the top view. Attach the end portions to the pattern to complete the construction.

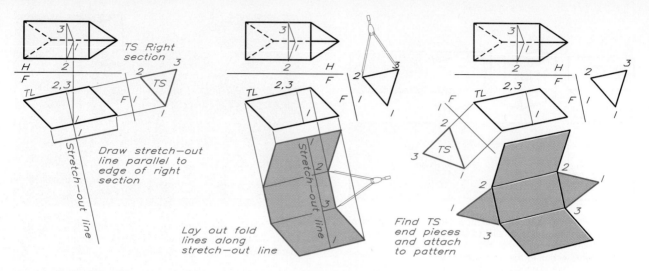

31.33 Development of an oblique prism.

Step 1 Draw the edge view of the right section perpendicular to the true-length axis in the front view. Find the true-size view of the right section in the auxiliary view. Draw the stretch-out line parallel to the edge of the right section. The line through point 1 is the first line of the development.

Step 2 Because the pattern is to be laid out to the right from line 1, the next point is line 2 (from the auxiliary view). Transfer true-length lines 1–2, 2–3, and 3–1 from the right section to the stretch-out line to locate fold lines. Determine the lengths of bend lines by projection.

Step 3 Find true-size views of the end pieces by projecting auxiliary views from the front view. Connect these ends to the development to form the completed pattern. Draw fold lines as thin, dark lines and outside lines as thicker, visible object lines.

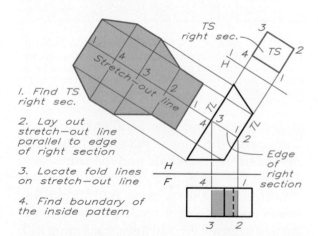

1. Find TS right sec.

2. Lay out stretch—out line parallel to edge of right section

3. Locate fold lines on stretch—out line

4. Find boundary of the inside pattern

31.34 To develop this oblique chute, locate the right-section true size in the auxiliary view. Draw the stretch-out line parallel to its right section. Find fold lines by transferring their spacing from the true-size right section to the stretch-out line.

A prism that does not project true length in either view may be developed, as shown in **Fig. 31.35**. The fold lines are true length in an auxiliary view projected from the front view. The right section appears as an edge perpendicular to the fold lines in the auxiliary view and true size in a secondary auxiliary view.

Draw the stretch-out line parallel to the edge view on the right section. Locate the fold lines on the stretch-out line by measuring around the right section in the secondary auxiliary view, beginning with the shortest one. Then project the lengths of the fold lines to the development from the primary auxiliary view.

31.13 Development of Cylinders

Figure 31.36 illustrates how to develop a flat pattern of a right cylinder. The elements of the cylinder are true length in the front view, so

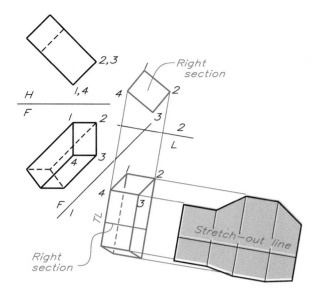

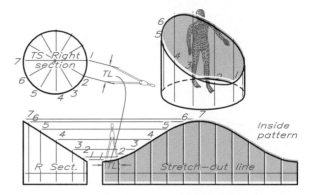

31.36 To develop an inside pattern for a truncated right cylinder, draw the stretch-out line parallel to the right section. Transfer points 1–7 from the top view to the stretch-out line that is parallel to the right section. Point 2 is to the right of point 1 for an inside pattern.

31.35 To develop an oblique prism, draw a primary auxiliary view in which the fold lines are true length and a secondary auxiliary view in which the right section appears true size. Use these views to develop the pattern the same way as in Fig. 31.34.

31.37 This large transition piece makes a 90° turn, enabling the cylinder to join with a square-end duct.

the right section appears as an edge in this view and true size in the top view. The stretch-out line is parallel to the edge view of the right section, and point 1 is the beginning point because it lies on the shortest element.

Let's assume that you are standing inside the cylinder in the top view and are looking at point 1: You will see that point 2 is to the right of point 1. Therefore, lay off point 2 to the right of point 1 to develop an inside pattern.

By drawing radial lines at 15° or 30° intervals, you can equally space the elements in the top view and conveniently lay them out along the stretch-out line as equal measurements. To complete the pattern, find the lengths of the elements by projecting from the front view. An application of a large developed cylinder joined with a transition piece is shown in **Fig. 31.37**.

31.14 Development of Oblique Cylinders

The pattern for an oblique cylinder (**Fig. 31.38**) involves the same determinations as the preceding cases but with the additional step of finding a true-size view of the right section in an auxiliary view. You should first locate a series of equally spaced elements around the right section in the auxiliary view and then project them back to the true-length view. Draw the stretch-out line parallel to the

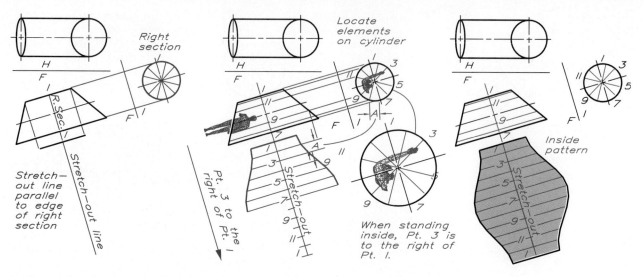

31.38 Development of an oblique cylinder.

Step 1 Draw the right section perpendicular to the true-length axis in the front view. Draw an auxiliary view to find the right-section true size; divide it into equal chords. Draw a stretch-out line parallel to the edge of the right section. Locate the shortest line at 1.

Step 2 Project elements from the right section to the front view. Transfer the chordal measurements in the auxiliary view to the stretch-out line to locate cylindrical elements and determine their lengths by projection.

Step 3 Locate the remaining elements to complete layout as was begun in step 2. Connect the ends of the elements with a smooth curve. This is an inside pattern.

edge view of the right section in the front view. Lay out the spacing between the elements along the stretch-out line, and draw the elements through these points perpendicular to the stretch-out line. Find the lengths of the elements by projecting from the front view and complete the pattern.

A more general case is the oblique cylinder shown in **Fig. 31.39**, where the elements are not true length in the given views. A primary auxiliary view gives the elements true length, and a secondary auxiliary view yields a true-size view of the right section. Draw the stretch-out line parallel to the edge view of the right section in the primary auxiliary view. Transfer the elements to the stretch-out line from the true-size right section.

Draw the elements perpendicular to the stretch-out line and find their lengths by projecting from the primary auxiliary view. Connect the endpoints with a smooth curve to complete the pattern.

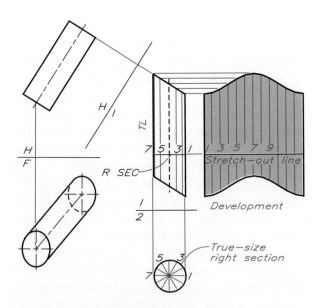

31.39 To develop an oblique cylinder, construct a primary auxiliary view in which the elements appear true length. Find the right-section true size in a secondary auxiliary view and complete the development, as shown in Fig. 31.38.

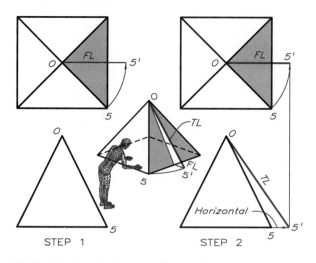

31.40 True length of a pyramid by revolution.

Step 1 To find the true length of corner line 0–5 of a pyramid, revolve it into the frontal plane in the top view, to 0–5'.

Step 2 Project point 5' to the front view, where frontal line 0–5' is true length.

31.41 This sheet-metal hopper is an application of a design involving development of a pyramid. (Courtesy of Gar-Bro.)

31.15 Development of Pyramids

All lines used to draw patterns must be true length, but pyramids have few lines that are true length in the given views. For this reason, you must find the sloping corner lines true length before drawing a development.

Figure 31.40 shows the method of finding the corner lines of a pyramid true length by revolution. Revolve line 0–5 into the frontal plane to line 0–5' in the top view so that it will be true length in the front view. An application of the development of a pyramid is the sheet-metal hopper shown in **Fig. 31.41**.

Figure 31.42 shows the development of a right pyramid. Line 0–1 is revolved into the frontal plane in the top view to find its true length in the front view. Because it is a right pyramid, all corner lines are equal in length. Line 0–1' is the radius for the base circle of the development. When you transfer distance 1–2 from the base in the top view to the development, it forms a chord on the base circle. Find

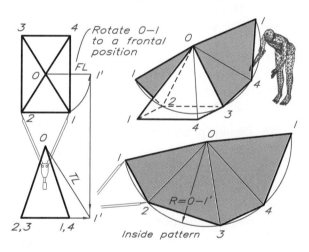

31.42 To develop this right pyramid, lay out an arc by using the true length of corner line 0–1' as the radius. Transfer true-length distances around the base in the top view to the arc and darken the lines.

lines 2–3, 3–4, and 4–1 in the same manner and in sequence. Draw the fold lines as thin lines from the base to the apex, point 0.

A variation of this case is the truncated pyramid (**Fig. 31.43**). Development of the inside pattern occurs as in the preceding case, but establishing the upper lines of the devel-

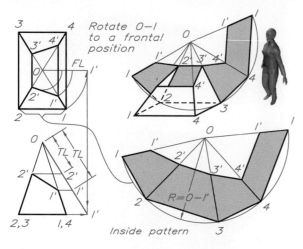

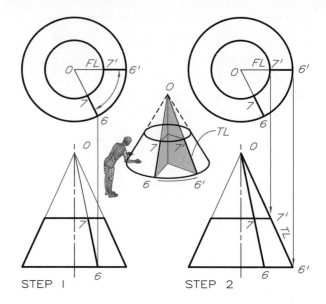

31.43 To develop an inside pattern of a truncated right pyramid, use the method shown in Fig. 31.40. Find the true lengths of elements 0–1′, 0–2′, 0–3′ and 0–4′ in the front view by revolution. Lay them off along their respective elements to find the upper boundary of the pattern.

31.44 True length by revolution (cone).

Step 1 Revolve an element of a cone, 0–6, into a frontal plane in the top view.

Step 2 Project point 6′ to the front view, where it is a true-length outside element of the cone. Find the true length of line 0–7′ by projecting point 7′ to the outside element in the front view.

opment requires an additional step. Revolution yields the true-length lines from the apex to points 1′, 2′, 3′, and 4′. Lay off these distances along their respective lines on the pattern to find the upper boundary of the pattern.

31.16 Development of Cones

All elements of a right cone are equal in length, as shown in **Fig. 31.44**. Revolving element 0–6 into its frontal position at 0–6′ gives its true length. When projected to the front view, line 0–6′ is true length and is the outside element of the cone. Projecting point 7 horizontally to element 0–6′ locates point 7′.

To develop the right cone depicted in **Fig. 31.45**, divide the base into equally spaced elements in the top view and project them to the front view, where they radiate to the apex at 0. The outside elements in the front view, 0–10 and 0–4, are true length.

Using element 0–10 as a radius, draw the base arc of the development. The spacing of the elements along the base circle is equal to

the chordal distances between them on the base in the top view. Inspection of the top view from the inside, where point 2 is to the right of point 1, indicates that this is an inside pattern. The fuselage of the Premier in Fig. 31.31 is an example of cone developed in sheet metal.

Figure 31.46 shows the development of a truncated cone. To find its pattern, lay out the entire cone by using the true-length element 0-1 as the radius, ignoring the portion removed from it. Locate the hyperbolic section formed by the inclined plane through the front view of the cone in the top view by projecting points on each element of the cone to the top view of these elements. For example, determine the true length of line 0–3′ by projecting point 3′ horizontally to the true-length element 0–1 in the front view. Lay off these distances, and others, along their respective elements to establish a smooth curve.

31.17 Development of Transition Pieces

A transition piece changes the shape of a section at one end to a different shape at the other end. In **Fig. 31.47**, you can see examples of transition pieces that convert one cross-sectional shape to another. In industrial applications, transition pieces may be huge (**Fig. 31.48**) or relatively small.

Figure 31.49 shows the steps in the development of a transition piece. Radial elements are extended from each corner to the equally spaced points on the circular end of the piece. Revolution is used to find the true length of each line. True-length lines 2–D, 3–D, and 2–3 yield the inside pattern of 2–3–D.

The true-length radial lines, used in combination with the true-length chordal distances in the top view, give a series of abutting triangles to form the pattern beginning with element D2. Adding the triangles A–1–2 and G–3–4 at each end of the pattern completes the development of a half-pattern. Only a half-pattern is shown in this example.

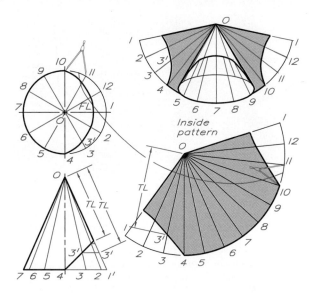

31.46 To develop a truncated cone with a side opening, begin by laying it out as in Fig. 31.45. Find true-length elements by revolution in the front view and transfer them to their respective elements in the pattern.

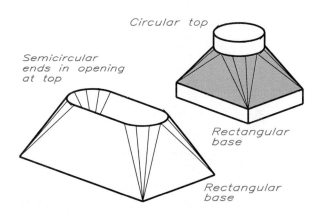

31.47 Examples of transition pieces that connect parts having different cross-sections.

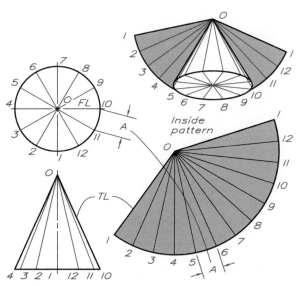

31.45 To develop an inside pattern of a right cone, use a true-length element (0–4 or 0–10 in the front view) as the radius. Transfer chordal distances from the true-size base in the top view and mark them off along the arc.

Problems

The scale of these problems allows you to fit two solutions on a size A sheet for a grid size of 0.20 in. (5 mm). For a grid size of 0.40 in. (10 mm), you can fit only one solution on a size A sheet.

31.48 Large transition pieces were used to join a circular shape with a rectangular section in this application. (Courtesy of Western Precipitation Group, Joy Manufacturing Company.)

Intersections

1–24. (**Fig. 31.50**) Lay out the given views of the problems and find the intersections between the shapes that are necessary to complete the views.

Developments

25–48. (**Fig. 31.51**) Lay out the problems and draw their developments. Orient the long side of the sheet horizontally to allow space at the right of the given views for the development.

49. Select an appropriate scale, draw two orthographic views of the corner bracket, and draw a developed, flat pattern of it on a size A sheet. Label your construction.

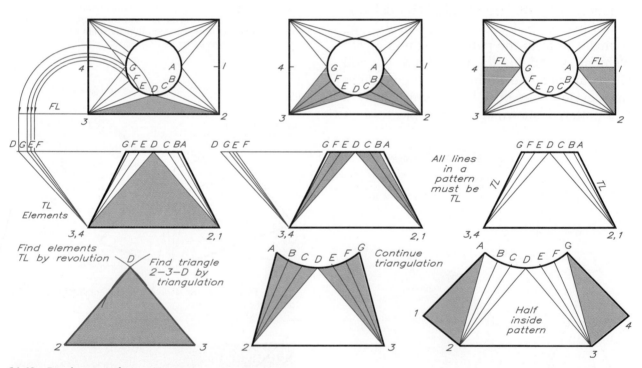

31.49 Development of a transition piece.

Step 1 Divide the circular end into equal parts in the top view and connect these points with lines to corner points 2 and 3. Find the true length of these lines by revolving and projecting them to the front view. Using true-length lines, draw triangle 2–3–D.

Step 2 Using other true-length lines and the chord distances on the circular end in the top view, draw a series of triangles joined at common sides. For example, draw arcs 2–C from point 2. To find point C, draw arc DC from D. Chord DC is true length in the top view.

Step 3 Construct the remaining planes, A–1–2 and G–3–4, by triangulation to complete the inside half-pattern of the transition piece. Draw the fold lines, where the surface is to be bent, as thin lines. The seam line for the pattern is line A–1, the shortest line.

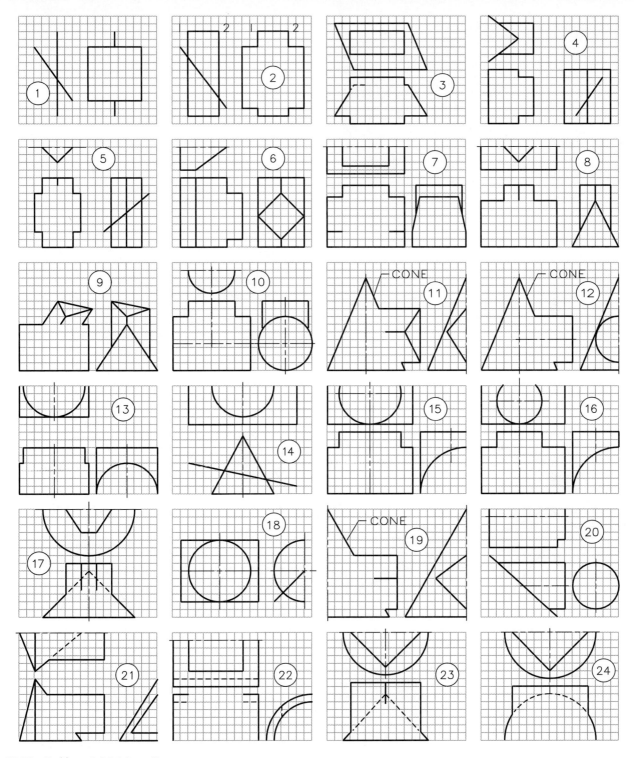

31.50 Problems 1–24. Intersections.

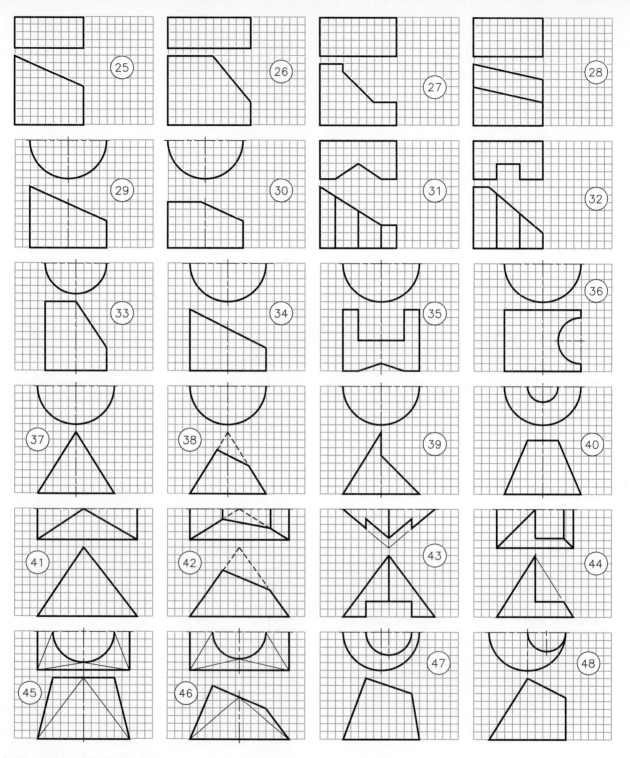

31.51 Problems 25–48. Developments.

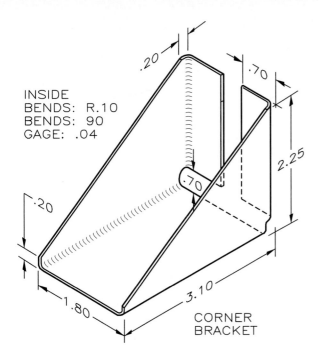

INSIDE
BENDS: R.10
BENDS: 90
GAGE: .04

.20

.70

2.25

.20

.70

1.80

3.10

CORNER
BRACKET

50–52. Four sheet-metal parts are shown here drawn on a 0.40 grid. Select an appropriate scale, draw the orthographic views of the parts, and draw a developed flat pattern of each; one per size-A sheet.

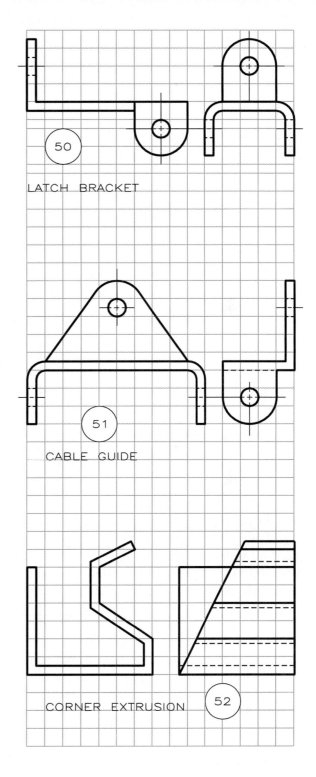

50

LATCH BRACKET

51

CABLE GUIDE

CORNER EXTRUSION   52

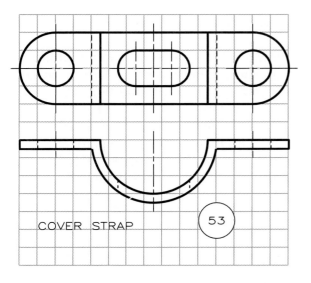

COVER STRAP   53

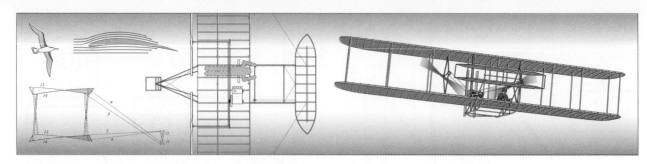

32

Graphs

32.1 Introduction

Data and information expressed as numbers and words are usually difficult to analyze or evaluate unless transcribed into graphical form, or as a **graph**. The term **chart** is an acceptable substitute for graph, but it is more appropriate when applied to maps, a specialized form of graphs. Graphs are especially useful in presenting data at briefings where the data must be interpreted and communicated quickly to those in attendance. Graphs are a convenient way to condense and present data visually, allowing the data to be grasped much more easily than when presented as tables of numbers or verbally.

Several different types of graphs are widely used. Their application depends on the data and the nature of the presentation required. There are six common types of graphs:

1. Pie graphs
2. Bar graphs
3. Linear coordinate graphs
4. Logarithmic coordinate graphs
5. Semilogarithmic coordinate graphs
6. Schematics and diagrams

Proportions

Graphs are used on large display boards, and in technical reports, as slides for a projector, or as transparencies for an overhead projector. Consequently, the proportion of the graph must be determined before it is constructed to match the page, slide, or transparency.

A graph that is to be photographed with a 35-mm camera must be drawn to the proportions of the film, or approximately 3 × 2 (**Fig. 32.1**). This area may be enlarged or reduced proportionally by using the diagonal-line method.

The proportions of an overhead projector transparency are approximately 10 × 8. The image size should not exceed 9.5 in. × 7.5 in.

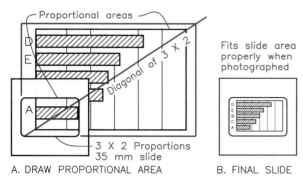

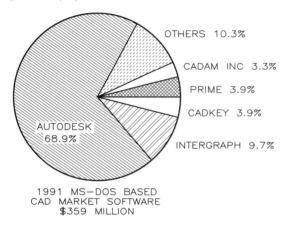

32.1 This diagonal-line method may be used to lay out drawings that are proportional to the area of a 35-mm slide.

32.2 A pie graph shows the relationship of parts to a whole. It is most effective when there are only a few parts. (Courtesy of Autodesk.)

to allow adequate margin for mounting the transparency on a frame (usually of cardboard).

32.2 Pie Graphs

Pie graphs compare the relationship of parts to a whole. For example, **Fig. 32.2** shows a pie graph that compares industry's use of various types of computer graphics software as percentages of the total, or their market share.

Figure 32.3 illustrates the steps involved in drawing a pie graph. Note how the data in this example, as simple as it is, are not as easily compared in numerical form. The pie graph illustrates relationships much more clearly. Position thin sectors of a pie graph as nearly horizontal as possible to provide more space for labeling. When space is not available within the sectors, place labels outside the pie graph and, if necessary, use leaders (see Fig. 32.2). Showing the percentage represented by each sector is important and giving the actual numbers or values as part of the label is also desirable.

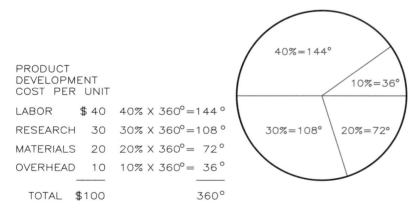

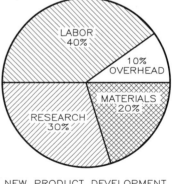

PRODUCT DEVELOPMENT COST PER UNIT		
LABOR	$ 40	40% X 360°=144°
RESEARCH	30	30% X 360°=108°
MATERIALS	20	20% X 360°= 72°
OVERHEAD	10	10% X 360°= 36°
TOTAL	$100	360°

32.3 Pie graphs.

Step 1 Find the sum of the parts and the percentage that each is of the total. Multiply each percentage by 360° to obtain the angle of each sector.

Step 2 Draw the circle and construct each sector using the degrees of each from step 1. Place small sectors as nearly horizontal as possible.

Step 3 Label sectors with their proper names and percentages. Exact numbers may also be included in each sector to add more clarity.

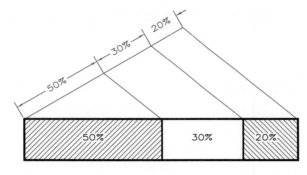

32.4 The diagonal-line method may be used to find the percentages of the parts to the whole, where the total bar represents 100 percent.

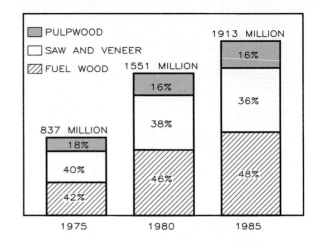

32.5 In this graph, each bar represents 100 percent of the total amount and shows the percentages of the parts to the total.

DIVIDENDS PAID
BY THE
SMITH COMPANY

YEAR	DIVIDEND
1996	$0.40
1997	0.60
1998	0.90

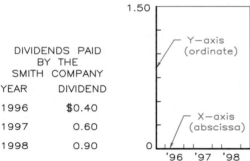

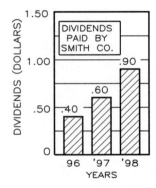

32.6 Drawing a bar graph.

Given These numerical data are to be plotted as a bar graph for better presentation in a report.

Step 1 Scale the vertical and horizontal axes so that the data will fit on the grid. Begin the bars at zero.

Step 2 The width of the bars should be greater than the space between them. Lines should not cross the bars.

Step 3 Strengthen lines, place a title in the graph, label the axes, and cross-hatch the bars.

32.3 Bar Graphs

Bar graphs are widely used for comparing values because the general public understands them. A bar graph may be a single bar (**Fig. 32.4**), where the length of the bar representing 100 percent is divided into lengths proportional to the percentages of its three parts. In **Fig. 32.5**, the bars not only show the overall production of timber (the total heights of the bars), but also the percentages of the total devoted to three uses of the timber.

Figure 32.6 shows how to convert data into a bar graph that can be used in a report or briefing. The axes of the graph carry labels, and its title appears inside the graph where space is available.

The bars of a bar graph should be sorted in ascending or descending order unless there is an overriding reason not to, such as a chronological sequence. An arbitrary arrangement of the bars, such as in alphabetical or numerical order, makes a graph difficult to evaluate (**Fig. 32.7A**). However, ranking the categories by bar length allows easier comparisons from smallest to largest (**Fig. 32.7B**). If the data are sequential and involve time, such as sales per

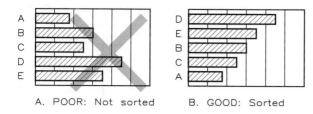

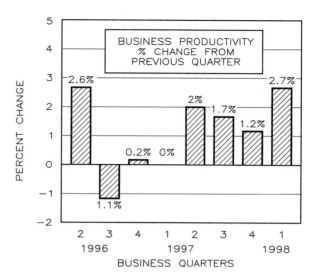

A. POOR: Not sorted   B. GOOD: Sorted

32.7 Sorting bars by length.

A When bars are arbitrarily sorted, as they are here alphabetically, the bar graph is difficult to interpret.

B When the bars are sorted by length, the graph is much easier to interpret.

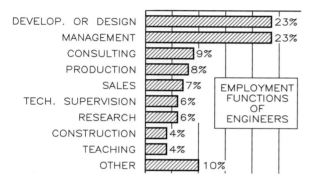

32.8 The horizontal bars of this graph are arranged in descending order to show the employment functions of engineers.

32.9 Bar graphs may illustrate both negative and positive directions.

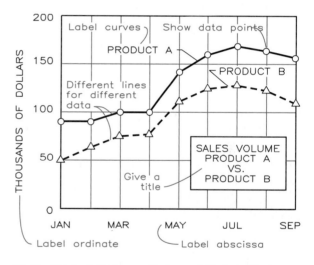

32.10 This basic linear coordinate graph illustrates the important features on a graph.

month, a better arrangement of the bars is chronologically, to show the effect of time.

Bars in a bar graph may be horizontal (**Fig. 32.8**) or vertical. Data cannot be compared accurately unless each bar is full length and originates at zero. Also, bars should not extend beyond the limits of the graph (giving the impression that the data were "too hot" to hold). **Figure 32.9** shows another form of bar graph, with plus as well as minus changes extending from a base value.

32.4 Linear Coordinate Graphs

Figure 32.10 shows a typical linear coordinate graph, with notes explaining its important features. Divided into equal divisions, the axes are referred to as linear scales. Data points are plotted on the grid by using measurements, called coordinates, along each

axis from zero. The plotted points are marked with symbols such as circles or squares that may be easily drawn with a template. The horizontal scale of the graph is called the **abscissa**, or *x* axis. The vertical scale is called the **ordinate**, or *y* axis.

When the points have been plotted, a curve is drawn through them to represent the data. The line drawn to represent data points

32.4 LINEAR COORDINATE GRAPHS • 571

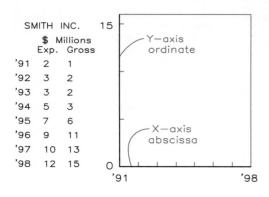

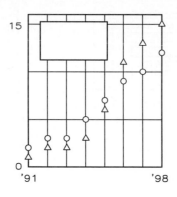

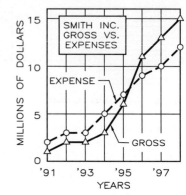

32.11 A broken-line graph.

Given A record of expenses and gross income for Smith Inc.

Step 1 Lay off the vertical (ordinate) and horizontal (abscissa) axes to provide space for the largest values.

Step 2 Draw division lines and plot the data, using different symbols for each set of data.

Step 3 Connect points with straight lines, label the axes, title the graph, darken the lines, and label the curves.

is called a **curve**, even if it is a straight line, smooth curve, or broken line. The curve should not extend through the plotted points; rather, the points should be left as open circles or other symbols.

The curve is the most important part of the graph, so it should be drawn as the most prominent (thickest) line. If there are two curves in a graph, they should be drawn as different line types and labeled. The title of the graph is placed in a box inside the graph and units are given along the *x* and *y* axes with labels identifying the scales of the graph.

Broken-Line Graphs

The steps required to draw a linear coordinate graph are shown in **Fig. 32.11**. Because the data points represent sales, which have no predictable pattern, the data do not give a smooth progression from point to point. Therefore, the points are connected with a **broken-line curve** drawn as an angular line from point to point.

Again, leave the symbols used to mark the data points open rather than extending grid lines or the data curve through them (**Fig. 32.12**). Each circle or symbol used to plot

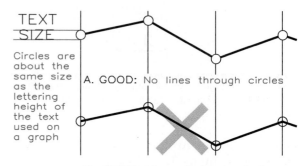

32.12 The curve of a graph drawn from point to point should not extend through the symbols used to represent data points.

points should be about 1/8 in. (3 mm) in diameter. **Figure 32.13** shows typical data-point symbols and lines.

Computer Method A data point may be produced as a CIRCLE, DONUT (open and closed), or POLYGON. The grid lines and curves that pass through the open symbols may be removed easily with the TRIM command (**Fig. 32.14**).

Titles The title of a graph may be located in any of the positions shown in **Fig. 32.15**. A graph's title should never be as meaningless as

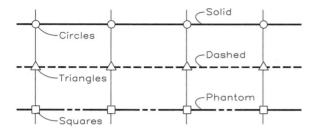

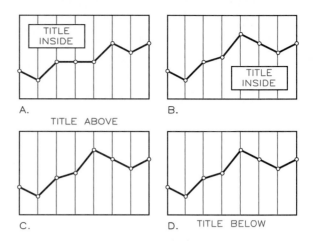

32.13 Symbols and lines such as these may be used to represent different curves on a graph. The data-point symbols should be drawn about the same size as the letter height being used, 1/8 in. (3 mm) in diameter.

32.15 Placement of titles on a graph.

A and **B** The title of a graph may be placed inside a box within the graph. Box perimeter lines should not coincide with grid lines.

C Titles may be placed above the graph.

D Titles may be placed below the graph.

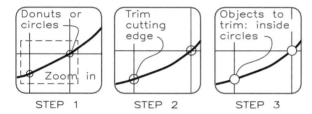

STEP 1   STEP 2   STEP 3

32.14 Editing data points.

Step 1 An open data point may be plotted on a graph as a CIRCLE, DONUT, or POLYGON. To remove lines from inside the open points, ZOOM in on several points.

Step 2 Command: TRIM (Enter)

Select cutting edges (s):...

Select objects: (Select the circle.) (Enter)

Step 3 Select object to trim: (Select the lines inside the circle, and they will be removed.)

Continue this process for all points.

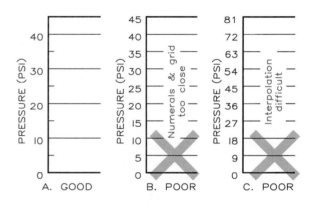

32.16 Calibrating graph scales.

A The scale is properly labeled and calibrated. It has about the right number of grid lines and divisions, and the numbers are well spaced and easy to interpolate.

B The numbers are too close together, and there are too many grid lines.

C The increments selected make interpolation difficult.

"graph" or "coordinate graph." Instead, it should identify concisely what the graph shows.

Calibration and Labeling The calibration and labeling of the axes affects the appearance and readability of a graph. In **Fig. 32.16**, part A shows a properly calibrated and labeled axis. Parts B and C illustrate common mistakes: placing the grid lines too close together and labeling too many divisions along the axis. In part C, the choice of the interval between the labeled values (9 units) makes interpolation

between them difficult. For example, locating the value 22 by eye is more difficult on this scale than on the one shown in part A.

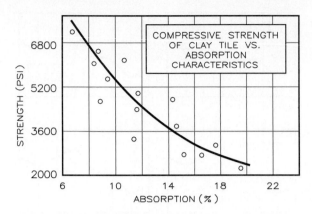

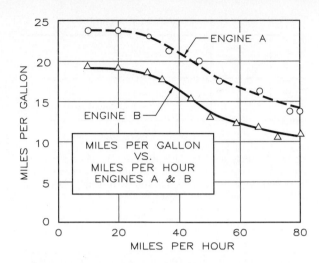

32.17 Since the compressive strength of clay tile varies gradually with respect to its absorption characteristics, this relationship is plotted as a smooth curve.

32.18 Smooth, best-fit curves are drawn to approximate the data without necessarily passing through each point. Since fuel consumption changes gradually with respect to speed, smooth-line curves represent the data better than broken-line curves.

Smooth-Line Graphs

When the data being graphed involve gradual, continuous changes in relationships, the curve is drawn as a smooth line. In **Fig. 32.17**, the strength of clay tile related to its absorption is plotted. The strength of clay tile changes gradually and continuously in relation to absorption time. Therefore, the data points are connected with a smooth-line rather than a broken-line curve. These relationships are represented by the **best-fit curve**, a smooth curve that is an average representation of the points. There is a smooth-line curve relationship between miles per gallon and the speed at which a car is driven. **Figure 32.18** compares the results for two engines.

A smooth-line curve on a graph implies that interpolations between data points can be made to estimate other values. Data points connected by a broken-line curve imply that interpolations between the plotted points cannot be made.

Straight-Line Graphs

Some graphs have neither broken-line curves nor smooth-line curves but straight-line

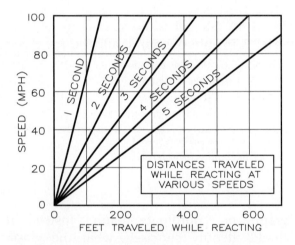

32.19 This graph may be used to determine a third value from the other two variables. For example, select a speed of 70 mph and a time of 5 seconds to find a distance traveled of 550 ft.

curves (**Fig. 32.19**). On this type of graph, a third value can be determined from the two given values. For example, if you are driving 70 miles per hour and you take 5 seconds to react and apply your brakes, you will have traveled 550 feet in that time.

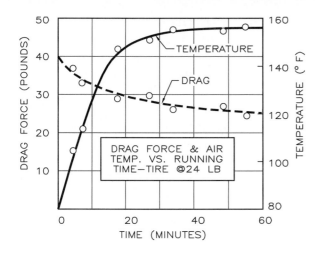

32.20 A two-scale graph has different scales along each *y* axis; labels identify which scale applies to which curve.

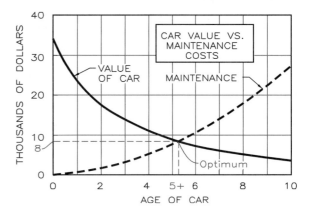

32.21 This graph shows the optimum time to sell a car based on the intersection of curves representing the depreciating value of the car and its increasing maintenance costs.

Two-Scale Coordinate Graphs

Graphs may contain different scales in combination with a common horizontal scale, as shown in **Fig. 32.20**, where the vertical scale at the left is in units of pounds and the one at the right is in degrees of temperature. Both curves are drawn with respect to their *y* axes and each curve is labeled. Two-scale graphs of this type may be confusing unless they are clearly labeled. Two-scale graphs are effective for comparing related variables, as shown here.

Optimization Graphs

The act of getting the most for the least is called **optimization**. Examples of optimization are the most miles per gallon, the most square footage per dollar, the most production per day, and so forth. An **optimization graph** is shown in **Fig. 32.21**, where an automobile's depreciation is compared with its maintenance costs to show the optimum time for replacing the car. These two sets of data cross at an *x*-axis value of slightly more than five years, or the optimum point. At that time, the cost of maintenance is equal to the value of the car, indicating that it might be a desirable time to buy a new car.

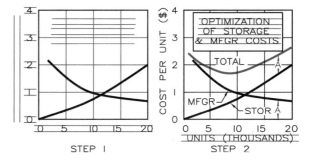

32.22 Constructing an optimization graph.

Step 1 Lay out the graph and plot the curves from the data given.

Step 2 Add the two curves to find the third curve. For example, transfer distance A to locate a point on the third curve. The lowest point of the "total" curve is the optimum point, or 8000 units.

The steps involved in drawing an optimization graph are illustrated in **Fig. 32.22**. Here, the manufacturing cost per unit reduces as more units are made, causing warehousing costs to increase. Adding the two curves to get a third (total) curve indicates that the optimum number to manufacture at a time is about 8000 units (the low point on the total curve). When more or fewer units are manufactured, the total cost per unit is greater.

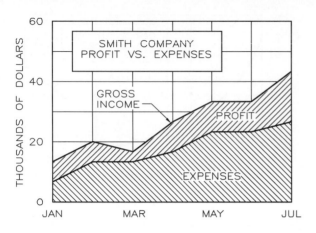

32.23 This graph is a composite of a coordinate graph and an area graph. The upper area represents the difference between the two plotted curves.

Composite Graphs

A **composite graph** combines the attributes of a linear graph with those of an area graph to present sets of data. An example of a composite graph is shown in **Fig. 32.23**, where a company's profit and expenses are compared. The upper curve is a plot of the company's gross income. The lower curve is a plot of the com-

pany's expenses. The difference between the two is the company's profit. Cross-hatching the areas emphasizes the relative sizes of expenses and profits.

Break-Even Graphs

Break-even graphs help in evaluating marketing and manufacturing costs to determine the selling price of a product. As **Fig. 32.24** shows, if the desired break-even point for a product is 10,000 units, it must sell for $3.50 each to cover the costs of manufacturing and development.

32.5 Semilogarithmic-Coordinate Graphs

Semilogarithmic coordinate graphs are also called ratio graphs because they represent ratios graphically. One scale, usually the vertical scale, is logarithmic, and the other is linear (divided into equal divisions).

The same data plotted on a linear grid (arithmetic scale) and on a semilogarithmic grid are compared in **Fig. 32.25**. The semilogarithmic graph reveals that the percentage change from 0 to 5 is greater for curve B than

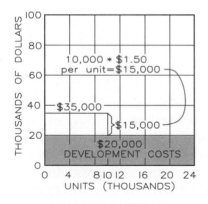

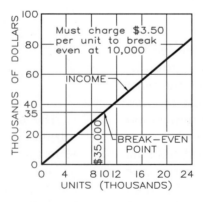

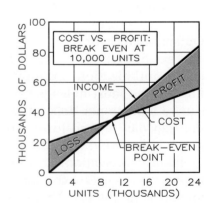

32.24 Drawing a break-even graph.

Step 1 Plot the development cost ($20,000). At $1.50 per unit to make, the total cost would be $35,000 for 10,000 units, the break-even point.

Step 2 To break even at 10,000, the manufacturer must sell each unit for $3.50. Draw a line from zero through the break-even point of $35,000 to represent income.

Step 3 There is a loss of $20,000 at zero units, but progressively less until the break-even point is reached. Profit is the difference between curves at the right of the break-even point.

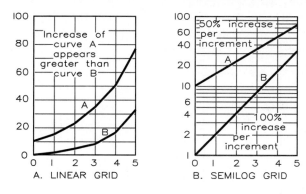

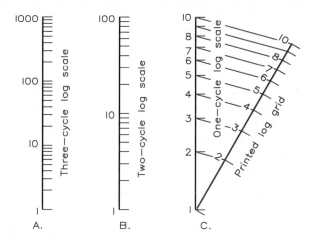

32.27 Logarithmic scales may have several cycles: (A) three-cycle scales; (B) two-cycle scales, and (C) one-cycle scales. Calibrations may be projected to a scale of any length from a printed scale, as shown here.

32.25 When plotted on a linear grid, curve A appears to be increasing at a greater rate than curve B. However, plotting the data on a semilogarithmic grid reveals the true rate of change.

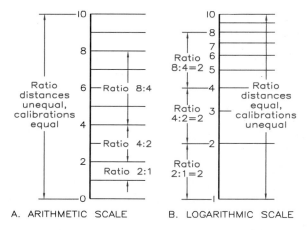

32.26 The divisions on an linear scale are equal and represent unequal ratios between points. The divisions on logarithmic scales are unequal and represent equal ratios.

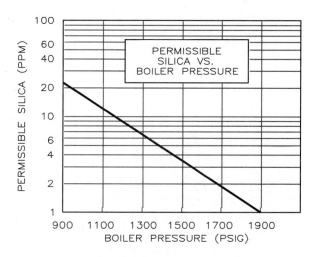

32.28 This semilogarithmic graph relates permissible silica (parts per million) to boiler pressure.

for curve A because here curve B is steeper. The plot on the linear grid appears to show the opposite result since linear grids do not reveal ratios as well.

Figure 32.26 shows the relationship between the linear scale and the logarithmic scale. Equal divisions along the linear scale have unequal ratios, but equal divisions along the log scale have equal ratios.

Log scales may have one or many cycles. Each cycle increases by a factor of 10. For example, the scale shown in **Fig. 32.27A** is a

three-cycle scale, and the one shown in **Fig. 32.27B** is a two-cycle scale. When scales must be drawn to a certain length, commercially printed log scales may be used to transfer graphically the calibrations to the scale being used (**Fig. 32.27C**).

An application of a semilogarithmic graph for presenting industrial data is illustrated in **Fig. 32.28**. People who do not realize that semilog graphs are different from linear coor-

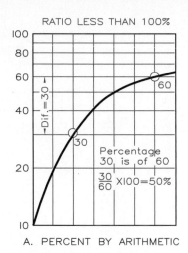

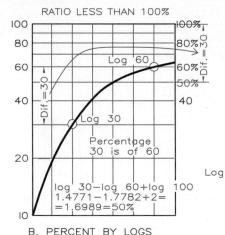

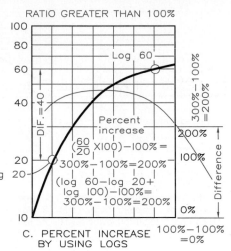

A. PERCENT BY ARITHMETIC   **B. PERCENT BY LOGS**   **C. PERCENT INCREASE BY USING LOGS**

32.29   Percentage graphs.

A Find the percentage that one data point is of another point (the percentage that 30 is of 60, for example). You may calculate this mathematically: $(30/60)(100) = 50\%$.

B Find the percentage that 30 is of 60 mathematically by using the logarithms of the numbers. You can also find it graphically by transferring the distance between 30 and 60 to the scale at the right, which shows that 30 is 50%.

C To find a percentage increase greater than 100%, divide the smaller number into the larger number. Find the difference between the logs of 60 and 20 with dividers and measure upward from 100% to find the increase of 200%.

dinate graphs may misunderstand them. Also, zero values cannot be shown on log scales.

Percentage Graphs

The percentage that one number is of another, or the percentage increase of one number to a greater number can be determined on a semilogarithmic graph (**Fig. 32.29**). Data plotted in part A are used to find the percentage that 30 is of 60 (two points on the curve) by arithmetic. The vertical distance between them is the difference between their logarithms, so the percentage can be found graphically in part B. The distance from 30 to 60 is transferred to the log scale at the right of the graph and subtracted from the log of 100 to find the value of 50 percent as a direct reading of percentage.

In part C, the percentage increase between two points is transferred from the grid to the lower end of the log scale and measured upward because the increase is greater than zero. These methods may be used to find percentage increases or decreases for any set of points on the grid.

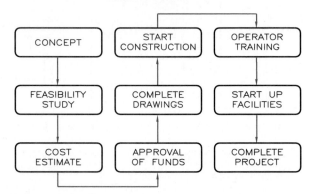

32.30 A schematic diagram showing the steps required to complete a construction project.

32.7 Schematics

The schematic diagram shown in **Fig. 32.30** depicts the steps required to complete a construction project. Each step is blocked in and connected with arrows to give the sequence.

The organization of a group of people can be represented in an organizational graph (usually called an **organization chart**), as shown in **Fig. 32.31**. This particular example was drawn by a computer program especially

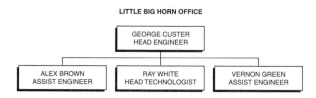

32.31 A computer-drawn organization chart outlining lines of responsibility.

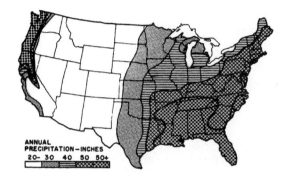

32.32 A map showing the annual precipitation for various parts of the United States. (Courtesy of the Structural Clay Products Institute.)

designed for drawing these types of graphs. The positions represented by the blocks in the lower part of the graph are responsible to those represented by the blocks above them. Lines of authority connecting the blocks give the routes for communication both upward and downward.

Geographical charts are used to combine maps and other relationships, such as weather. In **Fig. 32.32**, different hatching symbols represent the annual precipitation in various parts of the United States.

32.8 Graphs by Computer

Many computer programs are available for converting numerical data into various types of graphs to improve comprehension and interpretation. These programs vary from data representation as bar graphs, coordinate graphs, and pie graphs to three-dimensional mathematical models.

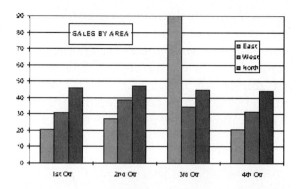

32.33 A computer-drawn three-dimensional bar graph.

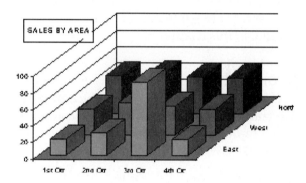

32.34 A three-dimensional plot of the data shown in Fig. 32.33, drawn by a computer program.

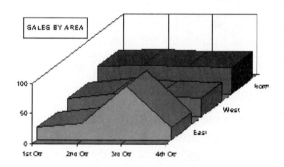

32.35 Another version of the data from Fig. 32.33.

Computer-produced graphs are especially useful for preparing visual aids for projection on a screen for a presentation and for technical reports. **Figure 32.33** is an example of a 3D bar graph printed from data that was input in tabular form. The same data was also plotted in the 3D forms shown in **Figs. 32.34** and **Fig. 32.35**.

It is up to you, the user, to determine which format is best in communicating a particular type of data to your audience, whether by an illustration in a written report or by a visual aid during an oral presentation. Whether drawn by hand or by computer, the principles of preparing effective graphs are the same.

Problems

Draw your solutions to these problems on size A sheets. Apply the techniques and principles covered in this chapter.

Pie Graphs

1. Draw a pie graph that shows the comparative sources of retirees' income: investments, 34%; employment, 24%; social security, 21%; pensions, 19%; other, 2%.

2. Draw a pie graph that shows the number of members of the technological team: engineers, 985,000; technicians, 932,000; scientists, 410,000.

3. Construct a pie graph of the employment status of graduates of two-year technician programs a year after graduation: employed, 63%; continuing full-time study, 23%; considering job offers, 6%; military, 6%; other, 2%.

4. Draw a pie graph showing the types of degrees held by aerospace engineers: bachelor's, 65%; master's, 29%; Ph.D.'s, 6%.

Bar Graphs

5. Draw a bar graph that shows expected job growth by city: Austin, 1.3%; San Antonio, 0.8%; Houston, 1.6%; Fort Worth, 0%; Dallas, 0.4%; all of Texas, 0.8%.

6. Draw a single-bar bar graph that represents 100% of a die casting alloy. The propor-

tional parts of the alloy are tin, 16%; lead, 24%; zinc, 38.8%; aluminum, 16.4%; copper, 4.8%.

7. Draw a bar graph that compares the number of skilled workers employed in various occupations. Use the following data and arrange the graph for ease of comparing occupations:

carpenters	82,000
all-round machinists	310,000
plumbers	350,000
bricklayers	200,000
appliance servicers	185,000
automotive mechanics	760,000
electricians	380,000
painters	400,000.

8. Draw a bar graph that shows the characteristics of a typical U.S. family's spending:

housing	29.6%
food	15.1%
transportation	16.7%
clothing	5.9%
retirement	8.6%
entertainment	4.9%
insurance	5.2%
health care	3.0%
charity	3.1%
other	7.9%.

9. Draw a bar graph that compares the corrosion resistance of the materials listed in the following table.

	Loss in Weight (%)	
	In Atmosphere	In Sea Water
Common steel	100	100
10% nickel steel	70	80
25% nickel steel	20	55

10. Draw a bar graph of the data from Problem 1.

11. Draw a bar graph of the data from Problem 2.

12. Draw a bar graph of the data from Problem 3.

13. Construct a bar graph comparing sales and earnings of Apricot Computers from 1990 through 1999. Data are by year for sales and earnings (profit) in billions of dollars: 1990, 0 and 0; 1991, 0.33 and 0.05; 1992, 0.60 and 0.07; 1993, 1.00 and 0.09; 1994, 1.51 and 0.08; 1995, 1.90 and 0.07; 1996, 1.85 and 0.12; 1997, 2.70 and 0.25; 1998, 4.15 and 0.40; 1999, 5.50 and 0.45.

Linear Coordinate Graphs

14. Draw a linear coordinate graph to show the estimated U.S. population growth from 1992 through 2050 in millions of people: 1992, 255; 2000, 270; 2010, 295; 2020, 325; 2030, 348; 2040, 360; 2050, 383.

15. Construct a linear coordinate graph that shows the relationship of energy costs (mills per kilowatt-hour) on the y axis to the percent capacity of a nuclear power plant and a gas-or oil-fired power plant on the x axis. Gas- or oil-fired plant data: 17 mills, 10%; 12 mills, 20%; 8 mills, 40%; 7 mills, 60%; 6 mills, 80%; 5.8 mills, 100%. Nuclear plant data: 24 mills, 10%; 14 mills, 20%; 7 mills, 40%; 5 mills, 60%; 4.2 mills, 80%; 3.7 mills, 100%.

16. Plot the data from Problem 13 as a linear coordinate graph.

17. Construct a linear coordinate graph to show the relationship between the transverse resilience in inch-pounds (ip) on the y axis and the single-blow impact in foot-pounds (fp) on the x axis of gray iron. Data: 21 fp, 375 ip; 22 fp, 350 ip; 23 fp, 380 ip; 30 fp, 400 ip; 32 fp, 420 ip; 33 fp, 410 ip; 38 fp, 510 ip; 45 fp, 615 ip; 50 fp, 585 ip; 60 fp, 785 ip; 70 fp, 900 ip; 75 fp, 920 ip.

18. Draw a linear coordinate graph to illustrate the trends in the export of U.S. services and products from 1986 through 1999:

1986	+8.5%
1987	+2.5%
1988	+7.5%
1989	+4.5%
1990	+7%
1991	+2%
1992	+7%
1993	+12.5%
1994	+17.5%
1995	+10%
1996	+6%
1997	+3%
1998	+3%
1999	+7%.

19. Draw a linear coordinate graph that shows the voltage characteristics for a generator as given in the following table of values: abscissa, armature current in amperes, (Ia); ordinate, terminal voltage in volts, (Et):

Ia	Et	Ia	Et	Ia	Et
0	288	31.1	181.8	41.5	68
5.4	275	35.4	156	40.5	42.5
11.8	257	39.7	108	39.5	26.5
15.6	247	40.5	97	37.8	16
22.2	224.5	40.7	90	13.0	0
26.2	217	41.4	77.5		

20. Draw a linear coordinate graph for the centrifugal pump test data in the table below. The units along the x axis are to be gallons per minute. Use four curves to represent the variables given.

Gallons per Minute	Discharge Pressure	Water HP	Electric HP	Efficiency (%)
0	19.0	0.00	1.36	0.0
75	17.5	0.72	2.25	32.0
115	15.0	1.00	2.54	39.4
154	10.0	1.00	2.74	36.5
185	5.0	0.74	2.80	26.5
200	3.0	0.63	2.83	22.2

21. Draw a linear coordinate graph that compares two of the values shown in **Table 32.1**—ultimate strength and elastic limit—with degrees of temperature (*x* axis).

Table 32.1

F°	Ultimate Strength	Elastic Limit
400	257,500	208,000
500	247,000	224,500
600	232,500	214,000
700	207,500	193,500
800	180,500	169,000
900	159,500	146,500
1000	142,500	128,500
1100	126,500	114,000
1200	114,500	96,500
1300	108,000	85,500

Break-Even Graphs

22. Draw a break-even graph that shows the earnings for a new product that has a development cost of $12,000. The break-even point is at 8000 units, and each unit costs $0.50 to manufacture. What would be the profit at volumes of 20,000 and 25,000?

23. Repeat Problem 22 but make the following changes: the development costs are $80,000, the manufacturing cost of the first 10,000 units is $2.30 each, and the desired break-even point is 10,000 units. What is the profit at volumes of 20,000 and 30,000?

Logarithmic Graphs

24. Use the data given in **Table 32.2** to construct a logarithmic graph. Plot the vibration amplitude (A) as the ordinate and the vibration frequency (F) as the abscissa. The data for curve 1 represent the maximum limits of machinery in good condition with no danger from vibration. The data for curve 2 are the lower limits of machinery that is being vibrated excessively to the danger point. The

vertical scale is three cycles, and the horizontal scale is two cycles.

25. Plot this data on a two-cycle log graph to show the current in amperes (*y* axis) versus the voltage in volts (*x* axis) of precision temperature-sensing resistors. Data: 1 volt, 1.9 amps; 2 volts, 4 amps; 4 volts, 8 amps; 8 volts, 17 amps; 10 volts, 20 amps; 20 volts, 30 amps; 40 volts, 36 amps; 80 volts, 31 amps; 100 volts, 30 amps.

26. Plot the data in Problem 14 as a logarithmic graph.

27. Plot the data in Problem 20 as a logarithmic graph.

Semilogarithmic Graphs

28. Construct a semilogarithmic graph making the *y* axis a two-cycle log scale from 1 to 100 and the *x* axis a linear scale from 1 to 7. The object of the graph is to show the survivability of a shelter at varying distances from the atmospheric detonation of a one-megaton thermonuclear bomb. Plot overpressure in psi along the *y* axis, and distance from ground zero in miles along the *x* axis. The data points represent an 80-percent chance of survival of the shelter. Data: 1 mile, 55 psi; 2 miles, 11 psi; 3 miles, 4.5 psi; 4 miles, 2.5 psi; 5 miles, 2.0 psi; 6 miles, 1.3 psi.

29. The growth of Division A and Division B of a company is shown by the following data. Plot the data on a semilog graph with a one-cycle log scale on the *y* axis for sales in thousands of dollars and a linear scale on the *x* axis for years. Data: first year, A = $11,700 and B = $44,000; second year, A = $19,500 and B = $50,000; third year, A = $25,000 and B = $55,000; fourth year, A = $32,000 and B = $64,000; fifth year, A = $42,000 and B = $66,000; sixth year, A = $48,000 and B = $75,000. Which division has the better growth rate?

Table 32.2

F	100	200	500	1000	2000	5000	10,000
A(1)	0.0028	0.002	0.0015	0.001	0.0006	0.0003	0.00013
A(2)	0.06	0.05	0.04	0.03	0.018	0.005	0.001

30. Draw a semilog chart showing probable engineering progress based on the following indices:

40,000 B.C.	21
30,000 B.C	21.5
20,000 B.C.	22
16,000 B.C.	23
10,000 B.C.	27
6000 B.C.	34
4000 B.C.	39
2000 B.C.	49
500 B.C.	60
1900 A.D.	100

Use a horizontal scale of 1 in. = 10,000 years, a height of about 5 in., and two-cycle printed paper, if available.

31. Plot the data in Problem 19 as a semilogarithmic graph.

32. Plot the data in Problem 21 as a semilogarithmic graph.

Percentage Graphs

33. Plot the data from Problem 14 as a semilog graph. What is the percentage increase in the demand for water from 1890 to 1920? What percentage of demand is the supply for 1900, 1930, and 2000?

34. Using the graph plotted in Problem 30, determine the percentage of increase of Division A and Division B growth from year 1 to year 4. What percentage of sales of Division A are the sales of Division B at the end of year 2? At the end of year 6?

35. Plot the values for water horsepower and electric horsepower from Problem 22 on semilog paper against gallons per minute along the x axis. What percentage of electric horsepower is water horsepower when 75 gallons per minute are being pumped? What percentage increase in electric horsepower corresponds to an increase in pumping volume from 0 to 185 gallons per minute?

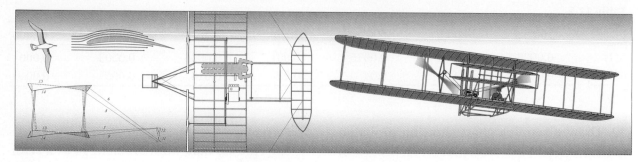

33

Nomography

33.1 Introduction

A helpful aid in interpreting data is the "number chart" that is called a **nomograph** or **nomogram**. A nomograph is a graphical combination of specially arranged and calibrated scales constructed to permit the repetitive solution of equations in an easy and rapid manner.

A nomogram frequently denotes a specific type of scale arrangement called an **alignment graph**. Two examples of alignment graphs are shown in **Fig. 33.1**. Other types have curved scales or different scale arrangements for use with more complex problems.

An alignment graph usually is constructed to solve for one or more unknowns in a formula or empirical relationship between two or more quantities. For example, an alignment graph can be used to convert degrees Celsius to degrees Fahrenheit or to find the

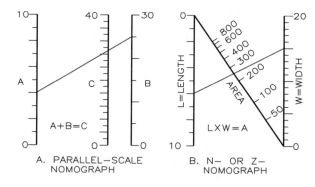

33.1 These two types of alignment graphs are typical nomographs.

size of a structural member to sustain a certain load. To read an alignment graph, place a straightedge or draw a line, called an **isopleth**, across the scales of the graph and read corresponding values from the scale on this line. **Figure 33.2** shows readings for the formula $W = U + V$.

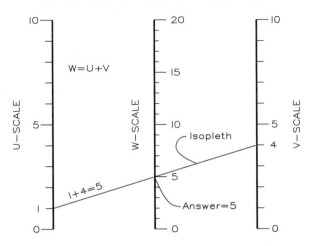

33.2 The use of an isopleth to solve graphically for unknowns in equation $W = U + V$.

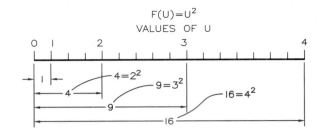

33.3 A functional scale containing units of measurement proportional to $F(U) = U2$.

33.2 Alignment Graph Scales

To construct any alignment graph, you must first determine the graduations of the scales that give the desired relationships. Alignment graph scales, called **functional scales**, are graduated according to values of some function of a variable. **Figure 33.3** illustrates a functional scale for $F(U) = U^2$. Substituting a value of $U = 2$ into the equation gives the position of U on the functional scale 4 units ($2^2 = 4$) from zero. Repeating this procedure for as many values of U as required yields a scale for finding all corresponding values of the function.

The Scale Modulus
Because the graduations on a functional scale are spaced in proportion to values of the function, a proportionality, or scaling factor, is needed. This constant of proportionality is called the scale modulus, m:

$$m = \frac{L}{F(U2) - F(U1)} \qquad (1)$$

where

m = scale modulus, in inches per functional unit,

L = desired length of the scale in inches,

$F(U2)$ = functional value at the end of the scale, and

$F(U1)$ = functional value at the start of the scale.

For example, suppose that you are to construct a functional scale for $F(U) = sin\ U$ from $0°$ to $45°$ and a scale 6 in. in length. Thus, $L = 6$ in., $F(U2) = \sin 45° = 0.707$, $F(U1) = \sin 0° = 0$. Substituting these values into Eq. (1) gives

$$m = \frac{6}{0.707 - 0} = 8.49 \text{ in.}$$
$$\text{per (sine) unit.}$$

The Scale Equation
A scale equation makes possible graduation and calibration of functional scales. The general form of this equation is a variation of Eq. (1):

$$X = m[F(U) - F(U1)], \qquad (2)$$

where

X = distance from the measuring point of the scale to any graduation point,

m = scale modulus,

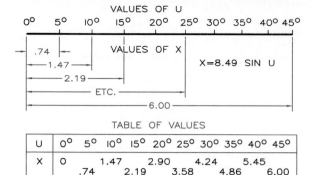

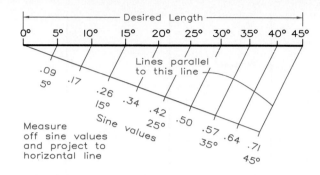

33.4 A functional scale calibrated with values from the table that were derived from the scale equation $X = 8.49 \sin U$.

33.5 A functional scale showing the sine of the angles from 0° to 45° drawn by the proportional-line method. Draw the scale to a desired length and lay off the sine values of angles at 5° intervals along a construction line passing through the 0° end of the scale.

$F(U)$ = functional value at the graduation point, and

$F(U1)$ = functional value at the measuring point of the scale.

For example, let's construct a functional scale for the equation $F(U) = \sin U$ (0° to 45°). We have already determined that $m = 8.49$, $F(U) = \sin U$, and $F(U1) = \sin 0° = 0$. By substitution, Eq. (2) becomes

$$X = 8.49(\sin U - 0) = 8.49 \sin U.$$

Using this equation, we can substitute values of U and construct a table of positions at 5° intervals, as shown in **Fig. 33.4**. The values of X give the positions in inches for the corresponding graduations measured from $U = 0°$. The initial measuring point does not have to be at one end of the scale, but an end is usually the most convenient point, especially if the functional value is zero at that point.

Figure 33.5 shows how to locate functional values along a scale with the proportional-line method. Measure the sine functions along a line at 5° intervals, with the end of the line passing through the 0° end of the scale. Transfer these functions from the inclined line with parallel lines back to the scale and label the functions.

33.3 Concurrent Scales

Concurrent scales aid in the rapid conversion of terms in one system of measurement into terms of a second system of measurement. Formulas of the type $F1 = F2$, which relate two variables, may be adapted to the concurrent scale format. An example is the Fahrenheit–Celsius temperature relationship:

$$°F = \frac{9}{5} °C + 32.$$

Another is the area of a circle:

$$A = \pi R^2.$$

Construction of a concurrent scale chart involves determining a functional scale for each side of the mathematical formula so that the position and lengths of each scale coincide. To construct a conversion chart 5 inches long that gives the areas of circles whose radii range from 1 to 10, we first write $F1(A) = A$, $F2(R) = \pi R^2$, $R1 = 1$, and $R2 = 10$. The scale modulus for R is

$$m_R = \frac{L}{F2(R2) - F2(r1)}$$

$$= \frac{5}{\pi(10)^2 - p(1)^2} = 0.0161$$

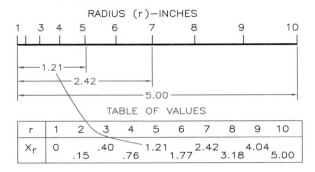

RADIUS (r)—INCHES

TABLE OF VALUES

r	1	2	3	4	5	6	7	8	9	10
X_r	0	.15	.40	.76	1.21	1.77	2.42	3.18	4.04	5.00

33.6 Calibrate one scale of a concurrent scale chart by using values from the table that have been calculated.

Thus, the scale equation for r becomes

$$X_r = m_R[F2(R) - F2(r1)]$$
$$= 0.0161[\pi R^2 - \pi(1)^2]$$
$$= 0.0161\,\pi(R^2 - 1)$$
$$= 0.0505\,(R^2 - 1).$$

Figure 33.6 shows a table of values for X_r and R. The R-scale values come from this table. From the original formula, $A = \pi R^2$, the limits of A are found to be $A1 = \pi = 3.14$ and $A2 = 100\,\pi = 314$. The scale modulus for concurrent scales is always the same for equal-length scales; therefore, $m_A = m_R = 0.0161$, and the scale equation for A becomes

$$X_A = m_A[F1(A) - F1(A1)]$$
$$= 0.0161(A - 3.14).$$

We then compute the corresponding table of values for selected values of A, as shown in **Fig. 33.7.** We superimposed the A scale on the R scale and placed its calibrations on the other side of the line to facilitate reading.

To expand or contract one of the scales, use the technique shown in **Fig. 33.8.** Draw the scales parallel at any distance apart and calibrate them in opposite directions. If they have different lengths, a different scale modulus and scale equation must be calculated for each scale.

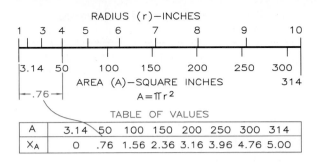

RADIUS (r)—INCHES

AREA (A)—SQUARE INCHES
$A = \pi r^2$

TABLE OF VALUES

A	3.14	50	100	150	200	250	300	314
X_A	0	.76	1.56	2.36	3.16	3.96	4.76	5.00

33.7 A completed concurrent scale chart for the formula $A = \pi R^2$. Values for the A scale are from table shown.

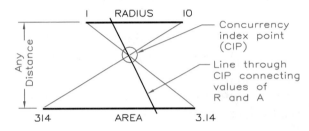

33.8 A construction used to draw a concurrent graph with unequal scales.

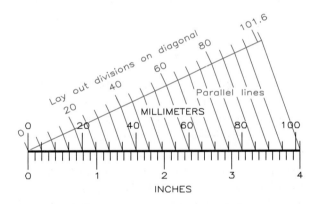

33.9 The proportional-line method can be used to construct an alignment graph that converts inches to millimeters. The units at each end of the scales must be known—101.6 mm and 4 inches in this case.

To draw concurrent scales, use the proportional-line method shown in **Fig. 33.9.** There are 101.6 mm in 4 inches. Project millimeters to the upper side of the inch scale with a series of parallel projectors.

33.3 CONCURRENT SCALES • 587

33.4 Alignment Graphs: Three Variables

For a formula containing three functions (of one variable each), draw a nomograph by selecting the lengths and positions of two scales according to the size of the graph desired. Calibrate these scales by using the scale equations presented in Section 33.3. Mathematical relationships may be used to locate the third scale, but graphical methods are simpler and less subject to error. Examples of the various forms of nomographs are discussed in the following sections.

33.5 Parallel Scale Graphs: Linear Scales

Any formula of the type $F3 = F1 + F2$ may be represented as a parallel scale alignment graph (**Fig. 33.10**). For addition, the three scales increase (functionally) in the same direction, and the function of the middle scale represents the sum of the other two. Reversing the direction of any scale changes the sign of its function in the formula, as for $F1 - F2 = F3$.

The formula $Z = X + Y$ is used to illustrate this type of alignment graph (**Fig. 33.11**).

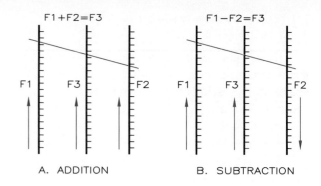

33.10 Two common forms of parallel-scale alignment nomographs showing the directions in which the scales increase for addition and subtraction.

Draw and calibrate the outer scales for X and Y, then use two sets of data that yield a Z of 8 to locate the parallel Z scale. Divide the Z scale into 16 units. Add various values of X and Y with an isopleth to find their sums along the Z scale.

Figure 33.12 illustrates how to calibrate the outer scales of a parallel-scale nomograph for the equation $U + 2V = 3W$. The scales are placed any distance apart and are divided into linear divisions from 0 to 14 for U and 0 to 8

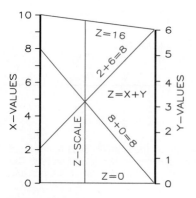
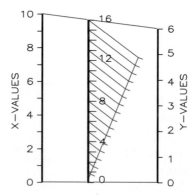
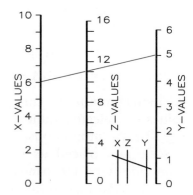

33.11 Constructing a parallel scale nomograph (linear scales).

Step 1 Draw and calibrate two parallel scales of any length. Locate the parallel Z scale by using two sets of values that give the same value (8 in this case). The ends of the Z scale are 0 and 16, the sum of the end values of X and Y.

Step 2 Draw the Z scale through the point located in step 1 parallel to the other scales. Calibrate the scale from 0 to 16 by using the proportional line method.

Step 3 Calibrate and label the Z scale. Draw a key to show how to use the nomograph. If the Y scale were calibrated with 0 at the upper end instead of at the bottom, a different Z scale could be computed and the nomograph could be used for $Z = X - Y$.

for *V*. These scales are used to complete the nomograph that is explained in **Fig. 33.13**.

Obtain the end calibrations for the middle scale by connecting the endpoints of the outer scales and substituting these values into the formula. *W* is 0 and 10 at its ends. Select two pairs of corresponding values of *U* and *V* that give the same value of *W*; for example, *U* = 0 and *V* = 7.5 give *W* = 5; also, *W* = 5 where *U* = 14 and *V* = 0.5. Because the *W* scale is linear (3*W* is a linear function), it can be subdivided into uniform intervals of equal parts. For a nonlinear scale, find the scale modulus and equation by substituting length and end values into Eq. (1).

Logarithmic Scales

Problems involving equations of the type *F*3 = (*F*1)(*F*2) can be solved in the manner that was described in Fig. 33.11 by using logarithmic scales instead of linear scales. The first step in drawing a nomograph with logarithmic scales is to transfer the logarithmic functions to the scales. **Figure 33.14** shows the graphical method where units are projected from a

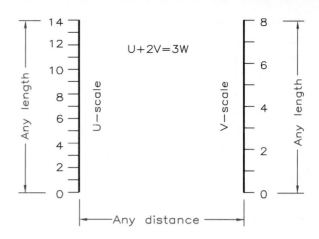

33.12 A calibration of the outer scales for the equation *U* + 2*V* = 3*W*, where *U* = 0 to 12 and *V* = 0 to 8.

printed logarithmic scale to the nomographic scale.

Figure 33.15 illustrates the conversion of the formula *Z* = *XY* into a nomograph. The desired end values of the *X* and *Y* scales are 1 and 10. Sets of values of *X* and *Y* that give the same value of *Z* (10 in this case) are used to locate the *Z* axis with end values of 1 and 100.

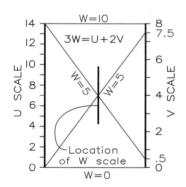

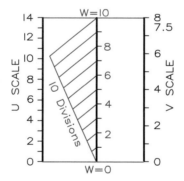

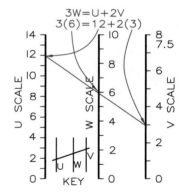

33.13 Constructing a parallel scale nomograph (linear scales).

Step 1 Substitute the end values of *U* and *V* into the formula to find the end values of the *W* scale: *W* = 0 and *W* = 10. Use any two sets of *U* and *V* that give the same *W* (*U* = 0 and *V* = 7.5, *W* = 5, and *U* = 14 and *V* = 0.5, *W* = 5) to locate the *W* scale.

Step 2 Draw the *W* scale parallel to the outer scales to the limit lines of *W* = 10 and *W* = 0. This scale is 10 linear divisions long, so divide it graphically into 10 units. The *W* scale is a linear scale; construct it as shown in Fig. 33.11.

Step 3 Connect any values of *U* and *V* with an isopleth to determine the resulting value of *W*. Draw a key that shows how to use the nomograph. The example values of *U* = 12 and *V* = 3 verify the graph's accuracy.

The Z axis is drawn and calibrated as a two-cycle log scale. A key explains how to use an isopleth to add the logarithms of X and Y to give the log of Z. The addition of logarithms performs multiplication. Had the Y axis been calibrated in the opposite direction with 1 at the upper end and 10 at the lower end, a new Z axis could have been calibrated for the formula $Z = Y/X$ for division by subtracting logarithms.

33.6 N or Z Nomographs

Whenever $F2$ and $F3$ are linear functions, we can avoid using logarithmic scales for formulas of the type

$$F1 = \frac{F2}{F3}$$

We use an **N** graph (**Fig. 33.16**) where the outer scales are functional scales and are linear if $F2$ and $F3$ are linear. If a parallel-scale graph were used for the same formula, all its scales would be logarithmic. Four main features are common to N graphs:

1. The outer scales are parallel functional scales of $F2$ and $F3$.

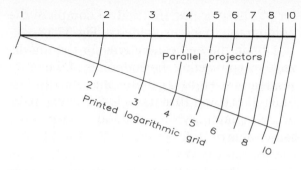

33.14 A scale calibrated graphically by projection from a printed logarithmic scale.

2. The outer scales increase functionally in opposite directions.

3. The diagonal scale connects the functional zeros of the outer scales.

4. The diagonal scale is not a functional scale for the function $F1$, and it is nonlinear.

Construction of an N nomograph is simplified because locating the middle (diagonal) scale is usually less of a problem than it is for a parallel scale graph. Calibration of the diagonal scale is most easily accomplished graphically.

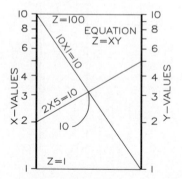 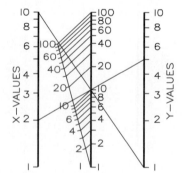 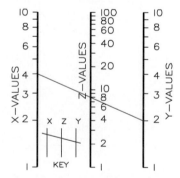

33.15 Parallel scale nomograph (logarithmic scales).

Step 1 For the equation $Z = XY$, draw parallel log scales. Construct sets of X and Y points that give the same value of Z (10 in this case) to locate the Z scale with end values of 1 and 100.

Step 2 Graphically calibrate the Z axis as a two-cycle logarithmic scale from 1 to 100 by projecting from a printed log scale. The Z scale is parallel to the X and Y scales.

Step 3 Draw a key showing how to use nomograph. By reversing the Y value scale from 1 to 10 downward and computing a different Z scale, the nomograph could be used for $Z = Y/X$.

Figure **33.17** shows how to construct a basic N graph of the equation $Z = Y/X$. Draw the diagonal to connect the zero ends of each scale, then locate whole values along the diagonal by using combinations of X and Y values. Use whole-value units along the diagonal that are easy to interpolate between. Label the diagonal and provide a key explaining how to use the nomograph. A sample isopleth verifies the correctness of the graphical relationship of the scales.

A more advanced N graph can be drawn for the equation

$$A = \frac{B + 2}{C + 5},$$

where $B=0$ and $C= 15$. This equation takes the form

$$F1 = \frac{F2}{F3},$$

where $F1(A) = A$, $F2(B) = B + 2$, and $F3(C) = C + 5$.

Thus, the outer scales will represent $B + 2$ and $C + 5$ and the diagonal scale will be for A.

Begin the construction in the same manner as for a parallel scale graph by selecting

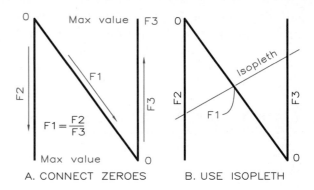

33.16 An N graph solving an equation of the form $F1 = F2/F3$.

the layout of the outer scales (**Fig. 33.18**). Determine the limits of the diagonal scale by connecting the endpoints on the outer scales, giving $A = 0.1$ for $B = 0$, $C = 15$ and $A = 2.0$ for $B = 8$, and $C = 0$. **Figure 33.19** shows these relationships and gives the remainder of the construction.

Locate the diagonal scale by finding the functional zeros of the outer scales (that is, the points where $B + 2 = 0$ or $B = -2$ and $C + 5 = 0$ or $C = -5$), then draw the diagonal scale by connecting these points. Calibrating the diagonal scale is most easily accomplished by sub-

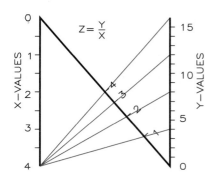 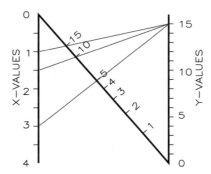 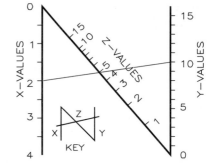

33.17 Constructing an N graph.

Step 1 Draw an N graph for the equation of $Z = Y/X$ and connect the zero ends of each scale with a diagonal scale. Draw isopleths to locate units along the diagonal scale.

Step 2 Draw additional isopleths to locate other units along the diagonal. The units on the diagonal should be whole units to make interpolation between them easy.

Step 3 Label the diagonal scale and draw a key. An isopleth confirms that $10/2 = 5$. The accuracy of the N graph is greatest at the 0 end of the diagonal; the other end approaches infinity.

stituting into the formula. Select the upper limit of an outer scale, say, $B = 8$, which will give the equation

$$A = \frac{10}{C+5}.$$

Values on the C scale can be found that correspond to the desired values on the A scale by writing the equation as:

$$C = \frac{10}{A} - 5.$$

Use this equation to make a table of A values that are easy to interpolate between (2, 1.5, 1, etc.) and their corresponding values of C up to the limit of 15, as shown in step 2 of Fig. 33.19. Connect isopleths from $B = 8$ to the tabulated values of C. Their intersections with the diagonal scale give the required calibrations for approximately half the diagonal scale. Calibrate the rest of the diagonal scale by substituting the end value of the other outer scale ($C = 15$) into the formula, giving

$$A = \frac{B+2}{20}.$$

Solving for B yields $B = 20A - 2$.

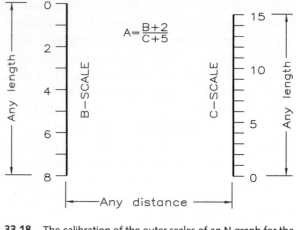

33.18 The calibration of the outer scales of an N graph for the equation $A = (B + 2)/(C + 5)$.

Construct a table for the desired values of A (Fig. 33.19) with isopleths connecting $C = 15$ with values of B, calibrating the rest of the A scale.

Problems

Solve the following problems on size A sheets. Show both the construction and calculations in the solutions involving both. If the calculations are extensive, use a separate sheet.

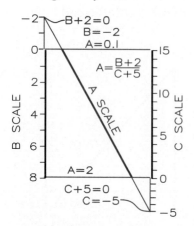

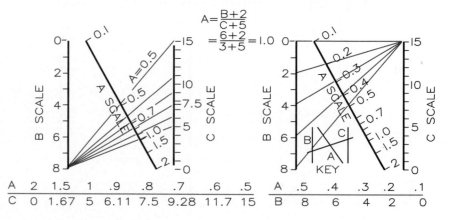

33.19 Constructing an N graph.

Step 1 Locate the diagonal scale by finding the functional zeros of the outer scales. Set $B + 2 = 0$ and $C + 5 = 0$, which gives a zero value for A.

Step 2 Select the upper limit of an outer scale ($B = 8$ in this case), substitute it into the equation, and obtain values of C for the whole values of A. Draw isopleths from $B = 8$ to the values of C to calibrate the A-scale.

Step 3 Calibrate the rest of the A scale by substituting $C = 15$ into the equation to determine values on the B scale for whole values on the A scale. Draw isopleths from $C = 15$ to calibrate the A scale.

Concurrent Scales

Draw concurrent scales for converting one type of unit to the other using the ranges given.

1. Kilometers and miles:

   1.609 km = 1 mile; from 10 to 100 miles.

2. Liters and U.S. gallons:

   1 L = 0.2692 U.S. gal; from 1 to 10 L.

3. Knots and miles per hour:

   1 knot = 1.15 mph; from 0 to 45 knots.

4. Horsepower and British thermal units:

   1 hp = 42.4 Btu; from 0 to 1200 hp.

5. Radius and area of a circle:

   $A = \pi r^2$; from $r = 0$ to 10.

6. Inches and millimeters:

   1 in. = 25.4 mm; from 0 to 5 in.

7. Numbers and their logarithms:

   Use logarithm tables; numbers from 1 to 10.

Addition and Subtraction Nomographs

Construct parallel scale nomographs to solve the following addition and subtraction problems.

8. $A = B + C$, where $B = 0$ to 10 and $C = 0$ to 5.

9. $Z = X + Y$, where $X = 0$ to 8 and $Y = 0$ to 12.

10. $Z = Y - X$, where $X = 0$ to 6 and $Y = 0$ to 24.

11. $A = C - B$, where $C = 0$ to 30 and $B = 0$ to 6.

12. $W = 2V + U$, where $U = 0$ to 12 and $V = 0$ to 9.

13. $W = 3U + V$, where $U = 0$ to 10; $V = 0$ to 10.

Multiplication and Division: Parallel Scales

Construct parallel scale nomographs with logarithmic scales for performing the following multiplication and division operations.

14. Area of a rectangle: $A = HXW$, where $H = 1$ to 10 and $W = 1$ to 12.

15. Area of a triangle: $A = 1/2B \, X \, H$, where $B = 1$ to 10, and $H = 1$ to 5.

16. Pythagorean theorem:

$$C^2 = A^2 + B^2,$$

where C = hypotenuse of a right triangle in (cm), A = one leg of the right triangle, 5 to 50 cm, and B = second leg of the triangle, 20 to 80 cm.

17. Miles per gallon (mpg) an automobile gets: Miles vary from 1 to 500; gallons from 1 to 24.

18. Cost per mile (cpm) of an automobile: Miles vary from 1 to 500; cost varies from $1 to $28.

N Nomographs

Construct N graphs that will solve the following equations.

19. **Stress** $= P/A$, where P varies from 0 to 1000 psi and A varies from 0 to 15 in^2.

20. Volume of a cylinder: $V = \pi r^2 h$, where V = volume in in^3; r = radius (5 to 10 ft); and h = height (2 to 20 in.).

21. Repeat Problem 14.

22. Repeat Problem 15.

23. Repeat Problem 16.

24. Repeat Problem 17.

25. Repeat Problem 18.

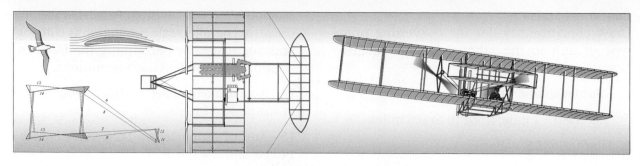

34

Empirical Equations and Calculus

34.1 Introduction

Graphical methods are useful supplements to mathematical techniques of solving problems dealing with experimental data. Graphics can be used to determine the mathematical equation of data obtained from laboratory testing and experimentation. Data from laboratory experiments and field tests are called **empirical data**. Empirical data often are expressed as one of three types of equations: (1) **linear**, (2) **power**, or (3) **exponential**.

Analysis of empirical data begins by plotting the data on three types of standard grids: (1) **rectangular (linear)**, (2) **logarithmic**, or (3) **semilogarithmic**. When the data plots as a straight line on one of these grids, its mathematical equation can be determined by using the characteristics of that particular grid (**Fig. 34.1**).

To find the equation of a straight-line plot of data, you will need to know the slope (M) of

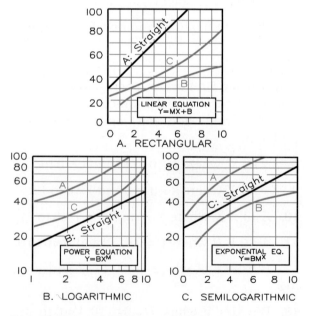

34.1 Plotting empirical data on the three types of standard grids. A straight-line plot on one of these grids means that a mathematical equation can be calculated to describe the data.

594

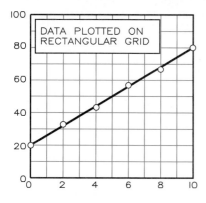

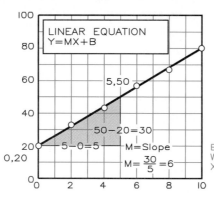

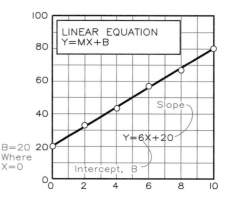

34.2 The linear equation Y = MX + B.

Step 1 When plotted on a linear grid, the data form a straight line represented by the equation Y = MX + B.

Step 2 Two points (0,20) and (4,60) are selected on the data curve to find the slope, M, which is Y/X. The slope is found to be 6.

Step 3 The intercept, B, where the curve crosses the Y axis where X = 0, is found to be 20. The equation of the data is Y = 6X + 20.

the curve, and its intercept, B, and apply it to the slope-intercept equation of Y = MX + B.

Slope is found by taking any two widely separated points on the curve and drawing a triangle to find the ratio of the height (ΔY) to the width (ΔX) which is the tangent of the curve's angle with the horizontal. The slope in this construction in **Fig. 34.2** is 6.

The intercept (B) is the point where the curve intersects the Y axis at X = 0. The intercept in Fig. 34.2 is B = 20. The values of slope and intercept can be combined with variables X and Y to find the equations of the data, as will be shown in the following examples.

34.2 Linear Equation: Y = MX + B

The curve representing empirical data in Fig. 34.2 plots as a straight line on a rectangular graph, which identifies the data as linear. Each measurement along the Y axis is directly proportional to the measurement along the X axis.

By selecting two points on the curve, the slope can be found as the first step in determining its equation. The vertical and horizontal differences between the coordinates of each point establish the adjacent sides of a

right triangle in step 2. In the slope-intercept equation, Y = MX + B, the slope, M, is the tangent of the angle between the curve and the horizontal, B is the Y intercept of the curve (where X = 0), and X and Y are variables. Here, M = 30/5 = 6, and the intercept is 20.

Substituting these values into the slope-intercept equation, we obtain Y = 6X + 20, from which we may determine values of Y by substituting any value of X into the equation. If the curve had sloped downward to the right, the slope would have been negative.

Figure 34.3 shows the relationship between the transverse strength and impact resistance of gray iron obtained from empirical data. The data plot as a straight line on a linear grid, so its equation takes the linear form.

34.3 Power Equation: Y = BX^M

Data plotted on a logarithmic grid that yields a straight line (**Fig. 34.4**) may be expressed in the power equation form in which Y is a function of X raised to a power, or Y = BX^M. We obtain the equation of the data by using the Y intercept as B and the slope of the curve as M.

Select two points on the curve to form the slope triangle and use an engineer's scale to measure its slope. If you draw the horizontal side of the right triangle as 1 or a multiple of 10, the vertical distance can be read directly. Here, the slope M (tangent of the angle) is 0.54 and the Y intercept, B, is 7; thus the equation is $Y = 7X^{0.54}$, which in logarithmic form is

$$\log Y = \log B + M \log X,$$

$$= \log 7 + 0.54 \log X.$$

We used base-10 logarithms in these examples, but natural logs may be used with base e (2.718). In **Fig. 34.5** the intercept, B, lies on the Y axis where $X = 1$ because the curve is plotted on a logarithmic grid. The Y intercept ($B = 80$) is found where $X = 1$. Recall that the intercept at this point is analogous to the linear form of the equation because the log of 1 is 0. The curve slopes downward to the right, making the slope, M, negative.

Figure 34.6 shows plots of empirical data that relate the specific weight (pounds per horsepower) of generators and hydraulic pumps to horsepower. The plots of the data are represented by straight lines on a logarith-

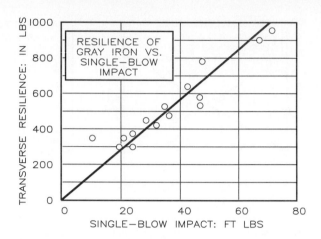

34.3 The relationship between the transverse strength of gray iron and impact resistance plots as a straight line yielding an equation of the form $Y = MX + B$ or $Y = 13.3X$.

mic grid, which means that their equations take the power form.

34.4 Exponential Equation: $Y = BM^X$

When data plotted on a semilogarithmic grid (**Fig. 34.7**) yield a straight line, the equation takes the form $Y = BM^X$, where B is the Y intercept and M is the slope of the curve. Select two points along the curve, draw a right triangle, and find its slope, M. The slope of the curve is

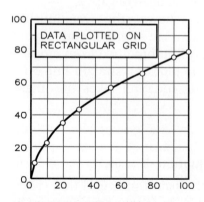

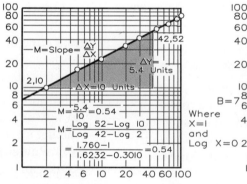

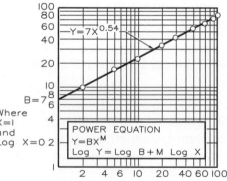

34.4 Power equation, $Y = BX^M$.

Step 1 Data plotted on a rectangular (linear) grid give a parabolic curve. Because the curve is not a straight line on this grid, its equation is not linear.

Step 2 The data plots as a straight line on a logarithmic grid. Find the slope, M, graphically with an engineer's scale by setting ΔX at 10 units and measuring ΔY as 5.4 units. $M = 5.4/10$, or 0.54.

Step 3 The intercept $B = 7$ lies at $X = 1$. Substitute the slope (0.54) and intercept (7) into the equation form of $Y = BX^M$ to obtain the equation of the data: $Y = 7X^{0.54}$.

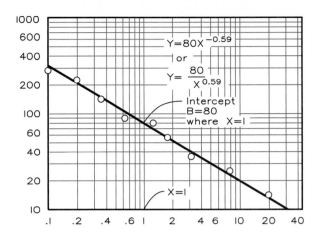

34.5 When using the slope-intercept method on a logarithmic grid, the intercept must be selected where X = 1. In this case, the intercept lies at 80 (near the middle of the graph).

$$\log M = \frac{\log 40 - \log 6}{8 - 3} = 0.1648,$$

or

$$M = (10)^{0.1648} = 1.46.$$

Substitute this value of M into the exponential equation:

$$Y = BM^X \qquad \text{or} \qquad Y = 2(1.46)^X,$$

$$Y = B(10)^{MX} \quad \text{or} \quad Y = 2(10)^{0.1648X},$$

where X is a variable that can be substituted into the equation to give infinitely many values for Y. In logarithmic form, the equation is

$$\log Y = \log B + X \log M,$$

or

$$\log Y = \log 2 + X \log 1.46.$$

The same methods give the negative slope of a curve. The curve shown in **Fig. 34.8** slopes downward to the right and has a negative slope. M, is the antilog of −.0274. The intercept of 70 and the slope yield the equation for the curve.

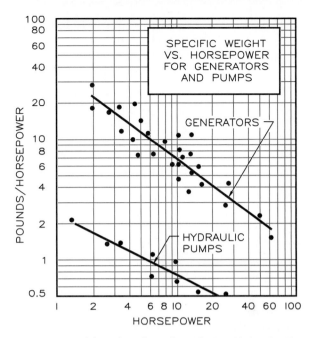

34.6 Empirical data plotted on a logarithmic grid showing the specific weight versus horsepower of electric generators and hydraulic pumps are straight lines. Their equations take the power form, Y = BX^M.

The half-life decay of radioactivity plotted in **Fig. 34.9** compares decay to time. The half-life of different isotopes varies, so different values would be assigned along the X axis for them. However, the curves for all isotopes would be straight lines with the exponential equation form.

34.5 Graphical Calculus

If the equation of a curve is known, calculus may be used to perform various types of calculations. However, experimental data often do not fit standard mathematical equations, making impossible the mathematical application of calculus. In these cases, graphical calculus can be used. The two basic forms of calculus are (1) **differential calculus** and (2) **integral calculus**.

Differential calculus is used to determine the rate of change of one variable with respect to another (**Fig. 34.10A**). The rate of change at

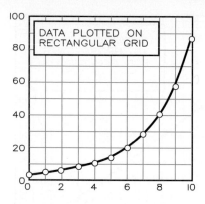
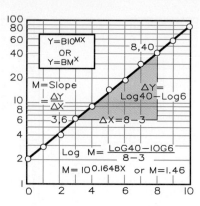
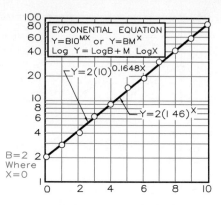

34.7 Exponential equation, $Y = BM^X$.

Step 1 When plotted on a rectangular (linear) grid, the data forms and a straight line on a semilogarithmic grid. It takes the equation form $Y = BM^X$.

Step 2 The slope must be found mathematically, because the X and Y scales are unequal. Write the slope equation in either of the forms shown above. The slope is found to be M = 1.46.

Step 3 Find the intercept, B = 2 on the Y axis, where X = 0. Substitute the values for M (1.46) and B (2) into the equation to get the equation:

$$Y = 2(10)^{0.1648X} \text{ or } Y = 2(1.46)^X.$$

any instant along the curve is the slope of a line tangent to the curve at that point. Constructing a chord at any interval allows approximation of this slope. The tangent, $\Delta Y/\Delta X$, may represent miles per hour, weight versus length, or various other rates of change important in the analysis of data.

Integral calculus is the reverse of differential calculus. Integration is used to find the area under a curve (the product of the variables plotted on the X and Y axes). The area under a curve is approximated by dividing one of the variables into a number of very small rectangular bars under the curve (**Fig. 34.10B**). Each bar is drawn so that equal areas lie above and below the curve and the average height of the bar is near its midpoint.

34.6 Graphical Differentiation

Graphical differentiation is used to determine the rate of change of two variables with respect to each other at any given point. **Figure 34.11** illustrates the preliminary construction of a derivative scale and the pole point for plotting a derivative curve.

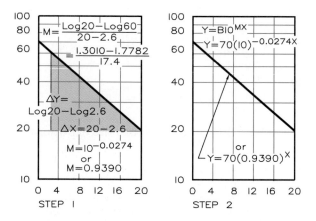

34.8 Negative slope (M).

A When a curve slopes downward to the right, its slope is negative.

B Substitution gives these two forms of the equation (shown above) for the curve.

The graphical differentiation process is illustrated in **Fig. 34.12**. The maximum slope of the data curve is estimated to be slightly less than 12; an ordinate scale long enough to accommodate the maximum slope is selected. To locate the pole point, a line is drawn from point 12 on the ordinate axis of the derivative grid parallel to the known slope on the given curve grid and the line is extended to the X axis.

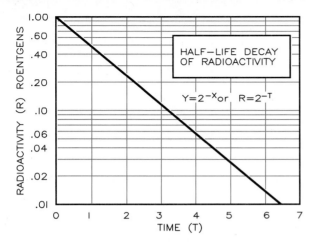

34.9 The decay of radioactivity is represented by a straight line on a semilog grid, indicating that its equation takes the exponential form, $Y = BM^X$ or $R = 2^{-T}$.

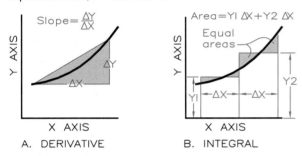

A. DERIVATIVE   B. INTEGRAL

34.10 Derivatives and integrals.

A The derivative of a curve is the rate of change at any point on the curve, or its slope, $\Delta Y/\Delta X$.

B The integral of a curve is the cumulative area enclosed by the curve, or the summation of the incremental areas comprising the whole.

A series of chords is constructed on the given curve. The interval between 0 and 1, where the curve is steepest, is divided in half to obtain a more accurate plot. After other bars are found, a smooth curve is drawn through the top of them so that the area above and under the top of each bar is equal. The rate of change, $\Delta Y/\Delta X$, can be found for any value of X in the derivative graph.

Applications

The mechanical handling shuttle shown in **Fig. 34.13** converts rotational motion into lin-

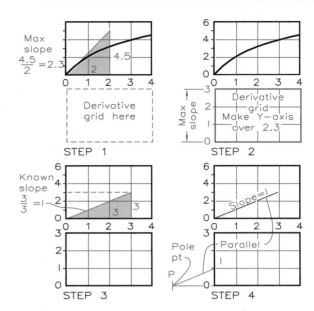

34.11 Scales for graphical differentiation.

Step 1 Estimate the maximum slope of the curve (here, 2.3) by drawing a line tangent to the curve where it is steepest.

Step 2 Draw the derivative grid with a maximum ordinate of 3.0 to accommodate the slope of 2.3.

Step 3 Find a known slope of 1 on the given grid. The slope has no relationship to the data curve.

Step 4 Draw a line from 1 on the Y axis of the derivative grid parallel to the slope of the triangle drawn in the given grid. This locates the pole point on the extension of the X axis.

ear motion. The drawing of the linkage shows the end positions of point P, which is the zero point for plotting travel versus degrees of revolution. Rotation is constant at one revolution every three seconds, so the degrees of revolution may be converted to time (**Fig. 34.14**). The drive crank, R1, is revolved at 30° intervals, and the distance that point P travels from its end position is plotted on the graph to give the distance–time relationship.

The ordinate scale of the derivative grid is scaled with an end value of 100 in./sec, or slightly larger than the estimated maximum slope of the curve. A slope of 40 is drawn on the given data grid. Pole point P is found by

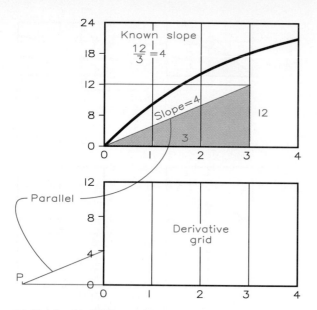

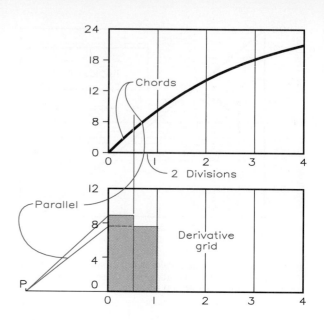

34.12 Graphical differentiation.

Required Find the derivative curve of the given data.

Step 1 Find the derivative grid and the pole point by using the steps introduced in Fig. 34.11.

Step 2 Construct chords between intervals on the given curve and draw lines parallel to them through point P on the derivative grid. These lines locate the heights of bars in their respective intervals. Divide the first interval into two bars where the curve is sharpest.

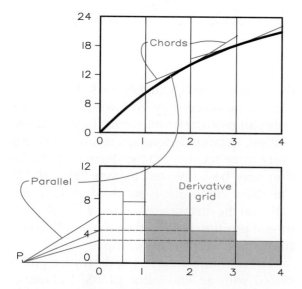

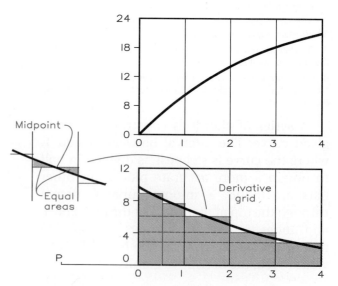

Step 3 Draw additional chords on the curve in the last three intervals. Draw lines parallel to the chords through point P to the Y axis. This construction locates the remainder of the bars needed to draw the derivative curve.

Step 4 The vertical bars represent the slopes of the curve at different intervals. Draw the derivative curve through the midpoints of the bars so that the areas below and above the bars are equal.

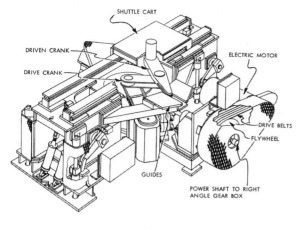

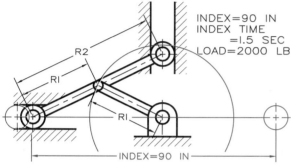

INDEX=90 IN
INDEX TIME
=1.5 SEC
LOAD=2000 LB

34.13 These are drawings of a mechanical handling shuttle used to move automobile parts on an assembly line. (Courtesy of General Motors Corporation.)

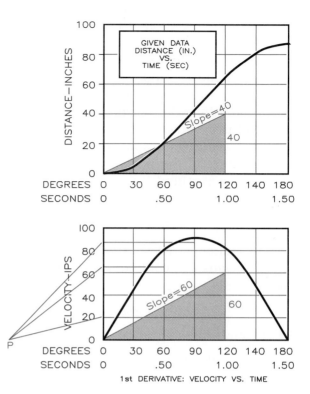

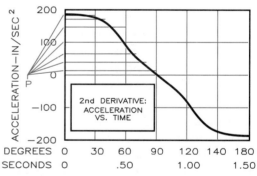

34.14 Velocity and acceleration of the mechanical handling shuttle may be obtained with graphical differential calculus.

drawing a line from 40 on the derivative ordinate scale parallel to the slope of 40 in the given data graph to the extension of the X axis of the derivative graph.

Chords drawn on the curve approximate the slope at various points. Draw lines from point P of the derivative scale parallel to the chordal lines to the ordinate axis. These intersections on the Y axis are projected horizontally to their respective intervals to form vertical bars. A curve is drawn through the tops of the bars to give an average of the bars. This curve shows the velocity of the shuttle in inches per second at any time interval.

The second derivative curve, acceleration versus time, is drawn in the same manner as the first derivative curve. Inspection of the first derivative curve shows that the maximum slope is about 200 in./sec/sec. An easily measured scale for the ordinate is chosen. A pole, P, is found in the same manner as for the previous pole.

Chords are drawn at intervals on the first derivative curve. Lines drawn from P parallel to these chords intersect the Y axis of the second derivative graph, where they are projected horizontally to their respective intervals, establishing the heights of the bars. A smooth curve is drawn through the tops of the bars to gives a close approximation of the average areas of the bars. The minus scale indicates deceleration.

These velocity and acceleration graphs show that parts being handled by the shuttle accelerate at a rapid rate until the maximum velocity is attained at 90°, at which time deceleration begins and continues until the parts come to rest.

34.7 Graphical Integration

Graphical integration is used to determine the area (product of two variables) under a curve. For example, if the Y axis represented pounds and the X axis represented feet, the integral curve would give the product of the variables, foot-pounds, at any interval along the X axis. The steps in determining the pole point and the scales for integration are illustrated in **Fig. 34.15**.

In **Fig. 34.16**, the total area under the curve is estimated to be less than 80 units. Therefore, the maximum height of the Y axis on the integral curve is 80, drawn at a convenient scale. Pole P is found by the steps shown in Fig. 34.15.

A series of vertical bars is constructed to approximate the areas under the curve. The narrower the bars, the more accurate will be the resulting plotted curve. The interval between 2 and 3 is divided into two bars to obtain more accuracy where the curve is sharpest. The tops of the bars are extended horizontally to the Y axis and are connected to point P.

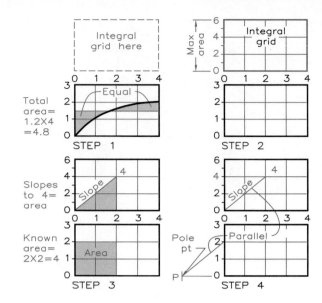

34.15 Scales for graphical integration.

Step 1 To determine the maximum value on the Y axis, draw a line to approximate the area under the given curve (4.8 square units here).

Step 2 Draw the integral with a Y axis of 6 to accommodate the maximum area.

Step 3 Find a known area of 4 on the given grid. Draw a slope from 0 to 4 on the integral grid directly above the known area to establish the integral.

Step 4 Draw a line from 2 on the Y axis of the given grid parallel to the slope line in the integral grid. Locate the pole point on the extension of the X axis.

Lines are drawn parallel to AP, BP, CP, DP, and EP in the integral grid to correspond to the respective intervals in the given grid. The intersection points of the chords are connected by a smooth curve—the integral curve—to give the cumulative product of the X and Y variables along the X axis.

Problems

Use size A sheets for solutions to the following problems. Solutions involving mathematical calculations should be shown on separate sheets if space is not available on the sheet with the graphical solutions.

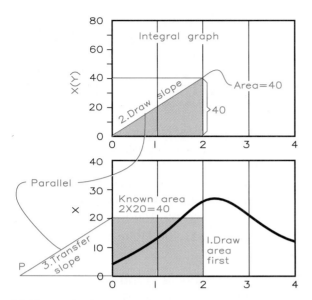

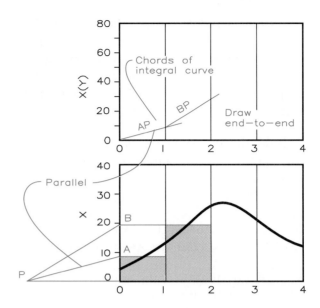

34.16 Graphical integration.

Required Plot the integral curve of the given data.

Step 1 Find pole point P by using the technique described in Fig. 34.15.

Step 2 Construct bars to approximate the areas under the curve. Project the heights of the bars to the Y axis and draw lines to pole P. Draw sloping lines AP and BP at their respective intervals parallel to the lines drawn to P.

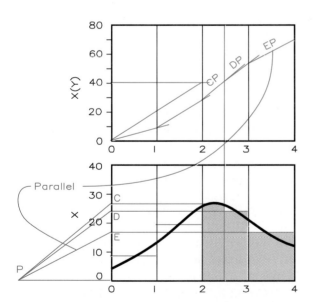

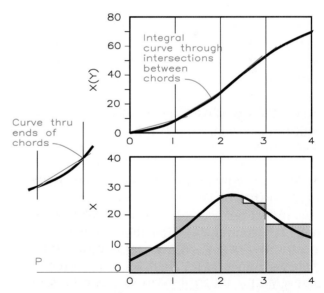

Step 3 Draw additional bars from 2 to 4 on the X axis. Project the heights of the bars to the Y axis and draw rays to pole P. Draw lines CP, DP, and EP at their respective intervals and parallel to their rays in the integral grid.

Step 4 The lines connected in the integral grid are chords of the integral curve. Draw the curve to pass through the points where the chords intersect. An ordinate value on the integral curve represents the cumulative area under the given curve from zero to that point on the X axis.

Empirical Equations: Logarithmic

1. Find the equation for empirical data that compares input voltage, V (Y axis), with input current in amperes, I (X axis), to a heat pump: V = 0.8, I = 20; V = 1.3, I = 30; Y = 1.75, I = 40; V = 1.85, I = 45.

2. Find the equation of the empirical data that gives the relationship between peak allowable current in amperes, I (Y axis), with the overload operating time in cycles at 60 cycles per second, C (X axis): I = 2000, C = 1; I = 1840, C = 2; I = 1640, C = 5; I = 1480, C = 10; I = 1300, C = 20; I = 1200, C = 50; I = 1000, C = 100.

3. Find the equation of the empirical data of a low-voltage circuit breaker used on a welding machine that give the maximum loading during welding in amperes, rms (Y axis), for the percentage of duty, pdc (X axis): rms = 7500, pdc = 3; rms = 5200, pdc = 6; rms = 4400, pdc = 9; rms = 3400, pdc = 15; rms = 2300, pdc = 30; rms = 1700, pdc = 60.

4. Construct a three-cycle by three-cycle logarithmic graph to find the equation of a machine's vibration displacement in mills (Y axis) and vibration frequency in cycles per minute, cpm (X axis). Data: 100 cpm, 0.80 mills; 400 cpm, 0.22 mills; 1000 cpm, 0.09 mills; 10,000 cpm, 0.009 mills; 50,000 cpm, 0.0017 mills.

5. Find the equation for data that compares the velocities of air moving over a plane surface in feet per second, V (X axis) at different heights in inches, Y (Y axis) above the surface: Y = 0.1, V = 18.8; Y = 0.2, V = 21.0; Y = 0.3, V = 22.6; Y = 0.4, V = 24.1; Y = 0.6, V = 26.0, Y = 0.8, V = 27.3; Y = 1.2, V = 29.2; Y = 1.6, V = 30.6; Y = 2.4, V = 32.4; Y = 3.2, V = 33.7.

6. Find the equation of the data below that shows the distance traveled in feet, S (Y axis), at various times in seconds, T (X axis), of a test vehicle: T = 1, S = 15.8; T = 2, S = 63.3; T = 3, S = 146; T = 4, S = 264; T = 5, S = 420; T = 6, S = 580.

Empirical Equations: Linear

7. Construct a linear graph to determine the equation for the annual cost of a compressor (Y axis) in relation to the compressor's size in horsepower (X axis). Data: 0 hp, $0; 50 hp, $2100; 100 hp, $4500; 150 hp, $6700; 200 hp, $9000; 250 hp, $11,400. Write the equation for these data.

8. Construct a linear graph on which the X axis is the mat depth from 0 to 4 in. and the Y axis is tons per hour per foot of width from 0 to 70 for a conveyor traveling at a rate of 50 feet per minute. The conveyor is a moving shaker that screens particles of coal by size. X = 0, Y = 0; X = 1, Y = 13; X = 2, Y = 25; X = 3, Y = 38; and X = 4, Y = 50.

9. Write the equation for the empirical data plotted in Fig. 34.7.

10. Plot the empirical data on a linear graph and determine the equation that shows the deflection in centimeters of a spring, D (Y axis), when it is loaded with different weights in kilograms, W (X axis): W = 0, D = 0.45; W = 1, D = 1.10; W = 2, D = 1.45; W = 3, D = 2.03; W = 4, D = 2.38; W = 2.38, D = 3.09.

11. Plot the empirical data on a linear graph and determine their equation. The data represent temperatures on a Fahrenheit thermometer, °F (Y axis), and a Celsius thermometer, °C (X axis):C = -6.8, F = 20; C = 6 F = 43; C = 16, F = 60.8; C = 32.2, F = 90; C = 52, F = 125.8; C = 76, F = 169.

Empirical Equations: Semilogarithmic

12. Draw a semilog graph of the data to find their equations. The Y axis is a two-cycle log scale for V (voltage) and the X axis a 10-unit linear scale for T (time) in sixteenths of a second. These data represent resistor voltage during capacitor charging. Data: 0 sec, 10 V; 2 sec, 6 V; 4 sec, 3.6 V; 6 sec, 2.2 V; 8 sec, 1.4 V; 10 sec, 0.8 V.

13. Write the equation for the data plotted in Fig. 34.9.

14. Construct a semilog graph of the data to find their equations. Make the Y axis a three-cycle log scale for the reduction factor, R, and the X axis a linear scale from 0 to 250 for mass thickness per square foot, MT, of a nuclear protection barrier. Data: 0MT, 1.0R; 100MT, 0.9R; 150MT, 0.028R; 200MT, 0.009R; 300MT, 0.0011R.

15. An engineering firm considering expansion is reviewing its past income. Their years of operation are represented by X (X axis), and their annual income (in tens of thousands of dollars) by N (Y-axis): X = 1, N = 0.05; X = 2, Y = 0.08; X = 3, N = 012; X = 4, N = 0.2; X = 5, Y = 0.32; X = 6, N = 0.51; X = 7, N = 0.8; X = 8, N = 1.3; X = 9, N = 2.05; X = 10, N3..25.

Calculus: Graphical Differentiation

16. Plot the equation $Y = X^3/6$ as a linear graph. Graphically determine the first and second derivatives. (See Fig. 34.14.)

17. Plot the equation $Y = 2X^2$ and find its derivative curve on a graph placed below the first.

18. Plot the equation $4Y = 8 - X^2$ and find its derivative curve on a graph placed below the first.

19. Plot the equation $3Y = X^2 + 16$ and find its derivative curve on a graph placed below the first.

20. Plot the equation $X = 3Y^2 - 5$ and find the derivative curve on a graph placed below the first.

Calculus: Graphical Integration

21. Plot the following equation and find its integral curve on a graph placed above the first: $Y = X^2$.

22. Plot the equation and find its integral curve on a graph placed above the first: $Y = 9 - X^2$.

23. Plot the equation on a graph and find the integral curve of the data on a graph placed above the first: $Y = X$.

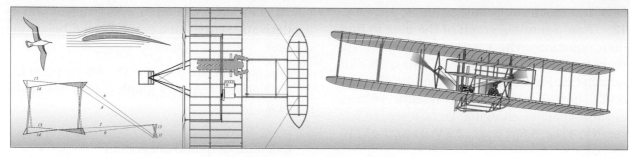

35

Pipe Drafting

35.1 Introduction

An understanding of pipe drafting begins with a familiarity with the most commonly used types of pipe: (1) steel pipe, (2) cast-iron pipe, (3) copper, brass, and bronze pipe and tubing, and (4) plastic pipe. The standards for the grades and weights for pipe and pipe fittings are specified by several organizations to ensure the uniformity of size and strength of inter-changeable components. Among these organizations are the American National Standards Institute (ANSI), the American Society for Test Materials (ASTM), the American Petroleum Institute (API), and the Manufacturers Standardization Society (MSS).

35.2 Welded and Seamless Steel Pipe

Traditionally, steel pipe has been specified in three weights: standard (STD), extra strong (XS), and double extra strong (XXS). These designations and their specifications are listed in the ANSI B 36.10 standards. However, additional designations for pipe, called schedules, have been introduced to provide the pipe designer with a wider selection of pipe to cover more applications.

There are ten such schedules: Schedule 10, Schedule 20, Schedule 30, Schedule 40, Schedule 60, Schedule 80, Schedule 100, Schedule 120, Schedule 140, and Schedule 160. The wall thicknesses of the pipes vary from the thinnest, in Schedule 10, to the thickest, in Schedule 160. The outside diameters are of a constant size for pipes of the same nominal size in all schedules. Schedule designations correspond to STD, XS, and XXS specifications in some cases, as shown in **Table 35.1**. This table has been abbreviated from the ANSI B 36.10 tables by omitting a number of the pipe sizes and schedules. The most often used schedules are 40, 80, and 120.

Table 35.1   Dimensions and Weights of Welded and Seamless Steel Pipe (ANSI B 36.10)

Inch Units				Identification		SI Units		
Inch Nominal Size	Outside Diameter (in.)	Wall Thickness (in.)	Weight lbs/ft	STD XS XXS	Schedule Number	Outside Diameter (mm)	Wall Thickness (mm)	Weight kg/m
1/2	0.84	0.11	0.85	STD	40	21.3	2.8	1.3
1	1.32	0.13	1.68	STD	40	33.4	3.4	2.5
1	1.3	0.18	2.17	XS	80	33.4	4.6	3.2
1	1.3	0.36	3.66	XXS		33.4	9.1	5.5
2	2.38	0.22	3.65	STD	40	60.3	3.9	5.4
2	2.38	0.22	5.02	XS	80	60.3	5.5	7.5
2	2.38	0.44	9.03	XXS		60.3	11.1	13.4
4	4.50	0.23	10.79	STD	40	114.3	6.0	16.1
4	4.50	0.34	14.98	XS	80	114.3	8.6	42.6
4	4.50	0.67	27.54	XXS		114.3	17.1	41.0
8	8.63	0.32	28.55	STD	40	219.1	8.2	42.6
8	8.63	0.50	43.39	XS	80	219.1	12.7	64.6
8	8.63	0.88	74.40	XXS		219.1	22.2	107.9
12	12.75	0.38	49.56	STD		323.0	9.5	67.9
12	12.75	0.50	65.42	XS		323.0	12.7	97.5
12	12.75	1.00	125.40	XXS	120	133.9	25.4	187.0
14	14.00	0.38	54.57	STD	30	355.6	9.5	87.3
14	14.00	0.50	72.08	XS		355.6	12.7	107.4
18	18.00	0.38	70.59	STD		457	9.5	106.2
18	18.00	0.50	93.45	XS		457	12.7	139.2
24	24.00	0.38	94.62	STD	20	610	9.5	141.1
24	24.00	0.50	125.49	XS		610	12.7	187.1
30	30.00	0.38	118.65	STD		762	9.5	176.8
30	30.00	0.50	157.53	XS	20	762	12.7	234.7
40	40.00	0.38	158.70	STD		1016	9.5	236.5
40	40.00	0.50	210.90	XS		1016	12.7	314.2

Pipes from the smallest size up to and including 12-in. pipes are specified by their inside diameter (ID), which means that the outside diameter (OD) is larger than the specified size. The inside diameters are the same size as the nominal sizes of the pipe for STD weight pipe. For XS and XXS pipe, the inside diameters are slightly different in size from the nominal size. Beginning with the 14-in. diameter pipes, the nominal sizes represent the outside diameters of the pipe.

The standard lengths for steel pipe are 20 ft and 40 ft. Seamless steel (SMLS STL) pipe is a smooth pipe with no weld seams along its length. Welded pipe is formed into a cylinder and is butt-welded (BW) at the seam, or it is joined with an electric resistance weld (ERW).

35.3 Cast-Iron Pipe

Cast-iron pipe is used for the transportation of liquids, water, gas, and sewerage. When used as a sewage pipe, cast-iron pipe is referred to as "soil pipe." Cast-iron pipe is available in diameter sizes from 3 in. to 60 in.

The standard lengths of cast-iron pipe are 5 ft and 10 ft. Cast iron is more brittle and more subject to cracking when loaded than is steel pipe. Therefore, cast-iron pipe should not be used where high pressures or weights will be applied to it.

35.4 Copper, Brass, and Bronze Piping

Copper, brass, and bronze are used to manufacture piping and tubing for use in applications where there must be a high resistance to corrosive elements, such as acidic soils and chemicals transmitted through the pipes. Copper pipe is used when the pipes are placed within or under concrete slab foundations of buildings to ensure that they will resist corrosion. The standard length of pipes made of these nonferrous materials is 12 ft.

Tubing is a smaller pipe that can be easily bent when it is made of copper, brass, or bronze. The term **piping** refers to rigid pipes that are larger than tubes, usually in excess of 2-in. diameters.

35.5 Miscellaneous Piping Materials

Other materials that are used to manufacture pipes are aluminum, asbestos-cement, concrete, polyvinyl chloride (PVC), and various other plastics. Each of these materials has its special characteristics that make it desirable or economical for certain applications. The method of designing and detailing piping systems by the pipe drafter is essentially the same, regardless of the piping material used.

35.6 Pipe Joints

The basic connection in a pipe system is the joint where two straight sections of pipe fit together. Three types of joints—screwed, welded, and flanged—are illustrated in **Fig. 35.1**.

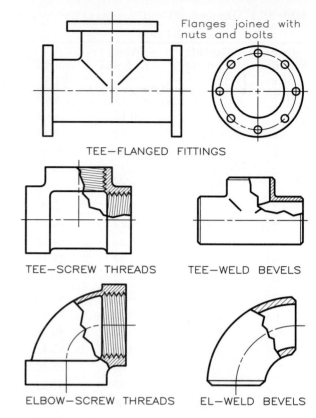

35.1 The three basic types of joints are flanged, screwed, and welded.

Screwed joints are joined by pipe threads of the type covered in Chapter 10 and Appendix 9. Pipe threads are tapered at a ratio of 1 to 16 along the outside diameter (**Fig. 35.2**). As the pipes are screwed together, the threads bind to form a snug, locking fit. A cementing compound is applied to the threads before joining to improve the seal.

Flanged joints, shown in **Fig. 35.3**, are welded to the straight sections of pipe, which are then bolted together around the perimeter of the flanges. Flanged joints form strong rigid joints that can withstand high pressure and permit disassembly of the joints when needed. Several types of flange faces are shown in **Fig. 35.4** and in Appendix 43.

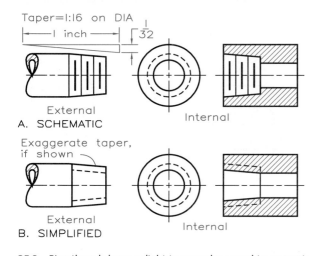

A. SCHEMATIC

Taper=1:16 on DIA

External   Internal

Exaggerate taper, if shown

External   Internal

B. SIMPLIFIED

35.2 Pipe threads have a slight taper and are used to connect screwed pipe fittings.

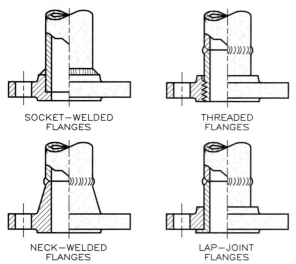

SOCKET—WELDED FLANGES

THREADED FLANGES

NECK—WELDED FLANGES

LAP—JOINT FLANGES

35.3 Types of flanged joints and the methods of attaching the flanges to the pipes.

Welded joints are joined by welded seams around the perimeter of the pipe to form butt welds. Welded joints are used extensively in "big inch" pipelines that are used for transporting petroleum products cross-country.

Bell and spigot (B&S) joints are used to join cast-iron pipes, (**Fig. 35.5**). The spigot is placed inside the bell and the two are sealed with molten lead or a sealing ring that snaps into position to form a sealed joint.

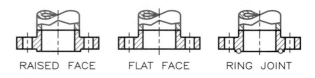

RAISED FACE   FLAT FACE   RING JOINT

35.4 Three types of flange faces: raised face (RF), flat face (FF), and ring joint (RJ).

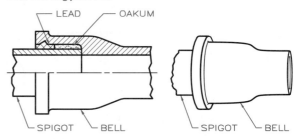

LEAD   OAKUM

SPIGOT   BELL   SPIGOT   BELL

35.5 A bell and spigot joint (B&S) is used to connect cast-iron pipes.

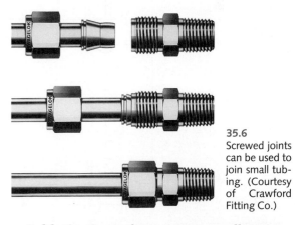

35.6 Screwed joints can be used to join small tubing. (Courtesy of Crawford Fitting Co.)

Soldering is used to connect smaller pipes and tubes, especially nonferrous tubing. Screwed fittings are available to connect tubing, as shown in **Fig. 35.6**.

35.7 Pipe Fittings

Pipe fittings are placed within a pipe system to join pipes at various angles, to transform the pipe diameter to a different size, or to control the flow and its direction within the system. Fittings are placed in the system using any of the previously covered joints. A pipe system can be drawn with single-line symbols or double-line symbols.

	FLANGED	SCREWED	BELL & SPIGOT	WELDED	SOLDERED
1. JOINT					
2. 90° ELBOW					
TOP VIEW					
FRONT VIEW					
BOTTOM VIEW					
3. ELBOW—LONG RADIUS					
4. ELBOW—REDUCING					
5. TEE					
TOP VIEW					
FRONT VIEW					
BOTTOM VIEW					
6. 45° ELBOW					
TOP VIEW					
FRONT VIEW					
BOTTOM VIEW					
7. 45° LATERAL					
TOP VIEW					
FRONT VIEW					
8. REDUCER					
9. GATE VALVE					
10. GLOBE VALVE					
11. CHECK VALVE					
12. UNION					

35.7 These single-line pipe fittings symbols are extracted from the ANSI Z32.2 standards.

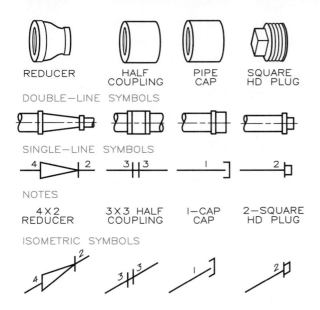

REDUCER   HALF COUPLING   PIPE CAP   SQUARE HD PLUG

DOUBLE—LINE SYMBOLS

SINGLE—LINE SYMBOLS

NOTES

4X2 REDUCER   3X3 HALF COUPLING   1—CAP CAP   2—SQUARE HD PLUG

ISOMETRIC SYMBOLS

35.8 Examples of standard fittings for screwed connections along with the single-line and double-line symbols that are used to represent them. Nominal pipe sizes can be indicated by numbers placed near the joints. The major flow direction is labeled first, with the branches labeled second. The large openings are labeled to precede the smaller openings.

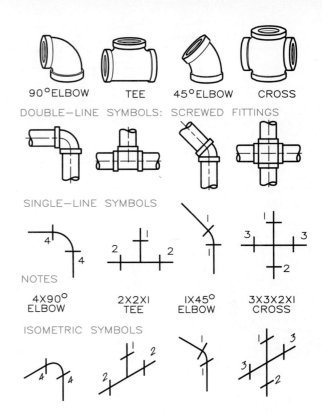

90°ELBOW   TEE   45°ELBOW   CROSS

DOUBLE—LINE SYMBOLS: SCREWED FITTINGS

SINGLE—LINE SYMBOLS

NOTES

4X90° ELBOW   2X2X1 TEE   1X45° ELBOW   3X3X2X1 CROSS

ISOMETRIC SYMBOLS

35.10 Further examples of standard fittings for screwed connections.

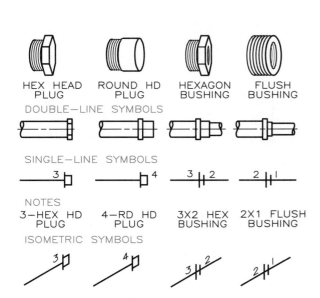

HEX HEAD PLUG   ROUND HD PLUG   HEXAGON BUSHING   FLUSH BUSHING

DOUBLE—LINE SYMBOLS

SINGLE—LINE SYMBOLS

NOTES

3—HEX HD PLUG   4—RD HD PLUG   3X2 HEX BUSHING   2X1 FLUSH BUSHING

ISOMETRIC SYMBOLS

35.9 A single-line piping system with the major valves represented as double-line symbols.

The fittings in **Fig. 35.7** are represented as single-line symbols with flanged, screwed, bell and spigot, welded, and soldered joints. Most symbols have been shown as they would be drawn to appear in various orthographic views—top, front, and side views. These symbols have been extracted from ANSI Z32.2.3 standards.

35.8 Screwed Fittings

A number of standard fittings are shown in **Fig. 35.8** through **Fig. 35.11**. The two types of graphical symbols that are used to represent fittings and pipe are **double-line symbols** and **single-line symbols**.

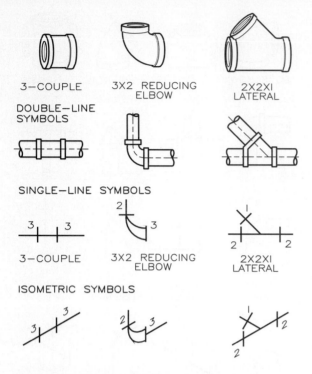

3—COUPLE   3X2 REDUCING   2X2XI
             ELBOW   LATERAL

DOUBLE—LINE SYMBOLS

SINGLE—LINE SYMBOLS

3—COUPLE   3X2 REDUCING   2X2XI
             ELBOW   LATERAL

ISOMETRIC SYMBOLS

35.11 Examples of standard fittings for screwed connections.

Double-line symbols are more descriptive of the fittings and pipes since they are drawn to scale with double lines. Single-line symbols are more symbolic since the pipe and fittings are drawn with single lines and schematic symbols.

Fittings are available in three weights: standard (STD), extra strong (XS), and double extra strong (XXS) to match the standard weights of the pipes with which they will be connected. Other weights of fittings are available, but these three weights are stocked by practically all suppliers.

A piping system of screwed fittings is shown in **Fig. 35.12**, with double-line symbols in a single-line system to call attention to them. These could just as well have been drawn using the single-line symbols.

35.9 Flanged Fittings

Flanges are used to connect fittings into a piping system when heavy loads are supported in

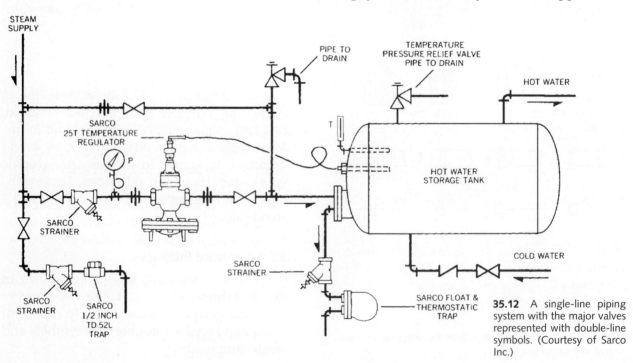

35.12 A single-line piping system with the major valves represented with double-line symbols. (Courtesy of Sarco Inc.)

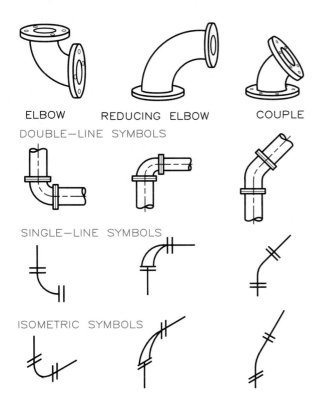

ELBOW   REDUCING ELBOW   COUPLE

DOUBLE—LINE SYMBOLS

SINGLE—LINE SYMBOLS

ISOMETRIC SYMBOLS

35.13   Examples of standard flanged fittings along with the singleline and single-line and double-line symbols that are used to represent them.

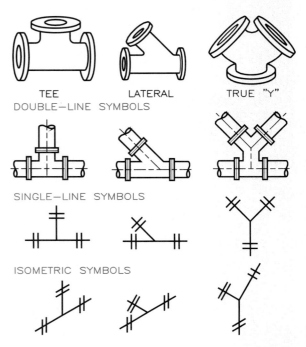

TEE   LATERAL   TRUE "Y"

DOUBLE—LINE SYMBOLS

SINGLE—LINE SYMBOLS

ISOMETRIC SYMBOLS

35.14   Examples of standard flanged fittings along with the single-line and double-line symbols that are used to represent them.

large pipes and where pressures are great. Since flanges are expensive, their usage should be kept to a minimum if other joining methods can be used. Flanges are welded to straight pipe sections so that they can be bolted together.

Examples of several fittings are drawn as double-line and single-line symbols in **Fig. 35.13** and **Fig. 35.14**. The elbow is commonly referred to as an "ell," and it is available in angles of turn of 90° and 45° in both short and long radii. The radius of a short-radius ell is equal to the diameter of the larger end. The long-radius (LR) ells have radii that are approximately 1.5 times the nominal diameter of the large end of the ell. A table of dimensions for 125-lb and 250-lb cast-iron fittings is given in Appendixes 42 and 43.

35.10 Welded Fittings

Welding is used to join pipes and fittings for permanent, pressure-resistant joints. Examples of double-line and single-line fittings connected by welding are shown in **Figs. 35.15** and **35.16**. Fittings are available with beveled edges that are ready for welding. The piping layout in **Fig. 35.17** illustrates a series of welded joints with a double-line drawing. The location of the welded joints has been dimensioned. Several flanged fittings have been welded into the system in order for the flanges to be used.

35.11 Valves

Valves are used to regulate the flow of gas and liquid transported within pipelines or to turn off the flow completely. The three basic types

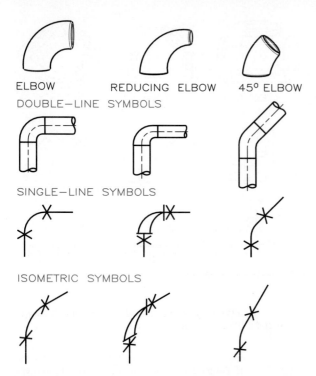

ELBOW   REDUCING ELBOW   45° ELBOW

DOUBLE—LINE SYMBOLS

SINGLE—LINE SYMBOLS

ISOMETRIC SYMBOLS

35.15   Examples of welded fittings along with the single-line and double-line symbols that are used to represent them.

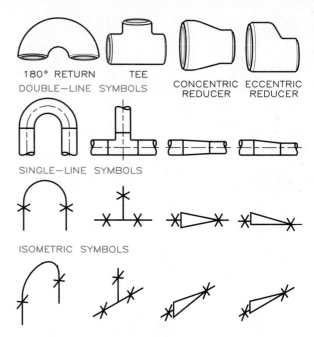

180° RETURN   TEE

DOUBLE—LINE SYMBOLS   CONCENTRIC REDUCER   ECCENTRIC REDUCER

SINGLE—LINE SYMBOLS

ISOMETRIC SYMBOLS

35.16   Additional examples of standard welded fittings along with the single-line and double-line symbols that are used to represent them.

of valves—**gate, globe, and check**—are shown in **Fig. 35.18** drawn with single-line symbols. Other types include **angle, safety, diaphragm, float,** and **relief** valves.

Gate valves: Used to turn the flow within a pipe on or off with the least restriction of flow through the valve. Gate valves are not meant to be used to regulate the degree of flow.

Globe valves: Used to not only turn the flow on and off but also to regulate the flow to a desired level. **Angle valves:** Types of globe valves that turn at 90° angles at bends in the piping system. They have the same controlling features as the straight globe valves.

Check valves: Restrict the flow in the pipe to only one direction. A backward flow is prevented by either a movable piston or a swinging washer activated by a reverse in the flow. Orthographic symbols for types of valves are

shown in **Fig. 35.19**. These symbols can be converted to isometric views, as shown in **Fig. 35.21**.

35.12 Fittings in Orthographic Views

Fittings and valves must be shown from any view in orthographic projection. Two and three views of typical fittings were shown in Fig. 35.7 as single-line screwed fittings, but the same general principles are used to represent similar joints as double-line drawings. Observe the various views of the fittings and notice how the direction of an elbow can be shown by a slight variation in the different views. The same techniques are used to represent tees and laterals.

A piping system is shown in a single orthographic view in **Fig. 35.20**, where a combination of double-line and single-line symbols are drawn. Note that arrows are used to give

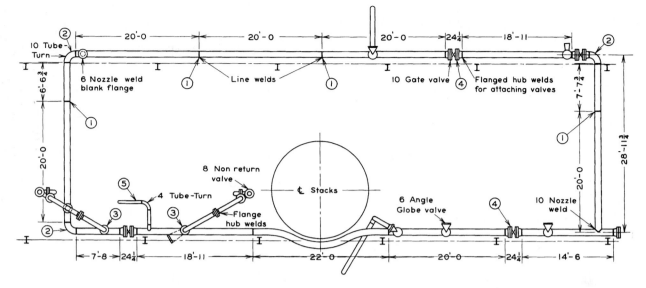

35.17 This piping layout uses a series of welded and flanged joints and is drawn using double-line symbols.

the direction of flow in the system. Joints are screwed, welded, and flanged.

Horizontal elevation lines are given to dimension the heights of each horizontal pipe. Station 5 + 12 − 0–1/4″ represents a distance of 500 feet plus 12′ − 0–1/4″, or 512′ − 0–1/4″ from the beginning station point of 0 + 00.

The dimensions in Fig. 35.20 are measured from the centerlines of the pipes; this is indicated by the C_L symbols. In some cases, the elevations of the pipes are dimensioned to the bottom of the pipe, abbreviated as **BOP**.

35.13 Piping Systems in Pictorial

Isometric drawings of piping systems are very helpful in the representation of three-dimensional installations that would be difficult to interpret if drawn in orthographic projection. Isometric and axonometric drawings of piping systems, called "**spool drawings**," can be drawn using either single-line or double-line symbols.

A three-dimensional piping system is drawn orthographically in **Fig. 35.21** with top and front views. This pipe system is also

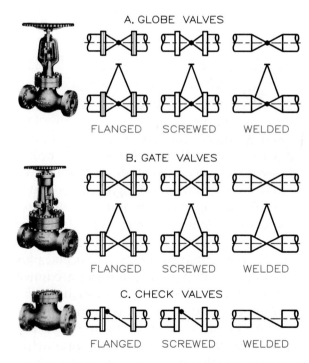

35.18 The three basic types of valves: (A) globe, (B) gate, and (C) check. (Photographs courtesy of Walworth/Aloyco.)

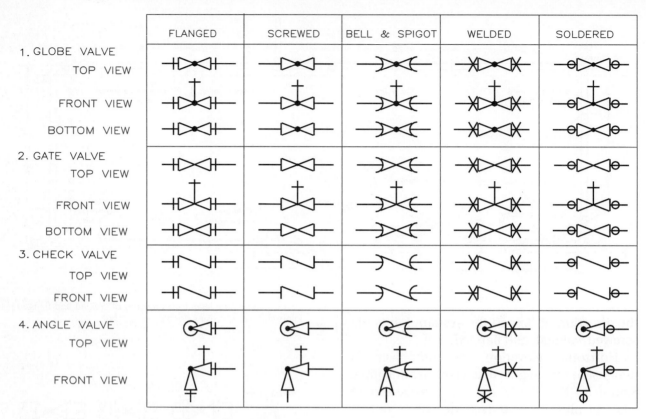

	FLANGED	SCREWED	BELL & SPIGOT	WELDED	SOLDERED
1. GLOBE VALVE TOP VIEW					
FRONT VIEW					
BOTTOM VIEW					
2. GATE VALVE TOP VIEW					
FRONT VIEW					
BOTTOM VIEW					
3. CHECK VALVE TOP VIEW					
FRONT VIEW					
4. ANGLE VALVE TOP VIEW					
FRONT VIEW					

35.19 Examples of orthographic views of valves drawn as single-line symbols.

drawn in isometric in this figure. Although this is a relatively simple three-dimensional system, a thorough understanding of orthographic projection is required to read the drawing. The piping system in **Fig. 35.22** is drawn with all of the pipes revolved into the same horizontal plane. You will notice that the vertical pipes and their fittings are drawn true size in the top view. This is called a **developed pipe drawing**. The fittings and pipe sizes are noted on this preliminary sketch from which the finished drawing will be made in **Fig. 35.23**.

The isometric schematic shown in **Fig. 35.21** explains the three-dimensional relationship of the parts of the system. The rounded bends in the elbows in an isometric

drawing can be constructed with ellipses using the isometric ellipse template, or the corners can be drawn square to reduce the effort and time required.

A north arrow is drawn on the plan view of the piping system in Fig. 35.21; this can be used to orient the isometric pictorial. This north direction is not necessarily related to compass north, which is a direction that is selected parallel to a major set of pipes within the system. In the isometric drawing, it is preferred that the north arrow point to the upper-left or upper-right corner of the pictorial.

35.14 Dimensioned Isometrics

An isometric drawing can be rendered as a fully dimensioned and specified drawing from

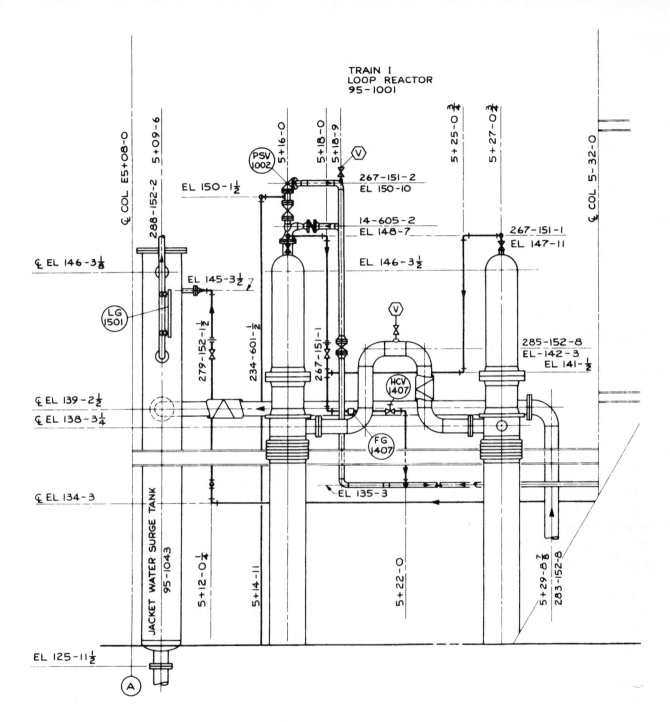

35.20 A piping system using a combination of single-line and double-line symbols. The connections are shown as screwed, welded, and flanged. (Courtesy of Bechtel Corporation.)

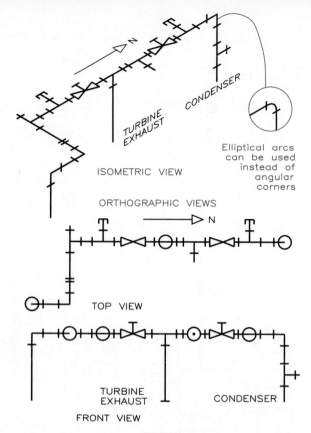

ISOMETRIC VIEW

Elliptical arcs
can be used
instead of
angular
corners

ORTHOGRAPHIC VIEWS

TOP VIEW

TURBINE
EXHAUST

CONDENSER

FRONT VIEW

35.21 Top and front orthographic views are used as the basis for drawing a three-dimensional piping system using single-line symbols.

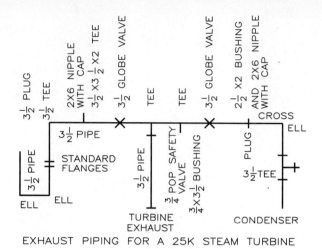

EXHAUST PIPING FOR A 25K STEAM TURBINE

35.22 The vertical pipes shown in Fig. 28.21 are revolved into the horizontal plane to form a developed drawing. The fittings and valves are noted on the sketch.

which a piping system can be constructed. The spool drawing in **Fig. 35.24** is an example where the specifications for the pipe, fittings, flanges, and valves are noted on the drawing and are itemized in the bill of materials.

A number of abbreviations are used to specify piping components and fittings, as you can see by referring to the bill of materials. Many of the standard abbreviations associated with pipe drawings and specifications are given in **Table 35.2**. Part number 1, for example, is an 8-in. diameter pipe of a Schedule 40 weight that is made of seamless steel by the open hearth (OH) process. Instead of OH, the abbreviation EF (electric furnace) may be used.

Under the column "materials" in **Fig. 35.24**, you will notice a code that begins with the letter A, such as A-53. The letter A is used to represent a grade of carbon steel that is listed in Table A of the ANSI B31.3: Petroleum Refinery Piping Standards. The codes for fittings, flanges, and valves are taken from the manufacturers' catalogs of these products.

The suggested format for spool drawings given in **Fig. 35.25**. This is the one used by the Bechtel Corporation, a major construction company in designing and constructing pipelines and refineries.

35.15 Vessel Detailing

Vessels are containers—cylindrical, spherical, or ellipsoidal—that are connected within piping systems for the storage of liquids, such as petroleum, water, or chemicals. The most common vessels are cylinders; they may be installed in vertical or horizontal positions.

Vessel drawings are made and dimensioned by using the standard rules of ortho-

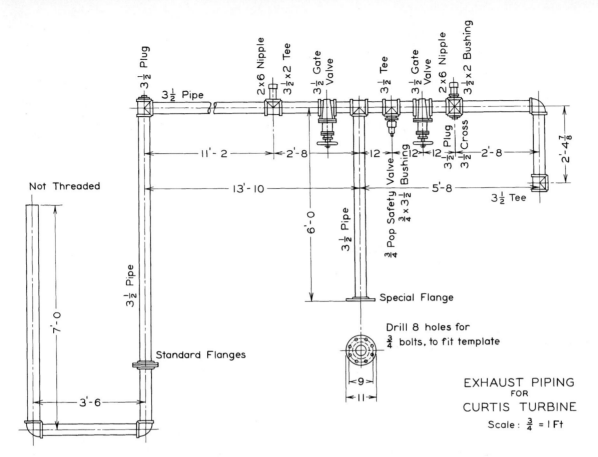

35.23 A finished developed drawing showing all of the components in the system true size with double-line symbols.

graphic projection and working drawings. Pipes connected to vessels and the types of joints required to withstand the pressures and loads must also be specified. Piping principles introduced in this chapter can be applied to describe systems of pipes and vessels in combination.

35.16 Computer Drawings

Pipe drawings range from renderings of plumbing installations of typical homes to offshore oil wells and similar large-scale applications. Due to the broadness of this field, several software packages have been developed for use by engineers and techni-

cians who are responsible for the preparation of pipe drawings.

Models have been extensively used for the design and layout of complex systems—refineries, processing plants, and similar installations—because of the difficulty in representing the intricate details of three-dimensional piping systems in two-dimensional drawings. Models are used to design, develop, and explain these projects with the supplementation of drawings (**Fig. 35.26**).

The ability to use three-dimensional models made by computer has alleviated some of the need for models. The computer permits the generation of views from any angle and viewpoint has greatly aided the in the under-

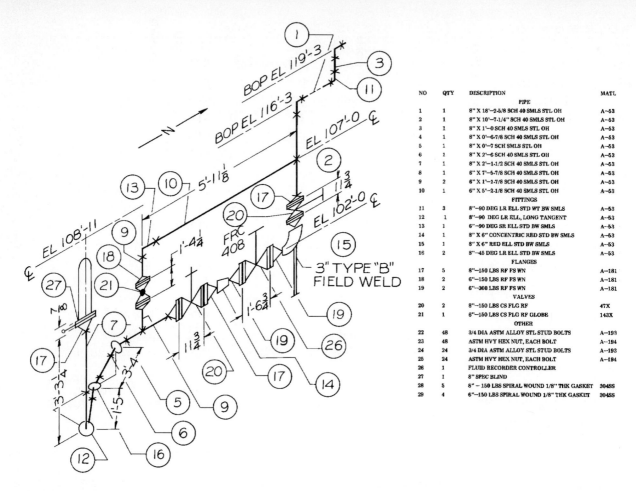

NO	QTY	DESCRIPTION	MATL
		PIPE	
1	1	8" X 18'-2-5/8 SCH 40 SMLS STL OH	A-53
2	1	8" X 10'-7-1/4" SCH 40 SMLS STL OH	A-53
3	1	8" X 1'-0 SCH 40 SMLS STL OH	A-53
4	1	8" X 0'-6-7/8 SCH 40 SMLS STL OH	A-53
5	1	8" X 0'-7 SCH SMLS STL OH	A-53
6	1	8" X 2'-6 SCH 40 SMLS STL OH	A-53
7	1	8" X 2'-1-1/2 SCH 40 SMLS STL OH	A-53
8	1	8" X 7'-5-7/8 SCH 40 SMLS STL OH	A-53
9	2	6" X 1'-1-7/8 SCH 40 SMLS STL OH	A-53
10	1	6" X 5'-2-1/8 SCH 40 SMLS STL OH	A-53
		FITTINGS	
11	3	8"-90 DEG LR ELL STD WT BW SMLS	A-53
12	1	8"-90 DEG LR ELL, LONG TANGENT	A-53
13	1	6"-90 DEG SR ELL STD BW SMLS	A-53
14	1	8" X 6" CONCENTRIC RED STD BW SMLS	A-53
15	1	8" X 6" RED ELL STD BW SMLS	A-53
16	2	8"-45 DEG LR ELL STD BW SMLS	A-53
		FLANGES	
17	5	8"-150 LBS RF FS WN	A-181
18	1	6"-150 LBS RF FS WN	A-181
19	2	6"-300 LBS RF FS WN	A-181
		VALVES	
20	2	8"-150 LBS CS FLG RF	47X
21	1	6"-150 LBS CS FLG RF GLOBE	143X
		OTHER	
22	48	3/4 DIA ASTM ALLOY STL STUD BOLTS	A-193
23	48	ASTM HVY HEX NUT, EACH BOLT	A-194
24	24	3/4 DIA ASTM ALLOY STL STUD BOLTS	A-193
25	24	ASTM HVY HEX NUT, EACH BOLT	A-194
26	1	FLUID RECORDER CONTROLLER	
27	1	8" SPEC BLIND	
28	5	8" - 150 LBS SPIRAL WOUND 1/8" THK GASKET	304SS
29	4	6"-150 LBS SPIRAL WOUND 1/8" THK GASKET	304SS

35.24 This dimensioned isometric pictorial is called a "spool drawing." It is sufficiently complete that it can serve as a working drawing when used with a bill of materials.

standing of pipe drawings. An example of a section of a pipe layout is shown in **Fig. 35.27**, where the hidden lines have been suppressed. The next step would be to render the pipes to give them a realistic appearance.

Summary

Pipe drafting is a complex topic in graphics and technology—in itself, worthy of a sizable textbook. The coverage here was limited to a brief introduction to the basics of piping. The standards of pipe drafting vary to a notable degree from company to company.

Problems

1. On a Size A sheet, draw five orthographic views of the fittings listed below. The views should include the front view, the top view, the bottom view, and the left and right views. Draw two fittings per page. Refer to Fig. 35.17 and Table 35.2 as guides in making these drawings. Use single-line symbols to draw the following screwed fittings: 90° ell, 45° ell, tee, lateral, cap reducing ell, cross, concentric reducer, check valve, union, globe valve, gate valve, and bushing.

Table 35.2 Standard Abbreviations Associated with Pipe Specifications

AVG	average	FS	forged steel	SPEC	specification
BC	bolt circle	FSS	forged stainless steel	SR	short radius
BE	beveled ends	FW	field weld	SS	stainless steel
BF	blind flange	GALV	galvanized	STD	standard
BM	bill of materials	GR	grade	STL	steel
BOP	bottom of pipe	ID	inside diameter	STM	steam
B&S	bell and spigot	INS	insulate	SW	socket weld
BWG	Birmingham wire gauge	IPS	iron pipe size	SWP	standard working pressure
CAS	cast allow steel	LR	long radius	TC	test connection
CI	cast iron	LW	lap weld	TE	threaded end
CO	clean out	MI	malleable iron	TEMP	temperature
CONC	concentric	MFG	manufacture	T&G	tongue and groove
CPLG	coupling	OD	outside diameter	TOS	top of steel
CS	carbon steel, cast steel	OH	open hearth	TYP	typical
DWG	drawing	PE	plain end—not beveled	VC	vitrified clay
ECC	eccentric	PR	pair	WE	weld end
EF	electric furnace	RED	reducer	WN	weld neck
EFW	electric fusion weld	RF	raised face	WB	welded bonnet
ELEV	elevation	RTG or RJ	ring type joint	WT	weight
ERW	electric resistance weld	SCH	schedule	XS	extra strong
FF	flat face	SCRD	screwed	XXS	double extra strong
FLG	flange	SMLS	seamless		
FOB	flat on bottom	SO	slip-on		

2. Same as Problem 1, but draw the fittings as flanged fittings.

3. Same as Problem 1, but draw the fittings as welded fittings.

4. Same as Problem 1, but draw the fittings as double-line screwed fittings.

5. Same as Problem 1, but draw the fittings as double-line flanged fittings.

6. Same as Problem 1, but draw the fittings as double-line welded fittings.

7. Convert the single-line sketch in **Fig. 35.28** into a double-line system that will fit on a Size A sheet.

8. Convert the single-line pipe system in **Fig. 35.29** into a double-line drawing that will fit on a Size A sheet, using the graphical scale given in the drawing to select the best scale for the system.

9. Convert the pipe system shown in isometric in **Fig. 35.30** into a double-line isometric drawing that will fit on a Size B sheet.

10. Convert the pipe system given in Fig. 35.20 into a two-view orthographic drawing using single-line symbols.

11. Convert the isometric drawing of the pipe system in Fig. 35.24 into a two-view orthographic drawing that will fit on a Size B sheet. Take the measurements from the given drawing, and select a convenient scale.

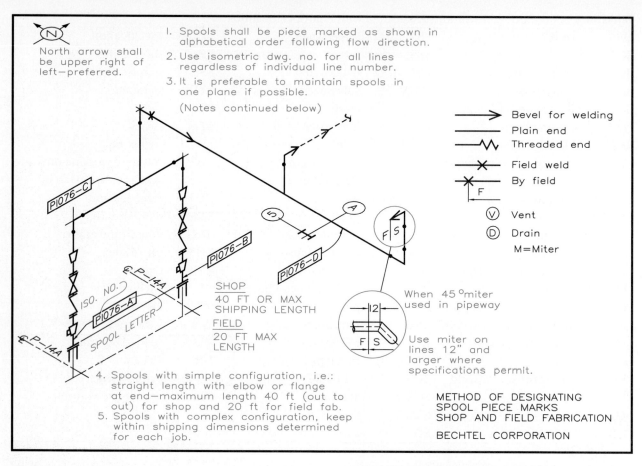

North arrow shall be upper right of left—preferred.

1. Spools shall be piece marked as shown in alphabetical order following flow direction.
2. Use isometric dwg. no. for all lines regardless of individual line number.
3. It is preferable to maintain spools in one plane if possible.

(Notes continued below)

→ Bevel for welding
— Plain end
—◇— Threaded end
—✕— Field weld
✕ By field

Ⓥ Vent
Ⓓ Drain
M=Miter

SHOP
40 FT OR MAX SHIPPING LENGTH

FIELD
20 FT MAX LENGTH

When 45° miter used in pipeway

Use miter on lines 12" and larger where specifications permit.

4. Spools with simple configuration, i.e.: straight length with elbow or flange at end—maximum length 40 ft (out to out) for shop and 20 ft for field fab.
5. Spools with complex configuration, keep within shipping dimensions determined for each job.

METHOD OF DESIGNATING SPOOL PIECE MARKS SHOP AND FIELD FABRICATION

BECHTEL CORPORATION

35.25 A suggested format for preparing spool drawings from a company standard. (Courtesy of the Bechtel Corporation.)

12. Convert the orthographic drawing of the pipe system in Fig. 35.19 into a double-line isometric drawing that will fit on a Size B sheet.

13. Convert the orthographic pipe system in Fig. 35.9 into a single-line isometric drawing that will fit on a Size B sheet. Estimate the dimensions.

14. Convert the orthographic pipe system in Fig. 35.9 into a double-line orthographic view that will fit on a Size B sheet.

35.26 Models are often necessary in the design, development, and explanation of projects involving piping. (Courtesy of the Bechtel Corporation.)

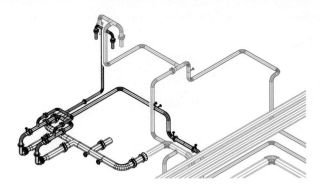

35.27 A piping system drawn by computer. (Courtesy of AutoDesk, Inc.)

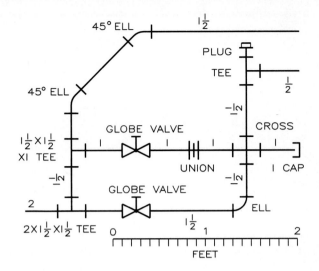

35.29 Problem 8.

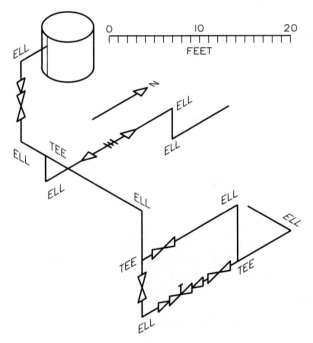

35.28 Problem 7.

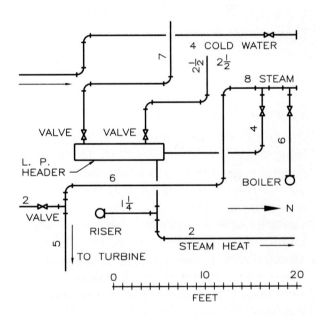

35.30 Problem 9.

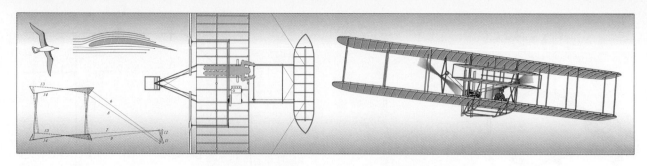

36

Electric/Electronic Graphics

36.1 Introduction

Electric/electronics graphics is a specialty area of the field of graphics technology. **Electrical graphics** is related to the transmission of electrical power that is used in large quantities in homes and industry for lighting, heating, and equipment operation. **Electronics graphics** deals with circuits constructed with transistors and electronic tubes, where power is used in much smaller quantities. Examples of electronic equipment are radios, televisions, computers, and similar products.

Electronics drafters are responsible for the preparation of drawings that will be used in fabricating the circuit and thereby bringing the product into being. They work from sketches and specifications developed by the engineer or electronics technologist. This chapter reviews the drafting practices that are necessary for the preparation of electronic diagrams.

A major portion of this chapter has been adapted from ANSI Y14.15, *Electrical and Electronics Diagrams,* the standards that regulate the graphics techniques used in this area. The symbols used were taken from ANSI Y32.2, *Graphic Symbols for Electrical and Electronics Diagrams.*

36.2 Types of Diagrams

Electronic circuits are classified and drawn according to the format in one of three types of diagrams:

1. Single-line

2. Schematic

3. Connection

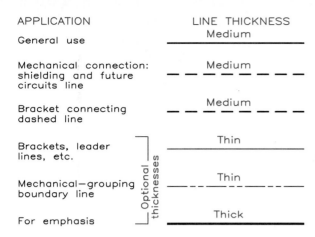

APPLICATION	LINE THICKNESS
General use	Medium
Mechanical connection: shielding and future circuits line	Medium
Bracket connecting dashed line	Medium
Brackets, leader lines, etc.	Thin
Mechanical—grouping boundary line	Thin
For emphasis	Thick

Optional thicknesses

36.1 The recommended line weights for drawing electronics diagrams.

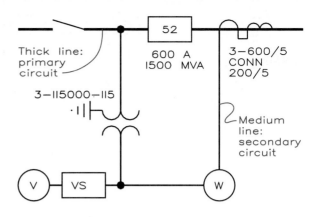

Thick line: primary circuit

3–115000–115

52

600 A
1500 MVA

3–600/5
CONN
200/5

Medium line: secondary circuit

V VS W

36.2 A portion of a single-line diagram where heavy lines represent the primary circuits and medium lines represent the connections to the current and potential sources.

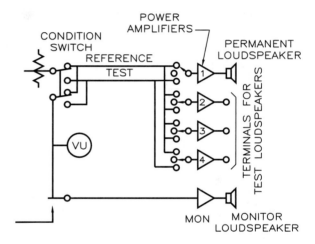

CONDITION SWITCH

POWER AMPLIFIERS

REFERENCE

TEST

PERMANENT LOUDSPEAKER

VU

1
2
3
4

TERMINALS FOR TEST LOUDSPEAKERS

MON MONITOR LOUDSPEAKER

36.3 A typical single-line diagram for illustrating a circuit in an audio system. Its basic functions are shown, but many of the details and components and the ratings that are shown in a schematic are omitted.

The suggested line weights for drawing these types of diagrams are shown in **Fig. 36.1**.

Single-Line Diagrams

Single-line diagrams are drawn with single lines and general symbols that are adequate to trace the flow of current through the circuit and obtain a basic understanding the parts and devices within it. Descriptions of the circuit components are not specified in detail. Single lines are used to represent both AC and DC systems, as illustrated in **Fig. 36.2**. An example of a single-line diagram of an audio system is shown in **Fig. 36.3**. Primary circuits are indicated by thick connecting lines, and medium lines are used to represent connections to the current and potential sources.

Single-line diagrams show the connections of meters, major equipment, and instruments. Ratings are often given to supplement the graphic symbols to provide information such as kilowatts, voltages, cycles and revolutions per minute, and generator ratings (**Fig. 36.4**).

Schematic Diagrams

Schematic diagrams use graphic symbols to show the electrical connections and functions of a specific circuit arrangement. Schematic diagrams provide more information and specifications that are necessary for the composition of a circuit than do single-line diagrams. Although a schematic diagram makes it possible to trace the circuit and its functions,

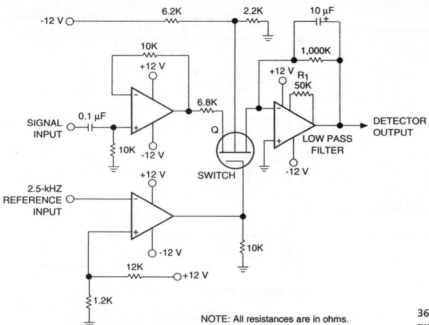

NOTE: All resistances are in ohms.

36.4 A single-line diagram illustrates a switching circuit complete with notation of device designations. (Courtesy of NASA.)

physical sizes, shapes, and locations of various components are not shown. The schematic diagram in **Fig. 36.5** can be referred to for many applications of the principles covered in this chapter. Another good reference example of a schematic diagram is given at the end of this chapter in Fig. 36.48.

Connection Diagrams

Connection diagrams show the connections and installations of the parts and devices of the system. In addition to showing the internal connections, external connections, or both, they show the physical arrangement of the parts, such as does the diagram shown in **Fig. 36.6**.

36.3 Schematic Diagram Connecting Symbols

The most basic symbols of a circuit are those that are used to represent connections of parts within the circuit. Connections, or junc-

tions, are indicated by using small black dots, as shown in part A of **Fig. 36.7**. The dots distinguish between connecting lines and those that simply pass over each other (part B). The use of dots to show connections is optional; it is preferable to omit them if clarity is not sacrificed and also preferable that connecting wires have single junctions wherever possible. When the layout of a circuit does not permit the use of single junctions, and lines within the circuit must cross, then dots must be used to distinguish between crossing and connecting lines (parts C and D).

Interrupted paths are breaks in lines within a schematic diagram that are interrupted to conserve space when this can be done without confusion. For example, the circuit in **Fig. 36.8** has been interrupted since the lines do not connect the left and the right sides of the circuit. Instead, the ends of the lines are labeled to correspond to the matching notes at the other side of the interrupted circuit.

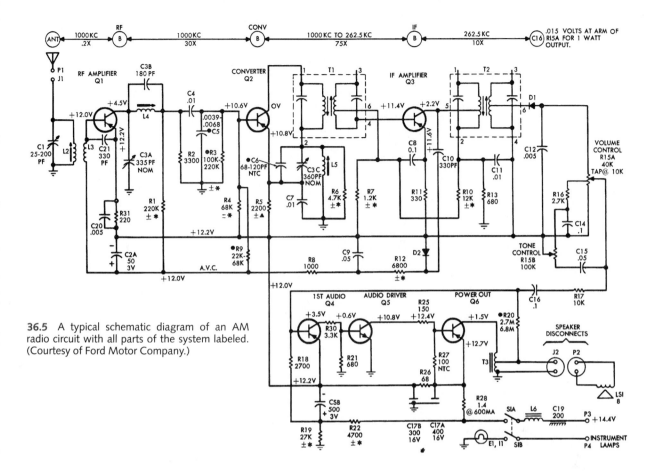

36.5 A typical schematic diagram of an AM radio circuit with all parts of the system labeled. (Courtesy of Ford Motor Company.)

There will be occasions where sets of lines in a horizontal or vertical direction will be interrupted (**Fig. 36.9**). Brackets will be used to interrupt the circuit and notes will be placed outside the brackets to indicate the destinations of the wires or their connections.

In some cases, a dashed line is used to connect brackets that interrupt circuits (**Fig. 36.10**). The dashed line should be drawn so that it will not be mistaken as a continuation of one of the lines within the bracket.

Mechanical linkages that are closely related to electronic functions may be shown as part of a schematic diagram (**Fig. 36.11**). An arrangement of this type helps clarify the relationship

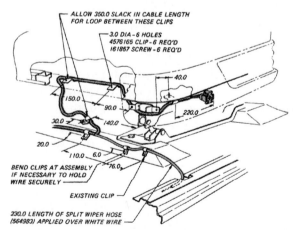

36.6 This three-dimensional connection diagram shows the circuit and its components with the necessary details to explain how it is connected or installed. (Courtesy of the General Motors Corporation.)

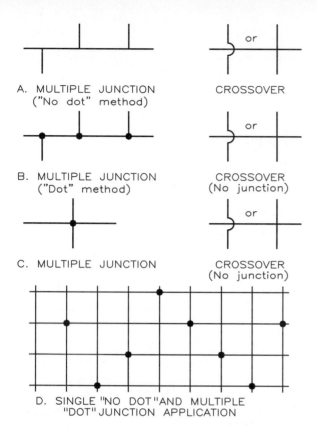

A. MULTIPLE JUNCTION ("No dot" method)   CROSSOVER

B. MULTIPLE JUNCTION ("Dot" method)   CROSSOVER (No junction)

C. MULTIPLE JUNCTION   CROSSOVER (No junction)

D. SINGLE "NO DOT" AND MULTIPLE "DOT" JUNCTION APPLICATION

36.7 Connections should be indicated with single-point junctions (A). Dots may be used to call attention to connections (B); dots must be used when there are multiples of the type shown at (C) and (D).

between the electronics circuit and the mechanical components.

36.4 Graphic Symbols

The electronics drafter must be familiar with the basic graphic symbols that are used to represent the parts and devices within electrical and electronics circuits, as shown in **Fig. 36.12–36.16**. These symbols, extracted from the ANSI Y32.2 standard, are adequate for practically all diagrams. However, when a highly specialized part needs to be shown and a symbol for it is not provided in these standards, it is permissible for drafters to develop their own symbols provided they are properly labeled and their meaning clearly conveyed.

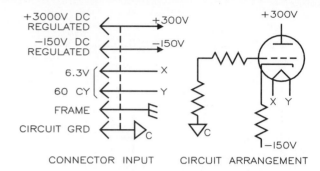

CONNECTOR INPUT   CIRCUIT ARRANGEMENT

36.8 Circuits may be interrupted and connections not shown by lines if they are properly labeled to clarify their relationship to the removed part of the circuit. The connections above are labeled to match those on the left and right sides of the illustration.

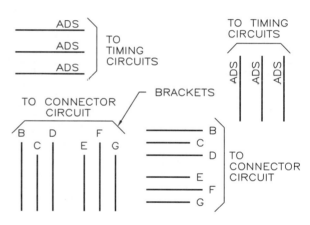

36.9 Brackets and notes may be used to specify the destinations of interrupted circuits, as shown in this illustration.

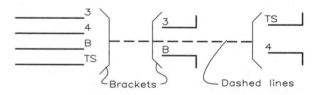

36.10 The connections of interrupted circuits may be indicated by using brackets and a dashed line in addition to labeling the lines. The dashed line should not be drawn to appear as an extension of one of the lines in the circuit.

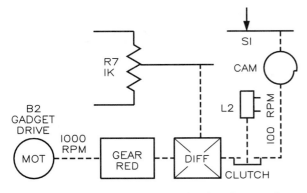

36.11 Mechanical linkages closely related to electronic functions are shown in this schematic diagram.

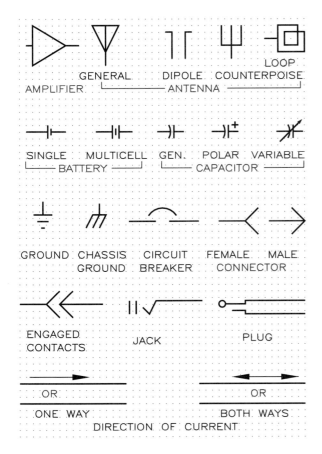

36.12 The proportions of these symbols are drawn on a 3-mm (0.13 in.) grid. When enlarged so the grid is full size, their suggested sizes can be found.

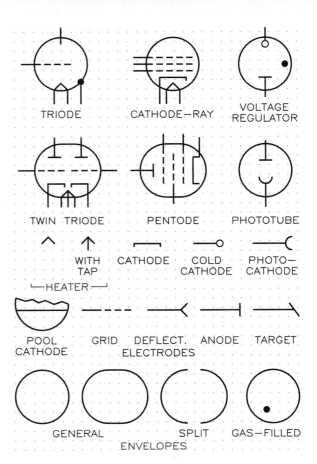

36.13 The upper six symbols are used to represent often-used types of electron tubes drawn on a 3-mm (0.13 in.) grid. An explanation of the parts that make up each symbol is given in the lower half of the figure. (Electron tubes have been mostly replaced by transistors.)

The symbols shown in Figs. 36.12–36.16 are drawn on a grid of 3-mm (0.13 in.) squares that have been reduced. The size of this grid is equal to the letter height used on the final drawing. It is general practice to size graphic symbols based on letter height since text and numerals cannot be enlarged or reduced as easily as graphic symbols without affecting readability. Symbols may be drawn larger or smaller to fit the size of your layout, provided the relative proportions of the symbols are kept about the same.

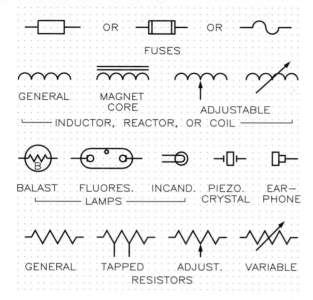

36.14 Graphics symbols of standard circuit components drawn on a 3-mm (0.13 in.) grid.

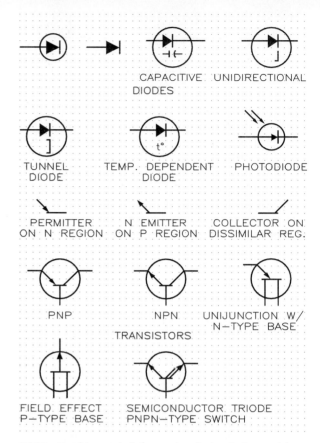

36.15 Graphics symbols for semiconductor devices and transistors drawn on a 3-mm (0.13 in.) grid. The arrows in the middle of the figure illustrate the meanings of the arrows used in the transistor symbols shown below them.

These symbols are but a few of the more commonly used ones. There are between five and six hundred different symbols in the ANSI standards for variations of the basic electrical/electronics symbols.

The preparation of a schematic diagram begins with drawing a freehand sketch to show the circuit and the placement of its components as shown in **Fig. 36.17A**. Using a printed grid makes sketching easier. When the sketch is completed, an instrument drawing can be made by hand or by computer to show the components at a proper scale and reference designations added (**Fig. 36.17B**).

Some of the symbols have been noted to provide designations of their sizes or ratings. The need for this additional information depends on the requirements and the usage of the schematic diagrams.

36.5 Terminals

Terminals are the ends of devices that are attached in a circuit with connecting wires. Examples of devices with terminals that are specified in circuit diagrams are switches, relays, and transformers. The graphic symbol for a terminal is an open circle that is the same size as the solid circle used to indicate a connection.

Switches are used to turn a circuit on or off, or else to actuate a certain part of it while turning another off. Examples of labeling switches are shown in Fig. 36.5 labeled as S1A and S1B.

When a group of parts is enclosed or shielded (drawn enclosed with dashed lines) and the terminal circles have been omitted, the terminal markings should be placed

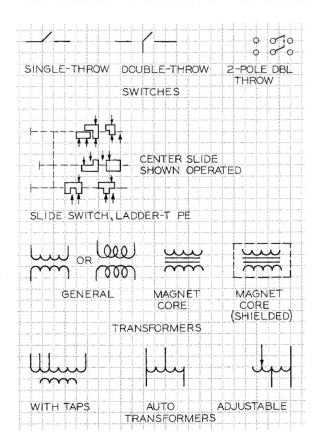

36.16 Graphics symbols for representing switches and transformers drawn on 3-mm (0.13 in.) grid.

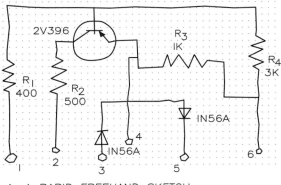

A. A RAPID FREEHAND SKETCH

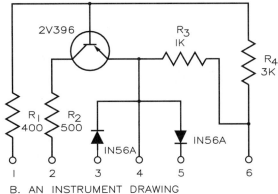

B. AN INSTRUMENT DRAWING

36.17 Drawing a diagram from a sketch.

A The circuit designer makes a freehand sketch of a circuit as a preliminary drawing.

B The final drawing of the circuit is made using the proper symbols and lines.

immediately outside the enclosure, as shown in Fig. 36.5 at T1 and T2. The terminal identifications should be added to the graphic symbols that correspond to the actual physical markings that appear on or near the terminals of the part (not given in this diagram). Several examples of notes and symbols that explain the parts of a circuit diagram are shown in **Figs. 36.18, 36.19,** and **36.20.**

When colored wires, numbers, or geometric symbols are used to identify the various leads or terminals of multilead parts, their identification is indicated near the connecting line adjacent to the symbol. Colored wires can be identified on a diagram with matching color abbreviations.

Rotary terminals are used to regulate the resistance in some circuits, and the direction of rotation of the dial is indicated on the schematic diagram. The abbreviations CW (clockwise) or CCW (counterclockwise) are placed adjacent to the movable contact when it is in its extreme clockwise position, as shown in **Fig. 36.21A.** The movable contact can be identified by an arrow at its end.

If the device terminals are not marked, numbers may be used with the resistor symbol and the number 2 assigned to the adjustable contact (**Fig. 36.21B**). Other fixed taps may be sequentially numbered and added, as shown in **Fig. 36.21C.**

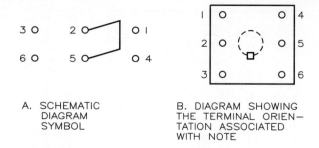

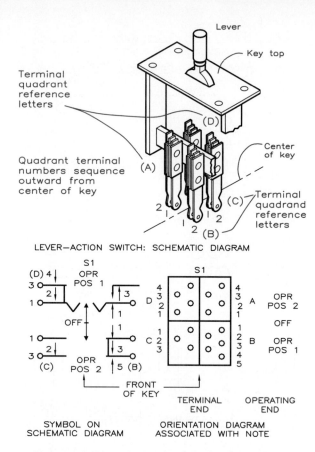

A. SCHEMATIC DIAGRAM SYMBOL

B. DIAGRAM SHOWING THE TERMINAL ORIENTATION ASSOCIATED WITH NOTE

36.18 An example of a method of labeling the terminals of a toggle switch on a schematic diagram (A) and a diagram that illustrates the toggle switch when it is viewed from the rear (B).

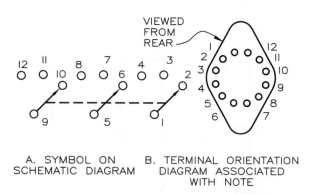

A. SYMBOL ON SCHEMATIC DIAGRAM

B. TERMINAL ORIENTATION DIAGRAM ASSOCIATED WITH NOTE

36.19 An example of a rotary switch as it would appear on a schematic diagram (A) and a diagram that shows the numbered terminals of the switch when viewed from the rear (B).

The position of a switch as it relates to the function of a circuit should be indicated on a schematic diagram. A method of showing functions of a variable switch is shown in **Fig. 36.22**. The arrow, representing the movable end of the switch, can be positioned to connect with several circuits. The different functional positions of the rotary switch are shown both by symbol and by table.

Another method of representing a rotary switch is shown in **Fig. 36.23** by symbol and table. The tabular form is preferred due to the complexity of this particular switch. The dashes between the numbers in the table indicate that the numbers have been connected. For example, when the switch is in

LEVER—ACTION SWITCH: SCHEMATIC DIAGRAM

SYMBOL ON SCHEMATIC DIAGRAM

ORIENTATION DIAGRAM ASSOCIATED WITH NOTE

36.20 An example of a typical lever switch as it would appear on a schematic diagram (left, part A) and an orientation diagram that shows the numbered terminals of the switch when viewed from its operating end (right, part A). A pictorial of the lever switch and its four quadrants is given in part B.

Position 2, the following terminals are connected: 1 and 3, 5 and 7, and 9 and 11. A table of this type should be placed at the bottom of a schematic diagram, if applicable.

Electron tubes have pins fitting into sockets that have terminals connecting into circuits. Pins are labeled with numbers placed outside the symbol used to represent the tube, as shown in **Fig. 36.24**, and are numbered in a clockwise direction with the tube viewed from its bottom.

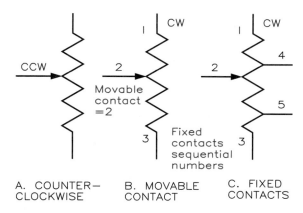

A. COUNTER-CLOCKWISE **B. MOVABLE CONTACT** **C. FIXED CONTACTS**

36.21 To indicate the direction of rotation of rotary switches on a schematic diagram, the abbreviations CW (clockwise) and CCW (counterclockwise) are placed near the movable contact (A). If the device terminals are not marked, numbers may be used with the resistor symbols and the number 2 assigned to the adjustable contact (B). Additional contacts may be labeled (C).

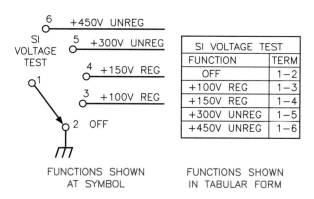

FUNCTIONS SHOWN AT SYMBOL

FUNCTIONS SHOWN IN TABULAR FORM

S1 VOLTAGE TEST	
FUNCTION	TERM
OFF	1-2
+100V REG	1-3
+150V REG	1-4
+300V UNREG	1-5
+450V UNREG	1-6

36.22 For more complex switches, position-to-position function relations may be shown using symbols on the schematic diagram, or in a table of values located elsewhere on the diagram.

36.6 Separation of Parts

In complex circuits, it is often advantageous to separate elements of a multielement part with portions of the graphic symbols drawn in different locations on the drawing. An example of this method of separation is the switch labeled S1A and S1B in Fig. 36.5. The switch is labeled S1 and the letter that follows, called a

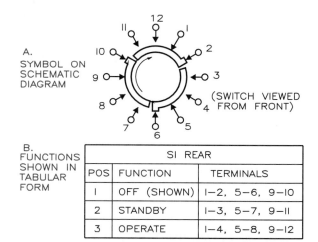

A. SYMBOL ON SCHEMATIC DIAGRAM

(SWITCH VIEWED FROM FRONT)

B. FUNCTIONS SHOWN IN TABULAR FORM

	S1 REAR		
POS	FUNCTION		TERMINALS
1	OFF (SHOWN)		1-2, 5-6, 9-10
2	STANDBY		1-3, 5-7, 9-11
3	OPERATE		1-4, 5-8, 9-12

36.23 A rotary switch may be shown on a schematic diagram with its terminals labeled (A), or its functions can be given in a table placed elsewhere on the drawing (B). Dashes are used to indicate the linkage of the numbered terminals. For example, 1–2 means that terminals 1 and 2 are connected in the "off" position.

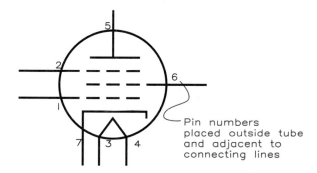

Pin numbers placed outside tube and adjacent to connecting lines

36.24 Tube pin numbers should be placed outside the tube envelope and adjacent to the connecting lines. (Courtesy of ANSI.)

suffix, is used to designate a different part of the same switch. Suffix letters may also be used to label subdivisions of an enclosed unit that is made up of a series of internal parts, such as the crystal unit shown in **Fig. 36.25**. These crystals are referred to as Y1A and Y1B.

Rotary switches of the type shown in **Fig. 36.26** are designated as S1A, S1B, etc. The suffix letters are labeled in sequence beginning with the knob and working away from it. Each end of the various sections of the switch

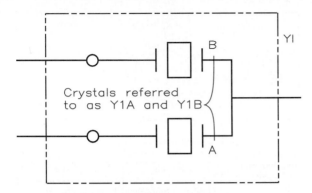

36.25 As subdivisions within the complete part, crystals A and B are referred to as Y1A and Y1B.

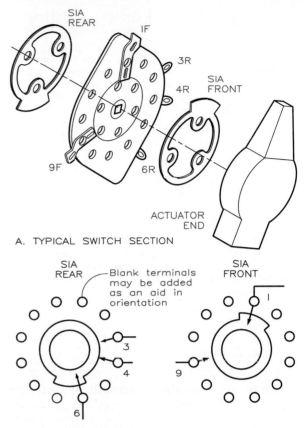

A. TYPICAL SWITCH SECTION

B. GRAPHICAL SYMBOL

36.26 (A) Parts of rotary switches are designated with suffix letters A, B, C, etc., and are referred to as S1A, S1B, S1C, etc. (B) The words FRONT and REAR are added to these designations when both sides of the switch are used. (Courtesy of ANSI.)

should be viewed from the same end. When the rear and front of the switches need to be used, the words FRONT and REAR are added to the designations.

Portions of items such as terminal boards, connectors, or rotary switches may be separated on a diagram. The words PART OF may precede the identification of the portion of the circuit of which it is a part, as shown in **Fig. 36.27A**. A second method of showing a part of a system is by using conventional break lines that make any additional note unnecessary (**Fig. 26.27B**).

36.7 Reference Designations

A combination of letters and numbers that identify items on a schematic diagram are called **reference designations**. These designations are used to identify the components not only on the drawing but in the related documents that refer to them. Reference designations should be placed close to the symbols that represent the replaceable items of a circuit on a drawing. Items that are not separately replaceable may be identified if this is considered necessary. Mounting devices for electron tubes, lamps, fuses, and so forth, are seldom identified on schematic diagrams.

It is standard practice to begin each reference designation with an uppercase letter that may be followed by a numeral with no hyphen between them. The number usually represents a portion of the part being represented. The lowest number of a designation should be assigned to begin at the upper left of the schematic diagram and proceed consecutively from left to right and top to bottom throughout the drawing.

Some of the standard abbreviations used to designate parts of an assembly are amplifier-A, battery-BT, capacitor-C, connector-J, piezoelectric crystal-Y, fuse-F, electron tube-V, generator-G, rectifier-CR, resistor-R, transformer-T, and transistor-Q.

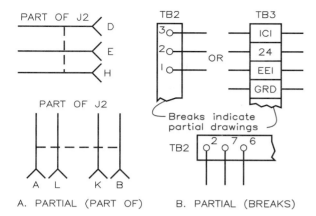

A. PARTIAL (PART OF)   B. PARTIAL (BREAKS)

36.27 The portions of connectors or terminal boards are functionally separated on a diagram; the words PART OF may precede the reference designation of the entire portion (A) or conventional breaks can be used to indicate graphically that the part is only a portion of the whole (B).

As the circuit is being designed, some of the numbered elements may be deleted from the circuit drawing. The numbered elements that remain should not be renumbered, even if there is a missing element within the sequence of numbers used to label the parts. Instead, a table of the type shown in **Fig. 36.28** can be used to list the parts that have been omitted from the circuit. The highest designations are also given in the table as a check to be sure that all parts have been considered.

Electron tubes are labeled not only with reference designations but with type designation and circuit function, as shown in **Fig. 36.29**. This information is labeled in three lines, such as V5 / 35C5 / OUTPUT, which are located adjacent to the symbol.

36.8 Numerical Units of Function

Functional units such as the values of resistance, capacitance, inductance, and voltage should be specified with the fewest number of zeros by using the multipliers in **Fig. 36.30A** as prefixes. Examples using this method of expression are shown in **36.30B** and **C**, where

HIGHEST REFERENCE DESIGNATIONS	
R72	C40
REFERENCE DESIGNATIONS NOT USED	
R8, R10, R61	C12, C15, C17
R64, R70	C20, C22

36.28 Reference designations are used to identify parts of a circuit. They are labeled in a numerical sequence from left to right beginning at the upper left of the diagram. Parts that are later deleted from the system should be listed in a table, along with the highest reference number designations.

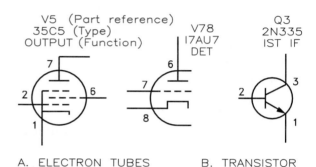

A. ELECTRON TUBES   B. TRANSISTOR

36.29 Three lines of notes can be used with electron tubes (A) and transistors (B) to specify reference designation, type designation, and function. This information should be located adjacent to and preferably above the symbol.

units of resistance and capacitance are given. When four-digit numbers are given, omit the commas; write one thousand as 1000, not as 1,000. You should recognize and use the lowercase or uppercase prefixes, as indicated in Fig. 36.30.

A general note can be used where certain units are repeated on a drawing to reduce time and effort:

UNLESS OTHERWISE SPECIFIED, RESISTANCE VALUES ARE IN OHMS; CAPACITANCE VALUES ARE IN MICROFARADS.

or

CAPACITANCE VALUES ARE IN PICOFARADS.

A. MULTIPLIERS

MULTIPLIER	PREFIX	SYMBOL METHOD 1	SYMBOL METHOD 2
10^{12}	TERA	T	T
10^9	GIGA	G	G
10^6 (1,000,000)	MEGA	M	M
10^3 (1,000)	KILO	k	K
10^{-3} (0.001)	MILLI	m	MILLI
10^{-6} (0.000,001)	MICRO	μ	U
10^{-9}	NANO	n	N
10^{-12}	PICO	p	P
10^{-13}	FEMTO	f	F
10^{-16}	ATTO	a	A

B. RESISTANCE

RANGE IN OHMS	EXPRESS AS	EXAMPLE
LESS THAN 1,000	OHMS	0.031 470
1,000 TO 99,999	OHMS OR KILOHMS	1800 15,853 10k
100,000 to 999,999	KILOHMS OR MEGOHMS	220k 0.22M
1,000,000 OR MORE	MEGOHMS	3.3M

C. CAPACITANCE

RANGE IN PICOFARADS	EXPRESS AS	EXAMPLE
LESS THAN 10,000	PICOFARADS	152.4pF 4700pF
10,000 OR MORE	MICROFARADS	0.015μF 30μF

36.30 Multipliers should be used to reduce the zeros in a number (A). Examples of units of resistance (B) and capacitance (C) are also shown.

The following note specifies capacitance values:

CAPACITANCE VALUES SHOWN AS NUMBERS EQUAL TO OR GREATER THAN UNITS ARE IN pF AND NUMBERS LESS THAN UNITY ARE IN μF.

Examples of the placement of the reference designations and the numerical values of resistors are shown in **Fig. 36.31**.

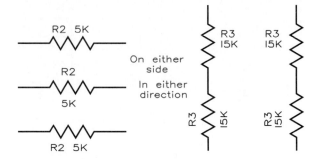

36.31 Methods of labeling the numerical values of resistors on a schematic diagram.

36.9 Functional Identification of Parts

The readability of a circuit is improved if parts are labeled to indicate their functions. Test points are labeled on drawings with the letters "Tp" and their suffix numbers. The sequence of the suffix numbers should be the same as the sequence of troubleshooting the circuit when it is defective. As an alternative, the test function can be indicated on the diagram below the reference designation.

Additional information may be included on a schematic diagram to aid in the maintenance of the system:

DC resistance of windings and coils.

Critical input and output impedance values.

Wave shapes (voltage or current) at significant points. Wiring requirements for critical ground points, shielding, pairing, etc.

Power or voltage ratings of parts.

Caution notation for electrical hazards at maintenance points.

Circuit voltage values at significant points (tube pins, test points, terminal boards, etc.).

Zones (grid system) on complex schematics.

Emphasis on signal flow direction in main signal paths.

36.32 Jack S. Kilby wrote in his notebook in 1958, "Extreme miniaturization of many electrical circuits could be achieved by making resistors, capacitors and transistors, and diodes on a single slice of silicon." (Courtesy of Texas Instruments.)

36.10 Printed Circuits

Printed circuits are universally used today for miniature electronic components and computer systems (**Fig. 36.32**). For years, the vacuum tube was the best means of controlling electrical current—but it was bulky, fragile, unreliable, and consumed large amounts of power. In 1947, the transistor replaced the vacuum tube, eliminating the disadvantages of the tube and reducing its size to fit on the head of pin.

In 1958, Jack Kilby advanced miniaturization further with his integrated circuit in which resistors, capacitors, transistors, and other components were merged on a slice of silicon called a **chip**. With each advancement in electronics have come even more breakthroughs, which have offered more productivity, at greater speeds, and at lower costs. The dramatic degree of miniaturization that has occurred can be seen in **Fig. 36.33**, where a complex circuit fits on a one-square-centimeter chip.

Printed circuit boards, on which chips and other devices are assembled, are drawn up to

36.33 An integrated circuit on a one-square-centimeter chip was a major breakthrough in electronics, but it has since been surpassed by more advanced technology. (Courtesy of Texas Instruments.)

36.34 A magnified view of a circuit that has been printed and etched on a board and the devices soldered in position. (Courtesy of Bishop Industries Corporation.)

four times their final size. The drawings are precisely drawn in ink on a highly stable acetate film and are photographically reduced to the desired size. The circuit is "printed" onto an insulated board made of plastic or ceramics, and the devices within the circuit are connected and soldered (**Fig. 36.34**).

Some circuits are printed on both sides of the circuit board, requiring two photographic negatives made from positive drawings (black lines on a white background), as shown in **Fig. 36.35**. The drawing for each side can be made on separate sheets of acetate that are laid over each other when the second diagram is drawn. However, a more efficient method uses red and blue tape to produce a single drawing (**Fig. 36.36**), from which two negatives are photographically made. Filters are used on the process camera to drop out the

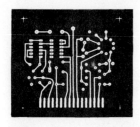 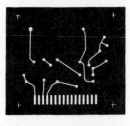

36.35 A printed circuit applied to both sides of the circuit board requires two drawings, one for each side, that are photographically converted to negatives for printing. (Courtesy of Bishop Industries Corporation.)

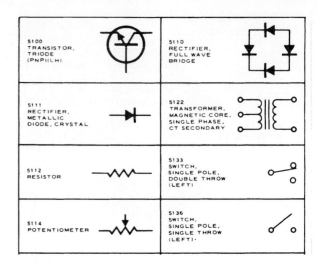

5100 TRANSISTOR, TRIODE (PNP)(LH)		5110 RECTIFIER, FULL WAVE BRIDGE	
5111 RECTIFIER, METALLIC DIODE, CRYSTAL		5122 TRANSFORMER, MAGNETIC CORE, SINGLE PHASE, CT SECONDARY	
5112 RESISTOR		5133 SWITCH, SINGLE POLE, DOUBLE THROW (LEFT)	
5114 POTENTIOMETER		5136 SWITCH, SINGLE POLE, SINGLE THROW (LEFT)·	

36.37 Preprinted symbols can be used for laying out schematic diagrams rather than drawing them. They give a higher contrast and sharpness that improves their reproducibility. (Courtesy of Bishop Industries Corporation.)

that can be burnished onto the surface of the drawing to form a permanent schematic diagram (**Fig. 36.37**).

The symbols can be connected with matching tape to represent wires between them instead of drawing the lines. Schematic diagrams made with these materials are of a very high quality that reproduces well when reduced for publication in specifications or in technical journals.

36.12 Installation Drawings

Many types of electric/electronics drawings are used to produce the finished installation, from the designer who visualized the system at the outset of the project to the contractor who builds it. Drawings are used to design the circuit, detail its parts for fabrication, specify the arrangement of the devices within the system, and instruct the contractor how to install the project.

A combination arrangement and wiring diagram drawing is shown in **Fig. 36.38**, where the system is shown in front and right-side views. The wiring diagram explains how the

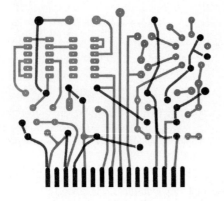

36.36 By using two colors, such as blue and red, one circuit drawing, with two negatives can be made from the same drawing by using camera filters that screen out one of the colors with each shot. The circuits are then printed on each side of the board. (Courtesy of Bishop Industries Corporation.)

red for one negative and a different filter for dropping out the blue for the second negative.

Printed circuits are usually coated with silicone varnish to prevent malfunction because of the collection of moisture or dust on the surface. They may also be enclosed in protective shells.

36.11 Shortcut Symbols

Preprinted symbols are available commercially that can be used for "drawing" high-quality electronic and printed circuits. The symbols are available on sheets or on tapes

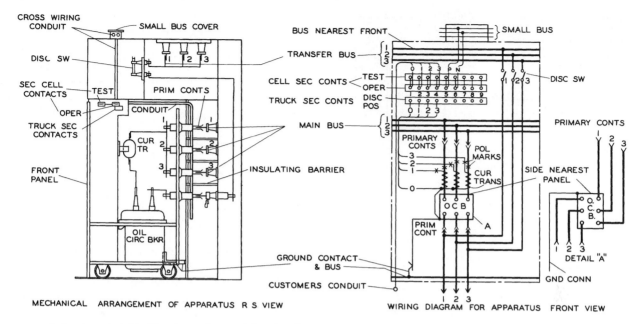

MECHANICAL   ARRANGEMENT OF APPARATUS R S VIEW

WIRING DIAGRAM FOR APPARATUS FRONT VIEW

36.38 A front and right-side view of a metal-enclosed switchgear. The diagram shows the arrangement of the apparatus and also gives details on the wiring for the unit.

wires and components within the system are connected for the metal-encased switchgear. Bus bars are conductors for the primary circuits.

The installation/circuit diagram in **Fig. 36.39** is a drawing used in a maintenance manual to show how to troubleshoot a defective circuit. It is a combination drawing that shows the physical arrangement as well as the circuit.

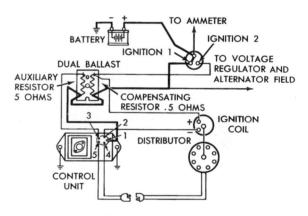

36.39 An installation/circuit diagram that describes how to troubleshoot a circuit. (Courtesy of Chrysler Corporation.)

Problems

1. On a Size A sheet, make a schematic diagram of the low-pass inductive-input filter shown in **Fig. 36.40**.

2. On a Size A sheet, make a schematic diagram of the quadruple-sampling processor shown in **Fig. 36.41**.

3. On a Size A sheet, make a schematic diagram of the temperature-compensating DC restorer circuit shown in **Fig. 36.42**.

4. On a Size A sheet, make a schematic diagram of the magnetic amplifier DC transducer shown in **Fig. 36.43**.

5. On a Size A sheet, make a schematic diagram of the improved power-factor controller shown in **Fig. 36.44**.

6. On a Size B sheet, make a schematic diagram of the "buck/boost" voltage regulator shown in **Fig. 36.45**.

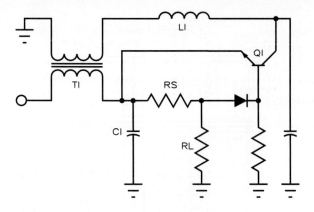

36.40 Problem 1: A low-pass inductive-input filter. (Courtesy of NASA.)

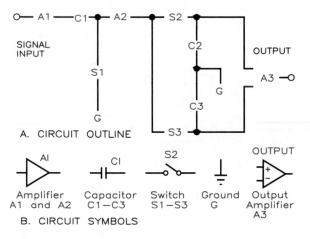

36.41 Problem 2: A quadruple-sampling processor. (Courtesy of NASA.)

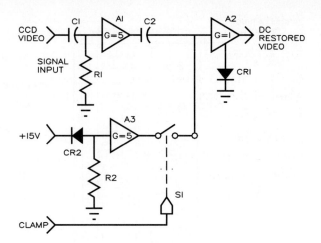

36.42 Problem 3: A temperature-compensating DC restorer circuit. (Courtesy of NASA.)

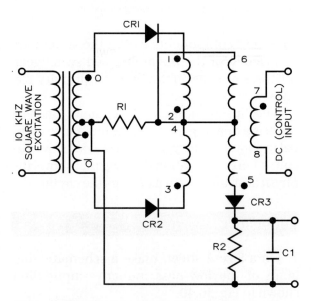

36.43 Problem 4: A magnetic amplifier DC transducer. (Courtesy of NASA.)

7. On a Size C sheet, make a schematic diagram of the overload protection circuit shown in **Fig. 36.46**.

8. On a Size B sheet, make a schematic diagram of the phase detector circuit shown in **Fig. 36.47**.

9. On a Size C sheet, draw a schematic diagram of the dual output manual station in **Fig. 36.48** and show a parts list.

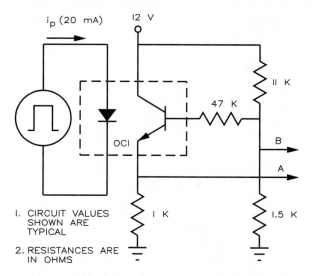

36.44 Problem 5: An improved power-factor controller. (Courtesy of NASA.)

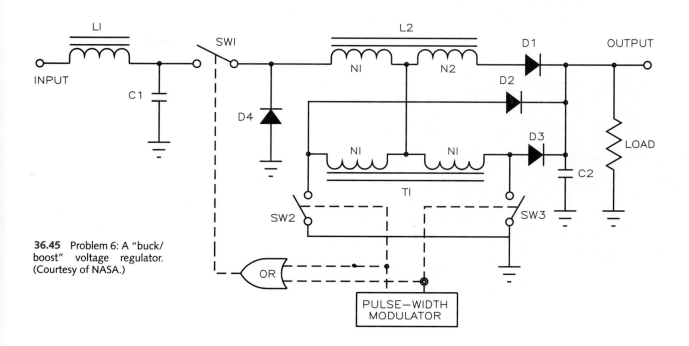

36.45 Problem 6: A "buck/boost" voltage regulator. (Courtesy of NASA.)

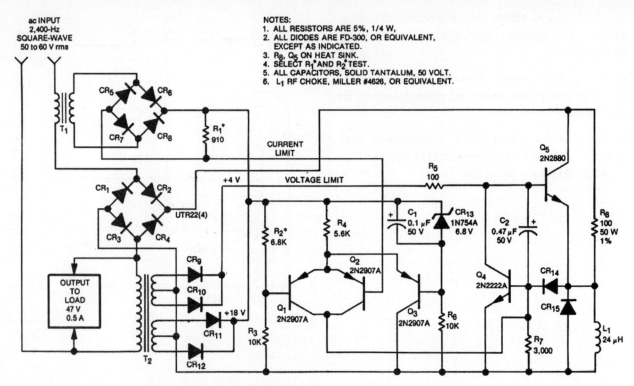

36.46 Problem 7: An overload-protection circuit. (Courtesy of NASA.)

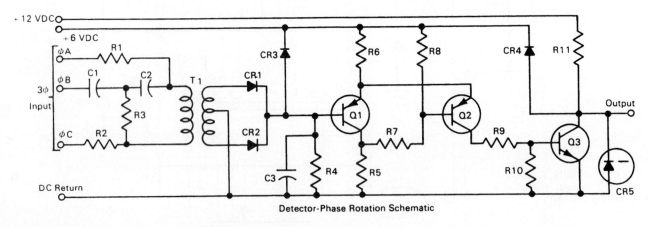

Detector-Phase Rotation Schematic

36.47 Problem 8: A schematic of a phase-detector circuit. (Courtesy of NASA.)

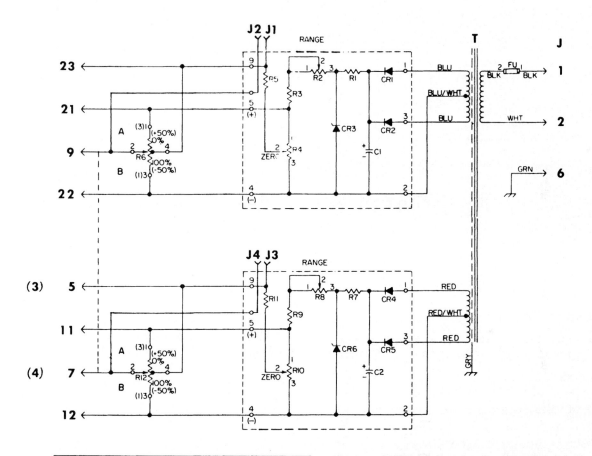

COMPONENT	DESCRIPTION	
CR1, CR2, CR3, CR4	DIODE	1N3193
CR3 & CR6	ZENER DIODE	24V 3W ±.5V TOL
R1 & R7	RESISTOR	560 Ω 5W 5%
R3 & R9	RESISTOR	56 Ω
R5 & R11	RESISTOR	2.7K 1/2W 5%
R2 & R8	TRIM POT	50 Ω
R4 & R10	TRIM POT	5K
R6 & R12	POT	500 Ω 2W DUAL
C1 & C2	CAPACITOR	60 MFD 60V
T	TRANSFORMER	
FU	FUSE 1 AMP	
M	EDGEWISE METER	
J	CONNECTOR	
J1, J2, J3, J4	TEST JACK	

TYPE	R6 SCALE		M SCALE	
	A	B	BOTTOM	TOP
RS1100C	0%	100%	0%	100%
RS1110C	0%	100%	PER ENGINEERING DATA	
RS3100C	+50%	-50%	-50%	+50%
RS4100C	0%	100%	0%	100%
RS2120C	100%	0%	OMIT	
RS2100C	100%	0%	0%	100%

NOTE: OUTPUT WIRED TO TERMINALS 3 AND 4 ON
TYPES RS1100C, RS1110C, AND RS3100C.

OUTPUT WIRED TO TERMINALS 5 AND 7 ON
TYPES RS2100C, RS2120C, AND RS4100C.

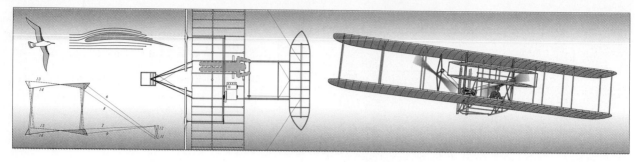

37

AutoCAD Computer Graphics

37.1 Introduction

This chapter provides an introduction to computer graphics using AutoCAD, Release 14, which runs on an Intel 486 or Pentium processor with 32 MB of RAM, at least 200 Mb of hard disk, a mouse (or tablet), an A-B plotter, and/or a printer. Windows 95, the industry standard, or Windows NT are recommended as the operating systems. AutoCAD was selected as the software for presenting computer graphics because it is the most widely used.

The coverage of AutoCAD in this book is brief and many operations have been omitted because of space limitations. Auto-CAD's concisely written *User's Guide* has 805 pages and its *Command Reference* has 932 pages. However, AutoCAD is covered here sufficiently well to guide you through applications typical of an engineering design graphics course.

You will find that the learning of computer graphics and its next upgrade will be a life-long experience.

37.2 Computer Graphics Overview

The major areas of computer graphics are *CAD* (computer-aided design), *CADD* (computer-aided design and drafting), *CIM* (computer-integrated manufacturing), and *CAD/CAM* (computer-aided manufacturing).

CAD (computer-aided design) is used to solve design problems, analyze design data, and store design information for easy retrieval. Many CAD systems perform these functions in an integrated manner, greatly increasing the designer's productivity.

CADD (computer-aided design drafting) is the computer process of making engineering drawings and technical documents more closely related to drafting that is CAD.

37.1 This automatic Chrysler assembly line uses computer-controlled robots as part of their CIM system for welding body parts together. (Courtesy of Chrysler Corporation.)

CAD/CAM (computer-aided design/computer-aided manufacturing) is a system that can be used to design a part or product, devise the production steps, and electronically communicate this data to control the operation of manufacturing equipment and robots.

CIM (computer-integrated manufacturing), a more advanced system than CAD/CAM, coordinates and operates all stages of manufacturing from design to finished product (**Fig. 37.1**).

Advantages of CAD and CADD

Computer-graphics systems offer the designer and drafter some or all of the following advantages:

1. **Increased accuracy**. CAD systems are capable of producing drawings that are 100-percent accurate.

2. **Increased drawing speed**. Engineering drawings and documents can be prepared more quickly, especially when standard details from existing libraries are incorporated in new drawings.

3. **Easy revisions**. Drawings can be more easily modified, changed, and revised than is possible by hand techniques.

4. **Better design analysis**. Alternative designs can be analyzed quickly and easily. Software is available to simulate a product's operation and test it under a variety of conditions, which thus lessens the need for models and prototypes.

5. **Better presentation**. Drawings can be presented in two or three dimensions, and rendered as technical illustrations to better communicate designs.

6. **Ongoing updates of libraries of drawing aids**. Databases of details, symbols, and figures that are used repeatedly can be archived for immediate used in making drawings.

7. **Improved filing**. Drawings can be conveniently filed, retrieved, and transmitted on disks and tapes.

37.3 Computer Hardware

The hardware of a computer graphics system includes the **computer**, **monitor**, **input devices** (keyboard, digitizers, and light pens), and **output devices** (plotters and printers).

Computer

The computer, with an installed **program**, receives input from the user from the keyboard, executes the instructions, and produces output. The part of the computer that follows the program's instructions is the **CPU** (**central processing unit**) (**Fig. 37.2**). The computer graphics computer should have at least 32 MB of RAM, and its hard disk storage should be large, preferably six to eight gigabytes.

The monitor is a **CRT** (**cathode-ray tube**) that has an electron gun that emits a beam that sweeps rows of raster lines on the screen.

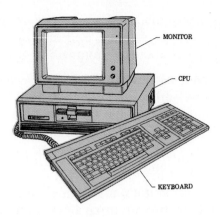

37.2 The basic components of a desktop computer system are the CPU (central processing unit), keyboard, and monitor.

Each line consists of dots called **pixels**. Raster-scanned CRTs refresh the picture display many times per second. A measure of monitor quality is **resolution**, which is the number of pixels per inch that can be produced on the screen. The greater the number of pixels, the greater will be the clarity of the image on the screen (**Fig. 37.3**).

Input Devices

Besides the standard keyboard, the **digitizer** is used to enter graphic data to the computer. The **mouse**, a hand-held *digitizer* that is moved about on a mouse pad or desktop to transmit information to the computer, is the most commonly used input device. Variations of the mouse are **thumbwheels** operated by fingertips, **joysticks** that let the user "steer" about the screen by tilting a lever, and **spherical balls** that can be rotated to input 3D data to the screen (**Fig. 37.4**).

A digitizer can also be used in combination with a **tablet** as an alternative to the mouse. The digitizer (stylus) can select commands from the menu attached to the tablet. Drawings can also be attached to tablets and "traced" with the stylus to convert it to x and y coordinates. The **light pen** enables the user to select points and lines that can be "drawn" on the screen.

37.3 Current technology enables computers to send three-dimensional graphics over networks for instantaneous communication. (Courtesy of Hewlett Packard Company.)

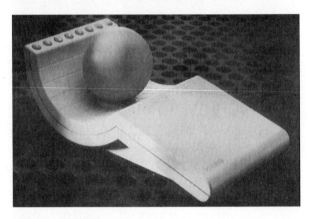

37.4 The Spaceball® digitizer is manipulated to move an object in three-dimensional space on the screen by pushing, pulling, and twisting it. (Courtesy of CalCom Corporation.)

Output Devices

The **plotter** makes a drawing on paper or film with a pen in the same manner a drawing is made by hand. The various types are **flatbed**, **drum**, and **sheet-fed plotters**. In the flatbed plotter, the drawing paper is held stationary while pens are moved about its surface. The drum plotter rolls the paper up and down over a cylinder while the pen moves left and right to make the drawing. The sheet-fed plotter holds the paper with grit wheels in a flat position as the sheet is moved forward and backward as the pen moves left and right to make the drawing.

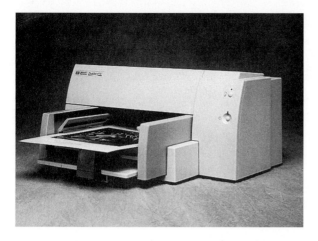

37.5 The DeskJet 690 series of printers produces photo-quality color images that are as close to traditional photographs as desktop printers have ever come. (Courtesy of Hewlett-Packard Company.)

Printers are of the impact type (much like typewriters) or nonimpact, where images are formed by sprays, laser beams, photography, or heat (**Fig. 37.5**). The laser printer gives excellent resolution of dense, accurately drawn lines in color as well as in black and white. Ink jet printers spray images onto a drawing surface in color or in black that approach the quality of the laser images. Larger, nonimpact printers (24 × 36 inches and larger) are most often ink jet printers since large lasers are much more expensive.

37.4 Your First AutoCAD Session

If this is your first AutoCAD session, you are probably anxious to turn the computer on, make a drawing on the screen, and plot it without reading the instructions. This section is what you're looking for.

Format of Presentation
In this chapter, the progression from one step of a command to the next level will be separated by an angle pointing to the right (>) in order to simplify presentation and reduce explanatory text. The commands and prompts that appear on the screen will be given in a distinct, lighter typeface to distinguish them from supplementary notes of explanation. Enter is the keyboard key with this name. Once a command is selected, additional prompts given at the Command line at the bottom of the screen must be followed.

Booting Turn on the computer and boot the system by typing ACAD14 (or the command used by your system) to activate the program (**Fig. 37.6**). Move the cursor around the screen with your mouse, select items, and try the pull-down menu.

Mouse Most interactions with the computer will be accomplished with a mouse (**Fig. 37.7**), but many commands can be entered at the keyboard, and maybe more quickly after you learn them. Press the left mouse button to click on, select, or pick a command or object; a double click is needed in some cases.

Creating a File To create a new file on a disk, place your formatted disk in its slot, pick File from the Menu bar and New from the dialog box (**Fig. 37.8**), and the Create New Drawing dialog box will appear on the screen (**Fig. 37.9**). Select the Use a Wizard button, Quick Setup, and OK. The Quick Setup box will appear where size Units can be specified (**Fig. 37.10**); select Decimal units, then select the Next button. When the Area box appears on the screen, insert the Width and Length (12 × 9) and then select the Done button (**Fig. 37.11**). The program returns the screen and its command menus ready for your drawing (**Fig. 37.12**).

Making a Drawing Since you don't have the menus and toolbars figured out, type L (for line), press Enter, and draw some lines on the screen with the mouse for the fun of it by selecting endpoints with the left button, as shown in **Fig. 37.12**. A line on the screen "rub-

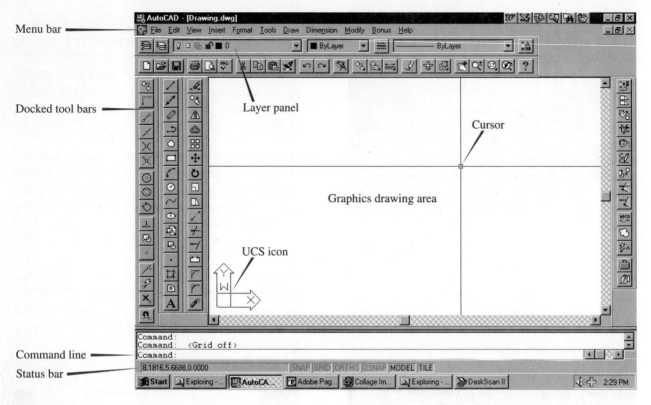

37.6 After booting the system, the AutoCAD, Release 14, screen will appear ready for drawing.

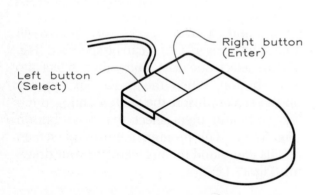

37.7 Press the left mouse button to pick points and to make selections from menus. Pressing the right button has the same effect as pressing the (Enter) key on the keyboard.

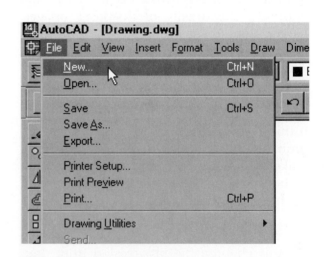

37.8 To open a new file, press `File> New` to obtain the `Create New Drawing` dialog box.

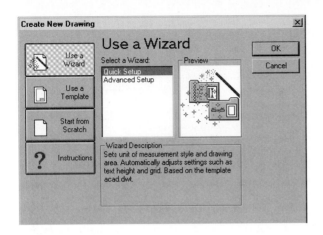

37.9 When you have entered the `Create New Drawing` box, select the `Use a Wizard` button and `OK`.

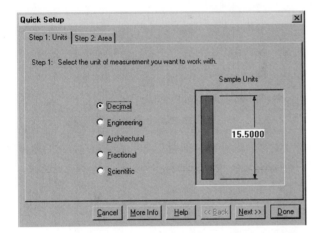

37.10 At the `Quick Setup` box (`Step 1: Units`), specify the type of units you wish to have, decimal units in this example.

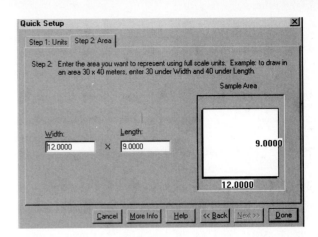

37.11 Select the second tab (`Step 2: Area`) to obtain this box and type in the width and length of the drawing that you wish to make and select `Done`.

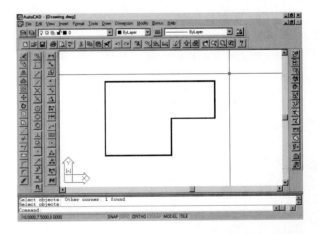

37.12 By typing L (for Line), or selecting `Line` (`Draw> Line`), endpoints of lines can be selected on the screen with the mouse.

berbands" from point to point. To disengage the rubberband, press the right mouse button, which is the same as pressing `Enter`.

Instead of typing L to enter the line command, pick the `Menu bar> Draw` and select the `Line` option to get the prompt line `From point:` in the `Command line` at the bottom of the screen. Now, use your mouse to draw on the screen. The `Line` command can also be selected from the `Draw toolbar` that you will learn about soon. Try drawing a circle and other objects on the screen by selecting icons from the `Draw` pull-down menu.

Repeating Commands By pressing `Enter` (the right mouse button) twice after the previous command, the command can be repeated. For example, if `Line` was the previous command, `Line` will appear in the `Command line` at the bottom of the screen after pressing Enter twice.

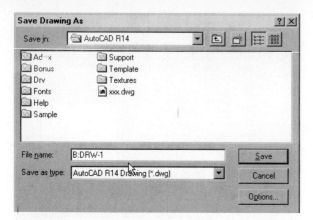

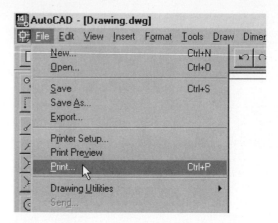

37.13 Select `Save As (File> Save As)` from the `Menu` bar, type the drawing drive (`B:`) and file name (`DRW-1`) in `File name` box, and select `Save`. Drawing `B:DRW-1.DWG` is saved.

37.14 Select `Print (File> Print)` to prepare the plotter.

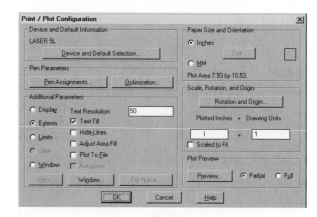

Saving Your File Click on `File` on the `Menu` bar and pick `Save As` from the pull-down menu to get the menu box (**Fig. 37.13**). Type `B:DRW-1` (if your disk is in Drive B) and select the `SAVE` button; the light over drive B: will blink briefly and the drawing, `DRW-1`, is saved on the disk in the B drive.

37.15 The `Print/Plot Configuration` box will appear on the screen for printing instructions. Select `Extents`, `Scaled to Fit`, and `OK` to plot the drawing.

Plotting Your Drawing Select `File` (**Fig. 37.14**) and the `Print` option from its pull-down menu to get the `Print/Plot Configuration` dialog box (**Fig. 37.15**). Select the `Extents` and `Scaled to Fit` buttons so the plotted drawing will fill the sheet. (The x and y origins, found under the `Rotation and Origin` box, must be set to 0, 0.) Load the A-size paper sheet in the plotter, which may be a a Hewlett-Packard 7475 that plots both A-size and B-size sheets with ink pens. Select the `OK` button to send the drawing data to the plotter.

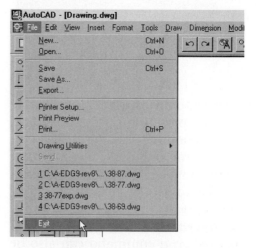

Quitting AutoCAD To be sure that your latest changes have been saved on disk, select `File` from the `Menu` bar and pick `Save` from the pull-down menu to update `B:DRW-1`. To quit AutoCAD, select `File` from the `Menu` bar and `Exit` from its pull-down dialog box (**Fig. 37.16**) to end the session and close the

37.16 If you want to quit AutoCAD, select `Menu bar> File> Exit`. It is good practice to save your drawing again just to be safe before exiting the program.

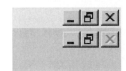

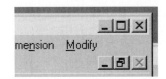

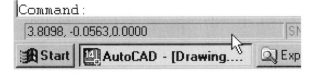

AutoCAD program. If you intend to continue drawing on AutoCAD, do not exit now; exit at the end your work session instead. When working from a floppy disk, do not remove it from its drive until it has been saved with either the `Save` or `Save As` options.

That's how AutoCAD works. Now, let's get into the details.

37.5 Introduction to Windows

The recommend operating system for Release 14 for a single user is Windows 95, which allows several programs to be open and running at the same time. For example, a word-processing program can be running in addition to AutoCAD.

A program can be minimized (reduced to a button in the `Windows` bar at the bottom of the screen) by pressing the dash button in the upper-right corner of the screen (**Fig. 37.17**). Maximize the program by selecting the button that represents the minimized program (**Fig. 37.18**), and it will return to the screen and cover the previously displayed program on the screen. When a program is maximized and fills the screen, two overlapping boxes appear in the window's corner (Fig. 37.17). By selecting this icon, the program window will be reduced to partial screen size so more than one program screen can be seen at the same

time. To make the program fill the screen, select the box icon (**Fig. 37.19**). By picking the X icon, you will exit the current program and it will close.

Windows can be resized by selecting the overlapping boxes icon to obtain a partial window with the arrow drawn with parallel lines in its corner (**Fig. 37.20**). Place the cursor over this arrow with the left button of the mouse, hold it down, and drag the window to the desired size. Toolbars can be moved and reshaped by selecting their edges, holding down the left mouse button, and dragging them to their new position.

When several overlapping program windows appear on the screen, select any point on a window to move it to the front, making it the current program.

37.6 Format of Presentation

In almost all cases, AutoCAD offers several alternative ways of executing a command. For example, a circle can be drawn by typing `C` at the `Command line`; with the `circle icon` from the `Draw toolbar`; or, by using the `Menu bar> Draw` drop-down menu, which includes the `Circle` command. In each of these examples, you must select from options—`Center, radius`; `Center, diameter`; and others—before drawing the circle.

Command Line When a circle is drawn by typing at the `Command line`, it will be presented as follows: `Command line> Circle (or C)> Center point>` **Select** `P1> Radius> Type` `4` (Enter). The underlined `P1` is a point on the screen and the underlined `4` is the radius typed at the command line in response to the prompt, `Radius`.

Menu Bar When the `Menu` bar is used to draw a circle, the sequence of steps is presented as `Menu bar> Draw> Circle> Center, radius> Center point> P1> Radius>` Drag to a radius of `4` .

Draw Toolbar When the `Draw toolbar` is used to draw a circle, the steps are presented as `Menu bar> View> Draw toolbar> Circle icon> Center, radius> Center point>` **Select** `P1> Radius>` **Type** `4`.

37.7 Using Dialog Boxes

Release 14 has many dialog boxes with names beginning with `DD` (`DD1modes`, for example) to interact with the user. The command `Filedia` can be used to turn off (0 = on and 1 = off) the dialog boxes if you prefer to type the commands without dialog boxes. When a command on a menu is followed by three dots (. . .) or an arrow in the right margin, additional dialog boxes will be displayed (**Fig. 37.21**), and some of these will have subdialog boxes.

Definition boxes are provided to identify the functions of each box on the screen. By clicking the cursor briefly on a box, a flyout will appear to define its function, as shown in **Fig. 37.22**.

Double clicking the mouse (quickly pressing the left button twice) selects a file from a file list and displays it. Single clicking followed by `OK` is an alternative for making the selec-

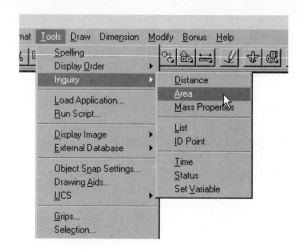

37.21 `Menu` bars have pull-down menus. Commands in the pull-down menu followed by black arrows have subdialog boxes. The sequence in this example is `Tools> Inquiry> Area`.

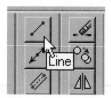

37.22 Definition boxes explain various icons in flyout boxes.

tion. By experimentation, you will soon learn where double clicking can be best applied.

The `Select File` box in **Fig. 37.23** has dialog boxes, lists, blanks, and buttons that can be selected by the cursor. When a file is selected, it is darkened by a gray bar and a reduced illustration of it is shown in the window. To obtain previews of drawing files in the `Browse/Search` screen from which to select, pick the `Find File` button (Fig. 37.23) to obtain the `Browse` screen (**Fig. 37.24**). The slider bar at the bottom of the **screen** lets you browse through the available files and select the one to load. The previews can be shown as small, medium, or large by using the `Size` box. Use the `Search` tab to obtain a window that will help you search for a file by type, name, pattern, and drive.

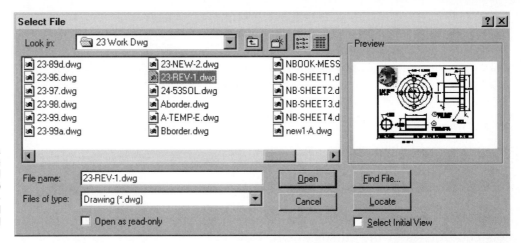

37.23 `File> Open` gives the `Select File` box that helps you find the file that you wish to open. A preview window gives a thumbnail view of any file selected before opening it.

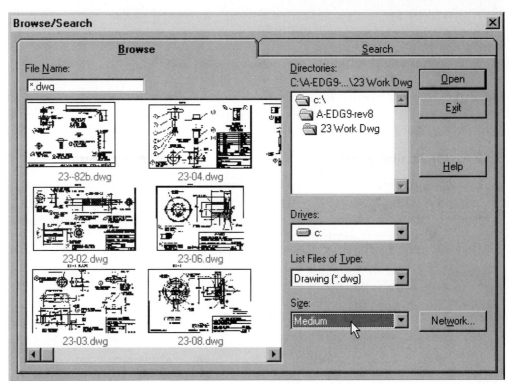

37.24 If you press `Menu bar> File> Open> Select File box> Find File> Browse/Search box> Browse tab`, you can browse preview illustrations of the files on your disk. The `Search tab` activates a screen from which you can insert specifications for finding a particular file on your disk.

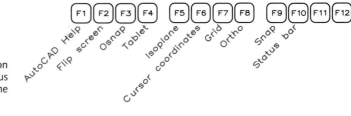

37.25 These function keys on the keyboard control various options, as indicated by the notes.

In many cases, speed is increased if you type commands at the `Command line` instead of using dialog boxes. What could be easier than typing `L` and pressing (Enter) for drawing a `Line`?

37.8 Drawing Aids

The most convenient drawing aids are available from the function keys that can be pressed to turn settings On and Off (**Fig. 37.25**). `F2` (flip screen) alternates between the graphics and text modes. `F6` (coordinates) sets coordinates of the cursor to On, which shows the numerical coordinates in the status bar at the bottom of the screen as the cursor is moved. `F7` (grid) turns the grid On or Off and refreshes the screen in the process, removing any blips or erasures. `F8` (ortho) forces all lines to be either horizontal or vertical. `F9` (snap), when On, makes all object points lie within an invisible grid. `F10` (tablet) activates a digitizing tablet, if one is attached to your computer.

To set the drawing area of the screen that will be filled with the `Grid` dots when `F7` is pressed, type `Limits`, or select `Drawing Limits` under `Format` on the `Menu bar`. A drawing that fills an A-size sheet (11 × 8.5 in.) has a plotting area of about 10.1 × 7.8 in. or 257 × 198 mm. The drawing size is specified with the `Limits` command as follows:

`Command line: LIMITS` (Enter)

`ON/OFF/<Lower-left corner> <0.00,0,00>:` (Enter) to accept default value.

`Upper-right corner <12.00,9.00>:`

`11,8.5` (Enter)

`Limits` can be reset at any time during the drawing session by repeating these steps.

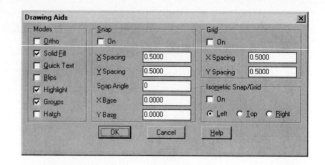

37.26 The `Drawing Aids` dialog box (`DDrmodes`) is used to make `Mode`, `Snap`, and `Grid` settings.

37.27 The `Status` bar beneath the `Command line` at the bottom of the screen displays the current settings. Double click on the `Drawing Aids` boxes to turn them on or off.

You can select the `Drawing Aids` box (`Menu bar> Tools> Drawing Aids`) or type `DDrmodes` to set `Mode`, `Snap`, and `Grid` (**Fig. 37.26**). By selecting the buttons or by filling in the blanks, you can activate these settings. Under `Modes`, the following settings can be made:

`Ortho`, which forces all lines to be either horizontal or vertical (not angular), can be set by picking the check box in **Fig. 37.26**, by pressing the `F6` key, by double clicking on `Ortho` in the `Status` bar (**Fig. 37.27**).

`Solid Fill` can be turned on or off to show areas drawn with the `Solid` command as filled or as outlined areas.

`Quick Text` is turned on to show text as rectangles to save regeneration time. Set to `off`, type `Regen`, and the text is restored as characters.

`Blips` are set `On` or `Off` by selecting `Blips` from the dialog box or by typing `Blipmode` and `On` or `Off`. Blips are temporary markers made on the screen when selections are made; they are removed by refreshing the screen (pressing F7).

Highlights shows selected objects with dashed lines to indicate which ones have been selected.

Groups can be turned on or off to select an object that is part of a group, which in turn selects the whole group of which the object is a part.

Hatch determines which objects will be hatched in associative hatching. When On, the hatch boundaries are selected also.

The SNAP panel of the Drawing Aids box is used to force the cursor to stop only at points on an imaginary grid of a specified spacing. The XBase and YBase options set the base position of the Snap grid. The x and y spacings can be set by typing the desired values in the text boxes. Snap can also be set by typing Snap at the Command line and typing the interval desired. When On, Snap is highlighted in the Status line at the bottom of the screen. Snap can be toggled on and off by pressing F9 or double clicking on its Status line icon. Select Aspect by typing A to assign x and y values if they are different. The Rotate option prompts for the angle of rotation of the Snap grid and its origin point, which is given by typing x and y values in the edit boxes, or by selecting Snap Angle from the Drawing Aids box.

Grid, in the Drawing Aids box, fills the Limits area with a dot spacing assigned by typing values in the x- and y-spacing boxes. By typing Grid and selecting the Snap option, the grid is set equal to the Snap spacing. By turning on Isometric/Snap Grid, the cursor and grid are set for drawing on the three planes of an isometric drawing—left, top, and right—as specified.

Running coordinates are displayed (by pressing F6) at the bottom of the screen in the status line to show the cursor's position.

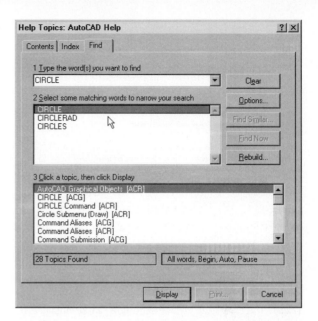

37.28 By selecting Help from the Menu bar and AutoCAD Help Topics, this screen with three tabs appears to help you find assistance with topics and commands. With experimentation, you will have a "manual" on the screen.

The Status Window at the bottom of the screen displays several of the settings discussed above as black if On or gray if Off (**Fig. 37.27**). Double clicking on these buttons toggles them off or on.

37.9 General Assistance Commands

Several examples of helpful commands that can be typed at the Command line are shown below.

Help (Menu bar> Help> AutoCAD Help Topics) gives menus Contents, Index, and Find to help you with all aspects of AutoCAD. Under Find, you will be able to insert keywords for commands and steps that you need assistance with, as shown in **Fig. 37.28**. When a topic such as Circle is selected, and Display is picked, a screen of instructions will appear to help you when using the Circle command (**Fig. 37.29**). Additional options are provided under Help that are self-explanatory as you experiment with them.

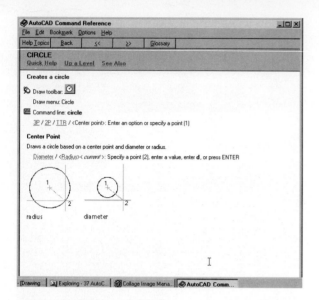

37.29 The `Command Reference` box can be obtained by `Menu bar> Help> AutoCAD Help Topics` to help you with a command. For example, type `CIRCLE` in all caps and select `Display` to get the screen shown here.

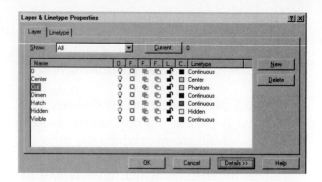

37.30 The `Layer & Linetype Properties` box (`Menu bar> Format Linetype` or type `DDlmodes`) lists the layers and their properties. The `Linetypes` tab lets you load line-types.

The `Purge` **command** (`Menu bar> File> Drawing Utilities> Purge> All`) can be used at any time to remove unused layers, blocks, and other attributes from files:

Command: <u>PURGE</u> (Enter)

Purge unused
Blocks/Dimstyles/LAyers/LTypes/
SHapes/ STyles/APpids/Mlinestyles/
All: <u>ALL</u> (Enter)

The `All` option is used to eliminate all unused references one at a time as prompted. The other options purge specific features of a drawing.

`List (Tools> Inquiry> List)` **enables** you to select any object drawn on the screen and obtain information about it—for example, the radius, circumference, and area of a circle plus the coordinates of its center point.

`Copy (Modify toolbar> Copy icon)` is used to select objects on the screen (single objects or groups of objects) with the cursor, pick a new position, and make a duplicate of the selection. The `Multiple (M)` option can be used to make more than a single copy.

`Unlock` is used to select `Files` to `Unlock`, then pick `OK` to execute.

37.10 Drawing Layers

An almost-infinite number of layers can be created, each assigned a name, color, and line type, on which to draw. For example, a yellow layer named `Hidden` with hidden linetypes assigned for drawing dashed lines may be created.

Architects use separate copies of the same floor plan for different applications: dimensions, floor finishes, electrical details, and so forth. The same basic plan is used for all of these applications by turning on the needed layers and turning others off.

Working with Layers The layers shown in the `Layer Control` **box** (`Menu bar> Format> Layer> Layer & Linetype Properties`) in **Fig. 37.30** are sufficient for most working drawings. Layers are assigned line types and different colors so they can be easily distin-

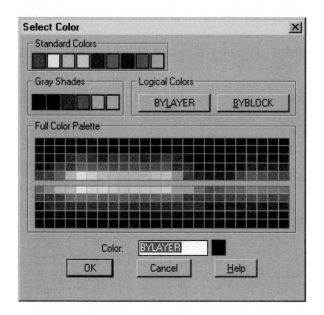

37.31 Select `Menu bar> Format> Color> Select Color` box or type `DDcolor` to get this box from which colors can be selected and assigned to layers.

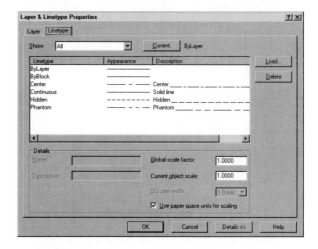

37.32 Select the `Linetype` tab of the `Layer & Linetype Properties` box, then press `Load` to select the available linetypes.

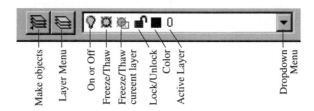

37.33 Selecting one of the icons shown on the `Layer Control` portion of the `Object Property` toolbar offers easy access to layers and their properties.

guished from each other. The 0 (zero) layer is the default layer, which can be turned off or frozen but not deleted.

Layers can be created by typing a layer name in the `Name` box (Fig. 37.30) and selecting `New`. The new layer will appear in the listing of layers with a default color of white and a continuous linetype, which can be changed to your specifications.

Color Pick a layer and select the `Color` button to assign it a color (Fig. 37.30). The `Select Color` menu appears on the screen from which a color can be chosen for that layer and pressing the `OK` button (**Fig. 37.31**).

Linetypes From the `Layer & Linetype Properties` menu (**Fig. 37.32**), select `Load` to obtain a list of linetypes to select from. Choose a linetype (`Hidden`, for example), OK, and it is loaded. To assign a linetype to a layer (a layer named Hidden, for example), pick this layer (**Fig. 37.30**), select the down arrow at the linetype box, pick the linetype from the list and pick `OK`.

Ltscale modifies the lengths of line segments of hidden lines and other noncontinuous lines.

A layer in the `Layer & Linetype Properties` box must be selected and made the current layer in order to draw on it with its assigned color and linetype. A layer is made current by selecting it, picking `Current`, and `OK` (**Fig. 37.30**). The current layer can be more easily picked by selecting it from the layer control portion of the `Object Properties` box (**Fig. 37.33**) at the top of the screen. Pick the down arrow and a listing of the layers will be shown (**Fig. 37.34**). Pick one to be the current layer and it's done.

This same portion of the Standard Toolbar can also be used to make other layer assignments: `On` and `Off`, `Freeze` and `Thaw`, and `Lock` and `Unlock`. By selecting the `Layers` icon, the `Layer & Linetype Properties`

37.10 DRAWING LAYERS • 657

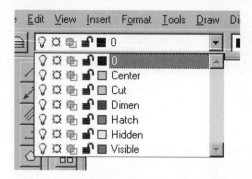

37.34 Select any point on the `Layer Control` toolbar where the name of the active layer's name appears, or select the down button, and this drop-down listing of the layers will appear.

box will be displayed from which the previously covered setting can be made. This box can also be obtained by typing `LA` at the command line.

Renaming a layer can be done from the `Layers & Linetypes Properties` box by dragging the cursor across the existing name and typing a new one in its place.

The `On` and `Off` settings of a layer are selected from the symbolic listing in the `Layer control` toolbar (Fig. 37.34). Layers can be turned on or off with buttons from the `Layer & Linetype Properties` box (Fig. 37.34), or by typing `LA` and `Off` at the command line. An `Off` layer that is selected as the current layer can be drawn on, but this is almost never done.

`Freeze` and `Thaw` options under the `Layer command` are used like the `On` and `Off` options (Fig. 37.34). `Freeze` a layer and it will (unlike an `Off` layer) be ignored by the computer until it has been `Thawed`, which makes regeneration faster than when `Off` is used.

37.11 Toolbars

Toolbars (`Menu bar> View> Toolbars`) are menus of icons representing commands that can be made active from the `Toolbar` box (**Fig. 37.35**). A section of the `Standard` toolbar (**Fig. 37.36**) gives a sequence of icons for commands from `New` to `Redo`. The remaining

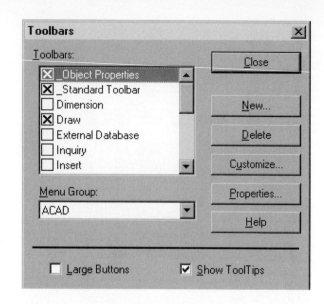

37.35 The commands, `Menu bar> View> Toolbars` can be used to obtain this listing of toolbars, which can be placed on or removed from the screen by checking a box.

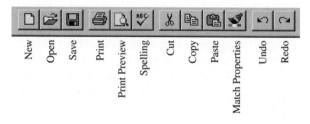

37.36 This section of a standard toolbar provides many routine operations that can be accessed with the cursor.

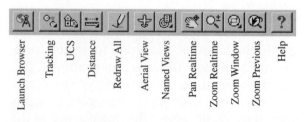

37.37 The remaining section of the standard toolbar has helpful commands needed for most drawings.

37.38 Toolbars can be arranged by dragging their corners to make single, double, or triple strips that are horizontal or vertical. They can be docked at the borders or left "floating" on the screen.

standard toolbar commands—from Launch Browser to Help—are given in **Fig. 37.37**.

Toolbars can be moved about the screen for convenience by selecting a point on one of their edges, holding the select button of the mouse down, and moving the cursor to a new position. Toolbars can be docked by moving them into contact with one of the perimeter borders of the screen (**Fig. 37.38**). They can be changed from single strips to double and triple blocks, by moving the corners of the toolbars, or into vertical or horizontal strips, depending on which border of the screen they are moved to. When located in the open area of the screen, toolbars will appear as floating menus. If you place the pointer on an icon, a flyout box will appear with its definition.

37.12 Creating a New Drawing

Setting Up a Title Block

To open a file for a new drawing, select Menu bar> File> New (**Fig. 37.39**). The AutoCAD box will ask if you wish to save the current drawing (**Fig. 37.40**). For a new session, pick No to obtain the Create New Drawing box, which offers four options: Use a Wizard, Use a Template, Start from Scratch, and Instructions. Pick Use a Wizard, followed

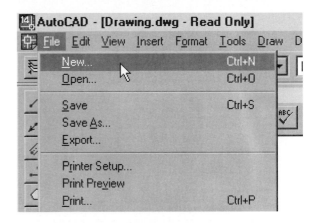

37.39 To open a new drawing, press Menu bar> File> New.

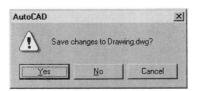

37.40 You will be asked if you want to save your current drawing before opening a new file.

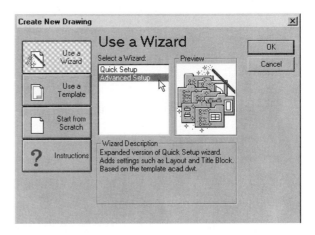

37.41 The Create New Drawing box offers three options for a new drawing. Select Use a Wizard> Advanced Setup > OK.

by Advanced Setup, and then pick OK (**Fig. 37.41**). The Advanced Setup box will appear with tabs with the following headings: Units, Angle, Angle measure, Angle direction, Area, Title block, and Layout. The Step 1: Units tab shown in **Fig. 37.42** lets you

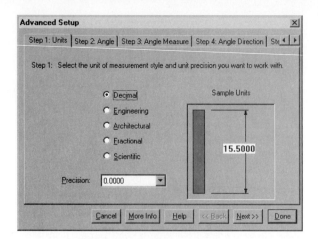

37.42 From the Step 1: Units tab, select Decimal units and set the Precision (0.00 for a number such as 2.00).

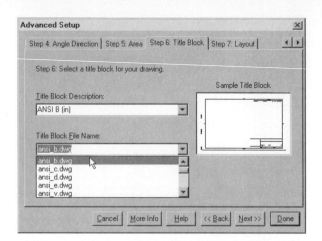

37.44 From Step 6: Title Block, select a title block (ANSI A.dwg in this example) and select the Next button.

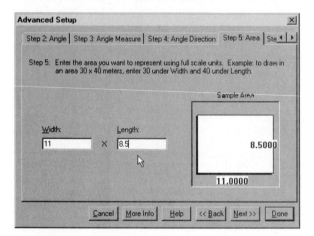

37.43 From the Step 2: Area tab, assign width and length to set up the drawing-area dimensions.

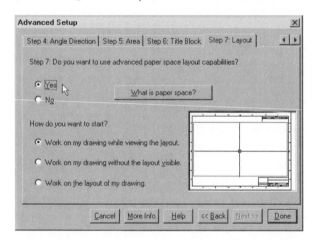

37.45 From Step 7: Layout, select No, since you do not need advanced layout capabilities for a simple, two-dimensional drawing, then press Done.

pick the units and precision (decimal units with 0.00 precision, for example). The tabs in Step 2 through Step 4 let angular units be specified. The Step 5: Area tab shown in **Fig. 37.43** prompts for the drawing area dimensions.

The Step 6: Title Block tab shown in **Fig. 37.44** gives you the option of selecting a title block from a listing in the Title Block File Name window. (In this example, ANSI A.dwg is picked.) After the sample image of the title block appears, press the Next button to go to the next tab.

Since you are working with a two-dimensional drawing, you do not need to use Step 7: Layout (**Fig. 37.45**), which offers the advanced paper space layout capabilities; select No, then press Done. The title block and border will appear on the screen (**Fig. 37.46**). The border and its contents is a Block, which is a collection of objects that are grouped as if they were a single object.

At this point, you have a defined area in which to do your drawing (**Fig. 37.47**). However, you may wish to remove all or parts of the title block within the outline border. To

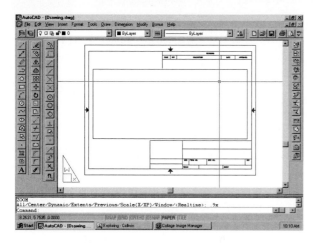

37.46 The title block appears in the graphics area of the screen ready for drawing.

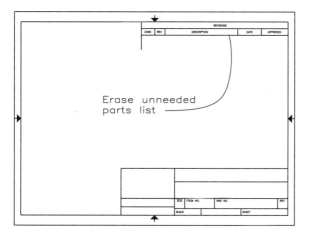

37.47 The title block inserted in the defined drawing area has a parts list and a title block that are larger than you will need for class assignments. You may wish to remove these.

do this, you must separate the `Block` into individual objects by typing `Explode` at the `Command line`, selecting any line on the title block, and pressing `Enter`. Now, the lines are individual objects, not a composite of objects. Type `Erase`, select the lines and text that you wish to remove, then press `Enter`. The title block and border will be modified to fit your needs (**Fig. 37.48**).

Section 37.68 covers the technique of customizing your own title block for classroom drawings that is more suitable than the one shown in the previous examples, but first, you

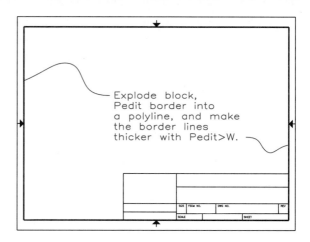

37.48 You may wish to `Explode` the title block into individual objects. `Erase` the parts list, convert the border into a polyline, and thicken the polyline with `Pedit`. (See Section 37.35.)

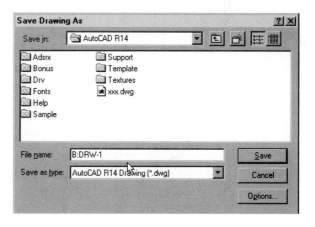

37.49 Name the file, `B:DRW-1`, select the drive to save it to, and save it by pressing `Menu bar> File> Save As`.

must learn more about the operation of AutoCAD. If this section seems a little advanced for you, leave it, and come back when you have learned a few more drawing principles.

Naming the File

Before making a drawing, you should name and save the file: press `File> Save As>` to get the `Save Drawing As` box shown in **Fig. 37.49**. Type the drive and the file name, `B:DRW-1`, in the `File Name` window and press `Save`. `B:DRW-1` becomes the current file. The graph-

37.12 CREATING A NEW DRAWING • 661

ics screen remains on ready for you to resume your drawing. As you progress with your drawing, type `Save` (`File> Save`) periodically and your current drawing, `B:DRW-1`, will be updated without having to specify its name.

In addition to making a drawing, you will need to set layers and linetypes, dimensioning variables, and, perhaps, many other variables. Once done, you may wish to use this file as a prototype file for similar drawings, which means you will not have to make these setting from scratch each time.

Making a Prototype File

To make a prototype file, save `B:DWR-1` (`File> Save As`) to a file with the name `B:PROTO1`. This will create a new file identical to `B:DWR-1` (linetype settings, layers, and all) and will make `B:PROTO1` the current file on the screen. Since this will be a prototype file with available settings and perhaps a title strip that you wish to keep as the start-up for the next drawing, erase the portion of the drawing that you do not wish to keep (maybe everything you see) and save it (`File> Save`) as `B:PROTO1`.

Now, when you start a new drawing (`File> Open`), select `B:PROTO1` and save it again (`File> Save As`) as `B:DRW-2` and press `OK` to adopt the settings of `B:PROTO1` while preserving `B:PROTO1`. You do not need to reset layers and other settings.

Making a Template File

A template file is the same as a `B:PROTO1` (with a `.DWG` extension) except when saving (`File> Save As`), the drive letter (B:) is omitted, the `Type of File` is specified as `Drawing Template File (*.DWT)`, and it is stored in the `Template directory`. The next time you create a new drawing, press the `Use a Template` button in the `Use a Wizard` box, and pick the `PROTO1.DWT`, which will be displayed in the list of template files.

37.13 Drawing Scale

It is best (and easiest) to work with a drawing at a full-size scale where 1 inch is equal to 1 inch. The previous examples of title blocks and template files were given and inserted as full-size layouts, which permits text size and all measurements to be easily handled.

Half-size drawings can be made by press the sequence (`Menu bar> Insert> Block> File> B:PROTO1> Open> Specify parameters or screen> Ok> Insertion point> 0,0> X- and Y-scale factor> 2` (Enter)`> Rotation angle:> 0` (Enter). The important aspect of these commands is the scale that was specified for the prototype file (and all its settings). It was set to a factor of 2, which is double size. (The factor would be 25.4 × 2 = 50.8 for millimeters.) When a full-size drawing is made within a double-size border, the drawing specifications and details can be handled conveniently. At plot time, the double-size border must be reduced to half size (0.5) so it will fit on its sheet, and the drawing within it reduced to half size.

Double-size drawings can be made by the same steps as half-size drawings except for the insertion of the prototype file with `X- and Y-scale factors` of `0.5` (half size). (The factor would be 25.4 × 0.5 = 12.7 for millimeters.) At plot time, the half-size border must be enlarged by a factor of 2 so it will fill the sheet, and the full-size drawing within it will be doubled.

Using this logic, other combinations of scale factors can be determined for drawings of any scale. The most important point to remember is that you are better off working with full-size drawings and scaling at plot time.

37.14 Saving and Exiting

The pull-down `File menu` (`Menu bar`) gives options for saving a drawing—`Save`, `Save As`, and `Exit` AutoCAD—which can be selected from the pull-down menu or the toolbar, or by typing at the command line.

Saving

Type the `Save` command to "quick save" to the current file's name if the file has been previously named and saved. If the file is unnamed, the `Save` command prompts for a file name by displaying the `Save Drawing As` dialog box in Fig. 37.49. Select the directory and name the file (`B:NEW-2`) to save it on the disk in drive B:. The new drawing, `B:NEW-2`, becomes the current file on the screen.

Ending the Session

To exit the drawing session and close AutoCAD, press `Menu bar> File> Exit`. If you have not saved the file immediately before selecting `Exit`, the dialog box shown in Fig. 37.40 asks if you want to save the updated drawing. Select `Yes` and the `Save Drawing As` menu box will appear for assigning the drive, directory, file name, and file type. To exit without saving, respond to `Save Changes?` with `No` and the latest changes made since the last `Save` will be discarded.

`End`, typed at the `Command line`, saves the current drawing and exits AutoCAD, whether or not any changes have been made. If the drawing has not been previously named, the `Create Drawing File` dialog box will prompt you for a file name. The previous version of the drawing is automatically saved as a backup file with a `.BAK` extension and the current drawing is saved with a `.DWG` extension.

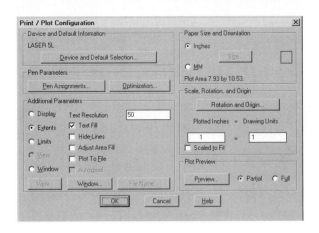

37.50 The printing options are given in the `Print/Plot Configuration` box (`Menu bar> File> Print`). Select the `Device and Default Selection` button.

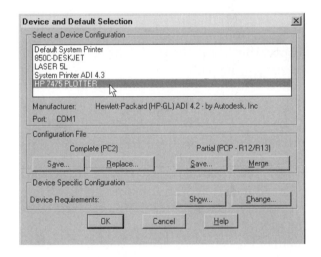

37.51 Select the printer or plotter that you wish to send your drawing to for making a paper copy.

37.15 Plotting Parameters

To plot a drawing before ending a drawing session, select `File> Print` to obtain the `Print/Plot Configuration` dialog box (**Fig. 37.50**). Select the `Device and Default Selection` button to obtain the subdialog box (**Fig. 37.51**) for selecting `HP 7475 plotter` to make a pen drawing.

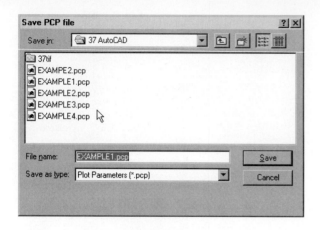

37.52 A `PCP` (Plot Configuration Parameters) file can be saved for repetitive usage for selection from `Merge Get Default from File` (Fig. 37.51).

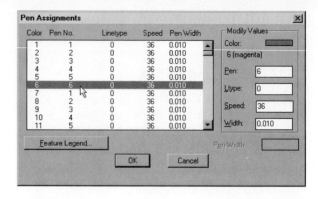

37.53 Select the `Pen Assignments` option from the `Print` menu, which shows color, pen number, linetypes, speed, and pen width.

Select the `Save` under `Partial` button to display the `Save PCP file` subdialog box for creating a file with a `.PCP` extension called a `Plot Configuration Parameters` file (**Fig. 37.52**). Name this file `B:DRW-1` (it is given a `.PCP` extension automatically) since it will be used in conjunction with the file `B:DRW-1.DWG`.

`Device Requirements> Show` is selected in Fig. 37.51 to obtain the dialog box for specifying how long the plotter will wait for plotter port time-outs and other requirements. Press `OK` to leave this box. Select `Device Requirements> Change` (Fig. 37.51) to change the plotting requirements, then `OK`. The `Pen Assignments` box shows the assignment of colors to pen numbers (slots in pen holder on the plotter) (**Fig. 37.53**). For example, a red line (`Visible` layer) will be plotted with the pen in slot 1, a P.7 pen.

`Linetype` is set to `0` (zero) since the layers have linetypes assigned to them. `Pen` assigns pens to the colors on the screen. `Speed` designates the rate at which the plotter pen moves. A speed of 9 in. per second is average and 36 is the maximum. `Pen Width` gives the line widths based on pen tip widths for drawing multistroke lines (polylines or traces). Use the default width of 0.01 for plotters. `Pen widths` are assigned when laser printers are used and this option is grayed out if not supported by a printer.

`Feature Legend`, a subdialog menu of the `Pen Assignments` box, shows linetypes for selection. Do not use this option when layers have been assigned linetypes as part of the drawing file.

`Optimization` of the `Print/Plot Configuration` menu in Fig. 37.50 shows a dialog box for reducing pen motion (**Fig. 37.54**). When a box is selected, those above it become active except for the `No` optimization box. The last two boxes pertain to three-dimensional figures and the elimination of multiple strokes where lines coincide.

The `Additional Parameters` box of the `Plot Configuration` menu (Fig. 37.54) offers the following settings:

- `Display` plots the drawing as shown on the screen. `Extents` plots a drawing to its extents of its boundary if the scale selected

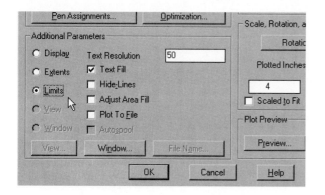

37.54 The `Additional Parameters` area is used to specify the area to be plotted, `Text Fill`, `Hide Lines` (when making a three-dimensional plot), and other options.

permits. It is good practice to apply `Zoom> Extents` to ready a drawing for plotting. `Limits` is used to plot the portion of the drawing lying within the grid pattern as defined by typing `Limits`.

- `View` lists saved `Views` that can be selected and plotted. This box is gray if no views are saved. `Window...` specifies the portion of a drawing to be plotted when `Pick` is chosen and the window is sized with the cursor, or coordinates type from the keyboard (**Fig. 37.55**).

- `Hide Lines` removes hidden lines from three-dimensional drawings that are plotted. `Adjust Area Fill` pulls in the boundaries of the filled area by one-half pen width for a more precisely drawn fill area.

- `Plot to File` sends the plot to a file with a `.PLT` extension for plotting with a utility program instead of to the plotter.

- `Paper Size and Orientation` is used to specify `Inches` or `MM` (millimeters) as units for a plot (**Fig. 37.56**). A rectangle in this area denotes whether the page has a portrait (vertical) or landscape (horizontal) orientation. `Paper Size` lists standard and user-specified plot sizes (**Fig. 37.57**).

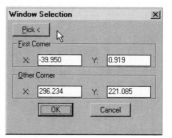

37.55 When `Window` is selected from the `Additional Parameters` box, the coordinates of the diagonal corners of the window to be plotted can be picked by the cursor or specified by numbers.

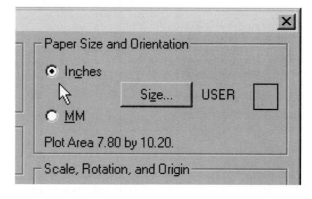

37.56 From the `Paper Size and Orientation` area, select `Inches` or `MM` as the units and select `Size` (size A here) to obtain a list of plot sizes.

Size	Width	Height
A	10.50	8.00
A4	11.20	7.80
MAX	15.64	9.96
USER	7.80	10.20

Size	Width	Height
USER:	7.80	10.20
USER1:		
USER2:		
USER3:		
USER4:		

Orientation is landscape

37.57 The `Paper Size` box gives a listing of the plot sizes that are available. User sizes for plots can be specified in the boxes at the right.

`MAX` is the largest size that the selected plotter can handle. When a size is selected and `OK` is picked, the dialog box disappears, and the size is shown in the `Plot Area` (Fig. 37.56).

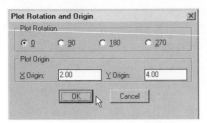

37.58 Select Plot Rotation and Origin to get this box for specifying rotation of the plot and its X and Y offsets.

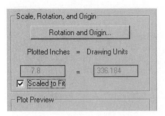

37.59 Type the desired scale (1 = 2, or half size, for example). If Scaled to Fit is checked (as in this case), the drawing will be enlarged to fill the sheet if the placement of the origin permits.

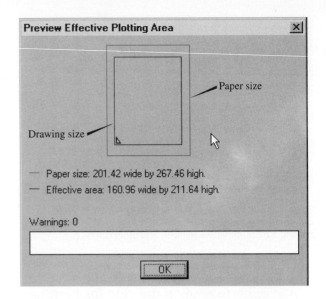

37.60 Select Partial Preview to get two rectangles on the screen that represent the paper size and the drawing size.

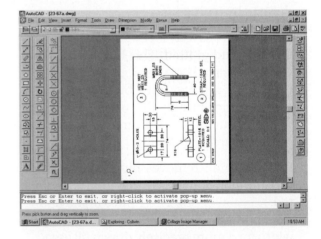

37.61 Full Preview will specify rotation of the plot and its X and Y offsets.

- Plot Rotation and Origin offers the plot-orientation options of rotation and origin (**Fig. 37.58**). Plot Rotation gives radio buttons for rotating plots 0, 90, 180, or 270 degrees. Plot Origin is usually specified as 0,0 (the lower-left corner of the sheet), but it can be offset by X and Y values.

- Scale is typed in the edit boxes, Plotted Inches = Drawing Units (**Fig. 37.59**). If metric units had been selected, the label would read Plotted MM with 1 = 1, being full size, 1 = 2 half size, and 2 = 1 double size. Scaled to Fit makes the drawing's extents fill the plotting area to as large a size as possible.

- Plot Preview (Fig. 37.50) gives partial and full previews. Partial Preview shows rectangles representing paper size and the drawing area (**Fig. 37.60**). The part of the drawing area that exceeds the paper size cannot be plotted unless the scale, origin, or both are adjusted. The triangular rotation icon is shown in the lower left-corner for zero rotation, upper left for 90° of rotation, upper right for 180° of rotation, and lower right for 270° of rotation.

- Full Preview (Fig. 37.50) shows the entire drawing on the screen and its relationship to the paper limits when plotted (**Fig. 37.61**). Press Enter to Pan and Zoom about the preview drawing. Press Enter to return to the Print/Plot Configuration dialog box.

To update the .PCP file, from the Print/Plot Configuration box> Device and Default Selection>, select Save in the Configuration File panel to get the dia-

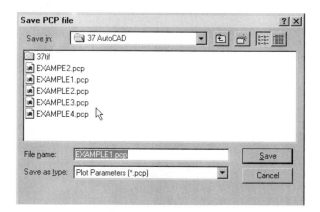

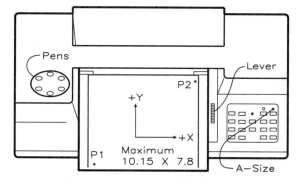

37.62 To keep the previous setting for future use, save the file as a .PCP file by using this dialog box: `File> Print> Print/Plot Configuration> Device and Default Selection> Save`.

1. Load paper, lower lever
2. Pens: P.7 (Slot 1); P.3 (Slot 2)
3. Set for A—Size
4. Press P1 and P2

37.63 The Hewlett-Packard 7575A plotter is often the plotter of choice for sizes A and B plots with pens.

log box shown in **Fig. 37.62**. These settings can be recalled from the `Get Defaults From Merge` in order to make a plot with the same settings.

37.16 Readying the Plotter

After the parameters are set and the preview is approved, press `OK`. The `Command` line will give the following messages at the bottom of the screen:

>   Effective plotting area: XX wide
>
>   by YY high (**Enter**)
>
>   Position paper in plotter. (**Enter**)
>
>   Press RETURN to continue or S to Stop
>
>   for hardware setup

Load paper in the plotter as shown in **Fig. 37.63** with the thick pen (P.7) in slot 1 and the thin pen (P.3) in slot 2, as specified by `Pen Assignments` in Fig. 37.53. Press `Enter` and the plot will begin. Plotting can be cancelled by pressing `Esc`, but it may take almost a minute for the command to take effect. When completed, select `File> Exit` to end the session.

Now that you know how to set a few drawing aids, save files, and plot, it is time to learn how to make drawings.

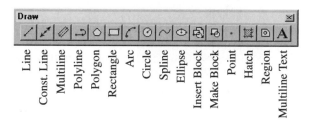

37.64 The `Draw` toolbar makes it easy for you to select commands with the cursor. Select `Menu bar> View> Toolbars> Draw` to obtain the above commands.

37.17 Two-Dimensional Lines (Draw Toolbar)

Open your prototype drawing, `Menu bar> File> Open> B:PROTO1` and use `Save As` to name the drawing <u>B:NO1</u>, which becomes the current drawing with the same settings of `B:PROTO`. Load the `Draw Toolbar` (`Menu bar> Toolbars> Draw`) to obtain the commands shown in **Fig. 37.64**.

A `Line` (called an object) can be drawn by using the command, `Draw toolbar`, or the `Menu bar` (`Menu bar> Draw> Line`). From the keyboard, type `Line` or `L` at the `Command Line`, and respond to the prompts, as shown in **Fig. 37.65**, to draw lines by picking endpoints with the left button of the mouse. The current line will rubber band from the last point, and you can draw lines in succession

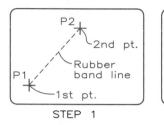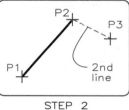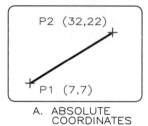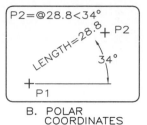

STEP 1	STEP 2

37.65 Drawing a LINE.

Step 1 Select Menu bar> Draw> Line. Specify first point <u>P1</u> with left mouse button and second point <u>P2</u>.

Step 2 Select third point <u>P3</u> and press right mouse button to leave command and rubber band.

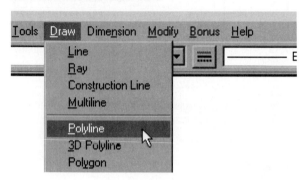

37.66 Various types of lines can be selected from the Draw drop-down menu or from the Menu toolbar.

until Enter or the right button on your mouse is pressed.

The Draw toolbar can be used to select the Line icon to get three flyout icons for types of lines: Lines and Construction Lines (Fig. 37.64). These types of lines can also be selected from Menu bar> Draw (**Fig. 37.66**). A Construction line is drawn across the entire screen, and a Ray is drawn from the selected point to the edge of the screen.

A comparison of absolute and polar coordinates is shown in **Fig. 37.67**. **Delta coordinates** can be typed as @2,4 to specify the end of a line 2 units in the x direction and 4 units in the y direction from the current end.

Polar coordinates are two-dimensional coordinates and are typed as @3.6<56 to draw a 3.6 long line from the current (and active) end of a line at an angle of 56° with the x axis.

A. ABSOLUTE COORDINATES	B. POLAR COORDINATES

37.67 Lines by coordinates.

A Absolute coordinates can be typed at the Command line to establish the ends of a line.

B Polar coordinates are relative to the current point and are specified with a length and the angle measured counter clockwise from the horizontal.

37.68 The Circle command (Menu bar> Draw> Circle) has a flyout menu with various options for drawing circles.

Last coordinates are found by typing @ while in the Line command. This causes the cursor to move to the last point.

World coordinates locate points in the World Coordinate System regardless of the User Coordinate System being used by preceding the coordinates with an asterisk (*). Examples are *4,3; *90<44; and @*1,3.

The Status line at the bottom of the screen shows the length of the line and its angle from the last point as it is rubber-banded from point to point. The Close command will close a continuous series of lines from the last to the first point selected.

37.18 Circles (Draw Toolbar)

The Circle command (Menu bar> Draw> Circle) draws circles when you select a center and radius, a center and diameter, or three points (**Fig. 37.68**). Typing C at the Command

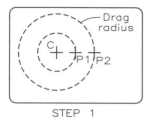

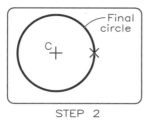

STEP 1   STEP 2

37.69 Drawing a circle

Step 1 Select `Menu bar> Draw> Circle> Center, Radius`, then select center C and drag radius to <u>P1</u> and <u>P2</u>; the circle is enlarged.

Step 2 Select the final end of the radius to complete the final circle, then click the left mouse button.

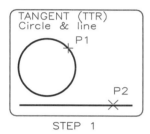

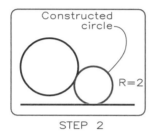

STEP 1   STEP 2

37.70 Drawing a circle tangent to two objects (TTR).

Step 1 Select `Menu bar> Draw> Circle> Tan, Tan, Radius`, then select point <u>P1</u> on the circle and point <u>P2</u> on the line.

Step 2 When prompted, type the length of the radius at the `Command line` (Enter). The circle will be drawn tangent to the line and circle.

`line` may be the fastest means of activating the `Circle` command, but the circle icon on the `Draw Toolbar` (Fig. 37.64) can also be selected. **Figure 37.69** illustrates how to draw a circle.

The `Tan, Tan, Radius` (tangent, tangent, radius) option of **Circle** draws a circle tangent to a circle and a line, three lines, three circles, or two lines and a circle. A circle is drawn tangent to a line and a circle by selecting circle and line and giving the radius (**Fig. 37.70**). The `Tan, Tan, Tan` option draws a circle tangent to three lines (**Fig. 37.71**).

37.19 Arcs (Draw Toolbar)

The `Arc` command (`Menu bar> Draw> Arc> options`) (**Fig. 37.72**) or the `Arc` icon in the `Draw toolbar` (Fig. 37.64) has eleven combinations of variables that use abbreviations for starting point, center, angle, ending point, length of

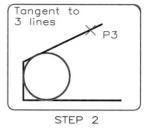

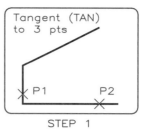

STEP 1   STEP 2

37.71 Drawing a circle, tangent to three lines (TTT).

Step 1 Select `Menu bar> Draw> Circle> Tan, Tan, Tan`, then select first line <u>P1</u>, select second line <u>P2</u>.

Step 2 Select the third line, <u>P3</u>; the length of the arcs will be calculated automatically and the arc will be drawn tangent to the three lines.

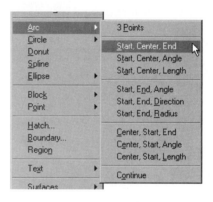

37.72 To draw an arc, select `Menu bar> Draw> Arc`. When using `Start, Center, End`, these elements must be specified in this order on the screen. This selection can also be made from the `Draw toolbar`.

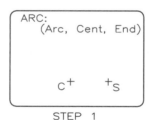

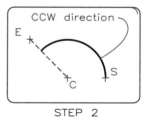

STEP 1   STEP 2

37.73 Drawing an arc: `Start, Center, End` (SCE) option.

Step 1 Select `Menu bar> Draw> Arc> Start, Center, End`, then select start point <u>S</u>, and the center <u>C</u>.

Step 2 With the cursor, drag the arc, select the final end of the radial line <u>CE</u>, and click the left mouse button. The arc is drawn.

chord, and radius. The `Start, Center, End` portion requires that you locate the starting point S, the center C, and the ending point E (**Fig. 37.73**). The arc begins at point S and is drawn counterclockwise by default to a point near E.

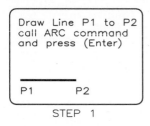
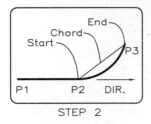

Draw Line P1 to P2 call ARC command and press (Enter)

P1   P2

STEP 1

End—
Chord—
Start—
                    P3

P1   P2   DIR.

STEP 2

37.74 Drawing an arc tangent to the end of a line.

Step 1 Select `Menu bar> Draw> Line>`, select ends of the line (P1 and P2) and press (Enter) or the right mouse button.

Step 2 Select `Arc` from the `Draw toolbar`, (Enter), select the endpoint of the arc (P3) with the cursor. An arc will be drawn from P2 tangent to line P1–P2.

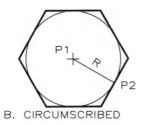

A. INSCRIBED   B. CIRCUMSCRIBED

37.75 Drawing inscribed and circumscribed polygons.

A Select `Menu bar> Draw> Polygon> Number of sides: 5` (Enter) `Edge/ <Center of polygon>: P1 Inscribed in circle/Circumscribed about circle (I/C)>: I` (Enter) `Radius of circle,` type `R-value` or draw with cursor (Enter).

B Select `Menu bar> Draw Polygon> Number of sides: 6` (Enter) `Edge/ <Center of polygon>: P1 Inscribed in circle/Circumscribed about circle (I/C)>: C` (Enter) `Radius of circle,` type `R-value` or draw with cursor (Enter).

A line can be continued with an arc drawn from the last point of a line and tangent to it, as shown in **Fig. 37.74** for drawing runouts of fillets and rounds. It can be used to draw a tangent line from an arc by applying the commands in reverse and dragging the line to its final length.

37.20 Polygon (Draw Toolbar)

An equilateral polygon can be drawn as shown in **Fig. 37.75** with `Menu bar> Draw> Polygon`, the **Polygon** icon can be selected from the **Draw toolbar** (**Fig. 37.76**) and the prompts responded to in the same manner. A

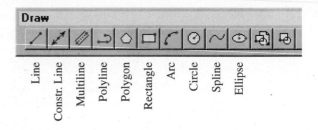

Draw

Line Constr. Line Multiline Polyline Polygon Rectangle Arc Circle Spline Ellipse

37.76 Select the `Polygon` icon and respond to the prompts in the same manner as described in Fig. 37.75.

polygon can be drawn by typing the command line as follows:

`Command:` `POLYGON` **(Enter)**

`Number of sides:` `5` **(Enter)**

`Edge/<Center of polygon>:` **(Locate center.)**

`Inscribed in circle\Circumscribed about circle (I/C):` `I` **(Enter)**

`Radius of circle:` **(Type length and** `(Enter)` **or select length with cursor.)**

When the `Edge` option is used, the next prompt asks:

`First endpoint of edge:` **(Select point.)**

`Second endpoint of edge:` **(Select point.)**

A polygon is drawn in a counterclockwise direction about the center point. Polygons can have a maximum of 1024 sides.

37.21 Ellipse (Draw Toolbar)

The `Ellipse` command in the `Draw Menu` gives icons for three methods drawing ellipses (**Fig. 37.77**). An ellipse is drawn by selecting the endpoints of the major axis and a third point, P3, to give the length of the minor radius (**Fig. 37.78**).

In **Fig. 37.79**, points are picked at the center of the ellipse at P1, one axis endpoint at P2, and the second axis length from P1 to P3. The ellipse is drawn through points P2 and the endpoint of the minor diameter specified by P3. The end-

37.77 Select `Menu bar> Draw> Ellipse> Axis, End` to draw ellipses.

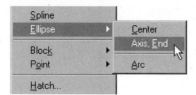

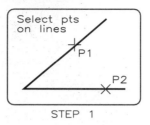

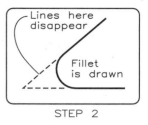

37.80 Rounding corners with the `Fillet` command.

Step 1 From the `Menu bar` select `Modify> Fillet> Polyline/Radius/Trim <Select first object>:` `R` (Enter) `Enter Fillet radius:` Type `R value` (Enter) (Enter) Select first object: `P1` Select second object: `P2`

Step 2 The fillet arc is drawn tangent to the two lines.

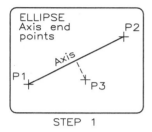

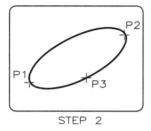

37.78 Ellipse: Endpoint option.

Step 1 Select `Menu bar> Draw> Ellipse> Axis, End` and axis ends `P1` and `P2`; drag to `P3`.

Step 2 Press (Enter) and the ellipse is drawn through P1, P2, and P3.

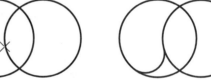

A. OBJECT SELECTIONS   RESULTING FILLETS

B. OBJECT SELECTIONS   RESULTING FILLETS

37.81 Tangent arcs: Fillets.

Step 1 Select `Menu bar> Modify>: Fillet>` type `R` for radius, (Enter), `Enter fillet radius`, type `1.2` (Enter).

Step 2 (Enter) `Select first object`, pick `P1`, `Select second object`, pick `P2` and the tangent arc is drawn.

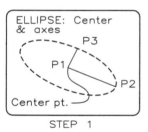

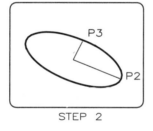

37.79 Ellipse: Center option.

Step 1 Select `Menu bar> Draw> Ellipse> Center` and pick center `P1` and axis end `P2`.

Step 2 Drag minor diameter to `P3` and select with left mouse button. The ellipse is drawn.

points of the axis can be located and the rotation angle specified. An angle of 0° gives a full circle, and an angle of 90° gives an edge.

37.22 Fillet (Modify Toolbar)

The corners of two lines can be rounded with the `Fillet` command (`Menu bar> Modify> Fillet`), whether or not they intersect. When the fillet is drawn, the lines are either trimmed or extended, as shown in **Fig. 37.80**. The assigned radius is remembered until it is

changed. By setting the radius to 0, lines will be extended to a perfect intersection. Fillets of a specified radius can be drawn tangent to circles or arcs, as shown in **Figs. 37.81** and **37.82**.

37.23 Chamfer (Modify Toolbar)

The `Chamfer` command (`Menu bar> Modify> Chamfer`) draws angular bevels at intersections of lines or polylines. After assigning chamfer distances (**D**), select two lines. They can be trimmed or extended, and the chamfer is drawn (**Fig. 37.83**). Press `Enter` to repeat this command using the previous settings.

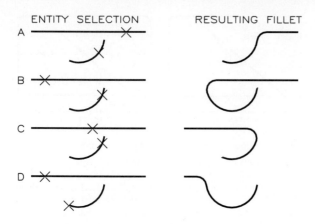

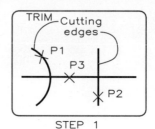

STEP 1   STEP 2

37.84 Trim: Cutting edges.

Step 1 `Menu bar> Modify> Trim>` (Enter) `Select cutting edge: (Projmode=UCS, Edgemode= Extend) Select objects`, pick P1 and P2 (Enter).

Step 2 `Select object to trim`, pick P3 and the line between the cutting edges is removed.

37.82 Each fillet that is applied to a line and an arc is determined by the location of the point selected.

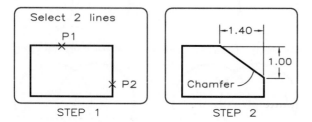

STEP 1   STEP 2

37.83 Chamfering a corner.

Step 1 `Menu bar> Modify> Chamfer> Distance>` `Enter first chamfer distance`, type 1.4, (Enter). `Enter second chamfer distance`, type 1.0 (Enter). (Enter) `Select first line`, pick P1.

Step 2 `Select first object`, pick P2. The chamfer is drawn and the corner is trimmed.

37.24 Trim (Modify Toolbar)

The `Trim` command (`Menu bar> Modify> Trim`) selects cutting edges for trimming selected lines, arcs, or circles that cross the edges at their crossing points (**Fig. 37.84**). The `Crossing` option of `Trim` is used to select four cutting edges in **Fig. 37.85** and the lines are trimmed.

37.25 Extend (Modify Toolbar)

The `Extend` command lengthens lines, plines, and arcs to intersect a selected boundary (**Fig. 37.86**). You are prompted to select the boundary object and the object to be

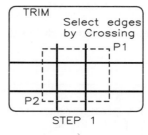

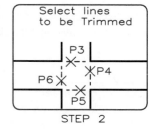

STEP 1   STEP 2

37.85 Trim: Crossing window.

Step 1 `Menu bar> Modify> Trim>` (Enter) `Select cutting edge: (Projmode=UCS, Edgemode=Extend) Select objects`, pick P1, hold down select button, drag to right to P2, making a crossing window to select the four lines.

Step 2 (Enter) `Select object to trim`, pick P3, P4, P5, P6. The four lines are trimmed.

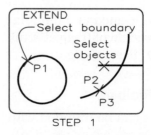

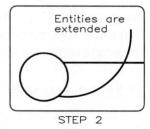

STEP 1   STEP 2

37.86 `Extend` command.

Step 1 Select from `Menu bar> Modify> Extend>` `Select boundary edge`, select circle with P1. (Enter)

Step 2 `Select object to extend`, pick line and arc with P2 and P3. The objects are extended to the circle.

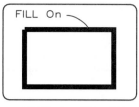

STEP 1: FILL ON   STEP 2: FILL OFF

37.87 Trace command.

Step 1 Select from Command line> type Fill> On (Enter). Command line> type Trace (Enter), Trace width: Type .2 (Enter), From point, select point with cursor, To point, continue selecting points to obtain filled lines.

Step 2 To obtain a Trace with unfilled lines, Command line> type Fill> Off (Enter) Regen (Enter) and draw the trace as in A above.

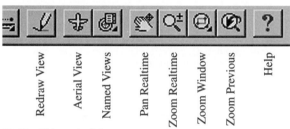

37.88 This part of the Standard toolbar contains many options for Panning and Zooming on the screen.

extended, which will extend the object to the boundary. More than one object can be extended. Extend will not work on "closed" polylines such as a polygon.

37.26 Trace (Command Line)

Wide lines made with multiple strokes can be drawn with the Trace command. At the Command line, type Fill and set to On. The Trace will be drawn as shown in **Fig. 37.87A**. When FILL is Off, the lines will be drawn as parallel lines with "mitered" angles (**Fig. 37.87B**).

37.27 Zoom and Pan (Standard Toolbar)

Parts of a drawing can be enlarged or reduced by the Zoom command (View>Standard toolbar> Zoom> options for Zooming), as shown in **Fig. 37.88**. An example of using a

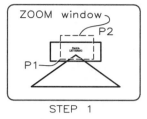

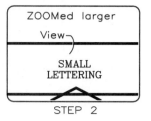

STEP 1   STEP 2

37.89 Select from Zoom command.

Step 1 Standard toolbar> Zoom realtime icon> right-click to activate pop-up menu> Right-click mouse; from menu, select Zoom Window. Hold down select button and place a box around the area to be zoomed (P1 and P2).

Step 2 The area in the window box is expanded to fill as much of the screen as possible.

Zoom window to enlarge part of a drawing is shown in **Fig. 37.89**.

Zoom (Z, when typing) has the following options:

All expands the drawing's Limits (dot pattern) to fill the screen.

Center is picked to specify the center of the zoomed image and specify its magnification or reduction.

Dynamic lets you Zoom and Pan by selecting points with the cursor.

Extents enlarges the drawing to its maximum size on the screen.

Previous displays the last zoomed view.

Scale X/XP magnifies a drawing relative to paper space. By typing 1/4XP, or .25XP, the drawing will be scaled so that 1/4 in. equals 1 in.

Window lets you pick the diagonal corners of a window to fill the screen.

Realtime (the default) lets you drag to the left to make a crossing window or to the right for the window to specify the area to be enlarged.

The Pan (P) command (**Fig. 37.90**) is used to pan the view across the screen by selecting two points. The drawing is not relocated, as in the Move command, but your viewpoint of it is changed.

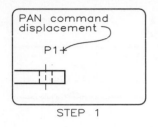
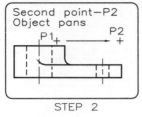

STEP 1   STEP 2

37.90   Pan command.

Step 1 Select `Standard toolbar> Pan icon>` right-click to activate pop-up menu> Right-click mouse; from menu select Pan. Select P1 and hold down select button.

Step 2 Drag to P2 and the drawing move to this point.

37.28 Selecting Objects

A recurring prompt, `Select objects:`, asks you to select an object or objects that are to be `Erased`, `Changed`, or modified in some way. Type `Select` when in a current command that requires a select (`Move`, for example) and the options will be displayed: `Window`, `Last`, `Crossing`, `Box`, `All Fence`, `Wpolygon`, `Cpolygon`, `Group`, `Add`, `Remove`, `Outline`, `Previous`, `Undo`, `Auto`, `Single`.

Figure 37.91 shows ways of selecting objects for applicable commands such as `Copy` to `Select objects`. The application of the options of `Single` and `Multiple`, `Window`, `Crossing`, and `Box` are shown in parts A–D.

`Window` lets you pick to diagonal corners of a box in which to select objects.

`Last` is used to pick the most recently drawn object.

`Crossing` is a window that selected any object it surrounds or crosses.

`Box` lets you make a window by selecting a point and dragging to the right. A crossing window can be made by dragging to the left.

`All` selects everything on the screen.

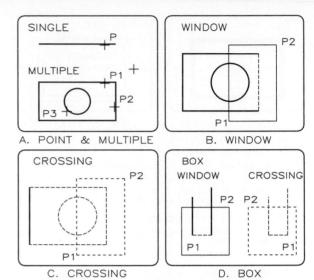

A. POINT & MULTIPLE   B. WINDOW

C. CROSSING   D. BOX

37.91   Selection options.

A When prompted, `Select objects:`, you can select objects one at a time or in multiples.

B A `Window` (W) can be used to select objects lying completely within the window.

C A `Crossing Window` (C) can be used to select objects within or crossed by the window.

D The `Box` option is used to make a `Window` or a `Crossing Window` determined by the direction the box is dragged.

`Fence` selects corner points of a polyline that will select any object it crosses, which is the same as a crossing window.

`WPolygon` (WP) forms a solid-line polygon that has the same effect as a window (**Fig. 37.92**).

`CPolygon` (CP) forms a dotted-line polygon that has the same effect as a crossing window (Fig. 37.92).

`Group` lets you recall a previously defined group of objects that were selected by the `Group` command.

`Remove` and `Add` are used while selecting objects to remove one by typing `R` (remove) or to add one by typing `A` (add). When finished, press `Enter` to end the `Select/remove` prompt.

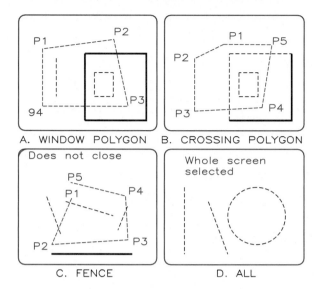

A. WINDOW POLYGON B. CROSSING POLYGON

C. FENCE D. ALL

37.92 Selection of objects.

A `Window Polygon` (`WP`) selects objects inside it.

B `Crossing Polygon` (`CP`) selects objects inside it and crossed by it.

C `Fence` (`F`), a nonclosing polyline, selects objects that it crosses.

D `All` selects all objects on the screen.

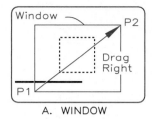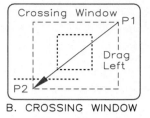

A. WINDOW B. CROSSING WINDOW

37.93 Selection of windows.

A A `Window` is formed by holding down the select button while dragging a diagonal corner to the right.

B A `Crossing Window` is formed in the same manner, but dragged to the left.

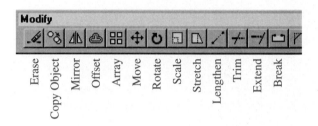

37.94 The `Modify Toolbar`, a portion of which is shown here, has options for making changes in a drawing.

`Multiple` selects multiple points without highlighting to speed up the selection process.

`Previous` recalls the previously selected set of objects. For example, enter `Move`, and type `P`, and the last selected objects are recalled.

`Undo` removes objects in reverse order one at a time by typing `U` (undo) repetitively.

`Auto` selects an object by pointing to it, and pointing to a blank area selects the first corner of a box defined by the `Box` option.

`Single` (`SI`) causes the program to act on the object or sets of objects without pausing for a response.

A `Window` or a `Crossing Window` can be obtained automatically by pressing the select button, holding it down, and selecting the diagonal of a window. By dragging it to the right, a window is obtained; by dragging it to the left, a crossing window is obtained (**Fig. 37.93**).

37.29 Erase and Break (Modify Toolbar)

The `Erase` command (`Modify toolbar> Erase icon`) (**Fig. 37.94**) deletes specified parts of a drawing. The selection techniques described previously can be used to select objects to be erased, as shown in **Figs. 37.95** and **37.96**. The default of the `Erase` command, `Select Objects`, allows you to pick one or more objects and delete them by pressing `Enter`. Type `Oops` to restore the last erasure, but only the last one.

The `Break` command (`Modify toolbar> Break icon`) removes part of a line, pline, arc, or circle (**Fig. 37.97**). To specify a break at an intersection with another line as shown in **Fig. 37.98**, select the line to be broken and select the `F` option. The endpoints

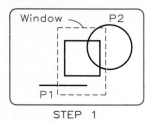
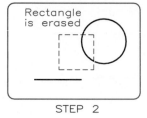

37.95 Erase: Window option.

Step 1 `Modify toolbar> Erase icon> Select objects`, pick P1, `Other corner`, pick P2.

Step 2 `Select objects`, (Enter) and the box completely within the erasing window is removed.

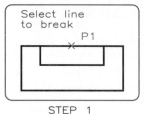
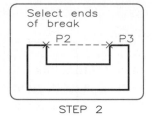

37.98 `Break: F` option.

Step 1 `Modify toolbar> Break icon> Select objects>` or F `for first point`, pick P1 on the line.

Step 2 `Enter first point`, pick P2, `Enter second point`, pick P3. Line P2–P3 is removed.

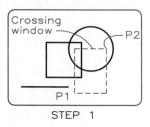
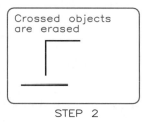

37.96 Erase: Crossing window option.

Step 1 `Modify toolbar> Erase icon> Select objects`, pick P1, `Other corner`, pick P2 (right to left for a crossing window).

Step 2 `Select objects`, (Enter), and the box completely within the erasing window is removed.

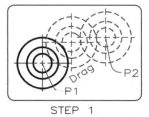
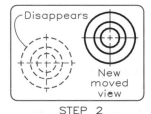

37.99 `Move` command.

Step 1 `Modify toolbar> Move icon> Select objects>` (Enter) `Base point>` Select 1st P1 (Enter), `Second point of displacement`, Pick 2nd, P2.

Step 2 Object is moved to new position.

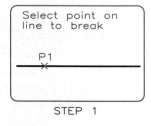
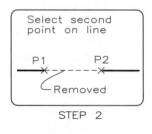

37.97 `Break` command.

Step 1 `Modify toolbar> Break icon> Select objects`, pick P1 on the line.

Step 2 `Enter second point`, and select P2. The line between P1–P2 is removed.

can be selected without fear of selecting the wrong lines.

37.30 Move and Copy (Modify Toolbar)

The `Move` command (`Modify toolbar> Copy icon`) repositions a drawing (**Fig. 36.99**) and the `Copy` command duplicates it, leaving the original in its original position. The `Copy` (C) command is applied in the same manner as the `Move` command. The `Copy` command has a `Multiple` option for locating multiple copies of drawings in different positions.

37.31 Undo (Standard Toolbar)

The Undo command (Command line> Type Undo or U) can reverse the previous commands one at a time back to the beginning of a session. The Redo command reverses the last Undo; Oops will not work. The Undo command has the following options: Auto, Control, BEgin, End, Mark, Back, and Number.

Command: <u>Undo</u> (Enter)

Auto/Cntrol/BEgin/End/Mark/Back/

<Number>: <u>4</u> (Enter)

Entering 4 has the same effect as using the U command four separate times.

Mark identifies a point in the drawing process to which subsequent additions can be undone by the Back option. Only the part of the drawing added after the placing the Mark will be undone at the prompt:

This will undo everything.

OK? <Y>: <u>Y</u> (Enter)

By responding Y, the Mark will be removed, making it possible for the next Undo (U) to proceed backward past the mark.

The BEgin and End options group a sequence of operations until End terminates the group. Undo treats the group as a single operation. The Control subcommand has three options: All, None, and One. All turns on the full features of the Undo command, None turns them off, and One uses Undo commands for single operations and requires the least disk space.

37.32 Change (Modify Toolbar)

Type Change to modify features: Lines, Circles, Text, Attribute Definitions, Blocks, Color, Layers, Linetypes, and

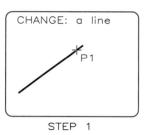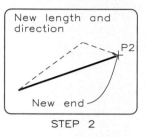

STEP 1   STEP 2

37.100 Change command: Line.

Step 1 Command line> Change> Select object> Pick point on line <u>P1</u>, Select objects> (Enter).

Step 2 Change point> select new endpoint <u>P2</u>. The end of the line moves to <u>P2</u>.

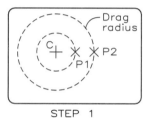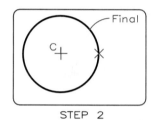

STEP 1   STEP 2

37.101 Change command: Circle.

Step 1 Command line> Change> Select object> Pick point on circle <u>P1</u>, Select objects> (Enter).

Step 2 Change point> drag to new radius endpoint <u>P2</u>. The new circle is drawn.

Thickness. The position of an endpoint of a Line is changed by selecting one end and locating a new endpoint (**Fig. 37.100**). Change varies the size of a circle by picking a point on its arc and dragging to a new size (**Fig. 37.101**).

The Text option of Change modifies text by pressing Enter until the prompts Insertion point, Style, Height, Rotation Angle, and Text String appear in sequence (**Fig. 37.102**). Attribute Definitions, including Tag, Prompt String, and Default Value, can be revised with the Change command. Blocks can be moved or rotated with the Change command, as shown in **Fig. 37.103**.

Property changes of the Change command, LAyer, Color, LType, and Thickness, are made by selecting objects and

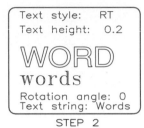

STEP 1   STEP 2

37.102 Change command: Text.

Step 1 Command line> Select object> **Select point on** text, Select objects> **(Enter).** Properties/Change point> **(Enter).** Enter text insertion point: **Select new point for text.**

Step 2 Text style: Outline, New style> RT, New height (.50): 0.20, New rotation angle (0): **(Enter),** New text (Word): Words **(Enter).**

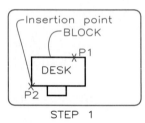
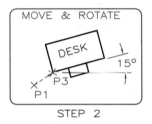

STEP 1   STEP 2

37.103 Change command: Block.

Step 1 Command line> Change> Select object> **Pick point P1,** Select objects> **(Enter).** Properties/Change point> **(Enter).**

Step 2 Enter block insertion point: **Select new insertion point for block,** P3. New rotation angle: 15 **(Enter).**

typing P (properties), as shown in **Fig. 37.104**. Type LA (Layer) and the name of the layer on which the text is to be changed.

Multiple Colors and LTypes can be assigned to objects on the same layer by the Change command, but it is better for each layer to have only one layer and linetype. The Thickness property is the height of an extrusion in the Z direction of a three-dimensional object drawn with the Elev and Thickness options (as covered in Chapter 38). Changing the thickness of a two-dimensional surface 0 to a nonzero value converts it to an extruded three-dimensional drawing.

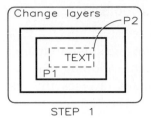
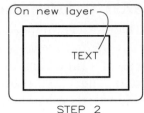

STEP 1   STEP 2

37.104 Change command: Layers.

Step 1 Command line> Change> Select object> **Pick points** P1 **and** P2 **to make window,** Select objects> **(Enter).**

Step 2 Properties/Change point> P, **(Enter)** (Color/Elev/LAyer/LType/ltScale/Thickness)? LA **(Enter).** New layer: VISIBLE **(An existing layer) (Enter).**

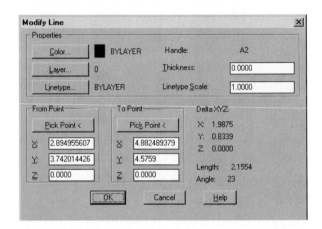

37.105 The Modify (DDmodify) box can be used to modify all objects in the same manner as when the Change command is used. Type DDmodify at the Command line and select the object to be modified to obtain this Modify Line box.

At the command line, type DDmodify (or Menu bar> Modify> Properties) and select an object when prompted to obtain a Modify box of this type to change colors, layers, linetypes, thickness, linetype scale, and ends of lines, as well as the properties of other objects (**Fig. 37.105**).

37.33 Grips (Tools Menu)

Grips are small squares that appear on selected objects at the midpoints and ends of lines, at the centers and quadrant points of

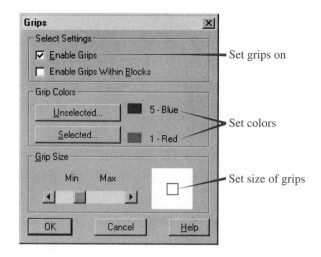

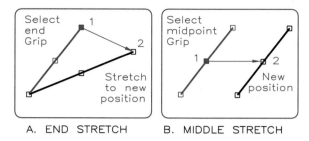

A. END STRETCH   B. MIDDLE STRETCH

37.107 Grips: Stretching.

Step 1 Select the line and grips will appear; click on end grip (1) and move to a new position 2.

Step 2 Select the midpoint grip 1, pick second point (2) to move the line to a new position.

37.106 Select Menu bar> Tools> Grips to obtain this box to Enable Grips and to set Grip size, and Grip color.

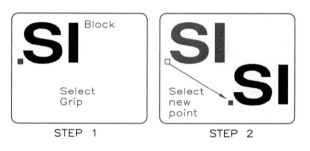

STEP 1   STEP 2

37.108 Grips: Moving.

Step 1 Select the block. A grip will appear at the insertion point. Pick this grip.

Step 2 Move the cursor to a new position to move the block.

circles, and at insertion points of text. Grips are used to Stretch, Move, Rotate, Scale, and Mirror.

The Grips dialog box (DDgrips) is found under Tools of the Standard menu (**Fig. 37.106**). The Enable Grips check box turns on grips for all objects. Enable Grips Within Blocks turns on grips for objects within a Block; when off, a single grip is given at the insertion point of the Block.

Grip Colors turns on the Color dialog box for assigning colors to selected and unselected grips; unselected grips are not filled in. Grip Size sets the size of grip boxes with a slider box.

Using Grips: By selecting an object with the cursor, grips will appear on it as open boxes. A grip that is picked and made a "hot" point is filled with color. By holding down the Shift key, more than one grip can be picked as a hot point, but the last grip of a series must be selected without pressing Shift. Press the Escape key to remove grips. By turning grips on and pressing (Enter), the options Stretch, Move, Rotate, Scale, and Move

will be shown in a menu box, each with its own subcommands.

Stretch lets the endpoint grip of a line be selected as a hot point and a second point as the new end of the line (**Fig. 37.107**). Select the midpoint grip to move the line to a new position.

To Move the Block in **Fig. 37.108**, select the insertion point as the hot point. Options of Base Point, Copy, Undo, and eXit can be used for these applications.

Rotate revolves an object about a selected Grip. The Reference option rotates an object about a selected grip by dragging or typing a number. The object in **Fig. 37.109** is rotated 60° from the reference line by typing 60.

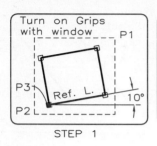

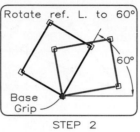

STEP 1   STEP 2

37.109 `Grips: Rotation.`

Step 1 Turn on grips by windowing the object; Select P3 as the base, (Enter) until `**Rotate**` appears. `<Rotation angle>/Base point/Copy/Undo/Reference/eXit:` **B** (Enter) `Reference angle <0>:` **10** (Enter).

Step 2 Select `<New angle>/Base point/Copy/Undo/ Reference/eXit:` **60** (Enter) Object is rotated 60° from its first position.

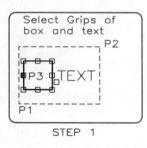

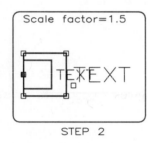

STEP 1   STEP 2

37.110 `Grips: Scaling.`

Step 1 Turn on grips by windowing object; select P3 as the base, (Enter) until `**Scale**` appears.

Step 2 `<Scale factor>/Base point/Copy/Undo/Refer- ence/eXit:` `1.5` (Enter) (object is scaled).

`Scale` uses the grip selected as a base point to size the object (**Fig. 37.110**). The scale factor is assigned by typing, dragging, or selecting a reference dimension and giving it a new dimension.

`Mirror` makes a mirror image of an object. The original object is removed when two grips are selected to specify a mirror line (**Fig. 37.111**). Hold down the `Shift` key while selecting the second grip point on the mirror line, and the initial drawing will not be removed.

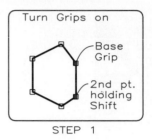

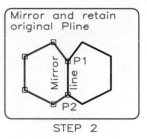

STEP 1   STEP 2

37.111 `Grips: Mirroring.`

Step 1 Turn on grips by windowing object; select P1 as the base, (Enter).

Step 2 Select `Mirror. <Second point>/Base point/ Copy/Undo/Reference/eXit:` P2 (While pressing Shift key) (Enter) (Object is mirrored and original is retained.)

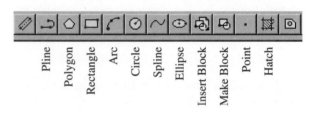

37.112 The `Pline` icon on the `Draw toolbar` is selected for drawing `Polylines`.

37.34 Polyline (Draw Toolbar)

The `Polyline` (PL) command from the `Draw toolbar` (**Fig. 37.112**) is used for drawing two-dimensional polylines, which are lines of continuously connected segments instead of separate segments as drawn by the `Line` command. The thickness of a `Pline` can be varied as well, which requires the pen to plot with multiple strokes when plotting (**Fig. 37.113**).

The Pline options are `Arc, Close, Halfwidth, Length, Undo, Width,` and `Endpoint of line.` Close automatically connects the last end of the polyline with its beginning point and ends the command. `Length` continues a `Pline` in the same direction by typing the length of the segment. If the first line was an arc, a line is drawn tangent to the arc. `Undo` erases the last segment of the polyline, and it can be repeated to continue

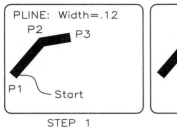
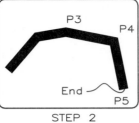

PLINE: Width=.12

STEP 1   STEP 2

37.113   `Pline:` Width **option.**

Step 1 `Draw toolbar> Select Pline icon> From point>` Select P1 (Enter), `Current line-width is 0.1.` `Arc/Close/Halfwidth/Length/Undo/Width/ <Endpoint of line>:` Width (Enter), `Starting width:` .5 (Enter).

`Ending width <.5>:` (Enter).

`Endpoint of line:` P2, `Endpoint of line:` P3.

Step 2 `Endpoint of line:` P4, to P5 (Enter).

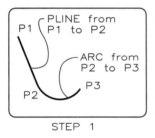
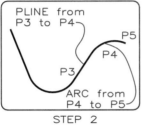

STEP 1   STEP 2

37.114   `Pline:` Lines **and** arcs.

Step 1 `Draw toolbar> Select Pline icon> From point>` P1 (Enter), `Current line-width is 0.1.` `Arc/Close/. . . <Endpoint of line>:` P2 (Enter) `Command:` Arc (Enter).

`Center/<Start point>:` (Enter) `Endpoint:` P3.

Step 2 `Command:` Line (Enter), `Line from point:` (Enter).

`Length of line:` P4, `To point:` (Enter) `Command:` Arc (Enter) `Center/<Start point>:` (Enter), `End point:` P5.

erasing segments. `Halfwidth` specifies the width of the line measured on both sides of its center line.

The `Arc` option of `Pline`, is selected to obtain the prompts shown in **Fig. 37.114**. The default draws the arc tangent from the endpoint of the last line and through the next selected point. `Angle` gives the prompt, `Included angle:`, to which a positive or negative value is given. The next prompt asks

for `Center/Radius/<End point>:` to draw an arc tangent to the previous line segment. Select `Center` and you will be prompted for the center of the next arc segment.

The next prompt is `Angle/Length/ <Endpoint>:`, where `Angle` is the included angle, and `Length` is the chordal length of the arc. `Close` causes the arc segment of the pline to close to its beginning point. `Direction` lets you override the default, which draws the next arc tangent to the last `Pline` segment. When prompted with `Direction from starting point:`, pick the beginning point and respond to the next prompt, `Endpoint`, by picking a second point to give the direction of the arc.

`Line` switches the `Pline` command back to the straight-line mode. `Radius` gives the prompt, `Radius:`, for specifying the size of the next arc. The next prompt, **Angle/Length/ <Endpoint>:**, lets you specify the included angle or the arc's chordal length. `Second Pt` gives two prompts, `Second point:` and `Endpoint:`, for selecting points on an arc.

37.35 Pedit (Draw Toolbar)

The `Pedit` command modifies `Plines` with the following options: `Close, Join, Width, Edit vertex, Fit, Spline, Decurve, Ltype generate, Undo,` and `eXit.` If the `Pline` is already closed, the `Close` command will be replaced by the `Open` option.

`Join` (J) gives the prompt, `Select objects, Window` or `Last:`, for selecting segments to join into a polyline. `Segments` must have exact meeting points to be joined.

`Width` (W) gives the prompt, `Enter new width for all segments:`, for assigning a new width to a `Pline`.

`Fit` (F) converts a polyline into a line composed of circular arcs that pass through each vertex (**Fig. 37.115**).

`Spline` (S) modifies the polyline, as did the `Fit` curve, but it draws cubic curves pass-

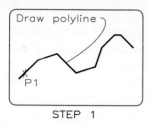

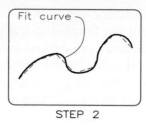

STEP 1   STEP 2

37.115 `Pedit: Fit` curve.

Step 1 `Modify II toolbar> Select Pedit icon> Select polyline:` **P1 (Enter).**

Step 2 `Close/Join/Width/Edit vertex/Fit/. . . Undo/eXit <X>:` `Fit` **(Enter) The curve is smoothed and passes through all points.**

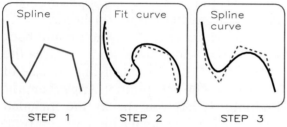

STEP 1   STEP 2   STEP 3

37.116 `Pedit: Fit` and `Spline`.

Step 1 Draw as polyline (Pline).

Step 2 `Modify II toolbar> Select Pedit icon> Select polyline:` **Select line.**

`Close/Join/Width/Edit vertex/Fit/. . .Undo/eXit <X>:` `Fit` (Enter) The curve is smoothed and passes through all points.

Step 3 Use `Pedit` in same way, but use the `Spline` option to obtain a "best curve" (which may not pass through the selected points).

ing through the first and last points, and not necessarily through the other points (**Fig. 37.116**). `Decurve (D)` converts `Fit` or `Spline` curves to their original straight-line forms.

`Ltype gen (L)` applies dashed lines (such as centerlines) in a continuous pattern on curved polylines. Without applying this option, dashed lines may omit gaps in curved lines (**Fig. 37.117**). By setting system variable `Plinegen On`, linetype generation will be applied as `Plines` are drawn.

`Undo (U)` reverses the most recent `Pedit` editing step. `Edit Vertex (E)` selects vertexes of the `Pline` for editing by placing an X on the

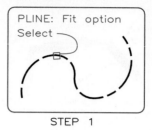

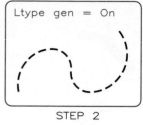

STEP 1   STEP 2

37.117 `Pedit: Ltype generate.`

Step 1 `Modify II toolbar> Select Pedit icon> Select polyline:` **Select polyline (Enter).**

Step 2 `Close/Join/Width/Edit vertex/Fit/. . . Ltype gen/Undo/eXit <X>:` **L (Enter).**

`Full PLINE line type ON/OFF:` `ON` (Enter). The dashes are uniform around all curves.

first vertex when the polyline is picked. The following options appear: `Next/Previous/ Break/Insert/Move/Regen/Straighten/ Tangent/Width/eXit/<N>:` `Next (Enter).`

`Next (N)` **and** `Previous (P)` options move the X marker to next or previous vertexes by pressing `Enter`. `Break (B)` prompts you to select a vertex with the X marker. You can then use `Next` or `Previous` to move to a second point and pick `Go` to remove the line between the vertexes. Select `eXit` to leave the `Break` command and return to `Edit Vertex`.

`Insert` adds a new vertex to the polyline between a selected vertex and the next vertex (**Fig. 37.118**). `Move (M)` relocates a selected vertex (**Fig. 37.119**). `Straighten (S)` converts the polyline into a straight line between two selected points. An X marker appears at the current vertex and the prompt, `Next/ Previous/ Go/eXit/<N>`, appears. Move the X marker to a new vertex with `Next` or `Previous`, select `Go`, and the line is straightened between the vertices (**Fig. 37.120**). Enter X to `eXit` and return to the `Edit Vertex` prompt.

`Tangent (T)` lets a tangent direction be selected at the vertex marked by the X for curve fitting by responding to the prompt,

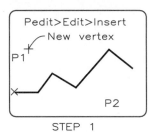

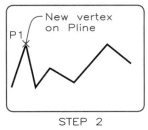

| STEP 1 | STEP 2 |

37.118 Pedit: Edit vertex, Insert.

Step 1 Modify II toolbar> Select Pedit icon> Select polyline: Select polyline (Enter).

Close/Join/Width/Edit vertex/Fit/. . . Ltype gen/Undo/eXit <X>: Edit (Enter).

Next/Previous/Break/Insert/. . . eXit: Insert (Enter) (X appears as end of polyline).

Enter location of new vertex: P1.

Step 2 Press (Enter); the new vertex is inserted, and the polyline passes through it.

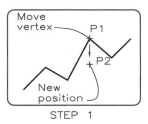

| STEP 1 | STEP 2 |

37.119 Pedit: Edit vertex, Move.

Step 1 Modify II toolbar> Select Pedit icon> Select polyline: Select polyline (Enter).

Close/Join/Width/Edit vertex/Fit/. . . Ltype gen/Undo/eXit <N>: E. Press (Enter) until X is on the vertex to be moved. Move (Enter).

Step 2 Enter new location: P2 (Enter) (Esc). The vertex is moved.

Direction of tangent. Enter the angle from the keyboard or by cursor.

Width (W) sets the beginning and ending widths of an existing line segment from the X-marked vertex. Use Next and Previous to confirm in which direction the line will be drawn from the X marker. The polyline will be changed to its new thickness when the screen is regenerated with Regen (R). Use eXit to escape from the Pedit command.

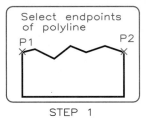

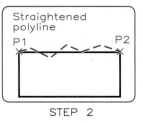

| STEP 1 | STEP 2 |

37.120 Pedit: Edit vertex, Straighten.

Step 1 Modify II toolbar> Select Pedit icon> Select polyline: Pick polyline (Enter).

Close/Join/Width/Edit vertex/Fit/. . . Ltype gen/Undo/eXit <N>: Press (Enter) until X is on the first vertex of the line to straighten, P1. Straighten (Enter).

Next/Previous/Go/eXit <N>: Press (Enter) until X is on the second vertex, P2.

Step 2 Next/Previous/Go/eXit: Go (Enter) (Esc). Line P1–P2 is straightened.

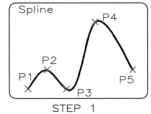

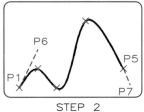

| STEP 1 | STEP 2 |

37.121 Spline.

Step 1 Draw toolbar> Select Spline icon> Object/<Enter first point>: P1, Enter point: P2, Close/Fit Tolerance/<Enter point>: P3. Continue this way to point P5.

Step 2 Close/Fit Tolerance/<Enter point>: (Enter).

Enter start tangent: Rubberband from P1 to P6.

Enter end tangent: Rubberband from P5 to P7. Spline appears.

37.36 Spline (Draw Toolbar)

The Spline command (Draw toolbar> Spline icon) draws a smooth curve with a sequence of points within a specified tolerance, as shown in **Fig. 37.121**. By setting Fit Tolerance to 0, the curve will pass through the points; when set to a value greater than 0, it will pass within a tolerance of each point. The Endpoint Tangents determine the angle of the spline at each end.

37.122 Select the hatch icon on the `Draw toolbar` to begin to hatch a sectioned area.

Hatch icon

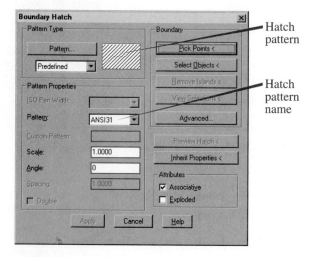

Hatch pattern

Hatch pattern name

37.123 When the hatch icon is selected, the `Boundary Hatch` box appears on the screen.

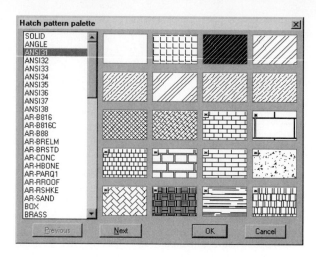

37.124 When the `Pattern` button in the `Boundary Hatch` box is selected, the hatch patterns and their names are displayed in this `Hatch pattern palette`.

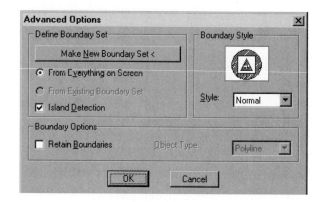

37.125 Select `Advanced` in the `Boundary Hatch` box (Fig. 37.123) to display this dialog box.

37.37 Hatching (Draw Toolbar)

Hatching is a pattern of lines that fills sectioned areas, bars on graphs, and similar applications. From the `Draw toolbar` (**Fig. 37.122**), the `Bhatch` (`Boundary hatch`) dialog box is displayed (**Fig. 37.123**). By selecting the down arrow at the `Pattern` box, a listing of pattern names is given. When one is selected, a view of the pattern will appear in one of the windows on the right. Examples of some predefined patterns are shown in **Fig. 37.124**.

The `Boundary Hatch` box (Fig. 37.123) lets you specify `Predefined`, `User-defined`, or `Custom` **patterns.** `Predefined` patterns are those provided by AutoCAD.

`Scale` sets the spacing between the lines of a pattern, and `Angle` assigns their angle. The `Advanced` button displays the `Advanced Options` dialog box (**Fig. 37.125**) from which

options of `Normal`, `Outer`, or `Ignore` can be selected. A circle with a square and a triangle inside it illustrates the effect of each choice. `Normal` hatches every other nested area beginning with the outside. `Outer` hatches the outside area, and `Ignore` hatches the entire area from the outer boundary. When selected, `Text` within hatching appears in an opening within the hatching area.

`Exploded Hatch` inserts a hatch pattern as a group of separate objects as if it were inserted as `*Name` (`*ANSI31`, for example). `ISO Pen Width` enables you to set the width of a hatch-

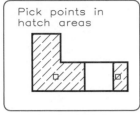
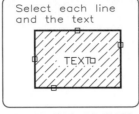

A. PICK POINTS   B. SELECT OBJECTS

37.126 Hatching areas.

A The `Pick Points` option (Fig. 37.123) of `Boundary Hatch` prompts for points inside the boundaries for hatching.

B The `Select Objects` option requires that boundary lines be selected.

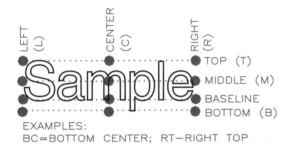

EXAMPLES:
BC=BOTTOM CENTER; RT—RIGHT TOP

37.127 Text can be added to a drawing by using any of the insertion points above. For example, BC means the bottom center of a word or sentence will be located at the cursor point.

ing line if a `Predefined ISO` pattern is selected (`ISO13W100`, for example). The `Pick Points` and `Select Objects` options (Fig. 37.123) are used to select areas inside of boundaries and then boundaries themselves (**Fig. 37.126**). When points are selected outside the boundary or if the boundary is not closed, error messages will appear.

37.38 Text and Numerals

`Justify` prompts for the insertion point for a string of text (**Fig. 37.127**). BC means bottom center, RT means right top, and so forth.

Figure 37.128 illustrates how multiple lines of `Dtext` are automatically spaced by pressing (Enter) at the end of each line. The special characters shown in **Fig. 37.129** can be

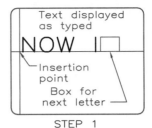
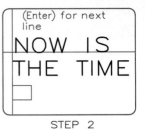

STEP 1   STEP 2

37.128 `Dtext` command.

Step 1 Command: <u>Dtext</u> (Enter).

`Justify/Style/<Start point>:` Select insertion point.

`Height:` **.125** (Enter), `Rotation angle <0>:` (Enter).

`Text:` <u>NOW IS</u>

Step 2 Press (Enter) to advance the box to the next line of text. <u>THE TIME</u> (Enter). Box spaces down with each (Enter).

%%O	Start or stop Overline of text
%%U	Start or stop Underline of text
%%D	Degree symbol: 45%%D =45°
%%P	Plus—minus: %%P0.05=±0.05
%%C	Diameter: %%C20=⌀20
%%%	Percent sign: 80%%% =80%
%%nnn	Special character number nnn

37.129 The special characters that begin with a double percent sign (%%) are used with `Dtext` to obtain these symbols.

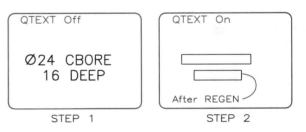

STEP 1   STEP 2

37.130 `Qtext` command.

Step 1 Command: <u>QTEXT</u> (Enter), `On/Off:` <u>On</u> (Enter).

Step 2 Command: <u>REGEN</u> (Enter) Text is shown as boxes.

inserted by typing a double percent sign (%%) in front of them.

Type <u>Qtext</u> and select On to reduce screen regeneration time by drawing text as boxes (**Fig. 37.130**). When `Qtext` is Off, the full text will be restored after regeneration (type <u>Regen</u>).

TXT	PRELIMINARY PLOTS
MONOTXT	FOR SPEED ONLY
	Simplex fonts
ROMANS	FOR WORKING DRAWINGS
SCRIPTS	*Handwritten Style, 1234*
GREEKS	ΓΡΕΕΚ ΣΙΜΠΛΕΞ, 12345
	Duplex fonts
ROMAND	**THICK ROMAN TEXT**
	Complex fonts
ROMANC	ROMAN WITH SERIFS
ITALICC	*ROMAN ITALICS TEXT*
SCRIPTC	*Thick–Stroke Script Text*
GREEKC	ΓΡΕΕΚ ΩΙΤΗ ΣΕΡΙΦΣ
	Triplex fonts
ROMANT	TRIPLE–STROKE ROMAN
ITALICT	*Triple–Stroke Italics*
	Gothic fonts
GOTHICE	𝕰nglish 𝕲othic 𝕿ext
GOTHICG	𝕲erman 𝕲othic 𝕿ext, 12
GOTHICI	𝕴talian 𝕲othic 𝖀ext, 12

37.131 Examples of some of the available AutoCAD fonts are shown here.

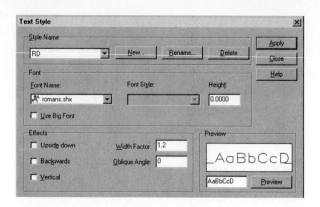

37.132 Menu bar> Format> Text Style (or type Style at the Command line) to display this Text Style box. From here, a New style can be named, Fonts can be assigned, Width Factors specified; and other assignments can also be made.

37.39 Text Style (Format Menu)

Many of AutoCAD's text fonts and their names are shown in **Fig. 37.131**. The default style, Standard, uses the Txt font. From the Format menu, select Text Style and type the Style name when prompted; the Select Font File dialog box will appear (**Fig. 37.132**). From here, you can assign a New Style name, pick a Font, select its Height, give a Width factor, specify an Oblique Angle, and other options. A sample of your choices is shown in the Preview window.

The Style names are listed in a drop-down menu in the window beneath the heading, Style Name of the Text Style dialog box. An example of the text font is displayed when a Style is selected (Fig. 37.132). A

define Style will retain its settings until they are changed. To select a font for a new style, pick a style from the Font Name window.

If you later create a Style with the same name as a previously defined style and select a different font, such as Romans, the text previously drawn with that style name will be redrawn with the Romans font. This technique is used to change the Txt and Monotxt fonts to more attractive fonts at plot time. Beforehand, time is saved by using Txt and Monotxt fonts because they regenerate quickly.

The DDedit command (or, select View> Toolbars> Modify> Text edit icon) is used to select a line of text to be displayed in a dialog box for editing (**Fig. 37.133**). Correct the text and select the OK button. The screen will be revised.

37.40 Multiline (Mtext): Draw Toolbar

From the View> Toolbars> Draw toolbar, select the A icon (Text), pick an insertion point, and specify the size of the text boundary by the diagonal window option by specifying the width with W, or by specifying

37.133 Press `Modify II toolbar>` Select `Ddedit icon` (or type `DDEDIT` at the `Command line`) to select a line of text on the screen. It will appear in an `Edit Text` box for editing.

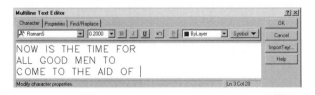

37.134 From the `Draw toolbar` select letter `A icon` for `Multiline text`, specify a window for the text with two diagonal corners when prompted, and this `Multiline Text Editor` appears. Type your lines of text in the box and pick `OK`. The text is drawn on the screen in the specified area. Three tabs—`Character`, `Properties`, and `Find/Replace`—of this box provide additional settings for the text.

two points with `2P`. The `Multiline Text Editor` dialog box will appear, and the paragraph of text can be typed in its window (**Fig. 37.134**). Press `OK` to place the paragraph on the drawing.

The following control keys can be used to edit the text:

`Ctrl+C` Copy text to `Clipboard`.

`Ctrl+V` Paste `Clipboard` contents over selection.

`Ctrl+X` Cut selection to the `Clipboard`.

`Ctrl+Z` Undo and `Redo`.

`Ctrl+Shift+Space` Insert nonbreaking space.

(Enter) End current paragraph and start new line.

Most of the options of the `Multiline Text Editor` box are obvious. `Stack`, under the `Character tab` lets a fraction (1/2, for example) be aligned vertically as fractions would

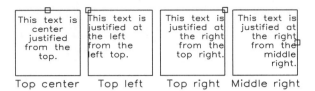

Top center   Top left   Top right   Middle right

37.135 Under the `Properties` tab, types of `Justification` can be selected. Several are shown here.

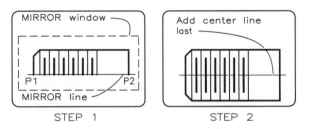

STEP 1   STEP 2

37.136 `Mirror` command.

Step 1 Draw the half to be mirrored. `Modify toolbar> Mirror icon> Select objects:` <u>W</u> (Enter) Window the drawing.

`Select objects:` (Enter).

`First point of mirror line:` <u>P1</u>, `Second point:` <u>P2</u>.

Step 2 `Delete old objects? <N>:` <u>No</u> (Enter). The drawing is mirrored.

be drawn. `Import Text` displays a `Text File` dialog box for importing files from other sources.

Select the `Properties` tab of the `Multiline text editor` box to obtain properties for editing the text displayed there. Examples of the text justification option are shown in **Fig. 37.135**. Attach controls the text justification and spill relative to the text boundary.

37.41 Mirror (Modify Toolbar)

Select the `Mirror (Modify toolbar> Mirror icon)` (or type `Mirror` at the `Command line`) to make mirror images about an axis (**Fig. 37.136**). A line that coincides with the `Mirror line` (`P1–P2`, for example) will be drawn twice when mirrored; therefore, parting lines should be drawn after the drawing has been mirrored.

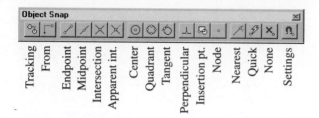

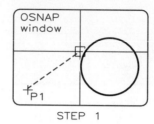 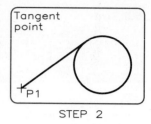

37.137 The Object Snap toolbar has these options for drawing object features on the screen.

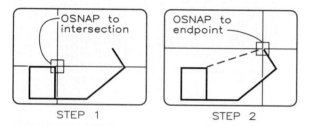

37.138 Osnap: Intersection and Endpoint.

Step 1 Draw toolbar> Line icon> Line from point: Pick Intersection icon, Int of: Pick intersection point.

Step 2 To point: Pick Endpoint icon on Osnap toolbar. To point: Select endpoint of line and the line is drawn.

A system variable, Mirrtext (Command line> Setvar> Mirrtext), is used for mirroring text. By setting Mirrtext to 0, Mirrtext is set to Off and text will not be mirrored. If it is set to 1 (On), the text will be mirrored along with the drawing.

37.42 Osnap (Object Snap Toolbar)

By using Osnap (Object Snap), you can snap to objects of a drawing rather than to the Snap grid. Osnap icons from the Object snap toolbar (**Fig. 37.137**) give the following options: Endpoint, Midpoint, Intersection, Apparent Intersection, Center, Quadrant, Tangent, Perpendicular, Node, Intersection, Nearest, Quick, and None. Osnap is used as an accessory to other commands: Line, Move, Break, and so forth.

Figure 37.138 shows how a line is drawn from an intersection to the endpoint of a line.

37.139 Osnap: Tangent.

Step 1 Draw toolbar> Line icon> Line from point: P1.

To point: Select Tangent icon, To point: Intersection icon.

Step 2 Tan to: Pick point on arc. The line is drawn from P1 tangent to the circle.

In **Fig. 37.139**, a line is drawn from P1 tangent to the circle by using the Tangent option of Osnap. This option can be used to draw a line tangent to two arcs.

The Node option snaps to a Point, the Quadrant option snaps to one of the four compass points on a circle, the Insert option snaps to the intersection point of a Block, and the None option turns off Osnap for the next selection. The Quick option reduces searching time by selecting the first object encountered rather than searching for the one closest to the aperture's center.

Osnap settings can be temporarily retained as "running" Osnaps for repetitive use. To set running Osnaps to Endpoints and circle Centers, do the following:

Command: OSNAP (Enter)

Object snap modes: Endpoint, Center (Enter)

Now, the cursor has an aperture target at its intersection for picking endpoints and centers of arcs. Remove running Osnap settings by typing Osnap and (Enter). Running Osnaps can be set by selecting the icon under the Object Snap toolbar. The Aperture command sets the size to the target box that appears at the cursor for Osnap applications. Its size can vary from 1 to 50 pixels square.

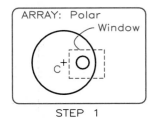

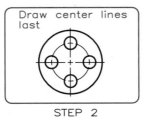

STEP 1 STEP 2

37.140 Array: Polar.

Step 1 Modify toolbar> Array icon> Select objects: W (Enter) (Window the hole.) Select objects: (Enter)

Step 2 Rectangular or Polar array (R/P): P (Enter)

Center point of array: C (Enter)

Number of items: 4 (Enter),

Angle to fill (+=ccw, -=cw) <360>: 360 (Enter)

Rotate objects as they are copied? <Y> (Enter)

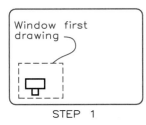

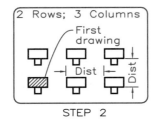

STEP 1 STEP 2

37.141 Array: Rectangle.

Step 1 Modify toolbar> Array icon> Select objects: W (Enter) (Window the desk.) Select objects: (Enter).

Step 2 Rectangular or Polar array (R/P): R (Enter).

Number of rows (—) <1>: 2 (Enter).

Number of columns (l l l) <1>: 3 (Enter).

Unit cell distance between rows (—): 4 (Enter).

Distance between columns (I I I) <1>: 3.5 (Enter).

37.43 Array (Modify Toolbar)

Modify toolbar> Array can be used to draw circular or rectangular patterns (rows and columns) of selected drawings. A series of holes can be drawn on a bolt circle by drawing the first hole and Arraying it as a Polar array (**Fig. 37.140**).

A rectangular array is begun by making the drawing in the lower-left corner and following the steps in **Fig. 37.141**. Rectangular arrays may be drawn at angles by using the Snap

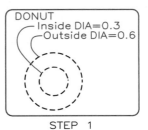

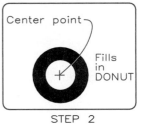

STEP 1 STEP 2

37.142 Donut.

Step 1 Menu bar> Draw> Donut> Inside diameter: 0.3 (Enter) Outside diameter: 0.6 (Enter).

Step 2 Center of donut: Select points and donuts are drawn.

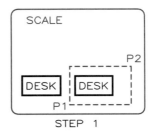

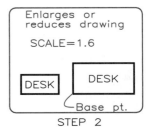

STEP 1 STEP 2

37.143 Scale.

Step 1 Modify toolbar> Scale icon> Select objects: W (Enter) Window the desk with P1 and P2.

Step 2 Base point: Select.

<Scale factor>/Reference: 1.6 (Enter) The desk is drawn 60% larger.

mode to Rotate the grid. The first object is drawn in the lower-left corner of the array and the number of rows, columns, and the cell distances are specified when prompted.

37.44 Donut (Draw Menu)

The Donut command, selected from the Circle flyout of the Draw toolbar, draws doughnuts by assigning outside and inside diameters (**Fig. 37.142**). Donuts can be drawn as solid circles by setting the inside diameter to 0.

37.45 Scale (Modify Toolbar)

The Scale command reduces or enlarges previously drawn objects. The desk in **Fig. 37.143** is enlarged by windowing it, selecting a

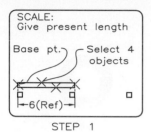 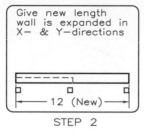 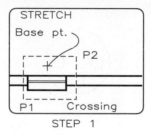 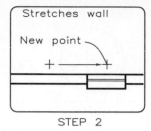

| STEP 1 | STEP 2 | STEP 1 | STEP 2 |

37.144 Scale: Reference.

Step 1 Modify toolbar> Scale icon> Select objects: **Select 4 objects.**

Base point: Select <Scale factor>/Reference: R (Enter).

Reference length <1>: 6 (Enter).

Step 2 New length: 12 (Enter). The drawing is enlarged in all directions.

37.145 Stretch.

Step 1 Modify toolbar> Stretch icon> Select objects: **Select with crossing window, P1 and P2.**

Base point: **Select point.**

Step 2 New point: **Select new point. The windowed portion of the drawing is repositioned.**

base point, and typing a scale factor of 1.6. The drawing and its text are enlarged in the x- and y-directions.

A second option of Scale, Reference, lets you select a length of a given object, specify its present length, as a reference, and assign a length as a ratio of the first dimension (**Fig. 37.144**). The lengths can be given by the cursor or typed at the keyboard in numeric values.

37.46 Stretch (Modify Toolbar)

The Stretch command (Modify toolbar> Stretch icon) lengthens or shortens a portion of a drawing while one end is left stationary. The window symbol in the floor plan (**Fig. 37.145**) is stretched to a new position, leaving the lines of the wall unchanged. A Crossing Window must be used to select lines that will be stretched.

37.47 Rotate (Modify Toolbar)

A drawing can be rotated about a base point by using the Rotate command (Modify toolbar> Rotate icon), as shown in **Fig. 37.146**. Window the drawing, select a base point, and type the rotation angle or select the angle with the cursor. Drawings made on multiple layers can also be rotated.

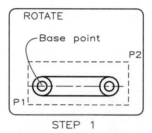 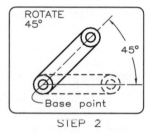

| STEP 1 | STEP 2 |

37.146 Rotate.

Step 1 Modify toolbar> Rotate icon> Select objects: **Select with window, P1 and P2.**

Step 2 Base point: **Select point.**

<Rotation angle>/Reference: 45 (Enter) Object is rotated 45° counterclockwise.

37.48 Setvar (Command Line)

Many system variables can be inspected by typing Setvar and ? at the Command line, and changed if they are not read-only commands. To change one or more variables (Textsize, for example), respond as follows:

Command: Setvar (Enter)

Variable name or ?: Textsize (Enter)

New value for TEXTSIZE <0.18>: 0.125 (Enter)

By entering the Setvar command with an apostrophe in front of it ('Setvar), it can be used transparently without leaving the command in progress.

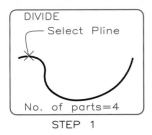

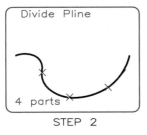

STEP 1   STEP 2

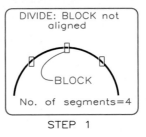

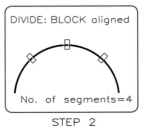

STEP 1   STEP 2

37.147 Divide.

Step 1 Command line: <u>Divide</u> (Enter).

Select object to divide: **Select Pline.**

Step 2 <Number of segments>/Block: <u>4</u> (Enter).

(Pdmode symbols are placed along the line, dividing it.)

37.148 Divide: Arc **with** Block.

Step 1 Command line: <u>Divide</u> (Enter).

Select object to divide: **Select arc.**

<Number of segments>/Block: <u>B</u> (Enter).

Block name to insert: <u>RECT</u> (Enter).

Align block with object? <Y> <u>N</u> (Enter).

Number of segments: <u>4</u> (Enter).

Step 2 Align block with object? <Y> (Enter) Blocks are drawn to radiate from the arc's center.

37.49 Divide (Draw Menu)

The Divide command (Menu bar> Draw> Divide) places markers on a line to show a specified number of equal divisions. The line in **Fig. 37.147** is selected by the cursor, the number of divisions is specified, and markers are equally spaced along it. The markers will be of the type and size currently set by the Pdmode and Pdsize variables under the Setvar command.

The Block option of Divide allows saved blocks (rectangles in this example) to be used as markers on the line (**Fig. 37.148**). Blocks can be either Aligned or Not Aligned, as shown.

37.50 Measure (Draw Menu)

The Measure command (Menu bar> Draw> Measure) repeatedly measures off a specified distance along an arc, circle, polyline, or line and places markers at these distances (**Fig. 37.149**). Respond to the Select object to measure prompt by picking a point near the end where measuring is to begin. When prompted, give the segment length; markers will be displayed along the line. The last segment is usually a shorter length.

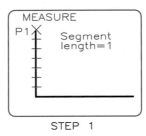

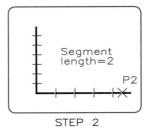

STEP 1   STEP 2

37.149 Measure.

Step 1 Command line: Measure (Enter).

Select object to measure: <u>P1</u>.

<Segment length>/Block: <u>1</u> (Enter) The line is divided into one segment starting at the point nearest P1.

Step 2 Command line: <u>Measure</u> (Enter).

Select object to measure: <u>P1</u>.

<Segment length>/Block: <u>2</u> (Enter) Divisions 2 apart are measured off along the line from the end at P2.

37.51 Offset (Modify Toolbar)

An object can be drawn parallel to and offset from another object, such as a polyline, by the Offset command (**Fig. 37.150**). Offset prompts for the distance or the point through which the offset line must pass, and then prompts for the side of the offset. The Offset command is helpful when drawing parallel lines to represent walls on a floor plan.

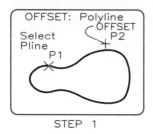
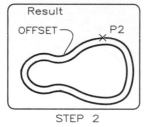

STEP 1 · STEP 2

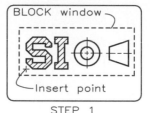
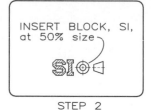

STEP 1 · STEP 2

37.150 Offset.

Step 1 Command: <u>Offset</u> (Enter).

Offset distance or Through <Through>: <u>T</u> (Enter).

Select object to offset: <u>P1</u> (Enter).

Step 2 Through point: <u>P2</u> (Enter). An enlarged Pline is drawn that passes through P2.

37.152 Block.

Step 1 Refer to the Block Definition box in Fig. 37.151. Pick Select Objects button. Window SI symbol on screen. Select objects: (Enter).

Step 2 Pick Select Point button. Select insertion point for block.

Block name: Type name, <u>SI</u>, in window. Select OK.

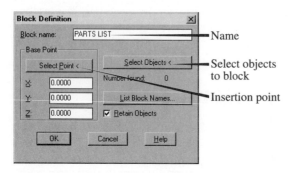

37.151 Draw toolbar> Make block icon, which will open this Block Definition box, can be used to make a Block.

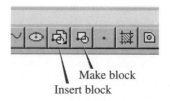

Make block
Insert block

37.153 The Make Block and Insert Block icons can be selected from the Draw toolbar.

37.52 Blocks (Draw Toolbar)

One of the more productive features of computer graphics is the capability of creating drawings called Blocks for repetitive use. Select Draw menu> Block> Block definition to make a block or use the Draw toolbar (**Fig. 37.151**). The SI symbol in **Fig. 37.152** is a typical drawing that is made into a Block and inserted into drawings using icons from the Draw toolbar shown in **Fig. 37.153**. The Insert heading on the Menu bar> Insert> Block) gives the Insert dialog box (**Fig. 37.154**) for inserting both Blocks and Files (Wblocks). Select Block, and the Defined Blocks dialog box will appear for selection of a Block (**Fig. 37.155**).

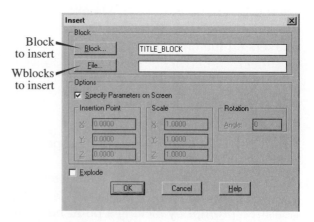

37.154 Select Menu bar> Insert> to get Insert box to specify Blocks or Files (Wblocks) to be inserted in a drawing.

When Insert is used, Defined Blocks are listed from which to select.

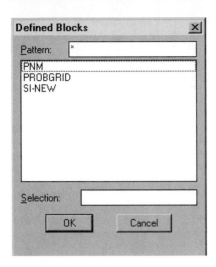

37.155 When `Insert` is selected, `Defined Blocks` will be listed.

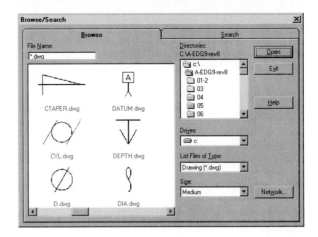

37.156 Select `Menu bar> Insert> File> Select Drawing File> Find File> Browse` tab and thumbnails of `Wblocks` are displayed.

When a `Block` is selected, it appears as a total unit. However, `Blocks` that have been `Inserted` by selecting the `Explode` box first, or by typing a star in front of the `Block` name (`*SI`, for example), can be selected one object at a time. An inserted `Block` can be separated into individual entities by typing `Explode` and selecting the `Block`.

`Blocks` can be used only in the current drawing file unless they are converted to `Wblocks` (`Write Blocks`), which become independent files, not parts of files. This conversion is performed as follows:

`Command: Wblock` (Enter)

`File Name: B:SI` (Enter) This assigns the name of the Wblock to drive A.

`Block Name: SI` (Enter) This is the name of the Block that is being changed to a Wblock.

`Blocks` can be redefined by selecting a previously used `Block` name to receive the prompt, `Redefine it? <N>:`. Type `Y` (Yes) and select the new drawing to be blocked. After doing so, the redefined `Blocks` automatically replace those in the current drawing with the same name. The `Browse` option can be used to locate thumbnail illustrations of `Wblocks`, as illustrated in **Fig. 37.156**.

37.53 Transparent Commands (Command Line)

Transparent commands can be used while another command is in progress by typing an apostrophe in front of the command name at the `Command line`. If you are dimensioning a part and wish to use `Pan`, type `'Pan`, (Enter), do the pan, and then complete the dimensioning command. Commands that can be used transparently are `'Graphscr`, `'Help`, `'Pan`, `'Redraw`, `'Resume`, `'Setvar`, `'Textscr`, `'View`, and `'Zoom`.

37.54 View (Command Line)

Portions of drawings can be saved as separate views with the `View` command (**Fig. 37.157**). The entire screen can be made into a `View` by the `Save` option and naming it when prompted. The `Window` option makes a `View` of the windowed portion of the drawing. Type `Restore`, and type the View's name, and press `Enter` to display it. Type `?` to list the saved views, and select `Delete` to remove a view from the list.

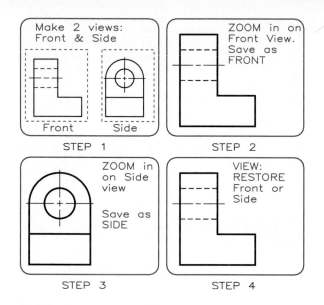

STEP 1 | STEP 2

STEP 3 | STEP 4

37.157 View.

Step 1 The two-view drawing can be saved as separate views.

Step 2 Zoom the front view to fill the screen.

Command line: View (Enter).

?/Delete/Restore/Save/Window: Save (Enter).

View name to save: FRONT (Enter).

Step 3 Zoom the side view to fill the screen.

Command line: View (Enter).

View name to save: SIDE (Enter).

Step 4 To display a view, Command line: View (Enter).

?/Delete/Restore/Save/Window: R (Enter).

View name to restore: SIDE (Enter) View is displayed.

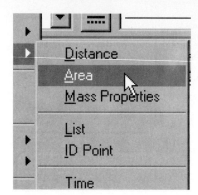

37.158 Menu bar> Tools> Inquiry menu offers these options to assist you in learning more about a particular drawing.

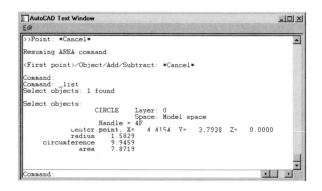

37.159 Inquiry toolbar> List icon> Select a circle on the screen. This box appears, and gives the center of the circle as well as its radius, circumference, and area.

37.55 Inquiry Commands (Tools)

From Tools on the Standard toolbar, select Inquiry or type it at the Command line, to obtain information about objects and files with Dist, Area, Mass Properties, List, ID Point, Time, Status, and Set Variables (Fig. 37.158). List is selected (or typed) and the circle (or any object) is selected when prompted to obtain information about it (**Fig. 37.159**).

Dist measures the distance, its angle, and its delta-x and delta-y distances between selected points without drawing a line. ID gives the x-, y-, and z-coordinates of a point that is picked on the screen. Area gives the perimeter and area space on the screen. Select the prompts First point:, Next point:, Next point:, and so on to pick all points, then press Enter. Areas can be added and removed when they are being selected, as shown in **Fig. 37.160**.

The Status option gives information about the settings, layers, coordinates, and disk space. The Time option displays information about the time spent on a drawing (**Fig. 37.161**). The timer can be Reset and turned On to record the time of a drawing session, but the cumulative time cannot be erased without deleting the drawing file. After Resetting, Display shows the time

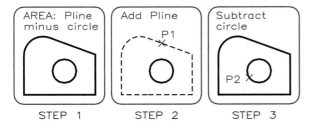

STEP 1   STEP 2   STEP 3

37.160 Area.

Step 1 The object is drawn with a `Pline` outline and a circular hole in it.

Step 2 Menu toolbar> Tools> Inquiry> Area> <First point>/Object/Add/Subtract: Add (Enter)

<First point>/Object/Subtract: Object (Enter)

(ADD mode) Select objects: P1

Area = 4.0300 Perimeter = 24.8560

Total area = 4.0300

(ADD mode) Select objects: (Enter)

Step 3 <First point>/Object/Subtract: Subtract (Enter)

<First point>/Object/Add: Object (Enter)

(SUBTRACT mode) Select objects: P2

Area = 0.4910 Circumference = 7.8540

Total area = 3.5400

```
Command: '_time
Current time:            Sunday, April 19, 1998 at 12:11:31:670 PM
Times for this drawing:
   Created:              Sunday, April 19, 1998 at 12:00:38.660 PM
   Last updated:         Sunday, April 19, 1998 at 12:00:38.660 PM
   Total editing time:   0 days 00:10:53.010
   Elapsed timer (on):   0 days 00:10:53.010
   Next automatic save in: 0 days 01:49:10.230

Display/ON/OFF/Reset :
```

37.161 The `Time` option under the `Inquiry` menu can be used to obtain the time spent on a drawing and for setting the time for an assignment.

of the current session opposite the heading, `Elapsed time:`.

37.56 Dimensioning

Figure 37.162 shows common types of dimensions that are applied to drawings. Drawings should be drawn full size on the screen since they are measured by the dimensioning commands; scaling should be done at plotting time.

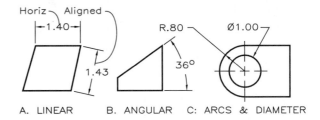

A. LINEAR   B. ANGULAR   C: ARCS & DIAMETER

37.162 The basic types of dimensions that appear on a drawing.

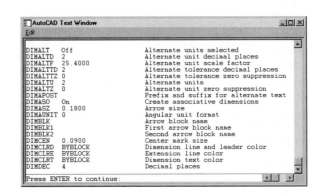

37.163 Type `Dimstyle` at the `Command` line, then `Status`, and (Enter) to obtain a listing of the dimension system variables.

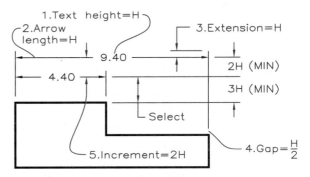

DIM VARS

1 DIMTXT (TEXT HT.)=H=.125
2 DIMASZ (ARROW)=H=.125
3 DIMEXE (EXTENSION)=H=.125
4 DIMEXO (OFFSET)=H/2=.06
5 DIMDLI (INCREMENT)=2H=.25 MIN.
6 DIMSCALE (SCALE)=1 for INCHES
   25.4 for MILLIMETERS

37.164 Dimensioning variables are based on the height of the lettering (text), usually .125-in. high.

Dimensions can be applied as associative or as nonassociative (Exploded) dimensions. Associative dimensions (when Dimaso is On) are inserted as if the dimension line, extension lines, text, and arrows were parts of a single block. Exploded dimensions are applied as individual objects that can be modified independently by setting Dimaso to Off. Except where noted, the examples that follow will be associative dimensions.

Many variables must be set before dimensioning is usable: Arrowheads and numerals must be sized, extension line offsets specified, text fonts assigned, and units adopted, to name a few.

37.57 Dimension Style (Dimstyle) Variables

To get a list of the current dimensioning variables, as shown in **Fig. 37.163**, type Dimstyle at the Command line, then type Status and press Enter. Sizes of dimensioning variables are based on the letter height, which is most often .125 in. (**Fig. 37.164**).

To set and save a few variables needed for basic applications, Open the file B:PROTO1. Each variable is set by typing Setvar and the name of the dimensioning variable (Dimtxt, text height, for example) and assigning a numerical value. A list of the 58 dimensioning variables is given in **Figs. 37.165** and **37.166**. Assign the basic variable values of Dimtxt, Dimasz, Dimexe, Dimexo, Dimdli, and Dimscale (Fig. 37.164) to B:PROTO1, since they apply to most applications. Type Units and set decimal fractions to two decimal places for inches and set Dimaso On.

Save these settings to B:PROTO1 with no drawings on it and use it as the prototype when creating a new file by Save As B:DWG-3, for example, which becomes the current file with the same settings as B:PROTO1.

Dim Vars	Default	Description
DIMADEC	−1	Decimal places for ang. dims.
DIMALT	OFF	Alternate units selected
DIMALTD	2	Alternate unit decimal places
DIMALTF	25.4	Alternate unit scale factor
DIMALTTD	2	Alternate tolerance dec. places
DIMALTTZ	0	Alternate tolerance zero suppress.
DIMALTU	2	Alternate units
DIMALTZ	0	Alternate unit zero suppression
DIMAPOST	—	Default suffix for alternate text
DIMASO	ON	Create associative dimensions
DIMASZ	.125	Arrow length
DIMAUNIT	0	Angular unit format
DIMBLK	—	Arrow block name
DIMBLK1	—	First arrow block name
DIMBLK2	—	Second arrow block name
DIMCEN	.09	Center mark size
DIMCLRD	BYLAYER	Dimension line color
DIMCLRE	BYLAYER	Extension line & leader color
DIMCLRT	BYLAYER	Dimension & extension color
DIMDEC	4	Decimal places for dimensions
DIMDLE	0	Dimension line extension
DIMDLI	.38	Dim. increment for continuation
DIMEXE	.125	Extension beyond dimension line
DIMEXO	.06	Extension line offset
DIMFIT	3	Fit text
DIMGAP	.06	Justification of text on dim. line
DIMJUST	0	Gap from dimension line to text
DIMLFAC	1	Length factor
DIMLIM	OFF	Gives tolerances in limit form
DIMPOST	—	Character suffix after dimensions

37.165 Type Dimstyle at the Command line and Status to obtain a listing of the dimension variables, their default settings, and definitions.

Dim Vars	Default	Description
DIMRND	0	Rounding value for distances
DIMSAH	OFF	Separate arrowheads at each end
DIMSCALE	1	Scale factor for all dim. vars.
DIMSD1	OFF	Suppress first dimension line
DIMSD2	OFF	Suppress second dimension line
DIMSE1	OFF	Suppress first extension line
DIMSE2	OFF	Suppress second extension line
DIMSHO	ON	Changes dimens. while dragging
DIMSOXD	OFF	Suppress outside dimension lines
DIMSTYLE	STANDARD	Current dimensioning stype
DIMTAD	0	Text placed above dimension line
DIMTDEC	4	Tolerance decimal places
DIMTFAC	1	Tolerance text scale factor
DIMTIH	ON	Text inside extension lines horiz.
DIMTIX	OFF	Text forced inside extension lines
DIMTM	0	Minus tolerance value
DIMTOFL	OFF	Forces dim. line inside, text out
DIMTOH	ON	Text outside ext. lines is horiz
DIMTOL	OFF	Applies tolerances to dimensions
DIMTOLJ	1	Tolerance vertical justification
DIMTP	0	Plus tolerance value
DIMTSZ	0	Tick size
DIMTVP	0	Text over or under dimen. line
DIMTXSTY	STANDARD	Text style
DIMTXT	.125	Text height
DIMTZIN	0	Tolerance zero suppression
DIMUNIT	2	Unit format
DIMUPT	OFF	User positioned text
DIMZIN	0	Zero suppression

37.166 Additional dimension variables.

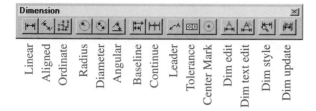

37.167 The Dimension toolbar and its options.

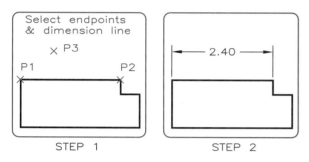

37.168 Dimension a line command.

Step 1 View> Toolbars> Dimension toolbar> Linear icon> First extension line origin or ENTER to select: P1.

Second extension line origin: P2.

Dimension line location (Text/Angle/Horizontal/Vertical/Rotated: P3.

Step 2 Dimension text <2.40>: (Enter) Dimension line is drawn.

Dimensioning variables can be set from dialog boxes also, instead of being typed; these techniques will be covered later. You will be more proficient by becoming familiar with both methods of assigning variables. For now, use the B:DWG-3 file to explore the fundamentals of dimensioning.

37.58 Linear (Dimension Toolbar)

The Dimension Toolbar (**Fig. 37.167**) is a convenient means of selecting dimensioning commands. Select the Linear icon, the points as prompted, and the horizontal option, as shown in **Fig. 37.168**.

A dimension is applied semiautomatically in **Fig. 37.169** by pressing (Enter) when prompted for Endpoints, selecting the line

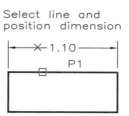
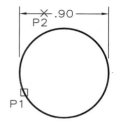

A. HORIZ. DIMEN.   B. HORIZ. DIMEN.

37.169 Semiautomatic Dimension: Line.

A Dimensioning toolbar> Linear> First extension line origin: (Enter).

Select object to dimension: P1.

Dimension line location: P2.

Dimension text <1.10>: (Enter).

B Use these same steps and select a point on the circle (P1) to dimension its diameter.

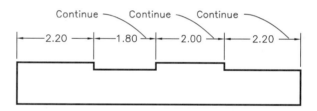

37.170 When dimensions are placed end to end, the Continue option is used to specify the second extension line origin after the first dimension line has been drawn.

or circle to be dimensioned, and locating its dimension line.

The Continue dimension command is selected from Dimension toolbar (or typed at the Command line) to continue a chain of linear, angular, or ordinate dimensions from the last extension line (**Fig. 37.170**). The Baseline option applies dimensions from a single endpoint and each dimension incrementally offset by the dimension line increment variable, Dimdli, that is set with Setvar command (**Fig. 37.171**).

Select the Aligned command from the Dimension toolbar, and you will be prompted

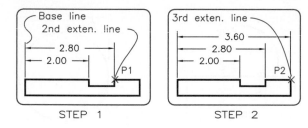

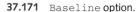

STEP 1 | STEP 2

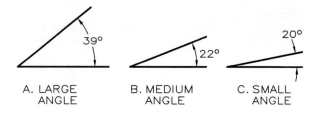

A. LARGE ANGLE | B. MEDIUM ANGLE | C. SMALL ANGLE

37.171 Baseline option.

Step 1 Dimension the first line as shown in Fig. 37.168. When prompted for the next dimension, select Baseline icon and you will be prompted for the Second extension line origin: P1

Step 2 The second dimension line is drawn. Select the Baseline icon again, Second extension line origin: P2. Continue in this manner for other baseline dimensions.

37.173 Angles are dimensioned in any of these three formats.

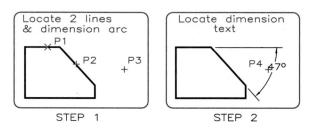

STEP 1 | STEP 2

37.174 Dimensioning Angles.

Step 1 Select Angular icon, Select arc, circle, line, or press ENTER: P1, Second line: P2. Dimension line arc location: P3.

Step 2 Dimension text <47>: (Enter).

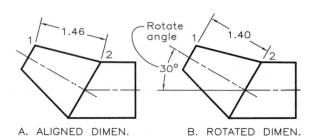

A. ALIGNED DIMEN. | B. ROTATED DIMEN.

37.172 Oblique dimension lines.

A Linear dimensions can be aligned by selecting the Align icon, and selecting endpoints 1 and 2.

B Linear dimensions can be rotated by selecting the Linear icon, the Rotate option (assign an angle of 30°), and endpoints 1 and 2.

the angle, and locate the dimension line arc, as shown in step 1 of **Fig. 37.174**. If space permits, the dimension text will be centered in the arc between the arrows (step 2).

An angle dimension can be applied by selecting its vertex and the endpoints of each line, as shown in step 1 of **Fig. 37.175**. For angles over 180°, select a point on the sides of the angle and move counterclockwise to the second point (step 2) and locate the dimension arc.

to select the first and second extension lines and the position of the dimension line (**Fig. 37.172A**). The dimension line will be inserted aligned with line 1–2. Extension lines can be automatically drawn by pressing (Enter) at the first prompt and selecting the line to be dimensioned. A rotated dimension can be applied with the Rotate option and an assigned angle of rotation (**Fig. 37.172B**).

37.59 Angular (Dimension Toolbar)

Figure 37.173 shows variations for dimensioning angles, depending on the space available. Select the Angular icon, select lines of

37.60 Diameter (Dimension Toolbar)

Diameters of circles can be placed as shown in **Fig. 37.176**, depending on the available space. By changing system variables Dimtix and Dimtofl, circles can be dimensioned as shown in **Figs. 37.177** and **37.178**. By setting the Dimtix system variable On, the text is forced inside the extension lines, regardless of the available space. The Dimtofl (On) dimension variable forces a dimension line to be

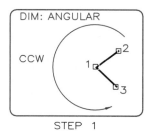
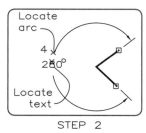

37.175 Dimensioning Angles: Endpoints.

Step 1 Select Angular icon> Select arc, circle, line, or press ENTER: (Enter) Angle vertex: 1

First angle endpoint: 2 Second angle endpoint: 3

Step 2 Dimension arc line location: 4

Dimension text <280>: (Enter) Dimension arc is drawn.

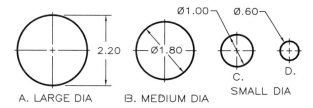

37.176 Examples of methods of dimensioning circles.

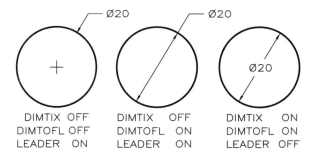

37.177 Examples of circle dimensions with various variable settings and their associated dimensioning variables.

drawn between the arrows when the text is located outside. Specifying whether or not a dimension has a Leader (Fit window of the Format box in Fig. 37.185) also affects the type dimension.

37.61 Radius (Dimension Toolbar)

Select Radius dimension from the Dimension toolbar to dimension areas with an R placed in front of the text (R1.00, for example),

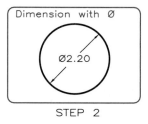
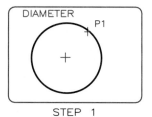

37.178 Dimensioning a circle.

Step 1 Pick Diameter icon> Select arc or circle: P1. Dimension text <2.20>: (Enter).

Step 2 Dimension line location (Text/Angle): (Enter).

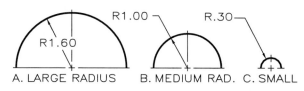

37.179 Arcs are dimensioned by one of the formats given here, depending on the size of the radius.

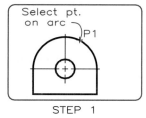
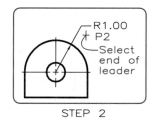

37.180 Radius: Leader option.

Step 1 Pick Radius icon> Select arc or circle: P1. Dimension text <2.20>: (Enter).

Step 2 Dimension line location (Text/Angle): (Enter).

as shown in **Fig. 37.179**. Dimensioning an arc with a radius and leader is shown in **Fig. 37.180**.

The Leader command is used to add a dimension or a note to a drawing, but it cannot measure the circle; it inserts the value of the last measurement made. The circle's diameter must be known and typed in to override this measurement (**Fig. 37.181**).

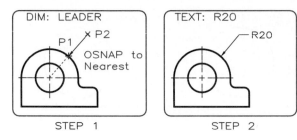

STEP 1   STEP 2

37.181 Dimensioning with leaders.

Step 1 `Dimension toolbar> Leader icon> From point:` P1.

`To point:` P2. `To point: (Format/Annotation/Undo) <Annotation>:` (Enter).

Step 2 `Annotation` (or press ENTER for options): R20 (Enter).

`Mtext:` (Enter) The leader is completed and the text is drawn.

37.62 Dimension Style (Dimension Toolbar)

Select `Dimension Style` icon (or type `DDim`) from the `Dimension toolbar` to display the `Dimension Styles` dialog box shown in **Fig. 37.182**, from which a number of variables can be assigned for three major categories: **Geometry, Format**, and **Annotation**. Each group of settings can be made and saved by style name (**NEW-4**, for example) for future use.

Geometry

Click on the `Geometry` box to get the `Geometry` menu shown in **Fig. 37.183**. The options under the `Dimension-Line` group controls the following variables: `Dimsd1`, `Dimsd2`, `Dimdle`, `Dimdli`, and `Dimclrd`. The `Suppress 1st` and `2nd` boxes turn `On` the `Dimsd1` and `Dimsd2` variables to suppress the first and second dimension lines. When `Oblique-Stroke` arrows are used, the value placed in the `Extension` box specifies the distance that the dimension line extends beyond the extension line. The `Spacing box` is used to set `Dimdli`, which controls the spacing between baseline dimensions. The `Color` button displays the color menu from

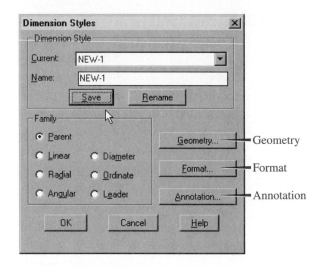

37.182 Select `Dimension toolbar> Dimstyle icon` to get this `Dimension Styles` box for setting `Geometry`, `Format`, and `Annotation` values.

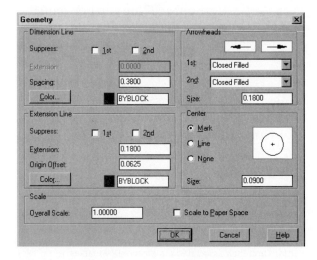

37.183 `Dimension Styles> Geometry` button opens this `Geometry` menu box.

which to select a color for the dimension line (`Dimclrd`).

The options of the `Extension-Line` group (Fig. 37.183) control the following variables: `Dimse1`, `Dimse2`, `Dimexe`, `Dimexo`, and `Dimclre`. The `Suppress 1st` and `2nd` boxes turn `On` the `Dimse1` and `Dimse2` variables to suppress the first and second exten-

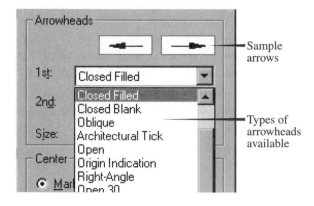

37.184 From the Geometry menu, Arrowheads can be selected from the list.

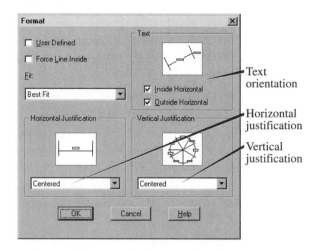

37.185 Dimension Styles> Format button opens this Format menu box for making a variety of settings.

sion lines. The value typed in the Extension box specifies the distance the extension line extends beyond the dimensioning arrow (Dimexe). The Origin Offset option is used to specify the size of the gap between the object and the extension line (Dimexo). The Color button lets you select the color of the extension lines (Dimclre).

The options of the Arrowheads group (**Fig. 37.184**) control the following variables: Dimasz, Dimtsz, Dimblk1, and Dimblk2. The value typed in the Size box gives the size of the arrowhead, the Dimasz variable. By selecting scroll arrows next to the 1st or 2nd boxes, the types of arrowheads for each end of the dimension are listed. If only the first arrow type is selected, it is automatically applied to the second end unless a second arrow type is specified. Tick marks are given when Oblique is selected, the Dimtsz variable. When User Arrow is selected, custom-made arrows can be inserted (Dimblk1 and Dimblk2).

Select Mark, Line, or None (equivalent to the Dimcenter command) to draw center marks, center lines, or nothing on arcs and circles when diameter or radius dimensions are placed outside by the Dimdiameter and Dimradius commands. The value placed in the Size box gives the size of the center mark

(a plus mark). A minus value gives center lines and a zero gives none.

The Overall Scale box specifies Dimscale, which controls the geometry of dimensioning variables—arrow size, text height, extension-line offsets, center size, and others. When set to 1, these variables are full size; when set to 2, they are twice as big. The Scale to Paper Space box (**Fig. 37.183**) is picked to adjust the dimensions in model space to the scale of paper space. A default of 1 is used if you are not working in paper space.

Format
Select the Format button (Fig. 37.182) to display the Format menu in **Fig. 37.185** to control the placement of text, arrows, leaders, and dimension lines. Select User Defined (Dimupt) to place the text of a dimension with the pointer, and disable this option for the text to be placed automatically. Select the scroll arrow at the Fit box (Dimfit) to obtain options of Text and Arrows, Text Only, Arrows Only, Best Fit, and Leader that specify what is placed inside extension lines if space is limited (**Fig. 37.186**).

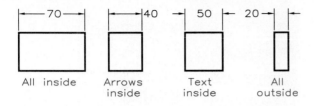

37.186 Examples of dimensions applied using the `Best Fit` option.

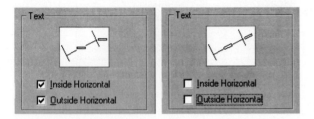

37.187 These examples show the results of having the `Dimtoh` and `Dimtih` set to `On` and `Off`.

37.188 Examples of settings of `Horizontal Justification`.

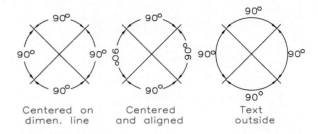

37.189 Examples of settings of `Vertical Justification`.

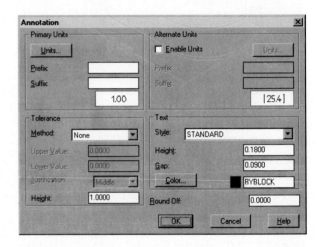

37.190 `Dimension Styles> Annotation` to get this `Annotation` menu box for setting text and numeral features.

When `Inside Horizontal (Dimtih)` and `Outside Horizontal (Dimtoh)` are picked, text inside extension lines and outside extension lines is placed horizontally (**Fig. 37.187**).

Select the scroll arrow under `Horizontal Justification` to obtain `Centered, 1st Extension Line, 2nd Extension Line, Over 1st Extension,` and `Over 2nd Extension` for placing text along the dimension line, as shown in **Fig. 37.188**. `Vertical Justification` has the options of `Centered, Above (Dimtad),` and `Outside` for positioning text relative to the dimension line (**Fig. 37.189**).

Annotation

Click on the `Annotation` box (Fig. 37.182) to get the menu shown in **Fig. 37.190**, which controls the dimensioning of text. `Select Units,` under `Primary,` and the `Primary Units` menu (**Fig. 37.191**) are displayed. Under `Units,` select the scroll arrows and a unit type. Using the scroll arrow under `Precision,` select the number of decimals places (or fractions) desired. From the `Angle` area, the `Units` of degrees and their `Precision` can be specified in the same manner as above.

In the `Primary Units` menu (Fig. 37.191), `Zero Suppression (Dimzin)` boxes suppress zeros that are leading or trailing dec-

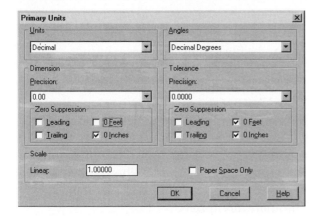

37.191 `Annotation menu box> Primary units box` lets you make changes to the dimensioning text settings.

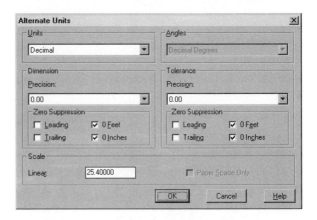

37.193 `Annotation Menu> Alternate Units` key gives this window for alternate and dual dimensions.

0 Ft 0 In	1/4"	4"	1'	1'−01/4"
No options	0'−01/4"	0'−6"	2'−0"	1'−01/4"
0 In	0'−01/4"	0'−4"	1'	1'−01/4"
0 Ft	1/4"	4"	1'−0"	1'−01/4"

37.192 The `Zero Suppression` (`Dimzin`) options control the leading and trailing zeros in dimensioning, especially when applied to architectural dimensions.

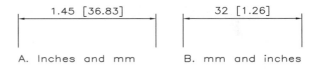

A. Inches and mm   B. mm and inches

37.194 Examples of alternate units (dual dimensions) made in inches and millimeters.

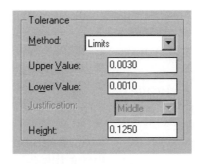

37.195 `Annotation menu> Tolerance` gives a menu for formatting dimensions with tolerances.

imal points. Select `0 Inches` and `Leading` to make 0.20 become .20. **Figure 37.192** shows the results of applying the four options to architectural dimensions. From under `Scale`, `Linear` (`Dimlfac,`) units for measuring dimensions can be given, and `OK` selected to return to the `Annotation` menu.

Under `Alternate Units`, select `Enable Units for dual dimensions` (`Dimalt`) to be used in dimensions; the scale of 25.4 gives millimeter equivalents for inches as an alternate dimension. When `Units` is selected, the `Alternate Units` menu is displayed (**Fig. 37.193**) and settings are made in the same way as `Primary Units`. Examples of dimensions with alternate units are shown in **Fig. 37.194**.

From the `Annotation` menu (**Fig. 37.195**), a `Prefix` and a `Suffix` (`Dimapost`) can be inserted to give text as R12 or as 26 mm for `Primary` and `Alternate Units` when `Enabled`. `Tolerance Method` can be selected from options of `None`, `Symmetrical`, `Deviation`, `Limits`, and `Basic`. The following values are assigned: `Upper Value` (`Dimtp`), `Lower Value` (`Dimtm`), `Justification` (`Dimtolj`), and

2.0000 ±.0020	2.0000 $\begin{array}{c}+.0030\\-.0020\end{array}$	2.0030 1.9980
DIMTP & DIMTM SAME	DIMTP & DIMTM DIFFERENT	DIMLIM

37.196 Examples of various formats for tolerancing.

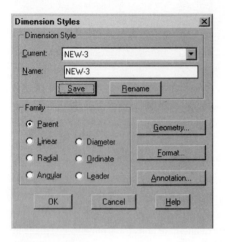

37.197 Select `Dimensioning toolbar> Dimension Styles> Parent> Save` to keep these settings.

`Height`. Some examples of these are shown in **Fig. 37.196**. `Height` is specified as a ratio of the primary text height, the basic dimension.

The `Text` group in Fig. 37.190 gives options for selecting a text `Style` (`Dimtxsty`) created with the `Style` command, assigning a text `Height` (`Dimtxt`), specifying the `Gap` (`Dimgap`) around the text, and picking a `Color` for the text. Set the `Round Off` value (`Dimrnd`) for all dimensions except angular measurements. Press the `OK` button to return to the `Dimension Styles` menu (**Fig. 37.197**).

37.63 Saving Dimension Styles

After setting the previous dimensioning variables, they can be saved as a parent `Dimension Style` (`Dimstyle`) (from the `Dimension toolbar`) by picking `Parent`, typing the name (`Style-1`, for example) in the `Name` box, and

selecting the `Save` button (Fig. 37.197). `Dimension Styles` should be saved to your prototype file to preserve these settings.

`Style families` can be created to include special applications of variables. If you want to `Force Line Inside` (`Dimtofl`) for diameter dimensions only, select the family name button, `Diameter`, select the `Parent Style` from the `Current Style` list, make modifications to the `Parent`, and `Save`. When this `Parent Style` is opened, the family member style is automatically applied.

37.64 Dimension Style Override (Dimension Toolbar)

If you want to `Override` (`Dimoverride`) one or more settings previously saved to a `Dimension Style`, from the `Dimension toolbar`, select `Dimension Style`, and pick the `Style` from the `Current Style` list. If you want to change `Dimscale`, for example, make this change to the style and choose the `OK` box. The style name is listed with a plus mark—for example, `+Style-1`.

To apply the `Override` settings, type `Dimstyle` at the `Command line`, enter `A` for `Apply`, and select the dimensions that are to be updated (**Fig. 37.198**). To `Restore` dimensions to a different set of variables, type `Dimstyle` at the `Command line`, enter `R` for `Restore`, and select the dimensions whose style is to be restored.

37.65 Editing Dimensions

When the dimensioning variable `Dimaso` is set to `On`, the dimensioning entities (arrows, text, extension lines, etc.) become a single unit (associative) once a dimension has been inserted into a drawing.

A closely related dimensioning variable is `Dimsho`, which can be set on or off. When on,

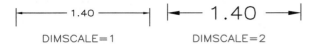

DIMSCALE=1   DIMSCALE=2

37.198 When `Dimscale` is changed from 1 to 2 and `Override` is used, the selected dimension is updated and doubled in size.

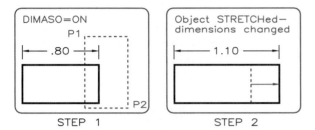

STEP 1   STEP 2

37.199 Associative dimensions: `Dimaso` and `Dimsho`.

Step 1 Set `Dimaso` and apply dimensions to the part. Use the `Stretch` command and a `Crossing` window at the end of the part.

Step 2 Drag to a new endpoint for the part, which will be lengthened, and a new dimension will be calculated and shown.

`Dimsho` causes the dimensioning numerals to dynamically change on the screen as the dimension line is `Stretched`. When `Dimaso` is on and `Dimsho` is off, the numerals will be changed after the `Stretch` (**Fig. 37.199**), but not dynamically during the `Stretch`.

When in the associative dimensioning mode, `Dimaso On, Dimension Text Edit` can be selected from `Dimension toolbar` to obtain the options of `Home, New, Rotate,` and `Oblique` for changing existing dimensions. `Home` repositions text to its standard position at the center of the dimension line after being changed by `Stretch` commands (**Fig. 37.200**). `New` changes text within a dimension line (**Fig. 37.201**) by pressing `Enter` when prompted and inserting the new text. `Rotate` positions dimension text at any specified angle (**Fig. 37.202**). `Oblique` converts extension lines to angular lines (**Fig. 37.203**).

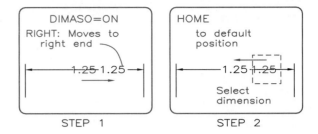

STEP 1   STEP 2

37.200 Associative dimensions: Left, right, home.

Step 1 `Dimension toolbar> Dimension text edit icon>` Select dimension P1, `Enter text location (Left/Right/Home/Angle):` <u>Right</u> **(Enter)** Numeral moves to right.

Step 2 Command: **(Enter)** `Select dimension:` <u>P2</u>.

`Enter text location (Left, Right/Home/Angle):` <u>Home</u>. **(Enter)**

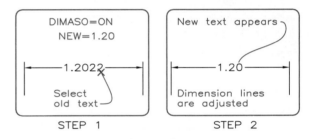

STEP 1   STEP 2

37.201 Associative dimensions: `New` text.

Step 1 `Dimension toolbar> Dimension Edit icon> New> Multiline Text Editor` appears. Type new text in the angles (`<NEW>`), pick `OK` to return to dimension.

Step 2 `Select objects:` <u>P1</u> **(Enter)**.

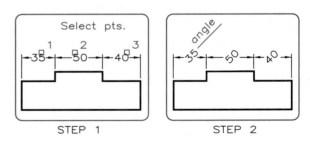

STEP 1   STEP 2

37.202 Associative dimensions: `Rotate` text.

Step 1 `Dimension toolbar> Dimension edit> (Home/New/Rotate/Oblique):` <u>Rotate</u> **(Enter)**, `Enter text angle:` <u>45</u> **(Enter)**.

Step 2 `Select objects:` <u>P1</u> **(Enter)**.

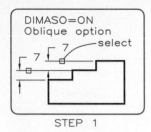
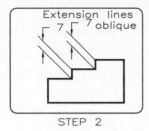

STEP 1   STEP 2

37.203 Associative dimensions: Oblique extension lines.

Step 1 `Dimension toolbar> Dimension edit icon>` `(Home/New/Rotate/Oblique):` <u>Oblique</u> (Enter).

`Select objects:` <u>P1</u>, `Enter text location (Left/` `Right/Home/Angle):` <u>Right</u> (Enter) `Numeral moves to right.`

Step 2 `Command:` (Enter) `Select objects:` (Enter).

`Enter obliquing angle:` <u>135</u> (Enter).

2.0000 ±.0020	2.0000 $^{+.0030}_{-.0020}$	2.0030 1.9980
DIMTP & DIMTM SAME	DIMTP & DIMTM DIFFERENT	DIMLIM

37.204 Dimensions can be toleranced in any of these three formats.

When `Dimension Text Edit` is selected from the `Dimension toolbar,` options of `Left/Right/Home/` and `Angle` are given for positioning and angling the text within the dimension lines.

37.66 Toleranced Dimensions

Dimensions can be toleranced automatically using the `Dimension Styles` defined previously in which the settings shown in **Fig. 37.204** are made: `Dimtol` (tolerance on), `Dimtp` (plus tolerance), and `Dimtm` (minus tolerance). When `Dimlim` is On, the upper and lower limits of the size are given (**Fig. 37.205**).

`Dimtfac` is a scale factor that controls the text height of the tolerance values, which is about 80 percent of the basic dimension.

37.67 Geometric Tolerances (Dimension Toolbar)

Geometric tolerances specify the permissible variations in form, profile, orientation, loca-

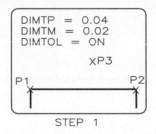
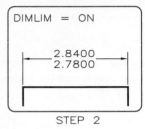

STEP 1   STEP 2

37.205 Tolerances: Limit form.

Step 1 Set the `Dimension variables` above to the values shown. Apply the dimension line by selecting the endpoints <u>P1</u> and <u>P2</u> and locate the dimension lines with <u>P3</u>.

Step 2 The dimension will be shown with its upper and lower limits based on the limits given by `Dimtp` and `Dimtm`.

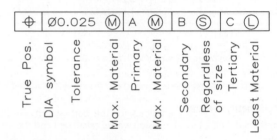

37.206 The parts of the feature control frame that give geometric tolerance specifications are defined here.

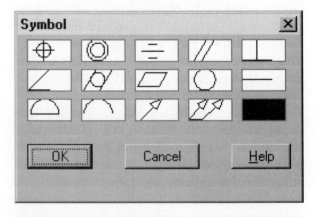

37.207 Various geometric tolerances can be selected from the `Symbol` menu.

tion, and runout. A typical geometric tolerance feature control frame is given in **Fig. 37.206**.

From the `Dimensioning toolbar,` select `Tolerance` to display the `Symbol` menu (**Fig. 37.207**), select the desired symbol,

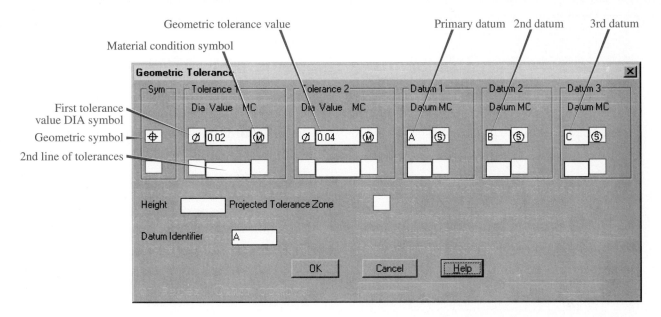

Geometric tolerance value

Material condition symbol

Primary datum 2nd datum 3rd datum

First tolerance value DIA symbol

Geometric symbol

2nd line of tolerances

37.208 The `Geometric Tolerance` box, (`View> Dimension toolbar> Tolerance`), is used to give specifications for geometric tolerances.

| ⊕ | Ø0.025 Ⓜ | P | S | T |

A. Tolerance of position

| ⊥ | 0.004 | A |

B. Tolerance of perpendicularity

37.209 Applications of `Geometric Tolerances` in the feature control frames are shown here.

and pick `OK` to display the `Geometric Tolerance` box (**Fig. 37.208**). In `Tolerance 1`, select `Dia` to insert the symbol; under `Value`, type the first tolerance `Value`, pick `MC` to give the material condition, then pick the material condition from the `Material Condition` box. Give the `Tolerance 2` value (optional) in the same way as `Tolerance 1`.

Under `Datum 1`, enter the letter for the primary reference datum, choose `MC`, and select your preference from the `Material Condition` box. Add `Datum 2` and `Datum 3` values in the same manner, select `OK`, and pick the location of the feature control frame (**Fig. 37.209**).

37.68 A Custom Title Block

Having covered most of the basics of two-dimensional drawing, you may wish to make your own customized title blocks with their own unique parameters that can be used to create new drawings as introduced in Section 37.11. Many problems in this format are given at the ends of the chapters. They can be laid out and solved by computer using your customized title block (**Fig. 37.210**).

To make a title block identical to the one used in these problems, draw the border that is 7.6 in. wide × 10.3 in. high, with its lower-left corner at 0,0, with a polyline that .016 in. thick. Draw a title strip across the bottom that has two rows of 1/8-in. text with margins of 1/8-in. above and below each line. Fill out the title strip using the **Romand** font (double-stroke Gothic) and fill in all blanks (name, date, etc.); they will be changed later. Set `Snap` and `Grid` to 0.2 in., set any other variables to

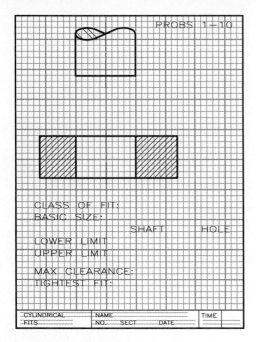

37.210 An example of a problem sheet title block and border that you can draw and save as a template file.

be based on a letter height of 1/8 in. Save the drawing (`Menu bar> Save As> File Name>`: <u>`PROB-1.DWT`</u> (by typing the file name and selecting `Save as Type` box and picking `Drawing Template File` (`*.DWT`).

When you next create a new drawing (`Menu bar> File> New`), pick `Use a Template`, select the `PROB-1.DWT` template, and you have a title block ready for drawing. But first use `DDedit` (`Modify II Toolbar> Text edit icon`), select the title strip text that needs to be edited, and update each entry. Save this setup to a file, (`Menu bar> File> Save As>`: <u>`ASSIGN-1`</u>), for example.

37.69 Digitizing with the Tablet

Drawings on paper can be taped to a digitizing tablet and digitized point-by-point. A drawing is calibrated in the following manner:

Command: <u>`Tablet`</u> (Enter)

Option (ON/OFF/CAL/CFG): <u>`CAL`</u> (Enter)

(Calibrate tablet for use.)

Digitize first known point: (`Digitize point.`)

Enter coordinates for first point: <u>`1,1`</u> (Enter)

Digitize second known point: (`Digitize point.`)

Enter coordinates for second point: <u>`10,1`</u> (Enter)

Digitize points from left to right, or from the bottom to the top of the drawing. Use `On` or `Off` to activate or deactivate the tablet mode. Function key `F10` also turns the tablet off so that the cursor can select from the screen menus. To draw lines, select the `Line` command from one of the menus (or type `L`), and pick points on the tablet with the stylus.

37.70 Sketch (Miscellaneous Toolbar)

The `Sketch` command can be used with the tablet for tracing drawings composed of irregular lines (**Fig. 37.211**). Tape the drawing to the tablet and calibrate it as discussed previously, following these steps:

Command: <u>`Sketch`</u> (Enter)

Record increment <0.1>: <u>`0.01`</u> (Enter)

Sketch. `Pen eXit Quit Record Erase Connect.`

The record increment specifies the distances between the endpoints of the connecting lines that are sketched. Other options are as follows:

`Pen`	Raises or lowers pen
`eXit`	Records lines and exits
`Quit`	Discards temporary lines and exits
`Record`	Records temporary lines
`Erase`	Deletes selected lines

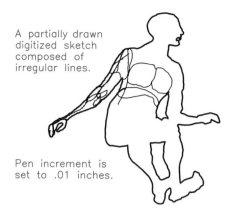

A partially drawn digitized sketch composed of irregular lines.

Pen increment is set to .01 inches.

37.211 This drawing was made with the `Sketch` command and a tablet instead of a mouse. The drawing was taped to the tablet and traced with a stylus in increments of .01 in.

`Connect` Joins current line to last endpoint

`.(period)` Draws line from current point to last endpoint

Begin sketching by moving your pointer to the first point, lower the pen (P), move the stylus over the line; the line is sketched on the screen. To erase, raise the pen (P), enter `Erase` (E), move the stylus backward from the current point, and select the point where the erasure is to stop. All lines are temporary until you select `Record` (R) or `eXit` (X) to save them. Begin new lines by repeating these steps.

The `Skpoly` variable can be set by `Command line> Setvar> Skpoly> 1 (On)`. When using the `Sketch` command with `Skpoly`, lines will be drawn as continuous polylines—as an alternative to lines composed of individual line segments.

37.71 Oblique Pictorials

An oblique pictorial can be constructed as shown in **Fig. 37.212**. The front orthographic view is drawn and `Copyed` behind the first view at the angle selected for the receding axis. The visible endpoints are connected with

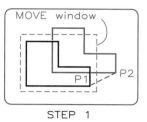
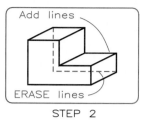

MOVE window

P1 P2

STEP 1

Add lines

ERASE lines

STEP 2

37.212 Oblique pictorial.

Step 1 Draw the front surface of the oblique and `Copy` the view from <u>P1</u> to <u>P2</u> using `Osnap`.

Step 2 Connect the corner points and erase the invisible lines to complete the oblique.

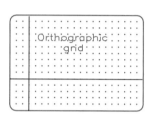
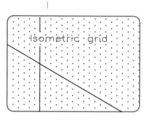

:Orthographic: :grid:

Isometric grid

A. ORTHOGRAPHIC GRID   B. ISOMETRIC GRID

37.213 Types of screen grids.

A The orthographic grid is called the `Standard(S)` style of the `Snap` mode command.

B The `isometric (I)` style of the `Snap` mode.

`Osnap` on and invisible lines are erased. Circles are drawn as true circles on the true-size front surface, but circular features should be avoided on the receding planes since their construction is complex.

37.72 Isometric Pictorials

The `Style` option of the `Snap` command is used to change the rectangular `Grid` from `Standard (S)` to `Isometric (I)` to show the grid dots in an isometric pattern (vertically and at 30° with the horizontal) (**Fig. 37.213**). The cursor will align with two of the isometric axes and can be snapped to the grid. The axes of the cursor are rotated 120° by pressing `Ctrl E` or by typing `Isoplane` at the `Command line`. When `Ortho` is `On`, lines are forced parallel to the isometric axes. **Figure 37.214** shows the steps for constructing an isometric pictorial.

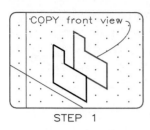

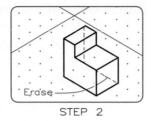

STEP 1   STEP 2

37.214 Isometric pictorial.

Step 1 Draw the front view of the isometric and `Copy` it at its proper depth.

Step 2 Connect the corner points and erase the invisible line. The cursor lines can be moved into three positions using `Ctrl E` for `Isoplane`.

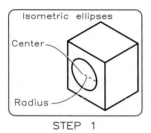

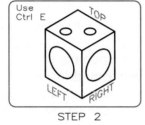

STEP 1   STEP 2

37.215 Ellipse: Isometric mode.

Step 1 When in isometric Snap mode:

Command: `Ellipse` (Enter).

`<Axis endpoint 1>/Center./Isocircle:` `I` (Enter).

`Center of circle:` Select center.

`<Circle radius>/Diameter:` Select radius.

Step 2 The isometric ellipse is drawn on the current Isoplane.

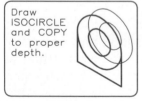

STEP 1   STEP 2

37.216 An isometric with partial ellipses.

Step 1 Draw an `Isocircle` (ellipse) as shown in step 1 of Fig. 37.215. Copy the `Isocircle` to establish the part's depth.

Step 2 Use the `Trim` command to remove the unneeded portions of the Isocircles. Add missing lines.

When the `Grid` is set to the isometric mode, the `Ellipse` command will give the following options: `<Axis endpoint 1>/Center/ Isocircle:` `I` (Enter). Select the `Isocircle` option to position ellipses in one of the isometric orientations, as shown in **Fig. 37.215** by using `Ctrl E`. An example of an isometric with partial ellipses is shown in **Fig. 37.216**.

The oblique and isometric drawings covered here are two-dimensional drawings that appear to be three-dimensional views, but they cannot be rotated on the screen to obtain different viewpoints of them.

Problems

1. Title block and border: Design a title block and border that can be used for an A-size sheet and/or a B-size sheet that could be used in your course. Begin by determining the essential information that belongs in a title block.

2. Setting up a drawing: Before drawing by computer or by hand, a certain amount of planning is necessary. Assume that you must make a drawing that would require a space of 50 in. × 30 in. (including dimensions and notes) if it were drawn full size. However, you must make it fit on a B-size sheet.

For it to be drawn and fill most of the B-size sheet, determine the following:

    A. Final scale on the B-size sheet. (This should be a standard scale.)

    B. Scale for drawing on the screen.

    C. Height of the lettering and the sizes of the dimensioning variables on the computer screen.

    D. The thicknesses of the polylines for visible, hidden, and dimension lines.

    E. What other helpful factors must be determined before beginning the drawing?

3. Advantages of graphical methods: On an 8.5 × 11 sheet of paper, make two vertical columns. In the column at the left, list the advantages of making an engineering drawing by hand. In the column at the right, list the advantages of making engineering drawings by computer. Compare your evaluation with a classmate's response.

4. Disadvantages of graphical methods: Same as problem 3, but list the disadvantages of each method: computer graphics and hand graphics.

5. Cost analysis: Will a computer graphics system pay for itself? If computer graphics is more productive than drawing by hand, you can build a case for the capital outlay. If it is not more productive, could you justify its expense?

You must begin this analysis by determining the productivity factor of computer graphics. If you think computer graphics is twice as fast as hand graphics, the productivity factor would be 2. This factor tells you that twice as much work can be done by computer, or half as many employees are necessary.

Next, you must determine the cost of a computer system: hardware, software, furniture, and printers. Visit computer stores and make an itemized list to determine a price range for a system.

Determine the average wage for drafters in your community by looking in the want ads and by making personal inquiries. (We can assume that computer drafters are paid the same as hand drafters.)

Begin with the productivity factor that you feel is correct, the wages of drafters, and cost of a computer system and determine the cost effectiveness of a computer system if it were to be paid for at the end of three years of ownership. Present your analysis in a clear manner so that your employer could make a decision based on your findings.

6. Self-employment study: If you wished to become self-employed and open an office in your community to offer graphical services, what opportunities can you recognize as being available? Is there a niche that needs filling that you could take advantage of? Brainstorm these possibilities: mapping services, visual aids, publication preparation, engineering drawings, 3D technical illustrations and similar applications.

Do a cost analysis to determine the expense for rent, furniture, equipment, 2 employees, air conditioning, utilities, and all other costs that you can think of. What is the bottom line expense and how much business must you generate per month to break even?

7. Want ad survey: Over a period of two weeks, collect want ad information from an area newspaper. Organize this data in a presentable form to determine the market for computer graphics skills. How much is the pay? How many jobs are available? What is your assessment of this field?

8. Work station: Design a computer graphics work station that conveniently accommodates the needed equipment: computer, monitor, printer, table space, and perhaps a scanner. Determine an effective method of consolidating the many connecting wires and power cables that must accompany computer systems. Can you design a station better than those on the market?

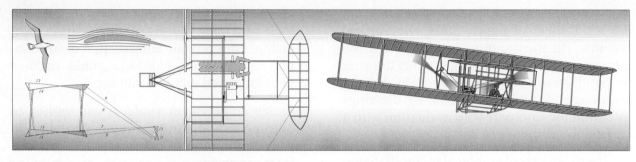

38

Three-Dimensional Modeling

38.1 Introduction

This chapter provides an introduction to the principles of making true three-dimensional pictorial drawings that can be rotated on the screen and viewed from any angle as if they were held in your hand. The three major divisions of this chapter focus on fundamentals of 3D drawing, solid modeling, and rendering.

AutoCAD provides several methods of executing most commands: from the Command line, the Menu bar, and Toolbars. Each of these techniques is used in the coverage that follows in this chapter, but the command-line technique is used most.

38.2 Paper Space and Model Space: An Overview

The two distinctly different ways of obtaining views of objects are with `Tilemode=1` (On) or `Tilemode=0` (Off). When `Tilemode`

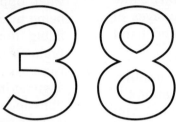

| A. TILEMODE=1 | B. TILEMODE=0 |

38.1 Effects of `Tilemode`.

A When `Tilemode` is set to 1 (On), the viewports are arranged to abut each other like flooring "tiles."

B When `Tilemode` is set to 0 (Off), 3D viewports are made with `Mview` (make view) and can overlap.

is on, the screen can be divided into viewports (`Vports`) that abut each other in standard arrangements as do flooring tiles (**Fig. 38.1A**). When `Tilemode` is off, the viewports are created with `Mview` (make view)in standard or nonstandard positions, as shown in **Fig. 38.1B**.

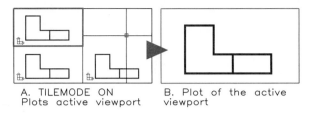

A. TILEMODE ON
Plots active viewport

B. Plot of the active
viewport

38.2 When `Tilemode` is set to 1 (On), only the active viewport can be plotted to paper.

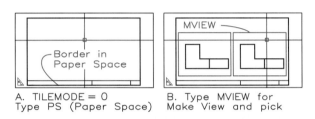

A. TILEMODE = 0
Type PS (Paper Space)

B. Type MVIEW for
Make View and pick

38.3 `Tilemode Off` and `On`.

A When `Tilemode` is set to zero and `PS` is typed, the screen is in paper space where a 2D border can be drawn.

B Use the `Mview` command to define the diagonal corner of a 3D viewport. Create a second viewport in the same manner.

Double Click

38.4 Double click on `Model` (or `Pspace`) in this box in the `Status bar` at the bottom of the screen to alternate between being in model space and paper space.

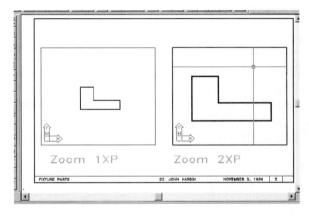

Zoom 1XP   Zoom 2XP

38.5 Type `MS` to enter `Model Space`, type `Zoom`, and `1XP` to show the drawing full-size in that viewport. Select the other viewport, then type `Zoom` and `2XP` to obtain a double-size view in that viewport.

Tilemode On

By setting `Tilemode=1` (On), the screen enters `Model Space` by default and drawings can be made in two dimensions (2D) and three dimensions (3D). The `Vports` command lets you create up to four tiled viewports at a time on the screen, and up to a total of 16 when viewports are subdivided. Drawings made before applying the `Vports` command will be duplicated in each viewport as if multiple monitors were wired to your computer (**Fig. 38.2**). Only the viewport in which the cursor appears is active and can be plotted.

Tilemode Off

Set `Tilemode=0` (Off), type `PS` (Paper Space) to enter paper space (2D space). The word

`Paper` appears in the command bar at the bottom of the screen, and a triangle icon appears at the lower left of the drawing area. Since drawings made in `Paper Space` (PS) are 2D drawings, now is a good time to insert a 2D border (**Fig. 38.3A**).

Type `Mview` and create two model-space windows by selecting their diagonal corners (**Fig. 38.3B**). Type `MS` (Model Space) or double click on the `Paper button` at the bottom of the screen in the `Status bar` to enter `Model Space` (`Model`) and the cursor will appear in the active viewport (**Fig. 38.4**). Move to a new viewport and select it with the cursor. The drawing in the model-space port is scaled by typing `Zoom` and `1XP` for a full-size drawing, `2XP` for a double-size drawing, and `0.5XP` for a half-size drawing (**Fig. 38.5**).

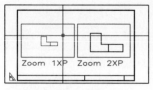 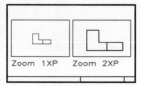

Zoom 1XP Zoom 2XP   Zoom 1XP Zoom 2XP

A. Plot from PS   B. All MS ports plot

38.6 Plotting from `paper space`.
A Select `PS` to enter paper space for plotting a drawing.
B Both paper space and model space are plotted.

To make a plot, return to paper space by typing `PS`; the triangular icon reappears, and `Paper` replaces `Model` at the bottom of the screen in the `Status bar`. Plot from paper space and both the 2D and 3D drawings will plot as they appear on the screen (**Fig. 38.6**).

38.3 Paper Space Versus Model Space

The following points summarize what you can do from paper space (`PS`):

1. `Mview` makes model space (`MS`) viewports.

2. `Stretch`, `Move`, and `Scale` `MS` viewports.

3. `Erase` `MS` viewports.

4. `Freeze` `MS` outlines.

5. `Insert` 2D drawings.

6. `Hideplot` removes invisible lines from selected viewports.

7. `Text` can be added across `MS` viewports.

The following points summarize what you can do from model space (`MS`):

1. `Modify` a 3D drawing.

2. `Rotate` the User Coordinate System (`UCS`).

3. `Pan`, `Zoom`, `Scale`, etc., `MS` drawings.

4. `Attach` dimensions to the `MS` drawing.

5. `Erase` the contents of an `MS` viewport.

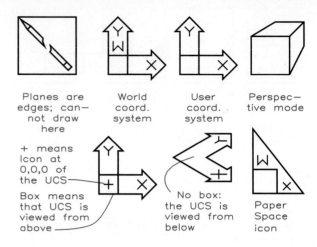

Planes are edges; cannot draw here

World coord. system

User coord. system

Perspective mode

+ means Icon at 0,0,0 of the UCS

Box means that UCS is viewed from above

No box: the UCS is viewed from below

Paper Space icon

38.7 The various icons that appear on the screen to show the X and Y axes when using the `Vport` and `UCS` commands.

38.4 Fundamentals of 3D Drawing

Experiment with 3D drawing; set `Tilemode=1 (On)`, type `UCSicon`, set to On, and the `XY` icon appears, as shown in **Fig. 38.7**. Select the `ORigin` option of the `UCSicon` command to make it appear at the origin. A plus sign appears in its corner, and a `W` appears on the icon when it represents the **World Coordinate System** (**WCS**). Without the `W`, you are in the **User Coordinate System** (**UCS**). The broken-pencil icon warns that the projection plane appears as an edge, making drawing in that viewpoint impractical. The oblique-box icon indicates that the current drawing is in a perspective mode. The triangular icon tells you that the screen is in 2D paper space. `UCSicon` never shows the `Z` axis; it is found by the right-hand rule shown in **Fig. 38.8**.

The standard `Tiled Vports` are shown in the `Tiled Viewport Layout` menu in **Fig. 38.9**. Use the `Save` option of the `Vports` command to save `Vport` arrangements.

38.5 Elementary Extrusions

The elementary extrusion technique is a method of drawing 3D objects with the `Elev`, `Thickness`, `Plan`, and `Hide` commands.

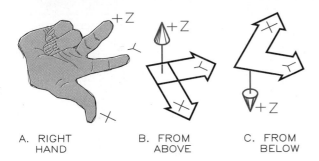

| A. RIGHT HAND | B. FROM ABOVE | C. FROM BELOW |

38.8 By pointing the thumb of your right hand in the positive X direction and your index finger in the positive Y direction, your middle finger points in the positive Z direction.

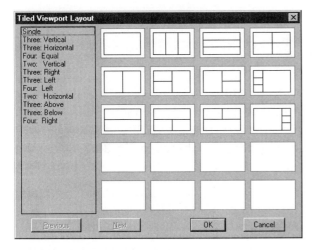

38.9 Select `Menu bar> View> Tile Viewports> Layout` to obtain the `Tiled Viewport Layout` dialogue box that is used for setting viewports on the screen.

`Elev` (elevation) sets the level of the base plane of the drawing.

`Thickness` is the distance of the extrusion parallel to the Z axis and perpendicular to the base plane.

`Plan` changes the UCS to give a true-size view of the XY icon and all surfaces parallel to it.

`Hide` removes invisible lines of the extruded surfaces.

Turn the `UCSicon` on to obtain the `XYicon`, type `Elev` and set it to 0, type `Thickness` and set it to 4, and draw the plan view of the object with the `Line` command (**Fig. 38.10**). The X

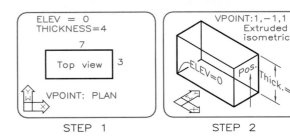

| STEP 1 | STEP 2 |

38.10 Extrusion of a box.
Step 1 Command: `Elev` (Enter)
New current elevation <0>: 0 (Enter)
New current thickness <0>: 4 (Enter)
Command: `Line` (Enter)
Line from point: (Draw 7 × 3 rectangle as a top view.)
Step 2 Command: `Vpoint` (Enter)
Rotate/<View> <current>: 1, -1, 1 An isometric view of the extruded box appears.

and Y axes are true size in the `Plan` view (top view), and the Z axis is perpendicular to them and points toward you.

Type `Vpoint` (viewpoint) and specify a line of sight with settings of 1, 21, 1 to obtain an isometric view of the extruded block in Fig. 38.10. When you type `Plan`, the UCS icon is shown true size and planes of the object that are parallel to it appear true size also. You may use all the regular `Draw` commands such as `Line`, `Circle`, and `Arc` to draw features, all of which will be extruded four units in the Z direction. The extrusion value of four units will remain in effect until reset.

`Hide` and `Enter` remove hidden lines from an extruded object and give an "empty-box" look (**Fig. 38.11**). Type `3Dface` and `Osnap` (object snap) to the corner points on the upper surface to make the top of the box opaque when the `Hide` command is applied (**Fig. 38.12**).

The `Solid` command can be used to make the top surface opaque by assigning it an `Elev` equal to the extruded `Thickness` (4), setting

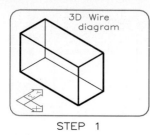
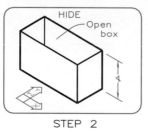
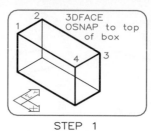
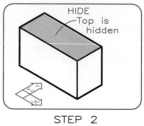

STEP 1 STEP 2 STEP 1 STEP 2

38.11 Hiding an extrusion.

Step 1 After the box is drawn, it appears as a wire frame on the screen.

Step 2 Select `Hide` and (Enter). The vertical surfaces are opaque (solid) planes, and the top is open.

38.12 Three-dimensional face.

Step 1 Command: `Osnap` (Enter)

Object snap modes: `End` (Enter)

Command: `3Dface` (Enter)

First point: 1 Second point: 2 Third point: 3

Fourth point: 4 Third point: (Enter)

Step 2 Command: `Hide` (Enter) The top surface appears as an opaque surface.

its `Thickness` to 0, and applying a solid area to the top by selecting the four corners. Type `Hide` to make the top appear opaque.

38.6 Coordinate Systems

Almost all drawing is done in the plane of the active coordinate system indicated by the `XY` icon. The two coordinate systems are the **World Coordinate System (WCS)** and the **User Coordinate System (UCS)**.

The WCS has an origin where X, Y, and Z are 0 and, usually, the X and Y axes are true length in the top, or plan, view. The `UCSicon` has a `W` and a plus sign on it when the icon is at the `WCS` origin.

The UCS can be located within the `WCS` with its X and Y axes positioned in any direction and its origin at any selected point. Type `UCSicon` and `On`, select `Origin` and move it to its origin, and establish a `User Coordinate System` in the following manner:

Command: `UCS` (Enter)

Origin/ZAxis/3point/OBject/View/X/ Y/Z/Prev/. . . Del/?/<World>: O (Enter)

Origin point <0,0,0>: (Select point) (Enter)

The options of the `UCS` command are as follows:

`World`: Set the system to the World Coordinate System.

`Origin`: Define an origin without changing the orientation of the X, Y, and Z axes by picking a point.

`ZAxis`: Select a new origin and pick a point on the positive portion of the new Z axis to locate a new `UCS`.

`3point`: Select a new origin, a point on the X axis, and a point on the Y axis to establish a new `UCS` (**Fig. 38.13**).

`OBject`: Pick an object (other than `3D Polyline` or a polygon mesh) and the UCS will have the same positive Z axis as the selected entity.

`View`: Select `View` to establish a `UCS` in which the XY plane is parallel to the screen, which allows the application of 2D text to a 3D drawing.

`X/Y/Z`: Specify X, Y, or Z as the axis about which to rotate the UCS and type the angle of

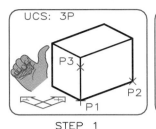 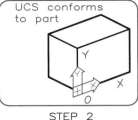

STEP 1 STEP 2

38.13 UCS: Three-point option.

Step 1 `Menu bar> Tools> UCS> 3 Point` (to rotate the UCS 90° about the x-axis.)

`Origin point <0,0,)>:` P1

`Point on positive portion of the x-axis:` P2

`Point on the positive portion of the UCS XY plane:` P3

Step 2 The `UCS icon` is transferred to the origin. The plus sign at its corner box indicates that it is at the origin.

rotation. (Example: Type X Enter, type 90 Enter, and the icon is rotated a positive 90° about the X axis.)

`Prev` (Previous): The previous UCS is recalled.

`Restore`: Select R and type the name of the saved UCS and it becomes the current UCS.

`Save`: Select S and name the current UCS to save it.

`Del` (Delete): Select D and give the name of the UCS to delete it.

`?: Pick ?` to get a listing of the current and saved coordinate systems. If unnamed, the current UCS is listed as *World* or *No Name*.

The `UCS Control dialogue box`, found by typing `DDUCS` or by selecting `View` and `Named UCS`, lists the saved coordinate systems (**Fig. 38.14A**). The *World* coordinate system is listed first. The current UCS is indicated by `Current`, but a new one can be selected with the cursor (`Setup2`, for example), or by picking the `Current` box. Delete a UCS by picking the `Delete` box, or rename one by picking the `Rename To` box and typing

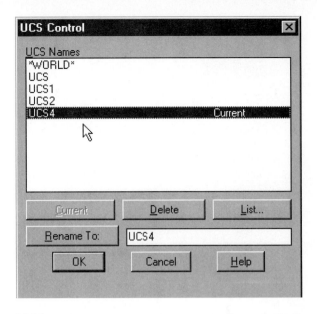

38.14 `Tools> UCS> Named UCS> UCS Control box` is displayed to list the saved UCS setups.

a new name. `OK` confirms any action made, and `Cancel` closes the dialogue box. When `List` is picked, the `UCS box` appears. The box shows the coordinates of the origin and the endpoints of the three axes. Select `OK` to return to the `UCS Control` box.

The `UCSicon` is turned on in the following manner:

`Command:` UCSICON `(Enter) On/Off/All/ Noorigan/ORigin<On>:` ON `(Enter)`

The functions of these `UCSicon` options are as follows:

`On/Off` turns the icon on and off.

`All` displays the icon in all viewports.

`Noorigin` displays the icon at the lower-left corner, regardless of the location of the UCS origin.

`Origin` places the icon at the origin of the current coordinate system if space permits, or at the lower left if space is unavailable.

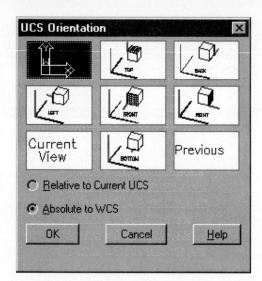

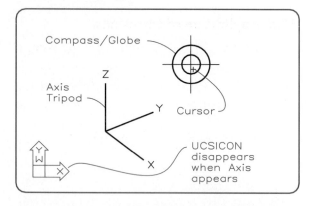

38.15 Select Menu bar> Tools> UCS> Preset UCS to get the UCS Orientation box, which presents views of objects.

38.16 When Vpoint is typed and (Enter) is pressed twice, a globe and axes appear on the screen for selecting a viewpoint.

38.7 Setting Views of Objects

Press UCS options (Menu bar> Tools> UCS> Preset UCS> UCS Orientation box) are available for specifying principal orthographic views of a 3D object (**Fig. 38.15**). Once a Preset UCS is selected, type Plan at the command line, press (Enter), and the specified view of the object is displayed. By using this method, the UCSicon is always true size when Plan is typed.

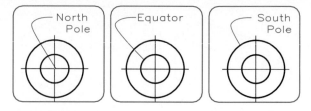

38.17 The compass globe.
A The north pole is at the crossing of the crosshairs.
B Points on the small circle locate viewpoints on the equator.
C Points on the large circle locate the viewpoint at the south pole.

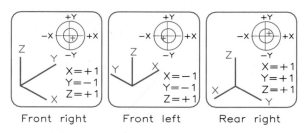

38.18 The relationship between the points on the Vpoint globe and the Vpoint values selected from the keyboard.

The Vpoint (viewpoint) command is used to obtain the direction of sight for viewing a UCS by typing coordinates at the command line, using the tripod axes, or by selecting Preset Viewpoints.

Type Vpoint and press (Enter) twice and the tripod axes will appear on the screen as shown in **Fig. 38.16**. A view of the object is found by selecting a point on the compass globe as described in **Fig. 38.17**. If necessary, repeat this command to select other views until the one desired is found.

Figure 38.18 compares the viewpoints specified on the compass globe with those types at the keyboard. By typing Vpoint and pressing (Enter) once, you will get the prompt, Rotate/<Viewpoint>: <0.000,0.000,1.000>. By typing a response of 1,−1,1, you specify that the origin (0,0,0) is viewed from a point that is 1 unit in the X direction, 1 unit in the negative Y direction, and 1 unit in the positive Z direction from 0,0,0 (**Fig. 38.19**).

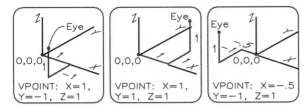

38.19 Graphical examples of a selection of Vpoints from the keyboard.

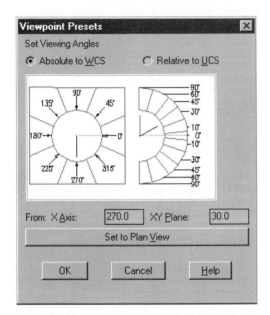

38.20 The Viewpoint Presets (Menubar> View> 3D Viewpoint> Select) dialogue box is used to select the point of view in the XY plane and with the XY plane by selecting intervals with the cursor or by typing.

Select from the Menu bar> View> 3D Viewpoint> Select to obtain the Viewpoint Presets box (**Fig. 38.20**) from which a line of sight for viewing an object from the X axis and from the XY plane can be chosen. Other options for viewpoints—top, bottom, left, right, front, back, SW isometric, SE isometric, NE isometric, and NW isometric—are available under 3D Viewpoint. Pick the desired viewpoint with your mouse from the menu and that view will be displayed on the screen.

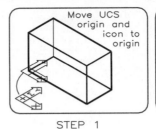
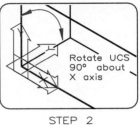

STEP 1 STEP 2

38.21 Setting the UCS.
Step 1 Command: UCSicon (Enter)
On/Off/All/Noorgin/ORigin <Off>: OR (Enter)
Command: UCS (Enter)
Origin, ZAxis/3point/Object/View/X/Y/Z/ . . . :
O (Enter)
Origin point: (Osnap to End at corner of box.)
Step 2 Command: UCS (Enter)
Origin/ZAxis . . . View/X/Y/Z . . . : X (Enter)
Rotation angle about X axis <0>: 90 (Enter) The icon is placed in the frontal plane of the box.

38.8 Application of Extrusions

An extruded box, similar to the one in Fig. 38.10, is shown in isometric by typing Vpoint (Enter), and giving coordinates of 1,21,1 in **Fig. 38.21**. The UCSicon is moved to the object's lower-left corner by the UCS command and rotated 90° about the X axis to lie in the frontal plane of the box. The right-hand rule is used to determine the direction of rotation by pointing your right thumb in the positive direction of the axis of rotation (**Fig. 38.22**). Type Plan, and the UCSicon and the front view will appear true size.

The circle is drawn as an extrusion by setting the Elev to 0 and Thickness to 23 (the depth of the box) and drawing a cylindrical hole (**Fig. 38.23**). Apply 3Dfaces to the upper and lower planes of the box and Hide invisible lines.

38.9 Dynamic View (Dview)

The Dview command is similar to the Vpoint command, but Dview changes the viewpoints

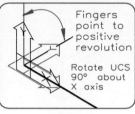

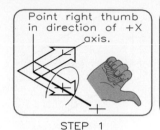

Point right thumb in direction of +X axis.

STEP 1

Fingers point to positive revolution

Rotate UCS 90° about X axis

STEP 2

38.22 Rotating the UCS.

Step 1 Command: <u>UCS</u> (Enter)

Origin/ZAxis . . ./<World>: <u>X</u> (Enter)

Step 2 Rotation and angle about the x axis <0>: <u>90</u> (Enter)

(The UCS rotates 90° about the X axis.)

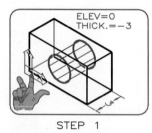

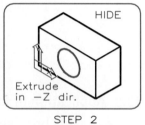

ELEV=0
THICK.=−3

STEP 1

HIDE

Extrude in −Z dir.

STEP 2

38.23 Extruding a hole.

Step 1 Command: <u>Elev</u> (Enter)

New current elevation <0>: <u>0</u> (Enter)

New current thickness <4>: <u>−3</u> (Enter)

Command: <u>Circle</u> (Enter) 3P/2P/TTR/<Center point>: (Enter)

Select center with cursor.

Diameter/<Radius>: <u>.5</u> (Enter) A 1 in. diameter cylinder is extruded 3 in. deep into the box.

Step 2 Command: <u>Hide</u> (Enter) The outline of the hole is shown, but it is not transparent.

of an object dynamically as they are changed by the cursor. In addition to axonometric views (parallel projections), three-point perspectives can be obtained for the most realistic pictorials (**Fig. 38.24**). The Dview command has the following options:

Command: <u>Dview</u> (Enter)

CAmera/TArget/Distance/POints/PAn/
Zoom/TWist/CLip/Hide/Off/Undo/
<eXit>:

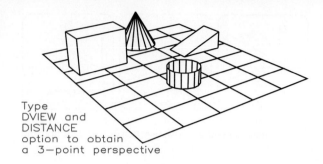

Type DVIEW and DISTANCE option to obtain a 3−point perspective

38.24 Select Menu bar> View> 3D Dynamic View> Select objects: (Window object) CAmera/TArget/. . . <eXit>: Distance to obtain a perspective view of the window objects.

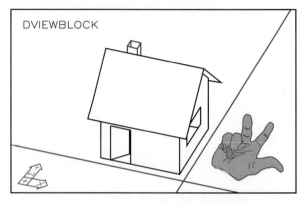

DVIEWBLOCK

38.25 Type Dview and press (Enter) twice to obtain the DviewBlock house, which can be used to experiment with various commands.

By typing Dview and pressing Enter twice, the top view of the DviewBlock house appears, which can be used for experimentation with the following options (**Fig. 38.25**):

CAmera rotates your viewpoint as if you were using a camera and were moving about the target (**Fig. 38.26**). As the cursor (the camera) is moved, the view is dynamically changed until a viewpoint is selected.

TArget is identical to the CAmera option, but the camera remains stationary as the target (and the drawing containing it) is rotated about the camera (**Fig. 38.27**).

Distance uses the current camera position and turns the view into a perspective. The XY icon is replaced with the perspective-box

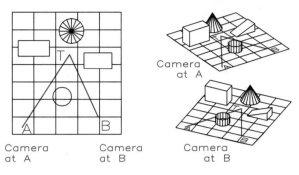

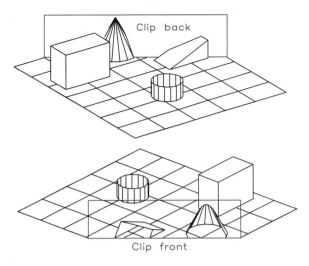

38.26 The CAmera option of the Dview command is used to obtain different views of a stationary target point by moving the position of the camera about it.

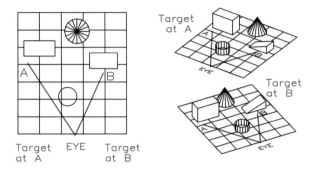

38.27 The TArget option of the Dview command is used to obtain views of a scene by moving the target to different locations about a stationary camera.

38.28 The Clip option of the Dview command removes the Back or Front portions of a drawing. The clipping plane is parallel to the drawing screen.

icon. When prompted, give a distance to the target by typing the value, or by using the slider bar, with a range from 0X to 16X; 1X is the current distance to the target.

POints specifies the target point and the camera position for viewing a drawing. This command is necessary to specify viewpoints for perspectives.

PAn moves the view of a drawing without changing its magnification or true position.

Zoom changes the magnification of a drawing in the same manner as the Zoom/Center command when perspective is off. When the perspective mode (Dist=On) is on, Zoom dynamically varies the magnification.

TWist rotates the drawing about an axis that is perpendicular to the screen.

CLip places cutting planes perpendicular to the line of sight to remove portions of a drawing either in front or in back of the plane by selecting Back/Front/<Off> (**Fig. 38.28**). Off exits from CLip. When Dist=On (perspective mode), the frontal clipping plane remains on at the camera position.

Hide suppresses the invisible lines.

OFF turns off the perspective mode enabled by Dist=On.

Undo reverses the previous Dview operations one at a time.

eXit ends the Dview command and displays the drawing.

38.10 Basic 3D Shapes (Surfaces)

From the Surfaces toolbar (**Fig. 38.29**), or by typing 3D, you can select a basic 3D shape or surface from the following: 2D solid 3D face, box, wedge, pyramid, cone, sphere, dome, dish, torus, edge, 3Dmesh, revolved surface, tabulated surface, ruled surface, and edge surface. These shapes are wire frames until Hide is used to make them

38.29 Menu bar> View> Toolbars> Surfaces gives this menu for drawing 3D surfaces (meshes).

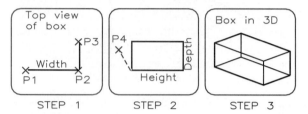

STEP 1   STEP 2   STEP 3

38.30 Box command.

Step 1 View> Toolbars> Surfaces> Box> Corner of box: P1

Length: P2 Cube/<Width>: P3 Height: P4

Step 2 Rotation angle about Z axis: 0 (Enter)
Type Vpoint of 1,−1,1 for an isometric of the 3D box.

appear as solids. They can be thought of as hollow shapes with meshes (or faces) applied to their surfaces. Remember that these shapes are drawn on the XY plane. The XY icon must be observed at all times:

Box draws a cube or a box when you select a corner, specify length, width, and height and give the angle of rotation about the Z axis (**Fig. 38.30**).

Wedge draws a wedge with the same steps used to draw the box except the Wedge command was typed at the Command line rather than using the Surfaces toolbar (**Fig. 38.31**).

Pyramid draws a pyramid extending to its apex (**Fig.38.32**), or it can be a truncated pyramid without an apex.

Cone draws a cone to its apex (**Fig. 38.33**) or it can be a truncated cone without an apex.

Sphere draws a ball by selecting its center and radius (**Fig. 38.34**). Its center lies on the XY plane of the UCS.

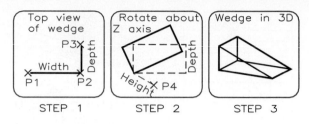

STEP 1   STEP 2   STEP 3

38.31 Wedge command.

Step 1 Command: Wedge (Enter)

Corner of wedge: P1 Length: P2 Width: P3

Step 2: Height: P4

Step 3: Rotation angle about Z axis: −15 (Enter) Type Vpoint and 1,−1,1 to get an isometric view of the wedge.

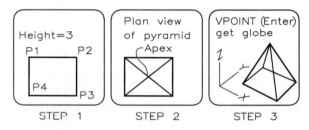

STEP 1   STEP 2   STEP 3

38.32 Pyramid command.

Step 1 Command: Pyramid (Enter)

First base point: P1 Second base point: P2

Third base point: P3 Tetrahedron/<Fourth base pt>: P4

Step 2 Ridge/Top <Apex point>: .XY (Enter)

of (Need Z): 2 (Enter)

Step 3 Set the Vpoint to 1,−1,1 to get a 3D view of the pyramid.

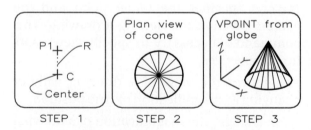

STEP 1   STEP 2   STEP 3

38.33 Cone command.

Step 1 Command: Cone (Enter) Base center point: C

Diameter/<radius> of base: P1

Diameter/<radius> of top <0>: (Enter)

Step 2 Height: 3 (Enter) Number of segments <16>: (Enter)

Step 3 Command: Vpoint (Enter) and set coordinates to 1,−1,1 to obtain an isometric of the cone.

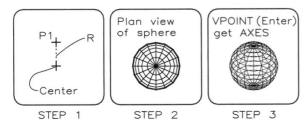

38.34 Sphere command.
Step 1 Command: <u>Sphere</u> (Enter) Center of sphere: <u>C</u>
Diameter/<radius>: <u>P1</u>
Number of longitudinal segments <16>: (Enter)
Number of latitudinal segments <16>: (Enter)
Step 2 The plan view of the sphere is generated.
Step 3 Set the <u>Vpoint</u> to <u>1,-1,1</u> to obtain an isometric view.

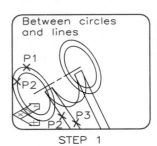

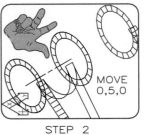

38.36 Rulesurf command (Part 1).
Step 1 Command: <u>Rulesurf</u> (Enter) Select object?: <u>P1</u>
Select object? <u>P2</u> Circles are faced. Select object? <u>P3</u>
Select object? <u>P4</u> Ends are faced. (Enter)
Step 2 Command: <u>Move</u> (Enter)
Select objects? **Select faces.** (Enter)
Base point or displacement: <u>0,5,0</u> (Enter) Faces are moved.

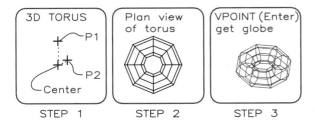

38.35 Torus command.
Step 1 Command: <u>Torus</u> (Enter) Center of Torus: <u>C</u>
Diameter/<radius> of torus: <u>P1</u>
Diameter/<radius> of tube: <u>P2</u>
Segments around tube circumference <16>: <u>8</u> (Enter)
Segments around torus circumference <16>: <u>8</u> (Enter)
Step 2 The plan view of the torus is drawn.
Step 3 Set the <u>Vpoint</u> to <u>1,-1,1</u> to obtain an isometric view.

Dome draws the upper hemisphere of a sphere by selecting its center and radius in the same manner as Dish.

Dish draws the lower hemisphere of a sphere by selecting its center and radius.

Torus draws a donut shape called a torus or toroid, as shown in **Fig. 38.35**.

Other operations available from the Surfaces toolbar are covered in Section 38.11.

38.11 Surface Modeling

Faces (or meshes) can be applied to cover wire frames with "skins" to make them look solid. Commands from the Surfaces toolbar for applying these meshes are Rulesurf, Tabsurf, Revsurf, Edgesurf, and 3Dmesh.

The application of these commands is illustrated by applying them to a 3D wire diagram, beginning with Fig. 38.36. Set Vpoint to 1,-1,1 to obtain an isometric view of the 3D frame. Type <u>Setvar</u>, (Enter), <u>Surftab1</u>, and set it to <u>20</u>, the number of faces to be applied.

Rulesurf (ruled surface) applies a surface (20 faces, as specified by Surftab1) between the circular ends (Fig. 38.36). The applied surface is Moved to 0,5,0 from the wire frame. Rulesurf can be used to place surfaces between two objects (curves, arcs, polylines, lines, or points), as shown in **Fig. 38.37**.

Tabsurf (tabulated surface) applies a surface from a curve that is parallel and equal in length to a directional vector (**Fig. 38.38**).

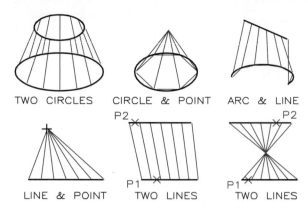

TWO CIRCLES CIRCLE & POINT ARC & LINE

LINE & POINT TWO LINES TWO LINES

38.37 The `Rulesurf` command connects objects with a series of 3D Faces as shown in these examples.

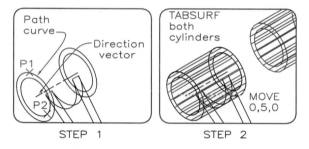

STEP 1 STEP 2

38.38 `Tabsurf` command (Part 2).

Step 1 Command: `Tabsurf` (Enter) Select path of curve: `P1`

Select direction of vector: `P2` The spacing between tabulated vectors is determined by the `Surftab1` variable.

Step 2 Command: `Move` (Enter) Select objects?: Select faces.

Base point or displacement: `0,5,0` (Enter) The faces are moved to a new position.

Select `Path Curve` (the circle) and `Select Direction Vector` and the cylinder is drawn. The `Surftab1` system variable (set to `20`) applies 20 faces.

`Revsurf` (revolved surface) revolves lines or shapes about an axis (**Fig. 38.39**) and it is `Moved` to `0,5,0` to add it to the hollow shell. Lines, polylines, arcs, or circles can be revolved to form a surface of revolution controlled by `Surftab1`. If a circle or a closed polyline is to be revolved (to make a torus, for example), system variable `Surftab2` controls

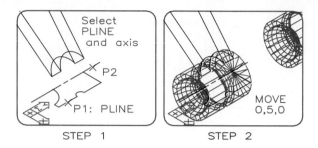

STEP 1 STEP 2

38.39 `Revsurf` command (Part 3).

Step 1 Command: `Revsurf` (Enter) Select path of curve: `P1`

Select axis of revolution: `P2` Start angle <0>: (Enter)

Included angle (+=ccw, -=cw) <Full circle>: `360` (Enter)

Step 2 Command: `Move` (Enter) Select objects?: Pick faces.

Base point or displacement: `0,5,0` (Enter) Faces are moved.

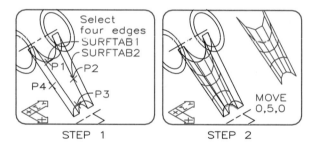

STEP 1 STEP 2

38.40 `Edgesurf` command (Part 4).

Step 1 Command: `Edgesurf` (Enter) Select edge 1: `P1`

Select edge 2: `P2` Select edge 3: `P3` Select edge 4: `P4` Mesh is drawn in boundaries. System variable `Surftab1` and `Surftab2` determine the density of the mesh.

Step 2 Command: `Move` (Enter) Select objects?: Pick faces.

Base point or displacement: `0,5,0` (Enter) Faces are moved.

the mesh density of the line, circle, arc, or polyline that is being revolved, and `Surftab1` controls the density of the path of revolution.

`Edgesurf` (edge surface) applies a mesh to four edges (joined boundary lines), as shown in **Fig. 38.40**. System variables `Surftab1` and `Surftab2` control the density of the first and second objects selected,

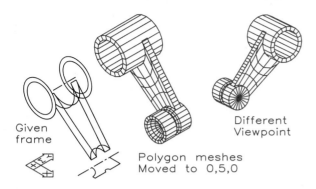

Given frame

Different Viewpoint

Polygon meshes Moved to 0,5,0

38.41 The given frame of the connector and the resulting 3D hollow shells from two viewpoints are shown here after using the Hide command to remove hidden lines.

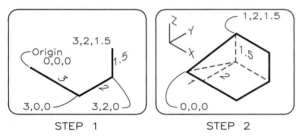

STEP 1

STEP 2

38.42 3Dpoly (absolute coordinates).

Step 1 Command: <u>3Dpoly</u> or <u>Line</u> (Enter)

From point: <u>0,0,0</u> (Enter)

Close/Undo/<Endpoint of line>: <u>3,0,0</u> (Enter)

Close/Undo/<Endpoint of line>: <u>3,2,0</u> (Enter)

Close/Undo/<Endpoint of line>: <u>3,2,1.5</u> (Enter)

Step 2 Close/Undo/<Endpoint of line>: <u>1,2,1.5</u> (Enter)

Close/Undo/<Endpoint of line>: <u>Close</u> Close figure. (Enter)

respectively. The surfaces are moved to <u>0,25,0</u> to complete the hollow shell of the connector.

The beginning frame and two final views of the meshed connector are shown in **Fig. 38.41** after the Hide command has been applied.

38.12 Line, Pline, and 3Dpoly

The Line command draws 2D lines when X and Y values are typed, and 3D lines when X, Y,

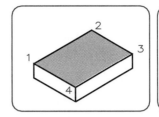 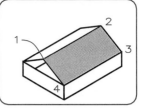

A. HORIZONTAL SURFACE   B. SLOPING SURFACE

38.43 3Dface on wire diagrams.

3Dface is used to opaque planes of wire frames by snapping to endpoints and to draw opaque faces without a wire diagram. The command can be applied to a horizontal surface (A) and to a sloping surface at (B).

and Z coordinates are given. Pline draws a 2D polyline and 3Dpoly draws 3D polylines with X, Y, and Z coordinates.

3Dpoly or Line commands draw lines with absolute coordinates from <u>0,0,0</u> located in 3D space in **Fig. 38.42**. Relative coordinates can be typed in the form of @xyz or <u>@3,0,0</u> to locate their endpoints with respect to the last point.

All Osnap modes apply to Line, Pline, and 3DFace. Three-dimensional objects can be stretched in the plane of the UCSicon (in the X and Y directions), but not in the Z direction. Height dimensions are modified by rotating the UCS so height lies in the X or Y directions where Stretch can be used.

38.13 The 3Dface Command

3Dface is used to apply faces that can be made opaque by the Hide command (**Fig. 38.43**). 3Dface is applied to wire frames by snapping to their endpoints with Osnap.

Figure 38.44 illustrates how corners of a 3Dface are used to form an opaque plane. After selecting four points with the 3Dface command, you will be prompted for points 3 and 4, using the previous two points as points 1 and 2. Successively added four-sided areas are connected with splice lines, yielding a "patchwork" area of faces. By typing I (for

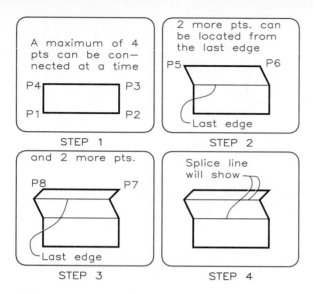

38.44 3Dface command.

Step 1 Command: <u>3Dface</u> (Enter) First point: <u>P1</u>
Second point: <u>P2</u> Third point: <u>P3</u> Fourth point: <u>P4</u>

Step 2 Third point: <u>P5</u> Fourth point: <u>P6</u>

Step 3 Third point: <u>P7</u> Fourth point: <u>P8</u>

Step 4 Third point: (Enter) Splice lines will show.

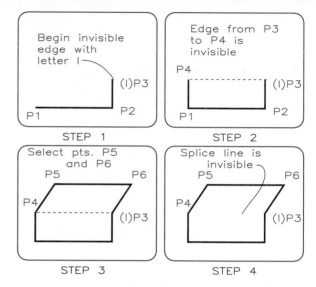

38.45 3Dface: Invisible seam.

Step 1 Command: <u>3Dface</u> (Enter) First point: <u>P1</u>
Second point: <u>P2</u> Third point: <u>I</u> (Enter) <u>P3</u>

Step 2 Fourth point: <u>P4</u>

Step 3 Third point: <u>P5</u> Fourth point: <u>P6</u>

Step 4 Third point: (Enter) Splice P3–P4 is invisible.

invisible) and Enter prior to selecting the beginning point of a splice line, the splice will be invisible (**Fig. 38.45**).

38.14 XYZ **Filters**

Filters are used for picking points that adopt the coordinates of 3D points (**Fig. 38.46**). When a command prompts for a point (as in the LINE command), type a period (.) followed by one- or two-letter coordinates (.X or .XY, for example) and select the point with the cursor. Respond to the prompt for the missing coordinate or coordinates with one or more numbers.

To locate the front view of a point projected from the left side and top views, select the .Y of the left point (Step 1) and the .X of the top point (Step 2); the front view of the point sharing these coordinates is found (Step 3).

A 3D drawing is made in **Fig. 38.47** using filters where the .XY coordinates of a given

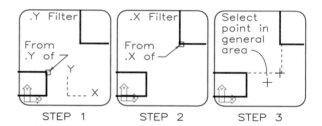

38.46 X and Y filters.

Step 1 Command: Line (Enter) Line from point: .Y (Enter) of (Select corner) of (Need XZ): (Enter)

Step 2 Close/Undo/<Endpoint of line>: .X of (Select corner) of (Need YZ)

Step 3 Select a point in the general area of the desired position. The point will appear at the intersection of the Y coordinate from the side view and the X coordinate from the top view.

point are selected and Z is specified as 2. Osnap was set to End for selecting endpoints. When the points are filtered in the XZ plane, a prompt will ask for the Y-coordinate to locate the point in 3D space.

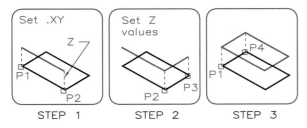

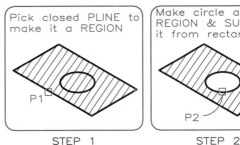

38.47 3D filters.

Step 1 Command: <u>3Dpoly</u> (or Line) (Enter)

From point: .<u>XY</u> of <u>P1</u> of (Need Z) <u>2</u> (Enter)

Close/Undo/<Endpoint of line>: .<u>XY</u> of <u>P2</u>

Step 2 Close/Undo/<Endpoint of line>: .<u>XY</u> of <u>P3</u>

Step 3 Close/Undo/<Endpoint of line>: .<u>XY</u> of (Select 1.) (Need Z) <u>2</u> (Enter)

38.49 2D Regions.

Step 1 Command: <u>Region</u> (Enter) Select objects: <u>P1</u> (Enter)

1 Region created. (Enter)

Step 2 Command: <u>Region</u> (Enter) Select objects: <u>P2</u> (Enter)

1 Region created. (Enter)

Command: <u>Subtract</u> (Enter) Select objects: <u>P1</u> (Enter) Select objects: <u>P2</u> (Enter) Region with a hole is created.

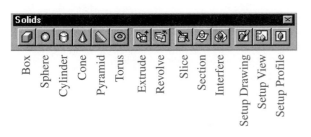

38.48 Select Menu bar> View> Toolbars> Solids to get this toolbar on your screen.

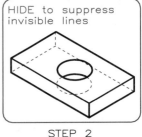

38.50 Extruding a region.

Step 1 View> Toolbars> Surfaces> Extrude icon> Select objects: <u>P1</u> (Enter)

Step 2 Path/<Height of Extrusion>: <u>1</u> (Enter)

Extrusion taper angle <0>: (Enter) Region is extruded.

38.15 Solid Modeling: Introduction

The AutoCAD modeler in Release 14 provides solid modeling capabilities of Regions (2D solids) and Solids (3D solids). The various commands for solid modeling found in the Solids toolbar (**Fig. 38.48**) are techniques for creating solid primitives—**boxes**, **cylinders**, **spheres**, etc.—that can be added or subtracted from each other to form a composite object.

Regions

Region modeling is a 2D version of solid modeling in which a closed surface can be converted into a solid plane (an object) (**Fig. 38.49**). The upper plane of a wire frame enclosed by a Pline is made into a 2D solid by typing Region and selecting the polyline.

The circle is made into a region and is removed from the rectangular Region by the Subtract command.

The Extrude **Command**

The Extrude command (from the Solids toolbar) is used with closed polylines, polygons, circles, ellipses, and 3D entities to extrude them to a specified height, with tapered sides, if desired. **Figure 38.50** shows the extrusion of the 2D Region developed in Fig. 38.49 to an assigned height of 1. Polylines

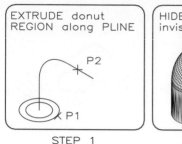

STEP 1 | STEP 2

38.51 Extruding a region.

Step 1 Command: <u>Extrude</u> (Enter) Select objects: <u>P1</u> (Enter)

Path/<Height of Extrusion>: <u>Path</u> (Enter)

Step 2 Select path: <u>P2</u> (Enter) Hide to show visibility.

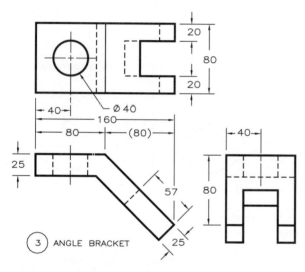

38.52 The angle bracket with an inclined surface is drawn as a 3D object in the five figures that follow.

with crossing or intersecting segments cannot be extruded.

A region can be an extruded path (usually a polyline) that forms a 3D shape, as illustrated in **Fig. 38.51**. The extruded shapes can be hidden and rendered.

38.16 Extrusion Example: `Tilemode=0`

Type `Tilemode` and set it to `0` for drawing the angle bracket with the inclined surface (**Fig. 38.52**). Begin by using `Mview` to create an MS

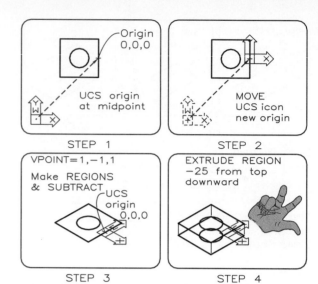

STEP 1 | STEP 2

STEP 3 | STEP 4

38.53 First extrusion (Part 1).

Step 1 Draw the top view of the bracket as a `Region` and set the UCS origin at the midpoint of the line.

Step 2 Use `UCSicon` and `Origin` to place the icon at the origin.

Step 3 Set a `Vpoint` of `1, -1, 1` to obtain an isometric view of the plane. Make the plane into a `Region` with a hole in it.

Step 4 `Extrude` the `Region` to a height of `-25`.

`Vport` that is about 200 × 180 to contain a single view of the bracket. Enter `Model Space` (MS) and begin drawing the bracket as follows:

Part 1 (**Fig. 38.53**): Draw the top view of the bracket as a polyline with a circle in it and set the UCS `origin` at the midpoint of one of its lines. Move the `UCSicon` to the UCS Origin, obtain an isometric view (`Vpoint=1, 21, 1`). Make the plane into a `Region` with a hole in it and `Extrude` the `Region` to `-25` mm below the upper surface.

Part 2 (**Fig. 38.54**): Type <u>PS</u> to return to `Paper Space` and define four `Vports` to fill the screen to obtain four views of the bracket. Type MS to enter `Model Space`, activate the lower-left port, and rotate the UCS 90° about the X axis, making the `UCSicon` parallel to the object's front view. Type `Plan` to obtain the

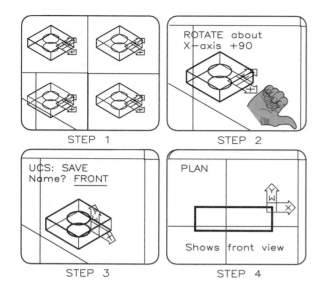

38.54 Setting Vports (Part 2).

Step 1 Command: <u>Vports</u> Select the 4-port option and the 3D drawing is shown in each.

Step 2 Type <u>MS</u> and pick the lower-left port. Command: <u>UCS</u> (Enter)

Select X option and rotate UCS 90° about X axis.

Step 3 Command: <u>UCS</u> (Enter) Type Save.

?/Desired UCS name: <u>Front</u> (Enter)

Step 4 Command: <u>Plan</u> (Enter) You will get front view; Zoom to the desired size.

front view and type UCS and SAVE to keep this UCS as Front.

Part 3 (Fig. 38.55): Restore the Front UCS and rotate the UCS 90° about the Y axis to align the XY icon parallel to the side plane of the bracket. Activate the lower-right port, type Plan to obtain the right-side view, and Save the UCS as Side. Obtain and Save the top view in the upper-left panel in the same manner that the two previous views were found.

Part 4 (Fig. 38.56): Activate the right-side port to Restore the UCS named Side and show it in isometric by setting Vpoint to <u>1,-1,1</u>. Set the UCS origin with coordinates of <u>0,-80,80</u> at the end of the sloping surface and the UCSicon will move to it. Rotate the UCSicon 45° about the X axis to make it lie in the plane of the inclined surface. Save this

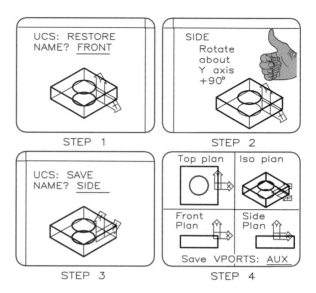

38.55 Drawing top and side views (Part 3).

Step 1 You can Restore the Front view by using the UCS command, but it is unnecessary here.

Step 2 Pick the lower-right port. Command: <u>UCS</u> (Enter)

Select <u>Y option</u> and rotate UCS 90° about the Y axis.

Step 3 Command: <u>UCS</u> (Enter) Type <u>Save</u>.

?/Desired UCS name: <u>Side</u> (Enter)

Step 4 Command: <u>Plan</u> (Enter) to get side view. Use similar steps to obtain and save a top view in the upper-left port. Leave the 3D view in the upper-right port.

UCS as Inclined and convert it into a Region. Draw the notch by using the dimensions from its given views (Fig. 38.52). You may draw these points with the cursor while observing the polar coordinates at the bottom of the screen because the plane of the inclined surface lies in the UCS.

Part 5 (Fig. 38.57): Use the Extrude command to make a solid shape of the inclined plane (Region) with a thickness (height) of <u>225</u>. The extruded Region appears in all ports. Use Union to join the two extruded shapes into a single bracket. Hide can be used to remove invisible lines at this point.

38.17 Solid Primitives

Solid primitives—box, sphere, wedge, cone, cylinder, and torus—can be selected from

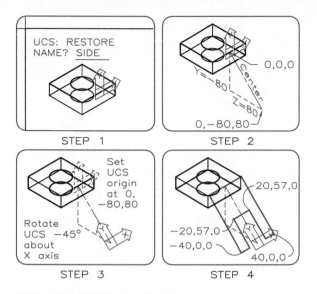

38.56 The inclined surface (Part 4).

Step 1 Select the side-view port. Command: UCS (Enter) Type Restore to recall the side view; show it in isometric.

?/Name of UCS to restore: Side (Enter) Command: Vports (Enter) Save this viewport as AUX1. Command: Vports (Enter) Type SIngle to obtain a full-screen image.

Step 2 Command: UCS (Enter) Origin/Zaxis/. . . <World>: Origin (Enter) Pick origin. Origin point: 0,-80,80 (Enter).

Step 3 Command: UCSicon (Enter) Type Origin to place icon at new origin. Command: UCS (Enter) Pick X and rotate UCS 45° about the X-axis. Use UCS command to Save as Inclined.

Step 4 Draw the inclined plane with Pline and convert it to a Region.

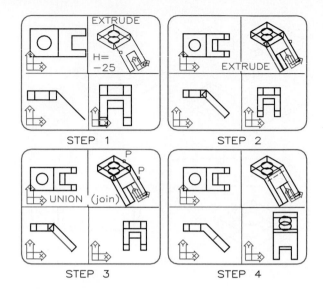

38.57 Completing the drawing (Part 5)

Step 1 Command: Vports (Enter) Restore the 4-port configuration, Aux1.) Extrude the inclined plane to a height of -25.

Step 2 The lower surface of the inclined plane is displayed.

Step 3 Use Union to join the two extruded shapes.

the Solids toolbar shown in Fig. 38.48. Hidden lines in solid primitives are suppressed by typing Hide.

The Box command is shown in **Fig. 38.58**, where the base of a box is drawn in the plane of the current UCS. The dimensions of the box can be created with separate widths and depths and the diagonal corners of the base, or as a cube by typing values at the keyboard or selecting them with the cursor.

Sphere creates a ball, by responding to the prompts with the center and radius, or center and diameter of the ball as shown in **Fig. 38.59**. The axis of the sphere connecting its

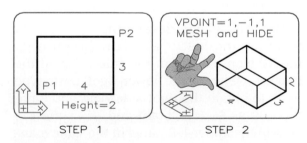

38.58 The Box command.

Step 1 Command: Box (Enter)

Center/<Corner of box>: P1 (Enter)

Cube/Length/<Other corner>: P2 (Enter)

Height <1>: 2 (Enter)

Step 2 Command: Vpoint (Enter)

Rotate/<View point> <0,0,0>: 1,-1,1 (Enter) The box is shown as a wire diagram; type Hide to suppress hidden lines.

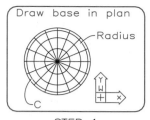

STEP 1   STEP 2

38.59   Sphere (solid).

Step 1 Command: <u>Sphere</u> (Enter)

Center of sphere <0,0,0>: <u>Center</u> (Enter)

Step 2 Command: <u>Vpoint</u> (Enter)

Enter Vpoint <0,0,0>: <u>1,-1,1</u> (Enter) A 3D view of the sphere is obtained; type Hide to suppress invisible lines.

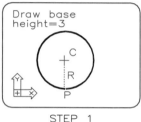

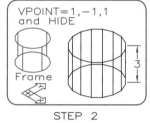

STEP 1   STEP 2

38.60   Cylinder (solid).

Step 1 Command: <u>Cylinder</u> (Enter)

Elliptical/<Center point>: <u>Center</u> (Enter)

Center of other end/<Height>: <u>3</u> (Enter)

Step 2 Command: <u>Vpoint</u> (Enter)

Rotate/<View point><0,0,0>: <u>1,-1,1</u> (Enter) 3D view of cylinder is obtained; type Hide to suppress invisible lines.

north and south poles is parallel to the Z axis of the current UCS, and its center is on the plane of the UCS.

Cylinder draws a cylinder in a manner similar to drawing a cone. **Figure 38.60** illustrates the steps of drawing a cylinder with a circular base. Cylinders with elliptical bases may also be drawn.

Cone draws cones with circular or elliptical bases, as shown in **Fig. 38.61**, with a circular base by locating its center, radius, and height.

Wedge draws the base of a wedge that lies in the plane of the current UCS with the upper

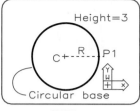

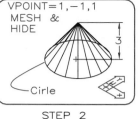

STEP 1   STEP 2

38.61   Cone (solid).

Step 1 Command: <u>Cone</u> (Enter)

Elliptical/<Center point><0,0,0>: <u>Center</u> (Enter)

Diameter/<Radius>: <u>P1</u>

Apex/<Height>: <u>3</u> (Enter)

Step 2 Command: <u>Vpoint</u> (Enter)

Rotate/<View point> <0,0,0>: <u>1,-1,1</u> (Enter) 3D view of the cone is drawn; type Hide to suppress invisible lines.

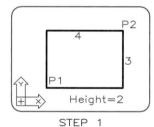

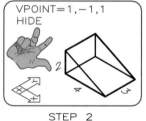

STEP 1   STEP 2

38.62   Wedge (solid).

Step 1 Command: <u>Wedge</u> (Enter)

Center/<Corner of wedge><0,0,0>: <u>P1</u> (Enter)

Cube/Length/<Other corner>: <u>P2</u> (Enter) Height: <u>2</u> (Enter)

Step 2 Command: <u>Vpoint</u> (Enter)

Rotate/<View point> <0,0,0>: <u>1,-1,1</u> (Enter) 3D view of wedge is drawn; type Hide to suppress invisible lines.

plane sloping toward the second point selected (**Fig. 38.62**). Prompts ask for the length, width, and height of the wedge.

Torus draws a donut solid by giving the center, diameter, or radius of the tube, and its diameter or radius of revolution (**Fig. 38.63**). The diameter of the torus will lie in the plane of the current UCS.

Revolve is used to sweep polylines, polygons, circles, ellipses, and 3D poly objects about an axis if they have at least 3, but less

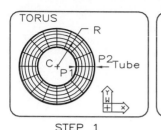

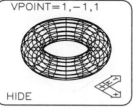

STEP 1	STEP 2

38.63 Torus (solid).

Step 1 Command: <u>Torus</u> (Enter) Center of torus: <u>C</u> (Enter)
Diameter/<Radius> of torus: <u>6</u> (Enter)
Diameter/<Radius> of tube: <u>2</u> (Enter)

Step 2 Command: <u>Vpoint</u> (Enter)

Rotate/<View point> <0,0,0>: <u>1,-1,1</u> (Enter) 3D view of torus is drawn; type Hide to suppress invisible lines.

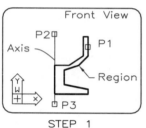

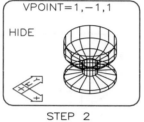

STEP 1	STEP 2

38.64 Revolved shape (solid).

Step 1 Command: <u>Revolve</u> (Enter) Select objects: <u>P1</u>

Axis of revolution-object/X/Y/<Start point of axis>: <u>P2</u>

<End point of axis>: <u>P3</u>

Angle of revolution <full circle>: (Enter) Surface is revolved.

Step 2 Command: <u>Vpoint</u> (Enter)

Rotate/<Viewpoint><0,0,0>: <u>1,-1,1</u> (Enter) 3D view of revolution is draw; type Hide (Enter) to suppress invisible lines.

than 300, vertices. In **Fig. 38.64**, a polyline is revolved a full 360° about an axis. The path of revolution can start and end at any point between 0 and 360°. Polylines that have been Fit or Splined will require extensive calculations.

38.18 Modifying Solids

Once Solids have been drawn, several modification commands can be used to refine them:

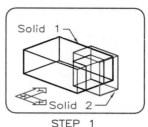

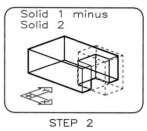

STEP 1	STEP 2

38.65 The Subtract command.

Step 1 Menu bar> Modify> Boolean> Subtract> Select solids and regions to subtract from . . .

Select objects: <u>P1</u> (Enter)

Select objects to subtract: <u>P2</u> (Enter)

Step 2 The notch is cut into the part. Type Hide to remove the hidden lines.

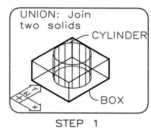

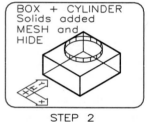

STEP 1	STEP 2

38.66 The Union command (solid).

Step 1 Command: <u>Union</u> (Enter)

Select objects: **Pick box** (Enter)

Select objects: **Pick cylinder** (Enter) Select objects: (Enter)

Step 2 The solids are unified (added) into a single part. Type Hide to remove the invisible lines.

Subtract, Union, Interfere, Explode, Fillet, Chamfer, Extend, and Trim.

Subtract is used to remove one intersecting solid from another, as illustrated in **Fig. 38.65**.

Union is used to join intersecting solids to form a single composite solid model. **Figure 38.66** shows how a box and cylinder are unified into a single solid.

Interfere is used to obtain a solid that is common to two intersecting solids. An inter-

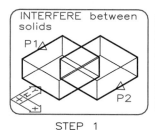

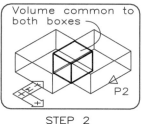

| STEP 1 | STEP 2 |

38.67 The `Interfere` command (solid).

Step 1 Command: <u>Interfere</u> (Enter)

Select first set of solids: Select objects: <u>P1</u> (Enter)

Select second set of solids:

Step 2 Select objects: <u>P2</u> (Enter)

Create interference solids? <N>: <u>Yes</u> (Enter) Interference solid is created; type Hide to suppress invisible lines.

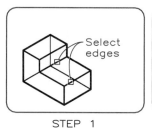

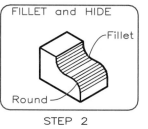

| STEP 1 | STEP 2 |

38.68 The `Fillet` command (solid).

Step 1 Command: <u>Fillet</u> (Enter) Select edges to be filleted (Press Enter when done): Select 2 edges to fillet. (Enter)

2 edges selected

Diameter/<Radius of fillet <default>: <u>3</u> (Enter)

Step 2 The `Fillets` are drawn. Type Hide to remove the hidden lines.

ference solid is found in **Fig. 38. 67** with this command.

`Explode` is used to separate solids or regions that were combined by the `Subtract` and `Union` commands to permit editing or correcting before redoing `Subtract` and `Union` commands.

`Fillet` is used to apply rounded intersections between planes by selecting the edges of solids and giving the diameter or radius of the fillet, as shown in **Fig. 38.68**. The `Fillet` command can also be used to fillet the edge of a cylindrical or curved feature.

`Chamfer` applies beveled edges by selecting the base surface, the adjoining surface, and the edges to be chamfered, giving the first and second chamfer distances (**Fig. 38.69**). When these prompts have been satisfied, the chamfers are automatically drawn.

38.19 Section

`Section` is used to pass a sectioning plane through a 3D solid to show a `Region` that outlines its internal features. `Bhatch` (hatch pattern) can be used to assign the hatching pattern to the `Region` if it lies in the plane of

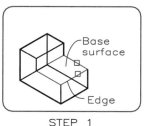

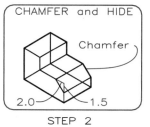

| STEP 1 | STEP 2 |

38.69 The `Chamfer` command (solid).

Step 1 Command: <u>Chamfer</u> (Enter) Select base surface: Pick <OK>/Next: (Enter) Select edges to be chamfered (Press Enter when done): Pick edge. (Enter) Select objects: (Enter)

Step 2 The `Chamfer` is drawn. Type Hide to remove the hidden lines.

the UCS. In Fig. 38.70, the hatch pattern is set to `ANSI31` for a cast iron symbol.

The `UCSicon` is placed on the object to establish the plane of the section, `Section` is typed, and the plane passing through the object appears. The section plane can be moved as shown in this example. Other sections through the object are found in this same manner by positioning the icon in the cutting plane or by selecting from one of the following options: `3point`, `Object`, `Zaxis`, `View`, `XY`, `YZ`, or `ZX`.

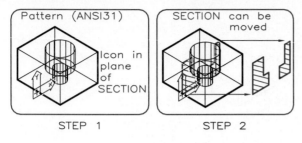

STEP 1   STEP 2

38.70   The `Section` command (solid).

Step 1 Position the UCS in the plane of the section.

Command: <u>Section</u> **(Enter)** Select objects: Pick box. **(Enter)**

Section plane by Object/. . ./XY/YZ/ZX: <u>.XY</u>
(Enter) The section establishes a Region through the plane of the icon.

Step 2 Command: <u>Move</u> **(Enter)** Move the Region outside the part; apply section lines to it if you like.

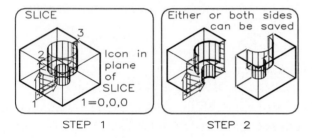

STEP 1   STEP 2

38.71   The `Slice` command (solid).

Step 1 Move the UCS to the plane of the `Slice`.

Command: <u>Slice</u> **(Enter)** Select objects: Pick the box.

Slicing plane by Object/. . . XY/YZ . . .: <u>XY</u>
(Enter)

Step 2 Point on XY plane <0,0,0>: <u>2</u> **(Enter)** Both sides/<point on desired side of plane: <u>B</u> **(Enter)**

38.20 Slice

With the `Slice` command, an object can be cut through and made into separate parts, either or both of which can be retained, as shown in **Fig. 38.71**. The `Slice` command has the following options:

`3points` defines three points on the slice plane.

`Object` aligns the cutting plane with a circle, ellipse, 2D spline, or 2D polyline element.

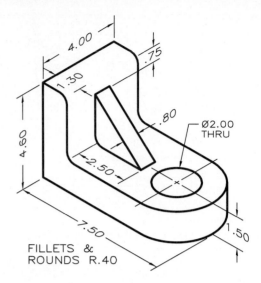

38.72   This bracket will be used as an example for illustrating solid modeling in the figures that follow.

`Zaxis` defines a plane by picking an origin point on the Z axis that is perpendicular to the selected points.

`View` makes the cutting plane parallel with the viewport's viewing plane when a single point is selected.

`XY`, `YZ`, or `ZX` aligns the cutting plane with the planes of the UCS by selecting only one point.

38.21 A Solid Model Example

The bracket shown in **Fig. 38.72** is to be drawn as a solid model by using the commands from the `Solids toolbar`. Create the plan view of the base as a `Pline` and obtain an isometric view of it by `Vpoint` <u>1,−1,1</u> (**Fig. 38.73**). You should then `Extrude` the base to a height of <u>1.50</u> units.

   A `Box` of <u>1.30 × 4.00 × 3.10</u> is drawn (**Fig. 38.74**) and moved to the base by snapping to end P1 of the box `End A`. Use `Wedge` to draw the bracket's rib, as shown in **Fig. 38.75**. Move the rib to join the base and the upright box by using the `Midpoint` option of `Snap`.

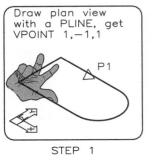
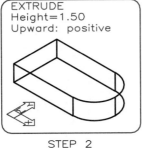

38.73 Extrude base (Part 1).

Step 1 Draw a the base with a closed `Pline`.

Command: <u>Vpoint</u> (Enter)

Rotate/<View point>: <u>1,-1,1</u> (Enter) Get isometric of base.

Command: <u>Extrude</u> (Enter) Select objects: <u>P1</u> (Enter)

Step 2 Path/<Height of Extrusion>: <u>1.50</u> (Enter)

Extrusion taper angle <0>: (Enter) Base is extruded 1.50".

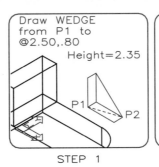
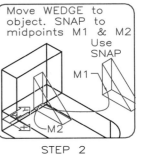

38.75 Draw wedge (solid) (Part 3).

Step 1 Command: <u>Wedge</u> (Enter)

Center/<Corner of wedge>: <u>P1</u> (Enter)

Length:

Step 2 Set Osnap to Midpoint. Command: <u>Move</u> (Enter)

Select objects: **Pick wedge.**

Base point or displacement: <u>M1</u> (Enter)

Second point or displacement: <u>M2</u> (Enter) Wedge is moved to midpoint of base.

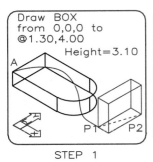
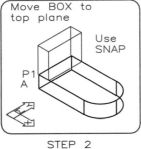

38.74 Draw box (Part 2)

Step 1 Command: <u>Box</u> (Enter)

Center/<Corner of box>: <u>0,0,0</u> (Enter)

Cube/Length/<Other corner>:

Height: <u>3.1</u> (Enter) Box is drawn.

Step 2 Set Osnap to End. Command: <u>Move</u> (Enter)

Select objects: **Pick box.** Base point or displacement: <u>P1</u> (Enter)

Second point of displacement: <u>A</u> (Enter) Box is moved.

In **Fig. 38.76**, use `Cylinder` to represent the hole and `Move` it to the center of the semicircular end of the base using the `Center` option of `Snap`. Use `Subtract` to create the hole in the base (**Fig. 38.77**). Use the `Union`

command to join the base, upright box, and wedge together into a composite solid.

`Fillet` is used in **Fig. 38.78** to select the edges to be rounded with a radius of <u>0.40</u> and `Hide` suppresses the invisible lines (**Fig. 38.79**). An infinite number of views of the bracket can be obtained with `Vpoint` or `Dview`.

38.22 Views of a Solid

Two commands, `Solview` and `Soldraw`, are used together to convert a 3D solid into orthographic views, auxiliary views, or sections. Once a 3D solid has been drawn, select the `Setup View` icon from the `Solids toolbar` (or type `Solview`). The screen will enter `Paper Space`, and you will be given the options of `UCS`, `Ortho`, `Auxiliary`, and `Section`. Select `UCS` and follow the prompts, as shown in **Fig. 38.80**.

You will be asked to select the `view center` and two `clip corners` (the diagonals of an `MS` viewport). When asked to name the

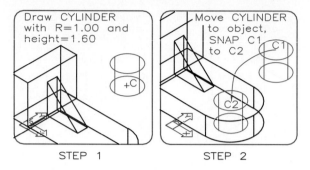

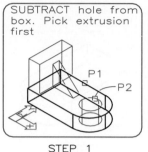

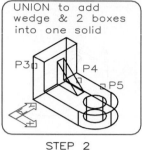

| STEP 1 | STEP 2 |

38.76 Draw cylinder (solid) (Part 4).

Step 1 Command: <u>Cylinder</u> (Enter)

Elliptical/<Center point> <0,0,0>: <u>C</u> (Enter)

Diameter/<Radius>: <u>1.00</u> (Enter) Center of other end/<Height>: <u>1.60</u> (Enter) Cylinder is drawn.

Step 2 Set Osnap to Center. Command: <u>Move</u> (Enter)

Select objects: Pick cylinder.

Base point or displacement: <u>P1</u> (Enter)

Second point or displacement: <u>P2</u> (Enter) Cylinder is moved to center of arc.

38.77 Subtract and union (solid) (Part 5).

Step 1 Command: <u>Subtract</u> (Enter) Select solids and regions to subtract from . . . Select objects: <u>P1</u> (Enter)

Select objects: <u>P2</u> (Enter) Hole is subtracted from base.

Step 2 Command: <u>Union</u> (Enter)

Select objects: <u>P3</u>, <u>P4</u>, <u>P5</u> (Enter) Select objects: (Enter)

The bracket is unified into one solid.

view, select <Current> Enter to use the XY plane of the current UCS to create an orthographic view.

The next prompt will ask for the type of view you want; choose Ortho for a side view and select the right edge of the window so it will be projected to the right of the front view. Select the view center when prompted to position the side view; several selections can be made to obtain the best location. Select the clip corner points to establish the MS window.

Since the object has an inclined plane (Step 3), the next view can be specified as an Auxiliary and two points selected on the inclined plane in the front view. Pick the side where the auxiliary is to be drawn, pick the view center, and pick the Clip corners; an auxiliary view is drawn. Solview automatically creates four layers for each named view. The name of the layer is followed by a dash and these abbreviations, −VIS, −HID, −DIM, and −Hat for visible, hidden, dimension, and hatch, respectively, on which these features can be drawn.

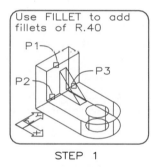

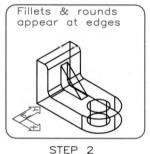

| STEP 1 | STEP 2 |

38.78 Fillet (solid) (Part 6).

Step 1 Command: <u>Fillet</u> (Enter) (Trim mode)

Current fillet radius=0.00 Polyline/Radius/Trim/<Select first object>: <u>R</u> (Enter) Enter fillet radius: <u>.40</u> (Enter)

Step 2 Command: <u>Fillet</u> (Enter)

Polyline/. . . <Select first object>: <u>P1</u> (Enter)

Enter radius <.40>: (Enter)

Chain/Radius/<Select edge>: <u>P1</u>, <u>P2</u>, <u>P3</u> (Enter)

In Step 4, select Solid toolbar> Setup Drawing icon (or type Soldraw at the command line), and you will be prompted for Viewports to draw. Select points on the MS windows and invisible lines will be converted to hidden (dashed) lines. It is necessary that linetype, Hidden, be loaded before using this command. From the Layer toolbar, Freeze

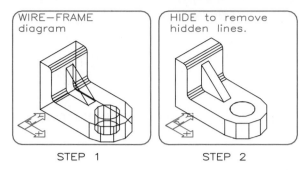

STEP 1   STEP 2

38.79 `Hide` invisible lines (solid) (Part 7).
Step 1 `Command:` <u>`Hide`</u> (Enter)
Step 2 The object is shown as solid with hidden lines suppressed.

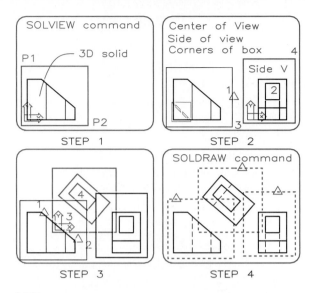

STEP 1   STEP 2

STEP 3   STEP 4

38.80 The `Solview` and `Soldraw` commands.
Step 1 `Command:` <u>`Solview`</u> (Enter)
`Ucs/Ortho/Auxiliary/Section/<eXit>:` <u>`UCS`</u> (Enter)
`Named/World/?/<Current>:` (Enter)
`Enter view scale <1.00>:` (Enter) `View Center:` <u>`P1`</u> (Enter)
`Clip first corner:` <u>`P2`</u> `Clip other corner:` <u>`P3`</u>
`View name:` <u>`Front`</u> (Enter)
Step 2 `Ucs/Ortho/Auxiliary/Section/<eXit>:` <u>`0`</u> (Enter)
`Pick side of viewport to be projected:` **Pick**
`View center:` **Pick** (Enter) `Clip first corner:` **Pick**
`Clip other corner:` **Pick** `View name:` <u>`RSIDE`</u> (Enter)
Step 3 `Ucs/Ortho/Auxiliary/Section/<eXit>:` <u>`A`</u> (Enter)
`Inclined Plane 1st point:` **Pick**
`Inclined Plane 2nd point:` **Pick** `Side view from:` **Pick**
`View center:` **Pick** (Enter) `Clip first corner:` **Pick**
`Clip other corner:` **Pick** `View name:` <u>`Aux1`</u> (Enter)
`Ucs/Ortho . . ./<eXit>:` (Enter)
Step 4 `Command:` <u>`Soldraw`</u> (Enter)
`Select view port to draw:` <u>`P1`</u>, <u>`P2`</u>, <u>`P3`</u>, (Enter) Solid lines that are invisible are converted to hidden (dashed lines). `Freeze` the `Vports` layer to remove the windows.

the layer `Vports` to remove the outline of the `MS` windows.

A sectional view is illustrated in **Fig. 38.81** where the same steps are followed, but instead of `Ortho` or `Aux`, `Section` is selected when prompted. The hatching is automatically applied to the sectional view and hidden lines are omitted when `Soldraw` is used to select the `MS` windows from paper space. Crosshatching can be varied by `HPname`, `HPscale`, and `HPangle` to set the name of the hatch pattern and its scale and angle, respectively.

Dimensions can be applied to the solid views by using the layers labeled with `-DIM` following the view's name (`Front-Dim`, for example).These dimensions are applied in much the same manner as will be discussed in Section 38.24.

38.23 Mass Properties (`Massprop`)

Various properties of regions and solids can be obtained with the `Massprop` command. An explanation of the options available from this command are as follows:

`Area` computes the area of a region or solid.

`Perimeter` calculates the perimeter of a region; not available for solids.

`Bounding Box` gives coordinates of the diagonal corners of a region's enclosing rectangle. For a solid, coordinates of the diagonal and opposite corners of a 3D box are given.

`Volume` gives the 3D space enclosed in a solid.

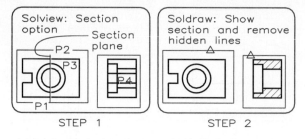

38.81 Solview: Section option.

Step 1 Draw a 3D solid and find the first view of it, the view to be sectioned, using the same steps as illustrated in Fig. 38.80. When prompted, Ucs/Ortho/Auxiliary/Section, select Section to get the prompts:

Cutting Plane's 1st point: P1

Cutting Plane's 2nd point: P2

Side view from: P3 Enter view scale <1.000>: (Enter)

View center: P4 Clip first corner: Pick

Clip other corner: Pick View name: Sect (Enter)

Ucs/Ortho/Auxiliary/Section/<eXit>: X (Enter)

Step 2 Command: Soldraw (Enter) Solid lines that are invisible are omitted and the cut surface of the section is hatched.

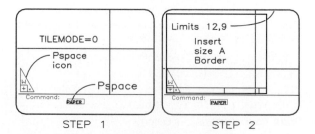

38.82 Tilemode=0: Paper space option.

Step 1 Command: Tilemode (Enter) New value for Tilemode <1>: 0 (Enter) Enters Pspace, 2D icon appears, and Pspace button shows in the Status line.

Step 2 Command: Insert (Enter)

Block name (or ?): Border-A (Enter)

Insertion point: 0,0 (Enter)

X scale factor <1>/Corner XYZ: 1 (Enter)

Y scale factor (default=X): 2 (Enter)

Rotation angle <0>: 0 (Enter) Border is inserted in Pspace. Set Limits to 12,9 and Zoom (Enter) (Enter).

Mass gives the weight of a solid.

Centroid gives the coordinates of the center of a region or the 3D center of a solid.

Moment of Inertia is given for regions and solids.

Product of Inertia is given for regions and solids.

Radius of Gyration is given for regions and solids.

MatlibB (materials library) assigns material types that can be assigned to solids that will have an effect on the mass properties listed above. The Materials Library box is found under the Render toolbar.

38.24 Paper Space and Model Space: Tilemode=0

By setting Tilemode=0 (Off), we can plot all Vports just as they appear on the screen, not just the active Vport. (An overview of Paper Space and Model Space was covered in Section 38.2.)

Type UCSicon and On so that the icon will appear. Type Tilemode and 0 (Off) to turn the screen into Paper Space (blank screen) with a Pspace icon (a triangle) in its lower-left corner. Paper appears in the Command line (**Fig. 38.82**). You should then set the paper space Limits large enough to contain the border, about 12″ × 9″, and insert a size A border. (You must design your own border and save it as a Wblock.)

Type Mview, select diagonal corners of a model-space viewport with the cursor, Copy it, type Mspace (MS) to enter Model space, and make one of the model spaces active (**Fig. 38.83**). Set the limits in this 3D viewport to 24,20, large enough to contain the drawing. Move between MS and PS by double clicking on the button in the Command line that will

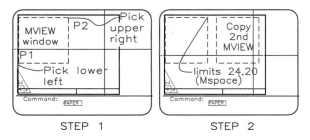

STEP 1   STEP 2

38.83 `Tilemode=0`: `Mview` option.

Step 1 Command: <u>Mview</u> (Enter) (Switch to Paper Space)

`On/Off/Hideplot/Fit/2/3/4/Restore/<First point>:` <u>P1</u> (Enter)

`Other corner:` <u>P2</u> (Enter) A model-space port is drawn.

Step 2 While still in paper space, make a copy of the 3D model-space port, return to `MS`, and set the view ports `Limits` to <u>24,20</u>.

be either `Model` or `Paper`. Other options under `Mview`, which operate from `Paper` space are `On`, `Off`, `Hideplot`, `Fit`, `2`, `3`, `4`, and `Restore`, which are explained below:

`On/Off`: Selects and turns off viewports to save regeneration time; leave at least one on.

`Hideplot`: When turned on, removes the hidden lines from the selected viewports.

`Fit`: Makes a viewport fill the current screen.

`2/3/4`: Lets you create an area and specify the number of viewports within it.

`Restore`: Recalls a viewport configuration saved by `Vports`.

Figure 38.84 shows a 3D object drawn by the `Extrude` command in one of the 3D viewports and shown simultaneously in a second port. Select an isometric `Vpoint` and type `PS` to enter paper space. The cursor spans the screen in paper space. Now the drawing must be scaled.

The paper-space `Limits` are `12,9`, inside of which are two model-space viewports each with `Limits` of `24,20`. These `MS` ports must be sized to fit, which requires calculations. The combined width of the two 24-in.-wide model spaces is 48. When scaled to half size,

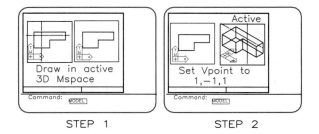

STEP 1   STEP 2

38.84 `Tilemode=0`: `model space`.

Step 1 Command: <u>Mspace</u> 3D icons appear in both ports and the cursor in the active port. Draw an extruded part; it will show in both views.

Step 2 Pick the right viewport to make it active.

Command: <u>Vport</u> (Enter) `Rotate/<View point>:` <u>1,-1,1</u> (Enter) An isometric is obtained.

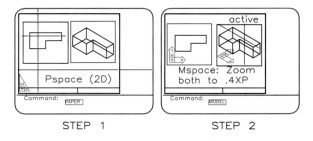

STEP 1   STEP 2

38.85 `Tilemode=0`: Zooming to scale.

Step 1 Command: <u>Pspace</u> (Enter) The 2D border reappears and the cursor spans the screen. Since `PS` had limits of 12,0 and `MS` had limits of 24,20, the `MS` ports must be `Zoomed` to about .20 size for both to fit in the border.

Step 2 Command: <u>Mspace</u> (Enter) One of the 3D ports becomes active. Command: <u>Zoom</u> (Enter)

`All/Center/. . ./<Scale (X/XP)>:` <u>.2XP</u> Active port is scaled to .2 size to fit in the 2D border. Select and `Zoom` the other port to .2XP.

their width is 24, too wide to fit. If scaled to 0.20 (two-tenths), their width is 9.6, small enough to fit inside the A size border.

From model space, type `Zoom` and use the `X/XP` (`2/10XP` or `0.2XP`) option to size the contents of each viewport (**Fig. 38.85**). This factor changes the width limit of each viewport from 24 to 4.8, with both drawings having the same scale on the screen. In other words, you may scale viewports with width limits of 24 inches to a full-size width of 4.8 in. of paper space.

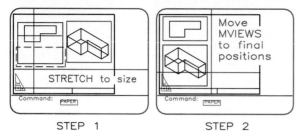

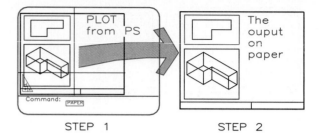

STEP 1 STEP 2

38.86 `Tilemode=0`: Moving ports.
Step 1 Command: `Pspace` (Enter) Switch to `PS` and use `Stretch` to size the 3D ports.
Step 2 Move `MS` ports to their final positions within the 2D border in paper space.

Type `PS` to enter paper space and use `Stretch` to reduce the size of the `Mspace` outlines (**Fig. 38.86**). Reposition the model-space views for plotting with `Move`.

Plotting must be performed from `Pspace`; both the 2D and 3D drawings are plotted at the same time, including the outlines of the model-space ports (**Fig. 38.87**). To remove the window outlines, `Change` them to a separate `Layer` (`Window`, for example), `Freeze` them, and `Plot` in the usual manner.

Remove hidden lines when plotting by selecting the `Hideplot` option of `Mview` from `PS`, and pick the `Remove hidden lines` button from the `Plot` dialogue box. When prompted by `Hideplot`, pick the outlines of the model-space viewports. Use `Vplayer` to select 3D viewports from paper space in which layers can be turned `Off` or `Frozen` while remaining `On` or `Thawed` in other viewports.

38.25 Dimensioning in 3D

Dimensions can be applied to objects in model space to match the dimensioning variables set for paper space by setting `Dimscale=0` (**Fig. 38.88**). Create two new layers, `H` and `F`, on which horizontal and frontal dimensions will be placed. Set associative dimensions on (`Dimaso=On`), set the layer `H` on, set the `UCS` to line in the top plane (see `XY` icon), and `Snap` to the endpoints of the object

38.87 `Tilemode=0`: Plotting.
Step 1 Command: `Plot` (Enter) Give specifications for plotting from paper space.
Step 2 The size A layout is plotted to show both 2D (paper space) and 3D (model space) in the same plot.

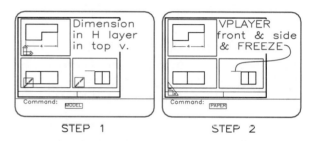

STEP 1 STEP 2

38.88 Dimensioning in model space.

Step 1 Create layer `H` and set it `On`. Set the `Dimscale` variable to 0 and the dimensioning variables in model space will match those in paper space. Move the `UCSicon` to the plane of the dimension and apply the width and depth dimensions by snapping to the endpoints of the part.

Step 2 When a dimension is applied in one viewport, it is shown in all 3D viewports and the dimensioning variables are the same size in `MS` as in `PS`. Dimensions are edges in the front and side views.

to attach width and depth dimensions to the top view. These dimensions appear as edges in the front and side views.

Select the front viewport and rotate the `XY` icon parallel to the frontal plane (`UCS/X/90°`) as shown in **Fig. 38.89**. Attach a vertical height dimension to the object by `Snapping` to the endpoints of the front view. The dimension appears as an edge in the top and side views.

Since the vertical dimensions cannot be seen in the top and side views, select these ports, type `Vplayer`, and `Freeze` the `F` layer (**Fig. 38.90**). Since the horizontal dimensions cannot be seen in the front and side views,

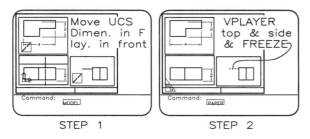

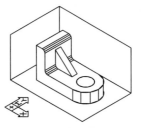

STEP 1   STEP 2

38.89   Dimensioning in 3D.

Step 1 Create layer F and set it On. Move the UCSicon to the plane of the frontal dimension. Apply the height dimension.

Step 2 The height dimension appears as an edge in the top and side views.

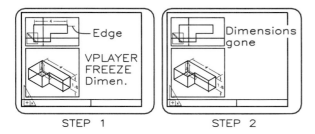

STEP 1   STEP 2

38.90   Dimensioning with Vplayer option.

Step 1 Since the frontal dimension is not readable in the top view, set Vplayer to On, select the border of the 3D viewport (while in PS) and Freeze layer F in that view.

Step 2 The dimensions are removed in the top view but remain in the 3D view. Freeze layer H in the front and side views also.

select these ports, type Vplayer, and Freeze the H layer. Other dimensions can be added and edited in this manner.

Options under Vplayer that operate from paper space are Freeze, Thaw, Reset, Newfrz, and Vpvisdflt.

Freeze/Thaw: Allows specified layers in selected VPORTS to be frozen or thawed. Thaw does not work on globally frozen layers.

Reset: Changes visibility of one or more layers in selected Vports to their current default setting.

Newfrz: Creates a new layer that is visible in the current viewport and is frozen in the rest.

Vpvisdflt: Used to set default visibility by viewport for any layer. The default setting makes layers frozen or thawed in new viewports.

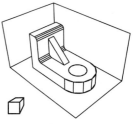

A. AXONOMETRIC   B. PERSPECTIVE

38.91   A bracket that will be used in illustrating rendering techniques in the examples that follow.

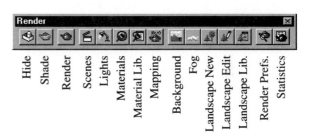

38.92   Select Menu bar> View> Toolbars> Render to obtain this toolbar for rendering 3D models.

38.26 The Render Process

Rendering is the process of adding color, lighting, and materials to 3D objects. The following examples illustrate how 3D drawings are rendered to give them a realistic appearance.

The bracket in Fig. 38.72 is displayed in two viewports as an axonometric and a perspective (**Fig. 38.91**). From the Render toolbar in Fig. 38.92, select Rendering Preferences to obtain the dialogue box shown in **Fig. 38.93**. Rendering options are Smooth Shade, Apply Materials, Shadows, and Render Cache. The More Options box gives Gouraud and Phong options to specify rendering quality (**Fig. 38.94**). The Phong option gives the smoother, more realistic rendering.

While a view of the bracket is on the screen, select Render from the Render toolbar, or type Render, and the Render

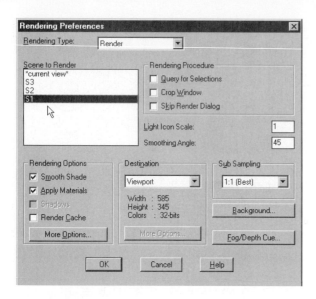

38.93 Select `Render toolbar> Render preferences icon` to obtain this dialogue box.

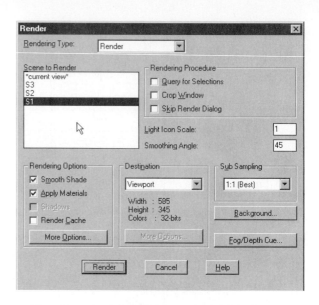

38.95 This `Render` dialogue box is selected from the `Render toolbar`.

38.94 The `More options` box of the `Rendering Preferences` menu gives this dialogue box.

`dialogue box` appears (**Fig. 38.95**). After making selections, the bracket will be rendered as shown in **Fig. 38.96**.

38.27 Lights

A rendering is enhanced by point lights, distant lights, spotlights, and ambient lighting that are under your control.

Point lights emit rays in all directions from a point source.

Distant lights emit parallel beams like those of sunlight.

Spotlights emit a cone of light to a target surface.

Ambient light comes from no particular source and provides a constant illumination to all surfaces of an object.

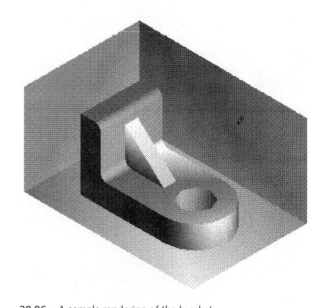

38.96 A sample rendering of the bracket.

New Lights

Light sources can be added to a drawing and indicated with the symbols shown in **Fig. 38.97**. From the `Render toolbar`, select `Lights` to get the dialogue box shown in **Fig. 38.98,** where several lights are listed. Select `Distant Light` from the pull-down

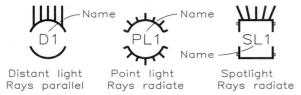

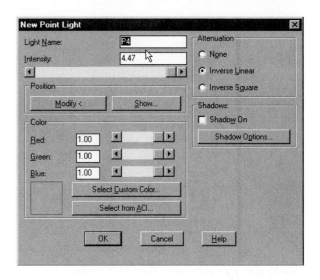

38.97 Light symbols.

Distant light, point light, and spotlight symbols are placed on drawings to indicate the positions for lighting.

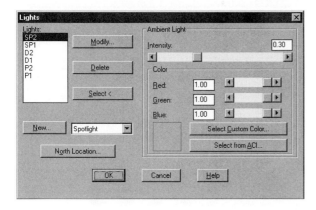

38.98 The `Lights` dialogue box is used to set and modify lights for rendering.

38.100 This menu is used to create `New Point Light`.

menu next to the `New` button, pick `New` to get the `New Distant Light` dialogue box (**Fig. 38.99**). From this menu, you can name the distant light, set its directions, and pick its colors. Select the `Distant Light` button and pick `OK` (Fig. 38.99). `Light Intensity` (`0=off`, `1=bright`) is set with the slider bar or by typing. Leave intensity set to `1`, select `OK` to return to `Lights` where `D2` is listed, and pick `OK` to exit. Select `Render` from the pull-down menu to obtain a new image of the bracket using distant light D2.

If you had selected `Point Light` and `New`, the `New Point Light` dialogue box would have appeared on the screen (**Fig. 38.100**). This dialogue box is different from the `New Distant Light` box since point lights have different characteristics.

Modifying Lights

Select `Lights` from the `Render toolbar` menu, select `Distant Light` to obtain the dialogue box shown in Fig. 38.99, and double click on `D1` (distant light) to obtain the `Modify Distant Light` box (**Fig. 38.101**). The `Select Custom Color` button can be picked to obtain the `Color` dialogue box (**Fig. 38.102**). Experiment with `Intensity` at

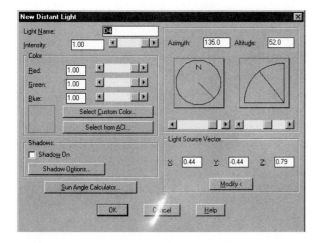

38.99 The `New Distant Light` dialogue box is used to create and name a new light source.

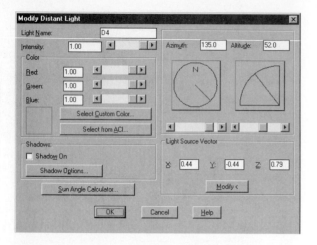

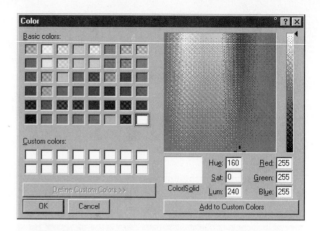

38.101 The `Modify Distant Light` dialogue box is used to make changes in an existing light.

38.102 Select `Render toolbar> Lights icon> Select Custom Color> Color menu` to obtain the color dialogue box.

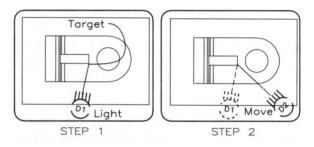

38.103 Moving a light.

Step 1 Select `Modify` from the `Modify Distant Light` dialogue box (Fig. 38.101), and you are transferred to the drawing.

`Enter light target <current>:` (Enter) Select current target.)

Step 2 `Enter light location <current>:` `.XY` (Enter) of (Need Z): `6` (Enter) The `Modify Distant Light` box reappears. Click on `OK` to exit and `Render` the bracket.

other settings (for example, `0.2` and `0.6`) and render an object to observe variations in the brightness of the image.

Moving Lights

To move distant light D2, type `Plan` to get a top view of the bracket, where the light sources are shown as symbols. Each light can be moved with the `Move` command (**Fig. 38.103**). You can also select `Modify` from the `Light` menu, pick distant light D1, and pick `Modify` to obtain the `Modify Distant Light` menu (Fig. 38.101). Select `Modify` under the heading, `Light Source Vector`, and the screen returns to the drawing where you are prompted for the direction of the light rays. The direction can also be changed by the `Azimuth` and `Altitude` adjustments (Fig. 38.101).

In Step 1 of Fig. 38.103, the prompt `Enter light direction TO <current>:` appears with a rubber-band line attached to the target. Press (`Enter`) to retain the current target and obtain a rubber-band line attached to it and the prompt `Enter light direction FROM <current>`. In Step 2, type `.XY`, pick the light location with the cursor and respond to the

prompt, `of (Need Z)` with 6 to locate the light source and establish the direction of the parallel light rays.

Type `VPOINT` and settings of `1,21,1` to obtain an isometric view of the bracket. Select `Render` from the `Render toolbar` to observe the original and new light position in **Fig. 38.104**.

Light Fall-off

The characteristic whereby light becomes dimmer as it travels farther from its source is called **fall-off**. `Point Lights` and `Spot-`

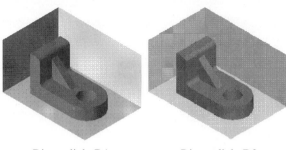

Distant light D1   Distant light D2

38.104 A comparison of the rendered views of a bracket with Distant Lights D1 and D2 in the positions shown in Fig. 38.103.

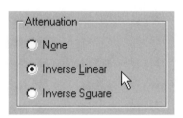

38.105 The type of point light fall-off can be specified as either Inverse Linear or Inverse Square.

A. Inverse Linear   B. Inverse Square

38.106 Light Fall-off options are None, Inverse Linear, and Inverse Square. By using Inverse Linear, an object 4 units away is one-fourth as bright as one that is 1 unit away from the light. By using Inverse Square, an object 4 units away is one-sixteenth as bright as one that is 1 unit away.

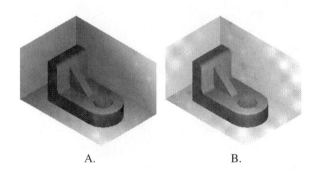

A.   B.

38.107 Ambient light.
(A) Ambient intensity set to 0.3. (B) Ambient intensity set to 0.80.

lights are affected by fall-off; Distant and Ambient Lights have uniform intensity at all distances. From the Lights menu, select P1 (Point Light), select Modify to get the Modify Point Light box, which has an Attenuation area. The options available here are None, Inverse Linear, and Inverse Square (**Fig. 38.105**):

None is a light with no fall-off, giving all objects the same brightness.

Inverse Linear makes a surface that is one-fourth as bright when 4 units from the light and one-eighth as bright when 8 units away.

Inverse Square makes an object one-sixteenth as bright when 4 units from the light and one-sixty-fourth as bright when 8 units away.

A comparison of lighting set to Inverse Linear and Inverse Square is shown in **Fig. 38.106**.

Ambient Light
So far, ambient light has been set to 0.30. From the Render toolbar obtain the Lights dialogue box and set ambient intensity to 0.80 by typing or using the slider bar. The variation in lighting the bracket is shown in **Fig. 38.107**.

Spotlights
Spotlights emit cones of light as defined in **Fig. 38.108** to highlight features. The creation of a Spotlight begins from the Lights dialogue box (Fig. 38.98), where Spotlight and New are selected to display the Create New Spotlight box. You can see the effects of a spotlight in **Fig. 38.109**.

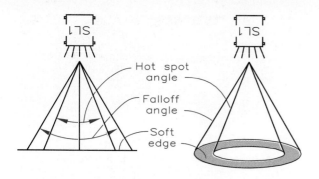

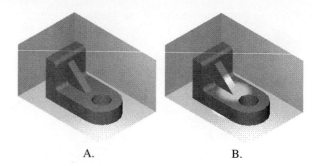

38.108 The characteristics of a spotlight are shown here.

38.110 Examples of Scenes, with (A) scene 1: two lights, and (B) scene 2: two lights and a spotlight.

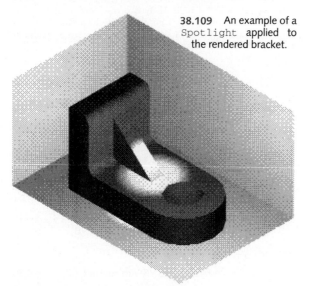

38.109 An example of a Spotlight applied to the rendered bracket.

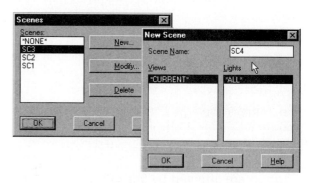

38.111 Render toolbar> Scene icon gives the Scenes dialogue box where the New button can be picked to obtain the New Scene box for naming a new scene.

38.28 Working with Scenes

Views and light settings can be saved in combinations as Scenes, which can have several lights (or no lights), but only one View. A viewpoint of a drawing can be saved as a View by typing View and giving it a name. Other viewpoints are saved in this manner. To recall a View, type View and Restore, and give its name when prompted.

Restored Views of the bracket illustrated in the previous examples are selected from Scenes of the Render toolbar menu to obtain the Scenes dialogue box (**Fig. 38.110**). Select New and the New Scene box appears in which all model-space views and lights are listed. Type the name of the Scene in the box (S2, for example). The name is trun-

cated to eight characters if it is longer. Select V2 as the View in S2, select point-light P1 and distant-light D1, select OK; the Scenes dialogue box reappears. Other scenes can be created by using these steps.

If *None* were selected as the scene to render, all lights would be on in the current view. If no lights existed, an over-the-shoulder distant light would be given. Select S2 from the dialogue box, followed by OK and Render to obtain a rendered S2. A comparison of S1 and S2 is given in Fig. 38.110 after they have been rendered.

A Scene can be modified by selecting it from the Scenes dialogue box (**Fig. 38.111**) and picking Modify to obtain the Modify Scene dialogue box, which is similar to the New Scene box. Different lights can be

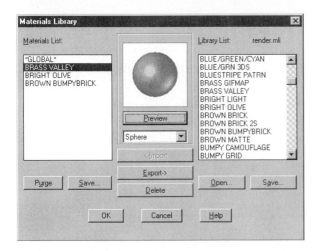

38.112 Select the `Materials  Library` icon from the `Render toolbar` to get this dialogue box, which lists the materials that can be assigned to 3D models. A preview of their application is shown on the sphere.

38.113 The `Materials` dialogue box is used to preview various materials as they are assigned to a part.

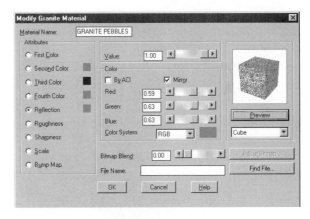

38.114 Selected materials can be modified by using this `Modify Material dialogue box`.

selected or deselected. Select `OK` to keep your modifications and return to the `Scenes` dialogue box; select `OK` to exit and `Render` the modified `Scene`.

38.29 Materials

The material of an object determines the reflective quality of its surfaces from dull to shiny. Select the `Materials Library` icon from the `Render toolbar` to get the `Materials Library` dialogue box (**Fig. 38.112**). Materials can be selected one at a time from the listing and previewed by selecting the `Preview` button. To add a material to the `Materials  List` for future use, pick `Import`. Click on `Save` to add the material to the `Materials List` file. Type `OK` to exit the `Materials Library` dialogue box and return to the drawing.

From the `Render toolbar`, select `Materials` to get the dialogue box shown in **Fig. 38.113**. To `Attach` a material finish to a part, select it from the list, select `Attach`, and the drawing reappears, with the prompt `Select objects to attach 'MATL' to:` Select the part and the `Materials` dialogue

box returns; click on `OK` to exit. `Render` the object to see the results.

By selecting `Modify` from the `Materials` dialogue box (Fig. 38.113), the `Modify Standard Materials` box (**Fig. 38.114**) lets you modify the selected material. These settings are `Color`, `Ambient light`, `Reflection`, and `Roughness`. Experiment with different settings and observe the results of these changes by picking `Preview`.

Problems

The problems at the ends of the previous chapters can be drawn and plotted using computer graphics techniques covered in this chapter as well as previous ones.

Appendix Contents

1 Decimal Equivalents and Temperature Conversions A-1
2 Metric and English Conversion Chart A-2
3 Logarithms of Numbers A-3
4 Trigonometric Functions A-5
5 Screw Threads: American National and Unified (inches) A-6
6 Screw Threads: American National and Unified (inches) Constant-Pitch Threads A-7
7 Screw Threads: American National and Unified (metric) A-8
8 Square and Acme Threads A-9
9 American Standard Taper Pipe Threads (NPT) A-9
10 Square Bolts (inches) A-10
11 Square Nuts A-10
12 Hexagon Head Bolts A-11
13 Hex Nuts and Hex Jam Nuts A-11
14 Round Head Cap Screws A-12
15 Flat Head Cap Screws A-12
16 Fillister Head Cap Screws A-13
17 Flat Socket Head Cap Screws A-13
18 Socket Head Cap Screws A-14
19 Round Head Machine Screws A-14
20 Set Screws A-15
21 Twist Drill Sizes A-16
22 Cotter Pins: American National Standard A-18
23 Straight Pins A-18
24 Woodruff Keys A-19
25 Standard Keys and Keyways A-20
26 Plain Washers (inches) A-21
27 Metric Washers (millimeters) A-22
28 Regular Helical Spring Lock Washers (inches) A-23
29 American Standard Running and Sliding Fits (hole basis) A-24
30 American Standard Clearance Locational Fits (hole basis) A-26
31 American Standard Transition Locational Fits (hole basis) A-28
32 American Standard Interference Locational Fits (hole basis) A-29
33 American Standard Force and Shrink Fits (hole basis) A-30
34 The International Tolerance Grades (ANSI B4.2) A-31
35 Preferred Hole Basis Clearance Fits— Cylindrical Fits (ANSI B4.2) A-32
36 Preferred Hole Basis Transition and Interference Fits—Cylindrical Fits (ANSI B4.2) A-34
37 Preferred Shaft Basis Clearance Fits— Cylindrical Fits (ANSI B4.2) A-36
38 Preferred Shaft Basis Transition and Interference Fits—Cylindrical Fits (ANSI B4.2) A-38
39 Hole Sizes for Nonpreferred Diameters (millimeters) A-40
40 Shaft Sizes for Nonpreferred Diameters (millimeters) A-42
41 Descriptive Geometry LISP Programs (ACAD.LSP) A-44
42 American National Standard 125-lb Cast Iron Screwed Fittings (inches) A-45
43 American Standard 250-lb Cast Iron Flanged Fittings (inches) A-46
44 Grading Graph A-47
45 Weights and Specific Gravities A-48

APPENDIX 1 • Decimal Equivalents and Temperature Conversions

DECIMAL EQUIVALENTS—INCH-MILLIMETER CONVERSION TABLE

1/2	1/4	1/8	1/16	1/32	1/64	Decimals	Millimeters
					1	.015625	.396875
				1		.031250	.793750
					3	.046875	1.190625
			1			.062500	1.587500
					5	.078125	1.984375
				3		.093750	2.381250
					7	.109375	2.778125
		1				.125000	3.175000
					9	.140625	3.571875
				5		.156250	3.968750
					11	.171875	4.365625
			3			.187500	4.762500
					13	.203125	5.159375
				7		.218750	5.556250
					15	.234375	5.953125
	1					.250000	6.350000
					17	.265625	6.746875
				9		.281250	7.143750
					19	.296875	7.540625
			5			.312500	7.937500
					21	.328125	8.334375
				11		.343750	8.731250
					23	.359375	9.128125
		3				.375000	9.525000
					25	.390625	9.921875
				13		.406250	10.318750
					27	.421875	10.715625
			7			.437500	11.112500
					29	.453125	11.509375
				15		.468750	11.906250
					31	.484375	12.303125
1						.500000	12.700000

1/2	1/4	1/8	1/16	1/32	1/64	Decimals	Millimeters
					33	.515625	13.096875
				17		.531250	13.493750
					35	.546875	13.890625
			9			.562500	14.287500
					37	.578125	14.684375
				19		.593750	15.081250
					39	.609375	15.478125
		5				.625000	15.875000
					41	.640625	16.271875
				21		.656250	16.668750
					43	.671875	17.065625
			11			.687500	17.462500
					45	.703125	17.859375
				23		.718750	18.256250
					47	.734375	18.653125
	3					.750000	19.050000
					49	.765625	19.446875
				25		.781250	19.843750
					51	.796875	20.240625
			13			.812500	20.637500
					53	.828125	21.034375
				27		.843750	21.431250
					55	.859375	21.828125
		7				.875000	22.225000
					57	.890625	22.621875
				29		.906250	23.018750
					59	.921875	23.415625
			15			.937500	23.812500
					61	.953125	24.209375
				31		.968750	24.606250
					63	.984375	25.003125
					64	1.00000	25.400000

LENGTH

1 millimeter (mm) = 0.03937 inch
1 centimeter (cm) = 0.39370 inch
1 meter (m) = 39.37008 inches
1 meter = 3.2808 feet
1 meter = 1.0936 yards
1 kilometer (km) = 0.6214 miles
1 inch = 25.4 millimeters
1 inch = 2.54 centimeters
1 foot = 304.8 millimeters
1 foot = 0.3048 meters
1 yard = 0.9144 meters
1 mile = 1.609 kilometers

AREA

1 square millimeter = 0.00155 square inch
1 square centimeter = 0.155 square inch
1 square meter = 10.764 square feet
1 square meter = 1.196 square yards
1 square kilometer = 0.3861 square mile
1 square inch = 645.2 square millimeters
1 square inch = 6.452 square centimeters
1 square foot = 929 square centimeters
1 square foot = 0.0929 square meter
1 square yard = 0.836 square meter
1 square mile = 2.5899 square kilometers

DRY CAPACITY

1 cubic centimeter (cm3) = 0.061 cubic inches
1 liter = 0.0353 cubic foot
1 liter = 61.023 cubic inches
1 cubic meter (m3) = 35.315 cubic feet
1 cubic meter = 1.308 cubic yards
1 cubic inch = 16.38706 cubic centimeters
1 cubic foot = 0.02832 cubic meter
1 cubic foot = 28.317 liters
1 cubic yard = 0.7646 cubic meter

LIQUID CAPACITY

1 liter = 1.0567 U.S. quarts
1 liter = 0.2642 U.S. gallons
1 liter = 0.2200 Imperial gallons
1 cubic meter = 264.2 U.S. gallons
1 cubic meter = 219.969 Imperial gallons
1 U.S. quart = 0.946 liters
1 Imperial quart = 1.136 liters
1 U.S. gallon = 3.785 liters
1 Imperial gallon = 4.546 liters

WEIGHT

1 gram (g) = 15.432 grains
1 gram = 0.03215 ounce troy
1 gram = 0.03527 ounce avoirdupois
1 kilogram (kg) = 35.274 ounces avoirdupois
1 kilogram = 2.2046 pounds
1000 kilograms = 1 metric ton (t)
1000 kilograms = 1.1023 tons of 2000 pounds
1000 kilograms = 0.9842 tons of 2240 pounds
1 ounce avoirdupois = 28.35 grams
1 ounce troy = 31.103 grams
1 pound = 453.6 grams
1 pound = 0.4536 kilogram
1 ton of 2240 pounds = 1016 kilograms
1 ton of 2240 pounds = 1.016 metric tons
1 grain = 0.0648 gram
1 metric ton = 0.9842 tons of 2240 pounds
1 metric ton = 2204.6 pounds

APPENDIX 3 • Logarithms of Numbers

N	0	1	2	3	4	5	6	7	8	9
1.0	.0000	.0043	.0086	.0128	.0170	.0212	.0253	.0294	.0334	.0374
1.1	.0414	.0453	.0492	.0531	.0569	.0607	.0645	.0682	.0719	.0755
1.2	.0792	.0828	.0864	.0899	.0934	.0969	.1004	.1038	.1072	.1106
1.3	.1139	.1173	.1206	.1239	.1271	.1303	.1335	.1367	.1399	.1430
1.4	.1461	.1492	.1523	.1553	.1584	.1614	.1644	.1673	.1703	.1732
1.5	.1761	.1790	.1818	.1847	.1875	.1903	.1931	.1959	.1987	.2014
1.6	.2041	.2068	.2095	.2122	.2148	.2175	.2201	.2227	.2253	.2279
1.7	.2304	.2330	.2355	.2380	.2405	.2430	.2455	.2480	.2504	.2529
1.8	.2553	.2577	.2601	.2625	.2648	.2672	.2695	.2718	.2742	.2765
1.9	.2788	.2810	.2833	.2856	.2878	.2900	.2923	.2945	.2967	.2989
2.0	.3010	.3032	.3054	.3075	.3096	.3118	.3139	.3160	.3181	.3201
2.1	.3222	.3243	.3263	.3284	.3304	.3324	.3345	.3365	.3385	.3404
2.2	.3424	.3444	.3464	.3483	.3502	.3522	.3541	.3560	.3579	.3598
2.3	.3617	.3636	.3655	.3674	.3692	.3711	.3729	.3747	.3766	.3784
2.4	.3802	.3820	.3838	.3856	.3874	.3892	.3909	.3927	.3945	.3962
2.5	.3979	.3997	.4014	.4031	.4048	.4065	.4082	.4099	.4116	.4133
2.6	.4150	.4166	.4183	.4200	.4216	.4232	.4249	.4265	.4281	.4298
2.7	.4314	.4330	.4346	.4362	.4378	.4393	.4409	.4425	.4440	.4456
2.8	.4472	.4487	.4502	.4518	.4533	.4548	.4564	.4579	.4594	.4609
2.9	.4624	.4639	.4654	.4669	.4683	.4698	.4713	.4728	.4742	.4757
3.0	.4771	.4786	.4800	.4814	.4829	.4843	.4857	.4871	.4886	.4900
3.1	4914	.4928	.4942	.4955	.4969	.4983	.4997	.5011	.5024	.5038
3.2	.5051	.5065	.5079	.5092	.5105	.5119	.5132	.5145	.5159	.5172
3.3	.5185	.5198	.5211	.5224	.5237	.5250	.5263	.5276	.5289	.5302
3.4	.5315	.5238	.5340	.5353	.5366	.5378	.5391	.5403	.5416	.5428
3.5	.5441	.5453	.5465	.5478	.5490	.5502	.5514	.5527	.5539	.5551
3.6	.5563	.5575	.5587	.5599	.5611	.5623	.5635	.5647	.5658	.5670
3.7	.5682	.5694	.5705	.5717	.5729	.5740	.5752	.5763	.5775	.5786
3.8	.5798	.5809	.5821	.5832	.5843	.5855	.5866	.5877	.5888	.5899
3.9	.5911	.5922	.5933	.5944	.5955	.5966	.5977	.5988	.5999	.6010
4.0	.6021	.6031	.6042	.6053	.6064	.6075	.6085	.6096	.6107	.6117
4.1	.6128	.6138	.6149	.6160	.6170	.6180	.6191	.6201	.6212	.6222
4.2	.6232	.6243	.6253	.6263	.6274	.6284	.6294	.6304	.6314	.6325
4.3	.6335	.6345	.6355	.6365	.6375	.6385	.6395	.6405	.6415	.6425
4.4	.6435	.6444	.6454	.6464	.6474	.6484	.6493	.6503	.6513	.6522
4.5	.6532	.6542	.6551	.6561	.6571	.6580	.6590	.6599	.6609	.6618
4.6	.6628	.6637	.6646	.6656	.6665	.6675	.6684	.6693	.6702	.6712
4.7	.6721	.6730	.6739	.6749	.6758	.6767	.6776	.6785	.6794	.6803
4.8	.6812	.6821	.6830	.6839	.6848	.6857	.6866	.6875	.6884	.6893
4.9	.6902	.6911	.6920	.6928	.6937	.6946	.6955	.6964	.6972	.6981
5.0	.6990	.6998	.7007	.7016	.7024	.7033	.7042	.7050	.7059	.7067
5.1	.7076	.7084	.7093	.7101	.7110	.7118	.7216	.7135	.7143	.7152
5.2	.7160	.7168	.7177	.7185	.7193	.7202	.7210	.7218	.7226	.7235
5.3	.7243	.7251	.7259	.7267	.7275	.7284	.7292	.7300	.7308	.7316
5.4	.7324	.7332	.7340	.7348	.7356	.7364	.7372	.7380	.7388	.7396
N	0	1	2	3	4	5	6	7	8	9

N	0	1	2	3	4	5	6	7	8	9
5.5	.7404	.7412	.7419	.7427	.7435	.7443	.7451	.7459	.7466	.7474
5.6	.7482	.7490	.7497	.7505	.7513	.7520	.7528	.7536	.7543	.7551
5.7	.7559	.7566	.7574	.7582	.7589	.7597	.7604	.7612	.7619	.7627
5.8	.7634	.7642	.7649	.7657	.7664	.7672	.7679	.7686	.7694	.7701
5.9	.7709	.7716	.7723	.7731	.7738	.7745	.7752	.7760	.7767	.7774
6.0	.7782	.7789	.7796	.7803	.7810	.7818	.7825	.7832	.7839	.7846
6.1	.7853	.7860	.7868	.7875	.7882	.7889	.7896	.7903	.7910	.7917
6.2	.7924	.7931	.7938	.7945	.7952	.7959	.7966	.7973	.7980	.7987
6.3	.7993	.8000	.8007	.8014	.8021	.8028	.8035	.8041	.8048	.8055
6.4	.8062	.8069	.8075	.8082	.8089	.8096	.8102	.8109	.8116	.8122
6.5	.8129	.8136	.8142	.8149	.8156	.8162	.8169	.8176	.8182	.8189
6.6	.8195	.8202	.8209	.8215	.8222	.8228	.8235	.8241	.8248	.8254
6.7	.8261	.8267	.8274	.8280	.8287	.8293	.8299	.8306	.8312	.8319
5.8	.8325	.8331	.8338	.8344	.8351	.8357	.8363	.8370	.8376	.8382
6.9	.8388	.8395	.8401	.8407	.8414	.8420	.8426	.8432	.8439	.8445
7.0	.8451	.8457	.8453	.8470	.8476	.8482	.8488	.8494	.8500	.8506
7.1	.8513	.8519	.8525	.8531	.8537	.8543	.8549	.8555	.8561	.8567
7.2	.8573	.8579	.8585	.8591	.8597	.8603	.8609	.8615	.8621	.8627
7.3	.8633	.8639	.8645	.8651	.8657	.8663	.8669	.8675	.8681	.8686
7.4	.8692	.8698	.8704	.8710	.8716	.8722	.8727	.8733	.8739	.8745
7.5	.8751	.8756	.8762	.8768	.8774	.8779	.8785	.8791	.8797	.8802
7.6	.8808	.8814	.8820	.8825	.8831	.8837	.8842	.8848	.8854	.8859
7.7	.8865	.8871	.8876	.8882	.8887	.8893	.8899	.8904	.8910	.8915
7.8	.8921	.8927	.8932	.8938	.8943	.8949	.8954	.8960	.8965	.8971
7.9	.8976	.8982	.8987	.8993	.8998	.9004	.9009	.9015	.9020	.9025
8.0	.9031	.9036	.9042	.9047	.9053	.9058	.9063	.9069	.9074	.9079
8.1	.9085	.9090	.9096	.9101	.9106	.9112	.9117	.9122	.9128	.9133
8.2	.9138	.9143	.9149	.9154	.9159	.9165	.9170	.9175	.9180	.9186
8.3	.9191	.9196	.9201	.9206	.9212	.9217	.9222	.9227	.9232	.9238
8.4	.9243	.9248	.9253	.9258	.9263	.9269	.9274	.9279	.9284	.9289
8.5	.9294	.9299	.9304	.9309	.9315	.9320	.9325	.9330	.9335	.9340
8.6	.9345	.9350	.9355	.9360	.9365	.9370	.9375	.9380	.9385	.9390
8.7	.9395	.9400	.9405	.9410	.9415	.9420	.9425	.9430	.9435	.9440
8.8	.9445	.9450	.9455	.9460	.9465	.9469	.9474	.9479	.9484	.9489
8.9	.9494	.9499	.9504	.9509	.9513	.9518	.9523	.9528	.9533	.9538
9.0	.9542	.9547	.9552	.9557	.9562	.9566	.9571	.9576	.9581	.9586
9.1	.9590	.9595	.9600	.9605	.9609	.9614	.9619	.9624	.9628	.9633
9.2	.9638	.9643	.9647	.9652	.9657	.9661	.9666	.9671	.9675	.9680
9.3	.9685	.9689	.9694	.9699	.9703	.9708	.9713	.9717	.9722	.9727
9.4	.9731	.9736	.9741	.9745	.9750	.9754	.9759	.9763	.9768	.9773
9.5	.9777	.9782	.9786	.9791	.9795	.9800	.9805	.9809	.9814	.9818
9.6	.9823	.9827	.9832	.9836	.9841	.9845	.9850	.9854	.9859	.9863
9.7	.9868	.9872	.9877	.9881	.9886	.9890	.9894	.9899	.9903	.9908
9.8	.9912	.9917	.9921	.9926	.9930	.9934	.9939	.9943	.9948	.9952
9.9	.9956	.9961	.9965	.9969	.9974	.9978	.9983	.9987	.9991	.9996
N	**0**	**1**	**2**	**3**	**4**	**5**	**6**	**7**	**8**	**9**

ANGLE in DEGREES	SINE	COSINE	TAN	COTAN	ANGLE in DEGREES
0	0.0000	1.0000	0.0000		90
1	.0175	.9998	.0175	57.290	89
2	.0349	.9994	.0349	28.636	88
3	.0523	.9986	.0524	19.081	87
4	.0698	.9976	.0699	14.301	86
5	.0872	.9962	.0875	11.430	85
6	.1045	.9945	.1051	9.5144	84
7	.1219	.9925	.1228	8.1443	83
8	.1392	.9903	.1405	7.1154	82
9	.1564	.9877	.1584	6.3138	81
10	.1736	.9848	.1763	5.6713	80
11	.1908	.9816	.1944	5.1446	79
12	.2079	.9781	.2126	4.7046	78
13	.2250	.9744	.2309	4.3315	77
14	.2419	.9703	.2493	4.0108	76
15	.2588	.9659	.2679	3.7321	75
16	.2756	.9613	.2867	3.4874	74
17	.2924	.9563	.3057	3.2709	73
18	.3090	.9511	.3249	3.0777	72
19	.3256	.9455	.3443	2.9042	71
20	.3420	.9397	.3640	2.7475	70
21	.3584	.9336	.3839	2.6051	69
22	.3746	.9272	.4040	2.4751	68
23	.3907	.9205	.4245	2.3559	67
24	.4067	.9135	.4452	2.2460	66
25	.4226	.9063	.4663	2.1445	65
26	.4384	.8988	.4877	2.0503	64
27	.4540	.8910	.5095	1.9626	63
28	.4695	.8829	.5317	1.8807	62
29	.4848	.8746	.5543	1.8040	61
30	.5000	.8660	.5774	1.7321	60
31	.5150	.8572	.6009	1.6643	59
32	.5299	.8480	.6249	1.6003	58
33	.5446	.8387	.6494	1.5399	57
34	.5592	.8290	.6745	1.4826	56
35	.5736	.8192	.7002	.14281	55
36	.5878	.8090	.7265	1.3764	54
37	.6018	.7986	.7536	1.3270	53
38	.6157	.7880	.7813	1.2799	52
39	.6293	.7771	.8098	1.2349	51
40	.6428	.7660	.8391	1.1918	50
41	.6561	.7547	.8693	1.1504	49
42	.6691	.7431	.9004	1.1106	48
43	.6820	.7314	.9325	1.0724	47
44	.6947	.7193	.9657	1.0355	46
45	.7071	.7071	1.0000	1.0000	45

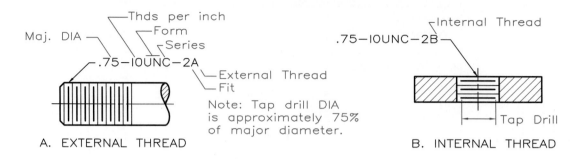

Note: Tap drill DIA is approximately 75% of major diameter.

A. EXTERNAL THREAD

B. INTERNAL THREAD

Nominal Diameter	Basic Diameter	Coarse NC & UNC		Fine NF & UNF		Extra Fine NEF/UNEF		Nominal Diameter	Basic Diameter	Coarse NC & UNC		Fine NF & UNF		Extra Fine NEF/UNEF	
		Thds per in.	Tap Drill DIA	Thds per in.	Tap Drill DIA	Thds per in.	Tap Drill DIA			Thds per in.	Tap Drill DIA	Thds per in.	Tap Drill DIA	Thds per in.	Tap Drill DIA
0	.060			80	.0469			1	1.000	8	.875	12	.922	20	.953
1	.073	64	No. 53	72	No. 53			1-1/16	1.063	...	...	...	...	18	1.000
2	.086	56	No. 50	64	No. 50			1-1/8	1.125	7	.904	12	1.046	18	1.070
3	.099	48	No. 47	56	No. 45			1-3/16	1.188	...	...	...	...	18	1.141
4	.112	40	No. 43	48	No. 42			1-1/4	1.250	7	1.109	12	1.172	18	1.188
5	.125	40	No. 38	44	No. 37			1-5/16	1.313	...	...	...	...	18	1.266
6	.138	32	No. 36	40	No. 33			1-3/8	1.375	6	1.219	12	1.297	18	1.313
8	.164	32	No. 29	36	No. 29			1-7/16	1.438	...	...	...	...	18	1.375
10	.190	24	No. 25	32	No. 21			1-1/2	1.500	6	1.344	12	1.422	18	1.438
12	.216	24	No. 16	28	No. 14	32	No. 13	1-9/16	1.563	...	...	...	...	18	1.500
1/4	.250	20	No. 7	28	No. 3	32	.2189	1-5/8	1.625	...	...	...	...	18	1.563
5/16	.3125	18	F	24	I	32	.2813	1-11/16	1.688	...	...	...	...	18	1.625
3/8	.375	16	.3125	24	Q	32	.3438	1-3/4	1.750	5	1.563	...	...	...	...
7/16	.4375	14	U	20	.3906	28	.4062	2	2.000	4.5	1.781	...	...	...	...
1/2	.500	13	.4219	20	.4531	28	.4688	2-1/4	2.250	4.5	2.031	...	...	...	...
9/16	.5625	12	.4844	18	.5156	24	.5156	2-1/2	2.500	4	2.250	...	...	...	...
5/8	.625	11	.5313	18	.5781	24	.5781	2-3/4	2.750	4	2.500	...	...	...	...
11/16	.6875	...	...	...	...	24	.6406	3	3.000	4	2.750	...	...	...	...
3/4	.750	10	.6563	16	.6875	20	.7031	3-1/4	3.250	4	...	...	...	...	...
13/16	.8125	...	...	...	...	20	.7656	3-1/2	3.500	4	...	...	...	...	...
7/8	.875	9	.7656	14	.8125	20	.8281	3-3/4	3.750	4	...	...	...	...	...
15/16	.9375	...	...	...	...	20	.8906	4	4.000	4	...	...	...	...	...

Source: ANSI/ASME B1.1—1989

Appendix 6 • Screw Threads: American National and Unified (inches)
Constant-Pitch Threads

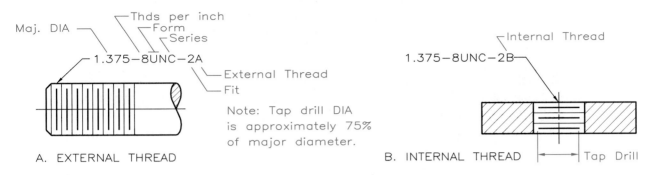

1.375—8UNC—2A
- Maj. DIA
- Thds per inch
- Form
- Series
- External Thread
- Fit

A. EXTERNAL THREAD

Note: Tap drill DIA is approximately 75% of major diameter.

1.375—8UNC—2B
- Internal Thread

B. INTERNAL THREAD — Tap Drill

Nominal Diameter	8 Pitch 8N & 8UN		12 Pitch 12N & 12UN		16 Pitch 16N & 16UN		Nominal Diameter	8 Pitch 8N & 8UN		12 Pitch 12N & 12UN		16 Pitch 16N & 16UN	
	Thds per in.	Tap Drill DIA	Thds per in.	Tap Drill DIA	Thds per in.	Tap Drill DIA		Thds per in.	Tap Drill DIA	Thds per in.	Tap Drill DIA	Thds per in.	Tap Drill DIA
.500	...	...	12	.422	...	...	2.063	...	...	...	...	16	2.000
.563	...	...	12	.484	...	...	2.125	...	...	12	2.047	16	2.063
.625	...	...	12	.547	...	...	2.188	...	...	...	...	16	2.125
.688	...	...	12	.609	...	...	2.250	8	2.125	12	2.172	16	2.188
.750	...	...	12	.672	16	.688	2.313	...	...	...	...	16	2.250
.813	...	...	12	.734	16	.750	2.375	...	...	12	2.297	16	2.313
.875	...	...	12	.797	16	.813	2.438	...	...	...	...	16	2.375
.934	...	...	12	.859	16	.875	2.500	8	2.375	12	2.422	16	2.438
1.000	8	.875	12	.922	16	.938	2.625	...	...	12	2.547	16	2.563
1.063	...	...	12	.984	16	1.000	2.750	8	2.625	12	2.717	16	2.688
1.125	8	1.000	12	1.047	16	1.063	2.875	...	...	12	...	16	...
1.188	...	...	12	1.109	16	1.125	3.000	8	2.875	12	...	16	...
1.250	8	1.125	12	1.172	16	1.188	3.125	...	...	12	...	16	...
1.313	...	...	12	1.234	16	1.250	3.250	8	...	12	...	16	...
1.375	8	1.250	12	1.297	16	1.313	3.375	...	...	12	...	16	...
1.434	...	...	12	1.359	16	1.375	3.500	8	...	12	...	16	...
1.500	8	1.375	12	1.422	16	1.438	3.625	...	...	12	...	16	...
1.563	...	...	...	...	16	1.500	3.750	8	...	12	...	16	...
1.625	8	1.500	12	1.547	16	1.563	3.875	...	...	12	...	16	...
1.688	...	...	...	...	16	1.625	4.000	8	...	12	...	16	...
1.750	8	1.625	12	1.672	16	1.688	4.250	8	...	12	...	16	...
1.813	...	...	...	...	16	1.750	4.500	8	...	12	...	16	...
1.875	8	1.750	12	1.797	16	1.813	4.750	8	...	12	...	16	...
1.934	...	...	...	...	16	1.875	5.000	8	...	12	...	16	...
2.000	8	1.875	12	1.922	16	1.938	5.250	8	...	12	...	16	...

Source: ANSI/ASME B1.1—1989.

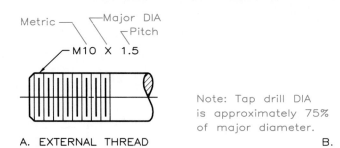

Metric — Major DIA — Pitch

M10 X 1.5

A. EXTERNAL THREAD

Note: Tap drill DIA is approximately 75% of major diameter.

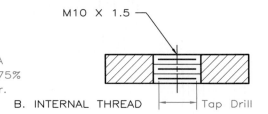

M10 X 1.5

B. INTERNAL THREAD — Tap Drill

COARSE		FINE		COARSE		FINE	
MAJ. DIA & THD PITCH	TAP DRILL	MAJ. DIA & THD PITCH	TAP DRILL	MAJ. DIA & THD PITCH	TAP DRILL	MAJ. DIA & THD PITCH	TAP DRILL
M1.6 × 0.35	1.25			M20 × 2.5	17.5	M20 × 1.5	18.5
M1.8 × 0.35	1.45			M22 × 2.5	19.5	M22 × 1.5	20.5
M2 × 0.4	1.6			M24 × 3	21.0	M24 × 2	22.0
M2.2 × 0.45	1.75			M27 × 3	24.0	M27 × 2	25.0
M2.5 × 0.45	2.05			M30 × 3.5	26.5	M30 × 2	28.0
M3 × 0.5	2.5			M33 × 3.5	29.5	M33 × 2	31.0
M3.5 × 0.6	2.9			M36 × 4	32.0	M36 × 3	33.0
M4 × 0.7	3.3			M39 × 4	35.0	M39 × 3	36.0
M4.5 × 0.75	3.75			M42 × 4.5	37.5	M42 × 3	39.0
M5 × 0.8	4.2			M45 × 4.5	40.5	M45 × 3	42.0
M6 × 1	5.0			M48 × 5	43.0	M48 × 3	45.0
M7 × 1	6.0			M52 × 5	47.0	M52 × 3	49.0
M8 × 1.25	6.8	M8 × 1	7.0	M56 × 5.5	50.5	M56 × 4	52.0
M9 × 1.25	7.75			M60 × 5.5	54.5	M60 × 4	56.0
M10 × 1.5	8.5	M10 × 1.25	8.75	M64 × 6	58.0	M64 × 4	60.0
M11 × 1.5	9.5			M68 × 6	62.0	M68 × 4	64.0
M12 × 1.75	10.3	M12 × 1.25	10.5	M72 × 6	66.0	M72 × 4	68.0
M14 × 2	12.0	M14 × 1.5	12.5	M80 × 6	74.0	M80 × 4	76.0
M16 × 2	14.0	M16 × 1.5	14.5	M90 × 6	84.0	M90 × 4	86.0
M18 × 2.5	15.5	M18 × 1.5	16.5	M100 × 6	94.0	M100 × 4	96.0

Source: ANSI/ASME B1.13

APPENDIX 8 • Square and Acme Threads

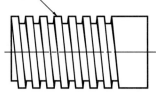

2.00−2.5 SQUARE

Typical thread note

Dimensions are in inches Size	Size	Thds per inch	Size	Size	Thds per inch	Size	Size	Thds per inch
3/8	.375	12	1-1/8	1.125	4	3	3.000	1-1/2
7/16	.438	10	1-1/4	1.250	4	3-1/4	3.125	1-1/2
1/2	.500	10	1-1/2	1.500	3	3-1/2	3.500	1-1/3
9/16	.563	8	1-3/4	1.750	2-1/2	3-3/4	3.750	1-1/3
5/8	.625	8	2	2.000	2-1/2	4	4.000	1-1/3
3/4	.75	6	2-1/4	2.250	2	4-1/4	4.250	1-1/3
7/8	.875	5	2-1/2	2.500	2	4-1/2	4.500	1
1	1.000	5	2-3/4	2.750	2	Larger		1

APPENDIX 9 • American Standard Taper Pipe Threads (NPT)

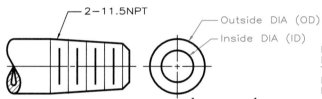

2−11.5NPT

Outside DIA (OD)

Inside DIA (ID)

PIPES THRU 12 INCHES IN DIA ARE SPECIFIED BY THEIR INSIDE DIAMETERS. LARGER PIPES ARE SPECI−FIED BY THEIR OD.

$\frac{1}{16}$ DIA to $1\frac{1}{4}$ DIA   Dimensions in inches

Nominal ID	$\frac{1}{16}$	$\frac{1}{8}$	$\frac{1}{4}$	$\frac{3}{8}$	$\frac{1}{2}$	$\frac{3}{4}$	1	1-1/4
Outside DIA	0.313	0.405	0.540	0.675	0.840	1.050	1.315	1.660
Thds/Inch	27	27	18	18	14	14	$11\frac{1}{2}$	$11\frac{1}{2}$

$1\frac{1}{2}$ DIA to 6 DIA

Nominal ID	$1\frac{1}{2}$	2	$2\frac{1}{2}$	3	$3\frac{1}{2}$	4	5	6
Outside DIA	1.900	2.375	2.875	3.500	4.000	4.500	5.563	6.625
Thds/Inch	$11\frac{1}{2}$	$11\frac{1}{2}$	8	8	8	8	8	8

8 DIA to 24 DIA

Nominal ID	8	10	12	14 OD	16 OD	18 OD	20 OD	24 OD
Outside DIA	8.625	10.750	12.750	14.000	16.000	18.000	20.000	24.000
Thds/Inch	8	8	8	8	8	8	8	8

Source: ANSI B2.1.

Appendix 10 • Square Bolts (inches)

DIA	E Max.	F Max.	G Avg.	H Max.	R Max.
1/4	.250	.375	.530	.188	.031
5/16	.313	.500	.707	.220	.031
3/8	.375	.563	.795	.268	.031
7/16	.438	.625	.884	.316	.031
1/2	.500	.750	1.061	.348	.031
5/8	.625	.938	1.326	.444	.062
3/4	.750	1.125	1.591	.524	.062
7/8	.875	1.313	1.856	.620	.062
1	1.000	1.500	2.121	.684	.093
1-1/8	1.125	1.688	2.386	.780	.093
1-1/4	1.250	1.875	2.652	.876	.093
1-3/8	1.375	2.625	2.917	.940	.093
1-1/2	1.500	2.250	3.182	1.036	.093

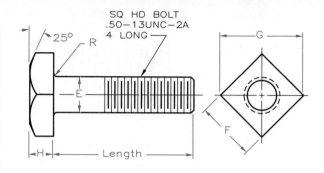

SQ HD BOLT
.50—13UNC—2A
4 LONG

25°   R
E
H — Length

STANDARD COMMERCIAL LENGTHS

DIAMETER	.50	.75	1.00	1.25	1.50	1.75	2.00	2.50	3.00	3.50	4.00	4.50	5.00	5.50	6.00	6.50	7.00	8.00 PLUS	
.25	•	•	•	•	•	•	•												
.31		•	•	•	•	•	•	•	•	•									
.375		•	•	•	•	•	•	•	•	•	•								
.500		•	•	•	•	•	•	•	•	•	•	•	•	•	•	•	•	•	*14
.625			•	•	•	•	•	•	•	•	•	•	•	•	•	•	•	•	*12
.750						•	•	•	•	•	•	•	•	•	•	•	•	•	*13
.875									•	•	•	•	•						
1.00															•	•	•	•	*12
1.25															•	•	•	•	

*14 MEANS THAT LENGTHS ARE AVAILABLE
AT 1 INCH INCREMENTS UP TO 10 INCHES.

Appendix 11 • Square Nuts

Dimensions are in inches.

DIA	DIA	F Max.	G Avg.	H Max.
1/4	.250	.438	.619	.235
5/16	.313	.563	.795	.283
3/8	.375	.625	.884	.346
7/16	.438	.750	1.061	.394
1/2	.500	.813	1.149	.458
5/8	.625	1.000	1.414	.569
3/4	.750	1.125	1.591	.680
7/8	.875	1.313	1.856	.792
1	1.000	1.500	2.121	.903
1-1/8	1.125	1.688	2.386	1.030
1-1/4	1.250	1.875	2.652	1.126
1-3/8	1.375	1.063	2.917	1.237
1-1/2	1.500	2.250	3.182	1.348

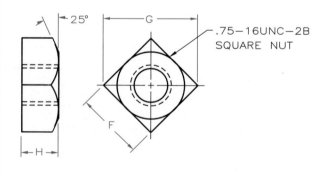

25°   G
.75—16UNC—2B
SQUARE NUT
F
H

Dimensions are in inches.

DIA	E Max.	F Max.	G Avg.	H Max.	R Max.
1/4	.250	.438	.505	.163	.025
5/16	.313	.500	.577	.211	.025
3/8	.375	.563	.650	.243	.025
7/16	.438	.625	.722	.291	.025
1/2	.500	.750	.866	.323	.025
9/16	.563	.812	.938	.371	.045
5/8	.625	.938	1.083	.403	.045
3/4	.750	1.125	1.299	.483	.045
7/8	.875	1.313	1.516	.563	.065
1	1.000	1.500	1.732	.627	.095
1-1/8	1.125	1.688	1.949	.718	.095
1-1/4	1.250	1.875	2.165	.813	.095
1-3/8	1.375	2.063	2.382	.878	.095
1-1/2	1.500	2.250	2.598	.974	.095
1-3/4	1.750	2.625	3.031	1.134	.095
2	2.000	3.000	3.464	1.263	.095
2-1/4	2.250	3.375	3.897	1.423	.095
2-1/2	2.500	3.750	4.330	1.583	.095
2-3/4	2.750	4.125	4.763	1.744	.095
3	3.000	4.500	5.196	1.935	.095

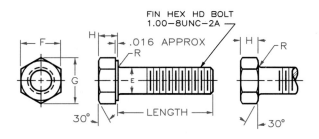

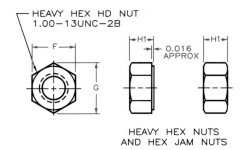

Appendix 13 • Hex Nuts and Hex Jam Nuts

MAJOR DIA		F Max.	G Avg.	H1 Max.	H2 Max.
1/4	.250	.438	.505	.226	.163
5/16	.313	.500	.577	.273	.195
3/8	.375	.563	.650	.337	.227
7/16	.438	.688	.794	.385	.260
1/2	.500	.750	.866	.448	.323
9/16	.563	.875	1.010	.496	.324
5/8	.625	.938	1.083	.559	.387
3/4	.750	1.125	1.299	.665	.446
7/8	.875	1.313	1.516	.776	.510
1	1.000	1.500	1.732	.887	.575
1-1/8	1.125	1.688	1.949	.899	.639
1-1/4	1.250	1.875	2.165	1.094	.751
1-3/8	1.375	2.063	2.382	1.206	.815
1-1/2	1.500	2.250	2.589	1.317	.880

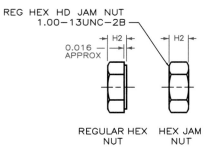

APPENDIX 14 • Round Head Cap Screws

Dimensions are in inches.

DIA	D Max.	A Max.	H Avg.	J Max.	T Max.
1/4	.250	.437	.191	.075	.117
5/16	.313	.562	.245	.084	.151
3/8	.375	.625	.273	.094	.168
7/16	.438	.750	.328	.094	.202
1/2	.500	.812	.354	.106	.218
9/16	.563	.937	.409	.118	.252
5/8	.625	1.000	.437	.133	.270
3/4	.750	1.250	.546	.149	.338

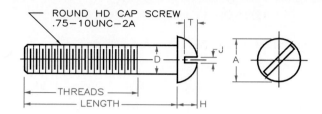

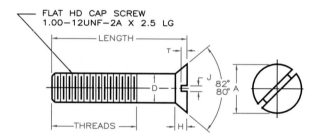

OTHER LENGTHS AND DIAMETERS ARE AVAILABLE, BUT THESE ARE THE MORE STANDARD ONES.

APPENDIX 15 • Flat Head Cap Screws

Dimensions are in inches.

DIA	D Max.	A Max.	H Avg.	J Max.	T Max.
1/4	.250	.500	.140	.075	.068
5/16	.313	.625	.177	.084	.086
3/8	.375	.750	.210	.094	.103
7/16	.438	.813	.210	.094	.103
1/2	.500	.875	.210	.106	.103
9/16	.563	1.000	.244	.118	.120
5/8	.625	1.125	.281	.133	.137
3/4	.750	1.375	.352	.149	.171
7/8	.875	1.625	.423	.167	.206
1	1.000	1.875	.494	.188	.240
1-1/8	1.125	2.062	.529	.196	.257
1-1/4	1.250	2.312	.600	.211	.291
1-3/8	1.375	2.562	.665	.226	.326
1-1/2	1.500	2.812	.742	.258	.360

OTHER LENGTHS AND DIAMETERS ARE AVAILABLE, BUT THESE ARE THE MORE STANDARD ONES.

APPENDIX 16 • Fillister Head Cap Screws

Dimensions are in inches.

DIA	D Max.	A Max.	H Avg.	J Max.	T Max.
1/4	.250	.375	.172	.075	.097
5/16	.313	.437	.203	.084	.115
3/8	.375	.562	.250	.094	.142
7/16	.438	.625	.297	.094	.168
1/2	.500	.750	.328	.106	.193
9/16	.563	.812	.375	.118	.213
5/8	.625	.875	.422	.133	.239
3/4	.750	1.000	.500	.149	.283
7/8	.875	1.125	.594	.167	.334
1	1.000	1.312	.656	.188	.371

Source: ANSI B18.6.2.

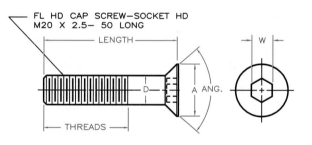

Appendix 17 • Flat Socket Head Cap Screws

Diameter mm	Diameter inches	Pitch	A	Ang.	W
M3	.118	.5	6	90	2
M4	.157	.7	8	90	2.5
M5	.197	.8	10	90	3
M6	.236	1	12	90	4
M8	.315	1.25	16	90	5
M10	.394	1.5	20	90	6
M12	.472	1.75	24	90	8
M14	.551	2	27	90	10
M16	.630	2	30	90	10
M20	.787	2.5	36	90	12

APPENDIX 18 • Socket Head Cap Screws

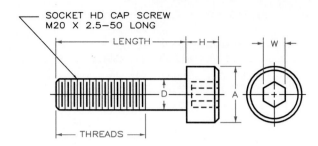

SOCKET HD CAP SCREW
M20 X 2.5—50 LONG

Diameter		Pitch	A	H	W
mm	inches				
M3	.118	.5	6	3	2
M4	.157	.7	8	4	3
M5	.187	.8	10	5	4
M6	.236	1	12	6	6
M8	.315	1.25	16	8	6
M10	.394	1.5	20	10	8
M12	.472	1.75	24	12	10
M14	.551	2	27	14	12
M16	.630	2	30	16	14
M20	.787	2.5	36	20	17

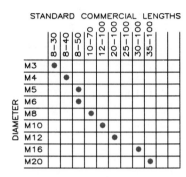

STANDARD COMMERCIAL LENGTHS

DIA 8–16: LENGTHS AT INTERVALS OF 2 MM
DIA 20–100: LENGTHS AT INTERVALS OF 5 MM

APPENDIX 19 • Round Head Machine Screws

Dimensions are in inches.

DIA	D Max.	A Max.	H Avg.	J Max.	T Max.
0	.060	.113	.053	.023	.039
1	.073	.138	.061	.026	.044
2	.086	.162	.069	.031	.048
3	.099	.187	.078	.035	.053
4	.112	.211	.086	.039	.058
5	.125	.236	.095	.043	.063
6	.138	.260	.103	.048	.068
8	.164	.309	.120	.054	.077
10	.190	.359	.137	.060	.087
12	.216	.408	.153	.067	.096
1/4	.250	.472	.175	.075	.109
5/16	.313	.590	.216	.084	.132
3/8	.375	.708	.256	.094	.155
7/16	.438	.750	.328	.094	.196
1/2	.500	.813	.355	.106	.211
9/16	.563	.938	.410	.118	.242
5/8	.625	1.000	.438	.133	.258
3/4	.750	1.250	.547	.149	.320

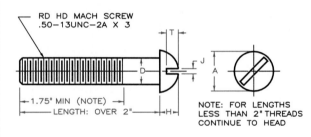

RD HD MACH SCREW
.50—13UNC—2A X 3

1.75" MIN (NOTE)
LENGTH: OVER 2"

NOTE: FOR LENGTHS LESS THAN 2" THREADS CONTINUE TO HEAD

STANDARD LENGTHS

OTHER LENGTHS AND DIAMETERS ARE AVAILABLE; THESE ARE THE MORE STANDARD ONES.

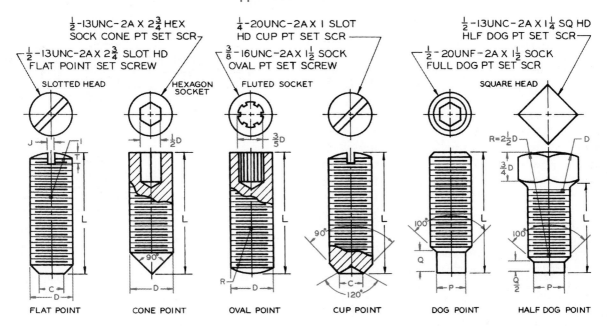

Dimensions for the set screws shown in ANSI Fig. 18.44 (dimensions in inches)

D	I	J	T	R	C		P		Q	q
					Diameter of Cup and Flat Points		Diameter of Dog Point		Length of Dog Point	
Nominal Size	Radius of Headless Crown	Width of Slot	Depth of Slot	Oval Point Radius	Max	Min	Max	Min	Full	Half
5 0.125	0.125	0.023	0.031	0.094	0.067	0.057	0.083	0.078	0.060	0.030
6 0.138	0.138	0.025	0.035	0.109	0.047	0.064	0.092	0.087	0.070	0.035
8 0.164	0.164	0.029	0.041	0.125	0.087	0.076	0.109	0.103	0.080	0.040
10 0.190	0.190	0.032	0.048	0.141	0.102	0.088	0.127	0.120	0.090	0.045
12 0.216	0.216	0.036	0.054	0.156	0.115	0.101	0.144	0.137	0.110	0.055
$\frac{1}{4}$ 0.250	0.250	0.045	0.063	0.188	0.132	0.118	0.156	0.149	0.125	0.063
$\frac{5}{16}$ 0.3125	0.313	0.051	0.076	0.234	0.172	0.156	0.203	0.195	0.156	0.078
$\frac{3}{8}$ 0.375	0.375	0.064	0.094	0.281	0.212	0.194	0.250	0.241	0.188	0.094
$\frac{7}{16}$ 0.4375	0.438	0.072	0.190	0.328	0.252	0.232	0.297	0.287	0.219	0.109
$\frac{1}{2}$ 0.500	0.500	0.081	0.125	0.375	0.291	0.270	0.344	0.344	0.250	0.125
$\frac{9}{16}$ 0.5625	0.563	0.091	0.141	0.422	0.332	0.309	0.391	0.379	0.281	0.140
$\frac{5}{8}$ 0.625	0.625	0.102	0.156	0.469	0.371	0.347	0.469	0.456	0.313	0.156
$\frac{3}{4}$ 0.750	0.750	0.129	0.188	0.563	0.450	0.425	0.563	0.549	0.375	0.188

Source: Courtesy of ANSI; B18.6.2.

Letter Size Drills

Size	Drill Diameter		Size	Drill Diameter		Size	Drill Diameter		Size	Drill Diameter	
	inches	mm		inches	mm		inches	mm		inches	mm
A	0.234	5.944	H	0.266	6.756	O	0.316	8.026	V	0.377	9.576
B	0.238	6.045	I	0.272	6.909	P	0.323	8.204	W	0.386	9.804
C	0.242	6.147	J	0.277	7.036	Q	0.332	8.433	X	0.397	10.084
D	0.246	6.248	K	0.281	7.137	R	0.339	8.611	Y	0.404	10.262
E	0.250	6.350	L	0.290	7.366	S	0.348	8.839	Z	0.413	10.490
F	0.257	6.528	M	0.295	7.493	T	0.358	9.093			
G	0.261	6.629	N	0.302	7.601	U	0.368	9.347			

Source: Courtesy of General Motors Corporation.

Number Size Drills

Size	Drill Diameter		Size	Drill Diameter		Size	Drill Diameter		Size	Drill Diameter	
	inches	mm		inches	mm		inches	mm		inches	mm
1	0.2280	5.7912	21	0.1590	4.0386	41	0.0960	2.4384	61	0.0390	0.9906
2	0.2210	5.6134	22	0.1570	3.9878	42	0.0935	2.3622	62	0.0380	0.9652
3	0.2130	5.4102	23	0.1540	3.9116	43	0.0890	2.2606	63	0.0370	0.9398
4	0.2090	5.3086	24	0.1520	3.8608	44	0.0860	2.1844	64	0.0360	0.9144
5	0.2055	5.2197	25	0.1495	3.7973	45	0.0820	2.0828	65	0.0350	0.8890
6	0.2040	5.1816	26	0.1470	3.7338	46	0.0810	2.0574	66	0.0330	0.8382
7	0.2010	5.1054	27	0.1440	3.6576	47	0.0785	19.812	67	0.0320	0.8128
8	0.1990	5.0800	28	0.1405	3.5560	48	0.0760	1.9304	68	0.0310	0.7874
9	0.1960	4.9784	29	0.1360	3.4544	49	0.0730	1.8542	69	0.0292	0.7417
10	0.1935	4.9149	30	0.1285	3.2639	50	0.0700	1.7780	70	0.0280	0.7112
11	0.1910	4.8514	31	0.1200	3.0480	51	0.0670	1.7018	71	0.0260	0.6604
12	0.1890	4.8006	32	0.1160	2.9464	52	0.0635	1.6129	72	0.0250	0.6350
13	0.1850	4.6990	33	0.1130	2.8702	53	0.0595	1.5113	73	0.0240	0.6096
14	0.1820	4.6228	34	0.1110	2.8194	54	0.0550	1.3970	74	0.0225	0.5715
15	0.1800	4.5720	35	0.1100	2.7940	55	0.0520	1.3208	75	0.0210	0.5334
16	0.1770	4.4958	36	0.1065	0.7051	56	0.0465	1.1684	76	0.0200	0.5080
17	0.1730	4.3942	37	0.1040	2.6416	57	0.0430	1.0922	77	0.0180	0.4572
18	0.1695	4.3053	38	0.1015	2.5781	58	0.0420	1.0668	78	0.0160	0.4064
19	0.1660	4.2164	39	0.0995	2.5273	59	0.0410	1.0414	79	0.0145	0.3638
20	0.1610	4.0894	40	0.0980	2.4892	60	0.0400	1.0160	80	0.0135	0.3428

Metric Drill Sizes   Decimal-inch equivalents are for reference only.

Drill Diameter		Drill Diameter		Drill Diameter		Drill Diameter		Drill Diameter		Drill Diameter		Drill Diameter	
mm	in.	mm	in.	mm	in.	mm	in.	mm	in.	mm	in.	mm	in.
.40	.0157	1.03	.0406	2.20	.0866	5.00	.1969	10.00	.3937	21.50	.8465	48.00	1.8898
.42	.0165	1.05	.0413	2.30	.0906	5.20	.2047	10.30	.4055	22.00	.8661	50.00	1.9685
.45	.0177	1.08	.0425	2.40	.0945	5.30	.2087	10.50	.4134	23.00	.9055	51.50	2.0276
.48	.0189	1.10	.0433	2.50	.0984	5.40	.2126	10.80	.4252	24.00	.9449	53.00	2.0866
.50	.0197	1.15	.0453	2.60	.1024	5.60	.2205	11.00	.4331	25.00	.9843	54.00	2.1260
.52	.0205	1.20	.0472	2.70	.1063	5.80	.2283	11.50	.4528	26.00	1.0236	56.00	2.2047
.55	.0217	1.25	.0492	2.80	.1102	6.00	.2362	12.00	.4724	27.00	1.0630	58.00	2.2835
.58	.0228	1.30	.0512	2.90	.1142	6.20	.2441	12.50	.4921	28.00	1.1024	60.00	2.3622
.60	.0236	1.35	.0531	3.00	.1181	6.30	.2480	13.00	.5118	29.00	1.1417		
.62	.0244	1.40	.0551	3.10	.1220	6.50	.2559	13.50	.5315	30.00	1.1811		
.65	.0256	1.45	.0571	3.20	.1260	6.70	.2638	14.00	.5512	31.00	1.2205		
.68	.0268	1.50	.0591	3.30	.1299	6.80	.2677	14.50	.5709	32.00	1.2598		
.70	.0276	1.55	.0610	3.40	.1339	6.90	.2717	15.00	.5906	33.00	1.2992		
.72	.0283	1.60	.0630	3.50	.1378	7.10	.2795	15.50	.6102	34.00	1.3386		
.75	.0295	1.65	.0650	3.60	.1417	7.30	.2874	16.00	.6299	35.00	1.3780		
.78	.0307	1.70	.0669	3.70	.1457	7.50	.2953	16.50	.6496	36.00	1.4173		
.80	.0315	1.75	.0689	3.80	.1496	7.80	.3071	17.00	.6693	37.00	1.4567		
.82	.0323	1.80	.0709	3.90	.1535	8.00	.3150	17.50	.6890	38.00	1.4961		
.85	.0335	1.85	.0728	4.00	.1575	8.20	.3228	18.00	.7087	39.00	1.5354		
.88	.0346	1.90	.0748	4.10	.1614	8.50	.3346	18.50	.7283	40.00	1.5748		
.90	.0354	1.95	.0768	4.20	.1654	8.80	.3465	19.00	.7480	41.00	1.6142		
.92	.0362	2.00	.0787	4.40	.1732	9.00	.3543	19.50	.7677	42.00	1.6535		
.95	.0374	2.05	.0807	4.50	.1772	9.20	.3622	20.00	.7874	43.50	1.7126		
.98	.0386	2.10	.0827	4.60	.1811	9.50	.3740	20.50	.0871	45.00	1.7717		
1.00	.0394	2.15	.0846	4.80	.1890	9.80	.3858	21.00	.8268	46.50	1.8307		

Appendix 22 • Cotter Pins: American National Standard

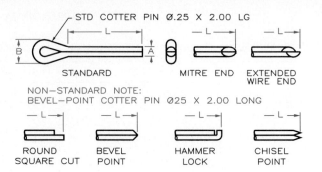

Nominal Diameter	Maximum DIA A	Minimum DIA B	Hole Size
0.031	0.032	0.063	0.047
0.047	0.048	0.094	0.063
0.062	0.060	0.125	0.078
0.078	0.076	0.156	0.094
0.094	0.090	0.188	0.109
0.109	0.104	0.219	0.125
0.125	0.120	0.250	0.141
0.141	0.176	0.281	0.156
0.156	0.207	0.313	0.172
0.188	0.176	0.375	0.203
0.219	0.207	0.438	0.234
0.250	0.225	0.500	0.266
0.312	0.280	0.625	0.313
0.375	0.335	0.750	0.375
0.438	0.406	0.875	0.438
0.500	0.473	1.000	0.500
0.625	0.598	1.250	0.625
0.750	0.723	1.500	0.750

Source: Courtesy of ANSI: B18.8.1—1983.

Appendix 23 • Straight Pins

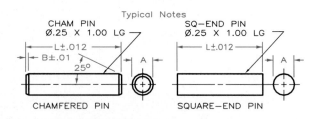

Nominal DIA	Diameter A		Chamfer B
	Max	Min	
0.062	0.0625	0.0605	0.015
0.094	0.0937	0.0917	0.015
0.109	0.1094	0.1074	0.015
0.125	0.1250	0.1230	0.015
0.156	0.1562	0.1542	0.015
0.188	0.1875	0.1855	0.015
0.219	0.2187	0.2167	0.015
0.250	0.2500	0.2480	0.015
0.312	0.3125	0.3095	0.015
0.375	0.3750	0.3720	0.030
0.438	0.4345	0.4345	0.030
0.500	0.4970	0.4970	0.030

Source: Courtesy of ANSI: B5.20.

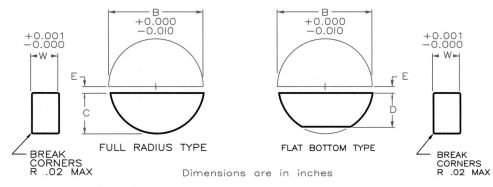

Dimensions are in inches

Key No.	W × B	C Max.	D Max.	E	Key No.	W × B	C Max.	D Max.	E
204	1/16 × 1/2	.203	.194	.047	506	5/32 × 3/4	.313	.303	.063
304	3/32 × 1/2	.203	.194	.047	606	3/16 × 3/4	.313	.303	.063
404	1/8 × 1/2	.203	.194	.047	507	5/32 × 7/8	.375	.365	.063
305	3/32 × 5/8	.250	.240	.063	607	3/16 × 7/8	.375	.365	.063
405	1/8 × 5/8	.250	.240	.063	807	1/4 × 7/8	.375	.365	.063
505	5/32 × 5/8	.250	.240	.063	608	3/16 × 1	.438	.428	.063
406	1/8 × 3/4	.313	.303	.063	609	3/16 × 1-1/8	.484	.475	.078

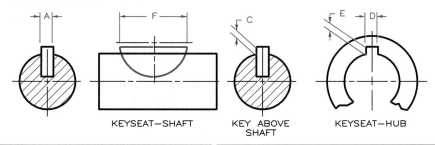

KEYSEAT—SHAFT   KEY ABOVE SHAFT   KEYSEAT—HUB

Key No.	A Min.	C +.005 −.000	F	D +.005 −.000	E +.005 −.000	Key No.	A Min.	C +.005 −.000	F	D +.005 −.000	E +.005 −.000
204	.0615	.0312	.500	.0635	.0372	506	.1553	.0781	.750	.1573	.0841
304	.0928	.0469	.500	.0948	.0529	606	.1863	.0937	.750	.1885	.0997
404	.1240	.0625	.500	.1260	.0685	507	.1553	.0781	.875	.1573	.0841
305	.0928	.0625	.625	.0948	.0529	607	.1863	.0937	.875	.1885	.0997
405	.1240	.0469	.625	.1260	.0685	807	.2487	.1250	.875	.2510	.1310
505	.1553	.0625	.625	.1573	.0841	608	.1863	.3393	1.000	.1885	.0997
406	.1240	.0781	.750	.1260	.0685	609	.1863	.3853	1.125	.1885	.0997

KEY SIZES VS. SHAFT SIZES

Shaft DIA	to .375	to .500	to .750	to 1.313	to 1.188	to 1.448	to 1.750	to 2.125	to 2.500
Key Nos.	204	304 305	404 405 406	505 506 507	606 607 608 609	807 808 809	810 811 812	1011 1012	1211 1212

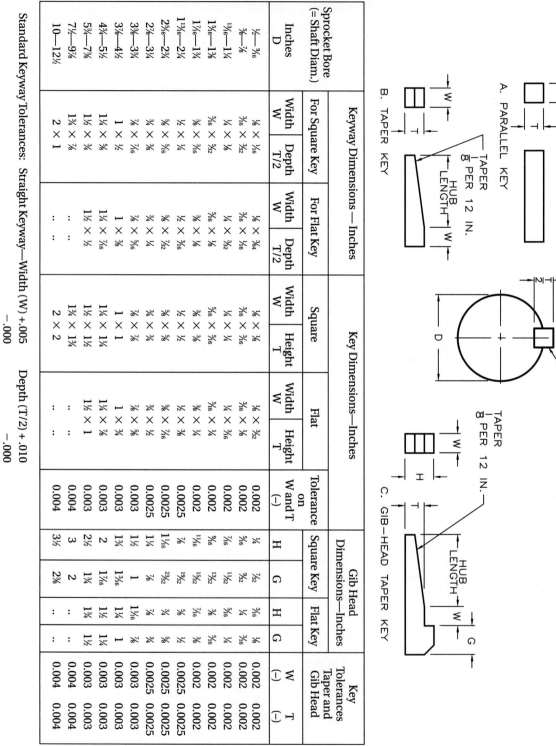

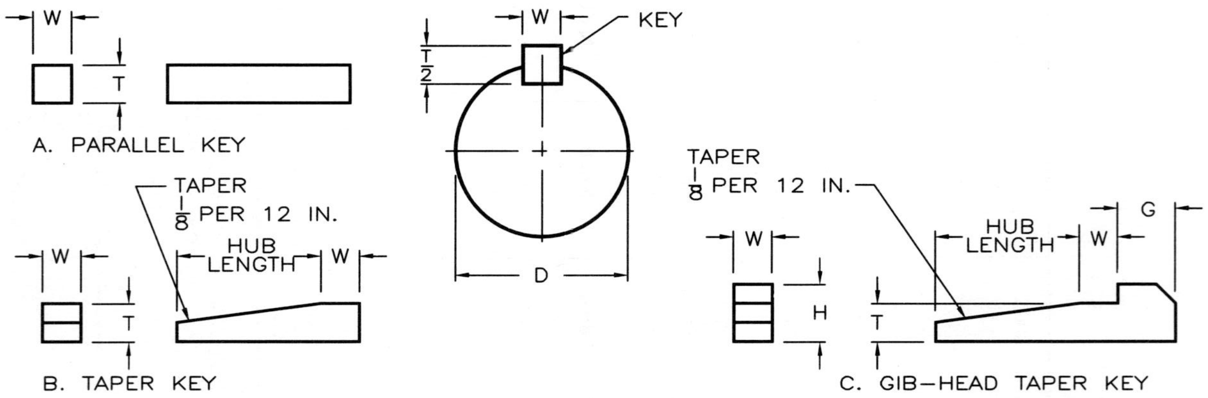
W
T
A. PARALLEL KEY
TAPER
1/8 PER 12 IN.
HUB LENGTH
W
T
B. TAPER KEY
W
KEY
T/2
D
TAPER
1/8 PER 12 IN.
W
H
T
HUB LENGTH
W
G
C. GIB—HEAD TAPER KEY

Sprocket Bore (= Shaft Diam.)	Keyway Dimensions — Inches		Key Dimensions—Inches			Gib Head Dimensions—Inches				Key Tolerances Taper and Gib Head	
	For Square Key	For Flat Key	Square	Flat	Tolerance on W and T	Square Key		Flat Key			
Inches D	Width W × Depth T/2	Width W × Depth T/2	Width W × Height T	Width W × Height T	(–)	H	G	H	G	W (–)	T (–)
1/2—9/16	1/8 × 1/16	1/8 × 3/64	1/8 × 1/8	1/8 × 3/32	0.002	1/4	7/32	3/16	1/8	0.002	0.002
5/8—7/8	3/16 × 3/32	3/16 × 1/16	3/16 × 3/16	3/16 × 1/8	0.002	5/16	9/32	1/4	3/16	0.002	0.002
13/16—1 1/4	1/4 × 1/8	1/4 × 3/32	1/4 × 1/4	1/4 × 3/16	0.002	7/16	11/32	5/16	1/4	0.002	0.002
1 5/16—1 3/8	5/16 × 5/32	5/16 × 1/8	5/16 × 5/16	5/16 × 1/4	0.002	9/16	13/32	3/8	5/16	0.002	0.002
1 7/16—1 3/4	3/8 × 3/16	3/8 × 1/8	3/8 × 3/8	3/8 × 1/4	0.002	11/16	15/32	7/16	3/8	0.002	0.002
1 13/16—2 1/4	1/2 × 1/4	1/2 × 3/16	1/2 × 1/2	1/2 × 3/8	0.0025	7/8	19/32	5/8	1/2	0.0025	0.0025
2 5/16—2 3/4	5/8 × 5/16	5/8 × 7/32	5/8 × 5/8	5/8 × 7/16	0.0025	1 1/16	23/32	3/4	5/8	0.0025	0.0025
2 7/8—3 1/4	3/4 × 3/8	3/4 × 1/4	3/4 × 3/4	3/4 × 1/2	0.0025	1 1/4	7/8	7/8	3/4	0.0025	0.0025
3 3/8—3 3/4	7/8 × 7/16	7/8 × 5/16	7/8 × 7/8	7/8 × 5/8	0.003	1 1/2	1	1 1/16	7/8	0.003	0.003
3 7/8—4 1/2	1 × 1/2	1 × 3/8	1 × 1	1 × 3/4	0.003	1 3/4	1 3/16	1 1/4	1	0.003	0.003
4 3/4—5 1/2	1 1/4 × 5/8	1 1/4 × 7/16	1 1/4 × 1 1/4	1 1/4 × 7/8	0.003	2	1 7/16	1 1/2	1 1/4	0.003	0.003
5 3/4—7 3/8	1 1/2 × 3/4	1 1/2 × 1/2	1 1/2 × 1 1/2	1 1/2 × 1	0.003	2 1/2	1 3/4	1 3/4	1 1/2	0.003	0.003
7 1/2—9 7/8	1 3/4 × 7/8	.. ..	1 3/4 × 1 3/4	.. ..	0.004	3	2	..	..	0.004	0.004
10—12 1/2	2 × 1	.. ..	2 × 2	.. ..	0.004	3 1/2	2 3/8	..	..	0.004	0.004

Standard Keyway Tolerances: Straight Keyway—Width (W) +.005 −.000 Depth (T/2) +.010 −.000
Taper Keyway—Width (W) +.005 −.000 Depth (T/2) +.000 −.010

APPENDIX 26 • Plain Washers (inches)

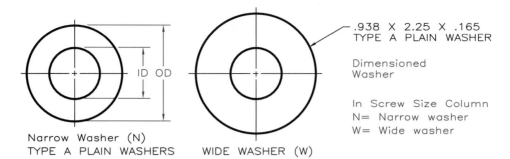

Narrow Washer (N)
TYPE A PLAIN WASHERS

WIDE WASHER (W)

.938 X 2.25 X .165
TYPE A PLAIN WASHER

Dimensioned
Washer

In Screw Size Column
N= Narrow washer
W= Wide washer

SCREW SIZE	ID SIZE	OD SIZE	THICK-NESS	SCREW SIZE	ID SIZE	OD SIZE	THICK-NESS
0.138	0.156	0.375	0.049	0.875 N	0.938	1.750	0.134
0.164	0.188	0.438	0.049	0.875 W	0.938	2.250	0.165
0.190	0.219	0.500	0.049	1.000 N	1.062	2.000	0.134
0.188	0.250	0.562	0.049	1.000 W	1.062	2.500	0.165
0.216	0.250	0.562	0.065	1.125 N	1.250	2.250	0.134
0.250 N	0.281	0.625	0.065	1.125 W	1.250	2.750	0.165
0.250 W	0.312	0.734	0.065	1.250 N	1.375	2.500	0.165
0.312 N	0.344	0.688	0.065	1.250 W	1.375	3.000	0.165
0.312 W	0.375	0.875	0.083	1.375 N	1.500	2.750	0.165
0.375 N	0.406	0.812	0.065	1.375 W	1.500	3.250	0.180
0.375 W	0.438	1.000	0.083	1.500 N	1.625	3.000	0.165
0.438 N	0.469	0.922	0.065	1.500 W	1.625	3.500	0.180
0.438 W	0.500	1.250	0.083	1.625	1.750	3.750	0.180
0.500 N	0.531	1.062	0.095	1.750	1.875	4.000	0.180
0.500 W	0.562	1.375	0.109	1.875	2.000	4.250	0.180
0.562 N	0.594	1.156	0.095	2.000	2.125	4.500	0.180
0.562 W	0.594	1.469	0.190	2.250	2.375	4.750	0.220
0.625 N	0.625	1.312	0.095	2.500	2.625	5.000	0.238
0.625 N	0.625	1.750	0.134	2.750	2.875	5.250	0.259
0.750 W	0.812	1.469	0.134	3.000	3.125	5.500	0.284
0.750 W	0.812	2.000	0.148				

APPENDIX 27 • Metric Washers (millimeters)

SCREW SIZE	ID SIZE	OD SIZE	THICK-NESS
3	3.2	9	0.8
4	4.3	12	1
5	5.3	15	1.5
6	6.4	18	1.5
8	8.4	25	2
10	10.5	30	2.5
12	13	40	3
14	15	45	3
16	17	50	3
18	19	56	4
20	21	60	4
2.6	2.8	5.5	0.5
3	3.2	6	0.5
4	4.3	8	0.5
5	5.3	10	1.0
6	6.4	11	1.5
8	8.4	15	1.5
10	10.5	18	1.5
12	13	20	2.0
14	15	25	2.0
16	17	27	2.0
18	19	30	2.5
20	21	33	2.5

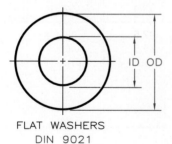

FLAT WASHERS
DIN 9021

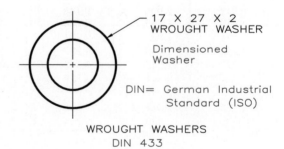

17 X 27 X 2
WROUGHT WASHER

Dimensioned
Washer

DIN= German Industrial
Standard (ISO)

WROUGHT WASHERS
DIN 433

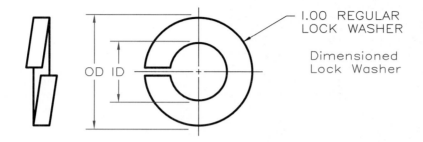

LOCK WASHERS—inches

SCREW SIZE	ID SIZE	OD SIZE	THICK-NESS
0.164	0.168	0.175	0.040
0.190	0.194	0.202	0.047
0.216	0.221	0.229	0.056
0.250	0.255	0.263	0.062
0.312	0.318	0.328	0.078
0.375	0.382	0.393	0.094
0.438	0.446	0.459	0.109
0.500	0.509	0.523	0.125
0.562	0.572	0.587	0.141
0.625	0.636	0.653	0.156
0.688	0.700	0.718	0.172
0.750	0.763	0.783	0.188
0.812	0.826	1.367	0.203
0.875	0.890	1.464	0.219
0.938	0.954	1.560	0.234
1.000	1.017	1.661	0.250
1.062	1.080	1.756	0.266
1.125	1.144	1.853	0.281
1.188	1.208	1.950	0.297
1.250	1.271	2.045	0.312
1.312	1.334	2.141	0.328
1.375	1.398	2.239	0.344
1.438	1.462	2.334	0.359
1.500	1.525	2.430	0.375

METRIC LOCK WASHERS—DIN 127 (Millimeters)

SCREW SIZE	ID SIZE	OD SIZE	THICK-NESS
4	4.1	7.1	0.9
5	5.1	8.7	1.2
6	6.1	11.1	1.6
8	8.2	12.1	1.6
10	10.2	14.2	2
12	12.1	17.2	2.2
14	14.2	20.2	2.5
16	16.2	23.2	3
18	18.2	26.2	3.5
20	20.2	28.2	3.5
22	22.5	34.5	4
24	24.5	38.5	5
27	27.5	41.5	5
30	30.5	46.5	6
33	33.5	53.5	6
36	36.5	56.5	6
39	39.5	59.5	6
42	42.5	66.5	7
45	45.5	69.5	7
48	49	73	7

Limits are in thousandths of an inch.
Limits for hole and shaft are applied algebraically to the basic size to obtain the limits of size for the parts.
Data in bold face are in accordance with ABC agreements.
Symbols H5, g5, etc., are Hole and Shaft designations used in ABC System.

Nominal Size Range Inches Over	To	Class RC1 Limits of Clearance	Class RC1 Standard Limits Hole H5	Class RC1 Standard Limits Shaft g4	Class RC2 Limits of Clearance	Class RC2 Standard Limits Hole H6	Class RC2 Standard Limits Shaft g5	Class RC3 Limits of Clearance	Class RC3 Standard Limits Hole H7	Class RC3 Standard Limits Shaft f6	Class RC4 Limits of Clearance	Class RC4 Standard Limits Hole H8	Class RC4 Standard Limits Shaft f7
0	−0.12	0.1	+0.2	−0.1	0.1	+0.25	−0.1	0.3	+0.4	−0.3	0.3	+0.6	−0.3
		0.45	0	−0.25	0.55	0	−0.3	0.95	0	−0.55	1.3	0	−0.7
0.12	−0.24	0.15	+0.2	−0.15	0.15	+0.3	−0.15	0.4	+0.5	−0.4	0.4	+0.7	−0.4
		0.5	0	−0.3	0.65	0	−0.35	1.12	0	−0.7	1.6	0	−0.9
0.24	−0.40	0.2	0.25	−0.2	0.2	+0.4	−0.2	0.5	+0.6	−0.5	0.5	+0.9	−0.5
		0.6	0	−0.35	0.85	0	−0.45	1.5	0	−0.9	2.0	0	−1.1
0.40	−0.71	0.25	+0.3	−0.25	0.25	+0.4	−0.25	0.6	+0.7	−0.6	0.6	+1.0	−0.6
		0.75	0	−0.45	0.95	0	−0.55	1.7	0	−1.0	2.3	0	−1.3
0.71	−1.19	0.3	+0.4	−0.3	0.3	+0.5	−0.3	0.8	+0.8	−0.8	0.8	+1.2	−0.8
		0.95	0	−0.55	1.2	0	−0.7	2.1	0	−1.3	2.8	0	−1.6
1.19	−1.97	0.4	+0.4	−0.4	0.4	+0.6	−0.4	1.0	+1.0	−1.0	1.0	+1.6	−1.0
		1.1	0	−0.7	1.4	0	−0.8	2.6	0	−1.6	3.6	0	−2.0
1.97	−3.15	0.4	+0.5	−0.4	0.4	+0.7	−0.4	1.2	+1.2	−1.2	1.2	+1.8	−1.2
		1.2	0	−0.7	1.6	0	−0.9	3.1	0	−1.9	4.2	0	−2.4
3.15	−4.73	0.5	+0.6	−0.5	0.5	+0.9	−0.5	1.4	+1.4	−1.4	1.4	+2.2	−1.4
		1.5	0	−0.9	2.0	0	−1.1	3.7	0	−2.3	5.0	0	−2.8
4.73	−7.09	0.6	+0.7	−0.6	0.6	+1.0	−0.6	1.6	+1.6	−1.6	1.6	+2.5	−1.6
		1.8	0	−1.1	2.3	0	−1.3	4.2	0	−2.6	5.7	0	−3.2
7.09	−9.85	0.6	+0.8	−0.6	0.6	+1.2	−0.6	2.0	+1.8	−2.0	2.0	+2.8	−2.0
		2.0	0	−1.2	2.6	0	−1.4	5.0	0	−3.2	6.6	0	−3.8
9.85	−12.41	0.8	+0.9	−0.8	0.8	+1.2	−0.8	2.5	+2.0	−2.5	2.5	+3.0	−2.5
		2.3	0	−1.4	2.9	0	−1.7	5.7	0	−3.7	7.5	0	−4.5
12.41	−15.75	1.0	+1.0	−1.0	1.0	+1.4	−1.0	3.0	+	−3.0	3.0	+3.5	−3.0
		2.7	0	−1.7	3.4	0	−2.0	6.6	0	−4.4	8.7	0	−5.2
15.75	−19.69	1.2	+1.0	−1.2	1.2	+1.6	−1.2	4.0	+1.6	−4.0	4.0	+4.0	−4.0
		3.0	0	−2.0	3.8	0	−2.2	8.1	0	−5.6	10.5	0	−6.5
19.69	−30.09	1.6	+1.2	−1.6	1.6	+2.0	−1.6	5.0	+3.0	−5.0	5.0	+5.0	−5.0
		3.7	0	−2.5	4.8	0	−2.8	10.0	0	−7.0	13.0	0	−8.0
30.09	−41.49	2.0	+1.6	−2.0	2.0	+2.5	−2.0	6.0	+4.0	−6.0	6.0	+6.0	−6.0
		4.6	0	−3.0	6.1	0	−3.6	12.5	0	−8.5	16.0	0	−10.0
41.49	−56.19	2.5	+2.0	−2.5	2.5	+3.0	−2.5	8.0	+5.0	−8.0	8.0	+8.0	−8.0
		5.7	0	−3.7	7.5	0	−4.5	16.0	0	−11.0	21.0	0	−13.0
56.19	−76.39	3.0	+2.5	−3.0	3.0	+4.0	−3.0	10.0	+6.0	−10.0	10.0	+10.0	−10.0
		7.1	0	−4.6	9.5	0	−5.5	20.0	0	−14.0	26.0	0	−16.0
76.39	−100.9	4.0	+3.0	−4.0	4.0	+5.0	−4.0	12.0	+8.0	−12.0	12.0	+12.0	−12.0
		9.0	0	−6.0	12.0	0	−7.0	25.0	0	−17.0	32.0	0	−20.0
100.9	−131.9	5.0	+4.0	−5.0	5.0	+6.0	−5.0	16.0	+10.0	−16.0	16.0	+16.0	−16.0
		11.5	0	−7.5	15.0	0	−9.0	32.0	0	−22.0	36.0	0	−26.0
131.9	−171.9	6.0	+5.0	−6.0	6.0	+8.0	−6.0	18.0	+8.0	−18.0	18.0	+20.0	−18.0
		14.0	0	−9.0	19.0	0	−11.0	38.0	0	−26.0	50.0	0	−30.0
171.9	−200	8.0	+6.0	−8.0	8.0	+10.0	−8.0	22.0	+16.0	−22.0	22.0	+25.0	−22.0
		18.0	0	−12.0	22.0	0	−12.0	48.0	0	−32.0	63.0	0	−38.0

Source: Courtesy of USASI; B4.1—1955.

| Class RC 5 | | | Class RC 6 | | | Class RC 7 | | | Class RC 8 | | | Class RC 9 | | | Nominal Size Range Inches | |
| Limits of Clearance | Standard Limits | | Limits of Clearance | Standard Limits | | Limits of Clearance | Sandard Limits | | Limits of Clearance | Standard Limits | | Limits of Clearance | Standard Limits | | | |
	Hole H8	Shaft e7		Hole H9	Shaft e8		Hole H9	Shaft d8		Hole H10	Shaft c9		Hole H11	Shaft	Over	To
0.6	+0.6	−0.6	0.6	+1.0	−0.6	1.0	+1.0	−1.0	2.5	+1.6	−2.5	4.0	+2.5	−4.0	0	− 0.12
1.6	−0	−1.0	2.2	−0	−1.2	2.6	0	−1.6	5.1	0	−3.5	8.1	0	−5.6		
0.8	+0.7	−0.8	0.8	+1.2	−0.8	1.2	+1.2	−1.2	2.8	+1.8	−2.8	4.5	+3.0	−4.5	0.12	− 0.24
2.0	−0	−1.3	2.7	−0	−1.5	3.1	0	−1.9	5.8	0	−4.0	9.0	0	−6.0		
1.0	+0.9	−1.0	1.0	+1.4	−1.0	1.6	+1.4	−1.6	3.0	+2.2	−3.0	5.0	+3.5	−5.0	0.24	− 0.40
2.5	−0	−1.16	3.3	−0	−1.9	3.9	0	−2.5	6.6	0	−4.4	10.7	0	−7.2		
1.2	+1.0	−1.2	1.2	+1.6	−1.2	2.0	+1.6	−2.0	3.5	+2.8	−3.5	6.0	+4.0	−6.0	0.40	− 0.71
2.9	−0	−1.9	3.8	−0	−2.2	4.6	0	−3.0	7.9	0	−5.1	12.8	−0	−8.8		
1.6	+1.2	−1.6	1.6	+2.0	−1.6	2.5	+2.0	−2.5	4.5	+3.5	−4.5	7.0	+5.0	−7.0	0.71	− 1.19
3.6	−0	−2.4	4.8	−0	−2.8	5.7	0	−3.7	10.0	0	−6.5	15.5	0	−10.5		
2.0	+1.6	−2.0	2.0	+2.5	−2.0	3.0	+2.5	−3.0	5.0	+4.0	−5.0	8.0	+6.0	−8.0	1.19	− 1.97
4.6	−0	−3.0	6.1	−0	−3.6	7.1	0	−4.6	11.5	0	−7.5	18.0	0	−12.0		
2.5	+1.8	−2.5	2.5	+3.0	−2.5	4.0	+3.0	−4.0	6.0	+4.5	−6.0	9.0	+7.0	−9.0	1.97	− 3.15
5.5	−0	−3.7	7.3	−0	−4.3	8.8	0	−5.8	13.5	0	−9.0	20.5	0	−13.5		
3.0	+2.2	−3.0	3.0	+3.5	−3.0	5.0	+3.5	−5.0	7.0	+5.0	−7.0	10.0	+9.0	−10.0	3.15	− 4.73
6.6	−0	−4.4	8.7	−0	−5.2	10.7	0	−7.2	15.5	0	−10.5	24.0	0	−15.0		
3.5	+2.5	−3.5	3.5	+4.0	−3.5	6.0	+4.0	−6.0	8.0	+6.0	−8.0	12.0	+10.0	−12.0	4.73	− 7.09
7.6	−0	−5.1	10.0	−0	−6.0	12.5	0	−8.5	18.0	0	−12.0	28.0	0	−18.0		
4.0	+2.8	−4.0	4.0	+4.5	−4.0	7.0	+4.5	−7.0	10.0	+7.0	−10.0	15.0	+12.0	−15.0	7.09	− 9.85
8.6	−0	−5.8	11.3	0	−6.8	14.3	0	−9.8	21.5	0	−14.5	34.0	0	−22.0		
5.0	+3.0	−5.0	5.0	+5.0	−5.0	8.0	+5.0	−8.0	12.0	+8.0	−12.0	18.0	+12.0	−18.0	9.85	− 12.41
10.0	0	−7.0	13.0	0	−8.0	16.0	0	−11.0	25.0	0	−17.0	38.0	0	−26.0		
6.0	+3.5	−6.0	6.0	+6.0	−6.0	10.0	+6.0	−10.0	14.0	+9.0	−14.0	22.0	+14.0	−22.0	12.41	− 15.75
11.7	0	−8.2	15.5	0	−9.5	19.5	0	13.5	29.0	0	−20.0	45.0	0	−31.0		
8.0	+4.0	−8.0	8.0	+6.0	−8.0	12.0	+6.0	−12.0	16.0	+10.0	−16.0	25.0	+16.0	−25.0	15.75	− 19.69
14.5	0	−10.5	18.0	0	−12.0	22.0	0	−16.0	32.0	0	−22.0	51.0	0	−35.0		
10.0	+5.0	−10.0	10.0	+8.0	−10.0	16.0	+8.0	−16.0	20.0	+12.0	−20.0	30.0	+20.0	−30.0	19.69	− 30.09
18.0	0	−13.0	23.0	0	−15.0	29.0	0	−21.0	40.0	0	−28.0	62.0	0	−42.0		
12.0	+6.0	−12.0	12.0	+10.0	−12.0	20.0	+10.0	−20.0	25.0	+16.0	−25.0	40.0	+25.0	−40.0	30.09	− 41.49
22.0	0	−16.0	28.0	0	−18.0	36.0	0	−26.0	51.0	0	−35.0	81.0	0	−56.0		
16.0	+8.0	−16.0	16.0	+12.0	−16.0	25.0	+12.0	−25.0	30.0	+20.0	−30.0	50.0	+30.0	−50.0	41.49	− 56.19
29.0	0	−21.0	36.0	0	−24.0	45.0	0	−33.0	62.0	0	−42.0	100	0	−70.0		
20.0	+10.0	−20.0	20.0	+16.0	−20.0	30.0	+16.0	−30.0	40.0	+25.0	−40.0	60.0	+40.0	−60.0	56.19	− 76.39
36.0	0	−26.0	46.0	0	−30.0	56.0	0	−40.0	81.0	0	−56.0	125	0	−85.0		
25.0	+12.0	−25.0	25.0	+20.0	−25.0	40.0	+20.0	−40.0	50.0	+30.0	−50.0	80.0	+50.0	−80.0	76.39	− 100.9
45.0	0	−33.0	57.0	0	−37.0	72.0	0	−52.0	100	0	−70.0	160	0	−110		
30.0	+16.0	−30.0	30.0	+35.0	−30.0	50.0	+25.0	−50.0	60.0	+40.0	−60.0	100	+60.0	−100	100.9	− 131.9
56.0	0	−40.0	71.0	0	−46.0	91.0	0	−66.0	125	0	−85.0	200	0	−140		
35.0	+20.0	−35.0	35.0	+30.0	−35.0	60.0	+30.0	−60.0	80.0	+50.0	−80.0	130	+80.0	−130	131.9	− 171.9
57.0	0	−47.0	85.0	0	−55.0	110.0	0	−80.0	160	0	−110	260	0	−180		
45.0	+25.0	−45.0	45.0	+40.0	−45.0	80.0	+40.0	−80.0	100	+60.0	−100	150	+100	−150	171.9	− 200
86.0	0	−61.0	110.0	0	−70.0	145.0	0	−105.0	200	0	−140	310	0	−210		

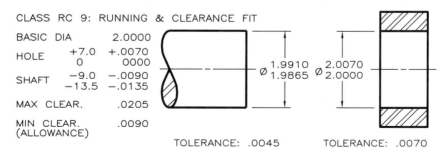

CLASS RC 9: RUNNING & CLEARANCE FIT

BASIC DIA	2.0000	
HOLE	+7.0 0	+.0070 0000
SHAFT	−9.0 −13.5	−.0090 −.0135
MAX CLEAR.	.0205	
MIN CLEAR. (ALLOWANCE)	.0090	

Ø 1.9910 / 1.9865   Ø 2.0070 / 2.0000

TOLERANCE: .0045   TOLERANCE: .0070

Limits are in thousandths of an inch.
Limits for hole and shaft are applied algebraically to the basic size to obtain the limits of size for the parts.
Data in bold face are in accordance with ABC agreements.
Symbols H9, f8, etc., are Hole and Shaft designations used in ABC System.

Nominal Size Range Inches Over	To	Class LC 1 Limits of Clearance	Standard Limits Hole H6	Shaft h5	Class LC2 Limits of Clearance	Standard Limits Hole H7	Shaft h6	Class LC 3 Limits of Clearance	Standard Limits Hole H8	Shaft h7	Class LC 4 Limits of Clearance	Standard Limits Hole H10	Shaft h9	Class LC 5 Limits of Clearance	Standard Limits Hole H7	Shaft g6
0	−0.12	0	+0.25	+0	0	+0.4	+0	0	+0.6	+0	0	+1.6	+0	0.1	+0.4	−0.1
		0.45	−0	−0.2	0.65	−0	−0.25	1	−0	−0.4	2.6	−0	−1.0	0.75	−0	−0.35
0.12	−0.24	0	+0.3	+0	0	+0.5	+0	0	+0.7	+0	0	+1.8	+0	0.15	+0.5	−0.15
		0.5	−0	−0.2	0.8	−0	−0.3	1.2	−0	−0.5	3.0	−0	−1.2	0.95	−0	−0.45
0.24	−0.40	0	+0.4	+0	0	+0.6	+0	0	+0.9	+0	0	+2.2	+0	0.2	+0.6	−0.2
		0.65	−0	−0.25	1.0	−0	−0.4	1.5	−0	−0.6	3.6	−0	−1.4	1.2	−0	−0.6
0.40	−0.71	0	+0.4	+0	0	+0.7	+0	0	+1.0	+0	0	+2.8	+0	0.25	+0.7	−0.25
		0.7	−0	−0.3	1.1	−0	−0.4	1.7	−0	−0.7	4.4	−0	−1.6	1.35	−0	−0.65
0.71	−1.19	0	+0.5	+0	0	+0.8	+0	0	+1.2	+0	0	+3.5	+0	0.3	+0.8	−0.3
		0.9	−0	−0.4	1.3	−0	−0.5	2	−0	−0.8	5.5	−0	−2.0	1.6	−0	−0.8
1.19	−1.97	0	+0.6	+0	0	+1.0	+0	0	+1.6	+0	0	+4.0	+0	0.4	+1.0	−0.4
		1.0	−0	−0.4	1.6	−0	−0.6	2.6	−0	−1	6.5	−0	−2.5	2.0	−0	−1.0
1.97	−3.15	0	+0.7	+0	0	+1.2	+0	0	+1.8	+0	0	+4.5	+0	0.4	+1.2	−0.4
		1.2	−0	−0.5	1.9	−0	−0.7	3	−0	−1.2	7.5	−0	−3	2.3	−0	−1.1
3.15	−4.73	0	+0.9	+0	0	+1.4	+0	0	+2.2	+0	0	+5.0	+0	0.5	+1.4	−0.5
		1.5	−0	−0.6	2.3	−0	−0.9	3.6	−0	−1.4	8.5	−0	−3.5	2.8	−0	−1.4
4.73	−7.09	0	+1.0	+0	0	+1.6	+0	0	+2.5	+0	0	+6.0	+0	0.6	+1.6	−0.6
		1.7	−0	−0.7	2.6	−0	−1.0	4.1	−0	−1.6	10	−0	−4	3.2	−0	1.6
7.09	−9.85	0	+1.2	+0	0	+1.8	+0	0	+2.8	+0	0	+7.0	+0	0.6	+1.8	−0.6
		2.0	−0	−0.8	3.0	−0	−1.2	4.6	−0	−1.8	11.5	−0	−4.5	3.6	−0	−1.8
9.85	−12.41	0	+1.2	+0	0	+2.0	+0	0	+3.0	+0	0	8.0	+0	0.7	+2.0	−0.7
		2.1	−0	−0.9	3.2	−0	−1.2	5	−0	−2.0	13	−0	−5	3.9	−0	−1.9
12.41	−15.75	0	+1.4	+0	0	+2.2	+0	0	+3.5	+0	0	+9.0	+0	0.7	+2.2	−0.7
		2.4	−0	−1.0	3.6	−0	−1.4	5.7	−0	−2.2	15	−0	−6	4.3	−0	−2.1
15.75	−19.69	0	+1.6	+0	0	+2.5	+0	0	+4	+0	0	+10.0	+0	0.8	+2.5	−0.8
		2.6	−0	−1.0	4.1	−0	−1.6	6.5	−0	−2.5	16	−0	−6	4.9	−0	−2.4
19.69	−30.09	0	+2.0	+0	0	+3	+0	0	+5	+0	0	+12.0	+0	0.9	+3.0	−0.9
		3.2	−0	−1.2	5.0	−0	−2	8	−0	−3	20	−0	−8	5.9	−0	−2.9
30.09	−41.49	0	+2.5	+0	0	+4	+0	0	+6	+0	0	+16.0	+0	1.0	+4.0	−1.0
		4.1	−0	−1.6	6.5	−0	−2.5	10	−0	−4	26	−0	−10	7.5	−0	−3.5
41.49	−56.19	0	+3.0	+0	0	+5	+0	0	+8	+0	0	+20.0	+0	1.2	+5.0	−1.2
		5.0	−0	−2.0	8.0	−0	−3	13	−0	−5	32	−0	−12	9.2	−0	−4.2
56.19	−76.39	0	+4.0	+0	0	+6	+0	0	+10	+0	0	+25.0	+0	1.2	+6.0	−1.2
		6.5	−0	−2.5	10	−0	−4	16	−0	−6	41	−0	−16	11.2	−0	−5.2
76.39	−100.9	0	+5.0	+0	0	+8	+0	0	+12	+0	0	+30.0	+0	1.4	+8.0	−1.4
		8.0	−0	−3.0	13	−0	−5	20	−0	−8	50	−0	−20	14.4	−0	−6.4
100.9	−131.9	0	+6.0	+0	0	+10	+0	0	+16	+0	0	+40.0	+0	1.6	+10.0	−1.6
		10.0	−0	−4.0	16	−0	−6	26	−0	−10	65	−0	−25	17.6	−0	−7.6
131.9	−171.9	0	+8.0	+0	0	+12	+0	0	+20	+0	0	+50.0	+0	1.8	+12.0	−1.8
		13.0	−0	−5.0	20	−0	−8	32	−0	−12	8	−0	−30	21.8	−0	−9.8
171.9	−200	0	+10.0	+0	0	+16	+0	0	+25	+0	0	+60.0	+0	1.8	+16.0	−1.8
		16.0	−0	−6.0	26	−0	−10	41	−0	−16	100	−0	−40	27.8	−0	−11.8

Source: Courtesy of USASI; B4.1—1955.

Class LC 6			Class LC 7			Class LC 8			Class LC 9			Class LC 10			Class LC 11			Nominal Size Range Inches	
Limits of Clearance	Standard Limits		Limits of Clearance	Standard Limits		Limits of Clearance	Standard Limits		Limits of Clearance	Standard Limits		Limits of Clearance	Standard Limits		Limits of Clearance	Standard Limits			
	Hole H9	Shaft f8		Hole H10	Shaft e9		Hole H10	Shaft d9		Hole H11	Shaft c10		Hole H12	Shaft		Hole H13	Shaft	Over	To
0.3	+1.0	−0.3	0.6	+1.6	−0.6	1.0	+0.6	−1.0	2.5	+2.5	−2.5	4	+4	−4	5	+6	−5	0 −	0.12
1.9	0	−0.9	3.2	0	−1.6	3.6	−0	−2.0	6.6	−0	−4.1	12	−0	−8	17	−0	−11		
0.4	+1.2	−0.4	0.8	+1.8	−0.8	1.2	+1.8	−1.2	2.8	+3.0	−2.8	4.5	+5	−4.5	6	+7	−6	0.12 −	0.24
2.3	0	−1.1	3.8	0	−2.0	4.2	−0	−2.4	7.6	−0	−4.6	14.5	−0	−9.5	20	−0	−13		
0.5	+1.4	−0.5	1.0	+2.2	−1.0	1.6	+2.2	−1.6	3.0	+3.5	−3.0	5	+6	−5	7	+9	−7	0.24 −	0.40
2.8	0	−1.4	4.6	0	−2.4	5.2	−0	−3.0	8.7	−0	−5.2	17	−0	−11	25	−0	−16		
0.6	+1.6	−0.6	1.2	+2.8	−1.2	2.0	+2.8	−2.0	3.5	+4.0	−3.5	6	+7	−6	8	+10	−8	0.40 −	0.71
3.2	0	−1.6	5.6	0	−2.8	6.4	−0	−3.6	10.3	−0	−6.3	20	−0	−13	28	−0	−18		
0.8	+2.0	−0.8	1.6	+3.5	−1.6	2.5	+3.5	−2.5	4.5	+5.0	−4.5	7	+8	−7	10	+12	−10	0.71 −	1.19
4.0	0	−2.0	7.1	0	−3.6	8.0	−0	−4.5	13.0	−0	−8.0	23	−0	−15	34	−0	−22		
1.0	+2.5	−1.0	2.0	+4.0	−2.0	3.0	+4.0	−3.0	5	+6	−5	8	+10	−8	12	+16	−12	1.19 −	1.97
5.1	0	−2.6	8.5	0	−4.5	9.5	−0	−5.5	15	−0	−9	28	−0	−18	44	−0	−28		
1.2	+3.0	−1.2	2.5	+4.5	−2.5	4.0	+4.5	−4.0	6	+7	−6	10	+12	−10	14	+18	−14	1.97 −	3.15
6.0	0	−3.0	10.0	0	−5.5	11.5	−0	−7.0	17.5	−0	−10.5	34	−0	−22	50	−0	−32		
1.4	+3.5	−1.4	3.0	+5.0	−3.0	5.0	+5.0	−5.0	7	+9	−7	11	+14	−11	16	+22	−16	3.15 −	4.73
7.1	0	−3.6	11.5	0	−6.5	13.5	−0	−8.5	21	−0	−12	39	−0	−25	60	−0	−38		
1.6	+4.0	−1.6	3.5	+6.0	−3.5	6	+6	−6	8	+10	−8	12	+16	−12	18	+25	−18	4.73 −	7.09
8.1	0	−4.1	13.5	0	−7.5	16	−0	−10	24	−0	−14	44	−0	−28	68	−0	−43		
2.0	+4.5	−2.0	4.0	+7.0	−4.0	7	+7	−7	10	+12	−10	16	+18	−16	22	+28	−22	7.09 −	9.85
9.3	0	−4.8	15.5	0	−8.5	18.5	−0	−11.5	29	−0	−17	52	−0	−34	78	−0	−50		
2.2	+5.0	−2.2	4.5	+8.0	−4.5	7	+8	−7	12	+12	−12	20	+20	−20	28	+30	−28	9.85 −	12.41
10.2	0	−5.2	17.5	0	−9.5	20	−0	−12	32	−0	−20	60	−0	−40	88	−0	−58		
2.5	+6.0	−2.5	5.0	+9.0	−5	8	+9	−8	14	+14	−14	22	+22	−22	30	+35	−30	12.41 −	15.75
12.0	0	−6.0	20.0	0	−11	23	−0	−14	37	−0	−23	66	−0	−44	100	−0	−65		
2.8	+6.0	−2.8	5.0	+10.0	−5	9	+10	−9	16	+16	−16	25	+25	−25	35	+40	−35	15.75 −	19.69
12.8	0	−6.8	21.0	0	−11	25	−0	−15	42	−0	−26	75	−0	−50	115	−0	−75		
3.0	+8.0	−3.0	6.0	+12.0	−6	10	+12	−10	18	+20	−18	28	+30	−28	40	+50	−40	19.69 −	30.09
16.0	0	−8.0	26.0	−0	−14	30	−0	−18	50	−0	−30	88	−0	−58	140	−0	−90		
3.5	+10.0	−3.5	7.0	+16.0	−7	12	+16	−12	20	+25	−20	30	+40	−30	45	+60	−45	30.09 −	41.49
19.5	0	−9.5	33.0	−0	−17	38	−0	−22	61	−0	−36	110	−0	−70	165	−0	−105		
4.0	+12.0	−4.0	8.0	+20.0	−8	14	+20	−14	25	+30	−25	40	+50	−40	60	+80	−60	41.49 −	56.19
24.0	0	−12.0	40.0	−0	−20	46	−0	−26	75	−0	−45	140	−0	−90	220	−0	−140		
4.5	+16.0	−4.5	9.0	+25.0	−9	16	+25	−16	30	+40	−30	50	+60	−50	70	+100	−70	56.19 −	76.39
30.5	0	−14.5	50.0	−0	−25	57	−0	−32	95	−0	−55	170	−0	110	270	−0	−170		
5.0	+20.0	−5	10.0	+30.0	−10	18	+30	−18	35	+50	−35	50	+80	−50	80	+125	−80	76.39 −	100.9
37.0	0	−17	60.0	−0	−30	68	−0	−38	115	−0	−65	210	−0	−130	330	−0	−205		
6.0	+25.0	−6	12.0	+40.0	−12	20	+40	−20	40	+60	−40	60	+100	−60	90	+160	−90	100.9 −	131.9
47.0	0	−22	67.0	−0	−27	85	−0	−45	140	−0	−80	260	−0	−160	410	−0	−250		
7.0	+30.0	−7	14.0	+50.0	−14	25	+50	−25	50	+80	−50	80	+125	−80	100	+200	−100	131.9 −	171.9
57.0	0	−27	94.0	−0	−44	105	−0	−55	180	−0	−100	330	−0	−205	500	−0	−300		
7.0	+40.0	−7	14.0	+60.0	−14	25	+60	−25	50	+100	−50	90	+160	−90	125	+250	−125	171.9 −	200
72.0	0	−32	114.0	−0	−54	125	−0	−65	210	−0	−110	410	−0	−250	625	−0	−375		

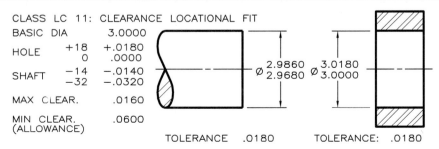

CLASS LC 11: CLEARANCE LOCATIONAL FIT

BASIC DIA		3.0000
HOLE	+18	+.0180
	0	.0000
SHAFT	−14	−.0140
	−32	−.0320
MAX CLEAR.		.0160
MIN CLEAR. (ALLOWANCE)		.0600

Ø 2.9860 / 2.9680   Ø 3.0180 / 3.0000

TOLERANCE .0180   TOLERANCE: .0180

APPENDIX 31 • American Standard Transition Locational Fits (hole basis)

Limits are in thousandths of an inch.
Limits for hole and shaft are applied algebraically to the basic size to obtain the limits of size for the mating parts.
Data in bold face are in accordance with ABC agreements.
"Fit" represents the maximum interference (minus values) and the maximum clearance (plus values).
Symbols H7, js6, etc., are Hole and Shaft designations used in ABC System.

In each cell below, "Fit" values are given as interference / clearance; "Standard Limits" values are given as upper / lower.

Nominal Size Range Inches Over	To	Class LT 1 Fit	LT 1 Hole H7	LT 1 Shaft js6	Class LT 2 Fit	LT 2 Hole H8	LT 2 Shaft js7	Class LT 3 Fit	LT 3 Hole H7	LT 3 Shaft k6	Class LT 4 Fit	LT 4 Hole H8	LT 4 Shaft k7	Class LT 5 Fit	LT 5 Hole H7	LT 5 Shaft n6	Class LT 6 Fit	LT 6 Hole H7	LT 6 Shaft n7
0	0.12	−0.10 / +0.50	+0.4 / −0	+0.10 / −0.10	−0.2 / +0.8	+0.6 / −0	+0.2 / −0.2							−0.5 / +0.15	+0.4 / −0	+0.5 / +0.25	−0.65 / +0.15	+0.4 / −0	+0.65 / +0.25
0.12	0.24	−0.15 / +0.65	+0.5 / −0	+0.15 / −0.15	−0.25 / +0.95	+0.7 / −0	+0.25 / −0.25							−0.6 / +0.2	+0.5 / −0	+0.6 / +0.3	−0.8 / +0.2	+0.5 / −0	+0.8 / +0.3
0.24	0.40	−0.2 / +0.8	+0.6 / −0	+0.2 / −0.2	−0.3 / +1.2	+0.9 / −0	+0.3 / −0.3	−0.5 / +0.5	+0.6 / −0	+0.5 / +0.1	−0.7 / +0.8	+0.9 / −0	+0.7 / +0.1	−0.8 / +0.2	+0.6 / −0	+0.8 / +0.4	−1.0 / +0.2	+0.6 / −0	+1.0 / +0.4
0.40	0.71	−0.2 / +0.9	+0.7 / −0	+0.2 / −0.2	−0.35 / +1.35	+1.0 / −0	+0.35 / −0.35	−0.5 / +0.6	+0.7 / −0	+0.5 / +0.1	−0.8 / +0.9	+1.0 / −0	+0.8 / +0.1	−0.9 / +0.2	+0.7 / −0	+0.9 / +0.5	−1.2 / +0.2	+0.7 / −0	+1.2 / +0.5
0.71	1.19	−0.25 / +1.05	+0.8 / −0	+0.25 / −0.25	−0.4 / +1.6	+1.2 / −0	+0.4 / −0.4	−0.6 / +0.7	+0.8 / −0	+0.6 / +0.1	−0.9 / +1.1	+1.2 / −0	+0.9 / +0.1	−1.1 / +0.2	+0.8 / −0	+1.1 / +0.6	−1.4 / +0.2	+0.8 / −0	+1.4 / +0.6
1.19	1.97	−0.3 / +1.3	+1.0 / −0	+0.3 / −0.3	−0.5 / +2.1	+1.6 / −0	+0.5 / −0.5	−0.7 / +0.9	+1.0 / −0	+0.7 / +0.1	−1.1 / +1.5	+1.6 / −0	+1.1 / +0.1	−1.3 / +0.3	+1.0 / −0	+1.3 / +0.7	−1.7 / +0.3	+1.0 / −0	+1.7 / +0.7
1.97	3.15	−0.3 / +1.5	+1.2 / −0	+0.3 / −0.3	−0.6 / +2.4	+1.8 / −0	+0.6 / −0.6	−0.8 / +1.1	+1.2 / −0	+0.8 / +0.1	−1.3 / +1.7	+1.8 / −0	+1.3 / +0.1	−1.5 / +0.4	+1.2 / −0	+1.5 / +0.8	−2.0 / +0.4	+1.2 / −0	+2.0 / +0.8
3.15	4.73	−0.4 / +1.8	+1.4 / −0	+0.4 / −0.4	−0.7 / +2.9	+2.2 / −0	+0.7 / −0.7	−1.0 / +1.3	+1.4 / −0	+1.0 / +0.1	−1.5 / +2.1	+2.2 / −0	+1.5 / +0.1	−1.9 / +0.4	+1.4 / −0	+1.9 / +1.0	−2.4 / +0.4	+1.4 / −0	+2.4 / +1.0
4.73	7.09	−0.5 / +2.1	+1.6 / −0	+0.5 / −0.5	−0.8 / +3.3	+2.5 / −0	+0.8 / −0.8	−1.1 / +1.5	+1.6 / −0	+1.1 / +0.1	−1.7 / +2.4	+2.5 / −0	+1.7 / +0.1	−2.2 / +0.4	+1.6 / −0	+2.2 / +1.2	−2.8 / +0.4	+1.6 / −0	+2.8 / +1.2
7.09	9.85	−0.6 / +2.4	+1.8 / −0	+0.6 / −0.6	−0.9 / +3.7	+2.8 / −0	+0.9 / −0.9	−1.4 / +1.6	+1.8 / −0	+1.4 / +0.2	−2.2 / +2.6	+2.8 / −0	+2.2 / +0.2	−2.6 / +0.4	+1.8 / −0	+2.6 / +1.4	−3.2 / +0.4	+1.8 / −0	+3.2 / +1.4
9.85	12.41	−0.6 / +2.6	+2.0 / −0	+0.6 / −0.6	−1.0 / +4.0	+3.0 / −0	+1.0 / −1.0	−1.4 / +1.8	+2.0 / −0	+1.4 / +0.2	−2.2 / +2.8	+3.0 / −0	+2.2 / +0.2	−2.6 / +0.6	+2.0 / −0	+2.6 / +1.4	−3.4 / +0.6	+2.0 / −0	+3.4 / +1.4
12.41	15.75	−0.7 / +2.9	+2.2 / −0	+0.7 / −0.7	−1.0 / +4.5	+3.5 / −0	+1.0 / −1.0	−1.6 / +2.0	+2.2 / −0	+1.6 / +0.2	−2.7 / +3.3	+3.5 / −0	+2.7 / +0.2	−3.0 / +0.6	+2.2 / −0	+3.0 / +1.6	−3.8 / +0.6	+2.2 / −0	+3.8 / +1.6
15.75	19.69	−0.8 / +3.3	+2.5 / −0	+0.8 / −0.8	−1.2 / +5.2	+4.0 / −0	+1.2 / −1.2	−1.8 / +2.3	+2.5 / −0	+1.8 / +0.2	−3.4 / +3.8	+4.0 / −0	+3.4 / +0.2	−3.4 / +0.7	+2.5 / −0	+3.4 / +1.8	−4.3 / +0.7	+2.5 / −0	+4.3 / +1.8

Source: Courtesy of ANSI; B4.1–1955.

Limits are in thousandths of an inch.
Limits for hole and shaft are applied algebraically to the basic size to obtain the limits of size for the parts.
Data in bold face are in accordance with ABC agreements.
Symbols H7, p6, etc., are Hole and Shaft designations used in ABC System.

Nominal Size Range Inches		Class LN 1			Class LN 2			Class LN 3		
		Limits of Interference	Standard Limits		Limits of Interference	Standard Limits		Limits of Interference	Standard Limits	
Over	To		Hole H6	Shaft n5		Hole H7	Shaft p6		Hole H7	Shaft r6
0	−0.12	0	+0.25	+0.45	0	+0.4	+0.65	0.1	+0.4	+0.75
		0.45	−0	+0.25	0.65	−0	+0.4	0.75	−0	+0.5
0.12	−0.24	0	+0.3	+0.5	0	+0.5	+0.8	0.1	+0.5	+0.9
		0.5	−0	+0.3	0.8	−0	+0.5	0.9	0	+0.6
0.24	−0.40	0	+0.4	+0.65	0	+0.6	+1.0	0.2	+0.6	+1.2
		0.65	−0	+0.4	1.0	−0	+0.6	1.2	−0	+0.8
0.40	−0.71	0	+0.4	+0.8	0	+0.7	+1.1	0.3	+0.7	+1.4
		0.8	−0	+0.4	1.1	−0	+0.7	1.4	−0	+1.0
0.71	−1.19	0	+0.5	+1.0	0	+0.8	+1.3	0.4	+0.8	+1.7
		1.0	−0	+0.5	1.3	−0	+0.8	1.7	−0	+1.2
1.19	−1.97	0	+0.6	+1.1	0	+1.0	+1.6	0.4	+1.0	+2.0
		1.1	−0	+0.6	1.6	−0	+1.0	2.0	−0	+1.4
1.97	−3.15	0.1	+0.7	+1.3	0.2	+1.2	+2.1	0.4	+1.2	+2.3
		1.3	−0	+0.7	2.1	−0	+1.4	2.3	−0	+1.6
3.15	−4.73	0.1	+0.9	+1.6	0.2	+1.4	+2.5	0.6	+1.4	+2.9
		1.6	−0	+1.0	2.5	−0	+1.6	2.9	−0	+2.0
4.73	−7.09	0.2	+1.0	+1.9	0.2	+1.6	+2.8	0.9	+1.6	+3.5
		1.9	−0	+1.2	2.8	−0	+1.8	3.5	−0	+2.5
7.09	−9.85	0.2	+1.2	+2.2	0.2	+1.8	+3.2	1.2	+1.8	+4.2
		2.2	−0	+1.4	3.2	−0	+2.0	4.2	−0	+3.0
9.85	−12.41	0.2	+1.2	+2.3	0.2	+2.0	+3.4	1.5	+2.0	+4.7
		2.3	−0	+1.4	3.4	−0	+2.2	4.7	−0	+3.5
12.41	−15.75	0.2	+1.4	+2.6	0.3	+2.2	+3.9	2.3	+2.2	+5.9
		2.6	−0	+1.6	3.9	−0	+2.5	5.9	−0	+4.5
15.75	−19.69	0.2	+1.6	+2.8	0.3	+2.5	+4.4	2.5	+2.5	+6.6
		2.8	−0	+1.8	4.4	−0	+2.8	6.6	−0	+5.0
19.69	−30.09		+2.0		0.5	+3	+5.5	4	+3	+9
			−0		5.5	−0	+3.5	9	−0	+7
30.09	−41.49		+2.5		0.5	+4	+7.0	5	+4	+11.5
			−0		7.0	−0	+4.5	11.5	−0	+9
41.49	−56.19		+3.0		1	+5	+9	7	+5	+15
			−0		9	−0	+6	15	−0	+12
56.19	−76.39		+4.0		1	+6	+11	10	+6	+20
			−0		11	−0	+7	20	−0	+16
76.39	−100.9		+5.0		1	+8	+14	12	+8	+25
			−0		14	−0	+9	25	−0	+20
100.9	−131.9		+6.0		2	+10	+18	15	+10	+31
			−0		18	−0	+12	31	−0	+25
131.9	−171.9		+8.0		4	+12	+24	18	+12	+38
			−0		24	−0	+16	38	−0	+30
171.9	−200		+10.0		4	+16	+30	24	+16	+50
			−0		30	−0	+20	50	−0	+40

Source: Courtesy of ANSI; B4.1–1955.

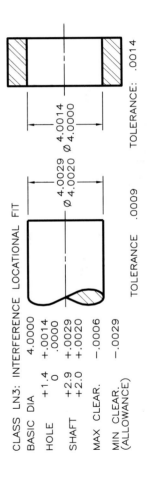

CLASS LN3: INTERFERENCE LOCATIONAL FIT

BASIC DIA.	4.0000	
HOLE	+1.4	+.0014
	0	.0000
SHAFT	+2.9	+.0029
	+2.0	+.0020
MAX CLEAR.		−.0006
MIN CLEAR. (ALLOWANCE)		−.0029

TOLERANCE .0009

∅ 4.0029 ∅ 4.0014
∅ 4.0020 ∅ 4.0000

TOLERANCE: .0014

Limits are in thousandths of an inch.
Limits for hole and shaft are applied algebraically to the basic size to obtain the limits of size for the parts.
Data in bold face are in accordance with ABC agreements.
Symbols H7, s6, etc., are Hole and Shaft designations used in ABC System.

Nominal Size Range Inches Over	To	Class FN 1 Limits of Interference	Class FN 1 Standard Limits Hole H6	Class FN 1 Standard Limits Shaft	Class FN 2 Limits of Interference	Class FN 2 Standard Limits Hole H7	Class FN 2 Standard Limits Shaft s6	Class FN 3 Limits of Interference	Class FN 3 Standard Limits Hole H7	Class FN 3 Standard Limits Shaft t6	Class FN 4 Limits of Interference	Class FN 4 Standard Limits Hole H7	Class FN 4 Standard Limits Shaft u6	Class FN 5 Limits of Interference	Class FN 5 Standard Limits Hole H8	Class FN 5 Standard Limits Shaft x7
0	0.12	0.05	+0.25	+0.5	0.2	+0.4	+0.85				0.3	+0.4	+0.95	0.3	+0.6	+1.3
		0.5	−0	+0.3	0.85	−0	+0.6				0.95	−0	+0.7	1.3	−0	+0.9
0.12	0.24	0.1	+0.3	+0.6	0.2	+0.5	+1.0				0.4	+0.5	+1.2	0.5	+0.7	+1.7
		0.6	−0	+0.4	1.0	−0	+0.7				1.2	−0	+0.9	1.7	−0	+1.2
0.24	0.40	0.1	+0.4	+0.75	0.4	+0.6	+1.4				0.6	+0.6	+1.6	0.5	+0.9	+2.0
		0.75	−0	+0.5	1.4	−0	+1.0				1.6	−0	+1.2	2.0	−0	+1.4
0.40	0.56	0.1	−0.4	+0.8	0.5	+0.7	+1.6				0.7	+0.7	+1.8	0.6	+1.0	+2.3
		0.8	−0	+0.5	1.6	−0	+1.2				1.8	−0	+1.4	2.3	−0	+1.6
0.56	0.71	0.2	+0.4	+0.9	0.5	+0.7	+1.6				0.7	+0.7	+1.8	0.8	+1.0	+2.5
		0.9	−0	+0.6	1.6	−0	+1.2				1.8	−0	+1.4	2.5	−0	+1.8
0.71	0.95	0.2	+0.5	+1.1	0.6	+0.8	+1.9				0.8	+0.8	+2.1	1.0	+1.2	+3.0
		1.1	−0	+0.7	1.9	−0	+1.4				2.1	−0	+1.6	3.0	−0	+2.2
0.95	1.19	0.3	+0.5	+1.2	0.6	+0.8	+1.9	0.8	+0.8	+2.1	1.0	+0.8	+2.3	1.3	+1.2	+3.3
		1.2	−0	+0.8	1.9	−0	+1.4	2.1	−0	+1.6	2.3	−0	+1.8	3.3	−0	+2.5
1.19	1.58	0.3	+0.6	+1.3	0.8	+1.0	+2.4	1.0	+1.0	+2.6	1.5	+1.0	+3.1	1.4	+1.6	+4.0
		1.3	−0	+0.9	2.4	−0	+1.8	2.6	−0	+2.0	3.1	−0	+2.5	4.0	−0	+3.0
1.58	1.97	0.4	+0.6	+1.4	0.8	+1.0	+2.4	1.2	+1.0	+2.8	1.8	+1.0	+3.4	2.4	+1.6	+5.0
		1.4	−0	+1.0	2.4	−0	+1.8	2.8	−0	+2.2	3.4	−0	+2.8	5.0	−0	+4.0
1.97	2.56	0.6	+0.7	+1.8	0.8	+1.2	+2.7	1.3	+1.2	+3.2	2.3	+1.2	+4.2	3.2	+1.8	+6.2
		1.8	−0	+1.3	2.7	−0	+2.0	3.2	−0	+2.5	4.2	−0	+3.5	6.2	−0	+5.0
2.56	3.15	0.7	+0.7	+1.9	1.0	+1.2	+2.9	1.8	+1.2	+3.7	2.8	+1.2	+4.7	4.2	+1.8	+7.2
		1.9	−0	+1.4	2.9	−0	+2.2	3.7	−0	+3.0	4.7	−0	+4.0	7.2	−0	+6.0
3.15	3.94	0.9	+0.9	+2.4	1.4	+1.4	+3.7	2.1	+1.4	+4.4	3.6	+1.4	+5.9	4.8	+2.2	+8.4
		2.4	−0	+1.8	3.7	−0	+2.8	4.4	−0	+3.5	5.9	−0	+5.0	8.4	−0	+7.0
3.94	4.73	1.1	+0.9	+2.6	1.6	+1.4	+3.9	2.6	+1.4	+4.9	4.6	+1.4	+6.9	5.8	+2.2	+9.4
		2.6	−0	+2.0	3.9	−0	+3.0	4.9	−0	+4.0	6.9	−0	+6.0	9.4	−0	+8.0
4.73	5.52	1.2	+1.0	+2.9	1.9	+1.6	+4.5	3.4	+1.6	+6.0	5.4	+1.6	+8.0	7.5	+2.5	+11.6
		2.9	−0	+2.2	4.5	−0	+3.5	6.0	−0	+5.0	8.0	−0	+7.0	11.6	−0	+10.0
5.52	6.30	1.5	+1.0	+3.2	2.4	+1.6	+5.0	3.4	+1.6	+6.0	5.4	+1.6	+8.0	9.5	+2.5	+13.6
		3.2	−0	+2.5	5.0	−0	+4.0	6.0	−0	+5.0	8.0	−0	+7.0	13.6	−0	+12.0
6.30	7.09	1.8	+1.0	+3.5	2.9	+1.6	+5.5	4.4	+1.6	+7.0	6.4	+1.6	+9.0	9.5	+2.5	+13.6
		3.5	−0	+2.8	5.5	−0	+4.5	7.0	−0	+6.0	9.0	−0	+8.0	13.6	−0	+12.0
7.09	7.88	1.8	+1.2	+3.8	3.2	+1.8	+6.2	5.2	+1.8	+8.2	7.2	+1.8	+10.2	11.2	+2.8	+15.8
		3.8	−0	+3.0	6.2	−0	+5.0	8.2	−0	+7.0	10.2	−0	+9.0	15.8	−0	+14.0
7.88	8.86	2.3	+1.2	+4.3	3.2	+1.8	+6.2	5.2	+1.8	+8.2	8.2	+1.8	+11.2	13.2	+2.8	+17.8
		4.3	−0	+3.5	6.2	−0	+5.0	8.2	−0	+7.0	11.2	−0	+10.0	17.8	−0	+16.0
8.86	9.85	2.3	+1.2	+4.3	4.2	+1.8	+7.2	6.2	+1.8	+9.2	10.2	+1.8	+13.2	13.2	+2.8	+17.8
		4.3	−0	+3.5	7.2	−0	+6.0	9.2	−0	+8.0	13.2	−0	+12.0	17.8	−0	+16.0
9.85	11.03	2.8	+1.2	+4.9	4.0	+2.0	+7.2	7.0	+2.0	+10.2	10.0	+2.0	+13.2	15.0	+3.0	+20.0
		4.9	−0	+4.0	7.2	−0	+6.0	10.2	−0	+9.0	13.2	−0	+12.0	20.0	−0	+18.0
11.03	12.41	2.8	+1.2	+4.9	5.0	+2.0	+8.2	7.0	+2.0	+10.2	12.0	+2.0	+15.2	17.0	+3.0	+22.0
		4.9	−0	+4.0	8.2	−0	+7.0	10.2	−0	+9.0	15.2	−0	+14.0	22.0	−0	+20.0
12.41	13.98	3.1	+1.4	+5.5	5.8	+2.2	+9.4	7.8	+2.2	+11.4	13.8	+2.2	+17.4	18.5	+3.5	+24.2
		5.5	−0	+4.5	9.4	−0	+8.0	11.4	−0	+10.0	17.4	−0	+16.0	24.2	+0	+22.0
13.98	15.75	3.6	+1.4	+6.1	5.8	+2.2	+9.4	9.8	+2.2	+13.4	15.8	+2.2	+19.4	21.5	+3.5	+27.2
		6.1	−0	+5.0	9.4	−0	+8.0	13.4	−0	+12.0	19.4	−0	+18.0	27.2	−0	+25.0
15.75	17.72	4.4	+1.6	+7.0	6.5	+2.5	+10.6	9.5	+2.5	+13.6	17.5	+2.5	+21.6	24.0	+4.0	+30.5
		7.0	−0	+6.0	10.6	−0	+9.0	13.6	−0	+12.0	21.6	−0	+20.0	30.5	−0	+28.0
17.72	19.69	4.4	+1.6	+7.0	7.5	+2.5	+11.6	11.5	+2.5	+15.6	19.5	+2.5	+23.6	26.0	+4.0	+32.5
		7.0	−0	+6.0	11.6	−0	+10.0	15.6	−0	+14.0	23.6	−0	+22.0	32.5	−0	+30.0

Courtesy of ANSI; B4.1

APPENDIX 34 • The International Tolerance Grades (ANSI B4.2)

Dimensions are in mm.

Basic sizes Over	Up to and including	IT01	IT0	IT1	IT2	IT3	IT4	IT5	IT6	IT7	IT8	IT9	IT10	IT11	IT12	IT13	IT14	IT15	IT16
0	3	0.0003	0.0005	0.0008	0.0012	0.002	0.003	0.004	0.006	0.010	0.014	0.025	0.040	0.060	0.100	0.140	0.250	0.400	0.600
3	6	0.0004	0.0006	0.001	0.0015	0.0025	0.004	0.005	0.008	0.012	0.018	0.030	0.048	0.075	0.120	0.180	0.300	0.480	0.750
6	10	0.0004	0.0006	0.001	0.0015	0.0025	0.004	0.006	0.009	0.015	0.022	0.036	0.058	0.090	0.150	0.220	0.360	0.580	0.900
10	18	0.0005	0.0008	0.0012	0.002	0.003	0.005	0.008	0.011	0.018	0.027	0.043	0.070	0.110	0.180	0.270	0.430	0.700	1.100
18	30	0.0006	0.001	0.0015	0.0025	0.004	0.006	0.009	0.013	0.021	0.033	0.052	0.084	0.130	0.210	0.330	0.520	0.840	1.300
30	50	0.0006	0.001	0.0015	0.0025	0.004	0.007	0.011	0.016	0.025	0.039	0.062	0.100	0.160	0.250	0.390	0.620	1.000	1.600
50	80	0.0008	0.0012	0.002	0.003	0.005	0.008	0.013	0.019	0.030	0.046	0.074	0.120	0.190	0.300	0.460	0.740	1.200	1.900
80	120	0.001	0.0015	0.0025	0.004	0.006	0.010	0.015	0.022	0.035	0.054	0.087	0.140	0.220	0.350	0.540	0.870	1.400	2.200
120	180	0.0012	0.002	0.0036	0.005	0.008	0.012	0.018	0.025	0.040	0.063	0.100	0.160	0.250	0.400	0.630	1.000	1.600	2.500
180	250	0.002	0.003	0.0045	0.007	0.010	0.014	0.020	0.029	0.046	0.072	0.115	0.185	0.290	0.460	0.720	1.150	1.850	2.900
250	315	0.0025	0.004	0.006	0.008	0.012	0.016	0.023	0.032	0.052	0.081	0.130	0.210	0.320	0.520	0.810	1.300	2.100	3.200
315	400	0.003	0.005	0.007	0.009	0.013	0.018	0.025	0.036	0.057	0.089	0.140	0.230	0.360	0.570	0.890	1.400	2.300	3.600
400	500	0.004	0.006	0.008	0.010	0.015	0.020	0.027	0.040	0.063	0.097	0.156	0.250	0.400	0.630	0.970	1.550	2.500	4.000
500	630	0.0045	0.006	0.009	0.011	0.016	0.022	0.030	0.044	0.070	0.110	0.175	0.280	0.440	0.700	1.100	1.750	2.800	4.400
630	800	0.005	0.007	0.010	0.013	0.018	0.025	0.035	0.050	0.080	0.125	0.200	0.320	0.500	0.800	1.250	2.000	3.200	5.000
800	1000	0.0055	0.008	0.011	0.015	0.021	0.029	0.040	0.056	0.090	0.140	0.230	0.360	0.560	0.900	1.400	2.300	3.600	5.600
1000	1250	0.0065	0.009	0.013	0.018	0.024	0.034	0.046	0.066	0.105	0.165	0.260	0.420	0.660	1.050	1.650	2.600	4.200	6.600
1250	1600	0.008	0.011	0.015	0.021	0.029	0.040	0.054	0.078	0.125	0.195	0.310	0.500	0.780	1.250	1.950	3.100	5.000	7.800
1600	2000	0.009	0.013	0.018	0.025	0.035	0.048	0.065	0.092	0.150	0.230	0.370	0.600	0.920	1.500	2.300	3.700	6.000	9.200
2000	2500	0.011	0.015	0.022	0.030	0.041	0.057	0.077	0.110	0.175	0.280	0.440	0.700	1.100	1.750	2.800	4.400	7.000	11.000
2500	3150	0.013	0.018	0.026	0.036	0.050	0.069	0.093	0.135	0.210	0.330	0.540	0.860	1.350	2.100	3.300	5.400	8.600	13.500

Tolerance grades[3]

[3]IT Values for tolerance grades larger than IT16 can be calculated by using the following formulas:
IT17 = IT12 × 10; IT18 = IT13 × 10; etc.

APPENDIX 35 • Preferred Hole Basis Clearance Fits—Cylindrical Fits (ANSI B4.2)

AMERICAN NATIONAL STANDARD PREFERRED METRIC LIMITS AND FITS ANSI B4.2—1978
Dimensions are in mm.

BASIC SIZE		LOOSE RUNNING Hole H11	Shaft c11	Fit	FREE RUNNING Hole H9	Shaft d9	Fit	CLOSE RUNNING Hole H8	Shaft f7	Fit	SLIDING Hole H7	Shaft g6	Fit	LOCATIONAL CLEARANCE Hole H7	Shaft h6	Fit
1	MAX	1.060	0.940	0.180	1.025	0.980	0.070	1.014	0.994	0.030	1.010	0.998	0.018	1.010	1.000	0.016
	MIN	1.000	0.880	0.060	1.000	0.955	0.020	1.000	0.984	0.006	1.000	0.992	0.002	1.000	0.994	0.000
1.2	MAX	1.260	1.140	0.180	1.225	1.180	0.070	1.214	1.194	0.030	1.210	1.198	0.018	1.210	1.200	0.016
	MIN	1.200	1.080	0.060	1.200	1.155	0.020	1.200	1.184	0.006	1.200	1.192	0.002	1.200	1.194	0.000
1.6	MAX	1.660	1.540	0.180	1.625	1.580	0.070	1.614	1.594	0.030	1.610	1.598	0.018	1.610	1.600	0.016
	MIN	1.600	1.480	0.060	1.600	1.555	0.020	1.600	1.584	0.006	1.600	1.592	0.002	1.600	1.594	0.000
2	MAX	2.060	1.940	0.180	2.025	1.980	0.070	2.014	1.994	0.030	2.010	1.998	0.018	2.010	2.000	0.016
	MIN	2.000	1.880	0.060	2.000	1.955	0.020	2.000	1.984	0.006	2.000	1.992	0.002	2.000	1.994	0.000
2.5	MAX	2.560	2.440	0.180	2.525	2.480	0.070	2.514	2.494	0.030	2.510	2.498	0.018	2.510	2.500	0.016
	MIN	2.500	2.380	0.060	2.500	2.455	0.020	2.500	2.484	0.006	2.500	2.492	0.002	2.500	2.494	0.000
3	MAX	3.060	2.940	0.180	3.025	2.980	0.070	3.014	2.994	0.030	3.010	2.998	0.018	3.010	3.000	0.016
	MIN	3.000	2.880	0.060	3.000	2.955	0.020	3.000	2.984	0.006	3.000	2.992	0.002	3.000	2.994	0.000
4	MAX	4.075	3.930	0.220	4.030	3.970	0.090	4.018	3.990	0.040	4.012	3.996	0.024	4.012	4.000	0.020
	MIN	4.000	3.855	0.070	4.000	3.940	0.030	4.000	3.978	0.010	4.000	3.988	0.004	4.000	3.992	0.000
5	MAX	5.075	4.930	0.220	5.030	4.970	0.090	5.018	4.990	0.040	5.012	4.996	0.024	5.012	5.000	0.020
	MIN	5.000	4.855	0.070	5.000	4.940	0.030	5.000	4.978	0.010	5.000	4.988	0.004	5.000	4.992	0.000
6	MAX	6.075	5.930	0.220	6.030	5.970	0.090	6.018	5.990	0.040	6.012	5.996	0.024	6.012	6.000	0.020
	MIN	6.000	5.855	0.070	6.000	5.940	0.030	6.000	5.978	0.010	6.000	5.988	0.004	6.000	5.992	0.000
8	MAX	8.090	7.920	0.260	8.036	7.960	0.112	8.022	7.987	0.050	8.015	7.995	0.029	8.015	8.000	0.024
	MIN	8.000	7.830	0.080	8.000	7.924	0.040	8.000	7.972	0.013	8.000	7.986	0.005	8.000	7.991	0.000
10	MAX	10.090	9.920	0.260	10.036	9.960	0.112	10.022	9.987	0.050	10.015	9.995	0.029	10.015	10.000	0.024
	MIN	10.000	9.830	0.080	10.000	9.924	0.040	10.000	9.972	0.013	10.000	9.986	0.005	10.000	9.991	0.000
12	MAX	12.110	11.905	0.315	12.043	11.950	0.136	12.027	11.984	0.061	12.018	11.994	0.035	12.018	12.000	0.029
	MIN	12.000	11.795	0.095	12.000	11.907	0.050	12.000	11.966	0.016	12.000	11.983	0.006	12.000	11.989	0.000
16	MAX	16.110	15.905	0.315	16.043	15.950	0.136	16.027	15.984	0.061	16.018	15.994	0.035	16.018	16.000	0.029
	MIN	16.000	15.795	0.095	16.000	15.907	0.050	16.000	15.966	0.016	16.000	15.983	0.006	16.000	15.989	0.000
20	MAX	20.130	19.890	0.370	20.052	19.935	0.169	20.033	19.980	0.074	20.021	19.993	0.041	20.021	20.000	0.034
	MIN	20.000	19.760	0.110	20.000	19.883	0.065	20.000	19.959	0.020	20.000	19.980	0.007	20.000	19.987	0.000
25	MAX	25.130	24.890	0.370	25.052	24.935	0.169	25.033	24.980	0.074	25.021	24.993	0.041	25.021	25.000	0.034
	MIN	25.000	24.760	0.110	25.000	24.883	0.065	25.000	24.959	0.020	25.000	24.980	0.007	25.000	24.987	0.000
30	MAX	30.130	29.890	0.370	30.052	29.935	0.169	30.033	29.980	0.074	30.021	29.993	0.041	30.021	30.000	0.034
	MIN	30.000	29.760	0.110	30.000	29.883	0.065	30.000	29.959	0.020	30.000	29.980	0.007	30.000	29.987	0.000

Source: American National Standard Preferred Metric Limits and Figs, ANSI B4.2—1978.

APPENDIX 35 • (continued)

BASIC SIZE		LOOSE RUNNING			FREE RUNNING			CLOSE RUNNING			SLIDING			LOCATIONAL CLEARANCE		
		Hole H11	Shaft c11	Fit	Hole H9	Shaft d9	Fit	Hole H8	Shaft f7	Fit	Hole H7	Shaft g6	Fit	Hole H7	Shaft h6	Fit
40	MAX	40.160	39.880	0.440	40.062	39.920	0.204	40.039	39.975	0.089	40.025	39.991	0.050	40.025	40.000	0.041
	MIN	40.000	39.720	0.120	40.000	39.858	0.080	40.000	39.950	0.025	40.000	39.975	0.009	40.000	39.984	0.000
50	MAX	50.160	49.870	0.450	50.062	49.920	0.204	50.039	49.975	0.089	50.025	49.991	0.050	50.025	50.000	0.041
	MIN	50.000	49.710	0.130	50.000	49.858	0.080	50.000	49.950	0.025	50.000	49.975	0.009	50.000	49.984	0.000
60	MAX	60.190	59.860	0.520	60.074	59.900	0.248	60.046	59.970	0.106	60.030	59.990	0.059	60.030	60.000	0.049
	MIN	60.000	59.670	0.140	60.000	59.826	0.100	60.000	59.940	0.030	60.000	59.971	0.010	60.000	59.981	0.000
80	MAX	80.190	79.850	0.530	80.074	79.900	0.248	80.046	79.970	0.106	80.030	79.990	0.059	80.030	80.000	0.049
	MIN	80.000	79.660	0.150	80.000	79.826	0.100	80.000	79.940	0.030	80.000	79.971	0.010	80.000	79.981	0.000
100	MAX	100.220	99.830	0.610	100.087	99.880	0.294	100.054	99.964	0.125	100.035	99.988	0.069	100.035	100.000	0.057
	MIN	100.000	99.610	0.170	100.000	99.793	0.120	100.000	99.929	0.036	100.000	99.966	0.012	100.000	99.978	0.000
120	MAX	120.220	119.820	0.620	120.087	119.880	0.294	120.054	119.964	0.125	120.035	119.988	0.069	120.035	120.000	0.057
	MIN	120.000	119.600	0.180	120.000	119.793	0.120	120.000	119.929	0.036	120.000	119.966	0.012	120.000	119.978	0.000
160	MAX	160.250	159.790	0.710	160.100	159.855	0.345	160.063	159.957	0.146	160.040	159.986	0.079	160.040	160.000	0.065
	MIN	160.000	159.540	0.210	160.000	159.755	0.145	160.000	159.917	0.043	160.000	159.961	0.014	160.000	159.975	0.000
200	MAX	200.290	199.760	0.820	200.115	199.830	0.400	200.072	199.950	0.168	200.046	199.985	0.090	200.046	200.000	0.075
	MIN	200.000	199.470	0.240	200.000	199.715	0.170	200.000	199.904	0.050	200.000	199.956	0.015	200.000	199.971	0.000
250	MAX	250.290	249.720	0.860	250.115	249.830	0.400	250.072	249.950	0.168	250.046	249.985	0.090	250.046	250.000	0.075
	MIN	250.000	249.430	0.280	250.000	249.715	0.170	250.000	249.904	0.050	250.000	249.956	0.015	250.000	249.971	0.000
300	MAX	300.320	299.670	0.970	300.130	299.810	0.450	300.081	299.944	0.189	300.052	299.983	0.101	300.052	300.000	0.084
	MIN	300.000	299.350	0.330	300.000	299.680	0.190	300.000	299.892	0.056	300.000	299.951	0.017	300.000	299.968	0.000
400	MAX	400.360	399.600	1.120	400.140	399.790	0.490	400.089	399.938	0.208	400.057	399.982	0.111	400.057	400.000	0.093
	MIN	400.000	399.240	0.400	400.000	399.650	0.210	400.000	399.881	0.062	400.000	399.946	0.018	400.000	399.964	0.000
500	MAX	500.400	499.520	1.280	500.155	499.770	0.540	500.097	499.932	0.228	500.063	499.980	0.123	500.063	500.000	0.103
	MIN	500.000	499.120	0.480	500.000	499.615	0.230	500.000	499.869	0.068	500.000	499.940	0.020	500.000	499.960	0.000

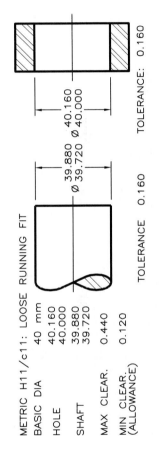

METRIC H11/c11: LOOSE RUNNING FIT

BASIC DIA	40 mm
HOLE	40.160 / 40.000
SHAFT	39.880 / 39.720
MAX CLEAR.	0.440
MIN CLEAR. (ALLOWANCE)	0.120

Ø 40.160 / Ø 40.000
Ø 39.880 / Ø 39.720
TOLERANCE 0.160
TOLERANCE 0.160
TOLERANCE: 0.160

APPENDIX 36 • Preferred Hole Basis Transition and Interference Fits—Cylindrical Fits (ANSI B4.2)

Dimensions are in mm.

BASIC SIZE		LOCATIONAL TRANSN.			LOCATIONAL TRANSN.			LOCATIONAL INTERF.			MEDIUM DRIVE			FORCE		
		Hole H7	Shaft k6	Fit	Hole H7	Shaft n6	Fit	Hole H7	Shaft p6	Fit	Hole H7	Shaft s6	Fit	Hole H7	Shaft u6	Fit
1	MAX	1.010	1.006	0.010	1.010	1.010	0.006	1.010	1.012	0.004	1.010	1.020	−0.004	1.010	1.024	−0.008
	MIN	1.000	1.000	−0.006	1.000	1.004	−0.010	1.000	1.006	−0.012	1.000	1.014	−0.020	1.000	1.018	−0.024
1.2	MAX	1.210	1.206	0.010	1.210	1.210	0.006	1.210	1.212	0.004	1.210	1.220	−0.004	1.210	1.224	−0.008
	MIN	1.200	1.200	−0.006	1.200	1.204	−0.010	1.200	1.206	−0.012	1.200	1.214	−0.020	1.200	1.218	−0.024
1.6	MAX	1.610	1.606	0.010	1.610	1.610	0.006	1.610	1.612	0.004	1.610	1.620	−0.004	1.610	1.624	−0.008
	MIN	1.600	1.600	−0.006	1.600	1.604	−0.010	1.600	1.606	−0.012	1.600	1.614	−0.020	1.600	1.618	−0.024
2	MAX	2.010	2.006	0.010	2.010	2.010	0.006	2.010	2.012	0.004	2.010	2.020	−0.004	2.010	2.024	−0.008
	MIN	2.000	2.000	−0.006	2.000	2.004	−0.010	2.000	2.006	−0.012	2.000	2.014	−0.020	2.000	2.018	−0.024
2.5	MAX	2.510	2.506	0.010	2.510	2.510	0.006	2.510	2.512	0.004	2.510	2.520	−0.004	2.510	2.524	−0.008
	MIN	2.500	2.500	−0.006	2.500	2.504	−0.010	2.500	2.506	−0.012	2.500	2.514	−0.020	2.500	2.518	−0.024
3	MAX	3.010	3.006	0.010	3.010	3.010	0.006	3.010	3.012	0.004	3.010	3.020	−0.004	3.010	3.024	−0.008
	MIN	3.000	3.000	−0.006	3.000	3.004	−0.010	3.000	3.006	−0.012	3.000	3.014	−0.020	3.000	3.018	−0.024
4	MAX	4.012	4.009	0.011	4.012	4.016	0.004	4.012	4.020	0.000	4.012	4.027	−0.007	4.012	4.031	−0.011
	MIN	4.000	4.001	−0.009	4.000	4.008	−0.016	4.000	4.012	−0.020	4.000	4.019	−0.027	4.000	4.023	−0.031
5	MAX	5.012	5.009	0.011	5.012	5.016	0.004	5.012	5.020	0.000	5.012	5.027	−0.007	5.012	5.031	−0.011
	MIN	5.000	5.001	−0.009	5.000	5.008	−0.016	5.000	5.012	−0.020	5.000	5.019	−0.027	5.000	5.023	−0.031
6	MAX	6.012	6.009	0.011	6.012	6.016	0.004	6.012	6.020	0.000	6.012	6.027	−0.007	6.012	6.031	−0.011
	MIN	6.000	6.001	−0.009	6.000	6.008	−0.016	6.000	6.012	−0.020	6.000	6.019	−0.027	6.000	6.023	−0.031
8	MAX	8.015	8.010	0.014	8.015	8.019	0.005	8.015	8.024	0.000	8.015	8.032	−0.008	8.015	8.037	−0.013
	MIN	8.000	8.001	−0.010	8.000	8.010	−0.019	8.000	8.015	−0.024	8.000	8.023	−0.032	8.000	8.028	−0.037
10	MAX	10.015	10.010	0.014	10.015	10.019	0.005	10.015	10.024	0.000	10.015	10.032	−0.008	10.015	10.037	−0.013
	MIN	10.000	10.001	−0.010	10.000	10.010	−0.019	10.000	10.015	−0.024	10.000	10.023	−0.032	10.000	10.028	−0.037
12	MAX	12.018	12.012	0.017	12.018	12.023	0.006	12.018	12.029	0.000	12.018	12.039	−0.010	12.018	12.044	−0.015
	MIN	12.000	12.001	−0.012	12.000	12.012	−0.023	12.000	12.018	−0.029	12.000	12.028	−0.039	12.000	12.033	−0.044
16	MAX	16.018	16.012	0.017	16.018	16.023	0.006	16.018	16.029	0.000	16.018	16.039	−0.010	16.018	16.044	−0.015
	MIN	16.000	16.001	−0.012	16.000	16.012	−0.023	16.000	16.018	−0.029	16.000	16.028	−0.039	16.000	16.033	−0.044
20	MAX	20.021	20.015	0.019	20.021	20.028	0.006	20.021	20.035	−0.001	20.021	20.048	−0.014	20.021	20.054	−0.020
	MIN	20.000	20.002	−0.015	20.000	20.015	−0.028	20.000	20.022	−0.035	20.000	20.035	−0.048	20.000	20.041	−0.054
25	MAX	25.021	25.015	0.019	25.021	25.028	0.006	25.021	25.035	−0.001	25.021	25.048	−0.014	25.021	25.061	−0.027
	MIN	25.000	25.002	−0.015	25.000	25.015	−0.028	25.000	25.022	−0.035	25.000	25.035	−0.048	25.000	25.048	−0.061
30	MAX	30.021	30.015	0.019	30.021	30.028	0.006	30.021	30.035	−0.001	30.021	30.048	−0.014	30.021	30.061	−0.027
	MIN	30.000	30.002	−0.015	30.000	30.015	−0.028	30.000	30.022	−0.035	30.000	30.035	−0.048	30.000	30.048	−0.061

Source: American National Standard Preferred Metric Limit and Fits, ANSI B4.2—1978.

APPENDIX 36 • (continued)

		LOCATIONAL TRANSN.			LOCATIONAL TRANSN.			LOCATIONAL INTERF.			MEDIUM DRIVE			FORCE		
BASIC SIZE		Hole H7	Shaft k6	Fit	Hole H7	Shaft n6	Fit	Hole H7	Shaft p6	Fit	Hole H7	Shaft s6	Fit	Hole H7	Shaft u6	Fit
0	MAX	40.025	40.018	0.023	40.025	40.033	0.008	40.025	40.042	-0.001	40.025	40.059	-0.018	40.025	40.076	-0.035
	MIN	40.000	40.002	-0.018	40.000	40.017	-0.033	40.000	40.026	-0.042	40.000	40.043	-0.059	40.000	40.060	-0.076
50	MAX	50.025	50.018	0.023	50.025	50.033	0.008	50.025	50.042	-0.001	50.025	50.059	-0.018	50.025	50.086	-0.045
	MIN	50.000	50.002	-0.018	50.000	50.017	-0.033	50.000	50.026	-0.042	50.000	50.043	-0.059	50.000	50.070	-0.086
60	MAX	60.030	60.021	0.028	60.030	60.039	0.010	60.030	60.051	-0.002	60.030	60.072	-0.023	60.030	60.106	-0.057
	MIN	60.000	60.002	-0.021	60.000	60.020	-0.039	60.000	60.032	-0.051	60.000	60.053	-0.072	60.000	60.087	-0.106
80	MAX	80.030	80.021	0.028	80.030	80.039	0.010	80.030	80.051	-0.002	80.030	80.078	-0.029	80.030	80.121	-0.072
	MIN	80.000	80.002	-0.021	80.000	80.020	-0.039	80.000	80.032	-0.051	80.000	80.059	-0.078	80.000	80.102	-0.121
100	MAX	100.035	100.025	0.032	100.035	100.045	0.012	100.035	100.059	-0.002	100.035	100.093	-0.036	100.035	100.146	-0.089
	MIN	100.000	100.003	-0.025	100.000	100.023	-0.045	100.000	100.037	-0.059	100.000	100.071	-0.093	100.000	100.124	-0.146
120	MAX	120.035	120.025	0.032	120.035	120.045	0.012	120.035	120.059	-0.002	120.035	120.101	-0.044	120.035	120.166	-0.109
	MIN	120.000	120.003	-0.025	120.000	120.023	-0.045	120.000	120.037	-0.059	120.000	120.079	-0.101	120.000	120.144	-0.166
160	MAX	160.040	160.028	0.037	160.040	160.052	0.013	160.040	160.068	-0.003	160.040	160.125	-0.060	160.040	160.215	-0.150
	MIN	160.000	160.003	-0.028	160.000	160.027	-0.052	160.000	160.043	-0.068	160.000	160.100	-0.125	160.000	160.190	-0.215
200	MAX	200.046	200.033	0.042	200.046	200.060	0.015	200.046	200.079	-0.004	200.046	200.151	-0.076	200.046	200.265	-0.190
	MIN	200.000	200.004	-0.033	200.000	200.031	-0.060	200.000	200.050	-0.079	200.000	200.122	-0.151	200.000	200.236	-0.265
250	MAX	250.046	250.033	0.042	250.046	250.060	0.015	250.046	250.079	-0.004	250.046	250.169	-0.094	250.046	250.313	-0.238
	MIN	250.000	250.004	-0.033	250.000	250.031	-0.060	250.000	250.050	-0.079	250.000	250.140	-0.169	250.000	250.284	-0.313
300	MAX	300.052	300.036	0.048	300.052	300.066	0.018	300.052	300.088	-0.004	300.052	300.202	-0.118	300.052	300.382	-0.298
	MIN	300.000	300.004	-0.036	300.000	300.034	-0.066	300.000	300.056	-0.088	300.000	300.170	-0.202	300.000	300.350	-0.382
400	MAX	400.057	400.040	0.053	400.057	400.073	0.020	400.057	400.098	-0.005	400.057	400.244	-0.151	400.057	400.471	-0.378
	MIN	400.000	400.004	-0.040	400.000	400.037	-0.073	400.000	400.062	-0.098	400.000	400.208	-0.244	400.000	400.435	-0.471
500	MAX	500.063	500.045	0.058	500.063	500.080	0.023	500.063	500.108	-0.005	500.063	500.292	-0.189	500.063	500.580	-0.477
	MIN	500.000	500.005	-0.045	500.000	500.040	-0.080	500.000	500.068	-0.108	500.000	500.252	-0.292	500.000	500.540	-0.580

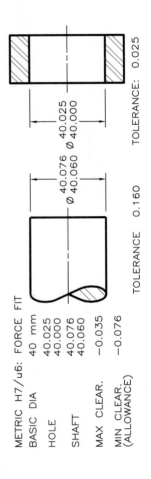

TOLERANCE 0.160

Ø 40.076
Ø 40.060

Ø 40.025
Ø 40.000

TOLERANCE: 0.025

METRIC H7/u6: FORCE FIT

BASIC DIA	40 mm
HOLE	40.025
	40.000
SHAFT	40.076
	40.060
MAX CLEAR.	-0.035
MIN CLEAR. (ALLOWANCE)	-0.076

APPENDIX 37 • Preferred Shaft Basis Clearance Fits—Cylindrical Fits (ANSI B4.2)

Dimensions are in mm.

BASIC SIZE		LOOSE RUNNING			FREE RUNNING			CLOSE RUNNING			SLIDING			LOCATIONAL CLEARANCE		
		Hole C11	Shaft h11	Fit	Hole D9	Shaft h9	Fit	Hole F8	Shaft h7	Fit	Hole G7	Shaft h6	Fit	Hole H7	Shaft h6	Fit
1	MAX	1.120	1.000	0.180	1.045	1.000	0.070	1.020	1.000	0.030	1.012	1.000	0.018	1.010	1.000	0.016
	MIN	1.060	0.940	0.060	1.020	0.975	0.020	1.006	0.990	0.006	1.002	0.994	0.002	1.000	0.994	0.000
1.2	MAX	1.320	1.200	0.180	1.245	1.200	0.070	1.220	1.200	0.030	1.212	1.200	0.018	1.210	1.200	0.016
	MIN	1.260	1.140	0.060	1.220	1.175	0.020	1.206	1.190	0.006	1.202	1.194	0.002	1.200	1.194	0.000
1.6	MAX	1.720	1.600	0.180	1.656	1.600	0.070	1.620	1.600	0.030	1.612	1.600	0.018	1.610	1.600	0.016
	MIN	1.660	1.540	0.060	1.620	1.575	0.020	1.606	1.590	0.006	1.602	1.595	0.002	1.600	1.594	0.000
2	MAX	2.120	2.000	0.180	2.045	2.000	0.070	2.020	2.000	0.030	2.012	2.000	0.018	2.010	2.000	0.016
	MIN	2.060	1.940	0.060	2.020	1.975	0.020	2.006	1.990	0.006	2.002	1.994	0.002	2.000	1.994	0.000
2.5	MAX	2.620	2.500	0.180	2.545	2.500	0.070	2.520	2.500	0.030	2.512	2.500	0.018	2.510	2.500	0.016
	MIN	2.560	2.440	0.060	2.520	2.475	0.020	2.506	2.490	0.006	2.502	2.494	0.002	2.500	2.494	0.000
3	MAX	3.120	3.000	0.180	3.045	3.000	0.070	3.020	3.000	0.030	3.012	3.000	0.018	3.010	3.000	0.016
	MIN	3.060	2.940	0.060	3.020	2.975	0.020	3.006	2.990	0.006	3.002	2.994	0.002	3.000	2.994	0.000
4	MAX	4.145	4.000	0.220	4.060	4.000	0.090	4.028	4.000	0.040	4.016	4.000	0.024	4.012	4.000	0.020
	MIN	4.070	3.925	0.070	4.030	3.970	0.030	4.010	3.988	0.010	4.004	3.992	0.004	4.000	3.992	0.000
5	MAX	5.145	5.000	0.220	5.060	5.000	0.090	5.028	5.000	0.040	5.016	5.000	0.024	5.012	5.000	0.020
	MIN	5.070	4.925	0.070	5.030	4.970	0.030	5.010	4.988	0.010	5.004	4.992	0.004	5.000	4.992	0.000
6	MAX	6.145	6.000	0.220	6.060	6.000	0.090	6.028	6.000	0.040	6.016	6.000	0.024	6.012	6.000	0.020
	MIN	6.070	5.925	0.070	6.030	5.970	0.030	6.010	5.988	0.010	6.004	5.992	0.004	6.000	5.992	0.000
8	MAX	8.170	8.000	0.260	8.076	8.000	0.112	8.035	8.000	0.050	8.020	8.000	0.029	8.015	8.000	0.024
	MIN	8.080	7.910	0.080	8.040	7.964	0.040	8.013	7.985	0.013	8.005	7.991	0.005	8.000	7.991	0.000
10	MAX	10.170	10.000	0.260	10.076	10.000	0.112	10.035	10.000	0.050	10.020	10.000	0.029	10.015	10.000	0.024
	MIN	10.080	9.910	0.080	10.040	9.964	0.040	10.013	9.985	0.013	10.005	9.991	0.005	10.000	9.991	0.000
12	MAX	12.205	12.000	0.315	12.093	12.000	0.136	12.043	12.000	0.061	12.024	12.000	0.035	12.018	12.000	0.029
	MIN	12.095	11.890	0.095	12.050	11.957	0.050	12.016	11.982	0.016	12.006	11.989	0.006	12.000	11.989	0.000
16	MAX	16.205	16.000	0.315	16.093	16.000	0.136	16.043	16.000	0.061	16.024	16.000	0.035	16.018	16.000	0.029
	MIN	16.095	15.890	0.095	16.050	15.957	0.050	16.016	15.982	0.016	16.006	15.989	0.006	16.000	15.989	0.000
20	MAX	20.240	20.000	0.370	20.117	20.000	0.169	20.053	20.000	0.074	20.028	20.000	0.041	20.021	20.000	0.034
	MIN	20.110	19.870	0.110	20.065	19.948	0.065	20.020	19.979	0.020	20.007	19.987	0.007	20.000	19.987	0.000
25	MAX	25.240	25.000	0.370	25.117	25.000	0.169	25.053	25.000	0.074	25.028	25.000	0.041	25.021	25.000	0.034
	MIN	25.110	24.870	0.110	25.065	24.948	0.065	25.020	24.979	0.020	25.007	24.987	0.007	25.000	24.987	0.000
30	MAX	30.240	30.000	0.370	30.117	30.000	0.169	30.053	30.000	0.074	30.028	30.000	0.041	30.021	30.000	0.034
	MIN	30.110	29.870	0.110	30.065	29.948	0.065	30.020	29.979	0.020	30.007	29.987	0.007	30.000	29.987	0.000

Source: American National Standard Preferred Metric Limits and Fits, ANSI B4.2—1978.

BASIC SIZE		LOOSE RUNNING Hole C11	Shaft h11	Fit	FREE RUNNING Hole D9	Shaft h9	Fit	CLOSE RUNNING Hole F8	Shaft h7	Fit	SLIDING Hole G7	Shaft h6	Fit	LOCATIONAL CLEARANCE Hole H7	Shaft h6	Fit
40	MAX	40.280	40.000	0.440	40.142	40.000	0.204	40.064	40.000	0.089	40.034	40.000	0.050	40.025	40.000	0.041
	MIN	40.120	39.840	0.120	40.080	39.938	0.080	40.025	39.975	0.025	40.009	39.984	0.009	40.000	39.984	0.000
50	MAX	50.290	50.000	0.450	50.142	50.000	0.204	50.064	50.000	0.089	50.034	50.000	0.050	50.025	50.000	0.041
	MIN	50.130	49.840	0.130	50.080	49.938	0.080	50.025	49.975	0.025	50.009	49.984	0.009	50.000	49.984	0.000
60	MAX	60.330	60.000	0.520	60.174	60.000	0.248	60.076	60.000	0.106	60.040	60.000	0.059	60.030	60.000	0.049
	MIN	60.140	59.810	0.140	60.100	59.926	0.100	60.030	59.970	0.030	60.010	59.981	0.010	60.000	59.981	0.000
80	MAX	80.340	80.000	0.530	80.174	80.000	0.248	80.076	80.000	0.106	80.040	80.000	0.059	80.030	80.000	0.049
	MIN	80.150	79.810	0.150	80.100	79.926	0.100	80.030	79.970	0.030	80.010	79.981	0.010	80.000	79.981	0.000
100	MAX	100.390	100.000	0.610	100.207	100.000	0.294	100.090	100.000	0.125	100.047	100.000	0.069	100.035	100.000	0.057
	MIN	100.170	99.780	0.170	100.120	99.913	0.120	100.036	99.965	0.036	100.012	99.979	0.012	100.000	99.979	0.000
120	MAX	120.400	120.000	0.620	120.207	120.000	0.294	120.090	120.000	0.125	120.047	120.000	0.069	120.035	120.000	0.057
	MIN	120.180	119.780	0.180	120.120	119.913	0.120	120.036	119.965	0.036	120.012	119.978	0.012	120.000	119.978	0.000
160	MAX	160.460	160.000	0.710	160.245	160.000	0.345	160.106	160.000	0.146	160.054	160.000	0.079	160.040	160.000	0.065
	MIN	160.210	159.750	0.210	160.145	159.900	0.145	160.043	159.960	0.043	160.014	159.975	0.014	160.000	159.975	0.000
200	MAX	200.530	200.000	0.820	200.285	200.000	0.400	200.122	200.000	0.168	200.061	200.000	0.090	200.046	200.000	0.075
	MIN	200.240	199.710	0.240	200.170	199.885	0.170	200.050	199.954	0.050	200.015	199.971	0.015	200.000	199.971	0.000
250	MAX	250.570	250.000	0.860	250.285	250.000	0.400	250.122	250.000	0.168	250.061	250.000	0.090	250.046	250.000	0.075
	MIN	250.280	249.710	0.280	250.170	249.885	0.170	250.050	249.954	0.050	250.015	249.971	0.015	250.000	249.971	0.000
300	MAX	300.650	300.000	0.970	300.320	300.000	0.450	300.137	300.000	0.189	300.069	300.000	0.101	300.052	300.000	0.084
	MIN	300.330	299.680	0.330	300.190	299.870	0.190	300.056	299.948	0.056	300.017	299.968	0.017	300.000	299.968	0.000
400	MAX	400.760	400.000	1.120	400.350	400.000	0.490	400.151	400.000	0.208	400.075	400.000	0.111	400.057	400.000	0.983
	MIN	400.400	399.640	0.400	400.210	399.860	0.210	400.062	399.943	0.062	400.018	399.964	0.018	400.000	399.964	0.000
500	MAX	500.880	500.000	1.280	500.385	500.000	0.540	500.165	500.000	0.228	500.083	500.000	0.123	500.063	500.000	0.103
	MIN	500.480	499.600	0.480	500.230	499.845	0.230	500.068	499.937	0.068	500.020	499.960	0.020	500.000	499.960	0.000

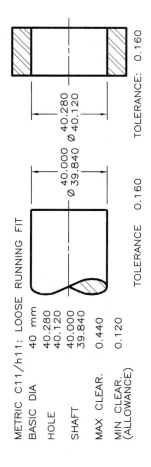

METRIC C11/h11: LOOSE RUNNING FIT

BASIC DIA	40 mm
HOLE	40.280 / 40.120
SHAFT	40.000 / 39.840
MAX CLEAR.	0.440
MIN CLEAR. (ALLOWANCE)	0.120

TOLERANCE: 0.160
TOLERANCE 0.160
TOLERANCE 0.160

Ø 40.280 Ø 40.120
Ø 40.000 Ø 39.840

APPENDIX 38 • Preferred Shaft Basis Transition and Interference Fits—Cylindrical Fits (ANSI B4.2)

Dimensions are in mm.

BASIC SIZE		LOCATIONAL TRANSN.			LOCATIONAL TRANSN.			LOCATIONAL INTERF.			MEDIUM DRIVE			FORCE		
		Hole K7	Shaft h6	Fit	Hole N7	Shaft h6	Fit	Hole P7	Shaft h6	Fit	Hole S7	Shaft h6	Fit	Hole U7	Shaft h6	Fit
1	MAX	1.000	1.000	0.006	0.996	1.000	0.002	0.994	1.000	0.000	0.986	1.000	−0.008	0.982	1.000	−0.012
	MIN	0.990	0.994	−0.010	0.986	0.994	−0.014	0.984	0.994	−0.016	0.976	0.994	−0.024	0.972	0.994	−0.028
1.2	MAX	1.200	1.200	0.006	1.196	1.200	0.002	1.194	1.200	0.000	1.186	1.200	−0.008	1.182	1.200	−0.012
	MIN	1.190	1.194	−0.010	1.186	1.194	−0.014	1.184	1.194	−0.016	1.176	1.194	−0.024	1.172	1.194	−0.028
1.6	MAX	1.600	1.600	0.006	1.596	1.600	0.002	1.594	1.600	0.000	1.586	1.600	−0.008	1.582	1.600	−0.012
	MIN	1.590	1.594	−0.010	1.586	1.594	−0.014	1.584	1.594	−0.016	1.576	1.594	−0.024	1.572	1.594	−0.028
2	MAX	2.000	2.000	0.006	1.996	2.000	0.002	1.994	2.000	0.008	1.986	2.000	−0.008	1.982	2.000	−0.012
	MIN	1.990	1.994	−0.010	1.986	1.994	−0.014	1.984	1.994	−0.016	1.976	1.994	−0.024	1.972	1.994	−0.028
2.5	MAX	2.500	2.500	0.006	2.496	2.500	0.002	2.494	2.500	0.000	2.486	2.500	−0.008	2.482	2.500	−0.012
	MIN	2.490	2.494	−0.010	2.486	2.494	−0.014	2.484	2.494	−0.016	2.476	2.494	−0.024	2.472	2.494	−0.028
3	MAX	3.000	3.000	0.006	2.996	3.000	0.002	2.994	3.000	0.000	2.986	3.000	−0.008	2.982	3.000	−0.012
	MIN	2.990	2.994	−.010	2.986	2.994	−0.014	2.984	2.994	−0.016	2.976	2.994	−0.024	2.972	2.994	−0.028
4	MAX	4.003	4.000	0.011	3.996	4.000	0.004	3.992	4.000	0.000	3.985	4.000	−0.007	3.981	4.000	−0.011
	MIN	3.991	3.992	−0.009	3.984	3.992	−0.016	3.980	3.992	−0.020	3.973	3.992	−0.027	3.969	3.992	−0.031
5	MAX	5.003	5.000	0.011	4.996	5.000	0.004	4.992	5.000	0.000	4.985	5.000	−0.007	4.981	5.000	−0.011
	MIN	4.991	4.992	−0.009	4.984	4.992	−0.016	4.980	4.992	−0.020	4.973	4.992	−0.027	4.969	4.992	−0.031
6	MAX	6.003	6.000	0.011	5.996	6.000	0.004	5.992	6.000	0.000	5.985	6.000	−0.007	5.981	6.000	−0.011
	MIN	5.991	5.992	−0.009	5.984	5.992	−0.016	5.980	5.992	−0.020	5.973	5.992	−0.027	5.969	5.992	−0.031
8	MAX	8.005	8.000	0.014	7.986	8.000	0.005	7.991	8.000	0.000	7.983	8.000	−0.008	7.978	8.000	−0.013
	MIN	7.990	7.991	−0.010	7.981	7.991	−0.019	7.976	7.991	−0.024	7.968	7.991	−0.032	7.963	7.991	−0.037
10	MAX	10.005	10.000	0.014	9.996	10.000	0.005	9.991	10.000	0.0000	9.983	10.000	−0.008	9.978	10.000	−0.013
	MIN	9.990	9.991	−0.010	9.981	9.991	−0.019	9.976	9.991	−0.024	9.968	9.991	−0.032	9.963	9.991	−0.037
12	MAX	12.006	12.000	0.017	11.995	12.000	0.006	11.989	12.000	0.000	11.979	12.000	−0.010	11.974	12.000	−0.015
	MIN	11.988	11.989	−0.012	11.977	11.989	−0.023	11.971	11.989	−0.029	11.961	11.989	−0.039	11.956	11.989	−0.044
16	MAX	16.006	16.000	0.017	15.995	16.000	0.006	15.989	16.000	0.000	15.979	16.000	−0.010	15.974	16.000	−0.015
	MIN	15.988	15.989	−0.012	15.977	15.989	−0.023	15.971	15.989	−0.029	15.961	15.989	−0.039	15.956	15.989	−0.044
20	MAX	20.006	20.000	0.019	19.993	20.000	0.006	19.986	20.000	−0.001	19.973	20.000	−0.014	19.967	20.000	−0.020
	MIN	19.985	19.987	−0.015	19.972	19.987	−0.028	19.965	19.987	−0.035	19.952	19.987	−0.048	19.946	19.987	−0.054
25	MAX	25.006	25.000	0.019	24.993	25.000	0.006	24.986	25.000	−0.001	24.973	25.000	−0.014	24.960	25.000	−0.027
	MIN	24.985	24.987	−0.015	24.972	24.987	−0.028	24.965	24.987	−0.035	24.952	24.987	−0.048	24.939	24.987	−0.061
30	MAX	30.006	30.000	0.019	29.993	30.000	0.006	29.986	30.000	−0.001	29.973	30.000	−0.014	29.960	30.000	−0.027
	MIN	29.985	29.987	−0.015	29.972	29.987	−0.028	29.965	29.987	−0.035	29.952	29.987	−0.048	29.939	29.987	−0.061

Source: American National Standard Preferred Metric Limits and Fits, ANSI B4.2—1978.

Dimensions are in mm.

BASIC SIZE		LOCATIONAL TRANSN. Hole K7	Shaft h6	Fit	LOCATIONAL TRANSN. Hole N7	Shaft h6	Fit	LOCATIONAL INTERF. Hole P7	Shaft h6	Fit	MEDIUM DRIVE Hole S7	Shaft h6	Fit	FORCE Hole U7	Shaft h6	Fit
40	MAX	40.007	40.000	0.023	39.992	40.000	0.008	39.983	40.000	-0.001	39.966	40.000	-0.018	39.949	40.000	-0.035
	MIN	39.982	39.984	-0.018	39.967	39.984	-0.033	39.958	39.984	-0.042	39.941	39.984	-0.059	39.924	39.984	-0.076
50	MAX	50.007	50.000	0.023	49.992	50.000	0.008	49.983	50.000	-0.001	49.966	50.000	-0.018	49.939	50.000	-0.045
	MIN	49.982	49.984	-0.018	49.967	49.984	-0.033	49.958	49.984	-0.042	49.941	49.984	-0.059	49.914	49.984	-0.086
60	MAX	60.009	60.000	0.028	59.991	60.000	0.010	59.979	60.000	-0.002	59.958	60.000	-0.023	59.924	60.000	-0.057
	MIN	59.979	59.981	-0.021	59.961	59.981	-0.039	59.949	59.981	-0.051	59.928	59.981	-0.072	59.894	59.981	-0.106
80	MAX	80.009	80.000	0.028	79.991	80.000	0.010	79.979	80.000	-0.002	79.952	80.000	-0.029	79.909	80.000	-0.072
	MIN	79.979	79.981	-0.021	79.961	79.981	-0.039	79.949	79.981	-0.051	79.922	79.981	-0.078	79.879	79.981	-0.121
100	MAX	100.010	100.000	0.032	99.990	100.000	0.012	99.976	100.000	-0.002	99.942	100.000	-0.036	99.889	100.000	-0.089
	MIN	99.975	99.978	-0.025	99.955	99.978	-0.045	99.941	99.978	-0.059	99.907	99.978	-0.093	99.854	99.978	-0.146
120	MAX	120.010	120.000	0.032	119.990	120.000	0.012	119.976	120.000	-0.002	119.934	120.000	-0.044	119.869	120.000	-0.109
	MIN	119.975	119.978	-0.025	119.955	119.978	-0.045	119.941	119.978	-0.059	119.899	119.978	-0.101	119.834	119.978	-0.166
160	MAX	160.012	160.000	0.037	159.988	160.000	0.013	159.972	160.000	-0.003	159.915	160.000	-0.060	159.825	160.000	-0.150
	MIN	159.972	159.975	-0.028	159.948	159.975	-0.052	159.932	159.975	-0.068	159.875	159.975	-0.125	159.785	159.975	-0.215
200	MAX	200.013	200.000	0.042	199.986	200.000	0.015	199.967	200.000	-0.004	199.895	200.000	-0.076	199.781	200.000	-0.190
	MIN	199.967	199.971	-0.033	199.940	199.971	-0.060	199.921	199.971	-0.079	199.849	199.971	-0.151	199.735	199.971	-0.265
250	MAX	250.013	250.000	0.042	249.986	250.000	0.015	249.967	250.000	-0.004	249.877	250.000	-0.094	249.733	250.000	-0.238
	MIN	249.967	249.971	-0.033	249.940	249.971	-0.060	249.921	249.971	-0.079	249.831	249.971	-0.169	249.687	249.971	-0.313
300	MAX	300.016	300.000	0.048	299.986	300.000	0.018	299.964	300.000	-0.004	299.850	300.000	-0.188	299.670	300.000	-0.298
	MIN	299.964	299.968	-0.036	299.934	299.968	-0.066	299.912	299.968	-0.088	299.798	299.968	-0.202	299.618	299.968	-0.382
400	MAX	400.017	400.000	0.053	399.984	400.000	0.020	399.959	400.000	-0.005	399.813	400.000	-0.151	399.586	400.000	-0.378
	MIN	399.960	399.964	-0.040	399.927	399.964	-0.073	399.902	399.964	-0.08	399.756	399.964	-0.244	399.529	399.964	-0.471
500	MAX	500.018	500.000	0.058	499.983	500.000	0.023	499.955	500.000	-0.005	499.771	500.000	-0.189	499.483	500.000	-0.477
	MIN	499.955	499.960	-0.045	499.920	499.960	-0.080	499.892	499.960	-0.1808	499.708	499.960	-0.292	499.420	499.960	-0.580

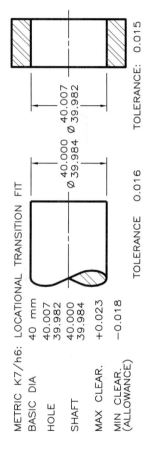

METRIC K7/h6: LOCATIONAL TRANSITION FIT

BASIC DIA	40 mm
HOLE	40.007 / 39.982
SHAFT	40.000 / 39.984
MAX CLEAR.	+0.023
MIN CLEAR. (ALLOWANCE)	-0.018

Ø 40.007 40.000
Ø 39.982 39.984

TOLERANCE 0.016

40.000 40.007
39.984 39.982

TOLERANCE: 0.015

Basic Size		C11	D9	F8	G7	H7	H8	H9	H11	K7	N7	P7	S7	U7
OVER	0	+0.120	+0.045	+0.020	+0.012	+0.010	+0.014	+0.025	+0.060	0.000	−0.004	−0.006	−0.014	−0.018
TO	3	+0.060	+0.020	+0.006	+0.002	0.000	0.000	0.000	0.000	−0.010	−0.014	−0.016	−0.024	−0.028
OVER	3	+0.145	+0.060	+0.028	+0.016	+0.012	+0.018	+0.030	+0.075	+0.003	−0.004	−0.008	−0.015	−0.019
TO	6	+0.070	+0.030	+0.010	+0.004	0.000	0.000	0.000	0.000	−0.009	−0.016	−0.020	−0.027	−0.031
OVER	6	+0.170	+0.076	+0.035	+0.020	+0.015	+0.022	+0.036	+0.090	+0.005	−0.004	−0.009	−0.017	−0.022
TO	10	+0.080	+0.040	+0.013	+0.005	0.000	0.000	0.000	0.000	−0.010	−0.019	−0.024	−0.032	−0.037
OVER	10	+0.205	+0.093	+0.043	+0.024	+0.018	+0.027	+0.043	+0.110	+0.006	−0.005	−0.011	−0.021	−0.026
TO	14	+0.095	+0.050	+0.016	+0.006	0.000	0.000	0.000	0.000	−0.012	−0.023	−0.029	−0.039	−0.044
OVER	14	+0.205	+0.093	+0.043	+0.024	+0.018	+0.027	+0.043	+0.110	+0.006	−0.005	−0.011	−0.021	−0.026
TO	18	+0.095	+0.050	+0.016	+0.006	0.000	0.000	0.000	0.000	−0.012	−0.023	−0.029	−0.039	−0.044
OVER	18	+0.240	+0.117	+0.053	+0.028	+0.021	+0.033	+0.052	+0.130	+0.006	−0.007	−0.014	−0.027	−0.033
TO	24	+0.110	+0.065	+0.020	+0.007	0.000	0.000	0.000	0.000	−0.015	−0.028	−0.035	−0.048	−0.054
OVER	24	+0.240	+0.117	+0.053	+0.028	+0.021	+0.033	+0.052	+0.130	+0.006	−0.007	−0.014	−0.027	−0.040
TO	30	+0.110	+0.065	+0.020	+0.007	0.000	0.000	0.000	0.000	−0.015	−0.028	−0.035	−0.048	−0.061
OVER	30	+0.280	+0.142	+0.064	+0.034	+0.025	+0.039	+0.062	+0.160	+0.007	−0.008	−0.017	−0.034	−0.051
TO	40	+0.120	+0.080	+0.025	+0.009	0.000	0.000	0.000	0.000	−0.018	−0.033	−0.042	−0.059	−0.076
OVER	40	+0.290	+0.142	+0.064	+0.034	+0.025	+0.039	+0.062	+0.160	+0.007	−0.008	−0.017	−0.034	−0.061
TO	50	+0.130	+0.080	+0.025	+0.009	0.000	0.000	0.000	0.000	−0.018	−0.033	−0.042	−0.059	−0.086
OVER	50	+0.330	+0.174	+0.076	+0.040	+0.030	+0.046	+0.074	+0.190	+0.009	−0.009	−0.021	−0.042	−0.076
TO	65	+0.140	+0.100	+0.030	+0.010	0.000	0.000	0.000	0.000	−0.021	−0.039	−0.051	−0.072	−0.106
OVER	65	+0.340	+0.174	+0.076	+0.040	+0.030	+0.046	+0.074	+0.190	+0.009	−0.009	−0.021	−0.048	−0.091
TO	80	+0.150	+0.100	+0.030	+0.010	0.000	0.000	0.000	0.000	−0.021	−0.039	−0.051	−0.078	−0.121
OVER	80	+0.390	+0.207	+0.090	+0.047	+0.035	+0.054	+0.087	+0.220	+0.010	−0.010	−0.024	−0.058	−0.111
TO	100	+0.170	+0.120	+0.036	+0.012	0.000	0.000	0.000	0.000	−0.025	−0.045	−0.059	−0.093	−0.146

Basic Size	c11	d9	f8	g7	h7	h8	h9	h11	k7	n7	p7	s7	u7
OVER 100	+0.400	+0.207	+0.090	+0.047	+0.035	+0.054	+0.087	+0.220	+0.010	-0.010	-0.024	-0.066	-0.131
TO 120	+0.180	+0.120	+0.036	+0.012	0.000	0.000	0.000	0.000	-0.025	-0.045	-0.059	-0.101	-0.166
OVER 120	+0.450	+0.245	+0.106	+0.054	+0.040	+0.063	+0.100	+0.250	+0.012	-0.012	-0.028	-0.077	-0.155
TO 140	+0.200	+0.145	+0.043	+0.014	0.000	0.000	0.000	0.000	-0.028	-0.052	-0.068	-0.117	-0.195
OVER 140	+0.460	+0.245	+0.106	+0.054	+0.040	+0.063	+0.100	+0.250	+0.012	-0.012	-0.028	-0.085	-0.175
TO 160	+0.210	+0.145	+0.043	+0.014	0.000	0.000	0.000	0.000	-0.028	-0.052	-0.068	-0.125	-0.215
OVER 160	+0.480	+0.245	+0.106	+0.054	+0.040	+0.063	+0.100	+0.250	+0.012	-0.012	-0.028	-0.093	-0.195
TO 180	+0.230	+0.145	+0.043	+0.014	0.000	0.000	0.000	0.000	-0.028	-0.052	-0.068	-0.133	-0.235
OVER 180	+0.530	+0.285	+0.122	+0.061	+0.046	+0.072	+0.115	+0.290	+0.013	-0.014	-0.033	-0.105	-0.219
TO 200	+0.240	+0.170	+0.050	+0.015	0.000	0.000	0.000	0.000	-0.033	-0.060	-0.079	-0.151	-0.265
OVER 200	+0.550	+0.285	+0.122	+0.061	+0.046	+0.072	+0.115	+0.290	+0.013	-0.014	-0.033	-0.113	-0.241
TO 225	+0.260	+0.170	+0.050	+0.015	0.000	0.000	0.000	0.000	-0.033	-0.060	-0.079	-0.159	-0.287
OVER 225	+0.570	+0.285	+0.122	+0.061	+0.046	+0.072	+0.115	+0.290	+0.013	-0.014	-0.033	-0.123	-0.267
TO 250	+0.280	+0.170	+0.050	+0.015	0.000	0.000	0.000	0.000	-0.033	-0.060	-0.079	-0.169	-0.313
OVER 250	+0.620	+0.320	+0.137	+0.069	+0.052	+0.081	+0.130	+0.320	+0.016	-0.014	-0.036	-0.138	-0.295
TO 280	+0.300	+0.190	+0.056	+0.017	0.000	0.000	0.000	0.000	-0.036	-0.066	-0.088	-0.190	-0.347
OVER 280	+0.650	+0.320	+0.137	+0.069	+0.052	+0.081	+0.130	+0.320	+0.016	-0.014	-0.036	-0.150	-0.330
TO 315	+0.330	+0.190	+0.056	0.017	0.000	0.000	0.000	0.000	-0.036	-0.066	-0.088	-0.202	-0.382
OVER 315	+0.720	+0.350	+0.151	+0.075	+0.057	+0.089	+0.140	+0.360	+0.017	-0.016	-0.041	-0.169	-0.369
TO 355	+0.360	+0.210	+0.062	+0.018	0.000	0.000	0.000	0.000	-0.040	-0.073	-0.098	-0.226	-0.426
OVER 355	+0.760	+0.350	+0.151	+0.075	+0.057	+0.089	+0.140	+0.360	+0.017	-0.016	-0.041	-0.187	-0.414
TO 400	+0.400	+0.210	+0.062	+0.018	0.000	0.000	0.000	0.000	-0.040	-0.073	-0.098	-0.244	-0.471
OVER 400	+0.840	+0.385	+0.165	+0.083	+0.063	+0.097	+0.155	+0.400	+0.018	-0.017	-0.045	-0.209	-0.467
TO 450	+0.440	+0.230	+0.068	+0.020	0.000	0.000	0.000	0.000	-0.045	-0.080	-0.108	-0.272	-0.530
OVER 450	+0.880	+0.385	+0.165	+0.083	+0.063	+0.097	+0.155	+0.400	+0.018	-0.017	-0.045	-0.229	-0.517
TO 500	+0.480	+0.230	+0.068	+0.020	0.000	0.000	0.000	0.000	-0.045	-0.080	-0.108	-0.292	-0.580

Appendix 40 • Shaft Sizes for Nonpreferred Diameters (millimeters)

Basic Size		c11	d9	f7	g6	h6	h7	h9	h11	k6	n6	p6	s6	u6
OVER	0	−0.060	−0.020	−0.006	−0.002	0.000	0.000	0.000	0.000	+0.006	+0.010	+0.012	+0.020	+0.024
TO	3	−0.120	−0.045	−0.016	−0.008	−0.006	−0.010	−0.025	−0.060	0.000	+0.004	+0.006	+0.014	+0.018
OVER	3	−0.070	−0.030	−0.010	−0.004	0.000	0.000	0.000	0.000	+0.009	+0.016	+0.020	+0.027	+0.031
TO	6	−0.145	−0.060	−0.022	−0.012	−0.008	−0.012	−0.030	−0.075	+0.001	+0.008	+0.012	+0.019	+0.023
OVER	6	−0.080	−0.040	−0.013	−0.005	0.000	0.000	0.000	0.000	+0.010	+0.019	+0.024	+0.032	+0.037
TO	10	−0.170	−0.076	−0.028	−0.014	−0.009	−0.015	−0.036	−0.090	+0.001	+0.010	+0.015	+0.023	+0.028
OVER	10	−0.095	−0.050	−0.016	−0.006	0.000	0.000	0.000	0.000	+0.012	+0.023	+0.029	+0.039	+0.044
TO	14	−0.205	−0.093	−0.034	−0.017	−0.011	−0.018	−0.043	−0.110	+0.001	+0.012	+0.018	+0.028	+0.033
OVER	14	−0.095	−0.050	−0.016	−0.006	0.000	0.000	0.000	0.000	+0.012	+0.023	+0.029	+0.039	+0.044
TO	18	−0.205	−0.093	−0.034	−0.017	−0.011	−0.018	−0.043	−0.110	+0.001	+0.012	+0.018	+0.028	+0.033
OVER	18	−0.110	−0.065	−0.020	−0.007	0.000	0.000	0.000	0.000	+0.015	+0.028	+0.035	+0.048	+0.054
TO	24	−0.240	−0.117	−0.041	−0.020	−0.013	−0.021	−0.052	−0.130	+0.002	+0.015	+0.022	+0.035	+0.041
OVER	24	−0.110	−0.065	−0.020	−0.007	0.000	0.000	0.000	0.000	+0.015	+0.028	+0.035	+0.048	+0.061
TO	30	−0.240	−0.117	−0.041	−0.020	−0.013	−0.021	−0.052	−0.130	+0.002	+0.015	+0.022	+0.035	+0.048
OVER	30	−0.120	−0.080	−0.025	−0.009	0.000	0.000	0.000	0.000	+0.018	+0.033	+0.042	+0.059	+0.076
TO	40	−0.280	−0.142	−0.050	−0.025	−0.016	−0.025	−0.062	−0.160	+0.002	+0.017	+0.026	+0.043	+0.060
OVER	40	−0.130	−0.080	−0.025	−0.009	0.000	0.000	0.000	0.000	+0.018	+0.033	+0.042	+0.059	+0.086
TO	50	−0.290	−0.142	−0.050	−0.025	−0.016	−0.025	−0.062	−0.160	+0.002	+0.017	+0.026	+0.043	+0.070
OVER	50	−0.140	−0.100	−0.030	−0.010	0.000	0.000	0.000	0.000	+0.021	+0.039	+0.051	+0.072	+0.106
TO	65	−0.330	−0.174	−0.060	−0.029	−0.019	−0.030	−0.074	−0.190	+0.002	+0.020	−0.032	+0.053	+0.087
OVER	65	−0.150	−0.100	−0.030	−0.010	0.000	0.000	0.000	0.000	+0.021	+0.039	+0.051	+0.078	+0.121
TO	80	−0.340	−0.174	−0.060	−0.029	−0.019	−0.030	−0.074	−0.190	+0.002	+0.020	+0.032	+0.059	+0.102
OVER	80	−0.170	−0.120	−0.036	−0.012	0.000	0.000	0.000	0.000	+0.025	+0.045	+0.059	+0.093	+0.146
TO	100	−0.390	−0.207	−0.071	−0.034	−0.022	−0.035	−0.087	−0.220	+0.003	+0.023	+0.037	+0.071	+0.124

Basic Size		c11	d9	f7	g6	h6	h7	h9	h11	k6	n6	p6	s6	u6
OVER	100	−0.180	−0.120	−0.036	−0.012	0.000	0.000	0.000	0.000	+0.025	+0.045	+0.059	+0.101	+0.166
TO	120	−0.400	−0.207	−0.071	−0.034	−0.022	−0.035	−0.087	−0.220	+0.003	+0.023	+0.037	+0.079	+0.144
OVER	120	−0.200	−0.145	−0.043	−0.014	0.000	0.000	0.000	0.000	+0.028	+0.052	+0.068	+0.117	+0.195
TO	140	−0.450	−0.245	−0.083	−0.039	−0.025	−0.040	−0.100	−0.250	+0.003	+0.027	+0.043	+0.092	+0.170
OVER	140	−0.210	−0.145	−0.043	−0.014	0.000	0.000	0.000	0.000	+0.028	+0.052	+0.068	+0.125	+0.215
TO	160	−0.460	−0.245	−0.083	−0.039	−0.025	−0.040	−0.100	−0.250	+0.003	+0.027	+0.043	+0.100	+0.190
OVER	160	−0.230	−0.145	−0.043	−0.014	0.000	0.000	0.000	0.000	+0.028	+0.052	+0.068	+0.133	+0.235
TO	180	−0.480	−0.245	−0.083	−0.039	−0.025	−0.040	−0.100	−0.250	+0.003	+0.027	+0.043	+0.108	+0.210
OVER	180	−0.240	−0.170	−0.050	−0.015	0.000	0.000	0.000	0.000	+0.033	+0.060	+0.079	+0.151	+0.265
TO	200	−0.530	−0.285	−0.096	−0.044	−0.029	−0.046	−0.115	−0.290	+0.004	+0.031	+0.050	+0.122	+0.236
OVER	200	−0.260	−0.170	−0.050	−0.015	0.000	0.000	0.000	0.000	+0.033	+0.060	+0.079	+0.159	+0.287
TO	225	−0.550	−0.285	−0.096	−0.044	−0.029	−0.046	−0.115	−0.290	+0.004	+0.031	+0.050	+0.130	+0.258
OVER	225	−0.280	−0.170	−0.050	−0.015	0.000	0.000	0.000	0.000	+0.033	+0.060	+0.079	+0.169	+0.313
TO	250	−0.570	−0.285	−0.096	−0.044	−0.029	−0.046	−0.115	−0.290	+0.004	+0.031	+0.050	+0.140	+0.284
OVER	250	−0.300	−0.190	−0.056	−0.017	0.000	0.000	0.000	0.000	+0.036	+0.066	+0.088	+0.190	+0.347
TO	280	−0.620	−0.320	−0.108	−0.049	−0.032	−0.052	−0.130	−0.320	+0.004	+0.034	+0.056	+0.158	+0.315
OVER	280	−0.330	−0.190	−0.056	−0.017	0.000	0.000	0.000	0.000	+0.036	+0.066	+0.088	+0.202	+0.382
TO	315	−0.650	−0.320	−0.108	−0.049	−0.032	−0.052	−0.130	−0.320	+0.004	+0.034	+0.056	+0.170	+0.350
OVER	315	−0.360	−0.210	−0.062	−0.018	0.000	0.000	0.000	0.000	+0.040	+0.073	+0.098	+0.226	+0.426
TO	355	−0.720	−0.350	−0.119	−0.054	−0.036	−0.057	−0.140	−0.360	+0.004	+0.037	+0.062	+0.190	+0.390
OVER	355	−0.400	−0.210	−0.062	−0.018	0.000	0.000	0.000	0.000	+0.040	+0.073	+0.098	+0.244	+0.471
TO	400	−0.760	−0.350	−0.119	−0.054	−0.036	−0.057	−0.140	−0.360	+0.004	+0.037	+0.062	+0.208	+0.435
OVER	400	−0.440	−0.230	−0.068	−0.020	0.000	0.000	0.000	0.000	+0.045	+0.080	+0.108	+0.272	+0.530
TO	450	−0.840	−0.385	−0.131	−0.060	0.040	0.063	−0.155	0.400	+0.005	+0.040	+0.068	+0.232	+0.490
OVER	450	−0.480	−0.230	−0.068	−0.020	0.000	0.000	0.000	0.000	+0.045	+0.080	+0.108	+0.292	+0.580
TO	500	−0.880	−0.385	−0.131	−0.060	−0.040	−0.063	−0.155	−0.400	+0.005	+0.040	+0.068	+0.252	+0.540

The following LISP programs were written by Professor Leendert Kersten of the University of Nebraska and are given here with his permission. These programs were introduced in Chapter 27, in which the principles of descriptive geometry are covered. These very valuable programs can be duplicated and added as supplements to your AutoCAD software.

```
)
(defun C:PARALLEL ()
  (setvar "aperture" 5)
    (setq sp (getpoint "\nSelect START point of
      parallel line:"))
    (setq ep (getpoint "\nSelect END point of
      parallel line:"))
    (setvar "osmode" 1)
    (setq sl (getpoint "\nSelect 1st point on line
      for parallelism:"))
    (setq el (getpoint "\nSelect 2nd point on line
      for parallelism:"))
    (setvar "osmode" 0)
    (setq pa (angle sl el ))
    (setq la (angle sp ep))
    (setq ll (distance sp ep))
    (setq m -1)
    (setq d 0)
    (if (> pa d) (setq m 1))
    (if (> la d) (setq d 1))
    (if (/= m d) (setq pa (+ pa 3.141593)))
    (setq ep (polar sp pa ll))
    (setvar "cmdecho" 0)
    (command line sp ep "" )
    (restore)
)
(defun C:PERPLINE ()
    (setvar "aperture" 5) (setvar "cmdecho" 0)
    (setq sp (getpoint "\nSelect START point of
      perpendicular line:"))
    (setvar "osmode" 128)
    (setq cc (getpoint sp "\nSelect ANY point on
      line to which perp'lr:"))
    (setq beta (angle sp cc)) (setvar "osmode" 0)
(setq ep
(getpoint "\nSelect END point of desired
 perpendicular (for length only): "))
    (setq length (distance sp ep))
    (setq ep (polar sp beta length))
    (command "line sp ep "")
    (restore)
)
(defun C:TRANSFER ()
    (setvar "aperture" 5) (setvar "cmdecho" 0)
    (setvar "osmode" 1)
    (setq aa (getpoint "\nSelect start of transfer
      distance:"))
    (setvar "osmode" 128)
    (setq bb (getpoint aa "\nSelect the reference
      plane:"))
    (setq length (distance aa bb))
    (setvar "osmode" 1)
    (setq cc (getpoint "\nSelect point to be
      projected:"))
    (setvar "osmode" 128)
    (setq dd (getpoint oc "\nSelect other reference
      plane:"))
(setvar "osmode" 0)
    (setq alpha (angle cc dd))
    (setq ep (polar dd alpha length))
    (COMMAND "CIRCLE" EP 0.05)
    (restore)
)
```

```
(defun RESTORE()
    (setvar "aperture" 10)
    (setvar "cmdecho" 1)
    (setvar "osmode" 0)
)
(DEFUN *ERROR* (MSG)
(SETVAR "OSMODE" 0)
(SETVAR "aperture" 10)
( setvar "cmdecho" 1)
(PRINC "error: ")
(princ msg)
(terpri)
)
(defun C:COPYDIST ()
    (setvar "aperture"5)
    (setvar "cmdecho" 0)
    (setvar "osmode" 1)
    (setq p1 (getpoint "\nSelect start point of
      line distance to be copied: "))
    (setq p2 (getpoint "\nEnd point?: "))
    (setvar "osmode" 0)
    (setq dist (distance p1 p2))
    (setq p1 (getpoint "\nStart point of new
      distance location:"))
    (setq ang (getangle p1 "\nWhich
      direction?: "))
    (setq p2 (polar p1 ang diet))
    (setvar "osmode" 0)
    (command "circle" p2 0.05)
    (restore)
)
(defun C:BISECT ()
    (setvar "aperture" 5)
    (setvar "osmode" 32)
    (setq sp (getpoint "\nSelect Corner of angle:"))
    (setvar "osmode" 2)
    (setq aa (getpoint "\nSelect first side
      (remember CCW):"))
    (setq alpha (angle sp aa))
    (setq bb (getpoint "\nSelect other side:"))
    (setvar "osmode" 0)
    (setq beta (angle sp bb))
    (setq m (/ (+ alpha beta) 2))
    (if (> alpha beta) (setq ang (+ pi m))
      (setq ang m))
(setq ep
    (getpoint "\nSelect endpoint of bisecting line
      (for length only): "))
    (setq length (distance sp ep))
    (setq ep (polar sp ang length))
    (setvar "cmdecho" 0)
    (command "line sp ep "")
    (restore)
}
```

APPENDIX 42 • American National Standard 125-lb Cast Iron Screwed Fittings (inches)

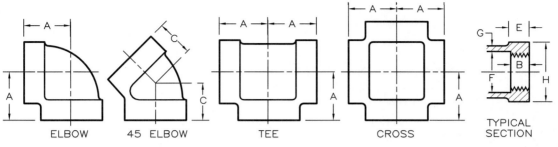

ELBOW   45 ELBOW   TEE   CROSS   TYPICAL SECTION

Nominal Pipe Size	A	C	B	E	F		G	H
			Min	Min	Min	Max	Min	Min
¼	0.81	0.73	0.32	0.38	0.540	0.584	0.110	0.93
⅜	0.95	0.80	0.36	0.44	0.675	0.719	0.120	1.12
½	1.12	0.88	0.43	0.50	0.840	0.897	0.130	1.34
¾	1.31	0.98	0.50	0.56	1.050	1.107	0.155	1.63
1	1.50	1.12	0.58	0.62	1.315	1.385	0.170	1.95
1¼	1.75	1.29	0.67	0.69	1.660	1.730	0.185	2.39
1½	1.94	1.43	0.70	0.75	1.900	1.970	0.200	2.68
2	2.25	1.68	0.75	0.84	2.375	2.445	0.220	3.28
2½	2.70	1.95	0.92	0.94	2.875	2.975	0.240	3.86
3	3.08	2.17	0.98	1.00	3.500	3.600	0.260	4.62
3½	3.42	2.39	1.03	1.06	4.000	4.100	0.280	5.20
4	3.79	2.61	1.08	1.12	4.500	4.600	0.310	5.79
5	4.50	3.05	1.18	1.18	5.563	5.663	0.380	7.05
6	5.13	3.46	1.28	1.28	6.625	0.725	0.430	8.28
8	6.56	4.28	1.47	1.47	8.625	8.725	0.550	10.63
10	8.08	5.16	1.68	1.68	10.750	10.850	0.690	13.12
12	9.50	5.97	1.88	1.88	12.750	12.850	0.800	15.47
14 O.D.	10.40	—	2.00	2.00	14.000	14.100	0.880	16.94
16 O.D.	11.82	—	2.20	2.20	16.000	16.100	1.000	19.30

Source: Extracted from American National Standards, "Cast-Iron Screwed Fittings, 125- and 250-lb" (ANSI B16.4), with the permission of the publisher, The American Society of Mechanical Engineers.

APPENDIX 43 • American Standard 250-lb Cast Iron Flanged Fittings (inches)

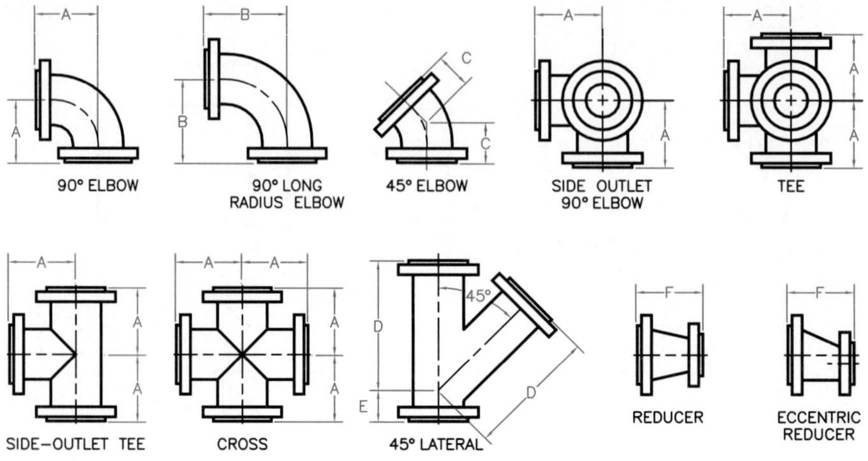
90° ELBOW
90° LONG RADIUS ELBOW
45° ELBOW
SIDE OUTLET 90° ELBOW
TEE
SIDE–OUTLET TEE
CROSS
45° LATERAL
45°
REDUCER
ECCENTRIC REDUCER
A
B
C
D
E
F

Nominal Pipe Size	Flanges: Dia of Flange	Flanges: Thickness of Flange (Min)	Flanges: Dia of Raised Face	Fittings: Inside Dia of Fittings (Min)	Fittings: Wall Thickness	Straight: Center to Face 90 Deg Elbow Tees, Crosses and True "Y" A	Straight: Center to Face 90 Deg Long Radius Elbow B	Straight: Center to Face 45 Deg Elbow C	Straight: Center to Face Lateral D	Straight: Short Center to Face True "Y" and Lateral E	Straight: Face to Face Reducer F
1	4 7/8	11/16	2 11/16	1	7/16	4	5	2	6 1/2	2	....
1 1/4	5 1/4	3/4	3 1/16	1 1/4	7/16	4 1/4	5 1/2	2 1/2	7 1/4	2 1/4	....
1 1/2	6 1/8	13/16	3 9/16	1 1/2	7/16	4 1/2	6	2 3/4	8 1/2	2 1/2	....
2	6 1/2	7/8	4 3/16	2	7/16	5	6 1/2	3	9	2 1/2	5
2 1/2	7 1/2	1	4 15/16	2 1/2	1/2	5 1/2	7	3 1/2	10 1/2	2 1/2	5 1/2
3	8 1/4	1 1/8	5 11/16	3	9/16	6	7 3/4	3 1/2	11	3	6
3 1/2	9	1 3/16	6 5/16	3 1/2	9/16	6 1/2	8 1/2	4	12 1/2	3	6 1/2
4	10	1 1/4	6 15/16	4	5/8	7	9	4 1/2	13 1/2	3	7
5	11	1 3/8	8 5/16	5	11/16	8	10 1/4	5	15	3 1/2	8
6	12 1/2	1 7/16	9 11/16	6	3/4	8 1/2	11 1/2	5 1/2	17 1/2	4	9
8	15	1 5/8	11 15/16	8	13/16	10	14	6	20 1/2	5	11
10	17 1/2	1 7/8	14 1/16	10	15/16	11 1/2	16 1/2	7	24	5 1/2	12
12	20 1/2	2	16 7/16	12	1	13	19	8	27 1/2	6	14
14	23	2 1/8	18 15/16	13 1/4	1 1/8	15	21 1/2	8 1/2	31	6 1/2	16
16	25 1/2	2 1/4	21 1/36	15 1/4	1 1/4	16 1/2	24	9 1/2	34 1/2	7 1/2	18
18	28	2 3/8	23 5/16	17	1 3/8	18	26 1/2	10	37 1/2	8	19
20	30 1/2	2 1/2	25 9/16	19	1 1/2	19 1/2	29	10 1/2	40 1/2	8 1/2	20
24	36	2 3/4	30 5/16	23	1 5/8	22 1/2	34	12	47 1/2	10	24
30	43	3	37 3/16	29	2	27 1/2	41 1/2	15	....	....	30

Source: Courtesy of ANSI; B16.1—1967.

This graph can be used to determine the individual grades of members of a team and to compute grade averages for those who do extra assignments.

The percent participation of each team member should be determined by the team as a whole (see Chapter 2 problems).

Example: written or oral report grades

Overall team grade: 82

Team members N=5	Contribution C=%	F=CN	Grade (graph)
J. Doe	20%	100	82.0
H. Brown	16%	80	75.8
L. Smith	24%	120	86.0
R. Black	20%	100	82.0
T. Jones	20%	100	82.0
	100%		

Example: quiz or problem sheet grades

Number assigned: 30
Number extra: 6
Total 36

Average grade for total (36): 82

$$F = \frac{\text{No. completed} \times 100}{\text{No. assigned}} = \frac{36 \times 100}{30} = 120$$

Final grade (from graph): 86.0

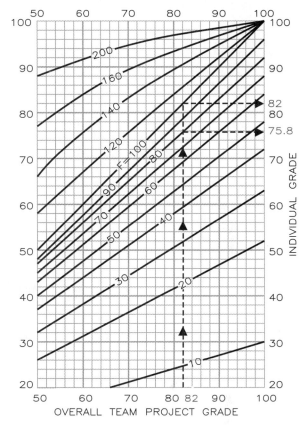

Figure A47-1 Grading graph.

APPENDIX 45 • Weights and Specific Gravities

SUBSTANCE	WEIGHT LB. PER CU. FT.	SPECIFIC GRAVITY	SUBSTANCE	WEIGHT LB. PER CU. FT.	SPECIFIC GRAVITY
METALS, ALLOYS, ORES			**TIMBER, U.S. SEASONED**		
Aluminum, cast, hammered	165	2.55–2.75	Moisture Content by Weight		
Brass, cast, rolled	534	8.4–8.7	Seasoned timber 15 t0 20%		
Bronze, 7.9 to 14% Sn	509	7.4–8.9	Green timber up to 50%		
Bronze, aluminum	481	7.7	Ash, white, red	40	0.62–0.65
Copper, cast, rolled	556	8.8–9.0	Cedar, white, red	22	0.32–0.38
Copper ore, pyrites	262	4.1–4.3	Chestnut	41	0.66
Gold, cast, hammered	1205	19.25–19.3	Cypress	30	0.48
Iron, cast, pig	450	7.2	Fir, Douglas spruce	32	0.51
Iron, wrought	485	7.6–7.9	Fir, eastern	25	0.40
Iron, spiegel-eisen	466	7.5	Elm, white	45	0.72
Iron, ferro-silicon	437	6.7–7.3	Hemlock	29	0.42–0.52
Iron ore, hematite	325	5.2	Hickory	49	0.74–0.84
Iron ore, hematite in bank	160–180	—	Locust	46	0.73
Iron ore, hematite loose	130–160	—	Maple, hard	43	0.68
Iron ore, limonite	237	3.6–4.0	Maple, white	33	0.53
Iron ore, magnetite	315	4.9–5.2	Oak, chestnut	54	0.86
Iron slag	172	2.5–3.0	Oak, live	59	0.95
Lead	710	11.37	Oak, red, black	41	0.65
Lead ore, galena	465	7.3–7.6	Oak, white	46	0.74
Magnesium, alloys	112	1.74–1.83	Pine, Oregon	32	0.61
Manganese	475	7.2–8.0	Pine, red	30	0.48
Manganese ore, pyrolusite	259	3.7–4.6	Pine, white	26	0.41
Mercury	849	13.6	Pine, yellow, long-leaf	44	0.70
Monel Metal	556	8.8–9.0	Pine, yellow, short-leaf	38	0.61
Nickel	565	8.9–9.2	Poplar	30	0.48
Platinum, cast, hammered	1330	21.1–21.5	Redwood, California	26	0.42
Silver, cast, hammered	656	10.4–10.6	Spruce, white, black	27	0.40–0.46
Steel, rolled	490	7.85	Walnut, black	38	0.61
Tin, cast, hammered	459	7.2–7.5			
Tin ore, cassiterite	418	6.4–7.0			
Zinc, cast, rolled	440	6.9–7.2			
Zinc ore, blends	253	3.9–4.2	**VARIOUS LIQUIDS**		
			Alcohol, 100%	49	0.79
VARIOUS SOLIDS			Acids, muriatic 40%	75	1.20
			Acids, nitric 91%	94	1.50
			Acids, sulphuric 87%	112	1.80
Cereals, oats ....bulk	32	—	Lye, soda	106	1.70
Cereals, barley ....bulk	39	—	Oils, vegetable	58	0.91–0.94
Cereals, corn, rye ....bulk	48	—	Oils, mineral, lubricants	57	0.90–0.93
Cereals, wheat ....bulk	48	—	Water, 4°C. max, density	62.428	1.0
Hay and Straw ....bales	20	—	Water, 100°C.	59.830	0.9584
Cotton, Flax, Hemp	93	1.47–1.50	Water, ice	56	0.88–0.92
Fats	58	0.90–0.97	Water, snow, fresh fallen	8	.125
Flour, loose	28	0.40–0.50	Water, sea water	64	1.02–1.03
Flour, pressed	47	0.70–0.80			
Glass, common	156	2.40–2.60			
Glass, plate or crown	161	2.45–2.72	**GASES**		
Glass, crystal	184	2.90–3.00			
Leather	59	0.86–1.02	Air, 0°C. 760 mm.	.08071	1.0
Paper	58	0.70–1.15	Ammonia	.0478	0.5920
Potatoes, piled	42	—	Carbon dioxide	.1234	1.5291
Rubber, caostchouc	59	0.92–0.96	Carbon monoxide	.0781	0.9673
Rubber goods	94	1.0–2.0	Gas, illuminating	.028–.036	0.35–0.45
Salt, granulated, piled	48	—	Gas, natural	.038–.039	0.47–0.48
Saltpeter	67	—	Hydrogen	.00559	0.0693
Starch	96	1.53	Nitrogen	.0784	0.9714
Sulphur	125	1.93–2.07	Oxygen	.0892	1.1056
Wool	82	1.32			

The specific gravities of solids and liquids refer to water at 4°C., those of gases to air at 0°C. and 760 mm. pressure. The weights per cubic foot are derived from average specific gravities, except where stated that weights are for bulk, heaped or loose materials, etc.

(Courtesy of the American Institute of Steel Construction.)

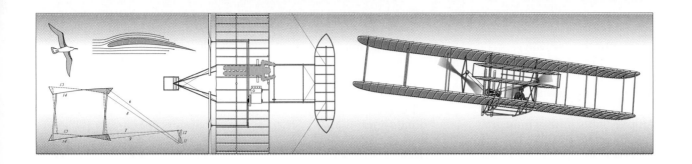

Index

A

Abbreviations
in notes, 332
ABC standards
for threads, 246
Abscissa (or x axis), 571
ACAD (LISP), 476, 477
Acceleration graphs, 602
Accessories
for exercise bench, 88–89
Acme threads, 248, 256
Activities Network (AN), 33
Activity sequence chart (ASC), 34
Actual size
English units, 339
Acute angles, 145
Addendum (A)
of bevel gear, 280
of gear, 277
Addendum angle (AA)
of bevel gear, 280
Addendum of tooth (AW), 283
Administrators, 74
Adjustable TV base, 110

Advanced Setup box, 659
Advertising, 42
and design analysis, 61
methods, 81
Aerodynamics, 6
Aerospace engineering, 6–7
Aesthetic awareness, 5
Aggregate conveyors
design applications for, 481
Agricultural engineering, 7
AH size sheet, 125
AIAA. See American Institute of
Aeronautics and
Astronautics
AIChE. See American Institute of
Chemical Engineers
AIIE. See American Institute of
Industrial Engineers
AIME. See American Institute of
Mining, Metallurgical, and
Petroleum Engineering
Airborne magnetometer, 13
AISI. See American Iron and Steel
Institute

Algebra, 136
ALIGNED command, 697
Aligned method
for obliques, 444
Aligned numerals
on isometric drawings, 454
Aligned vertical numeric dimen-
sions, 311
Alignment graphs, 584
three variables, 588
Alignment graph scales, 585–586
All option, 656, 674
Allowance
English units, 339
Alloys
of aluminum, 293
of copper, 292, 293
of magnesium, 293
Alloy steel, 292
Alphabet of lines, 188
and orthographic drawings,
180–181
for sketching, 162
Altitude adjustments (light), 744

Aluminum, 293, 298
  for pipes, 608
Ambient light, 742, 745
American Institute of Aeronautics
    and Astronautics (AIAA), 7
American Institute of Chemical
    Engineers (AIChE), 8
American Institute of Industrial
    Engineers (AIIE), 11
American Institute of Mining,
    Metallurgical, and
    Petroleum Engineering
    (AIME), 12, 13
American Iron and Steel Institute
    (AISI), 290
American National (N) form, 248
American National Standards
    Institute (ANSI), 260
  cylindrical fits of, 340
  cylindrical-fit tolerances of,
    338
  dimensioning standards of, 308
  pipe standards of, 606
  pipe thread notes abbrevia-
    tions of, 268
  rivet symbols of, 270
  and threaded parts, 246
  thread notes information from,
    250
  welding standards of, 370
American Nuclear Society, 12
American Petroleum Institute
    (API), 606
American Society for Agricultural
    Engineers, 7
American Society for Testing
    Materials (ASTM), 290, 606
American Society of Civil
    Engineers (ASCE), 9
American Standard thread, 246
American Welding Society, 370,
    378
Ames lettering guide, 129, 130,
    131, 132
AN. See Activities Network
Analysis, 19
  during design process, 20

with hanger bracket problem,
    23–25
Analytical geometry, 136, 150
  finding true length with, 479
Angle of taper, 295, 297
Angles, 136
  dihedral, 486, 488, 522
  dimensioning, 317
  drawing, 186
  of ellipse template, 221
  on isometric drawings,
    447–448
  line at specified, 526
  between lines and principal
    planes, 480
  on oblique drawings, 440–441
  slope of, 481
Angle valves, 614
Angular (Dimension toolbar),
    698
Angularity
  and orientation tolerancing,
    359–360
Angular surfaces
  dimensioning, 314
Angular tolerances, 390
  metric units, 348
Animation, 462
Annealing
  and heat treating, 294
Annotation box, 702
Annotation menu, 703
ANSI. See American National
    Standards Institute
ANSI Y14.5M-1994 Standards, 349
Aperture cards, 435
APERTURE command, 688
Apex to crown distance (AC)
  of bevel gear, 281
API. See American Petroleum
    Institute
Appearance
  and design analysis, 57
Appendix
  in technical report, 80
ARC command, 141, 199, 669, 715
Archery range, 104

Architects
  common fractions used by, 382
  dimensioning by, 309
  floor plan used by, 656
Architects' scale, 124
Architectural drawings
  aligned dimensions in, 311
Arc option, 681
Arcs, 184
  dimensioning, 314, 322, 327
  dimensioning with radius and
    leader, 699
  in perspective, 459
  revolving of, 724
  through three points, 142
Arcs (Draw toolbar), 669–670
Arc tangent
  to arc and line, 145
  to a line from point, 144
  to two arcs, 147–148
  to two lines, 145
Arc welding, 371, 377
Area option, 737
Arithmetic, 136
Arrangement and wiring diagram
    drawing, 638–639
Array (Modify toolbar), 689
Arrowheads, 309
Arrows
  directional, 523–524
Artwork
  for presentations, 76
ASAE. See American Society for
    Agricultural Engineers
Asbestos-cement
  for pipes, 608
ASC. See Activity sequence chart
ASCE. See American Society of
    Civil Engineers
Ascenders
  letter, 131
Assembly
  for presentations, 76
Assembly drawings, 387, 392–394,
    454–455
  for design problems, 99
  for exercise bench, 88

Associative dimensions, 696
ASTM. *See* American Society for
 Testing Materials
Attorney's fees
 for patents, 91
Attribute definitions, 677
ATTRIBUTES, 389
Audiocassette storage unit, 101
AutoCAD, 216
 to apply hatching, 229
 cross-sectional symbols with,
  228
 DIM command, 312
 dimensioning variables on, 313
 executing commands with, 712
 fonts within, 686
 isometric drawings with, 455
 lettering, 135
 predefined patterns within, 684
 raster images by, 382
 and representing human figures, 461
 ROMANS font, 133, 134
 SUPPORT directory of, 476–477
 surveyor's units within, 483
 three-dimension modeling
  with, 460
 toleranced dimensions in, 337
AutoCAD computer graphics
 Angular (Dimension toolbar),
  698
 Arc (Draw toolbar), 669–670
 Array (Modify toolbar), 689
 Blocks (Draw toolbar), 692–693
 Chamfer (Modify toolbar), 671
 Change (Modify toolbar),
  677–678
 Circles (Draw toolbar), 668–669
 computer graphics overview,
  644–645
 computer hardware, 645–647
 creating new drawing with,
  659–662
 custom title block, 707–708
 dialogue boxes, 652–654
 Diameter (Dimension toolbar),
  698–699

digitizing with tablet, 708
dimensioning with, 695–696
Dimension Style (Dimension
 toolbar), 700–704
Dimension style (Dimstyle)
 variables, 696–697
Dimension Style Override
 (Dimension toolbar), 704
Divide (Draw menu), 691
Donut (Draw menu), 689
drawing aids with, 654–655
drawing layers, 656–658
drawing scale, 662–663
editing dimensioning, 704–706
Ellipse (Draw toolbar),
 670–671
Erase and Break (Modify toolbar), 675–676
Extend (Modify toolbar),
 672–673
Fillet (Modify toolbar), 671
first session with, 647–651
format of presentation with,
 651–652
general assistance commands,
 655–656
Geometric tolerances
 (Dimension toolbar),
 706–707
Grips (Tools menu), 678–680
Hatching (Draw toolbar),
 684–685
Inquiry commands (Tools),
 694–695
isometric pictorials, 709–710
Linear (Dimension toolbar),
 697–698
Measure (Draw menu), 691
Mirror (Modify toolbar),
 687–688
Move and Copy (Modify toolbar), 676
Multiline (Mtext: Draw toolbar), 686–687
objects selection with, 674–675
oblique pictorials, 709
Offset (Modify toolbar), 691

Osnap (Object Snap toolbar),
 688
Pedit (Draw toolbar), 681–683
plotting parameters, 663–667
Polygon (Draw toolbar), 670
Polyline (Draw toolbar),
 680–681
Radius (Dimension toolbar),
 699
readying plotter, 667
Rotate (Modify toolbar), 690
saving and exiting, 663
saving dimension styles, 704
Scale (Modify toolbar), 689–690
Setvar (Command line), 690
Sketch (Miscellaneous tool-
 bar), 708–709
Spline (Draw toolbar), 683
Stretch (Modify toolbar), 690
text and numerals, 685
text style (Format menu), 686
toleranced dimensions, 706
toolbars, 658–659
Trace (Command line), 673
transparent commands, 693
Trim (Modify toolbar), 672
two-dimensional lines (Draw
 toolbar), 667–668
Undo (Standard toolbar), 677
View (Command line), 693
Windows introduction, 651
Zoom and Pan (Standard tool-
 bar), 673
Automobile stylists, 5
Auto option, 675
Auxiliary sections, 219–220,
 240–241
 drawn with horizontal refer-
  ence plane, 212
Auxiliary views, 208, 209. *See* also
  Primary auxiliary views;
  Secondary auxiliary views;
  Successive auxiliary views
 construction rules for, 212–213
 3D solids converted into, 735
Axonometric drawings
 of piping systems, 615

Axonometric pictorials, 438, 439
Axonometric projection, 456
Axonometric views (parallel pro-
jections), 720
Azimuth adjustments (light), 744
Azimuths, 481, 482
of lines, 481, 482

B

Baby stroller
comparative pricing of, 63–64
Background information
from consultants, 43
in magazines, 42
on patents, 43
Ball bearings, 231
Balloons
in assembly drawings, 454, 455
parts numbered with, 393, 394
Bar graphs, 568, 570–571
by computer, 579
3D, 579
Base circle (BC)
of gear, 277
Base flange, 381
Baseline dimensions, 325
Basic diameter
for tolerances and allowances,
340
Basic dimensions, 337, 355
Basic hole system, 340
Basic shaft system, 340
Basic size
English units, 339
metric units, 341
Basketball goal, 102
Bathing apparatus
product design problem, 110
Bearings
of lines, 481, 482–483
Bechtel Corporation, 618
Bed table, 108
Bell and spigot (B&S) joints, 609,
611
Best-fit curve, 574
Bevel gears, 276, 280–282

Bevel groove welds, 372, 373, 375
BHATCH, 229, 733
Bibliography
in technical report, 80
Bicycle child carrier, 107
Bicycle rental system, 104
"Big inch" pipelines, 609
Bilateral tolerancing, 336, 337,
358
Billets, 293, 296, 297
BISECT command, 476, 477
Bisecting angles, 141
Bisecting lines, 140–141
Blank hanger
working drawing for, 84
Blanks, 299, 394
Blips setting, 654
BLOCK command, 261
Blocks (Draw toolbar), 692–693
Blue-line print (or whiteprint),
433
Blue-line process
drawings reproduced by, 113,
114
Blueprinting, 434
Boat trailer, 111
Body
in technical report, 80
Body dimensions
and design analysis, 59–60
Bolt circle, 326
Bolts, 231, 258–259
Bonfire, 105
Bookholders, 101
Booting
of AutoCAD system, 647
Boot puller, 100
Boring
by drill press, 302
on lathe, 300
of machined holes, 328, 329
Bottom diameter of worm (BD),
283
Bottom of pipe (BOP), 615
Bottom tap, 267
Boundary hatch dialogue box,
684

Bounding Box option, 737
Bow compass, 116
Bow's notation, 534, 539
BOX command, 722, 730
Boxes
creating, 727
Box method
for sketching circles, 168
Box option, 674
Braddock-Rowe lettering triangle,
129, 131, 132, 312
Brainstorming, 19–20, 38–39
Brass pipe, 606, 608
Brazing
and welding, 378
BREAK command, 675
Break-even graphs, 576
Break (Modify toolbar), 675
Brittleness
of metal, 294
Broach, 303
Broaching machine, 300,
302–304
Broken-line curve, 572, 574
Broken-line graphs, 572
Broken-out sections, 234
Bronze pipe, 606, 608
Budgetary limitations
and design process, 16, 18
Buffing
and surface finishing, 304
Built-up welds, 377
application of symbols for, 377
Butt joint, 372
Buttress threads, 248
Butt-welded pipe, 607

C

Cabinet oblique drawing,
439, 440, 442
CAD. See Computer-aided design
CAD/CAM. See Computer-aided
design/computer-aided
manufacturing
CADD. See Computer-aided
design drafting

Calculus
   and empirical equations,
     595–602
   forms of, 597
   graphical, 597–598
Calibration
   of diagonal scale, 591–592
   and drawings on tablets, 708
   of functional scales, 585
   of outer scales of parallel-scale
     nomograph, 588
Camera option, 720
Cam followers
   types of, 285–286
Cams, 284–286
   designating plate, 286–287
   with offset follower, 287
   plate, 284
Can crusher, 109
Capacitance values, 635, 636
Cap screws, 259, 263
Carbon arc welding (CAW), 376
Carbon steel, 292
Car jack, 108
Car wash, 105
Car washer, 107
Case hardening
   and heat treating, 294
Cast aluminum, 293
Caster lock, 102
Casters, 53
Castings, 294, 324, 396
   working drawings for, 394–395
Cast-iron, 291
   bracket design analysis, 23–25
   pipe, 606, 607–608
   symbol, 228
Catalogs
   as design sources, 31
   pictorials in, 438
   product descriptions in, 67
Cathode-ray tube (CRT), 645
Caution notations
   in schematic diagrams, 636
Cavalier oblique, 169
Cavalier oblique drawings, 439,
    440, 442

Celsius measurements. *See*
    Fahrenheit/Celsius intercon-
    versions
Cement mixer, 110
Center
   finding center of, 142
Center distance (CD)
   of gear, 277
CENTER command, 721
Centerline (CL), 161, 172, 180,
    185, 309, 324
   for circular features, 167
   and removed sections, 235
CENTER option, 140
   of OSNAP, 145
Center-to-center distance (CD)
   between worm and spider,
    284
Central processing unit (CPU),
    645
Centroid option, 738
Ceramics, 290
Certification
   for drafters and engineering
    designers, 13–14
Chain dimensioning, 325
Chain vs datum dimensions,
    345–346
CHAMFER command, 671, 732,
    733
Chamfer (Modify toolbar), 671
Chamfers, 168, 330
   dimensioning of, 330
CHANGE command, 677
Change (Modify toolbar),
    677–678
Channel bracket, 104
Chart, 568. *See* also Graphs
   concurrent scale, 586
Checkers, 391, 433
Checking
   of working drawings, 391
Check valves, 614
Chemical engineering, 7–8
Chemical engineers, 13
Chemistry, 12
Chimney cover

product design problem, 109
Chip, silicon, 637
Chipping
   and surface contouring, 377
Chordal addendum (CA)
   of bevel gear, 281
Chordal thickness (CT)
   of gear, 277
"Chunnel"
   building of, 3
CIM. *See* Computer-integrated
    manufacturing
CIRCLE
   data point produced as, 572
CIRCLE command, 651, 655, 668,
    715
Circle of centers, 326
Circle(s), 137, 149
   area of, 137, 586
   degrees in, 136
   drawing with compass, 116
   in isometric drawings/pictori-
    als, 170–172, 448–450
   on oblique drawings, 442
   revolving of, 724
Circles (Draw toolbar), 668–669
Circuit diagram
   notes and symbols for, 631
Circuit function, 635
Circuits
   symbols of, 626, 628
Circuit voltage values, 636
Circular features
   sketching, 166–168
Circularity (roundness)
   and form tolerancing, 357
Circular pitch (CP)
   of gear, 277
Circular plates
   holes through, 325
Circular runout, 360
Circular thickness (CRT)
   of gear, 277
Civil engineering, 8–9
Civil engineers' scale, 121
Claims
   for patents, 91

Class
  of fit, 249
  thread, 247
Clearance fit (LC), 341, 343
  English units, 339
  metric units, 342
Clip option, 721
Clockwise (CCW), 631
CLOSE command, 668, 681
Close option, 680
Clothes hook, 102
Coarse-series fits
  thread, 252
Coarse thread, 248
Coarse tolerance, 247
Coarse tolerance class, 252
Coaxiality, 356
Cold-rolled steel (CRS), 299
Cold rolling (CR), 299
Colon
  for metric system, 390
Color
  adding to 3D objects, 741
  for layers, 657
  for presentations, 76, 77
  and working environment, 60
Color dialogue box, 743
Colored wires
  to identify leads and terminals,
    631
Color menu
  within Geometry box, 700
Comfort
  and working environment, 60
Command line, 652, 712
  with AutoCAD, 652
Commands
  general assistance with
    AutoCAD, 655–656
  repeating with AutoCAD, 649
Common fraction, 390
Communication
  of engineering data, 437
  through computer
    graphics/design, 645
  through graphs, 580
  through pictorials, 438

through sketching, 160
Communication systems
  and electrical engineering, 10
Company shareholders, 74
Comparative pricing, 62
  and design analysis, 63
Compass
  bearings, 481
  for drawing, 116–117
  sharpening point of, 113
  and triangulation, 139
Compass method
  for parallel lines, 143
Components, 531
Composite graphs, 576
Compression, 530, 531, 534, 536
Compression springs, 270, 271,
  273
Computer
  angle dimensioning by, 318
  arcs dimensioned by, 322
  arc tangent to two arcs on, 148
  arc tangent to two lines on, 145
  for auxiliaries from front view,
    215–216
  bisecting angles on, 141
  bisecting lines on, 141
  circle dimensioning by, 320
  constructing polygons on, 140
  detailed thread symbols drawn
    by, 254
  dimensioning by, 313–314
  dual dimensioning with, 310
  fillets and rounds drawn by,
    197–198
  finding line's true length by, 479
  geometry by, 476–477
  graphs by, 579–580
  hatching applied with, 229
  hexagon heads for bolt drawn
    by, 261
  isometric drawings by, 455–456
  laying out detail drawing by,
    387
  leaders produced by, 321–322
  linear dimensioning by,
    316–317

line tangent to arc on, 144
pipe drawings by, 619–620
plot plans by, 483–484
reproduction of drawings by,
  432
runouts on, 198–199
schematic thread symbols
  drawn by, 257
sectioning symbols on, 228
title block filled in by, 389
tolerances by, 337–338
variables dimensioned on,
  312–313
views by, 189
Computer-aided design (CAD),
  644
advantages of, 645
Computer-aided design/com-
  puter-aided manufacturing
  (CAD/CAM), 644
Computer-aided design drafting
  (CADD), 644
advantages of, 645
Computer drawings
  and diazo printing, 434
  transmission of, 436–437
Computer graphics, 2, 14, 112. *See
  also* AutoCAD computer
  graphics
  major areas of, 644
Computer graphics stations, 114
Computer hardware, 645
  input devices, 646
  output devices, 646–647
Computer images, 75
Computer-integrated manufac-
  turing (CIM), 645
Computer lettering, 133–135
Computer lines, 180–181
Computer mount, 107
Computer plotted transparencies
  for presentations, 77
Computers, 112
  and electrical engineers, 9
Computer visuals
  for presentations, 78
Computer wiring system, 105

Computer workstation, 110
Concave contour symbols, 377
Concentricity
    and location tolerancing, 354,
        356
Conceptual models, 64
    and design analysis, 64
Conclusion
    in technical report, 80
Conclusions
    to decisions, 73–74
Concrete
    for pipes, 608
Concurrent forces
    coplanar, 531–532
    noncoplanar, 532–533
Concurrent scales, 586–587
Concurrent system, 531
Conduit connector hanger, 103
CONE command, 722, 731
Cone of vision
    and perspectives, 459
Cones, 139
    development of, 562
    dimensioning of, 314, 320–321
    intersections between planes
        and, 551–552
    intersections between prisms
        and, 552
Conical helical spring, 271
Conical taper, 332
Conic sections, 149–152
Connection diagrams, 624, 626
Connectors, 634
Constant force springs, 270
Constant-pitch series, 248
Constant-pitch threads, 249
Construction
    and civil engineering, 8
Construction lines, 162, 172, 180,
        668
Contact area
    in surface texture, 361
Contact surfaces, 380
CONTINUE DIMENSION com-
        mand, 697
Contoured surfaces, 484–485

Contour lines, 484
Contour maps, 484, 485
Conventional breaks, 234,
        239–240
Conventional intersections,
        194–195
Conventional practices, 193–194
Conventional revolutions, 194,
        237–239
Convex contour symbols, 377
Coordinate graphs
    by computer, 579
Coordinate systems, 716–717
Coplanar system, 531
Coplanar truss analysis, 534–537
Copper, 292–293, 298
    pipe, 606, 608
COPY command, 254, 257, 674,
        676
COPYDIST command, 476, 477
Copy (Modify toolbar), 676
Copy option, 656
Corner joint, 372
Costs
    and refinement, 48
Cotter pin, 268
Counterboring
    by drill press, 302
    of machined holes, 329
Counterdrilling
    by drill press, 302
    of machined holes, 328
Countersinking
    by drill press, 302
    of machined holes, 328–329
Couplings, 249
Cover
    in technical report, 80
CPM. See Critical path method
CPolygon option (CP), 674
CPU. See Central processing unit
Create Drawing File dialogue box,
        663
Creativity
    and information accumula-
        tion, 36
Crest, 247, 491

Crest diameter tolerance, 252
Crest lines, 254, 257
Critical path
    in Activities Network, 34
Critical path method (CPM), 32
Criticism
    and brainstorming, 38
Critiques
    of presentations, 79
Crossing option, 674
Crossing Window, 675
CRT. See Cathode-ray tube
Crystal units
    labeling of, 633
Cup holder, 104
Cup point, 264
Curved parts
    dimensioning, 323–324
Curved shapes, 218–219
Curve plotting, 192–193
Curves, 572
    in isometric drawing, 452
    negative slope of, 597
    on oblique drawings, 443
Custom title block
    with AutoCAD, 707–708
Cut and fill process, 490–491
Cutting angle
    of bevel gear, 281
Cutting data, 179, 282
Cutting-plane method
    for intersections of two planes,
        495–496
Cutting planes, 226, 227, 547, 548,
        549, 554
    and auxiliary sections, 241
    and conventional revolutions,
        237
    and full sections, 229
    and half-sections, 233
    and offset sections, 234
    omitting, 231
    and phantom sections, 240
    and removed sections, 237
Cutting-tool holder, 336
CYLINDER command, 731, 735
Cylinders, 139

creating, 727
development of, 558–559
and half-sections, 233
intersections between, 549–550
intersections between planes
    and, 547–548
on isometric drawings,
    450–452
measuring, 302
on oblique drawings, 441–442
sketching, 172
standard views of, 226
Cylindrical datum features,
    351–353
Cylindrical features
dimensioning, 314
Cylindrical fits, 340–341
standard, 343–345
Cylindrical parts and holes
dimensioning, 318–320
one-view drawings of, 191
Cylindrical tolerancing, 338, 358

D

Dams
terminology for design of, 491
Data analysis, 2
and graphical methods, 20
Data collection
for problem identification, 29
Datum axis, 360
Datum dimensions
vs chain, 345–346
Datum K, 351, 352
Datum M, 351
Datum plane (or baseline), 346
DC resistance, 636
DDEDIT command, 389, 686, 708
Decagon, 137
Decimal fractions, 382, 390
Decimal inch (English system),
    309, 314
Decimals
fractional inches as, 309
spacing for, 132
Decision, 19, 72
within design process, 20

exercise bench, 73
with hanger bracket problem,
    25
presentations, 74–79
types of, 72
written reports, 80–82
Decision table, 25, 73
Decurve option, 682
Dedendum (D)
of bevel gear, 280
of gear, 277
Dedendum angle (DA)
of bevel gear, 280
Deer-hunting seat, 111
Defined blocks dialogue box, 692
Definition boxes, 652
Degree, 136
Delivery
of presentation, 78–79
Delta coordinates, 668
Depth
of thread, 247
Depth dimension
in auxiliary views, 213, 215
in six-view drawings, 182
in three-view drawings, 187
Derivative curve, 598, 601, 602
Derivative scale, 598, 601
Descenders
letter, 131
Descriptive geometry, 2–3, 20, 48,
    464, 504. *See also* Primary
    auxiliary views in descriptive
    geometry
application of, 49–51
and design analysis, 56
and problem refinement, 23
Design
and geometry, 504
ownership of, 390
quickie, 41–42
and working drawings, 381
Design analysis. *See also* Drawing
exercise bench, 65–67
graphics and analysis, 56
types of, 56–65
Design drafters
certification for, 14

Design engineering
aerospace, 6
Designers, 5, 20
computer-graphics systems
    used by, 644, 645
individual vs. team methods,
    37–38
and metals specifications, 290
problem identification by, 28
sketching by, 160
working drawings by, 85, 300
Design problems
examples of short, 100–104
individual approach to, 98
product, 106–111
scheduling team activities, 99
selection of, 99
specifications, 99
systems, 104–106
team approach to, 98–99
types of, 17–18
Design process, 16–17, 18–19
application of, 21–26
problem identification, 19–21
types of design problems, 17–18
Design Schedule & Progress
    Record (DS&PR), 33, 34
Design sketches, 41
Design worksheets, 30
Detail drawings, 178, 380
checking, 391
and geometry, 504
laying out of, 387–389
notes on, 332
Detailed symbols, 253, 254–256,
    258
Detail paper, 113–114
Detail views, 93
Developed pipe drawing, 616
Development
of cones, 562
of cylinders, 558–559
of oblique cylinders, 559–560
of oblique prisms, 557–558
principles and, 554–556
of pyramids, 561–562
of rectangular prisms, 556–557
of transition pieces, 563

Deviation
 metric units, 341
Diagonal scale, 591
Dialogue boxes
 use of, 652–654
Diameter (Dimension toolbar),
 698–699
Diameter symbol, 320
DIAMETR, 322
Diametral pitch (DP), 330
 of bevel gear, 280
 of gear, 277
Diametric symbol, 318
Diamond knurls, 330
Diazo (blue-line) print, 391
Diazo paper, 434
Diazo printer, 434
Diazo printing, 433–435
Diazo process, 113
Diazo-process machine, 434
Die casting, 294, 296
Dies, 396
Differential calculus, 597
Digitizers, 645, 646
Digitizing tablet
 with AutoCAD, 708
Dihedral angles, 486, 488,
 505–506, 522
DIMAD1, 700
DIMALT, 310, 703
DIMALTD, 310
DIMALTF, 310
DIMASO, 317, 704, 705
DIMASZ, 701
DIMBLK1, 701
DIMBLK2, 701
DIMCENTER command, 701
DIMCLRD, 700
DIMCLRE, 700
DIM command, 310, 312, 313,
 314, 316, 321, 337, 483
 ANGULAR option of, 318
 RADIUS option of, 322
DIMDIAMETER command, 701
DIMDLE, 700
DIMDLI, 700
Dimension and ranges
 and design analysis, 59–60

Dimensioning. *See also* Working
 drawings
 angles, 317
 arcs and radii, 322
 with AutoCAD, 695–696
 chamfers, 330
 by computer, 313–314
 curved and symmetrical parts,
 323–324
 cylindrical parts and holes,
 318–320
 dual, 310
 and English/metric conversions, 310
 finished surfaces, 324
 keyseats, 330
 location dimensions, 324–327
 machined holes, 328–329
 and metric units, 310–311
 miscellaneous, 332
 necks and undercuts, 331
 and numerals and symbols,
 311–313
 objects with rounded ends,
 326–327
 outline, 327–328
 prisms, 314
 pyramids, cones, and spheres,
 320–321
 rules, 314–323
 spacing in, 318
 tapers, 332
 terminology for, 308–309
 in 3D, 740–741
 and units of measurement, 309
Dimensioning and Tolerancing
 for Engineering Drawings
 (ANSI), 308
Dimension lines, 166, 308
Dimension numbers, 309
Dimensions, 178
 for bevel gears, 280–281
 editing, 704–706
 gear, 278, 279
 knurling, 330–331
 lettering, 128
 table of, 279
 tolerance, 336–338, 706

for working drawings, 381–387
for worm gear, 284
Dimension style (Dimension
 toolbar), 700–704
Dimension style (Dimstyle) vari-
 ables, 696–697
Dimension style override
 (Dimension toolbar), 704
Dimension styles
 saving, 704
Dimetric axonometric projection,
 456
DIMEXE, 700
DIMGAP, 704
DIMLIM, 337, 706
DIM1FAC, 703
DIMOVERRIDE, 704
DIMRADIUS command, 701
DIMRND, 704
DIMSCALE, 14, 312, 701, 704
DIMSHO, 317, 704, 705
DIMSTYLE, 704
DIMTFAC, 706
DIMTIH, 312, 702
DIMTM, 337, 703, 706
DIMTOH mode, 312
DIMTOH (Outside Horizontal),
 702
DIMTOL, 706
DIMTOLJ, 703
DIMTP, 337, 703, 706
DIMTSZ, 701
DIMTXSTY, 704
DIMTXT, 704
Dip, 492
Direction, 530
 determining, 523–524
Directional arrows, 523–524
Direct mail
 advertising through, 61
Directrix, 150
DISH command, 723
DISMD2, 700
Displacement diagram, 286, 287
Display booth
 product design problem, 111
Display setting, 664
Distance option, 720

Distant Light dialogue box, 743
Distant lights, 742
DIST command, 484
Dist option, 694
Distribution
    and implementation, 84
DIVIDE command, 691
Divide (Draw menu), 691
Dividers, 117, 210, 447, 452, 488
    with auxiliary views, 208, 210,
        211
    depth dimensions transferred
        with, 215
    measurements transferred
        with, 213
    and scale specifications, 390
    width dimensions transferred
        with, 218
Documentation drawings, 2
Documents
    electronic transmission of, 437
Dodecahedron, 138
Dog (or flat-point) set screw end,
    264
DOME command, 723
DONUT
    data point produced as, 572
DONUT command, 689
Donut (Draw menu), 689
Doorstop, 102
Dot-matrix printers, 180
Double-impression die, 296
Double-line drawings, 614
    of springs, 271, 273
Double-line symbols, 609, 611,
        612, 613, 614, 615
Double revolution, 521–522
Double-size drawings, 713
    making with AutoCAD, 662
Double thread, 250
Dowels, 231
Draft, 295, 297, 396
Draft angles
    and die castings, 296
Drafters, 5, 13, 391
    architects' scales used by, 119
    certification for, 13–14

computer-graphics systems
    used by, 644, 645
electronic, 624
logs kept by, 392
working drawings by, 380
Drafting, 13–14
Drafting machine, 114
DRAW command, 140, 715
Drawer handle, 103
Drawing aids
    with AutoCAD, 654–655
Drawing instruments
    drawing media, 112–114
    equipment, 114–117
    lines, 117–119
    measurement, 119–124
    and presentation of drawings,
        124–125
Drawing(s). See also
        Dimensioning; Geometric
        constructions; Orthographic
        drawings; Reproduction of
        drawings; Sketching;
        Working drawings
    aids within AutoCAD, 654
    analysis, 67
    architectural, 311
    assembly, 86, 99
    bevel gears, 282
    of concurrent scales, 587
    detail, 380
    developed pipe, 616
    dimensions applied to, 695
    double-size, 662
    with dual dimensioning, 310
    electronic transmission of, 437
    of ellipses, 448, 450, 451, 453,
        524
    forging, 297
    half-size, 662
    hexagon heads, 261
    hidden-line wire-frame, 460
    installation, 638–639
    isometric, 446–452
    layers with AutoCAD, 656–658
    light sources added to, 742–743
    making with AutoCAD, 647, 649

mechanical, 178
nomograph with logarithmic
    scales, 589
notes on detail, 332
nut and bolt combinations,
    262–263
oblique, 439–444
one-point perspectives, 457
one-view, 191
papers for, 113–114
partial views, 232–233
patent, 91–93
pie graphs, 569
plan-profile, 485
polygons with AutoCAD, 670
of section lines, 228
six-view, 181–182
small threads, 258
springs, 271–273
spur gears, 279
square heads, 261
square threads, 254
threads, 253
3D fundamentals, 774
three-view, 182
with triangles, 185–186
two-point perspectives, 457,
    459
two-view, 191
with welding specifications,
    370
wire-frame, 460
worm gears, 284
Drawing sets
    assembling, 435
Draw toolbar, 652
Drilling
    on lathe, 300
    of machined holes, 328
Drill press, 300, 302
Drive-through trash disposal sys-
    tem, 105
Drop forges, 296
Drum plotter, 646
Drum truck
    product design problem, 107
Dryseal threads, 268

DTEXT command, 133, 134, 135, 189
Dual dimensioning, 310
Ductile iron, 291
Ductility
of metal, 294
Durability
and design analysis, 57
DVIEW command, 719–721, 735
options of, 720–721

E

Economic factors
and design analysis, 56, 62
Edge joint, 372
EDGE option, 140
Edges
in orthographic projections, 185
in principal views, 210–211
EDGESURF command, 724
Edge views
of planes, 486, 488, 506, 512, 520, 521
of profile planes, 216–217
Edison, Thomas, 5
patent for electric lamp, 90
Editing
dimensions, 704–706
Education toy production, 106
EF. *See* Electric furnace
Einstein, Albert, 2
Elasticity
of metal, 294
Elastomers, 305, 306
Elbow (ell), 613
Electrical and Electronics
Diagrams (ANSI Y14.15), 624
Electrical engineering, 9–10
Electrical graphics, 624
Electrical power
and agricultural engineering, 7
Electric-arc welding
flux in, 371
Electric/electronic assembly
abbreviations designating
parts of, 634

Electric/electronic graphics
functional identification of
parts, 636
graphic symbols, 628–630
installation drawings, 638–639
numerical units of function, 635–636
printed circuits, 637–638
reference designations, 634–635
schematic diagram connecting
symbols, 626–628
separation of parts, 633–637
shortcut symbols, 638
terminals, 630–632
types of diagrams for, 624–626
Electric furnace (EF), 618
Electricians, 5
Electric resistance weld (ERW), 607
Electronic diagrams, 624
Electronic drawing instruments, 112
Electronics
and electrical engineering, 9–10
Electronics drafters, 624
graphic symbols used by, 628
Electronics graphics, 624
Electronic tubes, 624
Electron tubes, 632
labeling, 635
Elementary extrusions, 714–716
Elevation views, 93
ELEV command, 714, 715
Elev option, 678
ELLIPSE command, 670, 710
ISOCIRCLE option of, 455
Ellipse (Draw toolbar), 670–671
Ellipses, 149–150
circular shapes projected as, 221
drawing, 172, 448, 450, 451, 453, 524
in isometric pictorials, 170
on nonisometric planes, 452
plotting, 193

Ellipse templates, 149, 150, 442, 443, 448, 449, 452, 453, 454, 524
Elliptical curves
constructing, 218
Elliptical features
and primary auxiliary views, 221
Empire State Building, 3
Empirical data, 20, 56, 594
Empirical equations, 2
and calculus, 594–602
Employees
expenses with, 64
Enable Grips, 679
Enable Grips Within Blocks, 679
Endpoint option, 680
Engineering designers
certification for, 14
Engineering design graphics, 2
technical illustration, 13
Engineering drawings
on computer-graphics sys-
tems, 645
Engineering fields
aerospace, 6–7
agricultural, 7
chemical, 7–8
civil, 8–9
electrical, 9–10
industrial, 10–11
mechanical, 11–12
mining and metallurgical, 12
nuclear, 12
petroleum, 13
Engineering graphics, 2
descriptive geometry, 2–3
Engineering teams, 98
Engineering units, 314
Engineers, 4, 13, 74
dimensioning by, 309
graphics skills of, 128
pipe drawings prepared by, 619
plan-profile drawings used by, 485
reports prepared by, 79
reproduction of drawings by, 432

sketching by, 160
working drawings by, 380
Engineers' scale, 121–122
English (Imperial) system
    and specifications for threads,
        248–249
    and tolerancing, 339–340
    of units, 122
    for working drawings, 381–385
English/metric conversions
    and dimensioning, 310
English thread
    notes for, 249–250
Environment
    and design analysis, 60
Environmental engineering, 9
Environmental pollution control,
    8
Equal sign
    for English system, 390
Equations
    empirical data as, 594
Equilateral polygon
    drawing with AutoCAD, 670
Equilateral triangle, 137
Equilibrant, 531
Equilibrium
    forces in, 533–534
ERASE command, 675
Erase (Modify toolbar), 675
Erasers, 119
Erasing shield, 119
Ergonomics
    and design analysis, 57–58
ERW. *See* Electric resistance weld
Exercise bench
    analysis of, 65–67
    comparative pricing of, 63
    decision on, 73
    and implementation, 87–89
    preliminary ideas about, 44–45
    problem identification with,
        30–32
    refinement of, 51–52
Exiting
    drawing session in AutoCAD,
        663
EXit option, 721

Experimental data, 19
EXPLODE command, 732, 733
Exploded dimensions, 696
Exploded Hatch, 684
Exploded views, 93
Exponential equations, 594,
    596–597
EXTEND command, 672, 732
Extending spring, 271
Extend (Modify toolbar), 672–673
Extension lines, 308
    on computer, 316
    dimension lines as, 324
    drawing with AutoCAD, 698
Extension spring, 270
    working drawing of, 271
Extents setting, 665
External micrometer, 318
    calipers, 302
External threads, 246, 248,
    249–250, 252, 253
Extractive metallurgy, 12
Extra-fine threads, 248, 249
EXTRUDE command, 727, 729,
    739
Extrusions
    applications of, 719
    example: TILEMODE=0,
        728–729

F

Fabricators, 5
Face angle (FA)
    of bevel gear, 281
    of spider, 283
Face width (FW)
    of bevel gear, 281
    of gear, 277, 284
Facing
    on lathe, 300
Fahrenheit/Celsius interconver-
    sions
    alignment graph for, 584
    concurrent scales for, 586
FAO. *See* FINISHED ALL OVER
Farm structures
    and agricultural engineering, 7

Fastening
    advantages of welding over, 370
Fax machines
    drawings transmitted by, 436
Feature control frame, 357
    and true-position method, 355
Feature control frames
    priority of datum planes in,
        351, 352
    to specify geometric character-
        istics, 356
Fence option, 674
Ferrous metals, 291
Ferrules, 249
F-factor, 79
Fiberglass, 306
Fibers, 290
Field rivets, 270
"Figure-eight" breaks, 239
Files
    creating with AutoCAD, 647
    naming with AutoCAD,
        661–662
    prototype, 662
    template, 662
Filing fees
    for patents, 91, 94
FILLET command, 145, 148, 197,
    671, 732, 733, 735
Fillet (Modify toolbar), 671
Fillets, 195–197
    and castings, 295
    and die casting, 296
    and dimensioning, 322–323
    and forging, 297
    in isometric drawing, 454
Fillet welds, 373
    application of symbols for,
        374–375
Film
    for drawing, 114
Filters
    XYZ, 726
FIM. *See* Full indicator movement
Final reports, 79, 80
Fine-line holders, 113
Fine-series fits
    thread, 252

Fine threads, 248
Fine tolerance, 247
Fine tolerance class, 252
Finished head, 259
Finished surfaces
    and dimensioning, 324
Finish marks, 382
    drawing of, 324
    types of, 196–197
Finish mark symbol, 166
Fink truss, 534
FINISHED ALL OVER (FAO) note,
    324
Firewood rack, 107
First-angle projection, 199
First-angle system (European
    system), 124
Fit
    English units, 339
Fit option, 681
Fittings
    in orthographic views, 614–615
    and pipe threads, 268–269
    weights for, 612
Flame hardening
    and heat treating, 294
Flanged fittings, 612–613
Flanged joints, 608, 611, 615
Flanged pipe joints, 608
Flanges, 613
Flash welding, 371
Flatbed plotter, 646
Flatness
    and form tolerancing, 357
Flat patterns. See Developments
Flat springs, 270
Flat surface cam follower, 285
Flat taper, 332
Flaws
    in surface texture, 361
Flip charts
    for design problems, 99
    for presentations, 75
Flowcharts
    for design problems, 99
Flush contour symbols, 377
Flux
    in welding, 371

Folding
    drawing sets, 435
Folding-line method, 211
    for auxiliaries from front view,
        213–215
    for auxiliaries from profile
        view, 217–218
    for auxiliaries from top view,
        209
Fold lines, 185, 194, 213, 556, 557
    principles, 209
Fonts
    within AutoCAD, 686
Foot scraper
    short design problem, 101
Force, 530
Force fit (FN), 341
Foreshortened line, 185
Foreshortened plane, 185
    intersection between oblique
        prism and, 545
Forged parts
    working drawings for, 394–395
Forging, 294, 296–298, 396
    drawings, 297
Forklifts, 64
Form
    of thread, 247, 248
Formal presentations, 74
Format menu, 701
Form tolerancing, 357–358
Four-center ellipse construction,
    448, 449
Four-center ellipse method, 442,
    450, 451, 453
Fractional angles, 317
Fractional inches
    as decimals, 309
Fractions
    spacing for, 132
    on working drawings, 382
Freeboard, 491
Free-body diagram, 536
Freehand sketching. See also
    Sketching
    for design problems, 99
Freeze option, 658
French curves, 119

French National Assembly, 122
Frontal lines, 473
    principal, 466
Frontal orthographic planes, 209
Frontal planes, 178, 180, 465, 466
    of orthographic projection, 180
    principal, 468, 469
Frontal reference planes (FRP),
    216, 219
    and cylinders, 218
Front view
    auxiliaries from, 213–216
FRP. See frontal reference planes
Frustum
    of cone, 139
    of pyramid, 138
Full indicator movement (FIM),
    360
Full Preview setting, 666
Full sections, 226, 229–232, 232
Full-size drawing, 713
Full-size measurement, 121
Function
    and design analysis, 56, 57
    and product design, 17
    and refinement, 48
Functional scales, 585
Fundamental deviation
    metric units, 342
Funding
    and design process, 16
Funicular diagram, 539

G

Gas-tight joints
    and brazing, 378
Gas welding, 370–371
Gate hinge
    short design problem, 104
Gate opener
    product design problem, 109
Gate valves, 614
Gauge pin diameter, 356
Gears
    bevel, 280–282
    defined, 276

determining relationship
between meshing, 278
finding dimensions of, 281–282
number of teeth for, 279
pitch angle of, 280
ratios of, 278–279
spur gears, 276–278
tooth forms for, 277–278
worm, 282–284
General Information Concerning
Patents (PTO), 90
General Motors Corporation
dimensioning standards of, 308
General oblique drawings, 439,
440, 442
Generatrix, 139
Geologists, 13, 492
Geology
and strike and dip, 492
Geometric construction
angles, 136
arc through three points, 142
bisecting lines and angles,
140–141
circles, 137
conic sections, 149–151
constructing polygons,
139–140
division of lines, 141
geometric solids, 137–139
helixes, 152
parallel lines, 143
polygons, 136–137
quadrilaterals, 137
spirals, 152
tangents, 143–149
triangles, 137
Geometric solids
cones, 139
cylinders, 139
polyhedra, 137–138
prisms, 138
pyramids, 138
spheres, 139
Geometric symbols
to identify leads and terminals,
631
Geometric tolerances, 348–350

size limits for, 349–350
Geometric tolerances
(Dimension toolbar),
706–707
Geometry, 136
Geometry box, 700
Geotechnical engineering, 9
Ghost sections, 240
Glass, 306
Globe valves, 614
Golf driving range ball-return sys-
tem
systems design problem, 105
Gothic lettering
inclined, 131–133
vertical, 130–131
Gourad option, 741
Grades
thread tolerance, 252
Graph
alignment, 584
Graphical calculus, 597–598
Graphical differentiation,
598–602
Graphical integration, 602
Graphical mathematics, 2
Graphical methods
in analysis, 20
during implementation, 84
Graphics, 2, 13–14. *See* also
Computer graphics
and analysis, 56
during design process, 98
engineer's skill with, 128
future of three-dimensional,
462
Graphic symbols, 628–630
in schematic diagrams, 625
Graphic Symbols for Electrical
and Electronics Diagrams
(Y32.2), 624
Graphs
bar, 570–571
by computer, 579–580
linear coordinate, 571–576
optimization, 575
percentage, 578
pie, 569

with problem identification, 32
and proportions, 568–569
schematics, 578–579
semilogarithmic-coordinate,
576–578
types of, 568
Gravity motion (uniform acceler-
ation)
for cams, 285
for plate cams, 286
Gray iron, 291
Grease fittings, 269
Grid option, 655
Grids
isometric, 169
for sketching, 162
Grinding, 324
and surface contouring, 377
and surface finishing, 304
Grip Colors, 679
Grips dialogue box (DDgrips), 679
Grips (Tools menu), 678–680
Grocery store checkout system,
41
Groove welds
application of symbols for, 375
Ground line (GL), 457, 459
Ground plain ends
of springs, 271
Group option, 674
Groups setting, 655
Guide for Patent Draftsmen (U.S.
Government Printing
Office), 91
Guidelines, 180

H

Half sections, 233
on isometric drawings, 454
Half-size drawings, 713
making with AutoCAD, 662
Half-size measurement, 121
Halftones
reproduction of, 435
Half-view, 232
Halfwidth option, 680, 681
Hammering

and surface contouring, 377
Hammock support
  short design problem, 102
Handcrank of casement window
  and revolution techniques, 525
Hand truck
  product design problem, 109
Hanger bracket problem
  application of decision process
    with, 21–26
Hardening
  and heat treating, 294
Hardness
  of metal, 294
Harmonic motion
  for cams, 285, 287
  for plate cams, 286
HATCH command, 229
Hatching
  parts not requiring, 231
  symbols, 228
  webs and ribs, 231
Hatching (Draw toolbar), 684–685
Hatching lines
  for patent drawings, 92
Hatch setting, 655
Hauler, portable
  product design problem, 109
Headless set screw, 264
Heat treatment
  modifying metal properties by,
    294
Height dimension
  in auxiliary views, 209, 210
  in orthographic projection, 163
  in six-view drawings, 182
Helical curves, 254
Helical springs, 270
Helical torsion spring, 271
Helicopter frame
  descriptive geometry princi-
    ples for, 476
Helix angle (HA)
  of worm, 283
Helixes, 152
HELP command, 655
Heptagon, 137
Hexagon-head bolts, 260

Hexagon heads
  drawing, 261
Hexagon jam nut, 260
Hexagons
  constructing, 139
Hexahedron, 138
Hidden diameter
  minor diameter as drawn as,
    258
Hidden lines, 161, 172, 187, 262,
    324, 354, 468
  centerlines coinciding with,
    167
  dash lengths in, 180
  in exploded orthographic
    assembly drawing, 393
  invisible lines converted to, 736
  with LINE command, 189
  modified with Ltscale, 657
  omission of, 230, 233, 737
  in orthographic drawings, 184
  in orthographic views, 226, 229
  in precedence of lines, 185
  removed from extruded
    objects, 715
  removed when plotting, 740
  and selection of views, 183
  suppressed in pipe layout, 620
  suppressed in solid primitives,
    730
Hidden-line wire-frame drawing,
    460
HIDE command, 714, 715, 725
Hide Lines setting, 665
Hide option, 721
Highlights setting, 655
Historical records, 19
Hole basis
  metric units, 342
Hole-basis system
  preferred fits for, 343
Holes
  dimensioning cylindrical, 319
  dimensioning of, 325
  machined, 328–329
  in stampings, 299
  tapping, 266–267
Honing

and surface finishing, 304
Horizontal cutting planes, 551,
    552, 554
Horizontal lines, 118, 473
  drawing, 118
  principal, 466
Horizontal orthographic planes,
    209
Horizontal planes, 465, 466
  of orthographic projection, 180
  principal, 468, 469
Horizontal reference plane
    (HRP), 211–212
Hose spool
  short design problem, 104
Hot water supply
  systems design problem, 105
HRP. See Horizontal reference
    plane
Human factors
  and design analysis, 56, 57–58
Human figure, 461–462
Hunting blind
  product design problem, 106
Hunting seat
  comparative pricing of, 63
Hydraulics
  and civil engineering, 8
Hyperbolas, 149, 151

I

Icosahedron, 138
Ideographs, 373
  for fillet welds, 374
  for resistance welds, 376
IEEE. See Institute of Electrical
    and Electronic Engineers
Ignore hatches, 684
Illustrations
  in technical reports, 81
IMEXO, 700
Implementation, 19, 84
  and assembly drawings, 86
  of design process, 21
  for exercise bench, 87–89
  with hanger bracket problem,
    25–26

miscellaneous considerations, 86–87
of patent drawings, 91–93
and patents, 90–91
and patent searches, 93
specifications, 85–86
and working drawings, 84–85
Inch(es)
dimensions in millimeters converted to, 310
working drawings dimensioned in, 381
Inclined planes
in isometric drawings, 451–452
Inclined prism
intersection between vertical prism and, 546
intersection with vertical cylinder and, 549
India ink
for patent drawings, 92
for presentation lettering, 76
Individual approach
with design process, 37
Inductance values, 635
Industrial designers, 5
Industrial electronics, 9
Industrial engineering, 10–11
Industrial engineers, 10, 11
Informal presentations, 74
Information accumulation
and creativity, 36
Information center
systems design problem, 105
Ingots, 293
Ink
for patent drawings, 92
Ink drawings, 434
Ink jet printers, 432, 647
Ink jet printing
for reproduction of drawings, 432
Input devices, 645, 646
Input impedance values, 636
Inquiry commands (Tools), 694–695
INSERT command, 261, 389
Insert option, 682

Inside diameter (ID)
for pipe, 607
Inside pattern
in development, 556
Installation/circuit diagram, 639
Installation drawings, 638–639
Instant coffee spoon, 36
Institute of Electrical and Electronic Engineers (IEEE), 10
Instrumentation, 6
and chemical engineering, 8
and electrical engineering, 10
Instrument drawings, 51
for design problems, 99
triangles for, 185, 186
Instruments
orthographic drawing with, 178–199
Instrument set
for drawing, 116
Insurance
warehousing, 64
Integral calculus, 597, 598
Integral curve, 602
Integrated circuit, 637
INTERFERE command, 732
Interference
and patent applications, 95
Interference fit (LN), 341, 343
English units, 339
metric units, 342
Interior angles, 137
Interlocking dies, 297
Intermittent welds, 375
Internal micrometer, 318
calipers, 302
Internal threads, 247, 250, 252, 253, 266, 267
International Standards Organization (ISO), 122
threads, 246
thread table, 251
tolerancing system of, 341
International tolerance (IT)
grade, 347
metric units, 342
Interrupted paths, 626

Intersecting lines, 184, 467
Intersections
between cones and prisms, 552
between cylinders, 549–550
between cylinders and prisms, 549
of lines and planes, 544–545
between planes and cones, 551–552
between planes and cylinders, 547–548
between prisms, 546–547
between prisms and pyramids, 552–554
Inventors
joint, 95
patents applied for by, 90–91
sole, 95
Inverse linear option, 745
Inverse square option, 745
Investment casting, 294, 296
Investors, 74
Involute tooth, 277
Iron, 291, 298
Irregular curves, 119, 221, 442
in isometric drawings, 452
in oblique pictorials, 443
plotting of, 192
Irregular shapes
dimensioning, 323–324
ISO. See International Standards Organization
Isocircle option, 710
Isometric axonometric projection, 456
Isometric drawings, 446–452, 655
by computer, 455–456
dimensioning, 454, 616–618
of piping systems, 615
Isometric ellipse templates, 449–450, 451, 616
ISOMETRIC grid, 455
Isometric lines, 446
Isometric pictorials, 438, 439, 445–446, 709–710
sketching, 169–170
Isometric planes, 446
Isometric projection, 446

vs drawing, 445
ISO Metric Screw Threads, 253
ISOPLANE command, 456
Isopleth, 584, 590, 591, 592
Isosceles, 137
Issue fee
    for patents, 91, 94
Itemizing, 62–63
    and design analysis, 62–63

J

J-groove welds, 372, 373, 375
Jib crane brackets
    short design problem, 102
Jobs
    steps in planning of, 33
Join option, 681
Joint-by-joint analysis, 534, 536
Joysticks, 646

K

Kersten, Leendert, 216
Keyways, 330
Keyboard, 645
Keys, 246, 269
Keyseats, 330
Kilby, Jack, 637
Knife edge cam follower, 285
Knurling, 330–331
    diametral pitches for, 331

L

Labels and labeling
    of axes in graphs, 573
    in bar graphs, 570
    for datum targets, 354
    of electron tubes, 635
    of graphic symbols, 628
    of metric drawings, 311
    for orthographic views, 188
    for parts on working drawings,
        391
    for pins in electron tubes, 632
    of points, lines, and planes, 464
    for reference lines, 209

for sectional views, 227
in six-view drawings, 182
spacing of, 166
of switches, 630
in three-view drawings, 182
on working drawings, 384
Ladder attachment example
    problem identification with,
        28–29
Lamp bracket
    short design problem, 100
Lap joint, 372
Lapping, 324
    and surface finishing, 304
Lap-top computers
    slide shows shown on, 78
Laserjet printers, 437
Laser printers, 180, 432, 647
Laser printing
    for reproduction of drawings,
        432
Last option, 674
Lathes, 256, 300–301
Lay
    of surface texture, 361
    symbols, 363
LAYER command, 658
Layer Control box, 656
Layer & Linetype Properties box,
    657, 658
Layers
    drawing, 656–658
Layout
    rules for orthographic draw-
        ings, 187–188
Lead, 247
LEADER command, 699
Leaders, 250, 309
    in assembly drawings, 454
    and dimensioning, 321
    parts numbered by, 393
    team, 99
Lead (L)
    for worm, 283
Leaf cart
    short design problem, 102
Least material condition (LMC),
    350

Left-hand springs, 273
Left-hand threads, 247, 250
Legal documents
    working drawings as, 380–381
Length option, 680
Lengths of engagement, 252, 253
Lettering
    in auxiliary views, 213
    computer, 133–135
    Gothic, 130–133
    guidelines for, 129–130
    for patent drawings, 92
    for points, lines, and planes,
        464–465
    for presentations, 76
    tools for, 128
    for welding drawings, 373
    for working drawings, 388
Letter of transmittal
    for technical report, 80
Leveling devices, 53, 256
Light fall-off, 744–745
Lighting
    adding to 3D objects, 741
    and working environment, 60
Light pens, 645, 646
Lights
    modifying, 743–744
    moving, 744
    for renderings, 742–745
Lights dialogue box, 745
Limitations
    and problem identification, 29
Limit forms
    tolerances in, 336–337
LIMITS command, 654
Limits of tolerance
    English units, 339
Linear coordinate graphs, 568,
    571–576
    break-even graphs, 576
    broken-line graphs, 572
    calibration and labeling, 573
    composite graphs, 576
    computer method, 572–573
    optimization graphs, 575
    smooth-line graphs, 574
    straight-line graphs, 574

two-scale coordinate graphs, 575

Linear dimensions
metric units, 347–348

Linear (Dimension toolbar), 697–698

Linear equations, 594, 595

Linear pitch (P)
for worm, 283

Linear scales, 571

LINE command, 189, 199, 649, 715, 725, 726

Line fit, 339
English units, 340

Line methods
angle between line and plane, 513
shortest distance between skewed lines, 509
shortest distance from point to line, 508

Lines, 49, 465–467. *See also* Alphabet of lines
angular distance to, 511–512
arc tangent to two, 145
bearings and azimuths of, 481–483
computer, 180–181
crossing, 467–468
dimension, 308
division of, 141
erasing of, 119
intersection of planes and, 544–545
isometric, 446
locating point on, 467
in orthographic drawings, 185
parallelism of, 470
perpendicularity of, 472–473, 489
and planes, 468, 471
point view of, 504–505
precedence of, 167, 185
revolving of, 724
in sketches, 161
sloping, 480–482
at specified angles, 526
techniques for, 184–185
true-length, in primary auxiliary view, 478–479
types of, 118–119

visibility of, 467–468
widths on orthographic views, 180

Line tangent
to arc, 143

Linetypes
with AutoCAD, 657

Liquid-tight joints
and brazing, 378

LISP commands, 215–216

LISP program (ACAD), 476, 477

List option, 656

Locational fits (LC, LT, LN), 340–341

Location dimensions, 324–326

Location tolerancing, 348, 354–357

Lock washers, 267

Lofting, 485

Logarithmic coordinate graphs, 568

Logarithmic grids, 595, 596
empirical data plotted on, 594

Logarithmic scales, 589–590

Log splitter
product design problem, 110

Long (L) length of engagement, 252

Long-radius ells
radii of, 613

Lowercase letters
for external threads, 252
for fundamental deviations for shafts, 343
inclined Gothic, 132
vertical Gothic, 131

Lower deviation
metric units, 341

LTSCALE, 180

Ltscale, 657

Ltype gen (L) option, 682

Lubrication fittings, 268

M

Machined holes
and dimensioning, 328–329

Machinery's Handbook, 279

Machining

operations, 299–304
and surface contouring, 377

Machinists, 5

Machine screws, 259, 263–264

Magazines
for background information, 42

Magnesium, 293–294

Magnitude, 530

Mailbox
product design problem, 106

Mail questionnaire, 43
for opinion surveys, 44

Maintenance publications
pictorials in, 438

Major diameters, 263
of ellipse, 149
of thread, 247, 252

Malleability
iron, 291
of metal, 294

Managers, 74

Manned flying system (MFS), 537

MANNEQUIN, 461

Manufacturers
literature from, 53

Manufacturers' representatives, 43

Manufacturers Standardization Society (MSS), 606

Manufacturing
and mechanical engineering, 11
processes, 290
tolerances and costs of, 336

Map holder
product design problem, 108

Maps, 568. *See also* Graphs

Marketability
and accessories, 89
and design analysis, 57
and needs assessment, 17
projections, 25

Marketing
and design analysis, 61
and implementation, 87
surveys, 99

Mass option, 738

MASSPROP command

options of, 737
Mass properties, 737–738
Materials
adding to 3D objects, 741
Materials and processes, 290
commonly used metals, 291–294
forming metal shapes, 294–299
machining operations, 299–305
plastics and other materials, 305–306
properties of metals, 294
Materials dialogue box, 747
Materials Library dialogue box, 747
Mating parts, 338, 380, 455
MatlibB option, 738
Maximum material condition (MMC), 349–350
Maximum waviness height values, 363
Maxwell diagrams, 536–537
MEASURE command, 691
Measure (Draw menu), 691
Measurement
architects' scales, 119–121
dimensioning and units of, 309
engineers' scales, 121–122
English system of units, 122
metric scales, 122–123
metric symbols, 123–124
metric system of units, 122
metric units of, 199
scales, 119
Mechanical drawings, 178
Mechanical engineering, 11–12
Mechanical handling shuttle, 599
Mechanical linkages, 627
Mechanical power
and agricultural engineering, 7
Media
drawing, 112–114
Medicine, 12
Medium tolerance, 247
Medium tolerance class, 252
Menu bar, 652, 712
with AutoCAD, 652
Merchandise catalogs
standard parts specifications in, 53

Meshes
application of, 723
Metal
forming shapes from, 294–299
Metallurgical engineering. See Mining and metallurgical engineering
Metallurgy, 290
Metals
for arc welding, 371
commonly used, 291–294
ferrous, 291
properties of, 294
Meter, 122
Method
in technical report, 80
Metric drawings
dimensions on, 122
Metric/English conversions
and dimensioning, 310
Metric scales, 122–123
Metric system
for working drawings, 385–387
Metric thread
notes for, 251–253
standards, 246
Metric units, 531
and dimensioning, 310–311
expression of, 124
general tolerances, 347–348
of measurement, 199
system of, 122
tolerancing, 341–345
Microfilm enlarger-printer, 435
Microfilming
for reproduction of drawings, 433, 435
MIDPOINT mode, 141
Military electronics
and electrical engineering, 10
Military Standards, 349
Millimeter (mm), 122, 385, 390
dimensions in inches converted to, 310
fractions, 310
knurling calculations converted to, 331
in metric (SI) system, 309
Milling machine, 300, 304

Milliradian
finding, 348
Minimum length of worm (MLW), 283
Mining and metallurgical engineering, 12
Mining engineering
and strike and dip, 492
Mining engineers, 492
Minor diameter, 258
of ellipse, 149
of thread, 247, 252
Minutes, 136
Mirror (Modify toolbar), 687–688
Mirror option, 680
Mock-ups, 64
and design analysis, 64
Model airplane field
systems design problem, 104
Models, 75
and design analysis, 56, 64–65
for piping systems, 619
for presentations, 77–78
types of, 64
Model scale
and design analysis, 65
Model space (MS), 714
Moderator
of brainstorming session, 39
Modification of existing facility
systems design problem, 106
Modify Standard Materials box
settings within, 747
Moment of inertia option, 738
Monge, Gaspard, 2, 3
Monitor, 645
Monitor support
product design problem, 108
Monotxt fonts, 686
Motel, instant
systems design problem, 105
Motion analysis, 60, 62
Motor bracket
short design problem, 100
Motorized wheelchair, 306
Mountain lodge
systems design problem, 105
Mouse, 646, 647
double clicking, 652

Movable contact, 631
MOVE command, 388, 673, 676, 744
Move (Modify toolbar), 676
Move option, 679
MSS. *See* Manufacturers Standardization Society
Multiline (Mtext: Draw toolbar), 686–687
Multiline Text Editor dialogue box, 687
Multimedia
  in presentations, 78
MULTIPLE option, 254, 257, 675
Multiple-spindle bar machine, 276
Multiple threads, 249
MVIEW command, 189, 740
  options of, 739
MVSETUP, 191
Mylar, 114

N

Naming
  file with AutoCAD, 661–662
Napoleon, 3
NASA
  design analyses by, 58
Natural logs, 596
NEAREST, 321
Necks, 331
Negative slope, 597
New Distant Light box, 743
New Point Light dialogue box, 743
Newspapers
  advertising in, 61
  standard parts specifications in, 53
Next option, 682
N nomographs, 590–592
Nodular iron. *See* Ductile iron
Nominal size
  English units, 339
Nomograph, 584
Nomography, 2, 584
  alignment graphs: three variables, 588

alignment graph scales, 585–586
concurrent scales, 586–587
N or Z nomographs, 590–592
parallel scale graphs: linear scales, 588–590
Nonagon, 137
Nonassociative dimensions, 696
Noncoplanar vectors, 532
None option, 745
Nonintersecting lines, 467
Nonisometric planes
  ellipses on, 452
Nonperpendicular cylinders
  intersection between, 550
Nonprincipal plane, 208
Nonstandard fits
  nonpreferred sizes for, 345
Normal hatches, 684
Normalizing
  and heat treating, 294
Normal (N) length of engagement, 252
Note form
  in Activities Network, 33
Notes, 178
  ANSI abbreviations in pipe thread, 268
  for designating nuts, 261
  English thread, 249–250
  lettering, 128
  metric thread, 251–253
  miscellaneous dimensioning, 332
  patent rights, 390
  and preliminary ideas, 39
  tolerance, 346–347
  on working drawings, 84, 384, 389–391
Not to scale (NTS), 390
Nuclear energy
  and mechanical energy, 12
Nuclear engineering, 12
Nuclear power reactors
  and nuclear engineering, 12
Nuclear science, 8, 12
Number chart. *See* Nomograph
Numerals

in AutoCAD, 685
and dimensioning, 311
Gothic vertical, 131
inclined Gothic, 132
Numerically controlled manufacturing systems, 437
Nut, 259
Nutcracker
  product design problem, 110
Nuts, 231
  thickness of, 262

O

Object lines
  labeling for, 465
Objects
  material of, 747
  selecting with AutoCAD, 674–675
  setting views of, 718–719
  with rounded ends, 326–327
Oblique cylinders, 139
  development of, 559–560
Oblique drawings, 439–444
  advantages of, 441
  constructing, 440
  dimensioned, 444
  types of, 439–440
Oblique lines, 465
  finding true length of, 478–479
  line perpendicular to, 472
  slope of, 481
Oblique option, 705
Oblique pictorials, 438–439, 709
  sketching, 168–169
Oblique planes, 468, 470
  and double revolution, 521
  edge views of, 220
Oblique prisms, 138
  development of, 557–558
  intersection between foreshortened plane and, 545
Oblique projection
  vs oblique drawing, 444–445
  theory, 444–445
Oblique pyramid, 138
Observations, personal, 19

Obtuse angles, 145
Octagons
    constructing, 139
Octahedron, 138
OFFSET command, 691
Offset follower
    cam with, 287
Offset (Modify toolbar), 691
Ogee curve, 149
OH. *See* Open hearth
Omissions
    in student reports, 81
OmniShare conferencer, 437
One-point perspectives
    constructing, 457–459
One-view drawings, 191
Open hearth (OH) process, 618
Opinion surveys, 19
    and preliminary ideas, 43–44
Optimization, 575
Oral presentations, 72
    graphs for, 580
Ordinate (or y axis), 571
Ore-vein applications
    with descriptive geometry,
        492–494
Ore-vein outcrop, 493–494
Organization chart, 578
Organization of effort, 32
Orientation setting, 665
Orientation tolerancing, 348,
    358–360
Orthographic assembly drawings,
    385, 392
Orthographic drawings
    and alphabet of lines, 180–181
    arrangement of views in, 183
    conventional intersections,
        194–195
    conventional practices with,
        193–194
    curve plotting, 192–193
    fillets and rounds in, 195–199
    first-angle projection, 199
    with instruments, 178–199
    layout rules for, 187–188
    lines and planes in, 185
    line techniques in, 184–185

one-view drawing, 191
orthographic projection,
    178–180
partial views, 192
point numbering in, 185
readability of, 184
selection of views in, 183
simplified and removed views,
    191–192
six-view drawings, 181–182
three-view drawings, 182
three-view drawing layout,
    187–188
triangles for, 185–186
two-view drawings, 191
views by computer, 189–191
views by subtraction, 186–187
Orthographic projections, 161,
    178–180, 467, 476
    and auxiliary sections, 240
    axonometric projection, 456
    and conventional intersection,
        194
    fittings in, 614
    instrument-drawn, 23
    maximum number of principal
        views in, 163
    of piping systems, 615, 616
    plane represented in, 468
    principal views of, 165
    projection planes of, 180
    and readability, 183
    and vessel detailing, 618–619
    and working drawings, 380
Orthographic scale drawings, 51
Orthographic symbols
    for valves, 614
Orthographic views, 49, 226, 229,
    384, 454. *See* also Sections
    in assembly drawings, 86
    fillets and rounds on, 197
    fittings in, 611, 614–615
    point numbering in, 185
    readability of, 193
    and revolution, 518
    runouts on, 198
    with third-angle projection,
        387

of 3D object, 718
    3D solids converted into, 735
    in working drawings, 84
Ortho setting, 654
OSNAP command, 321, 688
    CENTER option of, 145, 479
    MIDPOINT mode of, 141
Osnap (Object snap toolbar), 688
OSNAP Tangent option, 144
Outer hatches, 684
Outline dimensioning, 327
Output devices, 645, 646
Output impedance values, 636
Outside circle, 279
Outside diameter (OD)
    of bevel gear, 281
    of gear, 277
    for pipes, 606, 607
    of worm, 283
Outside diameter of spider
    (ODS), 284
Outside pattern
    in development, 556
Overhead expenses, 81
Overhead projector transparen-
    cies
    for presentations, 77
    proportions of, 568
Overhead visuals, 75
Oxyacetylene welding method,
    371

P

Packaging
    and design analysis, 62
    for exercise bench, 88
    and implementation, 84, 86
Paint can holder
    short design problem, 101
Paint mixer
    product design problem, 109
PAN command, 673
Panel holder
    product design problem, 110
Pan option, 721
Pan (Standard toolbar), 673
Paper, 113–114

for patent drawings, 92
for presentations, 75
Paper dispenser
    short design problem, 103
Paper Size setting, 665
Paper space (PS), 189, 714
    and model space, 712–714
    and model space: TILE-
        MODE=0, 738–740
    vs model space, 714
Parabolas, 149, 150–151
PARALLEL command, 476, 477,
    479
Parallelepiped, 138
Parallelism, 470–471
    and orientation tolerancing,
        358–359
Parallel lines, 186
Parallelogram
    ellipse constructed inside of,
        150
Parallelogram method
    resultant, 531, 532
Parallel planes
    and ore-vein outcrop, 493
Parallel projectors, 178, 179
    and axonometric projection,
        456
    in isometric projection, 445
    and oblique projectors, 438
Parallel scale graphs
    linear scales, 588–589
Partial views, 192, 219, 232–233
Part modification
    short design problem, 100
Parts list, 388, 390
    on working drawings, 390
Parts manuals
    pictorials in, 438
Patentable ideas
    and design worksheets, 30
Patent agents, 93
"Patent applied for," 94
Patent attorneys, 93
Patent drawings, 91–93
    standards for, 92–93
Patent Office Search Room, 95, 96
"Patent pending," 94

Patent rights, 91
Patent rights note
    on working drawing, 390
Patents
    applying for, 90–91, 95
    for background information,
        42
    consideration for, 90
    expiration of, 93
    fees for, 91
    nature and duration of, 93–95
    petition and oath for, 91
    protection in foreign countries,
        97
    searches for, 93, 96
    technical knowledge available
        from, 96–97
Patio grill
    product design problem, 111
Patio table production
    systems design problem, 106
PEDIT command, 681, 683
Pedit (Draw toolbar), 681–683
Peening
    and surface contouring, 377
Pen Assignments box, 664
Pencils
    cast-iron symbols drawn by,
        228
    dimensioning lines with,
        308–309
    for drawing, 112–113
    for lettering, 128
    for sketching, 162
Pen plotter, 181
Pen plotting
    for reproduction of drawings,
        432
Pentagons
    constructing, 140
Pen Width, 684
Percentage graphs, 578
Percent grade, 481
Perimeter option, 737
Periodicals
    advertising in, 61
Permanent-mold casting, 294,
        295

Perpendicular centerlines, 167
Perpendicularity, 472–473
    and orientation tolerancing, 359
Perpendicular lines, 186
Perpend option, 145
PERPLINE command, 476, 477
Personal interview
    for opinion surveys, 43–44
Personal observations, 19
Perspective pictorials, 457
Perspective(s), 438
    arcs in, 459
    types of, 457
    views, 93
PERT. See Project evaluation and
    review technique
Petition and oath
    for patents, 91
Petroleum engineering, 13
Petroleum Refinery Piping
    Standards (ANSI B31.3), 618
Phantom (ghost) sections, 240
Phong option, 741
Photographic slides, 75
    for presentations, 76
Photographing
    layouts for presentations, 76
Photographs
    electronic transmission of, 437
Photostating
    for reproduction of drawings,
        433, 435
Physical metallurgy, 12
Physical properties
    and refinement, 48–49
Physical specifications
    and design analysis, 56, 61
Pick Points option, 685
Pickup truck hoist
    product design problem, 111
Pictorial assembly drawings, 392,
        394
Pictorials
    and axonometric projection,
        456
    for design problems, 99
    three-dimensional, 712
    types of, 438

Picture plane (PP), 457
Pie graphs, 568, 569
   by computer, 579
Piercing points, 488, 513, 544,
     545, 546, 547
Pinion, 276, 280
   finding dimensions of, 281–282
   pitch angle of, 280
   revolutions per minute (RPM)
     of, 278
Pins, 231, 246, 267, 268
Pipe
   commonly used, 606
   grade and weight standards for,
     606
Pipe aligner for welding
   short design problem, 100
Pipe clamp
   short design problem, 102
Pipe column support
   short design problem, 100
Pipe drafting
   cast-iron pipe, 607–608
   company standards for, 620
   computer drawings, 619–620
   copper, brass, and bronze pip-
     ing, 608
   dimensioned isometrics,
     616–618
   fittings in orthographic views,
     614–615
   flanged fittings, 612–613
   miscellaneous piping materi-
     als, 608
   pipe fittings, 609–611
   pipe joints, 608–609
   piping systems in pictorial, 615
   screwed fittings, 611–612
   valves, 613–614
   vessel detailing, 618–619
   welded and seamless steel
     pipe, 606–607
   welded fittings, 613
Pipe drawings
   by computer, 619–620
Pipe fittings, 609–611
   grade and weight standards for,
     606

Pipe joints, 608
Pipeline installation, 485
Pipe roll stand
   short design problem, 101
Pipe thread notes
   ANSI abbreviations for, 268
Pipe threads, 608
   and fittings, 268–269
Piping, 608
Pitch, 247, 254
Pitch angle, 280
Pitch circle (PC), 279
   of gear, 277
Pitch cone distance (PCD)
   of bevel gear, 280
Pitch diameter (PD), 247
   of bevel gear, 280
   of gear, 277
Pitch diameter of spider (PDS),
     283
Pitch diameter of worm (PDW),
     283
Pitch diameter tolerance, 252
Pixels, 646
PLACES (or PL) notes, 323
Plain ends
   of springs, 271
Plain washers, 267
Planar tolerance zone, 358
PLAN command, 714, 715
Plane
   angle between line and,
     512–513
   edge view of, 506, 512, 520, 521
   intersection between cylinder
     and, 548
   intersection between prism
     and, 545
   and line, 471
   line perpendicular to, 473
   nonprincipal, 208
   plane perpendicular to, 473
   point on, 469–470
   true size of, 506–507, 520–522
Plane geometry, 136
Plane method
   angle between line and plane,
     512–513

for shortest connector between
   lines, 511
   shortest distance between
     skewed lines, 509–510
Planer, 300, 304
Planes, 468–470
   angle between, 522–523
   edge views of, 486–488
   frontal, 178, 180
   intersections between,
     489–490, 495–496
   intersections between cones
     and, 551–552
   intersections between cylin-
     ders and, 547–548
   intersections of lines and,
     544–544
   isometric, 446
   line perpendicular to, 473
   and lines, 468
   lines of intersection between
     two, 544
   in orthographic drawings, 185
   of orthographic projection, 180
   parallelism of, 471
   picture, 457
   primary auxiliary, 209
   sloping, 490–492
Planning design
   activities for, 33–34
Plan of action
   for preliminary ideas, 38
Plan-profiles, 485–486
Plan views, 93
Plastic pipe, 606
Plastics, 290
   and other materials, 305–306
   for pipes, 608
Plate cams
   designating, 286–287
Playground, injury-proof
   systems design problem, 106
PLINE command, 681, 725
Pline options, 680
Plot
   making, 714
Plot Configuration menu, 664
Plot Origin setting, 666

Plot plans, 483–484
Plot Preview setting, 666
Plot Rotation setting, 666
Plotted ellipse, 448
Plotters, 112, 645, 646
  computer graphics, 180
  readying, 667
Plotting
  drawing with AutoCAD, 650
  parameters, 663–667
Plot to File setting, 665
Plug tap, 267
Plug welds, 373
Plus-and-minus tolerances, 337,
  339, 390
Point lights, 742
  and fall-off, 744
Point numbering
  in orthographic drawings, 185
Point of application, 530
Point plotting, 448
Points, 49
  arc tangent to line from, 144
  about axis, 524–526
  arc through three, 142
  labeling for, 464
  locating on line, 467
  on plane, 469–470
  point of tangency between line
    and, 143
  projection of, 465
Points on lines
  labeling for, 464
Points option, 721
Polar coordinates, 668
Polishing
  and surface finishing, 304
Polo-vault uprights
  product design problem, 107
Polyester film, 114
POLYGON
  data point produced as, 572
Polygon (Draw toolbar), 670
Polygon method, 531
  resultant, 532–533
POLYGON options, 140

Polygons, 136–137
  constructing, 139–140
  defined, 136
Polyhedra, 137
Polyline (Draw toolbar), 680–681
POLYLINE (PL) command, 680
Polylines
  revolving of, 724
Polymers, 305
Polyvinyl chloride (PVC)
  for pipes, 608
Portable bleachers
  systems design problem, 104
Poser (Fractal Design), 461
Position
  and location tolerancing, 354,
    355–356
Power
  and electrical engineering, 9
Power equations, 594, 595–596
Power generation
  and mechanical engineering,
    11
Power lawn-fertilizer attachment
  product design problem, 108
Power Point (Microsoft), 78
Practicality
  and refinement, 48
Preferred roughness width cutoff
  values, 362
Preliminary ideas
  background information for,
    42–44
  brainstorming, 38–39
  within design process, 19–20,
    36–37
  for exercise bench, 44–45
  with hanger bracket problem,
    21–23
  individual versus team meth-
    ods, 37–38
  plan of action, 38–39
  quickie design, 41–42
  refinement of, 48
  sketching and notes, 39–41
Presentations

  of drawings, 124–125
  making, 78–79
  oral, 72, 74, 77
  organizing, 74–75
  types of, 74
Press fits
  knurls for, 331
Press forges, 296
Pressure angle (PA)
  of gear, 277
Previous option, 675, 682
Pricing
  for exercise bench, 89
  methods of, 62
Primary auxiliary
  finding line's true length with,
    513
Primary auxiliary plane, 209
Primary auxiliary views, 208–209,
    504, 507, 508, 510, 560
  application for auxiliaries from
    top, 211–212
  auxiliary sections, 219–220
  construction rules for, 212–213
  curved shapes, 218–219
  elliptical views, 221
  finding line's true length by, 505
  folding-line principles, 109
  from front view, 213–216
  of intersections between cones
    and prisms, 552
  partial views, 219
  from profile view, 216–218
  secondary auxiliary views,
    220–221
  from top view, 209–211
Primary auxiliary views in
    descriptive geometry
  angles between lines and prin-
    cipal planes, 480–481
  application: plot plans,
    483–484
  bearings and azimuths of lines,
    481–483
  contour maps and profiles,
    484–485

edge views of planes, 486–488
geometry by computer, 476–478
intersections between planes, 495–496
ore-vein applications, 492–494
planes and lines, 488–490
plan-profiles, 485–486
sloping planes, 490–492
true-length lines, 478–480
Primary datum, 351
Primary datum plane, 354
Primary-diameter datums, 352
Primary internal/external parallel datums, 352–353
Principal lines, 466
line perpendicular to, 472
Principal planes, 468
angles between lines and, 480
Printed circuits, 637–638
Printers, 645, 647
and reproduction of drawing, 432
vs. plotters, 180–181
Prisms, 138
dimensioning, 314
intersections between, 546–547
intersections between cones and, 552
intersection between planes and, 545
intersections between pyramids and, 552–554
revolving about axis, 525–526
Problem identification, 19–21, 28
and design worksheets, 30
exercise bench example, 30–32
with hanger bracket problem, 21
ladder attachment example, 28–29
and organization of effort, 32–33
and planning design activities, 33–34
process of, 29–30
in technical report, 80

Problem refinement
with hanger bracket problem, 23
Problem statement, 29
Process control
and chemical engineering, 8
Product
sales of, 84
systems design problem for simple, 106
Product costs/pricing
and problem identification, 31
Product design, 17
problems, 106–111
Production costs, 17
Productivity
and computer graphics, 14
Product market
and design analysis, 56, 60, 61
Product of inertia option, 738
Product packaging, 81
and implementation, 86
Products
accessories for, 89
design analysis of, 56–60
market analysis of, 60–61, 62–64
marketing of, 87
prices of, 89
Profile lines
principal, 466, 467
Profile orthographic planes, 209
Profile planes, 465, 466
of orthographic projection, 180
principal, 468, 469
Profile reference plane (PRP), 218
Profiles, 484
Profile tolerancing, 348, 358
Profile view
auxiliaries from, 216–218
true-length lines, 519
Profit predictions, 17, 18
Progress reports, 79, 80
Project engineers
and working drawings, 391
Project evaluation and review technique (PERT), 32
Projection

first-angle, 199
maximum number of principal views in orthographic, 163
orthographic, 161, 178–180
principal views of ortho-graphic, 165
Projection lines
labeling for, 465
Projection planes, 465
Projector cabinet
product design problem, 108, 110
Property changes
of CHANGE command, 677–678
Proportional-line method
to draw concurrent scales, 587
Proposals, 79, 80
elements of, 80
Propulsion systems, 6
Prototype file
making, 662
Prototypes, 64
and design analysis, 64
Protractor, 115
PRP. See Profile reference plane
PSFACE command, 740
Pulley
dimensioned, 323
Pulley arm, 445
Pulley bracket clamp
short design problem, 100
Punching bag platform
product design problem, 111
PURGE command, 656
PYRAMID command, 722
Pyramids, 138
development of, 561–562
dimensioning of, 314, 320–321
intersections between prisms and, 552–554
Pythagorean theorem, 479

Q
Qtext, 685
Quadrilaterals, 137

Quenching
  and heat treating, 294
Questions and Answers About
    Patents (PTO), 93
Quickie design, 41–42
Quick Text setting, 654
Quitting
  AutoCAD, 650–651

R

Radial cutting planes, 551
Radial method
  for sketching circles, 168
Radiation
  and nuclear engineering, 12
Radii
  dimensioning, 322
Radio
  advertising on, 61
RADIUS, 322
Radius
  of circle, 137
Radius (Dimension toolbar), 699
Radius of gyration option, 738
Radius of spider throat (RST), 283
Railroad
  building of, 3
Ranges
  and design analysis, 61–62
Ratings
  in single-line diagrams, 625
Ratio
  and tapers, 332
Ratio graphs (or semilogarithmic
    graphs), 576
Ratios
  gear, 278
  and metric scales, 123
Rays
  in Draw toolbar, 668
Readability
  of circuits, 636
  of graphs, 573
  and orthographic projection,
    183
  of orthographic drawings, 184
  of orthographic views, 193

of simplified views, 192
of working drawings, 391
Reaming
  by drill press, 302
  on lathe, 300, 301
  of machined holes, 328, 329
Receding axes, 445
  of cabinet oblique, 440
  of cavalier oblique, 440
  of general oblique, 440
Recommendation summaries, 81
Recorder
  of brainstorming sessions, 39
Records
  of design activities, 30
Recreational facility
  systems design problem, 106
Rectangle
  ellipse constructed inside of,
    150
Rectangular arrays, 689
Rectangular (linear) grids
  empirical data plotted on, 594
Rectangular prisms
  development of, 556–557
REDO command, 677
Reference characters
  for patent drawings, 93
Reference designations, 634–635,
    636
Reference dimension, 327
Reference lines, 185, 213, 216
  labels and labeling for, 209, 464
Reference option, 679, 690
Reference-plane method
  for auxiliaries from front view,
    216
  for auxiliaries from profile
    view, 218
  for locating auxiliary view,
    211–212
Reference planes, 220
Refinement, 19
  considerations for, 51
  and descriptive geometry,
    49–51
  of exercise bench, 51–52
  and physical properties, 48–49

of preliminary ideas, 20
and standard parts, 53
Refinery installation, 510
Regardless of feature size (RFS),
    350
  datum features at, 352
  and tolerances of form, 359
Region
  extruding, 728
REGION command, 727
Region modeling, 727
Regular polygon, 137, 139
Regular polyhedron, 137–138
Relays, 630
Removed sections, 235–237
Removed views, 191–192
Remove option, 674
Rendered solids
  objects depicted as, 460
Rendering, 712
Render process, 741–742
Render toolbar menu, 746
Reports. *See also* Written reports
  for design problems, 99
Reproduction of drawings
  assembling drawing sets, 435
  computer drawing types, 432
  and transmittal, 436–437
  types of, 432–435
Requirements
  and problem identification, 29
Research engineering
  aerospace, 6
Resistance spot welding, 371–372,
    376–377
Resistance values, 635
Resistance welding, 371–372
Resolution, 646
Resultant, 531, 540
Resultants
  parallelogram method, 532
  polygon method, 532, 533
Retail outlets
  and design analysis, 61
Revolution, 561, 562, 563
  alternate points of, 519–520
  angle between planes, 522–523
  determining direction, 523–524

line at specified angles, 526
point about an axis, 524–526
principle of, 193
true-length lines: front view,
518–520
true size of plane, 520–522
REVOLVE command, 731
Revolved sections, 234–235
Revolving clamp assembly, 382
REVSURF command, 724
RFS. *See* Regardless of feature size
Ribs, 231–232
hatching, 231
revolution of, 237, 238
Right cone, 139
Right cylinder
altitude of, 139
Right-hand springs, 273
Right-hand thread, 247
Right pyramid, 138
development of, 561
Right triangles, 137
Rivets, 231, 246, 269–270
Robotics, 9
Rocket fuel development, 8
Roller bearings, 231
Roller cam follower, 285–286
Roll film, 435
Rolling, 298–299
and surface contouring, 377
Romand font, 707
ROMANS font, 133, 134, 686
Root, 247
Root circle, 279
Root diameter (RD)
of gear, 277
Root lines, 254, 258
Rotary switch, 6, 633
Rotary switches, 634
Rotary terminals, 631
ROTATE command, 690
Rotate (Modify toolbar), 690
Rotate option, 655, 679, 705
Rotation
of three-dimensional pictorial
drawings, 712
Roughness
of surface texture, 361

Roughness height
of surface texture, 361
Roughness height values, 362
Roughness width
of surface texture, 361
Roughness width cutoff
of surface texture, 361
Rounded corners
drawing, 451
Rounds, 168, 195–197
and castings, 295
and die casting, 296
and dimensioning, 322–323
and forging, 297
in isometric drawing, 454
RULESURF command, 723
Running coordinates, 655
Running (or sliding) clearance fits
(RC), 340
Runouts
on computer, 198–199
Runout tolerancing, 348, 360–361

S

SAE. *See* Society of Automotive
Engineers
Safety, 57
Sales
and product design, 17
Sales brochures
standard parts specifications
in, 53
Sales estimates, 81
Sales features
and design analysis, 61
Salespeople, 74
Sand castings, 294–295
Sanitary engineering
and civil engineering, 9
SAVE command, 663
Saving
with AutoCAD, 663
dimension styles, 704
files with AutoCAD, 650
views and light settings, 746
Sawhorse
product design problem, 108

SCALE command, 689
Scale drawings, 20
Scale equation, 585–586
Scale (Modify toolbar), 689–690
Scale modulus, 585, 589
for concurrent scales, 587
Scalene, 137
Scale option, 679, 680
Scale(s)
alignment graph, 585–586
architects', 119–121
concurrent, 586–587
conversion of, 124
drawing to, 662–663
engineers', 121–122
functional, 585
indicating, 390
lines and planes measured
with, 185
logarithmic, 589–590
metric, 122–123
for patent drawings, 93
refinement drawings drawn to,
52
setting, 666
Scanners, 112
Scenes
working with, 746–747
Scenes dialogue box, 746
Schedules
for pipe, 606
Schematic diagrams, 624,
625–626
connecting symbols, 626–627
for design problems, 99
maintenance information on,
636
preparation of, 630
reference designations on,
634–635
Schematics, 32, 578–579
Schematics and diagrams, 568
Schematic symbols, 253, 256–258,
258
Scientists, 4
Screwed fittings, 611–612
Screwed joints, 611, 615
Screwed pipe joints, 608

Screw jacks, 256
Screws, 246
  sheet metal, 249
  types of, 263–265
Seamless steel pipe, 606–607
Seam welds
  application of symbols for,
    376–377
Secondary auxiliary views, 208,
    220–221, 504, 505, 506, 507,
    508, 509, 510, 511, 524, 560
Secondary datum, 351, 353
Secondary datum plane, 354
Seconds, 136
SECTION command, 733
Sectioning symbols, 228–229
Sections
  assemblies of parts, 229
  auxiliary, 240–241
  basics of, 226–227
  broken-out, 234
  and conventional breaks,
    239–240
  and conventional revolutions,
    237–239
  full, 226, 229–232
  half, 233
  offset, 234
  partial views, 232–233
  phantom (ghost), 240
  removed, 235–237
  revolved, 234–235
  symbols of, 228–229
  on technical illustrations, 454
  3D solids converted into, 735
Section views, 93
Select File box, 652
Selective assembly
  English units, 340
Select Objects options, 685
Self-contained pipelayer, 40
Semifinished bolt heads and
    nuts, 259
Semilogarithmic-coordinate
    graphs, 568, 576–578
Semilogarithmic grids
  empirical data plotted on, 594
Semitransparent tracing paper, 434

Series
  thread, 247, 248
Set screws, 231, 264
SETVAR command, 690
Setvar (Command line), 690
Shading
  for patent drawings, 92
Shaft basis system
  metric units, 342
  preferred fits for, 343
Shafts, 231
Shape description, 160–161
Shaper, 300, 304
Shapes
  curved, 218–219
Shaping, 324
Sheet-fed plotter, 646
Sheet metal hopper, 561
Sheet metal screws, 249
Sheets
  standard sizes of, 125
Sheet sizes
  for working drawings, 388
Shelving
  warehousing, 64
Shipping
  and design analysis, 62
  of exercise bench, 88
  and implementation, 87
Shipping costs, 81
Shop bench
  product design problem, 111
Shopping caddy
  product design problem, 111
Shopping checkout system
  systems design problem, 106
Shop rivets, 270
Shortcut symbols, 638
Short-radius ell
  radius of, 613
Short (S) length of engagement,
    252
Shrinkage
  and castings, 296
SI. See Systeme International
    d'Unites (International
    System of Units)
Side-mounted mirror

short design problem, 100
Sight line, 213
Signatures
  for patent drawings, 93
Simplified symbols, 253, 258–261
Simplified views, 191–192
Single-impression die, 296
Single limits
  English units, 340
Single-line diagrams, 624, 625
Single-line drawings
  of springs, 271
Single-line screwed fittings, 614
Single-line symbols, 609, 611,
    612, 613, 614, 615
Single (SI) option, 675
Single threads, 249
SI symbol, 199, 385, 387, 388, 390
  made into Block, 692
Sit-up bench
  short design problem, 103
Six-view drawings, 181–182
Six-view sketching, 163
Size limits
  and geometric tolerances,
    349–350
Sizes and dimensions
  and design analysis, 61
Skateboard facility study
  systems design problem, 104
SKETCH command, 708, 709
Sketches
  with problem identification, 32
  for problem identification
    process, 29
Sketching
  circular features, 166–168
  isometric pictorial, 169–172
  oblique drawings, 443
  oblique pictorial, 168–169
  and preliminary ideas, 39
  shape description, 160–161
  six-view, 163
  techniques for, 162
  three-view, 163–166
Sketch (Miscellaneous toolbar),
    708–709
Skewed lines

shortest distance between (line method), 509

shortest distance between (plane method), 509–510

shortest grade distance between, 510–511

shortest level distance between, 510

SKPOLY, 709

SLICE command, 734

options of, 734

Slide projector elevator

short design problem, 101

Slides

making of, 76

Slide scripts

for presentations, 77

Slope, 595

defined, 480

negative, 597

Slope angle, 481

Slope-intercept equations, 595

Slope ratio, 481

Sloping lines, 480–481

Sloping planes, 490–492

Slots

dimensioning of, 327

Slot welds, 373

SME. See Society of Manufacturing Engineers

Smooth-line curve, 574

Smooth-line graphs, 574

SNAP command

Center option of, 735

STYLE option of, 455, 709

SNAP panel (Drawing Aids), 655

Society of Automotive Engineers, 290

Society of Manufacturing Engineers (SME), 11

Society of Petroleum Engineers (SPE), 13

Software

for pipe drawings, 619

for rendering human figure, 461

Soil and water control

and agricultural engineering, 7

Soil pipe, 608

Soldered joints, 611

Soldering

and welding, 378

Soldering iron, 378

SOLDRAW command, 735

SOLID command, 654, 715

Solid Fill setting, 654

Solid model and modeling, 461, 462, 712

example of, 734–735

introduction to, 727–728

Solid primitives, 729–732

Solids. See also Rendered solids

modifying, 732–733

three-dimensional, on computer, 189, 191

views of, 735–737

SOLIDS (3D solids) command, 727

Solids toolbar, 727

SOLVIEW command, 735

Sound

and design analysis, 60

effects, 462

Space diagram, 531

Space program

analysis of human factors in, 58

Spacing

in dimensioning, 318

for tolerance dimensions, 337

on working drawings, 38

Spatial relationships, 20

SPE. See Society of Petroleum Engineers

Specifications, 178. See also Drawings

design problem, 99

and implementation, 21, 85–86

lettering, 128

for patents, 91

for working drawings, 380

SPHERE command, 722, 730

Spheres, 139

creating, 727

dimensioning, 314, 320–321

Spherical balls, 646

Spheroidized iron. See Ductile iron

Spider, 282

terminology for, 283–284

Spirals, 152

SPLINE command, 683

Spline (Draw toolbar), 683

Spline option, 681

Spokes

and conventional revolution, 238

"Spool drawings," 615, 618

Spotfacing

by drill press, 302

of machined holes, 329

Spotlights, 742, 745

and fall-off, 744–745

Spot welds, 372

ideographs for, 376

Springs, 246, 270–271

drawing, 271

Spur gears, 276–278

drawing, 279

Spurs

defined, 276

revolutions per minute (RPM), 278

Squared ends

of springs, 271

Square groove welds, 372, 373

Square-head bolts, 260

Square heads

drawing, 261

Square nuts, 262

Square threads, 248

drawing, 254

Stadium expansion

systems design problem, 105–106

Stamping, 299

Standard parts

and refinement, 53

on working drawings, 87

Station point (SP), 457

Status option, 694

STD weight pipe, 606, 607

Steel, 291, 292

for forging, 298

Steel pipe, 606
  standard weights for, 606
Storage
  of drawings, 435
  for exercise bench, 88
  and implementation, 86
Straightedge
  for drawing, 118
Straighten option, 681
Straight knurls, 330
Straight-line graphs, 574
Straightness
  and form tolerancing, 357
Strength
  and design analysis, 56, 62
STRETCH command, 317, 690,
  705
Stretch (Modify toolbar), 690
Stretch option, 679
Stretch-out lines, 556, 557, 558,
  559, 560
Strike, 492
Strike and dip method
  for intersections of two planes,
  495
Structural design, 6
Structural engineers, 8
Stud, 259
Student design teams, 98
Stump remover
  product design problem, 109
STYLE command, 134, 135, 314
Stylists, 5
SUBTRACT command, 727, 732,
  733, 735
Subtraction
  views by, 186–187
Successive auxiliary views
  angular distance to line,
  511–512
  dihedral angles, 505–506
  line methods, 508–509, 513
  plane methods, 509–510,
  512–513
  point view of line, 504–505
  shortest distance between
  skewed lines, 510

shortest grade distance
  between skewed lines,
  510–511
true size of plane, 506–507
Surface contouring, 377
Surface finishing, 304–305
Surface modeling, 723–725
Surfaces, 49
Surfaces toolbar, 723
Surface texture
  defined, 361
Surface texture symbol, 197
SURFTAB1, 724
SURFTAB2, 724
Surgical light design, 50–51
Surveyors, 13, 485
Surveyor's units (AutoCAD), 483
Switches, 630
Symbol form
  in Activities Network, 33
Symbols
  ANSI for rivets, 270
  application of welding,
  374–377
  contour, 377
  datum target, 354
  detailed, 253, 254–256, 258
  diametric, 318, 320
  for dimensioning, 311, 313
  double-line, 609, 611, 612, 613,
  614, 615
  for geometric characteristics of
  dimensioned drawings, 349
  graphic, 628–630
  lay, 363
  for patent drawings, 93
  schematic, 253, 256–258
  schematic diagram connecting, 626–627
  sectioning, 228–229
  shortcut, 638
  SI, 385, 387, 388, 390
  simplified, 253, 258–261
  single-line, 609, 611, 612, 613,
  614, 615
  for surface texture, 362, 363
  tolerance, 342

welding, 373
Symmetrical parts
  dimensioning, 323–324
Symmetry
  and location tolerancing, 354,
  356–357
Systeme International d'Unites
  (International System of
  Units-SI), 122, 123, 124
System layout models, 64
  and design analysis, 64
System of forces, 531
Systems design, 17
  example of, 18
  problems, 104–106

T

Table leg design
  short design problem, 101
Table of contents
  of technical reports, 80
Table of dimensions
  and gear drawings, 279
Table of illustrations
  in technical reports, 80
Table of values, 390
Tablets
  and digitizer (stylus), 646
TABSURF command, 723
Tan, Tan, Radius option
  of CIRCLE command, 668
Tan, Tan, Tan option
  of CIRCLE command, 669
Tangency
  locating points of, 147
  marking points of, 143, 148
  points of, 145
Tangent option, 682
Tangents, 143–149
Tap, 301
Tapers, 332, 396
Taper tap, 267
Tapping
  by drill press, 302
  holes, 266–267
Tapping die, 301

Target option, 720
Team approach
    to design problems, 37, 98–99
Technical drawings
    preparation of, 112
Technical illustrations, 2, 13, 438,
    453–455
Technical magazines, 42
Technical reports
    elements of, 80–81
Technicians, 5
    pipe drawings prepared by, 619
    reports prepared by, 79
Technological and design team
    craftspeople, 5
    designers, 5
    engineers, 4
    scientists, 4
    stylists, 5
    technicians, 5
    technologists, 4
Technologists, 4–5
    reports prepared by, 79
    reproduction of drawings by,
    432
Technology, 3
Tee joint, 372
Teeth
    of broach, 303
Telephone interview, 43
    for opinion surveys, 44
Television
    advertising on, 61
Temperature
    warehousing and control of, 64
    and working environment, 60
Tempering
    and heat treating, 294
Template file
    making, 662
Templates
    for drawing, 117
    ellipse, 149, 150, 221
Tension, 530, 531, 534, 536
Terminal boards, 634
Terminals, 630–632
Tertiary datum, 351, 353

Tertiary datum plane, 354
Testing
    and design analysis, 65
Tetrahedron, 138
Text. See also Labels and labeling;
    Lettering
    in AutoCAD, 677, 685
    within hatching, 684
    mirroring, 688
Text option, 677
TEXT STYLE dialogue box, 134,
    135
Text style (Format menu), 686
Thaw option, 658
Thermonuclear engineering, 12
Thermoplastics, 305, 306
THICKNESS command, 714, 715
Thickness of tooth (TT)
    of bevel gear, 280
Thickness option, 678
Third-angle projections (U. S.
    system), 123–124, 387
    vs. first angle projections, 199
Thread angle, 247
Threaded fasteners, 246. See also
    Screws
Threading
    by drill press, 302
    on lathe, 300, 301
Thread notes, 258
    English, 249–250
    metric, 251–253
Thread relief, 331
Threads
    drawing, 253
    terminology for, 246–247
Thread tables, 250
Thread tolerance grades, 252
Three-cycle scale, 577
Three-datum plane concept,
    350–351
3D drawings
    fundamentals of, 714
    plotting, 740
3D face command, 725–726
Three-dimensional mathemati-
    cal models, 579

Three-dimensional orthographic
    views, 476
Three-dimensional pictorials, 48
    in assembly drawings, 86
    defined, 438
    isometric drawings, 446–452
    isometric pictorials, 445–446
    oblique drawings, 439–444
    oblique projection theory, 444
Three-dimensional solids
    on computer, 189, 191
Three-dimensional space
    parallelism of lines in, 470
Three-dimensional spatial geom-
    etry
    points, lines and planes in, 464
Three-dimension modeling,
    460–461
    application of extrusions, 719
    basic 3D shapes (surfaces),
    721–723
    coordinate systems, 716–717
    dimensioning in 3D, 740–741
    dynamic view (Dview),
    719–721
    elementary extrusions,
    714–716
    extrusion example: TILE-
    MODE=0, 728–729
    fundamentals of 3D drawing,
    714
    lights, 742–745
    LINE, PLINE, and 3DPOLY, 725
    mass properties, 737–738
    materials, 747
    modifying solids, 732–733
    paper space and model space:
    overview, 712–714
    paper space and model space:
    TILEMODE=0, 738–740
    paper space versus model
    space, 714
    render process, 741
    section, 733
    setting views of objects,
    718–719
    SLICE command, 734

solid model example, 734–735
solid modeling introduction,
 727–728
solid primitives, 729–732
surface modeling, 723–725
3Dface command, 725–726
views of solid, 735–737
working with scenes, 746–747
XYZ filters, 726
Three-dimension pictorials
axonometric projection, 456
future of, 462
and human figure, 461–462
isometrics by computer,
 455–456
perspective pictorials, 457–460
technical illustration, 453–455
three-dimensional modeling,
 460–461
3DPOLY command, 725
3D shapes (surfaces)
basic, 721–723
Three-point perspectives, 457
Three-view drawing layout, 187
Three-view drawings, 182
Three-view sketching, 163–166
Throat diameter (TD)
of spider, 283
Thumb screws, 265
Thumbwheels, 646
TILEMODE, 189
on, 713
off, 713
TILEMODE=0, 712, 738
TILEMODE=1, 712, 714
Title blocks, 389, 390, 660
with AutoCAD, 707–708
setting up, 659–661
Title page
of technical report, 80
Title strips
guidelines for lettering, 125
Toggle option
in ISOPLANE command, 456
Tolerance, 252
English units, 339
metric units, 342

Tolerance accumulation
elimination of, 346
Toleranced dimensions, 706
Tolerance grades, 252
Tolerance notes, 346–347
Tolerance positions, 252
Tolerances and tolerancing, 330,
 380
basic hole system, 340
chain vs. datum dimensions,
 345–346
by computer, 337–338
cylindrical, 338
cylindrical datum features,
 351–353
cylindrical fits, 340–341
datum targets, 354
definitions of, 339–340
dimensions, 336–338
form, 357–358
geometric, 348–350
location, 354–357
mating parts, 338
metric units, 341–345
metric units for general,
 347–348
notes, 346–347
orientation, 358–360
preferred sizes and fits for, 343
profile, 358
rules for, 350–351
runout, 360–361
and surface texture, 361–363
terms for, 339–340
on working drawings, 390
Tolerance symbols, 342
Tolerance values
applying, 341
Tolerance zone
metric units, 342
Toolbars, 651, 659, 712
with AutoCAD, 658–659
Tool table holder, 327
Toothbrush holder
short design problem, 102
Tooth forms
for gears, 277–278

Top views
auxiliaries from, 209–211
true-length lines, 519
in sections, 227
Torsion springs, 270
TORUS command, 723, 731
Total runout, 360–361
Toughness
of metal, 294
Towel bar
short design problem, 100
Toy design
product design problem, 111
Toy production, educational
systems design problem, 106
TRACE command, 673
Trace (Command line), 673
Tracing cloth, 114, 434
Tracing film, 432, 434
Tracing paper, 432
and transparencies, 77
Tracing paper (or tracing vellum),
 114
Tracing vellum, 443
Tracking antenna, 513
TRANSFER command, 476, 477,
 479
Transformers, 630
Transistors, 624, 637
Transition fits (LT), 341, 343
English units, 339
metric units, 342
Transition pieces
development of, 563
Transparencies
for presentations, 77
Transparent commands
(Command line), 693
Transportation
and civil engineering, 9
and mechanical engineering,
 11
Trash-can cover
short design problem, 103
Trash disposal system, drive-
through
systems design problem, 105

Triangle method
  for parallel lines, 143
Triangles, 114–115, 137
  constructing, 139
  for drawing, 114–115, 118,
    185–186
  for parallel lines, 186
Triangulation, 139
Trigonometry, 136
TRIM command, 572, 672, 732
Trimetric axonometric projec-
    tion, 456
Trim (Modify toolbar), 672
Triple thread, 250
True-length chordal distances,
    563
True-length diagram, 480, 539
True-length distances, 557
True-length lines, 185, 470, 563
  front view, 518–520
  labeling for, 465
True position
  of hole's center, 355
True-shape orthographic view,
    169
True size (TS), 213
  of plane, 520–522
  in primary auxiliary view, 216
  of width dimension in auxiliary
    view, 217
True-size planes, 185
  labeling for, 465
Truncated cone
  development of, 562
  with third-angle projection,
    387
Truncated prism, 138
Truncated pyramid, 138, 561
Tubing, 608
Turning
  on lathe, 300
Turret lathe, 301
Twist option, 721
Two-cycle log scale, 590
Two-cycle scale, 577
2D drawings, 714
  plotting, 740

Two-dimensional isometric
    drawings, 455
Two-dimensional lines (Draw
    toolbar), 667–668
Two-dimension polylines
    drawing with AutoCAD, 680
Two hole pattern
  gauging, 356
Two-point perspectives, 457
Two-scale coordinate graphs, 575
Two-scale graphs, 575
Two-view drawings, 191
Type designations, 635
TYPICAL (or TYP) notes, 323

U

U. S. Congress
  patent terms extended by, 93
UCS. See User Coordinate System
UCS command, 719
  options of, 716–717
UCSICON command, 714, 717,
    719, 728, 729, 733, 738
U-groove welds, 372, 373, 375
UN. See Unified National
Undercuts, 331
Undercutting
  on lathe, 300
Underground ore veins, 492–493
UNDO command, 677
Undo option, 675, 680, 682, 721
Undo (Standard toolbar), 677
Unfinished bolt heads and nuts,
    259–260
Unidirectional method
  for obliques, 444
Unidirectional numerals
  on isometric drawings, 454
Unidirectional vertical numeric
    dimensions, 311
Unified National Rolled form
    (UNR)
  for external threads, 248
Unified National (UN) thread
    form, 248
Unified National (UN) threads, 246

Uniform acceleration. See Gravity
    motion
Uniform motion
  for cams, 285
Unilateral tolerance zones, 358
Unilateral tolerancing, 336, 337
Uni-lodge, transportable, 40
UNION command, 732, 733
UNITS command, 483
Units of measurement
  and dimensioning, 309
UN/UNR-threads, 254
Uppercase letters
  for fundamental deviations for
    holes, 343
  inclined Gothic, 131
  for internal threads, 252
  vertical Gothic, 130–131
Upper deviation
  metric units, 341
U.S. Department of Agriculture, 7
U.S. Department of Defense, 349
U.S. Department of the Interior, 7
U.S. Government
  patents issued by, 93
U.S. Patent and Trademark Office
    (PTO), 43, 90, 91, 93, 94, 95, 97
User Coordinate System (UCS),
    668, 714, 716
Utility meter, multipurpose
  systems design problem, 104

V

Vacuum tubes, 637
Valves, 613–614
  types of, 613–614
Vanishing points (VPs)
  in perspective pictorials, 457
Variables
  dimensioning, 696–697
Variable switch, 632
Vector, 530
Vector diagram, 531
Vector graphics, 2
  coplanar, concurrent forces,
    531–532

coplanar truss analysis, 534–537
definitions for, 530–531
forces in equilibrium, 533–534
noncoplanar vector analysis, 537–539
noncoplanar, concurrent forces, 532–533
resultant of parallel, noncon-current forces, 539–540
Vector polygons, 539
solving for two unknown val-ues, 533
Velocity graphs, 602
Vendors
literature from, 53
Vertical cylinder
intersection with inclined prism, 549
Vertical dimensions, 311
Vertical letters
Gothic, 130–131
Vertical lines, 118
drawing, 118
Vertical prism
intersection between inclined prism and, 546
Vertical sections, 485
Vessel detailing, 618–619
V-groove welds, 372, 373, 375
Videotapes, 75
for presentations, 78
VIEW command, 693
Viewports (Vports), 712, 713
Views
arrangement in orthographic drawings, 183
by computer, 189
partial, 192, 232–233
for patent drawings, 93
primary auxiliary, 208–221
removed, 192
selection in orthographic drawings, 183
simplified and removed, 191–192
by subtraction, 186–187

View setting, 665
da Vinci, Leonardo, 58
Visibility, 467–468, 489
Visible lines, 161, 167
Visible object line, 185
Vision
and design analysis, 60
Visual aids, 75–78
for design problems, 99
preparation of, 75
Voice/data/fax mobile terminal, 36
Voltage
ratings, 636
Voltage values, 635
Vplayer option
dimensioning with, 741
VPOINT command, 718, 735
VPORTS command, 713, 714

W

Warehouses and warehousing
expenses with, 64
and implementation, 84
location factors with, 87
Washers, 231, 267–268
Washing machine
product design problem, 111
Water level, 491
Wave shapes, 636
Waviness (height and width)
of surface texture, 361
WCS. See World Coordinate System
Wearable data terminal, 36
Webs
hatching, 231
revolution of, 237
section lines through, 232
WEDGE command, 722, 731
Weight factors, 25
Welded joints, 609, 611, 615
Welded pipe, 606–607
Welded pipe joints, 608
Welded fittings, 613
Welders, 5

Welding
and brazing, 378
joints and welds, 372–373
processes, 370–372
and soldering, 378
surface contouring, 377
symbols, 373
Weld joints, 372–373
Welds, 373
White iron, 291
Whiteprint (or blue-line print), 433
Whitworth thread, 246
Whole depth (WD)
of bevel gear, 280
of gear, 277
Whole depth of tooth (WDT)
for worm, 283
Width dimensions
in auxiliary view, 218
in orthographic projection, 163, 165
in six-view drawings, 182
Width of thread at root (WT)
for worm, 283
Width option, 680, 681, 683
Window option, 674, 675
Windows
introduction to, 651
Windows 95, 651
Windows NT, 644
Wing screws, 264
Wire-frame drawing, 460
Women
with engineering careers, 6
Wood screws, 264–265
Workers' stilts
product design problem, 107
Working depth (WKD)
of gear, 277
Working drawings, 178
assembly drawings, 392–394
checking, 391
defined, 13
for design problems, 99
dimensions and notes on, 308
dimensions and units on, 381–387

drafter's log, 392
dual measurement dimensions for, 385, 387
for exercise bench, 87
for forged parts and castings, 394–395
freehand, 394
and implementation, 21, 25, 84–85
and laying out detail drawing, 387–388
as legal documents, 380–382
notes and other information on, 389–391
parts list on, 390
patent rights notes on, 390
scale specifications on, 390
sheet sizes for, 388
student, 391
tolerances on, 390
and vessel detailing, 619
Worksheets
for design problems and analysis, 67, 99
World Coordinate System (WCS), 668, 714, 716
World Wide Web

patent information on, 96
Worm gears, 276, 282–284
drawing, 284
Worms, 282
terminology for, 283
WPolygon option (WP), 674
Wright brothers
first flight of, 6
Writing table arm
product design problem, 106
Written reports, 72, 79–82
evaluating, 81–82
final reports, 80
format of, 80–81
graphs in, 580
illustrations in, 81
progress reports, 80
proposals, 80
Wrought aluminum alloys, 293
Wrought copper, 293

X

X axis
of linear coordinate graphs, 571, 572

Xerography
for reproduction of drawings, 433, 435
XS weight pipe, 606, 607
XXS weight pipe, 606, 607
X-Y icon, 189
XYZ filters, 726

Y

Yard helper
product design problem, 107
Y axis
of linear coordinate graphs, 571, 572

Z

Zero Suppression (DIMZIN) box, 702
Z nomographs, 590–591
Zones (grid system)
on schematic diagrams, 636
ZOOM command, 189, 191, 673, 721
Zoom option, 721
Zoom (Standard toolbar), 673

The following problem books are available from Creative Publishing Company.
They are compatible with this and all other texts authored by James H. Earle.

DESIGN GRAPHICS

Two college-level manuals for alternate semesters that include graphics, descriptive geometry, and design for solution by computer, sketching, instruments, or in combination.
Approximately 100 pages, 177 problems per book.

GRAPHICS FOR ENGINEERS

Three college-level manuals for alternate semesters. Includes design projects and computer graphics in addition to sketching and traditional graphics.
100 pages, 350 problems per book.

GRAPHICS AND GEOMETRY

Three college-level manuals for alternate semesters that include graphics, descriptive geometry, and design for solution by computer, sketching, or by instruments.
135 pages, 350 problems per book.

GEOMETRY FOR ENGINEERS

Three college-level manuals for alternate semesters. Coverage of spatial graphics, descriptive geometry, and design projects for a one-semester course.
90 pages, 150 problems

CREATIVE DRAFTING

Introductory graphics with a design flavor for solution by computer or by hand.
102 pages, 268 problems.

DRAFTING FUNDAMENTALS

A series of two problem books that can be used for alternate years for a two-semester high school course in beginning graphics and drafting.
98 pages, 282 problems per book.

DRAFTING TECHNOLOGY

Graphics and descriptive geometry problems for solution by computer, sketching, or by instruments.
130 pages, 400 problems.

DRAFTING & DESIGN

Beginning graphics and architectural drawing. Solution by computer or traditional instruments.
106 pages, 400 problems.

BASIC DRAFTING

An excellent high-school manual for beginners. Solution by hand or by computer.
58 pages, 232 problems.

TECHNICAL ILLUSTRATION

A text/problem manual for a one-semester course in 3D pictorial drawing, perspectives, and rendering techniques.
70 pages, 181 problems.

ARCHITECTURAL DRAFTING

A text/manual for a course in architectural drawing—house plans, detailing, design, and planning.
71 pages, 134 problems.

CREATIVE PUBLISHING COMPANY
Box 9292
College Station, Texas 77842
800-245-5841